COX RINGS

Cox rings are significant global invariants of algebraic varieties, naturally generalizing homogeneous coordinate rings of projective spaces. This book provides a largely self-contained introduction to Cox rings, with a particular focus on concrete aspects of the theory. Besides the rigorous presentation of the basic concepts, other central topics include the case of finitely generated Cox rings and its relation to toric geometry; various classes of varieties with group actions; the surface case; and applications in arithmetic problems, in particular Manin's conjecture. The introductory chapters require only basic knowledge of algebraic geometry. The more advanced chapters also touch on algebraic groups, surface theory, and arithmetic geometry.

Each chapter ends with exercises and problems. These comprise mini-tutorials and examples complementing the text, guided exercises for topics not discussed in the text, and, finally, several open problems of varying difficulty.

Ivan Arzhantsev received his doctoral degree in 1998 from Lomonosov Moscow State University and is a professor in its department of higher algebra. His research areas are algebraic geometry, algebraic groups, and invariant theory.

Ulrich Derenthal received his doctoral degree in 2006 from Universität Göttingen. He is a professor of mathematics at Leibniz Universität Hannover. His research interests include arithmetic geometry and number theory.

Jürgen Hausen received his doctoral degree in 1995 from Universität Konstanz. He is a professor of mathematics at Eberhard Karls Universität Tübingen. His field of research is algebraic geometry, in particular algebraic transformation groups, torus actions, geometric invariant theory, and combinatorial methods.

Antonio Laface received his doctoral degree in 2000 from Università degli Studi di Milano. He is an associate professor of mathematics at Universidad de Concepción. His field of research is algebraic geometry, more precisely linear systems, algebraic surfaces, and their Cox rings.

All the titles listed below can be obtained from good booksellers or from Cambridge University Press. For a complete series listing visit: www.cambridge.org/mathematics.

Cox Rings

IVAN ARZHANTSEV
Moscow State University

ULRICH DERENTHAL
Leibniz Universität Hannover

JÜRGEN HAUSEN
Eberhard Karls Universität Tübingen

ANTONIO LAFACE
Universidad de Concepción

CAMBRIDGE
UNIVERSITY PRESS

CAMBRIDGE
UNIVERSITY PRESS

University Printing House, Cambridge CB2 8BS, United Kingdom

One Liberty Plaza, 20th Floor, New York, NY 10006, USA

477 Williamstown Road, Port Melbourne, VIC 3207, Australia

314-321, 3rd Floor, Plot 3, Splendor Forum, Jasola District Centre, New Delhi - 110025, India

79 Anson Road, #06-04/06, Singapore 079906

Cambridge University Press is part of the University of Cambridge.

It furthers the University's mission by disseminating knowledge in the pursuit of
education, learning and research at the highest international levels of excellence.

www.cambridge.org
Information on this title: www.cambridge.org/9781107024625

First published 2015

A catalogue record for this publication is available from the British Library

Library of Congress Cataloging in Publication data
Arzhantsev, I. V. (Ivan Vladimirovich), 1972–
[Kol'tsa Koksa. English]
Cox rings / Ivan Arzhantsev, Department of Algebra, Faculty of Mechanics and
Mathematics, Moscow [and three others].
pages   cm. – (Cambridge studies in advanced mathematics)
Includes bibliographical references and index.
ISBN 978-1-107-02462-5 (hardback)
1. Algebraic varieties.   2. Rings (Algebra)   I. Title.
QA564.A7913   2015
516.3′53–dc23   2014005540

ISBN 978-1-107-02462-5 Hardback

Contents

Introduction

The basic principles regarding Cox rings become visible already in the classical example of the projective space $\mathbb{P}^n$ over a field $\mathbb{K}$, which we assume to be algebraically closed and of characteristic zero. The elements of $\mathbb{P}^n$ are the lines $\ell \subseteq \mathbb{K}^{n+1}$ through the origin $0 \in \mathbb{K}^{n+1}$. Such a line ℓ is concretely specified by its *homogeneous coordinates* $[z_0, \ldots, z_n]$, where $(z_0, \ldots, z_n)$ is *any* point on ℓ, different from the origin. Hence, this description comes with an intrinsic ambiguity. More formally speaking, that means that we should regard the projective space as a quotient by a group action

$$
\begin{array}{ccc}
\mathbb{K}^{n+1} \setminus \{0\} & \subseteq & \mathbb{K}^{n+1} \\
{\scriptstyle z \mapsto [z]} \Big\downarrow {\scriptstyle /\mathbb{K}^*} & & \\
\mathbb{P}^n & &
\end{array}
$$

where $\mathbb{K}^*$ acts on $\mathbb{K}^{n+1}$ via scalar multiplication. This presentation of the projective space $\mathbb{P}^n$ as the quotient of its *characteristic space* $\mathbb{K}^{n+1} \setminus \{0\}$ by the action of the *characteristic torus* $\mathbb{K}^*$ is the geometric way of thinking of Cox rings. In algebraic terms, the action of $\mathbb{K}^*$ on $\mathbb{K}^{n+1}$ is encoded by the associated decomposition of the polynomial ring into homogeneous parts

$$
\mathbb{K}[T_0, \ldots, T_n] = \bigoplus_{k \geq 0} \mathbb{K}[T_0, \ldots, T_n]_k,
$$

where $\mathbb{K}[T_0, \ldots, T_n]_k$ is the vector space of homogeneous polynomials f of degree k, which means that $f(tz) = t^k f(z)$ holds for all t and z. The polynomial ring together with this classical grading is the *Cox ring* of the projective space. Note that to construct $\mathbb{P}^n$ as a $\mathbb{K}^*$-quotient, we have to remove the origin, which is the vanishing locus of the *irrelevant ideal*, from the *total coordinate space* $\mathbb{K}^{n+1}$. The Cox ring can be seen in terms of algebraic geometry intrinsically

1

from the projective space as

$$\mathbb{K}[T_0, \ldots, T_n] = \bigoplus_{k \geq 0} \mathbb{K}[T_0, \ldots, T_n]_k \cong \bigoplus_{k \in \mathbb{Z}} \Gamma(\mathbb{P}^n, \mathcal{O}(kD)),$$

where the class of the hyperplane $D := V(T_0) \subseteq \mathbb{P}^n$ freely generates the *divisor class group* $\mathrm{Cl}(\mathbb{P}^n) \cong \mathbb{Z}$ and the sections $1, T_1/T_0, \ldots, T_n/T_0$ of degree 1 on the right-hand side correspond to the homogeneous generators $T_0, \ldots, T_n$ on the left-hand side. The brief discussion of this simple example shows us that Cox rings are located in the intersection of three fields: graded algebras, group actions and quotients, and divisors and their section rings.

Let us also take a brief look at the arithmetic aspects. The rational points $[z_0, \ldots, z_n]$ in the projective space $\mathbb{P}^n$ over $\mathbb{Q}$ are parameterized uniquely up to sign by primitive vectors, that is, tuples of coprime integers $z_0, \ldots, z_n$. This description of rational points is related to Cox rings via the diagram

$$
\begin{array}{ccc}
\mathbb{Z}_{\mathrm{prim}}^{n+1} \setminus \{0\} & \subseteq & \mathbb{Z}^{n+1} \\
{\scriptstyle z \mapsto [z]}\Big\downarrow {\scriptstyle /\mathbb{Z}^*} & & \\
\mathbb{P}^n(\mathbb{Q}) & &
\end{array}
$$

A typical problem is to estimate the number of rational points with bounded height $H([z_0, \ldots, z_n]) := \max\{|z_0|, \ldots, |z_n|\}$, which in our example essentially amounts to estimating the number of lattice points in an $(n+1)$-dimensional box. This is an instance of Manin's conjecture on the number of rational points of bounded height on Fano varieties.

The current interest in Cox rings has several sources. A first one dates back to the 1970s when Colliot-Thélène and Sansuc [96, 98] introduced universal torsors as a tool in arithmetic geometry in particular to investigate the existence of rational points on varieties. In the last few years, Salberger's approach [263] to study the distribution of rational points via universal torsors and their explicit representations in terms of Cox rings caused a considerable surge of research. Another source is the occurrence of characteristic spaces and Cox rings in toric geometry in the mid-1990s in work of Audin [23], Cox [104] and others, which had a tremendous impact on this field. Five years later, Hu and Keel [176] observed the fundamental connection between Mori theory and geometric invariant theory via Cox rings; one of the main insights is that, roughly speaking, finite generation of the Cox ring is equivalent to an optimal behavior with respect to the minimal model program. This put the toric case into a much more general framework and established Cox rings as an active field of research in algebraic geometry. For example, the explicit presentation of the Cox ring of a given variety in terms of generators and relations is a central question. The research in this direction was initiated by the work of Batyrev/Popov [33] and Hassett/Tschinkel [162] on (weak) del Pezzo surfaces.

The intention of this book is to provide an elementary access to Cox rings and their applications in algebraic and arithmetic geometry with a particular focus on the new, concrete aspects that Cox rings bring into these fields. The introductory part, consisting of the first three chapters, requires basic knowledge in algebraic geometry, and, in addition, some familiarity with toric varieties is helpful. The subsequent three chapters consider also more advanced topics such as algebraic groups, surface theory, and arithmetic questions.

Chapter 1 provides the mathematical framework for the ideas occurring in the preceding example discussion. We present the basics on graded algebras, quasitorus actions and their quotients, and divisors and sheaves of divisorial algebras. Building on this, we define an essentially unique Cox ring for any irreducible, normal variety with only constant invertible functions and a finitely generated divisor class group:

$$\mathcal{R}(X) := \bigoplus_{[D]\in \mathrm{Cl}(X)} \Gamma(X, \mathcal{O}(D)).$$

First results concern the algebraic properties of the Cox ring, in particular the divisibility properties: Cox rings are factorially graded rings in the sense that we have unique factorization for homogeneous elements. The further main results of the chapter elaborate the relations between the Cox ring and its geometric counterpart, the presentation of the underlying variety as a quotient of the characteristic space by the action of the characteristic quasitorus. For smooth varieties, this quotient presentation equals the universal torsor and in general it dominates the universal torsor.

Chapter 2 discusses the concepts provided in Chapter 1 for the example class of toric varieties; these come with an action of an algebraic torus having a dense open orbit. The basic feature of toric varieties is their complete description in terms of combinatorial data, so-called lattice fans. Approaching toric varieties via quotient presentations turns out to be combinatorial as well, and the describing data, which we call lattice bunches, correspond to lattice fans via linear Gale duality. Besides being illustrative examples, toric varieties are important in subsequent chapters as adapted ambient varieties.

Chapter 3 is devoted to varieties with a finitely generated Cox ring. In this setting, the varieties sharing the same Cox ring all occur as quotients of open subsets of their common total coordinate space, the spectrum of the Cox ring. Geometric invariant theory gives us a concrete combinatorial description of the possible characteristic spaces, which finally leads to the encoding of our varieties by "bunched rings." The resulting picture shares many combinatorial features with toric geometry. In fact, the varieties inherit many properties from a canonical toric ambient variety. We take a look from the combinatorial aspect to invariants such as Picard groups; local divisor class groups; and the cones of effective, movable, semiample, and ample divisors. Moreover, we treat singularities, intersection numbers, and the Mori chamber structure of the effective cone. A particularly interesting class are the rational varieties with a torus action of complexity 1, for example, $\mathbb{K}^*$-surfaces. Here, the bunched ring description leads to a very efficient approach to the geometry; for example, one obtains a concrete combinatorial resolution of singularities.

Chapter 4 begins with a study of Cox rings of embedded varieties and the effect of modifications, for example, blow-ups on the Cox ring. Then we investigate the various quotient presentations of a variety and show that they are all dominated by the characteristic space, playing here a similar role as the universal covering in topology. The problem of lifting group actions to quotient presentations and the automorphism group are investigated. We provide various criteria for finite generation of the Cox ring, for example, Knop's criterion for unirational varieties with a group action of complexity 1, a characterization via the multiplication map, and the characterization in terms of Mori theory due to Hu and Keel. Moreover, we relate Cox rings of blow-ups of the projective space to invariant rings of unipotent group actions following Nagata. For varieties coming with a torus action, we describe the Cox ring in terms of isotropy groups and a certain quotient; this generalizes the toric case and gives the foundation for the bunched ring approach to the more general case of complexity 1. Finally, we take a look at almost homogeneous varieties. After describing the Cox ring of a homogeneous space, we discuss embeddings with a small boundary in terms of bunched rings and then turn to Brion's description of Cox rings of spherical varieties and wonderful compactifications.

In Chapter 5, we take a close look at Cox rings of complex algebraic surfaces. A first general part is devoted to the classification of smooth Mori dream surfaces. We present a complete picture for surfaces with nonnegative anti-canonical Iitaka dimension, and study in detail the cases of elliptic rational surfaces, K3 surfaces, and Enriques surfaces. Then we turn to the explicit description of Cox rings by generators and relations. For del Pezzo surfaces, we show that the Cox ring is generated in anticanonical degree 1 and that the ideal of relations is generated by quadrics. A discussion of the relations between Cox rings of del Pezzo surfaces and flag varieties ends this part. Then we return to K3 surfaces. Here we provide a detailed study in the case of Picard number 2 and complete results are obtained for double covers of del Pezzo surfaces and of blow-ups of Hirzebruch surfaces in at most three points. Finally, we develop the theory of rational $\mathbb{K}^*$-surfaces. Here we allow singularities and show how their minimal resolution is encoded in the Cox ring. As an example class, we present the Gorenstein log del Pezzo $\mathbb{K}^*$-surfaces in terms of their Cox rings.

The aim of Chapter 6 is to indicate how Cox rings and universal torsors can be applied to arithmetic questions regarding rational points on varieties. We begin by discussing Colliot-Thélène's and Sansuc's theory of universal torsors over not necessarily algebraically closed fields, and explore the connection to characteristic spaces and Cox rings. Then we enter the problem of the existence of rational points on varieties over number fields. We discuss the Hasse principle and weak approximation. The failure of these principles is often explained by Brauer–Manin obstructions, and we indicate how they can be approached via universal torsors and give an overview of the existing results. Then we turn to Manin's conjecture. For del Pezzo surfaces, it is known in many cases, and a general strategy emerges. We discuss this strategy in detail and show how it can be applied to prove Manin's conjecture for a singular cubic surface.

Each chapter is followed by a choice of exercises and problems. The collections comprise small general background tutorials and examples complementing the text as well as guided exercises to topics going beyond the text including references and, finally, we pose several open problems (*) of varying presumed difficulty.

We are grateful to all people supporting the work on this text. In particular, we thank Carolina Araujo, Michela Artebani, Hendrik Bäker, Victor Batyrev, Benjamin Bechtold, Cinzia Casagrande, Jean-Louis Colliot-Thélène, Christopher Frei, Giuliano Gagliardi, Fritz Hörmann, Johannes Hofscheier, Elaine Huggenberger, Simon Keicher, Alvaro Liendo, Taras Panov, Marta Pieropan, Yuri Prokhorov, Fred Rohrer, Alexei Skorobogatov, Damiano Testa, Dmitri Timashev, Andrea Tironi, and Luca Ugaglia for helpful remarks and discussions.

1

Basic concepts

In this chapter we introduce the Cox ring and, more generally, the *Cox sheaf* and its geometric counterpart, the *characteristic space*. In addition, algebraic and geometric aspects are discussed. Section 1.1 is devoted to commutative algebras graded by monoids. In Section 1.2, we recall the correspondence between actions of quasitori (also called diagonalizable groups) on affine varieties and affine algebras graded by abelian groups and provide the necessary background on good quotients. Section 1.3 is a first step toward constructing Cox rings. Given an irreducible, normal variety X and a finitely generated subgroup $K \subseteq$ WDiv(X) of the group of Weil divisors, we consider the associated *sheaf of divisorial algebras*

$$\mathcal{S} = \bigoplus_{D \in K} \mathcal{O}_X(D).$$

We present criteria for local finite generation and consider the relative spectrum. A first result says that $\Gamma(X, \mathcal{S})$ is a unique factorization domain if K generates the divisor class group Cl(X). Moreover, we characterize divisibility in the ring $\Gamma(X, \mathcal{S})$ in terms of divisors on X. In Section 1.4, the Cox sheaf of an irreducible, normal variety X with finitely generated divisor class group Cl(X) is introduced; roughly speaking it is given as

$$\mathcal{R} = \bigoplus_{[D] \in \mathrm{Cl}(X)} \mathcal{O}_X(D).$$

The Cox ring then is the corresponding ring of global sections. In the case of a free divisor class group well-definedness is straightforward. The case of torsion needs some effort; the precise way to define $\mathcal{R}$ then is to take the quotient of an appropriate sheaf of divisorial algebras with respect to a certain ideal sheaf. Basic algebraic properties and divisibility theory of the Cox ring are investigated in Section 1.5. Finally, in Section 1.6, we study the characteristic space, that is, the relative spectrum $\widehat{X} = \mathrm{Spec}_X \mathcal{R}$ of the Cox sheaf. It comes with an action of the *characteristic quasitorus* $H = \operatorname{Spec} \mathbb{K}[\operatorname{Cl}(X)]$ and a good quotient $\widehat{X} \to X$. We relate geometric properties of X to properties of this action and describe the characteristic space in terms of geometric invariant theory.

1.1 Graded algebras

1.1.1 Monoid graded algebras

We recall basic notions on algebras graded by abelian monoids. In this subsection, R denotes a commutative ring with a unit element.

Definition 1.1.1.1 Let K be an abelian monoid. A *K-graded R-algebra* is an associative, commutative R-algebra A with a unit and a direct sum decomposition

$$A = \bigoplus_{w \in K} A_w$$

into R-submodules $A_w \subseteq A$ such that $A_w \cdot A_{w'} \subseteq A_{w+w'}$ holds for any two elements $w, w' \in K$. The R-submodules $A_w \subseteq A$ are the *(K-)homogeneous components* of A. An element $f \in A$ is *(K-)homogeneous* if $f \in A_w$ holds for some $w \in K$, and in this case w is called the *degree* of f. We write $A_\times \subseteq A$ for the multiplicative monoid of homogeneous elements.

We also speak of a K-graded R-algebra as a monoid graded algebra or just as a graded algebra. To compare R-algebras A and A' that are graded by different abelian monoids K and K', we work with the following notion of a morphism.

Definition 1.1.1.2 A *morphism* from a K-graded algebra A to a K'-graded algebra A' is a pair $(\psi, \widetilde{\psi})$, where $\psi \colon A \to A'$ is a homomorphism of R-algebras, $\widetilde{\psi} \colon K \to K'$ is a homomorphism of abelian monoids, and

$$\psi(A_w) \subseteq A'_{\widetilde{\psi}(w)}$$

holds for every $w \in K$. In the case $K = K'$ and $\widetilde{\psi} = \operatorname{id}_K$, we denote a morphism of graded algebras just by $\psi \colon A \to A'$ and also refer to it as a *(K-)graded homomorphism*.

Example 1.1.1.3 Given an abelian monoid K and $w_1, \ldots, w_r \in K$, the polynomial ring $R[T_1, \ldots, T_r]$ can be turned into a K-graded R-algebra by setting

$$R[T_1, \ldots, T_r]_w := \left\{ \sum_{\nu \in \mathbb{Z}_{\geq 0}^r} a_\nu T^\nu; \ a_\nu \in R, \ \nu_1 w_1 + \cdots + \nu_r w_r = w \right\}.$$

This K-grading is determined by $\deg(T_i) = w_i$ for $1 \le i \le r$. Moreover, $R[T_1, \ldots, T_r]$ comes with the natural $\mathbb{Z}^r_{\ge 0}$-grading given by

$$R[T_1, \ldots, T_r]_v := R \cdot T^v,$$

and we have a canonical morphism $(\psi, \tilde{\psi})$ from $R[T_1, \ldots, T_r]$ to itself, where $\psi = \mathrm{id}$ and $\tilde{\psi} \colon \mathbb{Z}^r_{\ge 0} \to K$ sends v to $v_1 w_1 + \cdots + v_r w_r$.

For any abelian monoid K, we denote by $K^{\pm}$ the associated group of differences and by $K_{\mathbb{Q}} := K^{\pm} \otimes_{\mathbb{Z}} \mathbb{Q}$ the associated rational vector space. Note that we have canonical maps $K \to K^{\pm} \to K_{\mathbb{Q}}$, where the first one is injective if and only if K admits cancellation and the second one is injective if and only if $K^{\pm}$ is torsion free. Given $w \in K$, we allow ourselves to write $w \in K^{\pm}$ and $w \in K_{\mathbb{Q}}$ for the respective images.

Definition 1.1.1.4 Let A be a K-graded R-algebra. The *weight monoid* of A is the submonoid $S(A) \subseteq K$ generated by all $w \in K$ with $A_w \ne 0$. The *weight group* of A is the subgroup $K(A) \subseteq K^{\pm}$ generated by $S(A) \subseteq K$. The *weight cone* of A is the convex cone $\omega(A) \subseteq K_{\mathbb{Q}}$ generated by $S(A) \subseteq K$.

By an *integral R-algebra*, we mean an R-algebra $A \ne 0$ without zero divisors. Note that for an integral R-algebra A graded by an abelian monoid K, the weight monoid of A is given as

$$S(A) = \{w \in K; \ A_w \ne 0\} \subseteq K.$$

We recall the construction of the algebra associated with an abelian monoid; it defines a covariant functor from the category of abelian monoids to the category of monoid graded algebras.

Construction 1.1.1.5 Let K be an abelian monoid. As an R-module, the associated *monoid algebra* over R is given by

$$R[K] := \bigoplus_{w \in K} R \cdot \chi^w$$

and its multiplication is defined by $\chi^w \cdot \chi^{w'} := \chi^{w+w'}$. If K' is a further abelian monoid and $\tilde{\psi} \colon K \to K'$ is a homomorphism, then we have a homomorphism

$$\psi := R[\tilde{\psi}] \colon R[K] \to R[K'], \qquad \chi^w \mapsto \chi^{\tilde{\psi}(w)}.$$

The pair $(\psi, \tilde{\psi})$ is a morphism from the K-graded algebra $R[K]$ to the K'-graded algebra $R[K']$, and this assignment is functorial.

Note that the monoid algebra $R[K]$ has K as its weight monoid, and $R[K]$ is finitely generated over R if and only if the monoid K is finitely generated. In general, if a K-graded algebra A is finitely generated over R, then its weight monoid is finitely generated and its weight cone is *polyhedral*, that is, the set of nonnegative linear combinations over a given finite collection of vectors.

Construction 1.1.1.6 (Trivial extension) Let $K \subseteq K'$ be an inclusion of abelian monoids and A a K-graded R-algebra. Then we obtain an K'-graded R-algebra A' by setting

$$A' := \bigoplus_{u \in K'} A'_u, \qquad A'_u := \begin{cases} A_u & \text{if } u \in K, \\ \{0\} & \text{else.} \end{cases}$$

Construction 1.1.1.7 (Lifting) Let $G \colon \widetilde{K} \to K$ be a homomorphism of abelian monoids and A a K-graded R-algebra. Then we obtain a $\widetilde{K}$-graded R-algebra

$$\widetilde{A} := \bigoplus_{u \in \widetilde{K}} \widetilde{A}_u, \qquad \widetilde{A}_u := A_{G(u)}.$$

Definition 1.1.1.8 Let A be a K-graded R-algebra. An ideal $I \subseteq A$ is called *(K-)homogeneous* if it is generated by (K-)homogeneous elements.

An ideal $I \subseteq A$ of a K-graded R-algebra A is homogeneous if and only if it has a direct sum decomposition

$$I = \bigoplus_{w \in K} I_w, \qquad I_w := I \cap A_w.$$

Construction 1.1.1.9 (Graded factor algebra) Let A be a K-graded R-algebra and $I \subseteq A$ a homogeneous ideal. Then the factor algebra A/I is K-graded by

$$A/I = \bigoplus_{w \in K} (A/I)_w \qquad (A/I)_w := A_w + I.$$

Moreover, for each homogeneous component $(A/I)_w \subseteq A/I$, one has a canonical isomorphism of R-modules

$$A_w/I_w \to (A/I)_w, \qquad f + I_w \mapsto f + I.$$

Construction 1.1.1.10 Let A be a K-graded R-algebra, and $\widetilde{\psi} \colon K \to K'$ be a homomorphism of abelian monoids. Then one may consider A as a K'-graded algebra with respect to the *coarsened grading*

$$A = \bigoplus_{u \in K'} A_u, \qquad A_u := \bigoplus_{\widetilde{\psi}(w)=u} A_w.$$

Example 1.1.1.11 Let $K = \mathbb{Z}^2$ and consider the K-grading of $R[T_1, \ldots, T_5]$ given by $\deg(T_i) = w_i$, where

$$w_1 = (-1, 2), \quad w_2 = (1, 0), \quad w_3 = (0, 1), \quad w_4 = (2, -1), \quad w_5 = (-2, 3).$$

Then the polynomial $T_1 T_2 + T_3^2 + T_4 T_5$ is K-homogeneous of degree $(0, 2)$, and thus we have a K-graded factor algebra

$$A = R[T_1, \ldots, T_5]/\langle T_1 T_2 + T_3^2 + T_4 T_5 \rangle.$$

The standard $\mathbb{Z}$-grading of the algebra A with $\deg(T_1) = \ldots = \deg(T_5) = 1$ may be obtained by coarsening via the homomorphism $\tilde{\psi} \colon \mathbb{Z}^2 \to \mathbb{Z}, (a, b) \mapsto a + b$.

Proposition 1.1.1.12 *Let A be a $\mathbb{Z}^r$-graded R-algebra satisfying $ff' \neq 0$ for any two nonzero homogeneous $f, f' \in A$. Then the following statements hold.*

(i) *The algebra A is integral.*
(ii) *If gg' is homogeneous for $0 \neq g, g' \in A$, then g and g' are homogeneous.*
(iii) *Every unit $f \in A^*$ is homogeneous.*

Proof Fix a lexicographic ordering on $\mathbb{Z}^r$. Given two nonzero $g, g' \in A$, write $g = \sum f_u$ and $g' = \sum f'_u$ with homogeneous f_u and f'_u. Then the maximal (minimal) component of gg' is $f_w f'_{w'} \neq 0$, where f_w and $f'_{w'}$ are the maximal (minimal) components of f and f' respectively. The first two assertions follow. For the third one observe that $1 \in A$ is homogeneous (of degree zero). □

1.1.2 Veronese subalgebras

We introduce Veronese subalgebras of monoid graded algebras and present statements relating finite generation of the algebra to finite generation of a given Veronese subalgebra and vice versa.

We begin with basic observations on finite generation of monoids. The first one is a generalization of the classical Gordan lemma which asserts that for any convex polyhedral cone $\sigma \subseteq \mathbb{Q}^r$, the monoid $\sigma \cap \mathbb{Z}^r$ is finitely generated.

Proposition 1.1.2.1 *Let K be a finitely generated abelian group and $L \subseteq M \subseteq K$ submonoids. If L is finitely generated and every $w \in M$ admits an $n \in \mathbb{Z}_{\geq 1}$ with $nw \in L$, then M is finitely generated.*

Proof First we prove Gordan's lemma. Consider $K = \mathbb{Z}^r$; let $L \subseteq K$ be generated by $w_1, \ldots, w_s \in \mathbb{Z}^r$ and $M = \sigma \cap \mathbb{Z}^r$ the monoid of integral points inside the convex cone $\sigma \subseteq \mathbb{Q}^r$ generated by $w_1, \ldots, w_s$. Then M is generated by the finite subset

$$([0, 1] \cdot w_1 + \cdots + [0, 1] \cdot w_s) \cap \mathbb{Z}^r \subseteq M.$$

We turn to the general case. Choose an epimorphism $\alpha \colon \mathbb{Z}^r \to K$. Then also $K' := \mathbb{Z}^r$ with $L' := \alpha^{-1}(L)$ and $M' := \alpha^{-1}(M)$ satisfy the assumptions. So, it suffices to show that M' is finitely generated. Take generators $w_1, \ldots, w_s \in \mathbb{Z}^r$

for L' and let $\sigma \subseteq \mathbb{Q}^r$ denote the convex cone generated by $w_1, \ldots, w_s$. By assumption, we have $M' \subseteq \sigma$. This leads to a tower of monoid algebras

$$\mathbb{Q} \subseteq \mathbb{Q}[M'] \subseteq \mathbb{Q}[\sigma \cap \mathbb{Z}^r].$$

By Gordan's lemma, proven earlier, $\mathbb{Q}[\sigma \cap \mathbb{Z}^r]$ is finitely generated over $\mathbb{Q}$. Moreover, for every $w \in \sigma \cap \mathbb{Z}^r$, some multiple kw with $k \in \mathbb{Z}_{\geq 1}$ belongs to L' and hence to M'. Thus, $\mathbb{Q}[\sigma \cap \mathbb{Z}^r]$ is integral and hence finite over $\mathbb{Q}[M']$. The Artin–Tate lemma [132, p. 144] tells us that $\mathbb{Q}[M']$ is finitely generated over $\mathbb{Q}$. Consequently, the weight monoid M' of $\mathbb{Q}[M']$ is finitely generated. $\qquad\square$

Proposition 1.1.2.2 *Let K, K', K'' be abelian monoids such that K admits cancellation and K', K'' are finitely generated.*

(i) *Let $\alpha\colon K' \to K$ be a homomorphism. Then $\ker(\alpha) \subseteq K'$ is a finitely generated submonoid.*

(ii) *For any two homomorphisms $\alpha\colon K' \to K$ and $\beta\colon K'' \to K$, the fiber product*

$$K' \times_K K'' := \{(w', w'') \in K' \times K''; \ \alpha(w') = \beta(w'')\}$$

is a finitely generated submonoid of the (finitely generated) product monoid $K' \times K''$.

(iii) *If $L, M \subseteq K$ are finitely generated submonoids, then $L \cap M \subseteq K$ is a finitely generated submonoid.*

(iv) *Let $\alpha\colon K' \to K$ be a homomorphism. If $L \subseteq K$ is a finitely generated submonoid, then $\alpha^{-1}(L) \subseteq K'$ is a finitely generated submonoid.*

Proof We prove (i). Because K' is finitely generated, there is an epimorphism $\pi\colon \mathbb{Z}_{\geq 0}^r \to K'$. It suffices to show that $\ker(\beta)$ is finitely generated for the composition $\beta := \alpha \circ \pi$. Consider the extension $\beta^{\pm}\colon \mathbb{Z}^r \to K^{\pm}$ to the groups of differences. Because K admits cancellation, we have $\ker(\beta) = \ker(\beta^{\pm}) \cap \mathbb{Z}_{\geq 0}^r$. The latter monoid is finitely generated due to Proposition 1.1.2.1.

We turn to (ii). Consider the group of differences $K^{\pm}$ and the diagonal subgroup $\Delta \subseteq K^{\pm} \times K^{\pm}$. Then we have a homomorphism

$$K' \times K'' \to (K^{\pm} \times K^{\pm})/\Delta, \qquad (w', w'') \mapsto (\alpha(w'), \beta(w'')) + \Delta.$$

According to (i), the kernel of this homomorphism is a finitely generated monoid. By construction, it equals the fiber product.

Assertion (iii) is obtained by applying (ii) to the inclusion homomorphisms $\alpha\colon L \to K$ and $\beta\colon M \to K$. Similarly, (iv) follows from (ii) by taking $\alpha\colon K' \to K$ and the inclusion $\beta\colon L \to K$. $\qquad\square$

We turn to Veronese subalgebras. In the sequel, R is a commutative ring with a unit element.

Definition 1.1.2.3 Let K be an abelian monoid and A a K-graded R-algebra. The *Veronese subalgebra* associated with a submonoid $L \subseteq K$ is

$$A(L) := \bigoplus_{w \in L} A_w \subseteq \bigoplus_{w \in K} A_w = A.$$

Proposition 1.1.2.4 *Let K be an abelian monoid admitting cancellation and A a finitely generated K-graded R-algebra. If $L \subseteq K$ is a finitely generated submonoid, then the associated Veronese subalgebra $A(L)$ is finitely generated over R.*

Proof We may assume that K is the weight monoid of A and thus finitely generated; note that L stays finitely generated due to Proposition 1.1.2.2 (iii). Let $f_1, \ldots, f_r \in A$ be homogeneous generators for A and $w_i \in K$ the degree of $f_i \in A$. Consider the canonical epimorphism (π, P) of graded algebras defined by

$$\pi \colon R[\mathbb{Z}_{\geq 0}^r] = R[T_1, \ldots, T_r] \to A, \quad T_i \mapsto f_i, \quad P \colon \mathbb{Z}_{\geq 0}^r \to K, \quad e_i \mapsto w_i.$$

By Proposition 1.1.2.2 (iv), the inverse image $M := P^{-1}(L) \subseteq \mathbb{Z}_{\geq 0}^r$ is a finitely generated monoid. Thus, the monoid algebra $R[M]$ is finitely generated over R. By construction, $\pi \colon R[\mathbb{Z}_{\geq 0}^r] \to A$ maps $R[M] \subseteq R[\mathbb{Z}_{\geq 0}^r]$ onto $A(L) \subseteq A$. This implies finite generation of $A(L)$. $\qquad\square$

Proposition 1.1.2.5 *Suppose that R is noetherian. Let K be a finitely generated abelian group, A a K-graded integral R-algebra, and $L \subseteq K$ a submonoid such that for every $w \in S(A)$ there exists an $n \in \mathbb{Z}_{\geq 1}$ with $nw \in L$. If the Veronese subalgebra $A(L)$ is finitely generated over R, then also A is finitely generated over R.*

Proof We may assume that L is the weight monoid of $A(L)$ and thus is finitely generated. According to Proposition 1.1.2.1, the following monoid is finitely generated as well:

$$\widetilde{L} := \{w \in K; \; kw \in L \text{ for some } k \in \mathbb{Z}_{\geq 1}\} \subseteq K.$$

Note that we have $L \subseteq S(A) \subseteq \widetilde{L}$. Choose $n \in \mathbb{Z}_{\geq 1}$ with $n\widetilde{L} \subseteq L$. Replacing L with $n\widetilde{L}$, we achieve $L = L^{\pm} \cap S(A)$ and Proposition 1.1.2.4 ensures that $A(L)$ is still finitely generated.

So, we may assume that $L \subseteq K$ is a subgroup of finite index. Coarsening the grading via $K \to K/L$, we reduce to the case that K is finite and $L = 0$ holds. In this situation, we only have to show that any homogeneous component $A_w \neq 0$ is finitely generated as an A_0-module. For this, take $n \in \mathbb{Z}_{\geq 1}$ with $nw = 0$ and $0 \neq f \in A_w$. Because A is integral, $g \mapsto f^{n-1}g$ embeds A_w as an A_0-submodule into A_0. Because A_0 is noetherian, A_w is finitely generated. $\quad\square$

Putting Propositions 1.1.2.4 and 1.1.2.5 together, we obtain the following well known statement on gradings by abelian groups.

Corollary 1.1.2.6 *Let R be noetherian, K a finitely generated abelian group, A an integral K-graded R-algebra, and $L \subseteq K$ a subgroup of finite index. Then the following statements are equivalent.*

(i) *The algebra A is finitely generated over R.*
(ii) *The Veronese subalgebra A(L) is finitely generated over R.*

Proposition 1.1.2.7 *Let L, K be abelian monoids admitting cancellation and (φ, F) a morphism from an L-graded R-algebra B to a K-graded R-algebra A. Assume that the weight monoid of B is finitely generated and $\varphi \colon B_u \to A_{F(u)}$ is an isomorphism for every $u \in L$. Then finite generation of A implies finite generation of B.*

Proof We may assume that K, L are the weight monoids of A, B respectively. Take homogeneous generators $f_1, \ldots, f_r \in A$ for A and let $w_i \in K$ be the degree of $f_i \in A$. Then we have a canonical epimorphism (π, P) of graded algebras given by

$$\pi \colon R[\mathbb{Z}_{\geq 0}^r] = R[T_1, \ldots, T_r] \to A, \quad T_i \mapsto f_i, \quad P \colon \mathbb{Z}_{\geq 0}^r \to K, \quad e_i \mapsto w_i.$$

Consider the fiber product $L \times_K \mathbb{Z}_{\geq 0}^r$ with respect to the homomorphisms $F \colon L \to K$ and $P \colon \mathbb{Z}_{\geq 0}^r \to K$. Then we obtain a commutative diagram

$$
\begin{array}{ccc}
R[L \times_K \mathbb{Z}_{\geq 0}^r] & \longrightarrow & R[\mathbb{Z}_{\geq 0}^r] \\
\downarrow & & \downarrow \\
B & \xrightarrow{\varphi} & A
\end{array}
$$

where the left-hand side downwards map sends $\chi^{(u,e)}$ to the element of B_u corresponding to $\chi^{P(e)} \in A_{F(u)}$. By Proposition 1.1.2.2 (ii), the monoid $L \times_K \mathbb{Z}_{\geq 0}^r$ is finitely generated. Consequently, $R[L \times_K \mathbb{Z}_{\geq 0}^r]$ is finitely generated over R. By assumption, it maps onto B. $\qquad\square$

1.2 Gradings and quasitorus actions

1.2.1 Quasitori

We recall the functorial correspondence between finitely generated abelian groups and quasitori (also called diagonalizable groups). Details can be found in the standard textbooks on algebraic groups; see, for example, [56, Sec. 8],

[178, Sec. 16], [236, Sec. 3.2.3] or [280, Sec. 2.5]. In this section, $\mathbb{K}$ is an algebraically closed field of characteristic zero.

Our language for algebraic geometry is mainly that of [178]. An *affine* ($\mathbb{K}$-)*variety* is a topological space X with a sheaf $\mathcal{O}$ of $\mathbb{K}$-valued functions such that X is isomorphic to a Zariski-closed subset of $\mathbb{K}^n$ endowed with the induced topology and the sheaf of regular functions, that is, functions being locally a quotient of polynomials. A ($\mathbb{K}$-)*prevariety* is a space X with a sheaf $\mathcal{O}_X$ of $\mathbb{K}$-valued functions covered by open subspaces $X_1, \dots, X_r$, each of which is an affine ($\mathbb{K}$-)variety. A ($\mathbb{K}$-)*variety* is a separated ($\mathbb{K}$-)prevariety, that is, a prevariety X with closed diagonal in $X \times X$. Thus, for readers preferring the language of schemes [160], the word prevariety refers to a reduced scheme of finite type over $\mathbb{K}$, and a variety is a separated reduced scheme of finite type over $\mathbb{K}$. Moreover, when we say point, we think of a closed point.

Recall that an *(affine) algebraic group* is an (affine) variety G together with a group structure such that the group laws are morphisms of varieties:

$$G \times G \to G, \quad (g_1, g_2) \mapsto g_1 g_2, \qquad G \to G, \quad g \mapsto g^{-1}$$

A *homomorphism* of two algebraic groups G and G' is a morphism $G \to G'$ of the underlying varieties that moreover is a homomorphism of groups.

A *character* of an algebraic group G is a homomorphism $\chi \colon G \to \mathbb{K}^*$ of algebraic groups, where $\mathbb{K}^*$ is the multiplicative group of the ground field $\mathbb{K}$. The *character group* of G is the set $\mathbb{X}(G)$ of all characters of G together with pointwise multiplication. Note that $\mathbb{X}(G)$ is an abelian group, and, given any homomorphism $\varphi \colon G \to G'$ of algebraic groups, one has a pullback homomorphism

$$\varphi^* \colon \mathbb{X}(G') \to \mathbb{X}(G), \qquad \chi' \mapsto \chi' \circ \varphi.$$

Definition 1.2.1.1 A *quasitorus* is an affine algebraic group H whose algebra of regular functions $\Gamma(H, \mathcal{O})$ is generated as a $\mathbb{K}$-vector space by the characters $\chi \in \mathbb{X}(H)$. A *torus* is a connected quasitorus.

Example 1.2.1.2 The *standard n-torus* $\mathbb{T}^n := (\mathbb{K}^*)^n$ is a torus in the sense of Definition 1.2.1.1. Its characters are precisely the Laurent monomials $T^\nu = T_1^{\nu_1} \cdots T_n^{\nu_n}$, where $\nu \in \mathbb{Z}^n$, and its algebra of regular functions is the Laurent polynomial algebra

$$\Gamma(\mathbb{T}^n, \mathcal{O}) = \mathbb{K}[T_1^{\pm 1}, \dots, T_n^{\pm 1}] = \bigoplus_{\nu \in \mathbb{Z}^n} \mathbb{K} \cdot T^\nu = \mathbb{K}[\mathbb{Z}^n].$$

We now associate with any finitely generated abelian group K in a functorial way a quasitorus, namely $H := \operatorname{Spec} \mathbb{K}[K]$; the construction will show that H is the direct product of a standard torus and a finite abelian group.

Construction 1.2.1.3 Let K be any finitely generated abelian group. Fix generators $w_1, \ldots, w_r$ of K such that the epimorphism $\pi \colon \mathbb{Z}^r \to K$, $e_i \mapsto w_i$ has the kernel

$$\ker(\pi) = \mathbb{Z} a_1 e_1 \oplus \ldots \oplus \mathbb{Z} a_s e_s$$

with $a_1, \ldots, a_s \in \mathbb{Z}_{\geq 1}$. Then we have the following exact sequence of abelian groups:

$$0 \longrightarrow \mathbb{Z}^s \xrightarrow{\ e_i \mapsto a_i e_i\ } \mathbb{Z}^r \xrightarrow{\ e_i \mapsto w_i\ } K \longrightarrow 0.$$

Passing to the respective spectra of group algebras we obtain with $H := \operatorname{Spec} \mathbb{K}[K]$ the following sequence of homomorphisms:

$$1 \longleftarrow \mathbb{T}^s \xleftarrow{\ (t_1^{a_1}, \ldots, t_s^{a_s}) \leftarrow\!\shortmid t\ } \mathbb{T}^r \xleftarrow{\quad \iota \quad} H \longleftarrow 1.$$

The ideal of $H \subseteq \mathbb{T}^r$ is generated by $T_i^{a_i} - 1$, where $1 \leq i \leq s$. Thus H is a closed subgroup of $\mathbb{T}^r$ and the sequence is an exact sequence of quasitori; note that

$$H \cong C(a_1) \times \ldots \times C(a_s) \times \mathbb{T}^{r-s}, \qquad C(a_i) := \{\zeta \in \mathbb{K}^*; \ \zeta^{a_i} = 1\}.$$

The group structure on $H = \operatorname{Spec} \mathbb{K}[K]$ does not depend on the choices made: the multiplication map is given by its comorphism

$$\mathbb{K}[K] \to \mathbb{K}[K] \otimes_{\mathbb{K}} \mathbb{K}[K], \qquad \chi^w \mapsto \chi^w \otimes \chi^w,$$

and the neutral element of $H = \operatorname{Spec} \mathbb{K}[K]$ is the ideal $\langle \chi^w - 1; \ w \in K \rangle$. Moreover, every homomorphism $\psi \colon K \to K'$ defines a homomorphism

$$\operatorname{Spec} \mathbb{K}[\psi] \colon \operatorname{Spec} \mathbb{K}[K'] \to \operatorname{Spec} \mathbb{K}[K].$$

Theorem 1.2.1.4 *We have contravariant exact functors being essentially inverse to each other:*

$$\{\textit{finitely generated abelian groups}\} \longleftrightarrow \{\textit{quasitori}\}$$

$$K \mapsto \operatorname{Spec} \mathbb{K}[K],$$

$$\psi \mapsto \operatorname{Spec} \mathbb{K}[\psi],$$

$$\mathbb{X}(H) \leftarrow\!\shortmid H,$$

$$\varphi^* \leftarrow\!\shortmid \varphi.$$

Under these equivalences, the free finitely generated abelian groups correspond to the tori.

This statement includes in particular the observation that closed subgroups as well as homomorphic images of quasitori are again quasitori. Note that homomorphic images of tori are again tori, but every quasitorus occurs as a closed subgroup of a torus.

Recall that a finite-dimensional *rational representation* of an affine algebraic group G is a homomorphism $\varrho\colon G \to \mathrm{GL}(V)$ to the (affine algebraic) group $\mathrm{GL}(V)$ of linear automorphisms of a finite-dimensional $\mathbb{K}$-vector space V. In terms of representations, one has the following characterization of quasitori, see, for example, [280, Thm. 2.5.2].

Proposition 1.2.1.5   *An affine algebraic group G is a quasitorus if and only if any finite-dimensional rational representation of G splits into one-dimensional subrepresentations.*

1.2.2 Affine quasitorus actions

Again we work over an algebraically closed field $\mathbb{K}$ of characteristic zero. Recall that one has contravariant equivalences between *affine algebras*, that is, finitely generated $\mathbb{K}$-algebras without nilpotent elements, and affine varieties:

$$A \;\mapsto\; \mathrm{Spec}\,A, \qquad\qquad X \;\mapsto\; \Gamma(X, \mathcal{O}),$$

where we interpret $\mathrm{Spec}\,A$ as the set of maximal ideals. We first specialize these correspondences to graded affine algebras and affine varieties with quasitorus action; here "graded" means graded by a finitely generated abelian group. Then we look at basic concepts such as orbits and isotropy groups from both sides.

A *G-variety* is a variety X together with a morphical action $G \times X \to X$ of an affine algebraic group G. A *morphism* from a G-variety X to G'-variety X' is a pair $(\varphi, \widetilde{\varphi})$, where $\varphi\colon X \to X'$ is a morphism of varieties and $\widetilde{\varphi}\colon G \to G'$ is a homomorphism of algebraic groups such that we have

$$\varphi(g\cdot x) \;=\; \widetilde{\varphi}(g)\cdot\varphi(x) \qquad \text{for all } (g, x) \in G \times X.$$

If G' equals G and $\widetilde{\varphi}$ is the identity, then we call $\varphi\colon X \to X'$ a *G-equivariant* morphism. We say that $f \in \Gamma(X, \mathcal{O})$ is *homogeneous* of weight $\chi \in \mathbb{X}(G)$ if

$$f(g\cdot x) \;=\; \chi(g)f(x) \qquad \text{for all } (g, x) \in G \times X.$$

Example 1.2.2.1   Let H be a quasitorus. Any choice of characters $\chi_1, \ldots, \chi_r \in \mathbb{X}(H)$ defines a *diagonal H-action* on $\mathbb{K}^r$ by

$$h\cdot z := (\chi_1(h)z_1, \ldots, \chi_r(h)z_r).$$

We now associate in a functorial manner with every affine algebra graded by a finitely generated abelian group an affine variety with a quasitorus action.

Construction 1.2.2.2 Let K be a finitely generated abelian group and A a K-graded affine algebra. Set $X = \operatorname{Spec} A$. If $f_i \in A_{w_i}$, $i = 1, \ldots, r$, generate A, then we have a closed embedding

$$X \rightarrow \mathbb{K}^r, \qquad x \mapsto (f_1(x), \ldots, f_r(x)),$$

and $X \subseteq \mathbb{K}^r$ is invariant under the diagonal action of $H = \operatorname{Spec} \mathbb{K}[K]$ given by the characters $\chi^{w_1}, \ldots, \chi^{w_r}$. Note that for any $f \in A$ homogeneity is characterized by

$$f \in A_w \iff f(h \cdot x) = \chi^w(h) f(x) \text{ for all } h \in H, \, x \in X.$$

This shows that the induced H-action on X does not depend on the embedding into $\mathbb{K}^r$: its comorphism is given by

$$A \rightarrow \mathbb{K}[K] \otimes_{\mathbb{K}} A, \qquad A_w \ni f_w \mapsto \chi^w \otimes f_w \in \mathbb{K}[K]_w \otimes_{\mathbb{K}} A_w.$$

This construction is functorial: given a morphism $(\psi, \widetilde{\psi})$ from a K-graded affine algebra A to K'-graded affine algebra A', we have a morphism $(\varphi, \widetilde{\varphi})$ from the associated H'-variety X' to the H-variety X, where $\varphi = \operatorname{Spec} \psi$ and $\widetilde{\varphi} = \operatorname{Spec} \mathbb{K}[\widetilde{\psi}]$.

For the other way around, that is, from affine varieties X with action of a quasitorus H to graded affine algebras, the construction relies on the fact that the representation of H on $\Gamma(X, \mathcal{O})$ is *rational*, that is, a union of finite-dimensional rational subrepresentations; see [280, Prop. 2.3.4] and [189, Lemma 2.5] for nonaffine X. Proposition 1.2.1.5 then shows that this representation of H splits into one-dimensional subrepresentations.

Construction 1.2.2.3 Let a quasitorus H act on a not necessarily affine variety X. Then $\Gamma(X, \mathcal{O})$ becomes a rational H-module by

$$(h \cdot f)(x) := f(h \cdot x).$$

The decomposition of $\Gamma(X, \mathcal{O})$ into one-dimensional subrepresentations turns it into a $\mathbb{X}(H)$-graded algebra:

$$\Gamma(X, \mathcal{O}) = \bigoplus_{\chi \in \mathbb{X}(H)} \Gamma(X, \mathcal{O})_\chi,$$

$$\Gamma(X, \mathcal{O})_\chi := \{f \in \Gamma(X, \mathcal{O}); \, f(h \cdot x) = \chi(h) f(x)\}.$$

Again this construction is functorial. If $(\varphi, \widetilde{\varphi})$ is a morphism from an H-variety X to an H'-variety X', then $(\varphi^*, \widetilde{\varphi}^*)$ is a morphism of the associated graded algebras.

Theorem 1.2.2.4 *We have contravariant functors being essentially inverse to each other:*

$$\{\text{graded affine algebras}\} \longleftrightarrow \{\text{affine varieties with quasitorus action}\}$$

$$A \mapsto \operatorname{Spec} A,$$
$$(\psi, \widetilde{\psi}) \mapsto (\operatorname{Spec} \psi, \operatorname{Spec} \mathbb{K}[\widetilde{\psi}]),$$

$$\Gamma(X, \mathcal{O}) \leftarrow\!\!\shortmid X,$$
$$(\varphi^*, \widetilde{\varphi}^*) \leftarrow\!\!\shortmid (\varphi, \widetilde{\varphi}).$$

Under these equivalences the graded homomorphisms correspond to the equivariant morphisms.

We use this equivalence of categories to describe some geometry of a quasitorus action in algebraic terms. The first basic observation is the following.

Proposition 1.2.2.5 *Let A be a K-graded affine algebra and consider the action of $H = \operatorname{Spec} \mathbb{K}[K]$ on $X = \operatorname{Spec} A$. Then for any closed subvariety $Y \subseteq X$ and its vanishing ideal $I \subseteq A$, the following statements are equivalent.*

(i) *The variety Y is H-invariant.*
(ii) *The ideal I is homogeneous.*

Moreover, if one of these equivalences holds, then one has a commutative diagram of K-graded homomorphisms

$$
\begin{array}{ccc}
\Gamma(X, \mathcal{O}) & \xleftarrow{\ \cong\ } & A \\
{\scriptstyle f \mapsto f|_Y} \downarrow & & \downarrow {\scriptstyle f \mapsto f + I} \\
\Gamma(Y, \mathcal{O}) & \xleftarrow{\ \cong\ } & A/I
\end{array}
$$

We turn to orbits and isotropy groups. First recall the following fact on general algebraic group actions, see, for example, [178, Sec. II.8.3].

Proposition 1.2.2.6 *Let G be an algebraic group, X a G-variety, and let $x \in X$. Then the isotropy group $G_x \subseteq G$ is closed, the orbit $G \cdot x \subseteq X$ is* *locally closed, and one has a commutative diagram of equivariant morphisms of G-varieties*

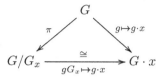

Moreover, the orbit closure $\overline{G \cdot x}$ *is the union of* $G \cdot x$ *and orbits of strictly lower dimension and it contains a closed orbit.*

Definition 1.2.2.7 Let A be a K-graded affine algebra and consider the action of $H = \operatorname{Spec} \mathbb{K}[K]$ on $X = \operatorname{Spec} A$.

(i) The *orbit monoid* of $x \in X$ is the submonoid $S_x \subseteq K$ generated by all $w \in K$ that admit a function $f \in A_w$ with $f(x) \neq 0$.
(ii) The *orbit group* of $x \in X$ is the subgroup $K_x \subseteq K$ generated by the orbit monoid $S_x \subseteq K$.

Proposition 1.2.2.8 *Let A be a K-graded affine algebra. Consider the action of $H = \operatorname{Spec} \mathbb{K}[K]$ on $X = \operatorname{Spec} A$ and let $x \in X$. Then there is a commutative diagram with exact rows*

$$
\begin{array}{ccccccccc}
0 & \longrightarrow & K_x & \longrightarrow & K & \longrightarrow & K/K_x & \longrightarrow & 0 \\
& & \downarrow{\cong} & & \downarrow{\cong}^{w \mapsto \chi^w} & & \downarrow{\cong} & & \\
0 & \longrightarrow & \mathbb{X}(H/H_x) & \xrightarrow{\ \pi^*\ } & \mathbb{X}(H) & \xrightarrow{\ \imath^*\ } & \mathbb{X}(H_x) & \longrightarrow & 0
\end{array}
$$

where $\imath \colon H_x \to H$ denotes the inclusion of the isotropy group and $\pi \colon H \to H/H_x$ the projection. In particular, we obtain $H_x \cong \operatorname{Spec} \mathbb{K}[K/K_x]$.

Proof Replacing X with $\overline{H \cdot x}$ does not change K_x. Moreover, take a homogeneous $f \in A$ vanishing along $\overline{H \cdot x} \setminus H \cdot x$ but not at x. Then replacing X with X_f does not affect K_x. Thus, we may assume that $X = H \cdot x$ holds. Then the weight monoid of the H-variety $H \cdot x$ is K_x and by the commutative diagram

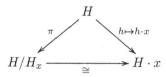

we see that $\pi^*(\mathbb{X}(H/H_x))$ consists precisely of the characters χ^w with $w \in K_x$, which gives the desired diagram. $\qquad\square$

Proposition 1.2.2.9 *Let A be a K-graded affine algebra. Consider the action of $H = \operatorname{Spec} \mathbb{K}[K]$ on $X = \operatorname{Spec} A$ and let $x \in X$. Then the orbit closure $\overline{H \cdot x}$ comes with an action of H/H_x, and there is an isomorphism $\overline{H \cdot x} \cong \operatorname{Spec} \mathbb{K}[S_x]$ of H/H_x-varieties.*

Proof Write for short $Y := \overline{H \cdot x}$ and $V := H \cdot x$. Then $V \subseteq Y$ is an affine open subset, isomorphic to H/H_x, and we have a commutative diagram

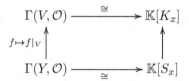

of graded homomorphisms, where the horizontal arrows send a homogeneous f of degree w to $f(x)\chi^w$. The assertion is part of this. $\square$

Proposition 1.2.2.10 *Let A be an integral K-graded affine algebra and consider the action of $H = \operatorname{Spec} \mathbb{K}[K]$ on $X = \operatorname{Spec} A$. Then there is a nonempty invariant open subset $U \subseteq X$ with*

$$S_x = S(A), \qquad K_x = K(A) \qquad \text{for all } x \in U.$$

Proof Let $f_1, \ldots, f_r$ be homogeneous generators for A. Then the set $U \subseteq X$ obtained by removing the zero sets $V(X, f_i)$ from X for $i = 1, \ldots, r$ is as wanted. $\square$

Recall that an action of a group G on a set X is said to be *effective* if $g \cdot x = x$ for all $x \in X$ implies $g = e_G$.

Corollary 1.2.2.11 *Let A be an integral K-graded affine algebra and consider the action of $H = \operatorname{Spec} \mathbb{K}[K]$ on $X = \operatorname{Spec} A$. Then the action of H on X is effective if and only if $K = K(A)$ holds.*

1.2.3 Good quotients

We summarize the basic facts on good quotients. Everything takes place over an algebraically closed field $\mathbb{K}$ of characteristic zero. Besides varieties, we consider more generally possibly nonseparated prevarieties.

Let an algebraic group G act on a prevariety X, where, here and later, we always assume that this action is given by a morphism $G \times X \to X$. Recall that a morphism $\varphi \colon X \to Y$ is said to be *G-invariant* if it is constant along the orbits. Moreover, a morphism $\varphi \colon X \to Y$ is called *affine* if for any open affine $V \subseteq Y$ the preimage $\varphi^{-1}(V)$ is an affine variety. When we speak of a *reductive* algebraic group, we mean a not necessarily connected affine algebraic group G such that every rational representation of G splits into irreducible ones.

Definition 1.2.3.1 Let a reductive algebraic group G act on a prevariety X. A morphism $p\colon X \to Y$ of prevarieties is called a *good quotient* for this action if it has the following properties:

(i) $p\colon X \to Y$ is affine and G-invariant,
(ii) The pullback $p^*\colon \mathcal{O}_Y \to (p_*\mathcal{O}_X)^G$ is an isomorphism.

A morphism $p\colon X \to Y$ is called a *geometric quotient* if it is a good quotient and its fibers are precisely the G-orbits.

Remark 1.2.3.2 Let $X = \operatorname{Spec} A$ be an affine G-variety with a reductive algebraic group G. The finiteness theorem of classical invariant theory ensures that the algebra of invariants $A^G \subseteq A$ is finitely generated [197, Sec. II.3.2]. This guarantees the existence of a good quotient $p\colon X \to Y$, where $Y :=$ $\operatorname{Spec} A^G$. The notion of a good quotient is locally modeled on this concept, because for any good quotient $p'\colon X' \to Y'$ and any affine open $V \subseteq Y'$ the variety V is isomorphic to $\operatorname{Spec} \Gamma(p'^{-1}(V), \mathcal{O})^G$, and the restricted morphism $p'^{-1}(V) \to V$ is the morphism just described.

Example 1.2.3.3 Consider the $\mathbb{K}^*$-action $t \cdot (z, w) = (t^a z, t^b w)$ on $\mathbb{K}^2$. The following three cases are typical.

(1) We have $a = b = 1$. Every $\mathbb{K}^*$-invariant function is constant and the constant map $p\colon \mathbb{K}^2 \to \{\mathrm{pt}\}$ is a good quotient.

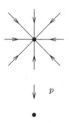

(2) We have $a = 0$ and $b = 1$. The algebra of $\mathbb{K}^*$-invariant functions is generated by z and the map $p\colon \mathbb{K}^2 \to \mathbb{K}$, $(z, w) \mapsto z$ is a good quotient.

(3) We have $a = 1$ and $b = -1$. The algebra of $\mathbb{K}^*$-invariant functions is generated by zw and $p\colon \mathbb{K}^2 \to \mathbb{K}$, $(z, w) \mapsto zw$ is a good quotient.

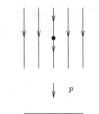

Note that the general p-fiber is a single $\mathbb{K}^*$-orbit, whereas $p^{-1}(0)$ consists of three orbits and is reducible.

Example 1.2.3.4   Let A be a K-graded affine algebra. Consider a homomorphism $\psi\colon K \to L$ of abelian groups and the coarsened grading

$$A = \bigoplus_{u \in L} A_u, \qquad A_u = \bigoplus_{w \in \psi^{-1}(u)} A_w.$$

Then the quasitorus $H := \operatorname{Spec} \mathbb{K}[L]$ acts on $X := \operatorname{Spec} A$, and for the algebra of invariants we have

$$A^H = \bigoplus_{w \in \ker(\psi)} A_w.$$

Note that in this special case, Proposition 1.1.2.4 ensures finite generation of the algebra of invariants.

Example 1.2.3.5 (Veronese subalgebras)   Let A be a K-graded affine algebra and $L \subseteq K$ a subgroup. Then we have the corresponding Veronese subalgebra

$$B = \bigoplus_{w \in L} A_w \subseteq \bigoplus_{w \in K} A_w = A.$$

By the preceding example, the morphism $\operatorname{Spec} A \to \operatorname{Spec} B$ is a good quotient for the action of $\operatorname{Spec} \mathbb{K}[K/L]$ on $\operatorname{Spec} A$.

We list basic properties of good quotients. The key to most of the statements is the following central observation.

Theorem 1.2.3.6   *Let a reductive algebraic group G act on a prevariety X. Then any good quotient $p\colon X \to Y$ has the following properties.*

(i) *G-closedness: If $Z \subseteq X$ is G-invariant and closed, then its image $p(Z) \subseteq Y$ is closed.*

(ii) *G-separation: If $Z, Z' \subseteq X$ are G-invariant, closed, and disjoint, then $p(Z)$ and $p(Z')$ are disjoint.*

Proof Because $p\colon X \to Y$ is affine and the statements are local with respect to Y, it suffices to prove them for affine X. This is done in [197, Sec. II.3.2], or [301, Thms. 4.6 and 4.7]. ☐

As an immediate consequence, one obtains basic information on the structure of the fibers of a good quotient.

Corollary 1.2.3.7 *Let a reductive algebraic group G act on a prevariety X, and let $p\colon X \to Y$ be a good quotient. Then p is surjective and for any $y \in Y$ one has:*

(i) *There is exactly one closed G-orbit $G \cdot x$ in the fiber $p^{-1}(y)$.*
(ii) *Every orbit $G \cdot x' \subseteq p^{-1}(y)$ has $G \cdot x$ in its closure.*

The first statement means that a good quotient $p\colon X \to Y$ parameterizes the closed orbits of the G-prevariety X.

Corollary 1.2.3.8 *Let a reductive algebraic group G act on a prevariety X, and let $p\colon X \to Y$ be a good quotient.*

(i) *The quotient space Y carries the quotient topology with respect to the map $p\colon X \to Y$.*
(ii) *For every G-invariant morphism of prevarieties $\varphi\colon X \to Z$, there is a unique morphism $\psi\colon Y \to Z$ with $\varphi = \psi \circ p$.*

Proof The first assertion follows from Theorem 1.2.3.6 (i). The second one follows from Corollary 1.2.3.7, Property 1.2.3.1 (ii) and the first assertion. ☐

An invariant morphism $p\colon X \to Y$ with Property 1.2.3.8 (ii) is also called a *categorical quotient*. The fact that a good quotient is categorical implies, in particular, that the good quotient space is unique up to isomorphy. This justifies the notation $X \to X /\!\!/ G$ for good and $X \to X/G$ for geometric quotients.

Proposition 1.2.3.9 *Let a reductive algebraic group G act on a prevariety X, and let $p\colon X \to Y$ be a good quotient.*

(i) *Let $V \subseteq Y$ be an open subset. Then the restriction $p\colon p^{-1}(V) \to V$ is a good quotient for the restricted G-action.*
(ii) *Let $Z \subseteq X$ be a closed G-invariant subset. Then the restriction $p\colon Z \to p(Z)$ is a good quotient for the restricted G-action.*

Proof The first statement is clear and the second one follows immediately from the corresponding statement on the affine case; see [197, Sec. II.3.2]. ☐

Example 1.2.3.10 (The Proj construction) Let $A = \oplus A_d$ be a $\mathbb{Z}_{\geq 0}$-graded affine algebra. The *irrelevant ideal* in A is defined as

$$A_{>0} := \langle f; \ f \in A_d \text{ for some } d > 0 \rangle \subseteq A.$$

For any homogeneous $f \in A_{>0}$ the localization A_f is a $\mathbb{Z}$-graded affine algebra; concretely, the grading is given by

$$A_f = \bigoplus_{d \in \mathbb{Z}} (A_f)_d, \qquad (A_f)_d := \{h/f^l \in A_f; \ \deg(h) - l \deg(f) = d\}.$$

In particular, we have the, again finitely generated, degree zero part of A_f; it is given by

$$A_{(f)} := (A_f)_0 = \{h/f^l \in A_f; \ \deg(h) = l \deg(f)\}.$$

Set $X := \mathrm{Spec}(A)$ and $Y_0 := \mathrm{Spec}(A_0)$, and, for a homogeneous $f \in A_{>0}$, set $X_f := \mathrm{Spec}\, A_f$ and $U_f := \mathrm{Spec}\, A_{(f)}$. Then, for any two homogeneous $f, g \in A_{>0}$, we have the commutative diagrams

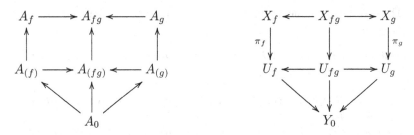

where the second one arises from the first one by applying the Spec-functor. The morphisms $U_{fg} \to U_f$ are open embeddings and gluing the U_f gives the variety $Y = \mathrm{Proj}(A)$. With the zero set $F := V(X, A_{>0})$ of the ideal $A_{>0}$, we have canonical morphisms, where the second one is projective:

$$X \setminus F \xrightarrow{\pi} Y \longrightarrow Y_0 \ .$$

Geometrically the following happened. The subset $F \subseteq X$ is precisely the fixed point set of the $\mathbb{K}^*$-action on X given by the grading. Thus, $\mathbb{K}^*$ acts with closed orbits on $W := X \setminus F$. The maps $X_f \to U_f$ are geometric quotients, and glue together to a geometric quotient $\pi \colon W \to Y$. Moreover, the $\mathbb{K}^*$-equivariant inclusion $W \subseteq X$ induces the morphism of quotients $Y \to Y_0$.

1.3 Divisorial algebras

1.3.1 Sheaves of divisorial algebras

We work over an algebraically closed field $\mathbb{K}$ of characteristic zero. We will deal with not only varieties over $\mathbb{K}$ but more generally also with possibly nonseparated prevarieties.

Let X be an irreducible prevariety. The group of *Weil divisors* of X is the free abelian group $\mathrm{WDiv}(X)$ generated by all *prime divisors*, that is, irreducible subvarieties $D \subseteq X$ of codimension 1. Via the *order* $\mathrm{ord}_D(f)$ along a prime divisor D, one associates with every nonzero rational function $f \in \mathbb{K}(X)^*$ the Weil divisor

$$\mathrm{div}(f) := \sum_{D \text{ prime}} \mathrm{ord}_D(f) \cdot D.$$

If f belongs to the local ring $\mathcal{O}_{X,D}$, then $\mathrm{ord}_D(f)$ is the length of the $\mathcal{O}_{X,D}$-module $\mathcal{O}_{X,D}/\langle f \rangle$, and otherwise one writes $f = g/h$ with $g, h \in \mathcal{O}_{X,D}$ and defines the order of f to be the difference of the orders of g and h. If the singular locus of X is of codimension at least 2, then $\mathcal{O}_{X,D}$ is a discrete valuation ring with valuation $f \mapsto \mathrm{ord}_D(f)$. This applies if X is *normal*, that is, every local ring $\mathcal{O}_{X,x}$ is integral and integrally closed in its quotient field.

The assignment $f \mapsto \mathrm{div}(f)$ is a homomorphism $\mathbb{K}(X)^* \to \mathrm{WDiv}(X)$, and its image $\mathrm{PDiv}(X) \subseteq \mathrm{WDiv}(X)$ is called the subgroup of *principal divisors*. We say that $D_1, D_2 \in \mathrm{WDiv}(X)$ are *linearly equivalent*, for short $D_1 \sim D_2$, if $D_2 - D_1$ is a principal divisor. The *divisor class group* of X is the factor group

$$\mathrm{Cl}(X) := \mathrm{WDiv}(X) \,/\, \mathrm{PDiv}(X).$$

A Weil divisor $D = a_1 D_1 + \cdots + a_s D_s$ with prime divisors D_i is called *effective*, denoted as $D \geq 0$, if $a_i \geq 0$ holds for $i = 1, \ldots, s$. With every divisor $D \in \mathrm{WDiv}(X)$, one associates a sheaf $\mathcal{O}_X(D)$ of $\mathcal{O}_X$-modules by defining its sections over an open $U \subseteq X$ as

$$\Gamma(U, \mathcal{O}_X(D)) := \{ f \in \mathbb{K}(X)^*; \ (\mathrm{div}(f) + D)|_U \geq 0 \} \cup \{0\},$$

where the restriction map $\mathrm{WDiv}(X) \to \mathrm{WDiv}(U)$ is defined for a prime divisor D as $D|_U := D \cap U$ if it intersects U and $D|_U := 0$ otherwise. Note that for any two functions $f_1 \in \Gamma(U, \mathcal{O}_X(D_1))$ and $f_2 \in \Gamma(U, \mathcal{O}_X(D_2))$ the product $f_1 f_2$ belongs to $\Gamma(U, \mathcal{O}_X(D_1 + D_2))$.

Definition 1.3.1.1 Let X be an irreducible, normal prevariety. The *sheaf of divisorial algebras* associated with a subgroup $K \subseteq \mathrm{WDiv}(X)$ is the sheaf of K-graded $\mathcal{O}_X$-algebras

$$\mathcal{S} := \bigoplus_{D \in K} \mathcal{S}_D, \qquad \mathcal{S}_D := \mathcal{O}_X(D),$$

where the multiplication in $\mathcal{S}$ is defined by multiplying homogeneous sections in the field of functions $\mathbb{K}(X)$.

Example 1.3.1.2 On the projective line $X = \mathbb{P}^1$, consider $D := \{\infty\}$, the group $K := \mathbb{Z}D$, and the associated K-graded sheaf of algebras $\mathcal{S}$. Then we have isomorphisms

$$\varphi_n \colon \mathbb{K}[T_0, T_1]_n \ \to \ \Gamma(\mathbb{P}^1, \mathcal{S}_{nD}), \qquad f \mapsto f(1, z),$$

where $\mathbb{K}[T_0, T_1]_n \subseteq \mathbb{K}[T_0, T_1]$ denotes the vector space of all polynomials homogeneous of degree n. Putting them together we obtain a graded isomorphism

$$\mathbb{K}[T_0, T_1] \cong \Gamma(\mathbb{P}^1, \mathcal{S}).$$

Fix an irreducible, normal prevariety X, a subgroup $K \subseteq \mathrm{WDiv}(X)$, and let $\mathcal{S}$ be the associated sheaf of divisorial algebras. We collect first properties.

Remark 1.3.1.3 If $V \subseteq U \subseteq X$ are open subsets such that $U \setminus V$ is of codimension at least 2 in U, then we have an isomorphism

$$\Gamma(U, \mathcal{S}) \to \Gamma(V, \mathcal{S}).$$

In particular, the algebra $\Gamma(U, \mathcal{S})$ equals the algebra $\Gamma(U_{\mathrm{reg}}, \mathcal{S})$, where $U_{\mathrm{reg}} \subseteq U$ denotes the set of smooth points.

Remark 1.3.1.4 Assume that $D_1, \ldots, D_s$ is a basis for $K \subseteq \mathrm{WDiv}(X)$ and suppose that $U \subseteq X$ is an open subset on which each D_i is principal, say $D_i = \mathrm{div}(f_i)$. Then, with $\deg(T_i) = D_i$ and $f_i^{-1} \in \Gamma(X, \mathcal{S}_{D_i})$, we have a graded isomorphism

$$\Gamma(U, \mathcal{O}) \otimes_{\mathbb{K}} \mathbb{K}[T_1^{\pm 1}, \ldots, T_s^{\pm 1}] \to \Gamma(U, \mathcal{S}),$$

$$g \otimes T_1^{\nu_1} \cdots T_s^{\nu_s} \mapsto g f_1^{-\nu_1} \cdots f_s^{-\nu_s}.$$

Remark 1.3.1.5 If K is of finite rank, say s, then the algebra $\Gamma(X, \mathcal{S})$ of global sections can be realized as a graded subalgebra of the Laurent polynomial algebra $\mathbb{K}(X)[T_1^{\pm 1}, \ldots, T_s^{\pm 1}]$. Indeed, let $D_1, \ldots, D_s$ be a basis for K. Then we obtain a monomorphism

$$\Gamma(X, \mathcal{S}) \to \mathbb{K}(X)[T_1^{\pm 1}, \ldots, T_s^{\pm 1}],$$

$$\Gamma(X, \mathcal{S}_{a_1 D_1 + \cdots + a_s D_s}) \ni f \mapsto f T_1^{a_1} \cdots T_s^{a_s}.$$

In particular, $\Gamma(X, \mathcal{S})$ is an integral ring and we have an embedding of the associated quotient fields

$$\mathrm{Quot}(\Gamma(X, \mathcal{S})) \to \mathbb{K}(X)(T_1, \ldots, T_s).$$

For quasiaffine X, we have $\mathbb{K}(X) \subseteq \mathrm{Quot}(\Gamma(X, \mathcal{S}))$ and for each variable T_i there is a nonzero function $f_i \in \Gamma(X, \mathcal{S}_{D_i})$. Thus, for quasiaffine X, one obtains

$$\mathrm{Quot}(\Gamma(X, \mathcal{S})) \cong \mathbb{K}(X)(T_1, \ldots, T_s).$$

Recall that the *support* $\mathrm{Supp}(D)$ of a Weil divisor $D = a_1 D_1 + \cdots + a_s D_s$ with prime divisors D_i is the union of those D_i with $a_i \neq 0$.

Definition 1.3.1.6 Let X be an irreducible, normal prevariety and D a Weil divisor on X. The *D-divisor* and the *D-localization* of a nonzero section $f \in \Gamma(X, \mathcal{O}_X(D))$ are

$$\mathrm{div}_D(f) := \mathrm{div}(f) + D \in \mathrm{WDiv}(X), \quad X_{D,f} := X \setminus \mathrm{Supp}(\mathrm{div}_D(f)) \subseteq X.$$

The D-divisor is always effective. Moreover, given sections $f \in \Gamma(X, \mathcal{O}_X(D))$ and $g \in \Gamma(X, \mathcal{O}_X(E))$, we have

$$\mathrm{div}_{D+E}(fg) = \mathrm{div}_D(f) + \mathrm{div}_E(g), \qquad f^{-1} \in \Gamma(X_{D,f}, \mathcal{O}_X(-D)).$$

For a projective X, the *complete linear system* $|D|$ of an effective divisor D on X is the collection of all D-divisors, coming with the structure of a projective space:

$$|D| = \{\mathrm{div}_D(f); \ f \in \Gamma(X, \mathcal{O}_X(D))\} = \mathbb{P}(\Gamma(X, \mathcal{O}_X(D))).$$

Remark 1.3.1.7 Let $D \in K$ and consider a nonzero homogeneous section $f \in \Gamma(X, \mathcal{S}_D)$. Then one has a canonical isomorphism of K-graded algebras

$$\Gamma(X_{D,f}, \mathcal{S}) \cong \Gamma(X, \mathcal{S})_f.$$

Indeed, the canonical monomorphism $\Gamma(X, \mathcal{S})_f \to \Gamma(X_{D,f}, \mathcal{S})$ is surjective, because for any $g \in \Gamma(X_{D,f}, \mathcal{S}_E)$, we have $gf^m \in \Gamma(X, \mathcal{S}_{mD+E})$ with some $m \in \mathbb{Z}_{\geq 0}$.

1.3.2 The relative spectrum

The geometric realization of a given sheaf of divisorial algebras is its relative spectrum. We begin with briefly recalling this construction in general. As before, we work over an algebraically closed field $\mathbb{K}$ of characteristic zero.

Construction 1.3.2.1 Let $\mathcal{S}$ be any quasicoherent sheaf of reduced $\mathcal{O}_X$-algebras on a prevariety X, and suppose that $\mathcal{S}$ is locally of finite type, that is, X is covered by open affine subsets $X_1, \ldots, X_r \subseteq X$ with $\Gamma(X_i, \mathcal{S})$ finitely generated. Cover each intersection $X_{ij} := X_i \cap X_j$ by open subsets $(X_i)_{f_{ijk}}$, where $f_{ijk} \in \Gamma(X_i, \mathcal{O})$. Set $\widetilde{X}_i := \mathrm{Spec}\, \Gamma(X_i, \mathcal{S})$ and let $\widetilde{X}_{ij} \subseteq \widetilde{X}_i$ be the union of the open subsets $(\widetilde{X}_i)_{f_{ijk}}$. Then we obtain commutative diagrams

$$
\begin{array}{ccccccc}
\widetilde{X}_i & \longleftarrow & \widetilde{X}_{ij} & \overset{\cong}{\longleftrightarrow} & \widetilde{X}_{ji} & \longrightarrow & \widetilde{X}_j \\
\downarrow & & \downarrow & & \downarrow & & \downarrow \\
X_i & \longleftarrow & X_{ij} & =\!=\!= & X_{ji} & \longrightarrow & X_j
\end{array}
$$

This allows us to glue together the $\widetilde{X}_i$ along the $\widetilde{X}_{ij}$, and we obtain a prevariety $\widetilde{X} = \mathrm{Spec}_X \mathcal{S}$ coming with a canonical morphism $p \colon \widetilde{X} \to X$. Note that

$p_*(\mathcal{O}_{\widetilde{X}}) = \mathcal{S}$ holds. In particular, $\Gamma(\widetilde{X}, \mathcal{O})$ equals $\Gamma(X, \mathcal{S})$. Moreover, the morphism p is affine and $\widetilde{X}$ is separated if X is so. Finally, the whole construction does not depend on the choice of the X_i.

Before specializing this construction to the case of our sheaf of divisorial algebras $\mathcal{S}$ on X, we provide two criteria for $\mathcal{S}$ being locally of finite type. The first one is an immediate consequence of Remark 1.3.1.7.

Proposition 1.3.2.2 *Let X be an irreducible, normal prevariety, $K \subseteq$ WDiv(X) a finitely generated subgroup, and $\mathcal{S}$ the associated sheaf of divisorial algebras. If $\Gamma(X, \mathcal{S})$ is finitely generated and X is covered by affine open subsets of the form $X_{D,f}$, where $D \in K$ and $f \in \Gamma(X, \mathcal{S}_D)$, then $\mathcal{S}$ is locally of finite type.*

A Weil divisor $D \in$ WDiv(X) on a prevariety X is called *Cartier* if it is locally a principal divisor, that is, locally of the form $D = \mathrm{div}(f)$ with a rational function f. The prevariety X is *locally factorial*, that is, all local rings $\mathcal{O}_{X,x}$ are unique factorization domains, if and only if every Weil divisor of X is Cartier. Recall that smooth prevarieties are locally factorial. More generally, an irreducible, normal prevariety is called $\mathbb{Q}$-*factorial* if for any Weil divisor some positive multiple is Cartier.

Proposition 1.3.2.3 *Let X be an irreducible, normal prevariety and $K \subseteq$ WDiv(X) a finitely generated subgroup. If X is $\mathbb{Q}$-factorial, then the associated sheaf $\mathcal{S}$ of divisorial algebras is locally of finite type.*

Proof By $\mathbb{Q}$-factoriality, the subgroup $K^0 \subseteq K$ consisting of all Cartier divisors is of finite index in K. Choose a basis $D_1, \ldots, D_s$ for K such that with suitable $a_i > 0$ the multiples $a_i D_i$, where $1 \le i \le s$, form a basis for K^0. Moreover, cover X by open affine subsets $X_1, \ldots, X_r \subseteq X$ such that for any $D \in K^0$ all restrictions $D|_{X_i}$ are principal. Let $\mathcal{S}^0$ be the sheaf of divisorial algebras associated with K^0. Then $\Gamma(X_i, \mathcal{S}^0)$ is the Veronese subalgebra of $\Gamma(X_i, \mathcal{S})$ defined by $K^0 \subseteq K$. By Remark 1.3.1.4, the algebra $\Gamma(X_i, \mathcal{S}^0)$ is finitely generated. Because $K^0 \subseteq K$ is of finite index, we can apply Proposition 1.1.2.5 and obtain that $\Gamma(X_i, \mathcal{S})$ is finitely generated. $\square$

Construction 1.3.2.4 Let X be an irreducible, normal prevariety; $K \subseteq$ WDiv(X) a finitely generated subgroup; and $\mathcal{S}$ the associated sheaf of divisorial algebras. We assume that $\mathcal{S}$ is locally of finite type. Then, in the notation of Construction 1.3.2.1, the algebras $\Gamma(X_i, \mathcal{S})$ are K-graded. This means that each affine variety $\widetilde{X}_i$ comes with an action of the torus $H := \mathrm{Spec}\,\mathbb{K}[K]$, and, because of $\mathcal{S}_0 = \mathcal{O}_X$, the canonical map $\widetilde{X}_i \to X_i$ is a good quotient for this action. Because the whole gluing process is equivariant, we end up with an H-prevariety $\widetilde{X} = \mathrm{Spec}_X \mathcal{S}$ and $p: \widetilde{X} \to X$ is a good quotient for the H-action.

Example 1.3.2.5 Consider once more the projective line $X = \mathbb{P}^1$; the group $K := \mathbb{Z}D$, where $D := \{\infty\}$; and the associated sheaf $\mathcal{S}$ of divisorial algebras. For the affine charts $X_0 = \mathbb{K}$ and $X_1 = \mathbb{K}^* \cup \{\infty\}$ we have the graded isomorphisms

$$\mathbb{K}[T_0^{\pm 1}, T_1] \to \Gamma(X_0, \mathcal{S}), \qquad \mathbb{K}[T_0^{\pm 1}, T_1]_n \ni f \mapsto f(1, z) \in \Gamma(X_0, \mathcal{S}_{nD}),$$

$$\mathbb{K}[T_0, T_1^{\pm 1}] \to \Gamma(X_1, \mathcal{S}), \qquad \mathbb{K}[T_0, T_1^{\pm 1}]_n \ni f \mapsto f(z, 1) \in \Gamma(X_1, \mathcal{S}_{nD}).$$

Thus, the corresponding spectra are $\mathbb{K}^2_{T_0}$ and $\mathbb{K}^2_{T_1}$. The gluing takes place along $(\mathbb{K}^*)^2$ and gives $\widetilde{X} = \mathbb{K}^2 \setminus \{0\}$. The action of $\mathbb{K}^* = \operatorname{Spec} \mathbb{K}[K]$ on $\widetilde{X}$ is the usual scalar multiplication.

The preceding example fits into the more general context of sheaves of divisorial algebras associated with groups generated by a *very ample divisor*, that is, the pullback of a hyperplane with respect to an embedding into a projective space.

Example 1.3.2.6 Let X be an irreducible, normal projective variety and $K :=$ $\mathbb{Z}D$ with a very ample divisor D on X. Then $\Gamma(X, \mathcal{S})$ is finitely generated and thus we have the affine cone $\overline{X} := \operatorname{Spec} \Gamma(X, \mathcal{S})$ over X. It comes with a $\mathbb{K}^*$-action and an attractive fixed point $\overline{x}_0 \in \overline{X}$, that is, $\overline{x}_0$ lies in the closure of any $\mathbb{K}^*$-orbit. The relative spectrum $\widetilde{X} = \operatorname{Spec}_X \mathcal{S}$ equals $\overline{X} \setminus \{\overline{x}_0\}$.

Remark 1.3.2.7 In the setting of Construction 1.3.2.4, let $U \subseteq X$ be an open subset such that all divisors $D \in K$ are principal over U. Then, using Remark 1.3.1.4, we obtain a commutative diagram of H-equivariant morphisms

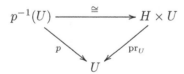

where H acts on $H \times U$ by multiplication on the first factor. In particular, if K consists of Cartier divisors, for example, if X is locally factorial, then $\widetilde{X} \to X$ is a locally trivial H-principal bundle.

Proposition 1.3.2.8 *Consider the situation of Construction 1.3.2.4. Then the prevariety $\widetilde{X}$ is irreducible and normal. Moreover, for any closed subset $A \subseteq X$ of codimension at least 2, the preimage $p^{-1}(A) \subseteq \widetilde{X}$ is as well of codimension at least 2.*

Proof First cover the set $X_{\operatorname{reg}} \subseteq X$ of smooth points by open subsets $U_1 \ldots, U_r$ as in Remark 1.3.2.7. Then $p^{-1}(X_{\operatorname{reg}})$ is the union of the irreducible open subsets $p^{-1}(U_i)$. Because the intersection over these sets is nonempty, also $p^{-1}(X_{\text{reg}})$ is irreducible. Moreover, $p^{-1}(X_{\text{reg}})$ is smooth. Now cover X by affine open subsets $V_1, \ldots, V_s$. Then Remark 1.3.1.3 gives

$$\Gamma(p^{-1}(V_i), \mathcal{O}) = \Gamma(p^{-1}(V_i) \cap p^{-1}(X_{\text{reg}}, \mathcal{O})).$$

Because $p^{-1}(X_{\text{reg}})$ is irreducible and smooth, the latter ring is integral and normal. We conclude that all $p^{-1}(V_i)$ are irreducible and normal and hence the same holds for X. The last assertion is an immediate consequence of Remark 1.3.1.3. $\qquad\square$

1.3.3 Unique factorization in the global ring

Here we investigate divisibility properties of the ring of global sections of the sheaf of divisorial algebras $\mathcal{S}$ associated with a subgroup $K \subseteq \operatorname{WDiv}(X)$ on an irreducible, normal prevariety X. The key statement is the following. The ground field $\mathbb{K}$ is algebraically closed and of characteristic zero.

Theorem 1.3.3.1 *Let X be an irreducible, smooth prevariety, $K \subseteq \operatorname{WDiv}(X)$ a finitely generated subgroup, $\mathcal{S}$ the associated sheaf of divisorial algebras and $\widetilde{X} = \operatorname{Spec}_X \mathcal{S}$. Then the following statements are equivalent.*

(i) *The canonical map $K \to \operatorname{Cl}(X)$ is surjective.*
(ii) *The divisor class group $\operatorname{Cl}(\widetilde{X})$ is trivial.*

We need a preparatory observation concerning the pullback of Cartier divisors. Recall that for any dominant morphism $\varphi \colon X \to Y$ of irreducible, normal prevarieties, there is a pullback of Cartier divisors: if a Cartier divisor E on Y is locally given as $E = \operatorname{div}(g)$, then the pullback divisor $\varphi^*(E)$ is the Cartier divisor locally defined by $\operatorname{div}(\varphi^*(g))$.

Lemma 1.3.3.2 *Consider the situation of Construction 1.3.2.4. Suppose that $D \in K$ is Cartier and consider a nonzero section $f \in \Gamma(X, \mathcal{S}_D)$. Then one has*

$$p^*(D) = \operatorname{div}(f) - p^*(\operatorname{div}(f)),$$

where on the right hand side f is first viewed as a homogeneous function on $\widetilde{X}$ and second as a rational function on X. In particular, $p^*(D)$ is principal.

Proof On suitable open sets $U_i \subseteq X$, we find defining equations f_i^{-1} for D and thus may write $f = h_i f_i$, where $h_i \in \Gamma(U_i, \mathcal{S}_0) = \Gamma(U_i, \mathcal{O})$ and $f_i \in \Gamma(U_i, \mathcal{S}_D)$. Then, on $p^{-1}(U_i)$, we have $p^*(h_i) = h_i$ and the function f_i is homogeneous of degree D and invertible. Thus, we obtain

$$p^*(D) = p^*(\mathrm{div}(f) + D) - p^*(\mathrm{div}(f))$$
$$= p^*(\mathrm{div}(h_i)) - p^*(\mathrm{div}(f))$$
$$= \mathrm{div}(h_i) - p^*(\mathrm{div}(f))$$
$$= \mathrm{div}(h_i f_i) - p^*(\mathrm{div}(f))$$
$$= \mathrm{div}(f) - p^*(\mathrm{div}(f)). \qquad \square$$

We are almost ready to prove the theorem. Recall that, given an action of an algebraic group G on an irreducible, normal prevariety X, we obtain an induced action of G on the group of Weil divisors by sending a prime divisor $D \subseteq X$ to $g \cdot D \subseteq X$. In particular, we can speak about invariant Weil divisors.

Proof of Theorem 1.3.3.1 Suppose that (i) holds. It suffices to show that every effective divisor $\widetilde{D}$ on $\widetilde{X}$ is principal. We work with the action of the torus $H = \mathrm{Spec}\,\mathbb{K}[K]$ on $\widetilde{X}$. Choosing an H-linearization of a line bundle associated with $\widetilde{D}$, see [189, Sec. 2.4] or 4.2.2, we obtain a representation of H on $\Gamma(\widetilde{X}, \mathcal{O}_{\widetilde{X}}(\widetilde{D}))$ such that for any section $\widetilde{f} \in \Gamma(\widetilde{X}, \mathcal{O}_{\widetilde{X}}(\widetilde{D}))$ one has

$$\mathrm{div}_{\widetilde{D}}(h \cdot \widetilde{f}) = h \cdot \mathrm{div}_{\widetilde{D}}(\widetilde{f}).$$

Taking a nonzero $\widetilde{f}$, which is homogeneous with respect to this representation, we obtain that $\widetilde{D}$ is linearly equivalent to the H-invariant divisor $\mathrm{div}_{\widetilde{D}}(\widetilde{f})$. This reduces the problem to the case of an invariant divisor $\widetilde{D}$; compare also [6, Thm. 4.2]. Now, consider any invariant prime divisor $\widetilde{D}$ on $\widetilde{X}$. Let $D := p(\widetilde{D})$ be the image under the good quotient $p \colon \widetilde{X} \to X$. Remark 1.3.2.7 gives $\widetilde{D} = p^*(D)$. By assumption, D is linearly equivalent to a divisor $D' \in K$. Thus, $\widetilde{D}$ is linearly equivalent to $p^*(D')$, which in turn is principal by Lemma 1.3.3.2.

Now suppose that (ii) holds. It suffices to show that any effective $D \in \mathrm{WDiv}(X)$ is linearly equivalent to some $D' \in K$. The pullback $p^*(D)$ is the divisor of some function $f \in \Gamma(\widetilde{X}, \mathcal{O})$. We claim that f is K-homogeneous. Indeed

$$F \colon H \times \widetilde{X} \to \mathbb{K}, \qquad (h, x) \mapsto f(h \cdot x)/f(x)$$

is an invertible function. By Rosenlicht's lemma [190, Sec. 1.1], or 4.2.2.2, we have $F(h, x) = \chi(h)g(x)$ with $\chi \in \mathbb{X}(H)$ and $g \in \Gamma(\widetilde{X}, \mathcal{O}^*)$. Plugging $(1, x)$ into F yields $g = 1$ and, consequently, $f(h \cdot x) = \chi(h)f(x)$ holds. Thus, we have $f \in \Gamma(X, \mathcal{S}_{D'})$ for some $D' \in K$. Lemma 1.3.3.2 gives

$$p^*(D) = \mathrm{div}(f) = p^*(D') + p^*(\mathrm{div}(f)),$$

where in the last term, f is regarded as a rational function on X. We conclude $D = D' + \mathrm{div}(f)$ on X. In other words, D is linearly equivalent to $D' \in K$. $\square$

As an immediate consequence, we obtain factoriality of the ring of global sections provided $K \to \mathrm{Cl}(X)$ is surjective; see [13, 42, 134] for the original references and note that combining Theorems 1.5.3.7 and 3.4.1.11 leads to another proof.

Theorem 1.3.3.3 *Let X be an irreducible, normal prevariety, $K \subseteq \mathrm{WDiv}(X)$ a finitely generated subgroup and $\mathcal{S}$ the associated sheaf of divisorial algebras. If the canonical map $K \to \mathrm{Cl}(X)$ is surjective, then the algebra $\Gamma(X, \mathcal{S})$ is a unique factorization domain.*

Proof According to Remark 1.3.1.3, the algebra $\Gamma(X, \mathcal{S})$ equals $\Gamma(X_{\mathrm{reg}}, \mathcal{S})$ and thus we may apply Theorem 1.3.3.1. $\square$

Divisibility and primality in the ring of global sections $\Gamma(X, \mathcal{S})$ can be characterized purely in terms of X.

Proposition 1.3.3.4 *Let X be an irreducible, normal prevariety, $K \subseteq \mathrm{WDiv}(X)$ a finitely generated subgroup projecting onto $\mathrm{Cl}(X)$ and let $\mathcal{S}$ be the associated sheaf of divisorial algebras.*

(i) *An element $0 \neq f \in \Gamma(X, \mathcal{S}_D)$ divides an element $0 \neq g \in \Gamma(X, \mathcal{S}_E)$ if and only if $\mathrm{div}_D(f) \leq \mathrm{div}_E(g)$ holds.*
(ii) *An element $0 \neq f \in \Gamma(X, \mathcal{S}_D)$ is prime if and only if the divisor $\mathrm{div}_D(f) \in \mathrm{WDiv}(X)$ is prime.*

Proof We may assume that X is smooth. Then $\widetilde{X} = \mathrm{Spec}_X \mathcal{S}$ exists, and Lemma 1.3.3.2 reduces (i) and (ii) to the corresponding statements on regular functions on $\widetilde{X}$, which in turn are well known. $\square$

1.3.4 Geometry of the relative spectrum

We collect basic geometric properties of the relative spectrum of a sheaf of divisorial algebras. We will use the following pullback construction for Weil divisors. The ground field $\mathbb{K}$ is algebraically closed and of characteristic zero.

Remark 1.3.4.1 Consider any morphism $\varphi \colon \widetilde{X} \to X$ of irreducible, normal prevarieties such that the closure of $X \setminus \varphi(\widetilde{X})$ is of codimension at least 2 in X. Then we may define a pullback homomorphism for Weil divisors

$$\varphi^* \colon \ \mathrm{WDiv}(X) \ \to \ \mathrm{WDiv}(\widetilde{X})$$

as follows: Given $D \in \mathrm{WDiv}(X)$, consider its restriction D' to X_{reg}, the usual pullback $\varphi^*(D')$ of Cartier divisors on $\varphi^{-1}(X_{\mathrm{reg}})$ and define $\varphi^*(D)$ to be the Weil divisor obtained by closing the support of $\varphi^*(D')$. Note that we always have

$$\mathrm{Supp}(\varphi^*(D)) \subseteq \varphi^{-1}(\mathrm{Supp}(D)).$$

If for any closed $A \subseteq X$ of codimension at least 2, $\varphi^{-1}(A) \subseteq \widetilde{X}$ is as well of codimension at least 2, then φ^* maps principal divisors to principal divisors, and we obtain a pullback homomorphism

$$\varphi^* \colon \ \mathrm{Cl}(X) \ \to \ \mathrm{Cl}(\widetilde{X}).$$

Example 1.3.4.2   Consider $X = V(\mathbb{K}^4; T_1 T_2 - T_3 T_4)$ and $\widetilde{X} = \mathbb{K}^4$. Then we have a morphism

$$p \colon \widetilde{X} \ \to \ X, \qquad z \mapsto (z_1 z_2, z_3 z_4, z_1 z_3, z_2 z_4).$$

For the prime divisor $D = \mathbb{K} \times 0 \times \mathbb{K} \times 0$ on X, we have

$$\mathrm{Supp}(p^*(D)) \ = \ V(\widetilde{X}; Z_4) \ \subsetneq \ V(\widetilde{X}; Z_4) \cup V(\widetilde{X}; Z_2, Z_3) \ = \ p^{-1}(\mathrm{Supp}(D)).$$

In fact, $p \colon \widetilde{X} \to X$ is the morphism determined by the sheaf of divisorial algebras associated with $K = \mathbb{Z}D$.

We say that a prevariety X is of *affine intersection* if for any two affine open subsets $U, U' \subseteq X$ the intersection $U \cap U'$ is again affine. For example, every variety is of affine intersection. Note that a prevariety X is of affine intersection if it can be covered by open affine subsets $X_1, \ldots, X_s \subseteq X$ such that all intersections $X_i \cap X_j$ are affine. Moreover, if X is of affine intersection, then the complement of any affine open subset $U \subsetneq X$ is of pure codimension 1.

Proposition 1.3.4.3 *In the situation of Construction 1.3.2.4, consider the pullback homomorphism $p^* \colon \mathrm{WDiv}(X) \to \mathrm{WDiv}(\widetilde{X})$ defined in Remark 1.3.4.1. Then, for every $D \in K$ and every nonzero $f \in \Gamma(X, \mathcal{S}_D)$, we have*

$$\mathrm{div}(f) = p^*(\mathrm{div}_D(f)),$$

where on the left-hand side f is a function on $\widetilde{X}$, and on the right-hand side a function on X. If X is of affine intersection and $X_{D,f}$ is affine, then we have moreover

$$\mathrm{Supp}(\mathrm{div}(f)) = p^{-1}(\mathrm{Supp}(\mathrm{div}_D(f))).$$

Proof   By Lemma 1.3.3.2, the first equation holds on $p^{-1}(X_{\mathrm{reg}})$. By Proposition 1.3.2.8, the complement $\widetilde{X} \setminus p^{-1}(X_{\mathrm{reg}})$ is of codimension at least 2 and thus the first equation holds on the whole $\widetilde{X}$. For the proof of the second one, consider

$$X_{D,f} \ = \ X \setminus \mathrm{Supp}(\mathrm{div}_D(f)), \qquad \widetilde{X}_f \ = \ \widetilde{X} \setminus V(\widetilde{X}, f).$$

Then we have to show that $p^{-1}(X_{D,f})$ equals $\widetilde{X}_f$. Because f is invertible on $p^{-1}(X_{D,f})$, we obtain $p^{-1}(X_{D,f}) \subseteq \widetilde{X}_f$. Moreover, Lemma 1.3.3.2 yields

$$p^{-1}(X_{D,f}) \cap p^{-1}(X_{\mathrm{reg}}) = \widetilde{X}_f \cap p^{-1}(X_{\mathrm{reg}}).$$

Thus the complement $\widetilde{X}_f \setminus p^{-1}(X_{D,f})$ of the affine subset $p^{-1}(X_{D,f}) \subseteq \widetilde{X}_f$ is of codimension at least 2. Because $p \colon \widetilde{X} \to X$ is affine, the prevariety $\widetilde{X}$ inherits the property of being of affine intersection from X and hence $\widetilde{X}_f \setminus p^{-1}(X_{D,f})$ must be empty. $\qquad\square$

Corollary 1.3.4.4 *Consider situation of Construction 1.3.2.4. Let $\widetilde{x} \in \widetilde{X}$ be a point such that $H \cdot \widetilde{x} \subseteq \widetilde{X}$ is closed, and let $0 \neq f \in \Gamma(X, \mathcal{S}_D)$. Then we have*

$$f(\widetilde{x}) = 0 \iff p(\widetilde{x}) \in \mathrm{Supp}(\mathrm{div}_D(f)).$$

Proof   Remark 1.3.4.1 and Proposition 1.3.4.3 show that $p(\mathrm{Supp}(\mathrm{div}(f)))$ is contained in $\mathrm{Supp}(\mathrm{div}_D(f))$. Moreover, they coincide along the smooth locus of X and Theorem 1.2.3.6 ensures that $p(\mathrm{Supp}(\mathrm{div}(f)))$ is closed. This gives

$$p(\mathrm{Supp}(\mathrm{div}(f))) = \mathrm{Supp}(\mathrm{div}_D(f)).$$

Thus, $f(\widetilde{x}) = 0$ implies $p(\widetilde{x}) \in \mathrm{Supp}(\mathrm{div}_D(f))$. If $p(\widetilde{x}) \in \mathrm{Supp}(\mathrm{div}_D(f))$ holds, then some $\widetilde{x}' \in \mathrm{Supp}(\mathrm{div}(f))$ lies in the p-fiber of $\widetilde{x}$. Because $H \cdot \widetilde{x}$ is closed, it is contained in the closure of $H \cdot \widetilde{x}'$; see Corollary 1.2.3.7. This implies $\widetilde{x} \in \mathrm{Supp}(\mathrm{div}(f))$. $\qquad\square$

Corollary 1.3.4.5 *Consider the situation of Construction 1.3.2.4. If X is of affine intersection and covered by affine open subsets of the form $X_{D,f}$, where $D \in K$ and $f \in \Gamma(X, \mathcal{S}_D)$, then $\widetilde{X}$ is a quasiaffine variety.*

Proof   According to Proposition 1.3.4.3, the prevariety $\widetilde{X}$ is covered by open affine subsets of the form $\widetilde{X}_f$ and thus is quasiaffine. $\qquad\square$

Corollary 1.3.4.6 *Consider the situation of Construction 1.3.2.4. If X is of affine intersection and $K \to \mathrm{Cl}(X)$ is surjective, then $\widetilde{X}$ is a quasiaffine variety.*

Proof   Cover X by affine open sets $X_1, \ldots, X_r$. Because X is of affine intersection, every complement $X \setminus X_i$ is of pure codimension 1. Because $K \to \mathrm{Cl}(X)$ is surjective, we obtain that $X \setminus X_i$ is the support of the D-divisor of some $f \in \Gamma(X, \mathcal{S}_D)$. The assertion thus follows from Corollary 1.3.4.5. $\qquad\square$

Proposition 1.3.4.7 *Consider the situation of Construction 1.3.2.4. For $x \in X$, let $K_x^0 \subseteq K$ be the subgroup of divisors that are principal near x and let $\widetilde{x} \in p^{-1}(x)$ be a point with closed H-orbit. Then the isotropy group $H_{\widetilde{x}} \subseteq H$ is given by $H_{\widetilde{x}} = \mathrm{Spec}\, \mathbb{K}[K/K_x^0]$.*

Proof   Replacing X with a suitable affine neighbourhood of x, we may assume that $\widetilde{X}$ is affine. By Proposition 1.2.2.8, the isotropy group $H_{\widetilde{x}}$ is $\mathrm{Spec}\, \mathbb{K}[K/K_{\widetilde{x}}]$ with the orbit group

$$K_{\widetilde{x}} = \langle D \in K;\ f(\widetilde{x}) \neq 0 \text{ for some } f \in \Gamma(X, \mathcal{S}_D) \rangle \subseteq K.$$

Using Corollary 1.3.4.4, we obtain that there exists an $f \in \Gamma(X, \mathcal{S}_D)$ with $f(\tilde{x}) \neq 0$ if and only if $D \in K_x^0$ holds. The assertion follows.   □

Corollary 1.3.4.8 *Consider the situation of Construction 1.3.2.4.*

(i) *If X is locally factorial, then H acts freely on $\tilde{X}$.*
(ii) *If X is $\mathbb{Q}$-factorial, then H acts with at most finite isotropy groups on $\tilde{X}$.*

1.4 Cox sheaves and Cox rings

1.4.1 Free divisor class group

As before, we work over an algebraically closed field $\mathbb{K}$ of characteristic zero. We introduce Cox sheaves and Cox rings for a prevariety with a free finitely generated divisor class group. As an example, we compute in Example 1.4.1.6 the Cox ring of a nonseparated curve, the projective line with multiplied points.

Construction 1.4.1.1 Let X be an irreducible, normal prevariety with free finitely generated divisor class group $\mathrm{Cl}(X)$. Fix a subgroup $K \subseteq \mathrm{WDiv}(X)$ such that the canonical map $c\colon K \to \mathrm{Cl}(X)$ sending $D \in K$ to its class $[D] \in \mathrm{Cl}(X)$ is an isomorphism. We define the *Cox sheaf* associated with K to be

$$\mathcal{R} := \bigoplus_{[D] \in \mathrm{Cl}(X)} \mathcal{R}_{[D]}, \qquad \mathcal{R}_{[D]} := \mathcal{O}_X(D),$$

where $D \in K$ represents $[D] \in \mathrm{Cl}(X)$ and the multiplication in $\mathcal{R}$ is defined by multiplying homogeneous sections in the field of rational functions $\mathbb{K}(X)$. The sheaf $\mathcal{R}$ is a quasicoherent sheaf of normal integral $\mathcal{O}_X$-algebras and, up to isomorphy, it does not depend on the choice of the subgroup $K \subseteq \mathrm{WDiv}(X)$. The *Cox ring* of X is the algebra of global sections

$$\mathcal{R}(X) := \bigoplus_{[D] \in \mathrm{Cl}(X)} \mathcal{R}_{[D]}(X), \qquad \mathcal{R}_{[D]}(X) := \Gamma(X, \mathcal{O}_X(D)).$$

Proof of Construction 1.4.1.1 Given two subgroups $K, K' \subseteq \mathrm{WDiv}(X)$ projecting isomorphically onto $\mathrm{Cl}(X)$, we have to show that the corresponding sheaves of divisorial algebras $\mathcal{R}$ and $\mathcal{R}'$ are isomorphic. Choose a basis $D_1, \ldots, D_s$ for K and define a homomorphism

$$\eta\colon K \to \mathbb{K}(X)^*, \qquad a_1 D_1 + \cdots + a_s D_s \mapsto f_1^{a_1} \cdots f_s^{a_s},$$

where $f_1, \ldots, f_s \in \mathbb{K}(X)^*$ are such that the divisors $D_i - \mathrm{div}(f_i)$ form a basis of K'. Then we obtain an isomorphism $(\psi, \tilde{\psi})$ of the sheaves of divisorial algebras $\mathcal{R}$ and $\mathcal{R}'$ by setting

$$\widetilde{\psi} \colon K \to K', \qquad\qquad\qquad D \mapsto -\operatorname{div}(\eta(D)) + D,$$

$$\psi \colon \mathcal{R} \to \mathcal{R}', \qquad \Gamma(U, \mathcal{R}_{[D]}) \ni f \mapsto \eta(D)f \in \Gamma(U, \mathcal{R}_{[\widetilde{\psi}(D)]}). \qquad \square$$

Example 1.4.1.2 Let X be the projective space $\mathbb{P}^n$ and $D \subseteq \mathbb{P}^n$ a hyperplane. The class of D generates $\mathrm{Cl}(\mathbb{P}^n)$ freely. We take K as the subgroup of $\mathrm{WDiv}(\mathbb{P}^n)$ generated by D, and the Cox ring $\mathcal{R}(\mathbb{P}^n)$ is the polynomial ring $\mathbb{K}[z_0, z_1, \ldots, z_n]$ with the standard grading.

Remark 1.4.1.3 If $X \subseteq \mathbb{P}^n$ is a closed, irreducible, normal subvariety whose divisor class group is generated by a hyperplane section, then $\mathcal{R}(X)$ coincides with $\Gamma(\overline{X}, \mathcal{O})$, where $\overline{X} \subseteq \mathbb{K}^{n+1}$ is the cone over X if and only if X is projectively normal; see also Corollary 4.1.1.7.

Remark 1.4.1.4 Let s denote the rank of $\mathrm{Cl}(X)$. Then Remark 1.3.1.5 realizes the Cox ring $\mathcal{R}(X)$ as a graded subring of the Laurent polynomial ring:

$$\mathcal{R}(X) \subseteq \mathbb{K}(X)[T_1^{\pm 1}, \ldots, T_s^{\pm 1}].$$

Using the fact that there are $f \in \mathcal{R}_{[D]}(X)$ with $X_{D,f}$ affine and Remark 1.3.1.7, we see that this inclusion gives rises to an isomorphism of the quotient fields

$$\mathrm{Quot}(\mathcal{R}(X)) \cong \mathbb{K}(X)(T_1, \ldots, T_s).$$

Proposition 1.4.1.5 *Let X be an irreducible, normal prevariety with free finitely generated divisor class group.*

(i) *The Cox ring $\mathcal{R}(X)$ is a unique factorization domain.*
(ii) *The units of the Cox ring are given by $\mathcal{R}(X)^* = \Gamma(X, \mathcal{O}^*)$.*

Proof The first assertion is a direct consequence of Theorem 1.3.3.3. To verify the second one, consider a unit $f \in \mathcal{R}(X)^*$. Then $fg = 1 \in \mathcal{R}_0(X)$ holds with some unit $g \in \mathcal{R}(X)^*$. This can happen only when f and g are homogeneous, say of degree $[D]$ and $-[D]$, and thus we obtain

$$0 = \operatorname{div}_0(1) = \operatorname{div}_D(f) + \operatorname{div}_{-D}(g) = (\operatorname{div}(f) + D) + (\operatorname{div}(g) - D).$$

Because the divisors $(\operatorname{div}(f) + D)$ and $(\operatorname{div}(g) - D)$ are effective, we conclude that $D = -\operatorname{div}(f)$. This means $[D] = 0$ and we obtain $f \in \Gamma(X, \mathcal{O}^*)$. $\square$

Example 1.4.1.6 Compare [169, Sec. 2]. Take the projective line $\mathbb{P}^1$, a tuple $A = (a_0, \ldots, a_r)$ of pairwise different points $a_i \in \mathbb{P}^1$ and a tuple $\mathfrak{n} = (n_0, \ldots, n_r)$ of integers $n_i \in \mathbb{Z}_{\geq 1}$. We construct a nonseparated smooth curve $\mathbb{P}^1(A, \mathfrak{n})$ mapping birationally onto $\mathbb{P}^1$ such that over each a_i lie precisely n_i points. Set

$$X_{ij} := \mathbb{P}^1 \setminus \bigcup_{k \neq i} a_k, \qquad 0 \leq i \leq r, \qquad 1 \leq j \leq n_i.$$

Gluing the X_{ij} along the common open subset $\mathbb{P}^1 \setminus \{a_0, \ldots, a_r\}$ gives an irreducible smooth prevariety $\mathbb{P}^1(A, \mathfrak{n})$ of dimension 1. The inclusion maps $X_{ij} \to \mathbb{P}^1$ define a morphism $\pi \colon \mathbb{P}^1(A, \mathfrak{n}) \to \mathbb{P}^1$, which is locally an isomorphism. Writing a_{ij} for the point in $\mathbb{P}^1(A, \mathfrak{n})$ stemming from $a_i \in X_{ij}$, we obtain the fiber over any $a \in \mathbb{P}^1$ as

$$\pi^{-1}(a) = \begin{cases} \{a_{i1}, \ldots, a_{in_i}\} & a = a_i \text{ for some } 0 \leq i \leq r, \\ \{a\} & a \neq a_i \text{ for all } 0 \leq i \leq r. \end{cases}$$

We compute the divisor class group of $\mathbb{P}^1(A, \mathfrak{n})$. Let K' denote the group of Weil divisors on $\mathbb{P}^1(A, \mathfrak{n})$ generated by the prime divisors a_{ij}. Clearly K' maps onto the divisor class group. Moreover, the group of principal divisors inside K' is

$$K_0' := K' \cap \mathrm{PDiv}(\mathbb{P}^1(A, \mathfrak{n})) = \left\{ \sum_{\substack{0 \leq i \leq r, \\ 1 \leq j \leq n_i}} c_i a_{ij}; \ c_0 + \cdots + c_r = 0 \right\}.$$

One directly checks that K' is the direct sum of K_0' and the subgroup $K \subseteq K'$ generated by $a_{01}, \ldots, a_{0n_0}$ and the $a_{i1}, \ldots, a_{in_i-1}$. Consequently, the divisor class group of $\mathbb{P}^1(A, \mathfrak{n})$ is given by

$$\mathrm{Cl}(\mathbb{P}^1(A, \mathfrak{n})) = \bigoplus_{j=1}^{n_0} \mathbb{Z} \cdot [a_{0j}] \ \oplus \ \bigoplus_{i=1}^{r} \left(\bigoplus_{j=1}^{n_i-1} \mathbb{Z} \cdot [a_{ij}] \right).$$

We are ready to determine the Cox ring of the prevariety $\mathbb{P}^1(A, \mathfrak{n})$. For every $0 \leq i \leq r$, define a monomial

$$T_i := T_{i1} \cdots T_{in_i} \in \mathbb{K}[T_{ij}; \ 0 \leq i \leq r, \ 1 \leq j \leq n_i].$$

Moreover, for every $a_i \in \mathbb{P}^1$ fix a presentation $a_i = [b_i, c_i]$ with $b_i, c_i \in \mathbb{K}$ and for every $0 \leq i \leq r - 2$ set $k = j + 1 = i + 2$ and define a trinomial

$$g_i := (b_j c_k - b_k c_j)T_i + (b_k c_i - b_i c_k)T_j + (b_i c_j - b_j c_i)T_k.$$

We claim that for $r \leq 1$ the Cox ring $\mathcal{R}(\mathbb{P}^1(A, \mathfrak{n}))$ is isomorphic to the polynomial ring $\mathbb{K}[T_{ij}]$, and for $r \geq 2$ it has a presentation

$$\mathcal{R}(\mathbb{P}^1(A, \mathfrak{n})) \cong \mathbb{K}[T_{ij}; \ 0 \leq i \leq r, \ 1 \leq j \leq n_i] \, / \, \langle g_i; \ 0 \leq i \leq r - 2 \rangle,$$

where, in both cases, the grading is given by $\deg(T_{ij}) = [a_{ij}]$. Note that all relations are homogeneous of degree

$$\deg(g_i) = [a_{i1} + \cdots + a_{in_i}] = [a_{01} + \cdots + a_{0n_0}].$$

Let us verify this claim. Set for short $X := \mathbb{P}^1(A, \mathfrak{n})$ and $Y := \mathbb{P}^1$. Let $K \subseteq \mathrm{WDiv}(X)$ be the subgroup generated by all $a_{ij} \in X$ different from $a_{1n_1}, \ldots, a_{rn_r}$, and let $L \subseteq \mathrm{WDiv}(Y)$ be the subgroup generated by $a_0 \in Y$. Then we may view the Cox rings $\mathcal{R}(X)$ and $\mathcal{R}(Y)$ as the rings of global sections of the sheaves of divisorial algebras $\mathcal{S}_X$ and $\mathcal{S}_Y$ associated with K and L. The canonical morphism $\pi \colon X \to Y$ gives rise to injective pullback homomorphisms

$$\pi^* \colon L \to K, \qquad \pi^* \colon \Gamma(Y, \mathcal{S}_Y) \to \Gamma(X, \mathcal{S}_X).$$

For any divisor $a_{ij} \in K$, let $T_{ij} \in \Gamma(X, \mathcal{S}_X)$ denote its canonical section, that is, the rational function $1 \in \Gamma(X, \mathcal{S}_{X,a_{ij}})$. Moreover, let $[z, w]$ be the homogeneous coordinates on $\mathbb{P}^1$ and consider the sections

$$S_i := \frac{b_i w - c_i z}{b_0 w - c_0 z} \in \Gamma(Y, \mathcal{S}_{Y,a_0}), \qquad 0 \le i \le r.$$

Finally, set $d_{in_i} := a_{01} + \cdots + a_{0n_0} - a_{i1} - \cdots - a_{in_i-1} \in K$ and define homogeneous sections

$$T_{in_i} := \pi^* S_i (T_{i1} \cdots T_{in_i-1})^{-1} \in \Gamma(X, \mathcal{S}_{X,d_{in_i}}), \qquad 1 \le i \le r.$$

We show that the sections T_{ij}, where $0 \le i \le r$ and $1 \le j \le n_i$, generate the Cox ring $\mathcal{R}(X)$. Note that we have

$$\mathrm{div}_{a_{ij}}(T_{ij}) = a_{ij}, \qquad \mathrm{div}_{d_{in_i}}(T_{in_i}) = a_{in_i}.$$

Consider $D \in K$ and $h \in \Gamma(X, \mathcal{S}_D)$. If there occurs an a_{ij} in $\mathrm{div}_D(h)$, then we may divide h in $\Gamma(X, \mathcal{S})$ by the corresponding T_{ij}, use Proposition 1.3.3.4 (i). Doing this as long as possible, we arrive at some $h' \in \Gamma(X, \mathcal{S}_{D'})$ such that $\mathrm{div}_{D'}(h')$ has no components a_{ij}. But then D' is a pullback divisor and hence h' is contained in

$$\pi^*(\Gamma(Y, \mathcal{S}_Y)) = \mathbb{K}[\pi^* S_0, \pi^* S_1] = \mathbb{K}[T_{01} \cdots T_{0n_0}, T_{11} \cdots T_{1n_1}].$$

Finally, we have to determine the relations among the sections $T_{ij} \in \Gamma(X, \mathcal{S}_X)$. For this, we first note that among the $S_i \in \Gamma(Y, \mathcal{S}_Y)$ we have the relations

$$(b_j c_k - b_k c_j) S_i + (b_k c_i - b_i c_k) S_j + (b_i c_j - b_j c_i) S_k = 0,$$

$$j = i+1, \quad k = i+2.$$

Given any nontrivial homogeneous relation $F = \alpha_1 F_1 + \cdots + \alpha_l F_l = 0$ with $\alpha_i \in \mathbb{K}$ and pairwise different monomials F_i in the T_{ij}, we achieve, by subtracting suitable multiples of pullbacks of the preceding relations, a homogeneous relation

$$F' = \alpha'_1 F''_1 \pi^* S_0^{k_1} \pi^* S_1^{l_1} + \cdots + \alpha'_m F''_m \pi^* S_0^{k_m} \pi^* S_1^{l_m} = 0$$

with pairwise different monomials F''_j, none of which has any factor $\pi^* S_i$. We show that F' must be trivial. Consider the multiplicative group M of Laurent monomials in the T_{ij} and the degree map

$$M \to K, \qquad T_{ij} \mapsto \deg(T_{ij}) = \begin{cases} a_{ij}, & i = 0 \text{ or } j \leq n_i - 1, \\ d_{in_i}, & i \geq 1 \text{ and } j = n_i. \end{cases}$$

The kernel of this degree map is generated by the Laurent monomials $\pi^* S_0 / \pi^* S_i$, where $1 \leq i \leq r$. The monomials of F' are all of the same K-degree and thus any two of them differ by a product of (integral) powers of the $\pi^* S_i$. It follows that all the F''_j coincide. Thus, we obtain the relation

$$\alpha'_1 \pi^* S_0^{k_1} \pi^* S_1^{l_1} + \cdots + \alpha'_m \pi^* S_0^{k_m} \pi^* S_1^{l_m} = 0.$$

This relation descends to a relation in $\Gamma(Y, \mathcal{S}_Y)$, which is the polynomial ring $\mathbb{K}[S_0, S_1]$. Consequently, we obtain $\alpha'_1 = \cdots = \alpha'_m = 0$.

1.4.2 Torsion in the divisor class group

We extend the definition of Cox sheaf and Cox ring to irreducible, normal prevarieties X having a finitely generated divisor class group $\mathrm{Cl}(X)$ with torsion. The idea is to take a subgroup $K \subseteq \mathrm{WDiv}(X)$ projecting onto $\mathrm{Cl}(X)$, to consider its associated sheaf of divisorial algebras $\mathcal{S}$ and to identify in a systematic manner homogeneous components $\mathcal{S}_D$ and $\mathcal{S}_{D'}$, whenever D and D' are linearly equivalent. Again we work over an algebraically closed field $\mathbb{K}$ of characteristic zero.

Construction 1.4.2.1 Let X be an irreducible, normal prevariety with $\Gamma(X, \mathcal{O}^*) = \mathbb{K}^*$ and finitely generated divisor class group $\mathrm{Cl}(X)$. Fix a subgroup $K \subseteq \mathrm{WDiv}(X)$ such that the map $c \colon K \to \mathrm{Cl}(X)$ sending $D \in K$ to its class $[D] \in \mathrm{Cl}(X)$ is surjective. Let $K^0 \subseteq K$ be the kernel of c, and let $\chi \colon K^0 \to \mathbb{K}(X)^*$ be a character, that is, a group homomorphism, with

$$\mathrm{div}(\chi(E)) = E \qquad \text{for all } E \in K^0.$$

Let $\mathcal{S}$ be the sheaf of divisorial algebras associated with K and denote by $\mathcal{I}$ the sheaf of ideals of $\mathcal{S}$ locally generated by the sections $1 - \chi(E)$, where 1 is homogeneous of degree zero, E runs through K^0, and $\chi(E)$ is homogeneous of degree $-E$. The *Cox sheaf* associated with K and χ is the quotient sheaf

$\mathcal{R} := \mathcal{S}/\mathcal{I}$ together with the $\mathrm{Cl}(X)$-grading

$$\mathcal{R} = \bigoplus_{[D]\in\mathrm{Cl}(X)} \mathcal{R}_{[D]}, \qquad \mathcal{R}_{[D]} := \pi\left(\bigoplus_{D'\in c^{-1}([D])} \mathcal{S}_{D'}\right),$$

where $\pi : \mathcal{S} \to \mathcal{R}$ denotes the projection. The Cox sheaf $\mathcal{R}$ is a quasicoherent sheaf of $\mathrm{Cl}(X)$-graded $\mathcal{O}_X$-algebras. The *Cox ring* is the ring of global sections

$$\mathcal{R}(X) := \bigoplus_{[D]\in\mathrm{Cl}(X)} \mathcal{R}_{[D]}(X), \qquad \mathcal{R}_{[D]}(X) := \Gamma(X, \mathcal{R}_{[D]}).$$

For any open set $U \subseteq X$, the canonical homomorphism $\Gamma(U, \mathcal{S})/\Gamma(U, \mathcal{I}) \to \Gamma(U, \mathcal{R})$ is an isomorphism. In particular, we have

$$\mathcal{R}(X) \cong \Gamma(X, \mathcal{S})/\Gamma(X, \mathcal{I}).$$

All the claims made in this construction will be verified as separate lemmas in the next subsection. The assumption $\Gamma(X, \mathcal{O}^*) = \mathbb{K}^*$ is crucial for the following uniqueness statement on Cox sheaves and rings.

Proposition 1.4.2.2 *Let X be an irreducible, normal prevariety with $\Gamma(X, \mathcal{O}^*) = \mathbb{K}^*$ and finitely generated divisor class group $\mathrm{Cl}(X)$. If K, χ and K', χ' are data as in Construction 1.4.2.1, then there is a graded isomorphism of the associated Cox sheaves.*

This is also proven in the next subsection. The construction of Cox sheaves (and thus also Cox rings) of a prevariety X can be made canonical by fixing a suitable point $x \in X$.

Construction 1.4.2.3 Let X be an irreducible, normal prevariety with $\Gamma(X, \mathcal{O}^*) = \mathbb{K}^*$ and finitely generated divisor class group $\mathrm{Cl}(X)$. Fix a point $x \in X$ with factorial local ring $\mathcal{O}_{X,x}$. For the subgroup

$$K^x := \{D \in \mathrm{WDiv}(X); \ x \notin \mathrm{Supp}(D)\}$$

let $\mathcal{S}^x$ be the associated sheaf of divisorial algebras and let $K^{x,0} \subseteq K^x$ denote the subgroup consisting of principal divisors. Then, for each $E \in K^{x,0}$, there is a unique section $f_E \in \Gamma(X, \mathcal{S}_{-E})$, which is defined near x and satisfies

$$\mathrm{div}(f_E) = E, \qquad f_E(x) = 1.$$

The map $\chi^x : K^x \to \mathbb{K}(X)^*$ sending E to f_E is a character as in Construction 1.4.2.1. We call the Cox sheaf $\mathcal{R}^x$ associated with K^x and χ^x the *canonical Cox sheaf of the pointed space* (X, x).

Example 1.4.2.4 (An affine surface with torsion in the divisor class group) Consider the two-dimensional affine quadric

$$X := V(\mathbb{K}^3; T_1 T_2 - T_3^2) \subseteq \mathbb{K}^3.$$

We have the functions $f_i := T_i|_X$ on X and with the prime divisors $D_1 := V(X; f_1)$ and $D_2 := V(X; f_2)$ on X, we have

$$\operatorname{div}(f_1) = 2D_1, \qquad \operatorname{div}(f_2) = 2D_2, \qquad \operatorname{div}(f_3) = D_1 + D_2.$$

The divisor class group $\operatorname{Cl}(X)$ is of order 2; it is generated by $[D_1]$. For $K := \mathbb{Z}D_1$, let $\mathcal{S}$ denote the associated sheaf of divisorial algebras. Consider the sections

$$g_1 := 1 \in \Gamma(X, \mathcal{S}_{D_1}), \qquad g_2 := f_3 f_1^{-1} \in \Gamma(X, \mathcal{S}_{D_1}),$$

$$g_3 := f_1^{-1} \in \Gamma(X, \mathcal{S}_{2D_1}), \qquad g_4 := f_1 \in \Gamma(X, \mathcal{S}_{-2D_1}).$$

Then g_1, g_2 generate $\Gamma(X, \mathcal{S}_{D_1})$ as a $\Gamma(X, \mathcal{S}_0)$-module, and g_3, g_4 are inverse to each other. Moreover, we have

$$f_1 = g_1^2 g_4, \qquad f_2 = g_2^2 g_4, \qquad f_3 = g_1 g_2 g_4.$$

Thus, g_1, g_2, g_3, and g_4 generate the $\mathbb{K}$-algebra $\Gamma(X, \mathcal{S})$. Setting $\deg(Z_i) := \deg(g_i)$, we obtain a K-graded epimorphism

$$\mathbb{K}[Z_1, Z_2, Z_3^{\pm 1}] \to \Gamma(X, \mathcal{S}), \qquad Z_1 \mapsto g_1, \; Z_2 \mapsto g_2, \; Z_3 \mapsto g_3,$$

which, by dimension reasons, is even an isomorphism. The kernel of the projection $K \to \operatorname{Cl}(X)$ is $K^0 = 2\mathbb{Z}D_1$ and a character as in Construction 1.4.2.1 is

$$\chi \colon K^0 \to \mathbb{K}(X)^*, \qquad 2nD_1 \mapsto f_1^n.$$

The ideal $\mathcal{I}$ is generated by $1 - f_1$, where $f_1 \in \Gamma(X, \mathcal{S}_{-2D_1})$; see Remark 1.4.3.2. Consequently, the Cox ring of X is given as

$$\mathcal{R}(X) \cong \Gamma(X, \mathcal{S})/\Gamma(X, \mathcal{I}) \cong \mathbb{K}[Z_1, Z_2, Z_3^{\pm 1}]/\langle 1 - Z_3^{-1}\rangle \cong \mathbb{K}[Z_1, Z_2],$$

where the $\operatorname{Cl}(X)$-grading on the polynomial ring $\mathbb{K}[Z_1, Z_2]$ is given by $\deg(Z_1) = \deg(Z_2) = [D_1]$.

1.4.3 Well-definedness

Here we prove the claims made in Construction 1.4.2.1 and Proposition 1.4.2.2. In particular, we show that, up to isomorphy, Cox sheaf and Cox ring do not depend on the choices made in their construction. The ground field $\mathbb{K}$ is algebraically closed and of characteristic zero.

Lemma 1.4.3.1 *Situation as in Construction 1.4.2.1. Consider the* $\operatorname{Cl}(X)$-*grading of the sheaf $\mathcal{S}$ defined by*

$$\mathcal{S} = \bigoplus_{[D] \in \operatorname{Cl}(X)} \mathcal{S}_{[D]}, \qquad \mathcal{S}_{[D]} := \bigoplus_{D' \in c^{-1}([D])} \mathcal{S}_{D'}.$$

Given $f \in \Gamma(U, \mathcal{I})$ and $D \in K$, the $\mathrm{Cl}(X)$-homogeneous component $f_{[D]} \in \Gamma(U, \mathcal{S}_{[D]})$ of f has a unique representation

$$f_{[D]} = \sum_{E \in K^0} (1 - \chi(E)) f_E, \quad \text{where } f_E \in \Gamma(U, \mathcal{S}_D) \text{ and } \chi(E) \in \Gamma(U, \mathcal{S}_{-E}).$$

In particular, the sheaf $\mathcal{I}$ of ideals is $\mathrm{Cl}(X)$-homogeneous. Moreover, if $f \in \Gamma(U, \mathcal{I})$ is K-homogeneous, then it is the zero section.

Proof To obtain uniqueness of the representation of $f_{[D]}$, observe that for every $0 \neq E \in K^0$, the product $-\chi(E) f_E$ is the K-homogeneous component of degree $D - E$ of $f_{[D]}$. We show existence. By definition of the sheaf of ideals $\mathcal{I}$, every germ $f_x \in \mathcal{I}_x$ can on a suitable neighborhood U_x be represented by a section

$$g = \sum_{E \in K^0} (1 - \chi(E)) g_E, \quad \text{where } g_E \in \Gamma(U_x, \mathcal{S}).$$

Collecting the $\mathrm{Cl}(X)$-homogeneous parts on the right-hand side represents the $\mathrm{Cl}(X)$-homogeneous part $h \in \Gamma(U_x, \mathcal{S}_{[D]})$ of degree $[D]$ of $g \in \Gamma(U_x, \mathcal{S})$ as follows:

$$h = \sum_{E \in K^0} (1 - \chi(E)) h_E, \quad \text{where } h_E \in \Gamma(U_x, \mathcal{S}_{[D]}).$$

Note that we have $h \in \Gamma(U_x, \mathcal{I})$ and h represents $f_{[D], x}$. Now, developing each $h_E \in \Gamma(U_x, \mathcal{S}_{[D]})$ according to the K-grading gives representations

$$h_E = \sum_{D' \in D + K^0} h_{E, D'}, \quad \text{where } h_{E, D'} \in \Gamma(U_x, \mathcal{S}_{D'}).$$

The section $h'_{E, D'} := \chi(D' - D) h_{E, D'}$ is K-homogeneous of degree D, and we have the identity

$$(1 - \chi(E)) h_{E, D'} = (1 - \chi(E + D - D')) h'_{E, D'} - (1 - \chi(D - D')) h'_{E, D'}.$$

Plugging this into the representation of h establishes the desired representation of $f_{[D]}$ locally. By uniqueness, we may glue the local representations. $\qquad\square$

Remark 1.4.3.2 Consider the situation of Construction 1.4.2.1. Then, for any two divisors $E, E' \in K^0$, one has the identities

$$1 - \chi(E + E') = (1 - \chi(E)) + (1 - \chi(E')) \chi(E),$$

$$1 - \chi(-E) = (1 - \chi(E))(-\chi(-E)).$$

Together with Lemma 1.4.3.1, this implies that for any basis $E_1, \ldots, E_s$ of K^0 and any open $U \subseteq X$, the ideal $\Gamma(U, \mathcal{I})$ is generated by $1 - \chi(E_i)$, where $1 \leq i \leq s$.

Lemma 1.4.3.3 *Consider the situation of Construction 1.4.2.1. If $f \in \Gamma(U, \mathcal{S})$ is $\mathrm{Cl}(X)$-homogeneous of degree $[D]$ for some $D \in K$, then there is a K-homogeneous $f' \in \Gamma(U, \mathcal{S})$ of degree D with $f - f' \in \Gamma(U, \mathcal{I})$.*

Proof Writing the $\mathrm{Cl}(X)$-homogeneous f as a sum of K-homogeneous functions $f_{D'}$, we obtain the assertion by means of the following trick:

$$f = \sum_{D' \in D + K^0} f_{D'} = \sum_{D' \in D + K^0} \chi(D' - D) f_{D'} + \sum_{D' \in D + K^0} (1 - \chi(D' - D)) f_{D'}.$$

$\square$

Lemma 1.4.3.4 *Consider the situation of Construction 1.4.2.1. Then, for every $D \in K$, we have an isomorphism of sheaves $\pi|_{\mathcal{S}_D} : \mathcal{S}_D \to \mathcal{R}_{[D]}$.*

Proof Lemma 1.4.3.1 shows that the homomorphism $\pi|_{\mathcal{S}_D}$ is stalkwise injective and from Lemma 1.4.3.3 we infer that it is stalkwise surjective. $\square$

Lemma 1.4.3.5 *Consider the situation of Construction 1.4.2.1. Then, for every open subset $U \subseteq X$, we have a canonical isomorphism*

$$\Gamma(U, \mathcal{S}) / \Gamma(U, \mathcal{I}) \cong \Gamma(U, \mathcal{S}/\mathcal{I}).$$

Proof The canonical map $\psi : \Gamma(U, \mathcal{S}) / \Gamma(U, \mathcal{I}) \to \Gamma(U, \mathcal{S}/\mathcal{I})$ is injective. To see that it is as well surjective, let $h \in \Gamma(U, \mathcal{S}/\mathcal{I})$ be given. Then there are a covering of U by open subsets U_i and sections $g_i \in \Gamma(U_i, \mathcal{S})$ such that $h|_{U_i} = \psi(g_i)$ holds and $g_j - g_i$ belongs to $\Gamma(U_i \cap U_j, \mathcal{I})$. Consider the $\mathrm{Cl}(X)$-homogeneous parts $g_{i,[D]} \in \Gamma(U_i, \mathcal{S}_{[D]})$ of g_i. By Lemma 1.4.3.1, the ideal sheaf $\mathcal{I}$ is homogeneous and thus also $g_{j,[D]} - g_{i,[D]}$ belongs to $\Gamma(U_i \cap U_j, \mathcal{I})$. Moreover, Lemma 1.4.3.3 provides K-homogeneous $f_{i,D}$ with $f_{i,D} - g_{i,[D]}$ in $\Gamma(U_i, \mathcal{I})$. The differences $f_{j,D} - f_{i,D}$ lie in $\Gamma(U_i \cap U_j, \mathcal{I})$ and hence, by Lemma 1.4.3.1, vanish. Thus, the $f_{i,D}$ fit together to K-homogeneous sections $f_D \in \Gamma(U, \mathcal{S})$. By construction, $f = \sum f_D$ satisfies $\psi(f) = h$. $\square$

Proof of Proposition 1.4.2.2 In a first step, we reduce to Cox sheaves arising from finitely generated subgroups of $\mathrm{WDiv}(X)$. So, let $K \subseteq \mathrm{WDiv}(X)$ and $\chi : K^0 \to \mathbb{K}(X)^*$ be any data as in Construction 1.4.2.1. Choose a finitely generated subgroup $K_1 \subseteq K$ projecting onto $\mathrm{Cl}(X)$. Restricting χ gives a character $\chi_1 : K_1^0 \to \mathbb{K}(X)^*$. The inclusion $K_1 \to K$ defines an injection $\mathcal{S}_1 \to \mathcal{S}$ sending the ideal $\mathcal{I}_1$ defined by χ_1 to the ideal $\mathcal{I}$ defined by χ. This gives a $\mathrm{Cl}(X)$-graded injection $\mathcal{R}_1 \to \mathcal{R}$ of the Cox sheaves associated with K_1, χ_1 and K, χ respectively. Lemma 1.4.3.3 shows that every $\mathrm{Cl}(X)$-homogeneous section of $\mathcal{R}$ can be represented by a K_1-homogeneous section of $\mathcal{S}$. Thus, $\mathcal{R}_1 \to \mathcal{R}$ is also surjective.

Next we show that for a fixed finitely generated $K \subseteq \mathrm{WDiv}(X)$, any two characters $\chi, \chi' \colon K^0 \to \mathbb{K}(X)^*$ as in Construction 1.4.2.1 give rise to isomorphic Cox sheaves $\mathcal{R}'$ and $\mathcal{R}$. For this, note that the product $\chi^{-1}\chi'$ sends K^0 to $\Gamma(X, \mathcal{O}^*)$. Using $\Gamma(X, \mathcal{O}^*) = \mathbb{K}^*$, we may extend $\chi^{-1}\chi'$ to a homomorphism $\vartheta \colon K \to \Gamma(X, \mathcal{O}^*)$ and obtain a graded automorphism (α, id) of $\mathcal{S}$ by

$$\alpha_D \colon \mathcal{S}_D \to \mathcal{S}_D, \qquad f \mapsto \vartheta(D)f.$$

By construction, this automorphism sends the ideal $\mathcal{I}'$ to the ideal $\mathcal{I}$ and induces a graded isomorphism from $\mathcal{S}/\mathcal{I}'$ onto $\mathcal{S}/\mathcal{I}$.

Now consider two finitely generated subgroups $K, K' \subseteq \mathrm{WDiv}(X)$, both projecting onto $\mathrm{Cl}(X)$. Then we find a homomorphism $\widetilde{\alpha} \colon K \to K'$ such that the following diagram is commutative:

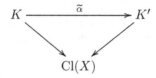

This homomorphism $\widetilde{\alpha} \colon K \to K'$ must be of the form $\widetilde{\alpha}(D) = D - \mathrm{div}(\eta(D))$ with a homomorphism $\eta \colon K \to \mathbb{K}(X)^*$. Choose a character $\chi' \colon K'^0 \to \mathbb{K}(X)^*$ as in Construction 1.4.2.1. Then, for $D \in K^0$, we have

$$D - \mathrm{div}(\eta(D)) = \widetilde{\alpha}(D) = \mathrm{div}(\chi'(\widetilde{\alpha}(D))).$$

Thus, D equals the divisor of the function $\chi(D) := \chi'(\widetilde{\alpha}(D))\eta(D)$. This defines a character $\chi \colon K^0 \to \mathbb{K}(X)^*$. Altogether, we obtain a morphism $(\alpha, \widetilde{\alpha})$ of the sheaves of divisorial algebras $\mathcal{S}$ and $\mathcal{S}'$ associated with K and K' by

$$\alpha_D \colon \mathcal{S}_D \to \mathcal{S}'_{\widetilde{\alpha}(D)}, \qquad f \mapsto \eta(D)f.$$

By construction, it sends the ideal $\mathcal{I}$ defined by χ to the ideal $\mathcal{I}'$ defined by χ'. Using Lemma 1.4.3.4, we see that the induced homomorphism $\mathcal{R} \to \mathcal{R}'$ is an isomorphism on the homogeneous components and thus it is an isomorphism. $\qquad\square$

1.4.4 Examples

For an irreducible, normal prevariety X with a free finitely generated divisor class group, we obtained in Proposition 1.4.1.5 that the Cox ring is a unique factorization domain having $\Gamma(X, \mathcal{O}^*)$ as its units. Here we provide two examples showing that these statements need not hold any more if there is torsion in the divisor class group. As usual, $\mathbb{K}$ is an algebraically closed field of characteristic zero.

Example 1.4.4.1 (An affine surface with nonfactorial Cox ring)   Consider the smooth affine surface

$$Z := V(\mathbb{K}^3; T_1^2 - T_2 T_3 - 1).$$

We claim that $\Gamma(Z, \mathcal{O}^*) = \mathbb{K}^*$ and $\mathrm{Cl}(Z) \cong \mathbb{Z}$ hold. To see this, consider $f_i := T_i|_Z$ and the prime divisors

$$D_+ := V(Z; f_1 - 1, f_2) = \{1\} \times \{0\} \times \mathbb{K},$$

$$D_- := V(Z; f_1 + 1, f_2) = \{-1\} \times \{0\} \times \mathbb{K}.$$

Then we have $\mathrm{div}(f_2) = D_+ + D_-$. In particular, D_+ is linearly equivalent to $-D_-$. Moreover, we have

$$Z \setminus \mathrm{Supp}(\mathrm{div}(f_2)) = Z_{f_2} \cong \mathbb{K}^* \times \mathbb{K}.$$

This gives $\Gamma(Z, \mathcal{O})^* = \mathbb{K}^*$, and shows that $\mathrm{Cl}(Z)$ is generated by the class $[D_+]$. Now suppose that $n[D_+] = 0$ holds for some $n > 0$. Then we have $nD_+ = \mathrm{div}(f)$ with $f \in \Gamma(Z, \mathcal{O})$ and $f_2^n = fh$ holds with some $h \in \Gamma(Z, \mathcal{O})$ satisfying $\mathrm{div}(h) = nD_-$. Look at the $\mathbb{Z}$-grading of $\Gamma(Z, \mathcal{O})$ given by

$$\deg(f_1) = 0, \qquad \deg(f_2) = 1, \qquad \deg(f_3) = -1.$$

Any element of positive degree is a multiple of f_2. It follows that in the decomposition $f_2^n = fh$ one of the factors f or h must be a multiple of f_2, a contradiction. This shows that $\mathrm{Cl}(Z)$ is freely generated by $[D_+]$.

Now consider the involution $Z \to Z$ sending z to $-z$ and let $\pi: Z \to X$ denote the quotient of the corresponding free $\mathbb{Z}/2\mathbb{Z}$-action. We claim that $\mathrm{Cl}(X)$ is isomorphic to $\mathbb{Z}/2\mathbb{Z}$ and is generated by the class of $D := \pi(D_+)$. Indeed, the subset

$$X \setminus \mathrm{Supp}(D) = \pi(Z_{f_2}) \cong \mathbb{K}^* \times \mathbb{K}$$

is factorial and $2D$ equals $\mathrm{div}(f_2^2)$. Moreover, the divisor D is not principal, because $\pi^*(D) = D_+ + D_-$ is not the divisor of a $\mathbb{Z}/2\mathbb{Z}$-invariant function on Z. This verifies our claim. Moreover, we have $\Gamma(X, \mathcal{O}^*) = \mathbb{K}^*$.

To determine the Cox ring of X, take $K = \mathbb{Z}D \subseteq \mathrm{WDiv}(X)$, and let $\mathcal{S}$ denote the associated sheaf of divisorial algebras. Then, as $\Gamma(X, \mathcal{S}_0)$-modules, $\Gamma(X, \mathcal{S}_D)$ and $\Gamma(X, \mathcal{S}_{-D})$ are generated by the sections

$$a_1 := 1, \ a_2 := f_1 f_2^{-1}, \ a_3 := f_2^{-1} f_3 \in \Gamma(X, \mathcal{S}_D),$$

$$b_1 := f_1 f_2, \ b_2 := f_2^2, \ b_3 := f_2 f_3 \in \Gamma(X, \mathcal{S}_{-D}).$$

Thus, using the fact that $f_2^{\pm 2}$ define invertible elements of degree $\mp 2D$, we see that $a_1, a_2, a_3, b_1, b_2, b_3$ generate the algebra $\Gamma(X, \mathcal{S})$. Now, take the character $\chi: K^0 \to \mathbb{K}(X)^*$ sending $2nD$ to f_2^{2n}. Then, by Remark 1.4.3.2, the associated ideal $\Gamma(X, \mathcal{I})$ is generated by $1 - f_2^2$. The generators of the factor algebra $\Gamma(X, \mathcal{S}) / \Gamma(X, \mathcal{I})$ are

$$Z_1 = a_2 + \mathcal{I} = b_1 + \mathcal{I}, \quad Z_2 = a_1 + \mathcal{I} = b_2 + \mathcal{I}, \quad Z_3 = a_3 + \mathcal{I} = b_3 + \mathcal{I}.$$

The defining relation is $Z_1^2 - Z_2 Z_3 = 1$. Thus the Cox ring $\mathcal{R}(X)$ is isomorphic to $\Gamma(Z, \mathcal{O})$. In particular, it is not a factorial ring.

Example 1.4.4.2 (An affine surface with only constant invertible global functions but nonconstant invertible elements in the Cox ring)   Consider the surface

$$X := V(\mathbb{K}^3; \ T_1 T_2 T_3 - T_1^2 - T_2^2 - T_3^2 + 4).$$

This is the quotient space of the torus $\mathbb{T}^2 := (\mathbb{K}^*)^2$ with respect to the $\mathbb{Z}/2\mathbb{Z}$-action defined by the involution $t \mapsto t^{-1}$; the quotient map is explicitly given as

$$\pi : \mathbb{T}^2 \ \to \ X, \qquad t \ \mapsto \ (t_1 + t_1^{-1}, t_2 + t_2^{-1}, t_1 t_2 + t_1^{-1} t_2^{-1}).$$

Because every $\mathbb{Z}/2\mathbb{Z}$-invariant invertible function on $\mathbb{T}^2$ is constant, we have $\Gamma(X, \mathcal{O}^*) = \mathbb{K}^*$. Moreover, using [190, Prop. 5.1], one verifies

$$\mathrm{Cl}(X) \ \cong \ \mathbb{Z}/2\mathbb{Z} \oplus \mathbb{Z}/2\mathbb{Z} \oplus \mathbb{Z}/2\mathbb{Z}, \qquad\qquad \mathrm{Pic}(X) \ = \ 0.$$

Let us see that the Cox ring $\mathcal{R}(X)$ has nonconstant invertible elements. Set $f_i := T_i|_X$ and consider the divisors

$$D_\pm := V(X; f_1 \pm 2, \ f_2 \pm f_3), \qquad D := D_+ + D_-.$$

Then, using the relations $(f_1 \pm 2)(f_2 f_3 - f_1 \pm 2) = (f_2 \pm f_3)^2$, one verifies $\mathrm{div}(f_1 \pm 2) = 2D_\pm$. Consequently, we obtain

$$2D = \mathrm{div}(f_1^2 - 4).$$

Moreover, D is not principal, because otherwise $f_1^2 - 4$ must be a square and hence also $\pi^*(f_1^2 - 4)$ is a square, which is impossible due to

$$\pi^*(f_1^2 - 4) \ = \ t_1^2 + t_1^{-2} - 4 \ = \ (t_1 + t_1^{-1} + 2)(t_1 + t_1^{-1} - 2).$$

Now choose Weil divisors D_i on X such that D, D_2, D_3 form a basis for a group $K \subseteq \mathrm{WDiv}(X)$ projecting onto $\mathrm{Cl}(X)$, and let $\mathcal{S}$ be the associated sheaf of divisorial algebras. As usual, let $K^0 \subseteq K$ be the subgroup consisting of principal divisors and fix a character $\chi : K^0 \to \mathbb{K}(X)^*$ with $\chi(2D) = f_1^2 - 4$. By Remark 1.4.3.2, the associated ideal $\Gamma(X, \mathcal{I})$ in $\Gamma(X, \mathcal{S})$ is generated by

$$1 - \chi(2D), \quad 1 - \chi(2D_2), \quad 1 - \chi(2D_3),$$

where $\chi(2D) = f_1^2 - 4$ lies in $\Gamma(X, \mathcal{S}_{-2D})$. Now consider $f_1 \in \Gamma(X, \mathcal{S}_0)$ and the canonical section $1_D \in \Gamma(X, \mathcal{S}_D)$. Then we have

$$(f_1 + 1_D)(f_1 - 1_D) = f_1^2 - 1_D^2 = 4 - 1_D^2 \cdot (1 - \chi(2D)) \in \mathbb{K}^* + \Gamma(X, \mathcal{I}).$$

Consequently, the section $f_1 + 1_D \in \Gamma(X, \mathcal{S})$ defines a unit in $\Gamma(X, \mathcal{R})$. Note that $f_1 + 1_D$ is not $\mathrm{Cl}(X)$-homogeneous.

1.5 Algebraic properties of the Cox ring

1.5.1 Integrity and normality

The aim is to prove that Cox rings are integral normal rings; recall that an integral ring R is called *normal* if it is integrally closed in its quotient field, that is, for every $h \in \mathrm{Quot}(R)$ one has $h \in R$ if $f(h) = 0$ holds with some monic polynomial $f \in R[T]$. As before, we work over an algebraically closed field $\mathbb{K}$ of characteristic zero.

Theorem 1.5.1.1 *Let X be an irreducible, normal prevariety with only constant invertible global functions, finitely generated divisor class group, and Cox sheaf $\mathcal{R}$. Then, for every open $U \subseteq X$, the ring $\Gamma(U, \mathcal{R})$ is integral and normal.*

The proof is based on the geometric construction 1.5.1.4, which is also used later and therefore occurs separately. We begin with two preparatory observations.

Lemma 1.5.1.2 *Consider the situation of Construction 1.4.2.1. For any two open subsets $V \subseteq U \subseteq X$ such that $U \setminus V$ is of codimension at least 2 in U, one has the restriction isomorphism*

$$\Gamma(U, \mathcal{R}) \rightarrow \Gamma(V, \mathcal{R}).$$

In particular, the algebra $\Gamma(U, \mathcal{R})$ equals the algebra $\Gamma(U_{\mathrm{reg}}, \mathcal{R})$, where $U_{\mathrm{reg}} \subseteq U$ denotes the set of smooth points.

Proof According to Remark 1.3.1.3, the restriction $\Gamma(U, \mathcal{S}) \rightarrow \Gamma(V, \mathcal{S})$ is an isomorphism. Lemma 1.4.3.1 ensures that $\Gamma(U, \mathcal{I})$ is mapped isomorphically onto $\Gamma(V, \mathcal{I})$ under this isomorphism. By Lemma 1.4.3.5, we have $\Gamma(U, \mathcal{R}) = \Gamma(U, \mathcal{S}) / \Gamma(U, \mathcal{I})$ and $\Gamma(V, \mathcal{R}) = \Gamma(V, \mathcal{S}) / \Gamma(V, \mathcal{I})$, which gives the assertion. $\square$

Lemma 1.5.1.3 *Consider the situation of Construction 1.4.2.1. Then for every open $U \subseteq X$, the ideal $\Gamma(U, \mathcal{I}) \subseteq \Gamma(U, \mathcal{S})$ is radical.*

Proof By Lemma 1.4.3.5, the ideal $\Gamma(U, \mathcal{I})$ is radical if and only if the algebra $\Gamma(U, \mathcal{R})$ has no nilpotent elements. Proposition 1.4.2.2 thus allows us to assume that $\mathcal{S}$ arises from a finitely generated group K. Moreover, by Remark 1.3.1.3, we may assume that X is smooth and it suffices to verify the assertion for affine $U \subseteq X$. We consider $\widetilde{U} = \operatorname{Spec} \Gamma(U, \mathcal{S})$ and the zero set $\widehat{U} \subseteq \widetilde{U}$ of $\Gamma(U, \mathcal{I})$. Note that $\widehat{U}$ is invariant under the action of the quasitorus $H_X = \operatorname{Spec} \mathbb{K}[\operatorname{Cl}(X)]$ on $\widetilde{U}$ given by the $\operatorname{Cl}(X)$-grading.

Now, let $f \in \Gamma(U, \mathcal{S})$ with $f^n \in \Gamma(U, \mathcal{I})$ for some $n > 0$. Then f and thus also every $\operatorname{Cl}(X)$-homogeneous component $f_{[D]}$ of f vanishes along $\widehat{U}$. Consequently, $f_{[D]}^m \in \Gamma(U, \mathcal{I})$ holds for some $m > 0$. By Lemma 1.4.3.3, we may write $f_{[D]} = f_D + g$ with $f_D \in \Gamma(U, \mathcal{S}_D)$ and $g \in \Gamma(U, \mathcal{I})$. We obtain $f_D^m \in \Gamma(U, \mathcal{I})$. By Lemma 1.4.3.1, this implies $f_D^m = 0$ and thus $f_D = 0$, which in turn gives $f_{[D]} \in \Gamma(U, \mathcal{I})$ and hence $f \in \Gamma(U, \mathcal{I})$. $\qquad\square$

Construction 1.5.1.4 Consider the situation of Construction 1.4.2.1. Assume that $K \subseteq \operatorname{WDiv}(X)$ is finitely generated and X is smooth. Consider $\widetilde{X} :=$ $\operatorname{Spec}_X \mathcal{S}$ with the action of the torus $H := \operatorname{Spec} \mathbb{K}[K]$ and the geometric quotient $p\colon \widetilde{X} \to X$ as in Construction 1.3.2.4. Then, with $\widehat{X} := V(\mathcal{I})$ and $H_X := \operatorname{Spec} \mathbb{K}[\operatorname{Cl}(X)]$, we have a commutative diagram

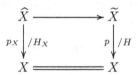

The prevariety $\widehat{X}$ is smooth, and, if X is of affine intersection, then it is quasiaffine. The quasitorus $H_X \subseteq H$ acts freely on $\widehat{X}$ and $p_X\colon \widehat{X} \to X$ is a geometric quotient for this action; in particular, it is an étale H_X-principal bundle. Moreover, we have a canonical isomorphism of sheaves

$$\mathcal{R} \cong (p_X)_*(\mathcal{O}_{\widehat{X}}).$$

Proof With the restriction $p_X\colon \widehat{X} \to X$ of $p\colon \widetilde{X} \to X$ we obviously obtain a commutative diagram as previously. Moreover, Lemma 1.5.1.3 gives us $\mathcal{R} \cong q_*(\mathcal{O}_{\widehat{X}})$. Because the ideal $\mathcal{I}$ is $\operatorname{Cl}(X)$-homogeneous, the quasitorus $H_X \subseteq H$ leaves $\widehat{X}$ invariant. Moreover, we see that $p_X\colon \widehat{X} \to X$ is a good quotient for this action, because we have the canonical isomorphisms

$$(p_X)_*(\mathcal{O}_{\widehat{X}})_0 \cong \mathcal{R}_0 \cong \mathcal{O}_X \cong \mathcal{S}_0 \cong p_*(\mathcal{O}_{\widetilde{X}})_0.$$

Freeness of the H_X-action on $\widehat{X}$ is due to the fact that H_X acts as a subgroup of the freely acting H; see Remark 1.3.2.7. As a consequence, we see that $p_X\colon \widehat{X} \to X$ is a geometric quotient. Luna's slice theorem [210] gives commutative diagrams

$$
\begin{array}{ccc}
H_X \times S & \longrightarrow & p_X^{-1}(U) \subseteq \widehat{X} \\
{\scriptstyle \mathrm{pr}_S} \downarrow & \quad {\scriptstyle p_X} \downarrow & \quad\quad \downarrow {\scriptstyle p_X} \\
S & \longrightarrow & U \quad \subseteq \quad X
\end{array}
$$

where $U \subseteq X$ are open sets covering X and the horizontal arrows are étale morphisms. By [219, Prop. I.3.17], étale morphisms preserve smoothness and thus $\widehat{X}$ inherits smoothness from X. If X is of affine intersection, then $\widetilde{X}$ is quasiaffine (see Corollary 1.3.4.6), and thus $\widehat{X}$ is quasiaffine. $\qquad\square$

Lemma 1.5.1.5 *Let $\mathbb{L}$ be a field of characteristic zero containing all roots of unity, and assume that $a \in \mathbb{L}$ is not a proper power. Then, for any $n \in \mathbb{Z}_{\geq 1}$, the polynomial $1 - at^n$ is irreducible in $\mathbb{L}[t, t^{-1}]$.*

Proof Over the algebraic closure of $\mathbb{L}$ we have $1 - at^n = (1 - a_1 t) \cdots (1 - a_n t)$, where $a_i^n = a$ and any two a_i differ by a nth root of unity. If $1 - at^n$ would split over $\mathbb{L}$ nontrivially into $h_1(t)h_2(t)$, then a_1^k must be contained in $\mathbb{L}$ for some $k < n$. But then also a_1^d lies in $\mathbb{L}$ for the greatest common divisor d of n and k. Thus a is a proper power, a contradiction. $\qquad\square$

Proof of Theorem 1.5.1.1 According to Proposition 1.4.2.2 and Lemma 1.5.1.2, we may assume that we are in the setting of Construction 1.5.1.4, where it suffices to prove that $\widehat{X}$ is irreducible. Because $p_X \colon \widehat{X} \to X$ is surjective, some irreducible component $\widehat{X}_1 \subseteq \widehat{X}$ dominates X. We verify that $\widehat{X}_1$ equals $\widehat{X}$ by checking that $p_X^{-1}(U)$ is irreducible for suitable open neighborhoods $U \subseteq X$ covering X.

Let $D_1, \ldots, D_s$ be a basis of K such that $n_1 D_1, \ldots, n_k D_k$, where $1 \leq k \leq s$, is a basis of K^0. Enlarging K, if necessary, we may assume that the D_i are primitive, that is, not proper multiples. We take affine open subsets $U \subseteq X$ such that on U every D_i is principal, say $D_i = \mathrm{div}(f_i)$. Then, with $\deg(T_i) := D_i$, Remark 1.3.1.4 provides a K-graded isomorphism

$$
\Gamma(U, \mathcal{O}) \otimes_{\mathbb{K}} \mathbb{K}[T_1^{\pm 1}, \ldots, T_s^{\pm 1}] \;\to\; \Gamma(U, \mathcal{S}),
$$

$$
g \otimes T_1^{\nu_1} \cdots T_s^{\nu_s} \;\mapsto\; g f_1^{-\nu_1} \cdots f_s^{-\nu_s}.
$$

In particular, this identifies $p^{-1}(U)$ with $U \times \mathbb{T}^s$, where $\mathbb{T}^s := (\mathbb{K}^*)^s$. According to Remark 1.4.3.2, the ideal $\Gamma(U, \mathcal{I})$ is generated by $1 - \chi(n_i D_i)$, where $1 \leq i \leq k$. Thus $p_X^{-1}(U)$ is given in $U \times \mathbb{T}^s$ by the equations

$$
1 - \chi(n_i D_i) f_i^{n_i} T_i^{n_i} \;=\; 0, \qquad 1 \leq i \leq k.
$$

To obtain irreducibility of $p_X^{-1}(U)$, it suffices to show that each $1 - \chi(n_i D_i) f_i^{n_i} T_i^{n_i}$ is irreducible in $\mathbb{K}(X)[T_i^{\pm 1}]$. With respect to the variable $S_i := f_i T_i$, this means to verify irreducibility of

$$1 - \chi(n_i D_i) S_i^{n_i} \in \mathbb{K}(X)[S_i^{\pm 1}].$$

In view of Lemma 1.5.1.5, we have to show that $\chi(n_i D_i)$ is not a proper power in $\mathbb{K}(X)$. Assume the contrary. Then we obtain $n_i D_i = k_i \mathrm{div}(h_i)$ with some $h_i \in \mathbb{K}(X)$. Because D_i is primitive, k_i divides n_i and thus, $n_i / k_i D_i$ is principal, a contradiction to the choice of n_i.

The fact that each ring $\Gamma(U, \mathcal{R})$ is normal follows directly from the fact that it is the ring of functions of an open subset of the smooth prevariety $\widehat{X}$. $\qquad \square$

1.5.2 Localization and units

We treat localization by homogeneous elements and consider the units of the Cox ring $\mathcal{R}(X)$ of an irreducible, normal prevariety X defined over an algebraically closed field $\mathbb{K}$ of characteristic zero. The main tool is the divisor of a homogeneous element of $\mathcal{R}(X)$, which we first have to define.

Construction 1.5.2.1 Consider the situation of Construction 1.4.2.1. Consider a divisor $D \in K$ and a nonzero element $f \in \mathcal{R}_{[D]}(X)$. According to Lemma 1.4.3.3, there is a (unique) element $\widetilde{f} \in \Gamma(X, \mathcal{S}_D)$ with $\pi(\widetilde{f}) = f$, where $\pi \colon \mathcal{S} \to \mathcal{R}$ denotes the projection. We define the *$[D]$-divisor* of f to be the effective Weil divisor

$$\mathrm{div}_{[D]}(f) := \mathrm{div}_D(\widetilde{f}) = \mathrm{div}(\widetilde{f}) + D \in \mathrm{WDiv}(X).$$

Proposition 1.5.2.2 *Consider the situation of Construction 1.5.2.1. The $[D]$-divisor depends neither on the representative $D \in K$ nor on the choices made in Construction 1.4.2.1. Moreover, the following statements hold.*

(i) *For every effective $E \in \mathrm{WDiv}(X)$ there are $[D] \in \mathrm{Cl}(X)$ and $f \in \mathcal{R}_{[D]}(X)$ with $E = \mathrm{div}_{[D]}(f)$.*

(ii) *Let $[D] \in \mathrm{Cl}(X)$ and $0 \neq f \in \mathcal{R}_{[D]}(X)$. Then $\mathrm{div}_{[D]}(f) = 0$ implies $[D] = 0$ in $\mathrm{Cl}(X)$.*

(iii) *For any two nonzero homogeneous elements $f \in \mathcal{R}_{[D_1]}(X)$ and $g \in \mathcal{R}_{[D_2]}(X)$, we have*

$$\mathrm{div}_{[D_1]+[D_2]}(fg) = \mathrm{div}_{[D_1]}(f) + \mathrm{div}_{[D_2]}(g).$$

Proof Let $f \in \mathcal{R}_{[D]}(X)$, consider any two isomorphisms $\varphi_i \colon \mathcal{O}_X(D_i) \to \mathcal{R}_{[D]}$ and let $\widetilde{f}_i$ be the sections with $\varphi_i(\widetilde{f}_i) = f$. Then $\varphi_2^{-1} \circ \varphi_1$ is multiplication with some $h \in \mathbb{K}(X)^*$ satisfying $\mathrm{div}(h) = D_1 - D_2$. Well-definedness of the $[D]$-divisor thus follows from

$$\operatorname{div}_{D_1}(\widetilde{f_1}) \;=\; \operatorname{div}(h\,\widetilde{f_1}) + D_2 \;=\; \operatorname{div}_{D_2}(\widetilde{f_2}).$$

If $\operatorname{div}_{[D]}(f) = 0$ holds as in (ii), then, for a representative $\widetilde{f} \in \Gamma(X, \mathcal{O}_X(D))$ of $f \in \mathcal{R}_{[D]}(X)$, we have $\operatorname{div}_D(\widetilde{f}) = 0$ and hence D is principal. Observations (i) and (iii) are obvious. $\qquad\square$

Note that the complete linear system associated with a divisor D on a projective variety X can as well be expressed in terms of the $[D]$-divisor: we have

$$|D| \;=\; \{\operatorname{div}_{[D]}(f);\ f \in \mathcal{R}_{[D]}(X)\} \;=\; \mathbb{P}(\mathcal{R}_{[D]}(X)).$$

Definition 1.5.2.3 Consider the situation of Construction 1.4.2.1. For every nonzero homogeneous element $f \in \mathcal{R}_{[D]}(X)$, we define the $[D]$-*localization* of X by f to be the open subset

$$X_{[D],f} \;:=\; X \setminus \operatorname{Supp}(\operatorname{div}_{[D]}(f)) \;\subseteq\; X.$$

Proposition 1.5.2.4 *Let X be an irreducible, normal prevariety with only constant invertible global functions, finitely generated divisor class group, and Cox ring $\mathcal{R}(X)$. Then, for every nonzero homogeneous $f \in \mathcal{R}_{[D]}(X)$, we have a canonical isomorphism*

$$\Gamma(X_{[D],f}, \mathcal{R}) \;\cong\; \Gamma(X, \mathcal{R})_f.$$

Proof Let the divisor $D \in K$ represent $[D] \in \operatorname{Cl}(X)$ and consider the section $\widetilde{f} \in \Gamma(X, \mathcal{S}_D)$ with $\pi(\widetilde{f}) = f$. According to Remark 1.3.1.7, we have

$$\Gamma(X_{D,\widetilde{f}}, \mathcal{S}) \;\cong\; \Gamma(X, \mathcal{S})_{\widetilde{f}}.$$

The assertion thus follows from Lemma 1.4.3.5 and the fact that localization is compatible with passing to the factor ring. $\qquad\square$

We turn to the units of the Cox ring $\mathcal{R}(X)$; the following result says in particular that for a complete irreducible, normal variety X they are all constant.

Proposition 1.5.2.5 *Let X be an irreducible, normal prevariety with only constant invertible global functions, finitely generated divisor class group, and Cox ring $\mathcal{R}(X)$.*

(i) *Every homogeneous invertible element of $\mathcal{R}(X)$ is constant.*
(ii) *If $\Gamma(X, \mathcal{O}) = \mathbb{K}$ holds, then every invertible element of $\mathcal{R}(X)$ is constant.*

Proof For (i), let $f \in \mathcal{R}(X)^*$ be homogeneous of degree $[D]$. Then its inverse $g \in \mathcal{R}(X)^*$ is homogeneous of degree $-[D]$, and $fg = 1$ lies in $\mathcal{R}(X)_0^* = \mathbb{K}^*$.

By Proposition 1.5.2.2 (iii), we have

$$0 = \mathrm{div}_0(fg) = \mathrm{div}_{[D]}(f) + \mathrm{div}_{[-D]}(g).$$

Because the divisors $\mathrm{div}_{[D]}(f)$ and $\mathrm{div}_{[-D]}(g)$ are effective, they both vanish. Thus, Proposition 1.5.2.2 (ii) yields $[D] = 0$. This implies $f \in \mathcal{R}(X)_0^* = \mathbb{K}^*$, as wanted.

For (ii), we have to show that any invertible $f \in \mathcal{R}(X)$ is of degree zero. Choose a decomposition $\mathrm{Cl}(X) = K_0 \oplus K_t$ into a free part and the torsion part, and consider the coarsened grading

$$\mathcal{R}(X) = \bigoplus_{w \in K_0} R_w, \qquad R_w := \bigoplus_{u \in K_t} \mathcal{R}(X)_{w+u}.$$

Then, as any invertible element of the K_0-graded integral ring $\mathcal{R}(X)$, also f is necessarily K_0-homogeneous of some degree $w \in K_0$. Decomposing f and f^{-1} into $\mathrm{Cl}(X)$-homogeneous parts we get representations

$$f = \sum_{u \in K_t} f_{w+u}, \qquad f^{-1} = \sum_{u \in K_t} f^{-1}_{-w+u}.$$

Because of $ff^{-1} = 1$, we have $f_{w+v} f^{-1}_{-w-v} \neq 0$ for at least one $v \in K_t$. Because $\Gamma(X, \mathcal{O}) = \mathbb{K}$ holds, $f_{w+v} f^{-1}_{-w-v}$ must be a nonzero constant. Using Proposition 1.5.2.2 we conclude $w + v = 0$ as before. In particular, $w = 0$ holds and thus each f_{w+u} has a torsion degree. For a suitable power f^n_{w+u} we have $n\mathrm{div}_{w+u}(f_{w+u}) = 0$, which implies $f_{w+u} = 0$ for any $u \neq 0$. $\qquad\square$

Remark 1.5.2.6 The affine surface X treated in Example 1.4.4.2 shows that requiring $\Gamma(X, \mathcal{O}^*) = \mathbb{K}^*$ is in general not enough to ensure that all units of the Cox ring are constant.

1.5.3 Divisibility properties

The ground field $\mathbb{K}$ is algebraically closed of characteristic zero. For irreducible, normal prevarieties X with a free finitely generated divisor class group, we saw that the Cox ring admits unique factorization. If we have torsion in the divisor class group this no longer needs to hold. However, restricting to homogeneous elements leads to a framework for a reasonable divisibility theory; the precise notions are the following.

Definition 1.5.3.1 Consider an abelian group K and a K-graded integral $\mathbb{K}$-algebra $R = \bigoplus_{w \in K} R_w$.

(i) A nonzero nonunit $f \in R$ is K-*prime* if it is homogeneous and $f | gh$ with homogeneous $g, h \in R$ implies $f | g$ or $f | h$.

(ii) We say that R is K-*factorial* or *factorially (K-)graded* if every homogeneous nonzero nonunit $f \in R$ is a product of K-primes.

(iii) By a K-*principal ideal* in R, we mean a principal ideal $\mathfrak{a} = \langle f \rangle$ with a homogeneous $f \in R$.

(iv) An ideal $\mathfrak{a} \lhd R$ is K-*prime* if it is homogeneous and for any two homogeneous $f, g \in R$ with $fg \in \mathfrak{a}$ one has either $f \in \mathfrak{a}$ or $g \in \mathfrak{a}$.

(v) A K-prime ideal $\mathfrak{a} \lhd R$ has K-*height d* if d is maximal admitting a chain $\mathfrak{a}_0 \subset \mathfrak{a}_1 \subset \ldots \subset \mathfrak{a}_d = \mathfrak{a}$ of K-prime ideals.

Let us look at these concepts also from the geometric point of view. Consider a prevariety Y with an action of an algebraic group H. Then H acts also on the group $\mathrm{WDiv}(Y)$ of Weil divisors via

$$h \cdot \sum a_D D := \sum a_D (h \cdot D).$$

Definition 1.5.3.2 Let an algebraic group H act on an irreducible, normal prevariety Y.

(i) An H-*prime divisor* on Y is a Weil divisor $0 \neq \sum a_D D$, where $a_D \in \{0, 1\}$, the D are prime, and those with $a_D = 1$ are transitively permuted by H.

(ii) An H-*homogeneous rational function* on Y is an element $f \in \mathbb{K}(Y)$, which is defined on an invariant open subset of Y and is H-homogeneous there.

(iii) We say that Y is H-*factorial* if every H-invariant Weil divisor on Y is the divisor of an H-homogeneous rational function.

Note that every H-invariant divisor is a unique sum of H-prime divisors. In the case of a quasiaffine variety, we see that the algebraic and the geometric notions correspond to each other.

Proposition 1.5.3.3 *Let $H = \mathrm{Spec}\,\mathbb{K}[K]$ be a quasitorus and W an irreducible, normal quasiaffine H-variety. Consider the K-graded algebra $R := \Gamma(W, \mathcal{O})$. Then the following statements are equivalent.*

(i) *Every K-prime ideal of K-height one in R is K-principal.*

(ii) *The variety W is H-factorial.*

(iii) *The ring R is K-factorial.*

Moreover, if one of these statements holds, then a homogeneous nonzero nonunit $f \in R$ is K-prime if and only if the divisor $\mathrm{div}(f)$ is H-prime, and every H-prime divisor is of the form $\mathrm{div}(f)$ with a K-prime $f \in R$.

Proof Assume that (i) holds and let D be an H-invariant Weil divisor on W. Write $D = a_1 D_1 + \cdots + a_r D_r$ with H-prime divisors D_i. Once we know that the vanishing ideal $\mathfrak{a}_i$ of D_i is of K-height one, we have $\mathfrak{a}_i = \langle f_i \rangle$ with some homogeneous f_i. We conclude $D_i = \mathrm{div}(f_i)$ and $D = \mathrm{div}(f_1^{a_1} \cdots f_r^{a_r})$, which in turn establishes (ii).

Thus, our task is to show that $\mathfrak{a}_i$ is of K-height one. Suppose we have $\mathfrak{a} \subset \mathfrak{a}_i$ with a nonzero K-prime ideal $\mathfrak{a}$. Choose an H-equivariant open embedding $W \subseteq Z$ into some irreducible, normal affine H-variety Z such that some nonzero homogeneous $h \in \mathfrak{a}$ and some homogeneous $f \in \mathfrak{a}_i \setminus \mathfrak{a}$ extend to functions on Z. Then $\mathfrak{b} := \mathfrak{a} \cap \Gamma(Z, \mathcal{O})$ is a nonzero K-prime ideal properly contained in the K-prime ideal $\mathfrak{b}_i := \mathfrak{a}_i \cap \Gamma(Z, \mathcal{O})$. Note that $\overline{D}_i = V_Z(\mathfrak{b}_i)$ is a proper subset of $V_Z(\mathfrak{b})$, because otherwise some power of f would belong to $V_Z(\mathfrak{b})$. Because $\mathfrak{b}$ is nonzero, we find a homogeneous $g \in \Gamma(Z, \mathcal{O})$ vanishing on $V_Z(\mathfrak{b}) \setminus \overline{D}_i$ but not on $\overline{D}_i$. Consequently, some power of fg lies in $\mathfrak{b}$, contradicting K-primeness of $\mathfrak{b}$.

Now, assume that (ii) holds. Given a homogeneous element $0 \neq f \in R \setminus R^*$, write $\operatorname{div}(f) = D_1 + \cdots + D_r$ with H-prime divisors D_i. Then $D_i = \operatorname{div}(f_i)$ holds with homogeneous f_i. One verifies directly that the f_i are K-prime. Thus we have $f = \alpha f_1 \cdots f_r$ with a homogeneous unit α as required in (iii).

If (iii) holds and $\mathfrak{a}$ is a K-prime ideal of K-height one, then we take any homogeneous $0 \neq f \in \mathfrak{a}$ and find a K-prime factor f_1 of f with $f_1 \in \mathfrak{a}$. This gives inclusions $0 \subsetneq \langle f_1 \rangle \subseteq \mathfrak{a}$ of K-prime ideals, which implies $\mathfrak{a} = \langle f_1 \rangle$. $\qquad \square$

Corollary 1.5.3.4 *In the setting of Proposition 1.5.3.3 assume that $R^* = \mathbb{K}^*$ holds. Then factoriality of R implies K-factoriality of R.*

We enter the divisibility theory of the Cox ring. As in the torsion-free case (see Proposition 1.3.3.4), divisibility and primality of homogeneous elements in the Cox ring $\mathcal{R}(X)$ can be characterized in terms of the corresponding divisors on X.

Proposition 1.5.3.5 *Let X be an irreducible, normal prevariety with only constant invertible global functions and finitely generated divisor class group.*

(i) *An element $0 \neq f \in \Gamma(X, \mathcal{R}_{[D]})$ divides $0 \neq g \in \Gamma(X, \mathcal{R}_{[E]})$ if and only if $\operatorname{div}_{[D]}(f) \leq \operatorname{div}_{[E]}(g)$ holds.*

(ii) *Two elements $0 \neq f_i \in \Gamma(X, \mathcal{R}_{[D_i]})$ are associated with each other if and only if $\operatorname{div}_{[D_1]}(f_1)$ equals $\operatorname{div}_{[D_2]}(f_2)$. In this case, we have $[D_1] = [D_2]$ and $f_2 = af_1$ holds with some $a \in \mathbb{K}^*$.*

(iii) *An element $0 \neq f \in \Gamma(X, \mathcal{R}_{[D]})$ is $\operatorname{Cl}(X)$-prime if and only if the divisor $\operatorname{div}_{[D]}(f) \in \operatorname{WDiv}(X)$ is prime.*

The second assertion yields in particular that the element $f \in \mathcal{R}_{[D]}(X)$ with $E = \operatorname{div}_{[D]}(f)$ for a given effective divisor E on X provided by Proposition 1.5.2.2 (i) is unique up to constants; we refer to such f as a *canonical section* of E.

Lemma 1.5.3.6 *In the situation of Construction 1.5.1.4, every nonzero element $f \in \Gamma(X, \mathcal{R}_{[D]})$ satisfies*

$$\operatorname{div}(f) = p_X^*(\operatorname{div}_{[D]}(f)),$$

where on the left-hand side f is a regular function on $\widehat{X}$ and on the right-hand side f is an element on $\mathcal{R}(X)$.

Proof In the notation of 1.5.1.4, let $D \in K$ represent $[D] \in \operatorname{Cl}(X)$, and let $\widetilde{f} \in \Gamma(X, \mathcal{S}_D)$ project to $f \in \Gamma(X, \mathcal{R}_{[D]})$. The commutative diagram of 1.5.1.4 yields

$$\operatorname{div}(f) = \imath^*(\operatorname{div}(\widetilde{f})) = \imath^*(p^*(\operatorname{div}_D(\widetilde{f}))) = p_X^*(\operatorname{div}_{[D]}(f)),$$

where $\imath \colon \widehat{X} \to \widetilde{X}$ denotes the inclusion and the equality $\operatorname{div}(\widetilde{f}) = p^*(\operatorname{div}_D(\widetilde{f}))$ was established in Lemma 1.3.3.2. □

Proof of Proposition 1.5.3.5 According to Lemma 1.5.1.2, we may assume that X is smooth. Then Construction 1.5.1.4 presents X as the geometric quotient of the smooth quasiaffine H_X-prevariety $\widehat{X}$ which has $\mathcal{R}(X)$ as its algebra of regular functions. For (i), note first that f divides g in $\mathcal{R}(X)$ if and only if $\operatorname{div}(f) \le \operatorname{div}(g)$ holds on $\widehat{X}$. By Lemma 1.5.3.6, the latter is equivalent to $\operatorname{div}_{[D]}(f) \le \operatorname{div}_{[E]}(g)$. The characterization of being associated in (ii) is clear by (i), and the additional assertion follows from the fact that the homogeneous invertible elements of $\mathcal{R}(X)$ are constant; see Proposition 1.5.2.5.

We prove (iii). If $\operatorname{div}_{[D]}(f)$ is prime, then Lemma 1.5.3.6 yields that $\operatorname{div}(f)$ is nontrivial and thus f is a nonzero nonunit in $\widehat{X}$. Moreover, if f divides a product $f_1 f_2$ of homogeneous elements $f_i \in \mathcal{R}(X)$, then, according to (i), it must divide one of the factors. Now suppose that f is $\operatorname{Cl}(X)$-prime but $\operatorname{div}_{[D]}(f)$ is not prime. Then $\operatorname{div}_{[D]}(f) = D_1 + D_2$ holds with nonzero divisors D_1, D_2 on X. Proposition 1.5.2.2 (i) yields homogeneous $f_i \in \mathcal{R}_{[D_i]}(X)$ with $\operatorname{div}_{[D_i]}(f_i) = D_1$. Using (i), we see that f divides $f_1 f_2$ but none of the factors f_1, f_2, a contradiction. □

Theorem 1.5.3.7 *Let X be an irreducible, normal prevariety with only constant invertible global functions and finitely generated divisor class group $\operatorname{Cl}(X)$. Then the Cox ring $\mathcal{R}(X)$ is $\operatorname{Cl}(X)$-factorial.*

Proof We have to show that every nonzero nonunit $f \in \mathcal{R}_{[D]}(X)$ is a product of $\operatorname{Cl}(X)$-primes. Write $\operatorname{div}_{[D]}(f) = D_1 + \cdots + D_r$ with prime divisors D_i on X. Then Proposition 1.5.2.2 (i) provides us with $f_i \in \mathcal{R}_{[D_i]}(X)$ such that $\operatorname{div}_{[D]}(f_i) = D_i$ holds. According to Proposition 1.5.3.5 (iii), every f_i is $\operatorname{Cl}(X)$-prime. Moreover, Proposition 1.5.3.5 (ii) shows that $f = a f_1 \cdots f_r$ holds with a constant $a \in \mathbb{K}^*$. □

Remark 1.5.3.8 Let X be an irreducible, normal prevariety with only constant invertible global functions, finitely generated divisor class group, and Cox ring $\mathcal{R}(X)$. Then the assignment $f \mapsto \mathrm{div}_{[D]}(f)$ induces an isomorphism from the multiplicative semigroup of homogeneous elements of $\mathcal{R}(X)$ modulo units onto the semigroup $\mathrm{WDiv}^+(X)$ of effective Weil divisors on X. The fact that $\mathcal{R}(X)$ is $\mathrm{Cl}(X)$-factorial reflects the fact that every effective Weil divisor is a unique nonnegative linear combination of prime divisors.

Remark 1.5.3.9 For the affine surface X considered in Example 1.4.4.1, the Cox ring $\mathcal{R}(X)$ is $\mathbb{Z}/2\mathbb{Z}$-factorial but not factorial.

1.6 Geometric realization of the Cox sheaf

1.6.1 Characteristic spaces

We study the geometric realization of a Cox sheaf, its relative spectrum, which we call a *characteristic space*. For locally factorial varieties, for example, smooth ones, this concept coincides with the universal torsor introduced by Colliot-Thélène and Sansuc in [96], see also [100, 279]. As soon as we have nonfactorial singularities, the characteristic space turns out to be no longer a torsor, that is, an étale principal bundle. As before, we work with irreducible, normal prevarieties defined over an algebraically closed field $\mathbb{K}$ of characteristic zero. First we provide two statements on local finite generation of Cox sheaves.

Proposition 1.6.1.1 *Let X be an irreducible, normal prevariety of affine intersection with only constant invertible functions and finitely generated divisor class group. If the Cox ring $\mathcal{R}(X)$ is finitely generated, then the Cox sheaf $\mathcal{R}$ is locally of finite type.*

Proof The assumption that X is of affine intersection guarantees that it is covered by open affine subsets $X_{[D],f}$, where $[D] \in \mathrm{Cl}(X)$ and $f \in \mathcal{R}_{[D]}(X)$. By Proposition 1.5.2.4, we have $\Gamma(X_{[D],f}, \mathcal{R}) = \mathcal{R}(X)_f$, which gives the assertion. $\qquad\square$

Proposition 1.6.1.2 *Let X be an irreducible, normal prevariety with only constant invertible functions and finitely generated divisor class group. If X is $\mathbb{Q}$-factorial, then any Cox sheaf $\mathcal{R}$ on X is locally of finite type.*

Proof By definition, the Cox sheaf $\mathcal{R}$ is the quotient of a sheaf of divisorial algebras $\mathcal{S}$ by some ideal sheaf $\mathcal{I}$. According to Proposition 1.4.2.2, we may assume that $\mathcal{S}$ arises from a finitely generated subgroup $K \subseteq \mathrm{WDiv}(X)$. Proposition 1.3.2.3 then tells us that $\mathcal{S}$ is locally of finite type, and Lemma 1.4.3.5 ensures that the quotient $\mathcal{R} = \mathcal{S}/\mathcal{I}$ can be taken at the level of sections. $\qquad\square$

We turn to the relative spectrum of a Cox sheaf. The following generalizes Construction 1.5.1.4, where the smooth case is considered.

Construction 1.6.1.3 Let X be an irreducible, normal prevariety with $\Gamma(X, \mathcal{O}^*) = \mathbb{K}^*$, finitely generated divisor class group $\mathrm{Cl}(X)$, and Cox sheaf $\mathcal{R}$. Suppose that $\mathcal{R}$ is locally of finite type, for example, X is $\mathbb{Q}$-factorial or $\mathcal{R}(X)$ is finitely generated. Taking the relative spectrum gives an irreducible, normal prevariety

$$\widehat{X} := \mathrm{Spec}_X(\mathcal{R}).$$

The $\mathrm{Cl}(X)$-grading of the sheaf $\mathcal{R}$ defines an action of the quasitorus $H_X := \mathrm{Spec}\,\mathbb{K}[\mathrm{Cl}(X)]$ on $\widehat{X}$, the canonical morphism $p_X \colon \widehat{X} \to X$ is a good quotient for this action, and we have an isomorphism of sheaves

$$\mathcal{R} \cong (p_X)_*(\mathcal{O}_{\widehat{X}}).$$

In particular, the Cox ring $\mathcal{R}(X)$ coincides, as a $\mathrm{Cl}(X)$-graded ring, with the algebra of functions $\Gamma(\widehat{X}, \mathcal{O})$. We call $p_X \colon \widehat{X} \to X$ the *characteristic space (over X)* associated with $\mathcal{R}$ and H_X the *characteristic quasitorus of X*.

Proof Everything is standard except irreducibility and normality, which follow from Theorem 1.5.1.1. □

The Cox sheaf $\mathcal{R}$ was defined as the quotient of a sheaf $\mathcal{S}$ of divisorial algebras by a sheaf $\mathcal{I}$ of ideals. Geometrically this means that the characteristic space is embedded into the relative spectrum of a sheaf of divisorial algebras; compare Construction 1.5.1.4 for the case of a smooth X. Before making this precise in the general case, we have to relate local finite generation of the sheaves $\mathcal{R}$ and $\mathcal{S}$ to each other.

Proposition 1.6.1.4 *Let X be an irreducible, normal prevariety with only constant invertible functions, finitely generated divisor class group, and Cox sheaf $\mathcal{R}$. Moreover, let $\mathcal{S}$ be the sheaf of divisorial algebras associated with a finitely generated subgroup $K \subseteq \mathrm{WDiv}(X)$ projecting onto $\mathrm{Cl}(X)$ and $U \subseteq X$ an open affine subset. Then the algebra $\Gamma(U, \mathcal{R})$ is finitely generated if and only if the algebra $\Gamma(U, \mathcal{S})$ is finitely generated.*

Proof Lemma 1.4.3.5 tells us that $\Gamma(U, \mathcal{R})$ is a factor algebra of $\Gamma(U, \mathcal{S})$. Thus, if $\Gamma(U, \mathcal{S})$ is finitely generated then the same holds for $\Gamma(U, \mathcal{R})$. Moreover, Lemma 1.4.3.4 says that the projection $\Gamma(U, \mathcal{S}) \to \Gamma(U, \mathcal{R})$ defines isomorphisms along the homogeneous components. Thus, Proposition 1.1.2.7 shows that finite generation of $\Gamma(U, \mathcal{R})$ implies finite generation of $\Gamma(U, \mathcal{S})$. □

Construction 1.6.1.5 Let X be an irreducible, normal prevariety with $\Gamma(X, \mathcal{O}^*) = \mathbb{K}^*$ and finitely generated divisor class group, and let $K \subseteq \mathrm{WDiv}(X)$ be a finitely generated subgroup projecting onto $\mathrm{Cl}(X)$. Consider the sheaf of divisorial algebras $\mathcal{S}$ associated with K and the Cox sheaf $\mathcal{R} = \mathcal{S}/\mathcal{I}$ as constructed in 1.4.2.1, and suppose that one of these sheaves is locally of finite type. Then the projection $\mathcal{S} \to \mathcal{R}$ of $\mathcal{O}_X$-algebras defines a commutative diagram

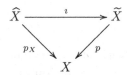

for the relative spectra $\widehat{X} = \mathrm{Spec}_X \mathcal{R}$ and $\widetilde{X} = \mathrm{Spec}_X \mathcal{S}$. We have the actions of $H_X = \mathrm{Spec}\, \mathbb{K}[\mathrm{Cl}(X)]$ on $\widehat{X}$ and $H = \mathrm{Spec}\, \mathbb{K}[K]$ on $\widetilde{X}$. The map $\imath : \widehat{X} \to \widetilde{X}$ is a closed embedding and it is H_X-invariant, where H_X acts on $\widetilde{X}$ via the inclusion $H_X \subseteq H$ defined by the projection $K \to \mathrm{Cl}(X)$. The image $\imath(\widehat{X}) \subseteq \widetilde{X}$ is precisely the zero set of the ideal sheaf $\mathcal{I}$.

Proposition 1.6.1.6 *Consider the situation of Construction 1.6.1.3; in particular, X is an irreducible, normal prevariety with characteristic space $p_X : \widehat{X} \to X$.*

(i) *The inverse image $p_X^{-1}(X_{\mathrm{reg}}) \subseteq \widehat{X}$ of the set of smooth points $X_{\mathrm{reg}} \subseteq X$ is smooth, the group H_X acts freely on $p_X^{-1}(X_{\mathrm{reg}})$, and the restriction $p_X : p_X^{-1}(X_{\mathrm{reg}}) \to X_{\mathrm{reg}}$ is an étale H_X-principal bundle.*

(ii) *For any closed $A \subseteq X$ of codimension at least 2, $p_X^{-1}(A) \subseteq \widehat{X}$ is as well of codimension at least 2.*

(iii) *The prevariety $\widehat{X}$ is H_X-factorial.*

(iv) *If X is of affine intersection, then $\widehat{X}$ is a quasiaffine variety.*

Proof For (i), we refer to the proof of Construction 1.5.1.4. To obtain (ii) consider an affine open set $U \subseteq X$ and $\widehat{U} := p_X^{-1}(U)$. By Lemma 1.5.1.2, the open set $\widehat{U} \setminus p_X^{-1}(A)$ has the same regular functions as $\widehat{U}$. Because $\widehat{X}$ is normal, we conclude that $\widehat{U} \cap p_X^{-1}(A)$ is of codimension at least 2 in $\widehat{U}$. Now, cover X by affine $U \subseteq X$ and Assertion (ii) follows.

We turn to (iii). According to (ii) we may assume that X is smooth. Let $\widehat{D}$ be an invariant Weil divisor on $\widehat{X}$. Using, for example, the fact that $p_X : \widehat{X} \to X$ is an étale principal bundle, we see that $\widehat{D} = p_X^*(D)$ holds with a Weil divisor D on X. Thus, we have to show that all pullback divisors $p_X^*(D)$ are principal. For this, it suffices to consider effective divisors D on X, and these are treated by Proposition 1.5.2.2 and Lemma 1.5.3.6. We show (iv). We may assume that we are in the setting of Construction 1.6.1.5. Corollary 1.3.4.6 then ensures that

$\widetilde{X}$ is quasiaffine and Construction 1.6.1.5 gives that $\widehat{X}$ is a closed subvariety of $\widetilde{X}$. $\qquad\square$

The characteristic space over a given X encodes among other things the geometric line bundles over X. Here is how to reconstruct them explicitly.

Proposition 1.6.1.7 *Situation as in Construction 1.6.1.3; in particular, X is an irreducible, normal prevariety with Cox sheaf $\mathcal{R}$ and characteristic space $p_X\colon \widehat{X} \to X$. Then, for any Cartier divisor $D \in \mathrm{WDiv}(X)$, the class $[D] \in \mathrm{Cl}(X)$ defines a character $\chi \in \mathbb{X}(H_X)$ and we have an action*

$$H_X \times (\widehat{X} \times \mathbb{K}) \ \to \ \widehat{X} \times \mathbb{K}, \qquad h\cdot(\widehat{x}, z) := (h^{-1}\cdot\widehat{x}, \chi(h)z),$$

The quotient space $L := (\widehat{X} \times \mathbb{K})/H_X$ is the total space of a line bundle $\varphi\colon L \to X$ over X, where the projection φ is induced by the H_X-invariant map

$$\psi\colon \widehat{X} \times \mathbb{K} \ \to \ X, \qquad (\widehat{x}, z) \mapsto p_X(\widehat{x}).$$

The isomorpism class of the line bundle $\varphi\colon L \to X$ in the Picard group $\mathrm{Pic}(X) \subseteq \mathrm{Cl}(X)$ equals the class of D in the divisor class group $\mathrm{Cl}(X)$.

Proof For the last statement, we have to show that $\mathcal{O}_X(D)$ equals the sheaf of sections of L. For this, it suffices to show that $\oplus_{n\geq 0}\mathcal{O}_X(-nD)$ is isomorphic to $\varphi_*(\mathcal{O}_L)$, the direct image of the sheaf of functions on L. We have canonical identifications

$$\varphi_*(\mathcal{O}_L) \ = \ \psi_*(\mathcal{O}_{\widehat{X}\times\mathbb{K}})^{H_X} \ = \ \psi_*(\mathcal{O}_{\widehat{X}} \otimes \mathcal{O}_{\mathbb{K}})^{H_X} \ = \ \bigoplus_{n\in\mathbb{Z}_{\geq 0}} \mathcal{R}_{[-nD]}.$$

For the last equality observe that the action of H_X on $\psi_*(\mathcal{O}_{\widehat{X}} \otimes \mathcal{O}_{\mathbb{K}})$ is given on an element of the form $f \otimes T^n$ with a homogeneous section f and $T^n \in \Gamma(\mathbb{K}, \mathcal{O}) = \mathbb{K}[T]$ as

$$h\cdot(f \otimes g) \ = \ \chi_f(h)f \otimes \chi(h^{-n})T^n,$$

where $\chi_f \in \mathbb{X}(H_X)$ is the weight of f. We conclude that the H_X-invariant sections of the form $f \otimes T^n$ are precisely those with f a section of $\mathcal{R}_{[-nD]}$. To obtain the assertion, we use Lemma 1.4.3.4, which gives us

$$\bigoplus_{n\in\mathbb{Z}_{\geq 0}} \mathcal{R}_{[-nD]} \ \cong \ \bigoplus_{n\in\mathbb{Z}_{\geq 0}} \mathcal{O}_X(-nD). \qquad\square$$

1.6.2 Divisor classes and isotropy groups

The aim of this subsection is to interpret properties of a prevariety in terms of its characteristic space and the characteristic quasitorus action there. As before, everything takes places over an algebraically closed field $\mathbb{K}$ of characteristic zero.

A first statement relates the divisor of a $[D]$-homogeneous function on a characteristic space to its $[D]$-divisor on the underlying prevariety; the smooth case was settled in Lemma 1.5.3.6.

Proposition 1.6.2.1 *In the situation of Construction 1.6.1.3, consider the pullback homomorphism* $p_X^*\colon \mathrm{WDiv}(X) \to \mathrm{WDiv}(\widehat{X})$ *defined in Remark 1.3.4.1. Then, for every* $[D] \in \mathrm{Cl}(X)$ *and every* $f \in \Gamma(X, \mathcal{R}_{[D]})$, *we have*

$$\mathrm{div}(f) = p_X^*(\mathrm{div}_{[D]}(f)),$$

where on the left-hand side f *is a function on* $\widehat{X}$, *and on the right-hand side a function on* X. *If* X *is of affine intersection and* $X \setminus \mathrm{Supp}(\mathrm{div}_{[D]}(f))$ *is affine, then we have moreover*

$$\mathrm{Supp}(\mathrm{div}(f)) = p_X^{-1}(\mathrm{Supp}(\mathrm{div}_{[D]}(f))).$$

Proof We may assume that we are in the setting of Construction 1.6.1.5. Let the divisor $D \in K$ represent the class $[D] \in \mathrm{Cl}(X)$, and let $\widetilde{f} \in \Gamma(X, \mathcal{S}_D)$ project to $f \in \Gamma(X, \mathcal{R}_{[D]})$. The commutative diagram of 1.6.1.5 yields

$$\mathrm{div}(f) = \iota^*(\mathrm{div}(\widetilde{f})) = \iota^*(p^*(\mathrm{div}_D(\widetilde{f}))) = p_X^*(\mathrm{div}_{[D]}(f)),$$

where $\iota\colon \widehat{X} \to \widetilde{X}$ denotes the inclusion, and the equality $\mathrm{div}(\widetilde{f}) = p^*(\mathrm{div}_D(\widetilde{f}))$ was established in Proposition 1.3.4.3. Similarly, we have

$$\mathrm{Supp}(\mathrm{div}(f)) = \iota^{-1}(\mathrm{Supp}(\mathrm{div}(\widetilde{f})))$$
$$= \iota^{-1}(p^{-1}(\mathrm{Supp}(\mathrm{div}_D(\widetilde{f}))))$$
$$= p_X^{-1}(\mathrm{Supp}(\mathrm{div}_{[D]}(f)))$$

provided that X is of affine intersection and $X \setminus \mathrm{Supp}(\mathrm{div}_{[D]}(f))$ is affine, because Proposition 1.3.4.3 then ensures $\mathrm{Supp}(\mathrm{div}(\widetilde{f})) = p^{-1}(\mathrm{Supp}(\mathrm{div}_D(\widetilde{f})))$. $\qquad\square$

Corollary 1.6.2.2 *Consider the situation of Construction 1.6.1.3. Let* $\widehat{x} \in \widehat{X}$ *be a point such that* $H_X \cdot \widehat{x} \subseteq \widehat{X}$ *is closed, and let* $f \in \Gamma(X, \mathcal{R}_{[D]})$. *Then we have*

$$f(\widehat{x}) = 0 \iff p_X(\widehat{x}) \in \mathrm{Supp}(\mathrm{div}_{[D]}(f)).$$

Proof The image $p_X(\mathrm{Supp}(\mathrm{div}(f)))$ is contained in $\mathrm{Supp}(\mathrm{div}_{[D]}(f))$. By the definition of the pullback and Proposition 1.6.2.1, the two sets coincide in X_{reg}. Thus, $p_X(\mathrm{Supp}(\mathrm{div}(f)))$ is dense in $\mathrm{Supp}(\mathrm{div}_{[D]}(f))$. By Theorem 1.2.3.6, the image $p_X(\mathrm{Supp}(\mathrm{div}(f)))$ is closed and thus we have

$$p_X(\mathrm{Supp}(\mathrm{div}(f))) = \mathrm{Supp}(\mathrm{div}_{[D]}(f)).$$

In particular, if $f(\widehat{x}) = 0$ holds, then $p_X(\widehat{x})$ lies in $\mathrm{Supp}(\mathrm{div}_{[D]}(f))$. Conversely, if $p_X(\widehat{x})$ belongs to $\mathrm{Supp}(\mathrm{div}_{[D]}(f))$, then some $\widehat{x}' \in \mathrm{Supp}(\mathrm{div}(f))$ belongs to the fiber of $\widehat{x}$. Because $H_x \cdot \widehat{x}$ is closed, Corollary 1.2.3.7 tells us that $\widehat{x}$ is contained in the orbit closure of $\widehat{x}'$ and hence belongs to $\mathrm{Supp}(\mathrm{div}(f))$. $\qquad\square$

Corollary 1.6.2.3 *Consider the situation of Construction 1.6.1.5 and suppose that X is of affine intersection. For $x \in X$, let $\widehat{x} \in p_X^{-1}(x)$ such that $H_X \cdot \widehat{x}$ is closed in $\widehat{X}$. Then $H \cdot \widehat{x}$ is closed in $\widetilde{X}$.*

Proof Assume that the orbit $H \cdot \widehat{x}$ is not closed in $\widetilde{X}$. Then there is a point $\widetilde{x} \in p^{-1}(x)$ having a closed H-orbit in $\widetilde{X}$, and $\widetilde{x}$ lies in the closure of $H \cdot \widehat{x}$. Because $\widetilde{X}$ is quasiaffine, we find a function $\widetilde{f} \in \Gamma(X, \mathcal{S}_D)$ with $\widetilde{f}(\widetilde{x}) = 0$ but $\widetilde{f}(\widehat{x}) \neq 0$. Corollary 1.3.4.4 gives $p(\widetilde{x}) \in \mathrm{Supp}(\mathrm{div}_D(\widetilde{f}))$. Because we have $p_X(\widehat{x}) = p(\widetilde{x})$, this contradicts Corollary 1.6.2.2. $\qquad\square$

For an irreducible, normal prevariety X and a point $x \in X$, let $\mathrm{PDiv}(X, x) \subseteq \mathrm{WDiv}(X)$ denote the subgroup of all Weil divisors that are principal on some neighborhood of x. We define the *local class group* of X in x to be the factor group

$$\mathrm{Cl}(X, x) := \mathrm{WDiv}(X)/\,\mathrm{PDiv}(X, x).$$

Obviously the group $\mathrm{PDiv}(X)$ of principal divisors is contained in $\mathrm{PDiv}(X, x)$. Thus, there is a canonical epimorphism $\pi_x \colon \mathrm{Cl}(X) \to \mathrm{Cl}(X, x)$. The *Picard group* of X is the factor group of the group $\mathrm{CDiv}(X)$ of Cartier divisors by the subgroup of principal divisors:

$$\mathrm{Pic}(X) \;=\; \mathrm{CDiv}(X)/\,\mathrm{PDiv}(X) \;=\; \bigcap_{x \in X} \ker(\pi_x).$$

Proposition 1.6.2.4 *Consider the situation of Construction 1.6.1.3. For $x \in X$, let $\widehat{x} \in p_X^{-1}(x)$ be a point with closed H_X-orbit. Define a submonoid*

$$S_x := \{[D] \in \mathrm{Cl}(X);\ f(\widehat{x}) \neq 0 \text{ for some } f \in \Gamma(X, \mathcal{R}_{[D]})\} \subseteq \mathrm{Cl}(X),$$

and let $\mathrm{Cl}_x(X) \subseteq \mathrm{Cl}(X)$ denote the subgroup generated by S_x. Then the local class groups of X and the Picard group are given by

$$\mathrm{Cl}(X, x) \;=\; \mathrm{Cl}(X)/\,\mathrm{Cl}_x(X), \qquad\qquad \mathrm{Pic}(X) \;=\; \bigcap_{x \in X} \mathrm{Cl}_x(X).$$

Proof First observe that Corollary 1.6.2.2 gives us the following description of the monoid S_x in terms of the $[D]$-divisors:

$$S_x = \{[D] \in \mathrm{Cl}(X);\ x \notin \mathrm{Supp}(\mathrm{div}_{[D]}(f)) \text{ for some } f \in \Gamma(X, \mathcal{R}_{[D]})\}$$

$$= \{[D] \in \mathrm{Cl}(X);\ D \geq 0,\ x \notin \mathrm{Supp}(D)\},$$

where the latter equation is due to the fact that the $[D]$-divisors are precisely
the effective divisors with class $[D]$. The assertions thus follow from

$$\mathrm{Cl}_x(X) = \{[D] \in \mathrm{Cl}(X); \ x \notin \mathrm{Supp}(D)\}$$

$$= \{[D] \in \mathrm{Cl}(X); \ D \text{ principal near } x\}. \qquad \square$$

Proposition 1.6.2.5 *Consider the situation of Construction 1.6.1.3.
Given $x \in X$, let $\widehat{x} \in p_X^{-1}(x)$ be a point with closed H_X-orbit. Then the inclusion $H_{X,\widehat{x}} \subseteq H_X$ of the isotropy group of $\widehat{x} \in \widehat{X}$ is given by the epimorphism $\mathrm{Cl}(X) \to \mathrm{Cl}(X, x)$ of character groups. In particular, we have*

$$H_{X,\widehat{x}} = \mathrm{Spec}\,\mathbb{K}[\mathrm{Cl}(X, x)], \qquad \mathrm{Cl}(X, x) = \mathbb{X}(H_{X,\widehat{x}}).$$

Proof Let $U \subseteq X$ be any affine open neighborhood of $x \in X$. Then U is of the
form $X_{[D],f}$ with some $f \in \Gamma(X, \mathcal{R}_{[D]})$ and $\widetilde{U} := p_X^{-1}(U)$ is affine. According
to Proposition 1.5.2.4, we have

$$\Gamma(\widetilde{U}, \mathcal{O}) = \Gamma(U, \mathcal{R}) = \Gamma(X, \mathcal{R})_f = \Gamma(\widehat{X}, \mathcal{O})_f.$$

Corollary 1.6.2.2 shows that the group $\mathrm{Cl}_x(X)$ is generated by the classes
$[E] \in \mathrm{Cl}(X)$ admitting a section $g \in \Gamma(U, \mathcal{R}_{[E]})$ with $g(\widehat{x}) \neq 0$. In other words,
$\mathrm{Cl}_x(X)$ is the orbit group of the point $\widehat{x} \in \widetilde{U}$. Now Proposition 1.2.2.8 gives the
assertion. $\qquad \square$

A point x of an irreducible, normal prevariety X is called *factorial* if its local
ring $\mathcal{O}_{X,x}$ admits unique factorization; this means exactly that near x every
Weil divisor is principal. Moreover, a point $x \in X$ is called $\mathbb{Q}$-*factorial* if near
x for every Weil divisor some nonzero multiple is principal.

Corollary 1.6.2.6 *Situation as in Construction 1.6.1.3.*

(i) *A point $x \in X$ is factorial if and only if the fiber $p_X^{-1}(x)$ is a single H_X-orbit with trivial isotropy.*
(ii) *A point $x \in X$ is $\mathbb{Q}$-factorial if and only if the fiber $p_X^{-1}(x)$ is a single H_X-orbit.*
(iii) *A point $x \in X$ is smooth if and only if the fiber $p_X^{-1}(x)$ is a single H_X-orbit with trivial isotropy and every $\widehat{x} \in p_X^{-1}(x)$ is smooth in $\widehat{X}$.*

Proof The point $x \in X$ is factorial if and only if $\mathrm{Cl}(X, x)$ is trivial, and it is
$\mathbb{Q}$-factorial if and only if $\mathrm{Cl}(X, x)$ is finite. Thus, the first two statements follow
from Proposition 1.6.2.5 and Corollary 1.2.3.7.

We prove (iii). The "only if" part follows directly from Proposition 1.6.1.6.
So, assume that $p_X^{-1}(x)$ is a single H_X-orbit and every $\widehat{x} \in p_X^{-1}(x)$ is smooth
in $\widehat{X}$. Then H acts freely over an open neighborhood $U \subseteq X$ of x and thus
p_X is an étale H-principal bundle over U. Because any $\widehat{x} \in p_X^{-1}(x)$ is smooth,
we can achieve by appropriate shrinking smoothness for the whole $p^{-1}(U)$.

Because étale morphisms preserve smoothness, we conclude that U is smooth and hence $x \in X$ is a smooth point. □

Corollary 1.6.2.7 *Consider the situation of Construction 1.6.1.3.*

(i) *The action of H_X on $\widehat{X}$ is free if and only if X is locally factorial.*
(ii) *The good quotient $p_X \colon \widehat{X} \to X$ is geometric if and only if X is $\mathbb{Q}$-factorial.*

Corollary 1.6.2.8 *Consider the situation of Construction 1.6.1.3. Let $\widehat{H}_X \subseteq H_X$ be the subgroup generated by all isotropy groups $H_{X,\widehat{x}}$, where $\widehat{x} \in \widehat{X}$. Then we have*

$$\ker\big(\mathbb{X}(H_X) \to \mathbb{X}(\widehat{H}_X)\big) = \bigcap_{\widehat{x} \in \widehat{X}} \ker\big(\mathbb{X}(H_X) \to \mathbb{X}(H_{X,\widehat{x}})\big)$$

and the projection $H_X \to H_X/\widehat{H}_X$ corresponds to the inclusion $\mathrm{Pic}(X) \subseteq \mathrm{Cl}(X)$ of character groups.

Corollary 1.6.2.9 *Consider the situation of Construction 1.6.1.3. If the variety $\widehat{X}$ contains an H_X-fixed point, then the Picard group $\mathrm{Pic}(X)$ is trivial.*

1.6.3 Total coordinate space and irrelevant ideal

Here we consider the situation that the Cox ring is finitely generated. This allows us to introduce the total coordinate space as the spectrum of the Cox ring. As before, we work over an algebraically closed field $\mathbb{K}$ of characteristic zero.

Construction 1.6.3.1 Let X be an irreducible, normal prevariety of affine intersection with $\Gamma(X, \mathcal{O}^*) = \mathbb{K}^*$ and finitely generated divisor class group $\mathrm{Cl}(X)$. Let $\mathcal{R}$ be a Cox sheaf and assume that the Cox ring $\mathcal{R}(X)$ is finitely generated. Then we have a diagram

$$\mathrm{Spec}_X \mathcal{R} = \widehat{X} \xrightarrow{\imath} \overline{X} = \mathrm{Spec}(\mathcal{R}(X))$$

$$p_X \downarrow$$

$$X$$

where the canonical morphism $\widehat{X} \to \overline{X}$ is an H_X-equivariant open embedding, the complement $\overline{X} \setminus \widehat{X}$ is of codimension at least 2, and $\overline{X}$ is an H_X-factorial affine variety. We call the H_X-variety $\overline{X}$ the *total coordinate space* associated with $\mathcal{R}$.

Proof Cover X by affine open sets $X_{[D],f} = X \setminus \mathrm{Supp}(\mathrm{div}_{[D]}(f))$, where $[D] \in \mathrm{Cl}(X)$ and $f \in \Gamma(X, \mathcal{R}_{[D]})$. Then, according to Proposition 1.6.2.1, the variety $\widehat{X}$ is covered by the affine sets $\widehat{X}_f = p_X^{-1}(X_{[D],f})$. Note that we have

$$\Gamma(\widehat{X}_f, \mathcal{O}) = \Gamma(\widehat{X}, \mathcal{O})_f = \Gamma(\overline{X}, \mathcal{O})_f = \Gamma(\overline{X}_f, \mathcal{O}).$$

Consequently, the canonical morphisms $\widehat{X}_f \to \overline{X}_f$ are isomorphisms. Gluing them together gives the desired open embedding $\widehat{X} \to \overline{X}$. □

Definition 1.6.3.2 Consider the situation of Construction 1.6.3.1. The *irrelevant ideal* of the prevariety X is the vanishing ideal of the complement $\overline{X} \setminus \widehat{X}$ in the Cox ring:

$$\mathcal{J}_{\text{irr}}(X) := \{f \in \mathcal{R}(X); \ f|_{\overline{X} \setminus \widehat{X}} = 0\} \subseteq \mathcal{R}(X).$$

Proposition 1.6.3.3 *Consider the situation of Construction 1.6.3.1.*

(i) *For any section $f \in \Gamma(X, \mathcal{R}_{[D]})$, membership in the irrelevant ideal is characterized as follows:*

$$f \in \mathcal{J}_{\text{irr}}(X) \iff \overline{X}_f = \widehat{X}_f \iff \widehat{X}_f \text{ is affine.}$$

(ii) *Let $0 \neq f \in \Gamma(X, \mathcal{R}_{[D]})$. If the $[D]$-localization $X_{[D],f}$ is affine, then we have $f \in \mathcal{J}_{\text{irr}}(X)$.*

(iii) *Let $0 \neq f_i \in \Gamma(X, \mathcal{R}_{[D_i]})$, where $1 \leq i \leq r$ be such that the sets $X_{[D_i],f_i}$ are affine and cover X. Then we have*

$$\mathcal{J}_{\text{irr}}(X) = \sqrt{\langle f_1, \ldots, f_r \rangle}.$$

Proof The first equivalence in (i) is obvious and the second one follows from the fact that $\overline{X} \setminus \widehat{X}$ is of codimension at least 2 in $\overline{X}$. Proposition 1.6.2.1 tells us that for affine $X_{[D],f}$ also $\widehat{X}_f$ is affine, which gives (ii). We turn to (iii). Proposition 1.6.2.1 and (ii) ensure that the functions $f_1, \ldots, f_r$ have $\overline{X} \setminus \widehat{X}$ as their common zero locus. Thus Hilbert's Nullstellensatz gives the assertion. □

Corollary 1.6.3.4 *Consider the situation of Construction 1.6.3.1. Then X is affine if and only if $\widehat{X} = \overline{X}$ holds.*

Proof Take $f = 1$ in the characterization 1.6.3.3 (i). □

Corollary 1.6.3.5 *Consider the situation of Construction 1.6.3.1 and assume that X is $\mathbb{Q}$-factorial. Then $0 \neq f \in \Gamma(X, \mathcal{R}_{[D]})$ belongs to $\mathcal{J}_{\text{irr}}(X)$ if and only if $X_{[D],f}$ is affine. In particular, we have*

$$\mathcal{J}_{\text{irr}}(X) = \lim_{\mathbb{K}}(f \in \Gamma(X, \mathcal{R}_{[D]}); \ [D] \in \text{Cl}(X), \ X_{[D],f} \text{ is affine}).$$

Proof We have to show that for any $[D]$-homogeneous $f \in \mathcal{J}_{\mathrm{irr}}(X)$, the $[D]$-localization $X_{[D],f}$ is affine. Note that $\widehat{X}_f$ is affine by Proposition 1.6.3.3 (i). The assumption of $\mathbb{Q}$-factoriality ensures that $p_X \colon \widehat{X} \to X$ is a geometric quotient; see Corollary 1.6.2.7. In particular, all H_X-orbits in $\widehat{X}$ are closed and thus Corollary 1.6.2.2 gives us $\widehat{X}_f = p_X^{-1}(X_{[D],f})$. Thus, as the good quotient space of the affine variety $\widehat{X}_f$, the set $X_{[D],f}$ is affine. $\qquad\square$

Recall that a Weil divisor D on a prevariety X is said to be *ample* if some multiple nD, where $n \in \mathbb{Z}_{\geq 1}$, admits sections $f_1, \ldots, f_r \in \Gamma(X, \mathcal{O}_X(nD))$ such that the sets X_{nD,f_i} are affine and cover X.

Corollary 1.6.3.6 *Consider the situation of Construction 1.6.3.1. If $[D] \in \mathrm{Cl}(X)$ is the class of an ample divisor, then we have*

$$\mathcal{J}_{\mathrm{irr}}(X) = \sqrt{\langle \Gamma(X, \mathcal{R}_{[D]}) \rangle}.$$

1.6.4 Characteristic spaces via GIT

As we saw, the characteristic space of a prevariety X of affine intersection is a quasiaffine variety $\widehat{X}$ with an action of the characteristic quasitorus H_X having X as a good quotient. Our aim is to characterize this situation in terms of geometric invariant theory (GIT). The crucial notion is the following. The ground field $\mathbb{K}$ is algebraically closed and of characteristic zero.

Definition 1.6.4.1 Let G be an affine algebraic group and W an irreducible G-prevariety. We say that the G-action on W is *strongly stable* if there is an open invariant subset $W' \subseteq W$ with the following properties:

(i) The complement $W \setminus W'$ is of codimension at least 2 in W.
(ii) The group G acts freely, that is, with trivial isotropy groups, on W'.
(iii) For every $x \in W'$ the orbit $G \cdot x$ is closed in W.

Remark 1.6.4.2 Let X be an irreducible, normal prevariety as in Construction 1.6.1.3 and consider the characteristic space $p_X \colon \widehat{X} \to X$ introduced there. Then Proposition 1.6.1.6 shows that the subset $p_X^{-1}(X_{\mathrm{reg}}) \subseteq \widehat{X}$ satisfies the properties of Definition 1.6.4.1.

Let X and $p_X \colon \widehat{X} \to X$ be as in Construction 1.6.1.3. In the sequel, we mean by a *characteristic space* for X more generally a good quotient $q \colon \mathcal{X} \to X$ for an action of a quasitorus H on a prevariety $\mathcal{X}$ such that there is an equivariant isomorphism $(\mu, \widetilde{\mu})$ making the following diagram commutative:

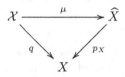

Recall that here $\mu \colon \mathcal{X} \to \widehat{X}$ is an isomorphism of prevarieties and $\widetilde{\mu} \colon H \to H_X$ is an isomorphism of algebraic groups such that we always have $\mu(h \cdot x) = \widetilde{\mu}(h) \cdot \mu(x)$. Note that a good quotient $q \colon \mathcal{X} \to X$ is a characteristic space if and only if we have an isomorphism of graded sheaves $\mathcal{R} \to q_*(\mathcal{O}_\mathcal{X})$, where $\mathcal{R}$ is a Cox sheaf on X.

Theorem 1.6.4.3 *Let a quasitorus H act on an irreducible, normal prevariety $\mathcal{X}$ with a good quotient $q \colon \mathcal{X} \to X$. Assume that $\mathcal{X}$ has only constant invertible global homogeneous functions, is H-factorial, and the H-action is strongly stable. Then X is an irreducible, normal prevariety, of affine intersection if $\mathcal{X}$ is so, $\Gamma(X, \mathcal{O}^*) = \mathbb{K}^*$ holds, $\mathrm{Cl}(X)$ is finitely generated, the Cox sheaf of X is locally of finite type, and $q \colon \mathcal{X} \to X$ is a characteristic space for X.*

The proof is given later in this section. First we also generalize the concept of the *total coordinate space* of a prevariety X of affine intersection with finitely generated Cox ring $\mathcal{R}(X)$: this is from now on any affine H-variety isomorphic to the affine H_X-variety $\overline{X}$ of Construction 1.6.3.1.

Corollary 1.6.4.4 *Let Z be an irreducible, normal affine variety with an action of a quasitorus H. Assume that every invertible homogeneous function on Z is constant, Z is H-factorial, and there exists an open H-invariant subset $W \subseteq Z$ with $\mathrm{codim}_Z(Z \setminus W) \geq 2$ such that the H-action on W is strongly stable and admits a good quotient $q \colon W \to X$. Then Z is a total coordinate space for X.*

A first step in the proof of Theorem 1.6.4.3 is to describe the divisor class group of the quotient space. Let us prepare the corresponding statement. Consider an irreducible prevariety $\mathcal{X}$ with an action of a quasitorus $H = \mathrm{Spec}\,\mathbb{K}[M]$. For any H-invariant morphism $q \colon \mathcal{X} \to X$ to an irreducible prevariety X, we have the pushforward homomorphism

$$q_* \colon \mathrm{WDiv}(\mathcal{X})^H \to \mathrm{WDiv}(X)$$

from the invariant Weil divisors of $\mathcal{X}$ to the Weil divisors of X sending an H-prime divisor $D \subseteq \mathcal{X}$ to the closure of its image $q(D)$ if the latter is of codimension 1 and to zero else. We denote the multiplicative group of nonzero homogeneous rational functions on $\mathcal{X}$ by $E(\mathcal{X})$ and the subset of nonzero rational functions of weight $w \in M$ by $E(\mathcal{X})_w$.

Proposition 1.6.4.5 *Let a quasitorus $H = \mathrm{Spec}\, \mathbb{K}[M]$ act on an irreducible, normal prevariety $\mathcal{X}$ with a good quotient $q \colon \mathcal{X} \to X$. Assume that $\mathcal{X}$ admits only constant invertible homogeneous functions and the H-action is strongly stable. Then X is an irreducible, normal prevariety, of affine intersection if $\mathcal{X}$ is so, and there is a homomorphism*

$$\delta \colon E(\mathcal{X}) \to \mathrm{WDiv}(X), \qquad f \mapsto q_*(\mathrm{div}(f)).$$

We have $\mathrm{div}(f) = q^*(\delta(f))$ for every $f \in E(\mathcal{X})$ and δ is surjective if $\mathcal{X}$ is H-factorial. Moreover, δ induces a well-defined monomorphism

$$M \to \mathrm{Cl}(X), \qquad w \mapsto [\delta(f)], \quad \text{with any } f \in E(\mathcal{X})_w$$

which is an isomorphism if $\mathcal{X}$ is H-factorial. Finally, for every $f \in E(\mathcal{X})_w$, and every open set $U \subseteq X$, we have an isomorphism of $\Gamma(U, \mathcal{O})$-modules

$$\Gamma(U, \mathcal{O}_X(\delta(f))) \to \Gamma(q^{-1}(U), \mathcal{O}_\mathcal{X})_w, \qquad g \mapsto fq^*(g).$$

Proof First note that the good quotient space X inherits normality from $\mathcal{X}$ and is of affine intersection if $\mathcal{X}$ is so.

Let $\mathcal{X}' \subseteq \mathcal{X}$ be as in Definition 1.6.4.1. Then, with $X' := q(\mathcal{X}')$, we have $q^{-1}(X') = \mathcal{X}'$. Consequently, $X' \subseteq X$ is open. Moreover, $X \setminus X'$ is of codimension at least 2 in X, because $\mathcal{X} \setminus \mathcal{X}'$ is of codimension at least 2 in $\mathcal{X}$. Thus, we may assume that $X = X'$ holds, which means in particular that H acts freely. Then we have homomorphisms of groups:

$$E(\mathcal{X}) \xrightarrow{\; f \mapsto \mathrm{div}(f) \;} \mathrm{WDiv}(\mathcal{X})^H \; \underset{q^*}{\overset{q_*}{\rightleftarrows}} \; \mathrm{WDiv}(X).$$

Clearly, the homomorphism from $E(\mathcal{X})$ to the group of H-invariant Weil divisors $\mathrm{WDiv}(\mathcal{X})^H$ is surjective if $\mathcal{X}$ is H-factorial. Moreover, q^* and q_* are inverse to each other, which follows from the observation that $q \colon \mathcal{X} \to X$ is an étale H-principal bundle. This establishes the first part of the assertion.

We show that δ induces a monomorphism $M \to \mathrm{Cl}(X)$. First we have to check that $[\delta(f)]$ does not depend on the choice of f. So, let $f, g \in E(\mathcal{X})_w$. Then f/g is H-invariant, and hence defines a rational function on X. We infer well-definedness of $w \mapsto [\delta(f)]$ from

$$q_*(\mathrm{div}(f)) - q_*(\mathrm{div}(g)) = q_*(\mathrm{div}(f) - \mathrm{div}(g)) = q_*(\mathrm{div}(f/g)) = \mathrm{div}(f/g).$$

To verify injectivity, let $\delta(f) = \mathrm{div}(h)$ for some $h \in \mathbb{K}(X)^*$. Then we obtain $\mathrm{div}(f) = \mathrm{div}(q^*(h))$. Thus, $f/q^*(h)$ is an invertible homogeneous function on $\mathcal{X}$ and hence is constant. This implies $w = \deg(f/q^*(h)) = 0$. If $\mathcal{X}$ is H-factorial, then $M \to \mathrm{Cl}(X)$ is surjective, because $E(\mathcal{X}) \to \mathrm{WDiv}(X)$ is surjective in this case.

We turn to the last statement. First we note that for every $g \in \Gamma(U, \mathcal{O}_X(\delta(f)))$ the function $f q^*(g)$ is regular on $q^{-1}(U)$, because we have

$$\mathrm{div}(f q^*(g)) = \mathrm{div}(f) + \mathrm{div}(q^*(g)) = q^*(\delta(f)) + \mathrm{div}(q^*(g))$$

$$= q^*(\delta(f) + \mathrm{div}(g)) \geq 0.$$

Thus, the homomorphism $\Gamma(U, \mathcal{O}_X(\delta(f))) \to \Gamma(q^{-1}(U), \mathcal{O}_X)_w$ sending g to $f q^*(g)$ is well defined. Note that $h \mapsto h/f$ defines an inverse. $\qquad\square$

Corollary 1.6.4.6 *Consider the characteristic space* $q \colon \widehat{X} \to X$ *obtained from a Cox sheaf* $\mathcal{R}$. *Then, for any nonzero* $f \in \Gamma(X, \mathcal{R}_{[D]})$ *the pushforward* $q_*(\mathrm{div}(f))$, *equals the* $[D]$-*divisor* $\mathrm{div}_{[D]}(f)$.

Proof Proposition 1.6.4.5 shows that $q^*(q_*(\mathrm{div}(f)))$ equals $\mathrm{div}(f)$ and Proposition 1.6.2.1 tells us that $q^*(\mathrm{div}_{[D]}(f))$ equals $\mathrm{div}(f)$ as well. $\qquad\square$

Proof of Theorem 1.6.4.3 Writing $H = \mathrm{Spec}\,\mathbb{K}[M]$ with the character group M of H, we are in the setting of Proposition 1.6.4.5. Choose a finitely generated subgroup $K \subseteq \mathrm{WDiv}(X)$ mapping onto $\mathrm{Cl}(X)$, and let $D_1, \ldots, D_s \in \mathrm{WDiv}(X)$ be a basis of K. By Proposition 1.6.4.5, we have $D_i = \delta(h_i)$ with $h_i \in E(\mathcal{X})_{w_i}$. Moreover, the isomorphism $M \to \mathrm{Cl}(X)$ given there identifies $w_i \in M$ with $[D_i] \in \mathrm{Cl}(X)$. For $D = a_1 D_1 + \cdots + a_s D_s$, we have $D = \delta(h_D)$ with $h_D = h_1^{a_1} \cdots h_s^{a_s}$.

Let $\mathcal{S}$ be the sheaf of divisorial algebras associated with K and for $D \in K$, let $w \in M$ correspond to $[D] \in \mathrm{Cl}(X)$. Then, for any open set $U \subseteq X$ and any $D \in K$, Proposition 1.6.4.5 provides an isomorphism of $\mathbb{K}$-vector spaces

$$\Phi_{U,D} \colon \Gamma(U, \mathcal{S}_D) \;\to\; \Gamma(q^{-1}(U), \mathcal{O})_w, \qquad g \mapsto q^*(g) h_D.$$

The $\Phi_{U,D}$ fit together to an epimorphism of graded sheaves $\Phi \colon \mathcal{S} \to q_*(\mathcal{O}_X)$. Once we know that Φ has the ideal $\mathcal{I}$ of Construction 1.4.2.1 as its kernel, we obtain an induced isomorphism $\mathcal{R} \to q_*\mathcal{O}_X$, where $\mathcal{R} = \mathcal{S}/\mathcal{I}$ is the associated Cox sheaf; this shows that $\mathcal{R}$ is locally of finite type and gives an isomorphism $\mu \colon \mathcal{X} \to \widehat{X}$.

Thus we are left with showing that the kernel of Φ equals $\mathcal{I}$. Consider a $\mathrm{Cl}(X)$-homogeneous element $f \in \Gamma(U, \mathcal{S})$ of degree $[D]$, where $D \in K$. Let K^0 be the kernel of the surjection $K \to \mathrm{Cl}(X)$. Then we have

$$f = \sum_{E \in K^0} f_{D+E}, \qquad \Phi(f) = \sum_{E \in K^0} q^*(f_{D+E}) h_{D+E}.$$

With the character $\chi \colon K^0 \to \mathbb{K}(X)^*$ defined by $q^*\chi(E) = h_E$, we may rewrite the image $\Phi(f)$ as

$$\Phi(f) = \sum_{E \in K^0} q^*(\chi(E) f_{D+E}) h_D = q^* \left(\sum_{E \in K^0} \chi(E) f_{D+E} \right) h_D.$$

So, f lies in the kernel of Φ if and only if $\sum \chi(E) f_{D+E}$ vanishes. Now observe that we have

$$f = \sum_{E \in K^0} (1 - \chi(E)) f_{D+E} + \sum_{E \in K^0} \chi(E) f_{D+E}.$$

The second summand is K-homogeneous, and thus we infer from Lemma 1.4.3.1 that $f \in \mathcal{I}$ holds if and only if $\sum \chi(E) f_{D+E} = 0$ holds. $\qquad \square$

Remark 1.6.4.7 Consider the isomorphism $(\mu, \widetilde{\mu})$ identifying the characteristic spaces $q \colon \mathcal{X} \to X$ and $p_X \colon \widehat{X} \to X$ in the above proof. Then the isomorphism $\widetilde{\mu}$ identifying the quasitori H and H_X is given by the isomorphism $M \to \mathrm{Cl}(X)$ of their character groups provided by Proposition 1.6.4.5.

Exercises to Chapter 1

Exercise 1.1 (Graded algebras) Let $\mathbb{K}$ be a field, K be an abelian monoid, and $A = \oplus_K A_w$ a K-graded $\mathbb{K}$-algebra.

(1) If K admits cancellation, then the unit element $1 \in A$ is K-homogeneous of degree zero. For arbitrary K, this need not hold.
(2) Assume that A is integral with $A^* = \mathbb{K}^*$. If $A_0 = \mathbb{K}$ holds, then the weight cone $\omega(A) \subseteq K_{\mathbb{Q}}$ contains no line. The converse is false in general.
(3) If K is a free finitely generated abelian group and the algebra A is normal, then $S(A) = \omega(A) \cap K(A)$ holds.
(4) If K is an abelian group and $L \subseteq K$ a subgroup, then normality of A implies normality of the Veronese subalgebra $A(L)$.
(5) Assume that $\mathbb{K}$ is algebraically closed of characteristic zero, K is a finitely generated abelian group, and A is finitely generated integral.
   (a) The K-grading of A extends to a K-grading of the integral closure $\overline{A} \subseteq \mathrm{Quot}(A)$.
   (b) A is normal if every element $f/g \in \mathrm{Quot}(A)$ with homogeneous f, $g \in A$ which is integral over A belongs to A.

Exercise 1.2 Let $\mathbb{K}$ be a field. All $\mathbb{K}$-algebras A in this exercise are graded by $K := \mathbb{Z}/2\mathbb{Z} = \{\bar{0}, \bar{1}\}$.

(1) Let $A := \mathbb{K}[T]/\langle T^2 - 1 \rangle$ be K-graded by $\deg(T) := \bar{1}$. Then A has zero-divisors, but no homogeneous ones.
(2) Assume that $\mathbb{K}$ is not of characteristic two. Consider $A := \mathbb{K}[T, T^{-1}]$ with the K-grading defined by $\deg(T + T^{-1}) := \bar{0}$ and $\deg(T - T^{-1}) := \bar{1}$. Then A has nonconstant invertible elements, but no homogeneous ones.

(3) Let $A := \mathbb{K}[T]$ with the K-grading given by $\deg(T) = \bar{1}$. Then the homogeneous element $1 - T^2 = (1 + T)(1 - T)$ is reducible but it is not a product of noninvertible homogeneous elements.

(4) Consider the $\mathbb{K}$-algebra $A := \mathbb{K}[T_i; \ i \in \mathbb{Z}_{>0}]/\langle T_i T_j; \ i, j \in \mathbb{Z}_{>0}\rangle$ with the grading defined by $\deg(T_i) = \bar{1}$. Then the subalgebra $A_{\bar{0}} = \mathbb{K}$ is finitely generated, but the algebra A is not.

Exercise 1.3 (Quasitori)   An affine algebraic group is a quasitorus if and only if it is isomorphic to a direct product of a torus $\mathbb{T}^n$ and a finite abelian group. Moreover,

  (i) Every closed subgroup of a quasitorus is a quasitorus.
 (ii) Images of quasitori under homorphisms of algebraic groups are quasitori.
(iii) If a quasitorus H is a finite group, then $\mathbb{X}(H) \cong H$ holds.

Exercise 1.4   Let G be an affine algebraic group and X be a G-variety. The G-action $(g \cdot f)(x) := f(g^{-1}x)$ defines on $\Gamma(X, \mathcal{O})$ a structure of a rational G-module.

Exercise 1.5 (Quasitorus actions)   Let K be a finitely generated abelian group and A a K-graded affine algebra. Consider the corresponding action of the quasitorus $H = \operatorname{Spec} \mathbb{K}[K]$ on $X = \operatorname{Spec} A$.

(1) Take $x \in X$. Then the orbit $H \cdot x \subseteq X$ is closed if and only if $K_x = S_x$ holds.

(2) Let $Y \subseteq X$ be a closed H-invariant subset. Then for any finite collection $x_1, \ldots, x_r \in X \setminus Y$, there is a homogeneous function $f \in A$ with $f|_Y = 0$ and $f(x_i) \neq 0$ for all $1 \leq i \leq r$.

(3) Give an example of a free $\mathbb{K}^*$-action on an affine variety X with $\Gamma(X, \mathcal{O}^*) = \mathbb{K}^*$.

Exercise 1.6* (Linearization problem)   Let $\mathbb{K}$ be an algebraically closed field, K a finitely generated abelian group, $H := \operatorname{Spec} \mathbb{K}[K]$ the corresponding quasitorus, and $A := \mathbb{K}[T_1, \ldots, T_r] = \Gamma(\mathbb{K}^r, \mathcal{O})$. Consider the following three statements.

(1) Given any K-grading on A, there exists an algebraically independent system of homogeneous generators of A.

(2) Any action $H \times \mathbb{K}^r \to \mathbb{K}^r$ is of the form $h \cdot w = (\chi_1(h)w_1, \ldots, \chi_r(h)w_r)$ for suitable coordinates $w_1, \ldots, w_r$ on $\mathbb{K}^r$.

(3) Any action $H \times \mathbb{K}^r \to \mathbb{K}^r$ is linear with respect to a suitable system of coordinates on $\mathbb{K}^r$.

Show that the three statements are equivalent. Prove that (2) holds provided that H acts effectively and $\dim(H) \geq r - 1$ holds. Prove that (1) holds if $A_0 = \mathbb{K}$ holds. Prove or disprove the statements in general. See [196] for more background.

Exercise 1.7 (Separatedness) Let X be a prevariety. Recall that X is separated, that is, a variety, if the diagonal $\{(x, x); \ x \in X\}$ is closed in $X \times X$.

(1) X is separated if and only if for every prevariety Y and any two morphisms $\varphi, \psi \colon Y \to X$ the subset $\{y \in Y; \ \varphi(y) = \psi(y)\}$ is closed in Y.
(2) Every locally closed subset of a variety is a variety.
(3) If X is a variety and $U, U' \subseteq X$ are open affine subsets, then the intersection $U \cap U'$ is affine.
(4) Let X be covered by affine open subsets $X_i \subseteq X$. Then X is separated if and only if all $X_i \cap X_j$ are affine and each algebra $\Gamma(X_i \cap X_j, \mathcal{O})$ is generated by the restrictions of $\Gamma(X_i, \mathcal{O})$ and $\Gamma(X_j, \mathcal{O})$.
(5) If for every two points $x_1, x_2 \in X$ there is an open affine subset $U \subseteq X$ with $x_1, x_2 \in U$, then X is separated.
(6) There are only finitely many separated open subsets of X which are maximal with respect to inclusion; see [45, Thm. 1] for a solution.

Exercise 1.8 (Prevarieties of affine intersection) Recall that a prevariety X is of affine intersection, if for any two affine open $U, U' \subseteq X$ the intersection $U \cap U'$ is affine.

(1) A prevariety X is of affine intersection if and only if the diagonal morphism $\Delta \colon X \to X \times X, x \mapsto (x, x)$ is affine.
(2) If a prevariety X is covered by affine open subsets $X_i \subseteq X$ such that all intersections $X_i \cap X_j$ are affine, then X is of affine intersection.
(3) The affine line with a doubled zero, that is, the gluing of two copies of $\mathbb{K}$ along $\mathbb{K}^*$, is a nonseparated prevariety of affine intersection. The affine plane with a doubled point is not of affine intersection.
(4) If a prevariety X is of affine intersection, then the complement of any affine open subset $U \subsetneq X$ is of pure codimension 1.
(5) Consider $Y := V(T_1 T_2 - T_3 T_4) \subseteq \mathbb{K}^4$ and the divisor $D := Y \cap V(T_1, T_3)$. Show that $Y \setminus D$ is not affine. Hint: Consider $Y \setminus D \cap V(T_2, T_4)$. Let X be the prevariety obtained from Y by doubling D. Then X is not of affine intersection, but every affine open $U \subseteq X$ has complement of pure codimension 1.
(6) Let $\widehat{X}$ be the relative spectrum of a sheaf of divisorial algebras on an irreducible, normal prevariety X. If the prevariety $\widehat{X}$ is separated, then X is of affine intersection.

Exercise 1.9 Let X be a prevariety.

(1) X is isomorphic to a quasiaffine variety if and only if there are elements $f_1, \ldots, f_r \in \Gamma(X, \mathcal{O})$ such that $X_{f_1}, \ldots, X_{f_r}$ is an affine cover of X.
(2) X is isomorphic to an affine variety if and only if there are elements $f_1, \ldots, f_r \in \Gamma(X, \mathcal{O})$ such that $X_{f_1}, \ldots, X_{f_r}$ is an affine cover of X and the ideal generated by $f_1, \ldots, f_r$ in $\Gamma(X, \mathcal{O})$ coincides with $\Gamma(X, \mathcal{O})$.

Exercise 1.10 (The Proj construction) Let $A = \oplus A_d$ be a $\mathbb{Z}_{\geq 0}$-graded affine algebra. For a homogeneous $f \in A_{>0}$ set $U_f = \operatorname{Spec} A_{(f)}$. Check that the morphisms $U_{fg} \to U_f$ are open embeddings. Use Exercise 1.7 (iv) to show that the gluing of the U_f along U_{fg} gives a variety.

Exercise 1.11 (Good quotients)

(1) Let G be a reductive group, X a G-prevariety, and $p\colon X \to X /\!\!/ G$ a good quotient. Then p is geometric if and only if $U = p^{-1}(p(U))$ holds for every open G-invariant subset $U \subseteq X$.
(2) If a torus T acts on a normal prevariety X with only finite isotropy groups, then this action admits a good quotient in the category of prevarieties; see [285, Cor. 3] for a solution. For arbitrary reductive group actions this statement need not hold in general.
(3) The action of $\mathbb{K}^*$ on $\mathbb{K}^2 \setminus \{(0, 0)\}$ given by $t \cdot z = (tz_1, t^{-1}z_2)$ admits a good quotient in the category of prevarieties but not in the category of varieties.

Exercise 1.12 (Unique factorization domains) (1) If X is an irreducible, normal prevariety with $\operatorname{Cl}(X) = 0$, then $\Gamma(X, \mathcal{O})$ is a unique factorization domain.
(2) If X is an irreducible, normal quasiaffine variety and $\Gamma(X, \mathcal{O})$ is a unique factorization domain, then $\operatorname{Cl}(X) = 0$ holds.

Exercise 1.13 Recall that an n-dimensional irreducible prevariety is rational if it contains an open subset isomorphic to some open subset of $\mathbb{K}^n$.

(1) If X is an irreducible, normal rational prevariety, then the group $\operatorname{Cl}(X)$ is finitely generated. Give an example where the converse is not true.
(2) If X is an irreducible, normal prevariety, which contains an open subset isomorphic to an affine space, then $\operatorname{Cl}(X)$ is finitely generated and torsion free.

Exercise 1.14 Let X be an irreducible, normal prevariety with finitely generated divisor class group $\operatorname{Cl}(X)$ of rank r. Prove that $\operatorname{Quot}(\mathcal{R}(X)) \cong \mathbb{K}(X)(T_1, \ldots, T_r)$ if and only if $\operatorname{Cl}(X)$ is torsion free.

Exercise 1.15 Describe the Cox ring of the projective plane with a doubled line.

Exercise 1.16 (Face subalgebras) Let K be a finitely generated abelian group and A a K-graded $\mathbb{K}$-algebra. Consider a face $\omega \preceq \omega(A)$ of the weight cone $\omega(A) \subseteq K_\mathbb{Q}$ and let $L \subseteq K$ be the subgroup of all weights mapped to ω. If A is factorially K-graded, then the Veronese subalgebra $A(L)$ is factorially L-graded.

Exercise 1.17 Let X be an irreducible, normal variety with $\Gamma(X, \mathcal{O}^*) = \mathbb{K}^*$, finitely generated divisor class group $\mathrm{Cl}(X)$, characteristic space $p_X \colon \widehat{X} \to X$, total coordinate space $\overline{X}$, and characteristic quasitorus H_X.

(1) Assume that X is locally factorial. Then $p_X \colon \widehat{X} \to X$ is a locally trivial H_X-principal bundle (with respect to the Zariski topology) if and only if H_X is connected.
(2) If X is a complete variety, then the quotient space of the total coordinate space $\overline{X}$ by H_X is a point.
(3) Give an example of a variety X with $\mathrm{Pic}(X) = 0$ such that the variety $\widehat{X}$ does not contain an H_X-fixed point.

Exercise 1.18 Let X be an irreducible, normal variety with $\Gamma(X, \mathcal{O}^*) = \mathbb{K}^*$ and characteristic space $p_X \colon \widehat{X} \to X$. Let D be a prime divisor on X such that $Y := X \setminus D$ satisfies $\Gamma(Y, \mathcal{O}^*) = \mathbb{K}^*$ and let $p_Y \colon \widehat{Y} \to Y$ be the characteristic space. Then there is a commutative diagram

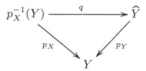

If the class $[D]$ has order n in $\mathrm{Cl}(X)$, then q is an n-fold covering; If the class $[D]$ has infinite order in $\mathrm{Cl}(X)$, then q is a trivial $\mathbb{K}^*$-bundle.

Exercise 1.19 Check that all three conditions in the definition of a strongly stable action are essential in Theorem 1.6.4.3.

Exercise 1.20 Compute the irrelevant ideal in Example 1.4.1.6.

Exercise 1.21 Let V be a finite-dimensional vector space and $G \subseteq \mathrm{GL}(V)$ a finite subgroup. Let $H \subseteq G$ be the (normal) subgroup generated by the pseudoreflections, that is, the elements having a hyperplane as fixed point set. Set $F := G/H$, let $\pi \colon G \to F$ be the canonical map and set $K := \pi^{-1}([F, F])$, where $[F, F] \subseteq F$ is the commutator subgroup. Then the affine variety $Z := V/K$ comes with an action of the finite abelian group $N := F/[F, F]$ and the canonical map $Z \to V/G$ is the quotient for the N-action. Show that $Z \to V/G$ is the characteristic space over V/G. Give an example where Z is not factorial. See [16] for a solution.

Exercise 1.22* Consider a representation of a reductive group G on a vector space V. Describe the Cox ring of $V /\!\!/ G$ in terms of the representation; note that for semisimple G, the quotient $V /\!\!/ G$ is factorial.

2

Toric varieties and Gale duality

Toric varieties form an important class of examples in algebraic geometry, as they admit a complete description in terms of combinatorial data, so-called lattice fans. In Section 2.1, we briefly recall this description and also some of the basic facts in toric geometry. Then we present Cox's construction of the characteristic space of a toric variety in terms of a defining fan and discuss the basic geometry around this. Section 2.2 is pure combinatorics. We introduce the notion of a "bunch of cones" and show that, in an appropriate setting, this is the Gale dual version of a fan. Under this duality, the normal fans of polytopes correspond to bunches of cones arising canonically from the chambers of the so-called Gelfand–Kapranov–Zelevinsky decomposition. In Section 2.3, we discuss the geometric meaning of bunches of cones: they encode the maximal separated good quotients for subgroups of the acting torus on an affine toric variety. In Section 2.4, we specialize these considerations to toric characteristic spaces, that is, to the good quotients arising from Cox's construction. This leads to an alternative combinatorial description of toric varieties in terms of "lattice bunches," which turns out to be particularly suitable for phenomena around divisors.

2.1 Toric varieties

2.1.1 Toric varieties and fans

We introduce toric varieties and their morphisms and recall that this category admits a complete description in terms of lattice fans. The details can be found in any textbook on toric varieties, for example [105], [141], or [234]. We work over an algebraically closed field $\mathbb{K}$ of characteristic zero.

Definition 2.1.1.1 A *toric variety* is an irreducible, normal variety X together with an algebraic torus action $T \times X \to X$ and a base point $x_0 \in X$ such that the orbit map $T \to X$, $t \mapsto t \cdot x_0$ is an open embedding.

In this setting, we refer to T as to the *acting torus* of the toric variety X. If we want to specify notation, we sometimes denote a toric variety X with acting torus T and base point x_0 as a triple (X, T, x_0).

Definition 2.1.1.2 Let X and X' be toric varieties. A *toric morphism* from X to X' is a pair $(\varphi, \widetilde{\varphi})$, where $\varphi \colon X \to X'$ is a morphism with $\varphi(x_0) = x'_0$ and $\widetilde{\varphi} \colon T \to T'$ is a homomorphism of the respective acting tori such that $\varphi(t \cdot x) = \widetilde{\varphi}(t) \cdot \varphi(x)$ holds for all $t \in T$ and $x \in X$.

Note that for a toric morphism $(\varphi, \widetilde{\varphi})$, the homomorphism $\widetilde{\varphi} \colon T \to T'$ of the acting tori is uniquely determined by the morphism $\varphi \colon X \to X'$ of varieties; we therefore often denote a toric morphism just by $\varphi \colon X \to X'$.

A first step in the combinatorial description of the category of toric varieties is to relate affine toric varieties to lattice cones. Recall that a *lattice cone* is a pair (σ, N), where N is a lattice and $\sigma \subseteq N_\mathbb{Q}$ a *pointed*, that is, σ contains no line, convex polyhedral cone in the rational vector space $N_\mathbb{Q} = N \otimes_\mathbb{Z} \mathbb{Q}$ associated with N; we also refer to this setting less formally as to a cone σ in a lattice N. By a *map of lattice cones* (σ, N) and (σ', N') we mean a homomorphism $F \colon N \to N'$ with $F(\sigma) \subseteq \sigma'$, where, as earlier, F also denotes the induced linear map $N_\mathbb{Q} \to N'_\mathbb{Q}$ of rational vector spaces. With any lattice cone we associate in a functorial way a toric variety.

Construction 2.1.1.3 Let N be a lattice and $\sigma \subseteq N_\mathbb{Q}$ a pointed cone. Set $M := \mathrm{Hom}(N, \mathbb{Z})$ and let $\sigma^\vee \subseteq M_\mathbb{Q}$ be the dual cone. Then we have the M-graded affine $\mathbb{K}$-algebra

$$A_\sigma := \mathbb{K}[\sigma^\vee \cap M] = \bigoplus_{u \in \sigma^\vee \cap M} \mathbb{K}\chi^u.$$

The corresponding affine variety $X_\sigma = \mathrm{Spec}\, A_\sigma$ comes with an action of the torus $T_N := \mathrm{Spec}\,\mathbb{K}[M]$ and is a toric variety with the base point $x_0 \in X$ defined by the maximal ideal

$$\mathfrak{m}_{x_0} = \langle \chi^u - 1;\ u \in \sigma^\vee \cap M \rangle \subseteq A_\sigma.$$

For every map $F \colon N \to N'$ of lattice cones (σ, N) and (σ', N'), the dual map $F^* \colon M' \to M$ sends $(\sigma')^\vee$ to $\sigma^\vee$. Hence $F \colon N \to N'$ induces a morphism of graded algebras from $A_{\sigma'}$ to A_σ, see Construction 1.1.1.5, and thus, finally, a toric morphism $(\varphi_F, \widetilde{\varphi}_F)$ from X_σ to $X_{\sigma'}$; see Theorem 1.2.2.4.

To go the other way round, that is, from affine toric varieties to lattice cones one works with the one-parameter subgroups of the acting torus. Recall that a *one-parameter subgroup* of a torus T is a homomorphism $\lambda \colon \mathbb{K}^* \to T$. The one-parameter subgroups of a torus T form a lattice $\Lambda(T)$ with respect to pointwise multiplication. Note that we have bilinear pairing

$$\mathbb{X}(T) \times \Lambda(T) \to \mathbb{Z}, \qquad (\chi, \lambda) \mapsto \langle \chi, \lambda \rangle,$$

where $\langle \chi, \lambda \rangle \in \mathbb{Z}$ is the unique integer satisfying $\chi \circ \lambda(t) = t^{\langle \chi, \lambda \rangle}$ for all $t \in \mathbb{K}^*$. For every homomorphism $\varphi \colon T \to T'$ of tori, we have a functorial pushforward of one-parameter subgroups:

$$\varphi_* \colon \Lambda(T) \to \Lambda(T'), \qquad \lambda \mapsto \varphi \circ \lambda.$$

Construction 2.1.1.4 Let X be an affine toric variety with acting torus T and base point $x_0 \in X$. We call a one-parameter subgroup $\lambda \in \Lambda(T)$ *convergent in X* if the orbit map $\mathbb{K}^* \to X$, $t \mapsto \lambda(t) \cdot x_0$ can be extended to a morphism $\mathbb{K} \to X$. In this case the image of $0 \in \mathbb{K}$ is denoted as

$$\lim_{t \to 0} \lambda(t) \cdot x_0 \in X.$$

The convergent one-parameter subgroups $\lambda \in \Lambda(T)$ in X generate a pointed convex cone $\sigma_X \subseteq \Lambda_{\mathbb{Q}}(T)$. Moreover, for every toric morphism $(\varphi, \widetilde{\varphi})$ from X to X', the pushforward of one-parameter subgroups $\widetilde{\varphi}_* \colon \Lambda(T) \to \Lambda(T')$ is a map of the lattice cones $(\sigma_X, \Lambda(T))$ and $(\sigma_{X'}, \Lambda(T'))$.

Proposition 2.1.1.5 *We have covariant functors being essentially inverse to each other:*

$$\{lattice\ cones\} \longleftrightarrow \{affine\ toric\ varieties\}$$

$$(\sigma, N) \mapsto (X_\sigma, T_N, x_0),$$

$$F \mapsto (\varphi_F, \widetilde{\varphi}_F),$$

$$(\sigma_X, \Lambda(T)) \leftmapsto (X, T, x_0),$$

$$\widetilde{\varphi}_* \leftmapsto (\varphi, \widetilde{\varphi}).$$

We are ready to describe general toric varieties. The idea is to glue the affine descriptions. On the combinatorial side this means considering lattice fans. We first recall this concept: A *quasifan* in a rational vector space $N_{\mathbb{Q}}$ is a finite collection Σ of convex, polyhedral cones in $N_{\mathbb{Q}}$ such that for any $\sigma \in \Sigma$ all faces $\tau \preceq \sigma$ belong to Σ, and for any two $\sigma, \sigma' \in \Sigma$ the intersection $\sigma \cap \sigma'$ is a face of both, σ and σ'. A quasifan is called a *fan* if it consists of pointed cones. A *lattice fan* is a pair (Σ, N), where N is a lattice and Σ is a fan in $N_{\mathbb{Q}}$. A *map of lattice fans* (Σ, N) and (Σ', N') is a homomorphism $F \colon N \to N'$ such that for every $\sigma \in \Sigma$, there is a $\sigma' \in \Sigma'$ with $F(\sigma) \subseteq \sigma'$.

Construction 2.1.1.6 Let (Σ, N) be a lattice fan. Then, for any two $\sigma_1, \sigma_2 \in \Sigma$, the intersection $\sigma_{12} := \sigma_1 \cap \sigma_2$ belongs to Σ, and there is a *separating linear form*, that is, an element $u \in M$ with

$$u|_{\sigma_1} \geq 0, \qquad u|_{\sigma_2} \leq 0, \qquad u^\perp \cap \sigma_1 = \sigma_{12} = u^\perp \cap \sigma_2.$$

The subset $X_{\sigma_{12}} \subseteq X_{\sigma_1}$ is the localization of X_{σ_1} by χ^u and $X_{\sigma_{12}} \subseteq X_{\sigma_2}$ is the localization of X_{σ_2} by χ^{-u}. This allows gluing the affine toric varieties X_σ, where $\sigma \in \Sigma$, together to a variety X_Σ. Because this gluing is equivariant and respects base points, X_Σ is a toric variety with acting torus T_N and well-defined base point x_0.

Moreover, given a map $F \colon N \to N'$ from a lattice fan (Σ, N) to a lattice fan (Σ', N'), fix for every $\sigma \in \Sigma$ a $\sigma' \in \Sigma'$ with $F(\sigma) \subseteq \sigma'$. Then the associated toric morphisms from X_σ to $X_{\sigma'}$ glue together to a toric morphism $(\varphi_F, \widetilde{\varphi}_F)$ from X_Σ to $X_{\Sigma'}$.

The key for the way from toric varieties to lattice fans is Sumihiro's theorem [285], see also [189, Thm. 1], which tells us that every irreducible, normal variety with torus action can be covered by invariant open affine subvarieties. Using finiteness of orbits for affine toric varieties, one concludes that any toric variety admits only finitely many invariant open affine subvarieties and is covered by them.

Construction 2.1.1.7 Let X be a toric variety with acting torus T. Consider the T-invariant affine open subsets $X_1, \ldots, X_r \subseteq X$ and let $\Sigma_X := \{\sigma_{X_1}, \ldots, \sigma_{X_r}\}$ be the collection of the corresponding cones of convergent one-parameter subgroups. Then $(\Sigma_X, \Lambda(T))$ is a lattice fan. Moreover, every toric morphism $(\varphi, \widetilde{\varphi})$ to a toric variety X' with acting torus T' defines a map $\widetilde{\varphi}_* \colon \Lambda(T) \to \Lambda(T')$ of the lattice fans $(\Sigma_X, \Lambda(T))$ and $(\Sigma_{X'}, \Lambda(T'))$.

Theorem 2.1.1.8 *We have covariant functors being essentially inverse to each other:*

$$\{lattice\ fans\} \longleftrightarrow \{toric\ varieties\},$$

$$(\Sigma, N) \mapsto (X_\Sigma, T_N, x_0),$$

$$F \mapsto (\varphi_F, \widetilde{\varphi}_F),$$

$$(\Sigma_X, \Lambda(T)), \leftarrow (X, T, x_0),$$

$$\widetilde{\varphi}_* \leftarrow (\varphi, \widetilde{\varphi}).$$

2.1.2 Some toric geometry

The task of toric geometry is to describe geometric properties of a toric variety in terms of its defining fan. We recall here some of the very basic observations. Again we refer to the textbooks [105], [141], or [234] for details and more. The ground field $\mathbb{K}$ is algebraically closed and of characteristic zero.

As any space with group action, also each toric variety is the disjoint union of its orbits. For an explicit description of this orbit decomposition, one introduces distinguished points as follows; for a cone σ in a rational vector space, we denote by σ° its relative interior.

Construction 2.1.2.1 Let X be the toric variety arising from a fan Σ in a lattice N. With every cone $\sigma \in \Sigma$, one associates a (well-defined) *distinguished point*:

$$x_\sigma := \lim_{t \to 0} \lambda_v(t) \cdot x_0 \in X, \qquad \text{where } v \in \sigma^\circ.$$

On every affine chart $X_\sigma \subseteq X$, where $\sigma \in \Sigma$, the distinguished point is the unique point with the property

$$\chi^u(x_\sigma) = \begin{cases} 1, & \text{where } u \in \sigma^\perp \cap M, \\ 0, & \text{where } u \in \sigma^\vee \cap M \setminus \sigma^\perp. \end{cases}$$

Note that the distinguished points are precisely the possible limits of the one-parameter subgroups of the acting torus passing through the base point. The following statement shows in particular that the distinguished points represent exactly the orbits of a toric variety.

Proposition 2.1.2.2 (Orbit decomposition) *Let X be the toric variety arising from a fan Σ and let T denote the acting torus of X. Then there is a bijection*

$$\Sigma \to \{T\text{-orbits of } X\}, \qquad \sigma \mapsto T \cdot x_\sigma.$$

Moreover, for any two $\sigma_1, \sigma_2 \in \Sigma$, we have $\sigma_1 \preceq \sigma_2$ if and only if $\overline{T \cdot \sigma_1} \supseteq \overline{T \cdot \sigma_2}$ holds. For the affine chart $X_\sigma \subseteq X$ defined by $\sigma \in \Sigma$, we have

$$X_\sigma = \bigcup_{\tau \preceq \sigma} T \cdot x_\tau.$$

The structure of the toric orbit corresponding to a cone of the defining fan is described as follows.

Proposition 2.1.2.3 (Orbit structure) *Let X be the toric variety arising from a fan Σ in an n-dimensional lattice N, and denote by T its acting torus. Then, for every $\sigma \in \Sigma$, the inclusion $T_{x_\sigma} \subseteq T$ of the isotropy group is given by the projection $M \to M/(\sigma^\perp \cap M)$ of character lattices. In particular, T_{x_σ} is a torus and we have*

$$\dim(T_{x_\sigma}) = \dim(\sigma), \qquad \dim(T \cdot x_\sigma) = n - \dim(\sigma).$$

To describe the fibers of a toric morphism, it suffices to describe the fibers over the distinguished points. This works as follows.

Proposition 2.1.2.4 (Fiber formula) *Let $(\varphi, \widetilde{\varphi})$ be the toric morphism from (X, T, x_0) to (X', T', x_0') defined by a map $F: N \to N'$ of fans Σ and Σ' in* lattices N and N', respectively. Then the fiber over a distinguished point x_τ, where $\tau \in \Sigma'$, is given by

$$\varphi^{-1}(x_\tau) = \bigcup_{\substack{\sigma \in \Sigma \\ F(\sigma)^\circ \subseteq \tau^\circ}} \widetilde{\varphi}^{-1}(T'_{x_\tau}) \cdot x_\sigma.$$

We turn to singularities of toric varieties. Recall that a cone σ in a lattice N is said to be *simplicial* if it is generated by linearly independent family $v_1, \ldots, v_r \in N$. Moreover, a cone σ in a lattice N is called *regular* if it is generated by a family $v_1, \ldots, v_r \in N$ that can be completed to a lattice basis of N.

Proposition 2.1.2.5 *Let X be the toric variety arising from a fan Σ in a lattice N, and let $\sigma \in \Sigma$.*

(i) *The point $x_\sigma \in X$ is $\mathbb{Q}$-factorial if and only if σ is simplicial.*
(ii) *The point $x_\sigma \in X$ is smooth if and only if σ is regular.*

The next subject is completeness and projectivity. The *support* of a quasifan Σ in a vector space $N_\mathbb{Q}$ is the union $\operatorname{Supp}(\Sigma) \subseteq N_\mathbb{Q}$ of its cones. A quasifan is *complete* if its support coincides with the space $N_\mathbb{Q}$. We say that a quasifan in a vector space $N_\mathbb{Q}$ is *normal* if it is the normal quasifan $\mathcal{N}(\Delta)$ of a polyhedron $\Delta \subseteq M_\mathbb{Q}$ in the dual vector space, that is, its cones arise from the faces of Δ via the bijection

$$\operatorname{faces}(\Delta) \;\rightarrow\; \mathcal{N}(\Delta),$$

$$\Delta_0 \;\mapsto\; \{v \in N_\mathbb{Q}; \; \langle u - u_0, v\rangle \geq 0 \text{ for all } u \in \Delta, \, u_0 \in \Delta_0\}.$$

Moreover, the support of $\mathcal{N}(\Delta)$ is the dual of the recession cone of Δ, that is, the unique cone $\sigma \subseteq M_\mathbb{Q}$ such that $\Delta = B + \sigma$ holds with a polytope $B \subseteq M_\mathbb{Q}$. We say that a quasifan is *polytopal* if it is normal and complete. In other words, a polytopal quasifan in $N_\mathbb{Q}$ is the normal quasifan of a polytope $\Delta \subseteq M_\mathbb{Q}$.

Proposition 2.1.2.6 *Let X be the toric variety arising from a fan Σ in a lattice N.*

(i) *X is complete if and only if Σ is complete.*
(ii) *X is projective if and only if Σ is polytopal.*

Now we take a look at the divisor class group $\operatorname{Cl}(X)$ of the toric variety X defined by a fan Σ in the lattice N. We assume that Σ is *nondegenerate*, that is, the primitive lattice vectors $v_1, \ldots, v_r \in N$ of the rays of Σ generate $N_\mathbb{Q}$ as a vector space; this just means that on X every globally invertible function is constant. Set $F := \mathbb{Z}^r$, consider the linear map $P \colon F \to N$ sending the ith canonical base vector $f_i \in F$ to $v_i \in N$ and the dual map $P^* \colon M \to E$. Then, with $L := \ker(P)$ and $K := E/P^*(M)$, we have the following two exact sequences of abelian groups:

$$0 \longrightarrow L \xrightarrow{\;Q^*\;} F \xrightarrow{\;P\;} N$$

$$0 \longleftarrow K \xleftarrow{\;Q\;} E \xleftarrow{\;P^*\;} M \longleftarrow 0$$

The lattice M represents the characters χ^u of the acting torus T of X, and each such character χ^u is a rational function on X; in fact, χ^u is regular on any affine chart X_σ with $u \in \sigma^\vee$. Moreover, the lattice E is isomorphic to the subgroup $\mathrm{WDiv}^T(X)$ of T-invariant Weil divisors via

$$E \;\to\; \mathrm{WDiv}^T(X), \qquad e \mapsto D(e) := \langle e, f_1 \rangle D_1 + \cdots + \langle e, f_r \rangle D_r,$$

where the $D_i := \overline{T \cdot x_{\varrho_i}}$ with $\varrho_i = \mathrm{cone}(v_i)$ are the T-invariant prime divisors. Along the open toric orbit, all Weil divisors are principal and hence every Weil divisor is linearly equivalent to a T-invariant one. Thus, denoting by $\mathrm{PDiv}^T(X) \subseteq \mathrm{WDiv}^T(X)$ the subgroup of invariant principal divisors, we arrive at the following description of the divisor class group.

Proposition 2.1.2.7 *Let X be the toric variety arising from a nondegenerate fan Σ in a lattice N. Then, in the above notation, there is a commutative diagram with exact rows*

$$
\begin{array}{ccccccccc}
0 & \longleftarrow & K & \xleftarrow{\;Q\;} & E & \xleftarrow{\;P^*\;} & M & \longleftarrow & 0 \\
& & \cong \downarrow & & {\scriptstyle e \mapsto D(e)}\downarrow\cong & & {\scriptstyle u \mapsto \mathrm{div}(\chi^u)}\downarrow\cong & & \\
0 & \longleftarrow & \mathrm{Cl}(X) & \longleftarrow & \mathrm{WDiv}^T(X) & \longleftarrow & \mathrm{PDiv}^T(X) & \longleftarrow & 0
\end{array}
$$

To compute intersection numbers, we first recall the basic notions from [142]. Let X be any n-dimensional variety. The group $Z_k(X)$ of k-cycles on X is the free abelian group over the set of *prime k-cycles* of X, that is, the irreducible k-dimensional subvarieties of X. The subgroup $B_k(X) \subseteq Z_k(X)$ of k-boundaries is generated by the divisors $\mathrm{div}(f)$ of rational functions $f \in \mathbb{K}(Y)^*$ living on $k+1$-dimensional subvarieties $Y \subseteq X$. The *kth Chow group* of X is the factor group $A_k(X) := Z_k(X)/B_k(X)$. There is a well-defined bilinear intersection map

$$\mathrm{Pic}(X) \times A_{k+1}(X) \;\to\; A_k(X), \qquad ([D], [Y]) \mapsto [D] \cdot [Y] := [\iota^* D],$$

where $\iota \colon Y \to X$ denotes the inclusion of a $(k+1)$-dimensional irreducible subvariety and D is a representative of $[D]$ such that Y is not contained in its support. Now suppose that X is complete and consider Cartier divisors $D_1, \ldots, D_n$

on X. Then the recursively obtained intersection $[D_1] \cdots [D_n]$ is represented by a divisor on a projective curve and has well defined degree $D_1 \cdots D_n$, called the *intersection number* of $D_1, \ldots, D_n$. Note that $(D_1, \ldots, D_n) \mapsto D_1 \cdots D_n$ is linear in every argument. If X is $\mathbb{Q}$-factorial, then one defines the intersection number of any n Weil divisors $D_1, \ldots, D_n$ to be the rational number $(a_1 D_1) \cdots (a_n D_n)/a_1 \cdots a_n$, where $a_i \in \mathbb{Z}_{\geq 0}$ is such that $a_i D_i$ is Cartier.

For computing intersection numbers on a $\mathbb{Q}$-factorial toric variety, one has to know the possible intersection numbers of toric prime divisors. To express these numbers in terms of the defining fan, we need the following notion for a cone σ in a lattice N: Let $v_1, \ldots, v_r$ be the primitive vectors on the rays of σ and set

$$\mu(\sigma) := [N \cap \operatorname{lin}_{\mathbb{Q}}(\sigma) : \operatorname{lin}_{\mathbb{Z}}(v_1, \ldots, v_r)].$$

Proposition 2.1.2.8 *Let X be an n-dimensional complete toric variety arising from a simplicial fan Σ in a lattice N. Let $D_1, \ldots, D_n$ be pairwise different invariant prime divisors on X corresponding to rays $\varrho_1, \ldots, \varrho_n \in \Sigma$ and set $\sigma := \varrho_1 + \cdots + \varrho_n$. Then the intersection number of $D_1, \ldots, D_n$ is given as*

$$D_1 \cdots D_n = \begin{cases} \mu(\sigma)^{-1}, & \sigma \in \Sigma, \\ 0, & \sigma \notin \Sigma. \end{cases}$$

2.1.3 The Cox ring of a toric variety

Roughly speaking, Cox's theorem says that, for a toric variety X with only constant invertible global functions, the Cox ring is given in terms of its invariant prime divisors $D_1, \ldots, D_r \subseteq X$ as

$$\mathcal{R}(X) \cong \mathbb{K}[T_1, \ldots, T_r], \qquad \deg(T_i) = [D_i] \in \operatorname{Cl}(X).$$

In fact, approaching this from the combinatorial side makes the statement more concrete and allows to determine the $\operatorname{Cl}(X)$-grading of the Cox ring and the characteristic space explicitly, see [104, Thm. 2.1] as well as [23, Sec. VI.2.2] for the simplicial case, [30, Sec. 2] for the regular case and [226, Thm. 1] for a similar result. As before, $\mathbb{K}$ is algebraically closed of characteristic zero.

We say that a fan Σ in a lattice N is *nondegenerate* if the primitive lattice vectors on the rays of Σ generate $N_{\mathbb{Q}}$ as a vector space. For the associated toric variety X this means precisely that $\Gamma(X, \mathcal{O}^*) = \mathbb{K}^*$ holds.

Construction 2.1.3.1 Let Σ be a nondegenerate fan in a lattice N and let X be the associated toric variety. Denote the primitive vectors on the rays of Σ by $v_1, \ldots, v_r \in N$, set $F := \mathbb{Z}^r$ and consider the linear map $P : F \to N$ sending the ith canonical base vector $f_i \in F$ to $v_i \in N$. There is a fan $\widehat{\Sigma}$ in F consisting of certain faces of the positive orthant $\delta \subseteq F_\mathbb{Q}$, namely

$$\widehat{\Sigma} := \{\widehat{\sigma} \preceq \delta; \ P(\widehat{\sigma}) \subseteq \sigma \text{ for some } \sigma \in \Sigma\}.$$

The fan $\widehat{\Sigma}$ defines an open toric subvariety $\widehat{X}$ of $\overline{X} := \operatorname{Spec}(\mathbb{K}[\delta^\vee \cap E])$, where $E := \operatorname{Hom}(F, \mathbb{Z})$. Note that all rays $\operatorname{cone}(f_1), \ldots, \operatorname{cone}(f_r)$ of the positive orthant $\delta \subseteq F_\mathbb{Q}$ belong to $\widehat{\Sigma}$ and thus we have

$$\Gamma(\widehat{X}, \mathcal{O}) = \Gamma(\overline{X}, \mathcal{O}) = \mathbb{K}[\delta^\vee \cap E].$$

As $P \colon F \to N$ is a map of the fans $\widehat{\Sigma}$ and Σ, it defines a toric morphism $p \colon \widehat{X} \to X$. Consider the dual map $P^* \colon M \to E$, where $M := \operatorname{Hom}(N, \mathbb{Z})$, set $K := E/P^*(M)$ and denote by $Q \colon E \to K$ the projection. Then, by Proposition 2.1.2.7, we have the following commutative diagram:

$$
\begin{array}{ccccccccc}
0 & \longleftarrow & \mathbb{X}(H) & \longleftarrow & \mathbb{X}(\mathbb{T}) & \overset{p^*}{\longleftarrow} & \mathbb{X}(T) & \longleftarrow & 0 \\
& & \cong \big\uparrow & & e \mapsto \chi^e \ \cong \big\uparrow & & u \mapsto \chi^u \ \cong \big\uparrow & & \\
0 & \longleftarrow & K & \underset{Q}{\longleftarrow} & E & \underset{P^*}{\longleftarrow} & M & \longleftarrow & 0 \\
& & \cong \big\downarrow & & e \mapsto D(e) \ \cong \big\downarrow & & u \mapsto \operatorname{div}(\chi^u) \ \cong \big\downarrow & & \\
0 & \longleftarrow & \operatorname{Cl}(X) & \longleftarrow & \operatorname{WDiv}^T(X) & \longleftarrow & \operatorname{PDiv}^T(X) & \longleftarrow & 0
\end{array}
$$

with the acting tori $T := \operatorname{Spec}(\mathbb{K}[M])$ of X and $\mathbb{T} := \operatorname{Spec}(\mathbb{K}[E])$ of $\widehat{X}$ and the quasitorus $H := \operatorname{Spec}(\mathbb{K}[K])$ of X. The map $Q \colon E \to K$ turns the polynomial ring $\mathbb{K}[E \cap \delta^\vee]$ into a K-graded algebra:

$$\mathbb{K}[E \cap \delta^\vee] = \bigoplus_{w \in K} \mathbb{K}[E \cap \delta^\vee]_w, \qquad \mathbb{K}[E \cap \delta^\vee]_w = \bigoplus_{e \in Q^{-1}(w) \cap \delta^\vee} \mathbb{K} \cdot \chi^e.$$

Theorem 2.1.3.2 *Consider the situation of Construction 2.1.3.1. Then the Cox ring of X is isomorphic to the K-graded polynomial ring $\mathbb{K}[E \cap \delta^\vee]$. Moreover, H is a characteristic quasitorus of X, the toric morphism $p \colon \widehat{X} \to X$ is a characteristic space over X, and $\overline{X}$ is a total coordinate space of X.*

Proof We follow the construction of the Cox ring performed in 1.4.2.1. The group $E = \operatorname{WDiv}^T(X)$ of invariant Weil divisors projects onto $K = \operatorname{Cl}(X)$ and the kernel of this projection is

$$E^0 = P^*(M) = \operatorname{PDiv}^T(X).$$

Let $\mathcal{S}$ be the divisorial sheaf associated with E. Consider $e \in E$, a cone $\sigma \in \Sigma$ and let $\widehat{\sigma} \in \widehat{\Sigma}$ be the cone with $P(\widehat{\sigma}) = \sigma$. Then we have

$$\Gamma(X_\sigma, \mathcal{S}_e) = \operatorname{lin}_\mathbb{K}\left(\chi^u; \ u \in M, \ \operatorname{div}(\chi^u) + D(e) \geq 0 \text{ on } X_\sigma\right)$$
$$= \operatorname{lin}_\mathbb{K}\left(\chi^u; \ u \in M, \ P^*(u) + e \in \widehat{\sigma}^\vee\right).$$

Using this identity, we define an epimorphism $\pi \colon \Gamma(X_\sigma, \mathcal{S}) \to \Gamma(\widehat{X}_{\widehat{\sigma}}, \mathcal{O})$ of K-graded algebras by

$$\Gamma(X_\sigma, \mathcal{S}_e) \ni \chi^u \mapsto \chi^{P^*(u)+e} \in \Gamma(\widehat{X}_{\widehat{\sigma}}, \mathcal{O})_{Q(e)}.$$

For $u \in M$, look at $\chi^u \in \Gamma(X_\sigma, \mathcal{S}_{-P^*(u)})$. Then $\pi(1 - \chi^u) = 0$ holds. Moreover, for $\chi^{u'} \in \Gamma(X_\sigma, \mathcal{S}_{e'})$ and $e \in E$ with $e' - e = P^*(u)$ we have

$$\chi^{u'} = \chi^u \chi^u + \chi^{u'}(1 - \chi^u), \qquad\qquad \chi^u \chi^u \in \Gamma(X_\sigma, \mathcal{S}_e).$$

It follows that $\ker(\pi)$ equals the ideal associated with the character $E^0 \to \mathbb{K}(X)^*$, $P^*(u) \mapsto \chi^u$ (see 1.4.2.1), and we obtain a K-graded isomorphism

$$\Gamma(X_\sigma, \mathcal{R}) \to \Gamma(\widehat{X}_{\widehat{\sigma}}, \mathcal{O}).$$

This shows that the K-graded sheaves $\mathcal{R}$ and $p_*\mathcal{O}_{\widehat{X}}$ are isomorphic. Consequently $p \colon \widehat{X} \to X$ is a characteristic space for X and the Cox ring of X is

$$\mathcal{R}(X) = \Gamma(\widehat{X}, \mathcal{O}) = \mathbb{K}[E \cap \delta^\vee]. \qquad\qquad \square$$

Example 2.1.3.3 Consider the projective plane $X = \mathbb{P}^2$. Its total coordinate space is $\overline{X} = \mathbb{K}^3$, we have $\widehat{X} = \mathbb{K}^3 \setminus \{0\}$, the characteristic torus $H = \mathbb{K}^*$ acts by scalar multiplication, and the quotient map is

$$p \colon \widehat{X} \to X, \qquad (z_0, z_1, z_2) \mapsto [z_0, z_1, z_2].$$

In terms of fans, the situation is the following. The fan Σ of X lies in $N = \mathbb{Z}^2$. With $v_0 = (-1, -1)$ and $v_1 = (1, 0)$ and $v_2 = (0, 1)$ its maximal cones are

$$\operatorname{cone}(v_1, v_2), \qquad \operatorname{cone}(v_2, v_0), \qquad \operatorname{cone}(v_0, v_1).$$

Thus, we have $F = \mathbb{Z}^3$ and the map $P \colon F \to N$ sends the canonical basis vector f_i to v_i, where $i = 0, 1, 2$. The fan $\widehat{\Sigma}$ of $\widehat{X}$ has the facets of the orthant $\operatorname{cone}(f_0, f_1, f_2)$ as its maximal cones.

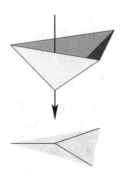

In general, the total coordinate space $\overline{X} = \operatorname{Spec} \mathbb{K}[E \cap \delta^\vee]$ is isomorphic to $\mathbb{K}^r$. More precisely, if $e_1, \ldots, e_r$ denote the primitive generators of $\delta^\vee$, then a concrete isomorphism $\overline{X} \to \mathbb{K}^r$ is given by the comorphism

$$\mathbb{K}[T_1, \ldots, T_r] \to \mathbb{K}[E \cap \delta^\vee], \qquad T_i \mapsto \chi^{e_i}.$$

We want to describe the *irrelevant ideal* of X in the Cox ring $\mathcal{R}(X) = \mathbb{K}[T_1, \ldots, T_r]$ in terms of the defining fan Σ; recall from Definition 1.6.3.2 that the irrelevant ideal $\mathcal{J}_{\mathrm{irr}}(X)$ is the vanishing ideal of $\overline{X} \setminus \widehat{X}$ in $\Gamma(\overline{X}, \mathcal{O}) = \mathcal{R}(X) = \mathbb{K}[T_1, \ldots, T_r]$. For any quasifan Σ, we denote by $\Sigma^{\max}$ the set of its maximal cones.

Proposition 2.1.3.4 *Let X be the toric variety arising from a nondegenerate fan Σ in a lattice N and let $v_1, \ldots, v_r$ be the primitive generators of Σ. For every $\sigma \in \Sigma$ define a vector*

$$v(\sigma) := (\varepsilon_1, \ldots, \varepsilon_r) \in \mathbb{Z}^r_{\geq 0} \qquad \varepsilon_i := \begin{cases} 1, & v_i \notin \sigma, \\ 0, & v_i \in \sigma. \end{cases}$$

Then the irrelevant ideal $\mathcal{J}_{\mathrm{irr}}(X)$ in the Cox ring $\mathcal{R}(X) = \mathbb{K}[T_1, \ldots, T_r]$ is generated by the monomials $T^{v(\sigma)} = T_1^{\varepsilon_1} \cdots T_r^{\varepsilon_r}$, where $\sigma \in \Sigma^{\max}$.

Proof The toric variety $\widehat{X}$ is the union of the affine charts $\widehat{X}_{\widehat{\sigma}}$, where $\sigma \in \Sigma^{\max}$. The complement of $\widehat{X}_{\widehat{\sigma}}$ in $\overline{X}$ is the zero set of $T^{v(\sigma)}$. Observing that the monomials $T^{v(\sigma)}$, where $\sigma \in \Sigma^{\max}$, generate a radical ideal gives the assertion. □

Now we give an explicit description of the complement $\overline{X} \setminus \widehat{X}$ as an arrangement of coordinate subspaces.

Proposition 2.1.3.5 *Let X be the toric variety arising from a nondegenerate fan Σ in a lattice N, let $v_1, \ldots, v_r$ be the primitive generators of Σ, and let $\mathcal{G}(\Sigma)$ be the collection of subsets $I \subseteq \{1, \ldots, r\}$ which are minimal with the property that the vectors v_i, where $i \in I$, are not contained in a common cone of Σ. Then one has*

$$\overline{X} \setminus \widehat{X} = \bigcup_{I \in \mathcal{G}(\Sigma)} V(T_i; \, i \in I) \subseteq \mathbb{K}^r.$$

Proof Take a point $z \in \mathbb{K}^r$ and set $I_z := \{i; \, z_i = 0\}$. By Proposition 2.1.3.4, the point z is not contained in $\widehat{X}$ if and only if the vectors v_i, where $i \in I_z$, are not contained in a cone of Σ. This means that $I \subseteq I_z$ for some element I of $\mathcal{G}(\Sigma)$, or that the point z is in $V(T_i; \, i \in I)$. □

Example 2.1.3.6 The first Hirzebruch surface is the toric variety X arising from the fan Σ in $N = \mathbb{Z}^2$ given as follows:

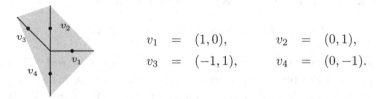

$$v_1 = (1,0), \qquad v_2 = (0,1),$$
$$v_3 = (-1,1), \qquad v_4 = (0,-1).$$

Thus, we have $F = \mathbb{Z}^4$ and, with respect to the canonical bases, the maps $P \colon F \to N$ and $Q \colon E \to K$ are given by the matrices

$$P = \begin{bmatrix} 1 & 0 & -1 & 0 \\ 0 & 1 & 1 & -1 \end{bmatrix}, \qquad Q = \begin{bmatrix} 0 & 1 & 0 & 1 \\ 1 & 0 & 1 & 1 \end{bmatrix}.$$

The total coordinate space of X is $\overline{X} = \mathbb{K}^4$ and, according to Proposition 2.1.3.5, the open subset $\widehat{X} \subseteq \overline{X}$ is obtained from $\overline{X}$ by removing

$$V(\mathbb{K}^4; T_1, T_3) \ \cup \ V(\mathbb{K}^4; T_2, T_4).$$

This describes the characteristic space $p \colon \widehat{X} \to X$ over X; the action of the characteristic torus $H = \mathbb{K}^* \times \mathbb{K}^*$ on $\widehat{X}$ is given by

$$h \cdot z = (h_2 z_1, h_1 z_2, h_2 z_3, h_1 h_2 z_4).$$

2.1.4 Geometry of Cox's construction

We give a self-contained discussion of basic geometric properties of Cox's quotient presentation of toric varieties, that is, without using general results on characteristic spaces. The ground field $\mathbb{K}$ is algebraically closed of characteristic zero. We begin with a general observation on lattices needed also later.

Let F and E be mutually dual lattices, and consider an epimorphism $Q \colon E \to K$ onto an abelian group K and a pair of exact sequences

$$0 \longrightarrow L \longrightarrow F \overset{P}{\longrightarrow} N,$$

$$0 \longleftarrow K \underset{Q}{\longleftarrow} E \underset{P^*}{\longleftarrow} M \longleftarrow 0.$$

Lemma 2.1.4.1 *Let $F_{\mathbb{Q}}^0 \subseteq F_{\mathbb{Q}}$ be a vector subspace and let $E_{\mathbb{Q}}^0 \subseteq E_{\mathbb{Q}}$ be the annihilating space of $F_{\mathbb{Q}}^0$. Then one has an isomorphism of abelian groups*

$$K/Q(E_{\mathbb{Q}}^0 \cap E) \cong (L \cap F_{\mathbb{Q}}^0) \oplus (P(F_{\mathbb{Q}}^0) \cap N)/P(F_{\mathbb{Q}}^0 \cap F).$$

Proof Set $F^0 := F_{\mathbb{Q}}^0 \cap F$ and $E^0 := E_{\mathbb{Q}}^0 \cap E$. Then E/E^0 is the dual lattice of F_0. Moreover, M/M^0 is the dual lattice of $N^0 := P(F_{\mathbb{Q}}^0) \cap N$, where $M^0 \subseteq M$ is the inverse image of $E^0 \subseteq E$ under P^*. These lattices fit into the exact sequences

$$0 \longrightarrow L^0 \longrightarrow F^0 \overset{P}{\longrightarrow} N^0$$

$$0 \longleftarrow K/Q(E^0) \overset{Q}{\longleftarrow} E/E^0 \overset{P^*}{\longleftarrow} M/M^0 \longleftarrow 0$$

where we set $L^0 := L \cap F^0$. In other words, $K/Q(E^0)$ is isomorphic to the cokernel of $M/M^0 \to E/E^0$. The latter is the direct sum of the cokernel and the kernel of the dual map $F^0 \to N^0$. $\qquad\square$

We now discuss actions of subgroups of the acting torus of an affine toric variety. Let $\delta \subseteq F_{\mathbb{Q}}$ be any pointed cone, and consider the associated affine toric variety X_δ. Then $Q \colon E \to K$ defines a K-grading of the algebra of regular functions

$$\Gamma(X_\delta, \mathcal{O}) \;=\; \mathbb{K}[E \cap \delta^\vee] \;=\; \bigoplus_{w \in K} \mathbb{K}[E \cap \delta^\vee]_w,$$

$$\mathbb{K}[E \cap \delta^\vee]_w \;:=\; \bigoplus_{e \in Q^{-1}(w) \cap \delta^\vee} \mathbb{K} \cdot \chi^e.$$

Thus, the quasitorus $H = \operatorname{Spec} \mathbb{K}[K]$ acts on X_δ. Note that H acts on X_δ as a subgroup of the acting torus T of X_δ, where the embedding $H \to T$ is given by the map $Q \colon E \to K$ of the character groups.

Proposition 2.1.4.2 *The inclusion $H_{x_\delta} \subseteq H$ of the isotropy group of $x_\delta \in X_\delta$ is given by the projection $K \to K_\delta$ of character lattices, where*

$$K_\delta \;:=\; K/Q(\delta^\perp \cap E) \;\cong\; (L \cap \operatorname{lin}_{\mathbb{Q}}(\delta)) \oplus (P(\operatorname{lin}_{\mathbb{Q}}(\delta)) \cap N)/P(\operatorname{lin}_{\mathbb{Q}}(\delta) \cap F).$$

Proof Use the characterization of the distinguished point $x_\delta \in X_\delta$ given in Construction 2.1.2.1 to see that $Q(\delta^\perp \cap E)$ is the orbit group of x_δ. Thus, Proposition 1.2.2.8 tells us that $K \to K_\delta$ gives the inclusion $H_{x_\delta} \subseteq H$. The alternative description of K_δ is obtained by applying Lemma 2.1.4.1 to $F_{\mathbb{Q}}^0 := \operatorname{lin}_{\mathbb{Q}}(\delta)$. $\qquad\square$

To determine a good quotient for the action of H on X_δ, consider the image $P(\delta)$ in $N_{\mathbb{Q}}$. This cone need not be pointed; we denote by $\tau \preceq P(\delta)$ its minimal face. Then, with $N_1 := N/(\tau \cap N)$, we have the projection $P_1 \colon F \to N_1$ and $\delta_1 := P_1(\delta)$ is pointed.

Proposition 2.1.4.3 *In the preceding notation, the toric morphism* $p_1 \colon X_\delta \to X_{\delta_1}$ *given by* $P_1 \colon F \to N_1$ *is a good quotient for the action of* H *on* X_δ.

Proof The algebra of H-invariant regular functions on X_δ is embedded into the algebra of functions as

$$\Gamma(X_\delta, \mathcal{O})^H \;=\; \mathbb{K}[\delta^\vee \cap M] \;\subseteq\; \mathbb{K}[\delta^\vee \cap E] \;=\; \Gamma(X_\delta, \mathcal{O}).$$

The sublattice M_1 generated by $\delta^\vee \cap M$ is the dual lattice of N_1 and $\delta_1 = P_1(\delta)$ in N_1 is the dual cone of $\delta^\vee \cap (M_1)_\mathbb{Q}$ in M_1. □

We specialize to Cox's Construction 2.1.3.1. That means that in the preceding setting, we have $F = \mathbb{Z}^r$ and the map $P \colon F \to N$ sends the canonical basis vectors $f_1, \ldots, f_r$ to the primitve generators $v_1, \ldots, v_r \in N$ of the rays of a nondegenerate fan Σ in N. Moreover, we have the othant $\delta = \operatorname{cone}(f_1, \ldots, f_r)$ and the fan $\widehat{\Sigma}$ in F defined by

$$\widehat{\Sigma} \;:=\; \{\widehat{\sigma} \preceq \delta;\; P(\widehat{\sigma}) \subseteq \sigma \text{ for some } \sigma \in \Sigma\}.$$

As before, we denote by $\widehat{X}$ the open toric subvariety of $\overline{X} = \operatorname{Spec}(\mathbb{K}[\delta^\vee \cap E])$ defined by the fan $\widehat{\Sigma}$. Moreover, $p \colon \widehat{X} \to X$ is the toric morphism defined by $P \colon F \to N$, and we set $H := \operatorname{Spec} \mathbb{K}[K]$. Then H acts on $\overline{X}$ and leaves $\widehat{X}$ invariant. Finally, let $\widehat{\varrho}_i := \operatorname{cone}(f_i)$, where $1 \leq i \leq r$, denote the rays of $\widehat{\Sigma}$ and set

$$\widehat{W} \;:=\; \overline{X}_{\widehat{\varrho}_1} \cup \ldots \cup \overline{X}_{\widehat{\varrho}_r} \;\subseteq\; \widehat{X}.$$

Proposition 2.1.4.4 *Consider the situation of Construction 2.1.3.1. The toric morphism* $p \colon \widehat{X} \to X$ *is the good quotient for the action of* H *on* $\widehat{X}$. *Moreover,* H *acts freely on the open subset* $\widehat{W} \subseteq \widehat{X}$ *defined above and every* H-*orbit on* $\widehat{W}$ *is closed in* $\widehat{X}$.

Proof We first show that $p \colon \widehat{X} \to X$ is affine. For $\sigma \in \Sigma$, let $\widehat{\sigma} \in \widehat{\Sigma}$ be the (unique) cone with $P(\widehat{\sigma}) = \sigma$. Then we have $p^{-1}(X_\sigma) = \widehat{X}_{\widehat{\sigma}}$ and Proposition 2.1.4.3 tells us that $p \colon \widehat{X} \to X$ is a good quotient for the H-action. Using Proposition 2.1.4.2 we see that H acts with trivial isotropy groups on $\widehat{W}$. To see the last assertion, note that the subset $\widehat{W}$ coincides with $p^{-1}(X')$, where $X' = X_{\varrho_1} \cup \ldots \cup X_{\varrho_r} \subseteq X$. □

Note that combining Proposition 2.1.4.4 with Theorem 1.6.4.3 shows that $p \colon \widehat{X} \to X$ is a characteristic space for X and hence provides another proof of Theorem 2.1.3.2. We conclude with two observations on the geometry of the H-action and the quotient $p \colon \widehat{X} \to X$; these two statements may also be obtained as a consequence of Corollary 1.6.2.7.

Proposition 2.1.4.5 *Consider the situation of Construction 2.1.3.1. For the action of H on X and the quotient $p \colon \widehat{X} \to X$, the following statements are equivalent.*

(i) *The fan Σ is simplicial.*
(ii) *One has $\dim(\widehat{\sigma}) = \dim(P(\widehat{\sigma}))$ for every $\widehat{\sigma} \in \widehat{\Sigma}$.*
(iii) *Any H-orbit in $\widehat{X}$ has at most finite isotropy group.*
(iv) *The quotient $p \colon \widehat{X} \to X$ is geometric.*
(v) *The variety X is $\mathbb{Q}$-factorial.*

Proof Statements (i) and (ii) are obviously equivalent. The equivalence of (ii) and (iii) is clear by Proposition 2.1.4.2. The equivalence of (ii) and (iv) follows from the Fiber Formula 2.1.2.4. Finally, (i) and (v) are equivalent by standard toric geometry, see Proposition 2.1.2.5. □

Proposition 2.1.4.6 *Consider the situation of Construction 2.1.3.1. For the action of H on X and the quotient $p \colon \widehat{X} \to X$, the following statements are equivalent.*

(i) *The fan Σ is regular.*
(ii) *$P \colon F \cap \mathrm{lin}_{\mathbb{Q}}(\widehat{\sigma}) \to N \cap \mathrm{lin}_{\mathbb{Q}}(P(\widehat{\sigma}))$ is an isomorphism for every $\widehat{\sigma} \in \widehat{\Sigma}$.*
(iii) *The action of H on $\widehat{X}$ is free.*
(iv) *The variety X is smooth.*

Proof Statements (i) and (ii) are obviously equivalent. The equivalence of (ii) and (iii) is clear by Proposition 2.1.4.2. Finally, (i) and (iv) are equivalent by standard toric geometry; see Proposition 2.1.2.5. □

2.2 Linear Gale duality

2.2.1 Fans and bunches of cones

We introduce the concept of a bunch of cones and state in Theorem 2.2.1.14 that, in an appropriate setting, this is the Gale dual version of the concept of a fan; the proof is given in Subsection 2.2.3. Our presentation is in the spirit of [235]; in particular, we make no use of the language of oriented matroids. A general reference is [107, Chap. 4]. In [133, Sec. 1] some historical aspects are discussed.

Definition 2.2.1.1 We say that a vector configuration $V = (v_1, \ldots, v_r)$ in a rational vector space $N_{\mathbb{Q}}$ and a vector configuration $W = (w_1, \ldots, w_r)$ in a rational vector space $K_{\mathbb{Q}}$ are *Gale dual* to each other if the following holds:

(i) We have $v_1 \otimes w_1 + \cdots + v_r \otimes w_r = 0$ in $N_{\mathbb{Q}} \otimes K_{\mathbb{Q}}$.

(ii) For any rational vector space U and any vectors $u_1, \ldots, u_r \in U$ with
$v_1 \otimes u_1 + \cdots + v_r \otimes u_r = 0$ in $N_{\mathbb{Q}} \otimes U$, there is a unique linear map
$\psi \colon K_{\mathbb{Q}} \to U$ with $\psi(w_i) = u_i$ for $i = 1, \ldots, r$.

(iii) For any rational vector space U and any vectors $u_1, \ldots, u_r \in U$ with
$u_1 \otimes w_1 + \cdots + u_r \otimes w_r = 0$ in $U \otimes K_{\mathbb{Q}}$, there is a unique linear map
$\varphi \colon N_{\mathbb{Q}} \to U$ with $\varphi(v_i) = u_i$ for $i = 1, \ldots, r$.

If we fix the first configuration in a Gale dual pair, then the second one
is, by the properties of Gale duality, uniquely determined up to isomorphism,
and therefore is also called the *Gale transform* of the first one. The following
characterization of Gale dual pairs is also used as a definition.

Remark 2.2.1.2 Consider vector configurations $V = (v_1, \ldots, v_r)$ and $W = (w_1, \ldots, w_r)$ in rational vector spaces $N_{\mathbb{Q}}$ and $K_{\mathbb{Q}}$ respectively, and let $M_{\mathbb{Q}}$ be
the dual vector space of $N_{\mathbb{Q}}$. Then Gale duality of V and W is characterized by
the following property: For any tuple $(a_1, \ldots, a_r) \in \mathbb{Q}^r$ one has

$$a_1 w_1 + \cdots + a_r w_r = 0 \iff u(v_i) = a_i \text{ for } i = 1, \ldots, r \text{ with some } u \in M_{\mathbb{Q}}.$$

The following construction shows existence of Gale dual pairs, and, up to
isomorphy, produces any pair of Gale dual vector configurations.

Construction 2.2.1.3 Consider a pair of mutually dual exact sequences of
finite-dimensional rational vector spaces

$$0 \longrightarrow L_{\mathbb{Q}} \longrightarrow F_{\mathbb{Q}} \overset{P}{\longrightarrow} N_{\mathbb{Q}} \longrightarrow 0$$

$$0 \longleftarrow K_{\mathbb{Q}} \underset{Q}{\longleftarrow} E_{\mathbb{Q}} \longleftarrow M_{\mathbb{Q}} \longleftarrow 0$$

Let $(f_1, \ldots, f_r)$ be a basis for $F_{\mathbb{Q}}$, let $(e_1, \ldots, e_r)$ be the dual basis for $E_{\mathbb{Q}}$ and
denote the image vectors by

$$v_i := P(f_i) \in N_{\mathbb{Q}}, \qquad w_i := Q(e_i) \in K_{\mathbb{Q}}, \qquad 1 \le i \le r.$$

Then the vector configurations $V = (v_1, \ldots, v_r)$ in $N_{\mathbb{Q}}$ and $W = (w_1, \ldots, w_r)$
in $K_{\mathbb{Q}}$ are Gale dual to each other.

Remark 2.2.1.4 Let $r = n + k$ with integers $n, k \in \mathbb{Z}_{>0}$. Consider matrices
$P \in \mathrm{Mat}(n, r; \mathbb{Q})$ and $Q \in \mathrm{Mat}(k, r; \mathbb{Q})$ such that the rows of Q form a basis
for the nullspace of P. Then the columns $(v_1, \ldots, v_r)$ of P in $\mathbb{Q}^n$ and the
columns $(w_1, \ldots, w_r)$ of Q in $\mathbb{Q}^k$ are Gale dual vector configurations. Note

that after fixing one of the matrices, Gale duality determines the other up to multiplicitation by an invertible matrix from the left.

Example 2.2.1.5 The columns of the following matrices P and Q are Gale dual vector configurations:

$$P = \begin{bmatrix} 1 & 0 & -1 & 0 \\ 0 & 1 & 1 & -1 \end{bmatrix}, \qquad Q = \begin{bmatrix} 0 & 1 & 0 & 1 \\ 1 & 0 & 1 & 1 \end{bmatrix}.$$

Now we turn to fans and bunches of cones. We work with convex polyhedral cones generated by elements of a fixed vector configuration. The precise notion is the following.

Definition 2.2.1.6 Let $N_{\mathbb{Q}}$ be a rational vector space and $\mathcal{V} = (v_1, \dots, v_r)$ a family of vectors in $N_{\mathbb{Q}}$ generating $N_{\mathbb{Q}}$. A $\mathcal{V}$-*cone* is a convex polyhedral cone generated by some of the $v_1, \dots, v_r$. The set of all $\mathcal{V}$-cones is denoted by $\Omega(\mathcal{V})$.

Definition 2.2.1.7 Let $N_{\mathbb{Q}}$ be a rational vector space, and $\mathcal{V} = (v_1, \dots, v_r)$ a family of vectors in $N_{\mathbb{Q}}$ generating $N_{\mathbb{Q}}$. A $\mathcal{V}$-*quasifan* is a quasifan in $N_{\mathbb{Q}}$ consisting of $\mathcal{V}$-cones, that is, a nonempty set $\Sigma \subseteq \Omega(\mathcal{V})$ such that

 (i) For all $\sigma_1, \sigma_2 \in \Sigma$, the intersection $\sigma_1 \cap \sigma_2$ is a face of both, σ_1 and σ_2.
 (ii) For every $\sigma \in \Sigma$, also all faces $\sigma_0 \preceq \sigma$ belong to Σ.

A $\mathcal{V}$-*fan* is a $\mathcal{V}$-quasifan consisting of pointed cones. We say that a $\mathcal{V}$-(quasi)fan is *maximal* if it cannot be enlarged by adding $\mathcal{V}$-cones. Moreover, we call a $\mathcal{V}$-(quasi)fan *true* if it contains all rays $\mathrm{cone}(v_i)$, where $1 \le i \le r$.

Example 2.2.1.8 Consider $N_{\mathbb{Q}} = \mathbb{Q}^2$ and let $\mathcal{V} = (v_1, \dots, v_4)$ be the family consisting of the columns of the matrix P given in Example 2.2.1.5.

    A true maximal $\mathcal{V}$-fan Σ_1 A maximal $\mathcal{V}$-fan Σ_2

Definition 2.2.1.9 Let $N_{\mathbb{Q}}$ be a rational vector space, and $\mathcal{V} = (v_1, \dots, v_r)$ a family of vectors in $N_{\mathbb{Q}}$ generating $N_{\mathbb{Q}}$. We say that a $\mathcal{V}$-quasifan $\Sigma_1 \subseteq \Omega(\mathcal{V})$ *refines* a $\mathcal{V}$-quasifan $\Sigma_2 \subseteq \Omega(\mathcal{V})$, written $\Sigma_1 \le \Sigma_2$, if for every $\sigma_1 \in \Sigma_1$ there is a $\sigma_2 \in \Sigma_2$ with $\sigma_1 \subseteq \sigma_2$.

Definition 2.2.1.10 Let $K_\mathbb{Q}$ be a rational vector space, and $W = (w_1, \ldots, w_r)$ a family of vectors in $K_\mathbb{Q}$ generating $K_\mathbb{Q}$. A W-*bunch* is a nonempty set $\Theta \subseteq \Omega(W)$ such that

(i) For all $\tau_1, \tau_2 \in \Theta$, one has $\tau_1^\circ \cap \tau_2^\circ \neq \emptyset$.

(ii) For every $\tau \in \Theta$, all W-cones τ_0 with $\tau^\circ \subseteq \tau_0^\circ$ belong to Θ.

We say that a W-bunch is *maximal* if it cannot be enlarged by adding W-cones. Moreover, we call a W-bunch Θ *true* if every cone $\vartheta_i = \mathrm{cone}(w_j; \ j \neq i)$, where $1 \leq i \leq r$, belongs to Θ.

Example 2.2.1.11 Consider $K_\mathbb{Q} = \mathbb{Q}^2$ and let $W = (w_1, \ldots, w_4)$ be the family consisting of the columns of the matrix Q given in Example 2.2.1.5.

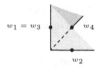   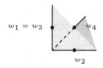

A true maximal W-bunch Θ_1 A maximal W-bunch Θ_2

Definition 2.2.1.12 Let $K_\mathbb{Q}$ be a rational vector space, and $W = (w_1, \ldots, w_r)$ a family of vectors in $K_\mathbb{Q}$ generating $K_\mathbb{Q}$. We say that a W-bunch $\Theta_1 \subseteq \Omega(W)$ *refines* a W-bunch $\Theta_2 \subseteq \Omega(W)$, written $\Theta_1 \leq \Theta_2$, if for every $\tau_2 \in \Theta_2$ there is a $\tau_1 \in \Theta_1$ with $\tau_1 \subseteq \tau_2$.

Definition 2.2.1.13 Let $V = (v_1, \ldots, v_r)$ in $N_\mathbb{Q}$ and $W = (w_1, \ldots, w_r)$ in $K_\mathbb{Q}$ be Gale dual vector configurations. Set $\mathfrak{R} := \{1, \ldots, r\}$. Then for any collection Σ of V-cones and any collection Θ of W-cones we set

$$\Sigma^\sharp := \{\mathrm{cone}(w_j; \ j \in \mathfrak{R} \setminus I); \ I \subseteq \mathfrak{R}, \ \mathrm{cone}(v_i; \ i \in I) \in \Sigma\},$$

$$\Theta^\sharp := \{\mathrm{cone}(v_i; \ i \in \mathfrak{R} \setminus J); \ J \subseteq \mathfrak{R}, \ \mathrm{cone}(w_j; \ j \in J) \in \Theta\}.$$

Theorem 2.2.1.14 *Let $V = (v_1, \ldots, v_r)$ in $N_\mathbb{Q}$ and $W = (w_1, \ldots, w_r)$ in $K_\mathbb{Q}$ be Gale dual vector configurations. Then we have an order reversing map*

$$\{W\text{-}bunches\} \ \rightarrow \ \{V\text{-}quasifans\}, \qquad \Theta \mapsto \Theta^\sharp.$$

Now assume that $v_1, \ldots, v_r$ generate pairwise different one-dimensional cones. Then there are mutually inverse order reversing bijections

$$\{true \ maximal \ W\text{-}bunches\} \ \longleftrightarrow \ \{true \ maximal \ V\text{-}fans\},$$

$$\Theta \ \mapsto \ \Theta^\sharp,$$

$$\Sigma^\sharp \ \leftarrow\!\shortmid \ \Sigma.$$

Under these bijections, the simplicial true maximal $\mathcal{V}$-fans correspond to the true maximal $\mathcal{W}$-bunches consisting of full-dimensional cones.

Example 2.2.1.15 For the true maximal $\mathcal{W}$-bunch Θ_1 and the true maximal $\mathcal{V}$-fan Σ_1 presented in Examples 2.2.1.11 and 2.2.1.8 one has

$$\Sigma_1 \;=\; \Theta_1^\sharp, \qquad\qquad \Theta_1 \;=\; \Sigma_1^\sharp.$$

Moreover, for the maximal $\mathcal{W}$-bunch Θ_2 and the maximal $\mathcal{V}$-fan Σ_2 presented there, we have $\Sigma_2 = \Theta_2^\sharp$ but $\Sigma_2^\sharp$ is not even a $\mathcal{W}$-bunch as it contains $\mathrm{cone}(w_4)$ and $\mathrm{cone}(w_2, w_4)$, contradicting property 2.2.1.10 (i).

2.2.2 The GKZ decomposition

Given a vector configuration $\mathcal{V}$, the normal ones among the possible $\mathcal{V}$-fans are obtained from the chambers of the so-called Gelfand–Kapranov–Zelevinsky (GKZ) decomposition of the dual vector configuration $\mathcal{W}$. We make this precise and study it in detail; the proofs are given in Subsection 2.2.4. A different treatment is given in [107, Chap. 5].

Construction 2.2.2.1 Let $K_\mathbb{Q}$ be a rational vector space and $\mathcal{W} = (w_1, \ldots, w_r)$ a family of vectors in $K_\mathbb{Q}$ generating $K_\mathbb{Q}$. For every $w \in \mathrm{cone}(\mathcal{W})$, we define its *chamber* to be

$$\lambda(w) \;:=\; \bigcap_{\substack{\tau \in \Omega(\mathcal{W}) \\ w \in \tau}} \tau \;=\; \bigcap_{\substack{\tau \in \Omega(\mathcal{W}) \\ w \in \tau^\circ}} \tau.$$

The *Gelfand–Kapranov–Zelevinsky decomposition (GKZ decomposition)* associated with $\mathcal{W}$ is the collection of all these chambers:

$$\Lambda(\mathcal{W}) := \{\lambda(w); \; w \in \mathrm{cone}(\mathcal{W})\}.$$

Note that for every $w \in \mathrm{cone}(\mathcal{W})$ one has $w \in \lambda(w)^\circ$. Moreover, for any $\lambda \in \Lambda(\mathcal{W})$ and $w \in \lambda^\circ$ one has $\lambda = \lambda(w)$. With every chamber $\lambda = \lambda(w)$, we associate a $\mathcal{W}$-bunch

$$\Theta(\lambda) := \{\tau \in \Omega(\mathcal{W}); \; w \in \tau^\circ\}.$$

Theorem 2.2.2.2 *Let $V = (v_1, \ldots, v_r)$ in $N_\mathbb{Q}$ and $\mathcal{W} = (w_1, \ldots, w_r)$ in $K_\mathbb{Q}$ be Gale dual vector configurations. Then $\Lambda(\mathcal{W})$ is a fan in $K_\mathbb{Q}$ with support $\mathrm{cone}(\mathcal{W})$. Moreover, the following statements hold.*

(i) *For every chamber $\lambda \in \Lambda(\mathcal{W})$, the associated $\mathcal{W}$-bunch $\Theta(\lambda)$ is maximal and $\Sigma(\lambda) := \Theta(\lambda)^\sharp$ is a normal maximal $\mathcal{V}$-quasifan.*

(ii) *Consider the situation of Construction 2.2.1.3 and set $\gamma :=$ $\mathrm{cone}(e_1, \ldots, e_r)$. Then the $\mathcal{V}$-quasifan $\Sigma(\lambda)$ associated with $\lambda \in \Lambda(\mathcal{W})$*

is the normal quasifan of any polyhedron $B_w \subseteq M_{\mathbb{Q}}$ obtained as follows:

$$B_w := (Q^{-1}(w) \cap \gamma) - e, \qquad w \in \lambda^\circ, \ e \in Q^{-1}(w).$$

(iii) *Let $\lambda_1, \lambda_2 \in \Lambda(\mathcal{W})$. Then $\lambda_1 \preceq \lambda_2$ is equivalent to $\Theta(\lambda_1) \leq \Theta(\lambda_2)$. In particular, if λ_1 is a face of λ_2 then $\Sigma(\lambda_2)$ refines $\Sigma(\lambda_1)$.*

(iv) *If Σ is a normal maximal $\mathcal{V}$-quasifan, then $\Sigma = \Sigma(\lambda)$ holds with some chamber $\lambda \in \Lambda(\mathcal{W})$.*

Recall that, given quasifans $\Sigma_1, \ldots, \Sigma_n$ in a rational vector space $N_{\mathbb{Q}}$ such that all Σ_i have the same support, the *coarsest common refinement* of $\Sigma_1, \ldots, \Sigma_n$ is the quasifan

$$\Sigma := \{\sigma_1 \cap \ldots \cap \sigma_n; \ \sigma_1 \in \Sigma_1, \ldots, \sigma_n \in \Sigma_n\}.$$

If each Σ_i is the normal quasifan of a polyhedron $B_i \subseteq M_{\mathbb{Q}}$ in the dual vector space, then the coarsest common refinement of $\Sigma_1, \ldots, \Sigma_n$ is the normal fan of the Minkowski sum $B_1 + \cdots + B_n$.

Theorem 2.2.2.3 *Let $\mathcal{V} = (v_1, \ldots, v_r)$ in $N_{\mathbb{Q}}$ and $\mathcal{W} = (w_1, \ldots, w_r)$ in $K_{\mathbb{Q}}$ be Gale dual vector configurations.*

(i) *The GKZ decomposition $\Lambda(\mathcal{V})$ is the coarsest common refinement of all quasifans $\Sigma(\lambda)$, where $\lambda \in \Lambda(\mathcal{W})$.*

(ii) *The GKZ decomposition $\Lambda(\mathcal{V})$ is the coarsest common refinement of all $\mathcal{V}$-quasifans having $\mathrm{cone}(\mathcal{V})$ as their support.*

(iii) *In the setting of Theorem 2.2.2.2 (iii), fix for each chamber $\lambda \in \Lambda(\mathcal{W})$ an element $w(\lambda) \in \lambda^\circ$. Then the Minkowski sum over the $B_{w(\lambda)}$ has $\Lambda(\mathcal{V})$ as its normal fan.*

Corollary 2.2.2.4 *Every complete quasifan in a rational vector space admits a refinement by a polytopal fan.*

Definition 2.2.2.5 *Let $K_{\mathbb{Q}}$ be a rational vector space and $\mathcal{W} = (w_1, \ldots, w_r)$ a family of vectors in $K_{\mathbb{Q}}$ generating $K_{\mathbb{Q}}$. The* moving cone *of $\mathcal{W}$ is*

$$\mathrm{Mov}(\mathcal{W}) := \bigcap_{i=1}^{r} \mathrm{cone}(w_j; \ j \neq i) \subseteq K_{\mathbb{Q}}.$$

Theorem 2.2.2.6 *Let $\mathcal{V} = (v_1, \ldots, v_r)$ in $N_{\mathbb{Q}}$ and $\mathcal{W} = (w_1, \ldots, w_r)$ in $K_{\mathbb{Q}}$ be Gale dual vector configurations.*

(i) *The cone $\mathrm{Mov}(\mathcal{W})$ is of full dimension in $K_{\mathbb{Q}}$ if and only if $v_1, \ldots, v_r$ generate pairwise different one-dimensional cones in $N_{\mathbb{Q}}$.*

(ii) *Assume that $\mathrm{Mov}(\mathcal{W})$ is of full dimension in $K_{\mathbb{Q}}$. Then the quasifan $\Sigma(\lambda)$ associated with $\lambda \in \Lambda(\mathcal{W})$ is a fan if and only if $\lambda^\circ \subseteq \mathrm{cone}(\mathcal{W})^\circ$ holds.*

(iii) *Assume that* Mov($\mathcal{W}$) *is of full dimension in* $K_{\mathbb{Q}}$. *Then we have mutually inverse order preserving bijections*

$$\{\lambda \in \Lambda(\mathcal{W}); \ \lambda^{\circ} \subseteq \text{Mov}(\mathcal{W})^{\circ}\} \longleftrightarrow \{\textit{true normal } \mathcal{V}\text{-fans}\},$$

$$\lambda \ \mapsto \ \Sigma(\lambda),$$

$$\bigcap_{\tau \in \Sigma^{\sharp}} \tau \ \leftharpoonup \ \Sigma.$$

Under these bijections, the simplicial true normal fans correspond to the full-dimensional chambers.

As an immediate consequence, one obtains the following; see [235, Cor. 3.8], [242, Thm. 4.1] and also [282, Thm. 8.3].

Corollary 2.2.2.7 *Assume that* $v_1, \ldots, v_r \in N_{\mathbb{Q}}$ *generate pairwise different one-dimensional cones. There exist normal true* $\mathcal{V}$*-fans and every such fan admits a refinement by a simplicial normal true* $\mathcal{V}$*-fan.*

Example 2.2.2.8 Consider the Gale dual vector configurations $\mathcal{V} = (v_1, \ldots, v_6)$ in $N_{\mathbb{Q}} = \mathbb{Q}^3$ and $\mathcal{W} = (w_1, \ldots, w_6)$ in $K_{\mathbb{Q}} = \mathbb{Q}^3$ given by

$$v_1 = (-1, 0, 0), \quad v_2 = (0, -1, 0), \quad v_3 = (0, 0, -1),$$

$$v_4 = (0, 1, 1), \quad v_5 = (1, 0, 1), \quad v_6 = (1, 1, 0)$$

and

$$w_1 = (1, 0, 0), \quad w_2 = (0, 1, 0), \quad w_3 = (0, 0, 1),$$

$$w_4 = (1, 1, 0), \quad w_5 = (1, 0, 1), \quad w_6 = (0, 1, 1).$$

Then there are 13 chambers $\lambda \in \Lambda(\mathcal{W})$ with $\lambda^{\circ} \subseteq \text{Mov}(\mathcal{W})^{\circ}$. These give rise to 13 polytopal true maximal $\mathcal{V}$-fans.

Moreover, one finds 14 true maximal $\mathcal{W}$-bunches not arising from a chamber. Thus we have in total 27 maximal $\mathcal{V}$-fans, 14 of them being nonpolytopal.

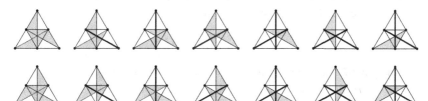

2.2.3 Proof of Theorem 2.2.1.14

We work in the setting of Construction 2.2.1.3. That means that we have a pair of mutually dual exact sequences of finite-dimensional rational vector spaces

$$0 \longrightarrow L_{\mathbb{Q}} \longrightarrow F_{\mathbb{Q}} \overset{P}{\longrightarrow} N_{\mathbb{Q}} \longrightarrow 0$$

$$0 \longleftarrow K_{\mathbb{Q}} \underset{Q}{\longleftarrow} E_{\mathbb{Q}} \longleftarrow M_{\mathbb{Q}} \longleftarrow 0$$

The idea is to decompose the $\sharp$-operation between collections in $K_{\mathbb{Q}}$ and collections in $N_{\mathbb{Q}}$ according to the following scheme of further operations on collections of cones:

$$
\begin{array}{ccc}
\{\text{collections in } E_{\mathbb{Q}}\} & \overset{*}{\longleftrightarrow} & \{\text{collections in } F_{\mathbb{Q}}\} \\
Q^{\uparrow} \big\uparrow \big\downarrow Q^{\downarrow} & & P^{\downarrow} \big\uparrow \big\downarrow P^{\uparrow} \\
\{\text{collections in } K_{\mathbb{Q}}\} & \underset{\sharp}{\longleftrightarrow} & \{\text{collections in } N_{\mathbb{Q}}\}
\end{array}
$$

We enter the detailed discussion. Let $(f_1, \ldots, f_r)$ be a basis for $F_{\mathbb{Q}}$, and let $(e_1, \ldots, e_r)$ be the dual basis for $E_{\mathbb{Q}}$. Then we have the image vectors

$$v_i := P(f_i) \in N_{\mathbb{Q}}, \qquad w_i := Q(e_i) \in K_{\mathbb{Q}}, \qquad 1 \le i \le r.$$

The vector configurations $V = (v_1, \ldots, v_r)$ in $N_{\mathbb{Q}}$ and $\mathcal{W} = (w_1, \ldots, w_r)$ in $K_{\mathbb{Q}}$ are Gale dual to each other. Moreover, we set

$$\delta := \text{cone}(f_1, \ldots, f_r), \qquad \gamma := \text{cone}(e_1, \ldots, e_r).$$

Then these cones are dual to each other, and we have the face correspondence, that is, mutually inverse bijections

$$\text{faces}(\delta) \longleftrightarrow \text{faces}(\gamma),$$

$$\delta_0 \mapsto \delta_0^* := \delta_0^{\perp} \cap \gamma,$$

$$\gamma_0^{\perp} \cap \delta =: \gamma_0^* \longleftarrow \gamma_0.$$

Definition 2.2.3.1 By an *$L_{\mathbb{Q}}$-invariant separating linear form* for two faces $\delta_1, \delta_2 \preceq \delta$, we mean an element $e \in E_{\mathbb{Q}}$ such that

$$e|_{L_{\mathbb{Q}}} = 0, \qquad e|_{\delta_1} \ge 0, \qquad e|_{\delta_2} \le 0, \qquad \delta_1 \cap e^{\perp} = \delta_1 \cap \delta_2 = e^{\perp} \cap \delta_2.$$

Lemma 2.2.3.2 (Invariant separation lemma) *Consider two faces $\delta_1, \delta_2 \preceq \delta$ and the corresponding faces $\gamma_i := \delta_i^* \preceq \gamma$. Then the following statements are equivalent.*

(i) *There is an $L_\mathbb{Q}$-invariant separating linear form for δ_1 and δ_2.*

(ii) *The relative interiors $Q(\gamma_i)^\circ$ satisfy $Q(\gamma_1)^\circ \cap Q(\gamma_2)^\circ \neq \emptyset$.*

Proof Let $\delta_1 = \operatorname{cone}(f_i;\ i \in I)$ and $\delta_2 = \operatorname{cone}(f_j;\ j \in J)$ with subsets I and J of $\mathfrak{R} := \{1, \ldots, r\}$. Condition (i) is equivalent to the existence of a linear form $u \in M_\mathbb{Q}$ with

$$u(v_i) > 0 \text{ for } i \in I \setminus J, \quad u(v_k) = 0 \text{ for } k \in I \cap J, \quad u(v_j) < 0 \text{ for } j \in J \setminus I.$$

According to Remark 2.2.1.2 such a linear form $u \in M_\mathbb{Q}$ exists if and only if there are $a_1, \ldots, a_r \in \mathbb{Q}$ with $a_1 w_1 + \cdots + a_r w_r = 0$ and

$$a_i > 0 \text{ for } i \in I \setminus J, \quad a_k = 0 \text{ for } k \in I \cap J, \quad a_j < 0 \text{ for } j \in J \setminus I.$$

This in turn is possible if and only if we find coefficients $b_m > 0$ for $m \in \mathfrak{R} \setminus I$ and $c_n > 0$ for $n \in \mathfrak{R} \setminus J$ such that $\sum_{m \in \mathfrak{R} \setminus I} b_m w_m$ equals $\sum_{n \in \mathfrak{R} \setminus J} c_n w_n$. The latter is condition (ii). $\qquad\square$

Definition 2.2.3.3 By a *δ-collection* we mean a collection of faces of δ. We say that a δ-collection $\mathfrak{A}$ is

(i) *separated* if any two $\delta_1, \delta_2 \in \mathfrak{A}$ admit an $L_\mathbb{Q}$-invariant separating linear form;

(ii) *saturated* if for any $\delta_1 \in \mathfrak{A}$ and any $\delta_2 \preceq \delta_1$, which is $L_\mathbb{Q}$-invariantly separable from δ_1, one has $\delta_2 \in \mathfrak{A}$;

(iii) *true* if all rays of δ belong to $\mathfrak{A}$;

(iv) *maximal* if it is maximal among the separated δ-collections.

Definition 2.2.3.4 By a *γ-collection* we mean a collection of faces of γ. We say that a γ-collection $\mathfrak{B}$ is

(i) *connected* if for any two $\gamma_1, \gamma_2 \in \mathfrak{B}$ the intersection $Q(\gamma_1)^\circ \cap Q(\gamma_2)^\circ$ is not empty;

(ii) *saturated* if for any $\gamma_1 \in \mathfrak{B}$ and any $\gamma \succeq \gamma_2 \succeq \gamma_1$ with $Q(\gamma_1)^\circ \subseteq Q(\gamma_2)^\circ$, one has $\gamma_2 \in \mathfrak{B}$;

(iii) *true* if all facets of γ belong to $\mathfrak{B}$;

(iv) *maximal* if it is maximal among the connected γ-collections.

Proposition 2.2.3.5 *We have mutually inverse bijections sending separated (saturated, true, maximal) collections to connected (saturated, true, maximal) collections:*

$$\{\text{separated } \delta\text{-collections}\} \longleftrightarrow \{\text{connected } \gamma\text{-collections}\},$$

$$\mathfrak{A} \mapsto \mathfrak{A}^* := \{\delta_0^*;\ \delta_0 \in \mathfrak{A}\},$$

$$\mathfrak{B}^* := \{\gamma_0^*;\ \gamma_0 \in \mathfrak{B}\} \leftarrow\!\shortmid \mathfrak{B}.$$

Proof The assertion is an immediate consequence of the invariant separation lemma. □

Definition 2.2.3.6 By a W-*collection* we mean a set of W-cones. We say that a W-collection Θ is

(i) *connected* if for any two $\tau_1, \tau_2 \in \Theta$ we have $\tau_1^\circ \cap \tau_2^\circ \neq \emptyset$,
(ii) *saturated* if for any $\tau \in \Theta$ also all W-cones σ with $\tau^\circ \subseteq \sigma^\circ$ belong to Θ,
(iii) *true* if every $\vartheta_i = \mathrm{cone}(w_j; \ j \neq i)$, where $1 \leq i \leq r$, belongs to Θ,
(iv) *maximal* if it is maximal among all connected W-collections.

Definition 2.2.3.7 Consider the set of all W-collections and the set of all γ-collections. We define the Q-*lift* and the Q-*drop* to be the maps

$$Q^\uparrow \colon \{W\text{-collections}\} \to \{\gamma\text{-collections}\},$$

$$\Theta \mapsto Q^\uparrow \Theta := \{\gamma_0 \preceq \gamma; \ Q(\gamma_0) \in \Theta\},$$

$$Q_\downarrow \colon \{\gamma\text{-collections}\} \to \{W\text{-collections}\},$$

$$\mathfrak{B} \mapsto Q_\downarrow \mathfrak{B} := \{Q(\gamma_0); \ \gamma_0 \in \mathfrak{B}\}.$$

Proposition 2.2.3.8 *The Q-lift is injective and sends connected (saturated, true, maximal) W-collections to Q-connected (saturated, true, maximal) γ-collections. Moreover, we have mutually inverse bijections sending true collections to true collections:*

$$\{maximal \ W\text{-}collections\} \longleftrightarrow \{maximal \ \gamma\text{-}collections\},$$

$$\Theta \mapsto Q^\uparrow \Theta,$$

$$Q_\downarrow \mathfrak{B} \leftarrow\!\shortmid \mathfrak{B}.$$

Proof By definition of the Q-lift and the Q-drop, we have $Q_\downarrow Q^\uparrow \Theta = \Theta$ for every W-collection Θ. In particular, $Q^\uparrow$ is injective. Moreover, $Q^\uparrow$ clearly preserves the properties connected, saturated, true, and maximal. If $\mathfrak{B}$ is a maximal γ-collection, then $Q_\downarrow \mathfrak{B}$ is a maximal W-collection and we have $Q^\uparrow Q_\downarrow \mathfrak{B} = \mathfrak{B}$. Thus, restricted to maximal collections, $Q^\uparrow$ and $Q_\downarrow$ are mutually inverse bijections. Obviously, $Q_\downarrow$ sends true collections to true collections. □

Definition 2.2.3.9 By a V-*collection* we mean a set of V-cones. We say that a V-collection Σ is

(i) *separated* if any two $\sigma_1, \sigma_2 \in \Sigma$ intersect in a common face;
(ii) *saturated* if for any $\sigma \in \Sigma$ also all faces $\sigma_0 \preceq \sigma$ belong to Σ;
(iii) *true* if every ray $\varrho_i = \mathrm{cone}(v_i)$, where $1 \leq i \leq r$, belongs to Σ;
(iv) *maximal* if it is maximal among all separated V-collections.

Definition 2.2.3.10 Consider the set of all $\mathcal{V}$-collections and the set of all δ-collections. We define the *P-lift* and the *P-drop* to be the maps

$$P^\uparrow \colon \{\mathcal{V}\text{-collections}\} \to \{\delta\text{-collections}\},$$

$$\Sigma \mapsto P^\uparrow \Sigma := \{\delta_0 \preceq \delta;\ P(\delta_0) \in \Sigma\},$$

$$P_\downarrow \colon \{\delta\text{-collections}\} \to \{\mathcal{V}\text{-collections}\},$$

$$\mathfrak{A} \mapsto P_\downarrow \mathfrak{A} := \{P(\delta_0);\ \delta_0 \in \mathfrak{A}\}.$$

Proposition 2.2.3.11 *The P-drop is surjective and sends separated (saturated, true, maximal) collections to separated (saturated, true, maximal) collections. If $v_1, \ldots, v_r$ generate pairwise different rays, then we have mutually inverse bijections sending saturated (maximal) collections to saturated (maximal) collections:*

$$\{\text{true separated } \delta\text{-collections}\} \longleftrightarrow \{\text{true separated } \mathcal{V}\text{-collections}\},$$

$$\mathfrak{A} \mapsto P_\downarrow \mathfrak{A},$$

$$P^\uparrow \Sigma \leftarrow\!\shortmid \Sigma.$$

Proof For every $\mathcal{V}$-collection Σ we have $\Sigma = P_\downarrow P^\uparrow \Sigma$. In particular, $P_\downarrow$ is surjective. The fact that $P_\downarrow$ preserves separatedness and saturatedness follows from the observation that an $L_{\mathbb{Q}}$-invariant separating linear form for two faces $\delta_1, \delta_2 \preceq \delta$ induces a separating linear form for the images $P(\delta_1)$ and $P(\delta_2)$. The fact that $P_\downarrow$ preserves the properties true and maximal is obvious.

Now assume that $v_1, \ldots, v_r$ generate pairwise different rays. Consider a true separated $\mathcal{V}$-collection Σ. Then, for every $\sigma \in \Sigma$ and every $1 \le i \le r$, we have $v_i \in \sigma$ if and only if $\mathbb{Q}_{\ge 0} \cdot v_i$ is an extremal ray of σ. Consequently, for every $\sigma \in \Sigma$ there is a unique $\delta_0 \preceq \delta$ with $P(\delta_0) = \sigma$. It follows that $P^\uparrow(\Sigma)$ is true and separated, and, if Σ is saturated (maximal), then also $P^\uparrow(\Sigma)$ saturated (maximal). Moreover, we conclude that $P_\downarrow$ restricted to the true separated collections is injective. □

Proof of Theorem 2.2.1.14 First observe that the (true, maximal) $\mathcal{W}$-bunches are precisely the (true, maximal) connected saturated $\mathcal{W}$-collections and the (true, maximal) $\mathcal{V}$-quasifans are precisely the (true, maximal) separated saturated $\mathcal{V}$-collections. Next observe that we have

$$\Theta^\sharp = P_\downarrow((Q^\uparrow \Theta)^*), \qquad \Sigma^\sharp = Q_\downarrow((P^\uparrow \Sigma)^*).$$

Then Propositions 2.2.3.8, 2.2.3.5, and 2.2.3.11 provide the statements made on $\Theta \mapsto \Theta^\sharp$ and $\Sigma \mapsto \Sigma^\sharp$; the fact that these assigments are order-reversing is obvious.

We still have to show that simplicial fans correspond to bunches of full-dimensional cones. A $\mathcal{V}$-fan Σ is simplicial exactly when for every cone$(v_i ; i \in I)$ in Σ and every subset $I_1 \subseteq I$ the cone$(v_i ; i \in I_1)$ is in Σ. For true Σ this means that for every $\tau = \mathrm{cone}(w_j ; j \in J)$ in the corresponding bunch Θ and every $J_1, J \subseteq J_1 \subseteq \mathfrak{R}$, one has $\tau^\circ \subseteq \mathrm{cone}(w_j ; j \in J_1)^\circ$. Because the vectors w_j generate $K_{\mathbb{Q}}$, this is exactly the case when all cones of Θ are of full dimension. $\qquad\qquad\square$

2.2.4 Proof of Theorems 2.2.2.2, 2.2.2.3, and 2.2.2.6

The setup is as in the preceding subsection, which means that we have the pair of mutually dual exact sequences

$$0 \longrightarrow L_{\mathbb{Q}} \longrightarrow F_{\mathbb{Q}} \overset{P}{\longrightarrow} N_{\mathbb{Q}} \longrightarrow 0,$$

$$0 \longleftarrow K_{\mathbb{Q}} \underset{Q}{\longleftarrow} E_{\mathbb{Q}} \longleftarrow M_{\mathbb{Q}} \longleftarrow 0,$$

fix a basis $(f_1, \ldots, f_r)$ for $F_{\mathbb{Q}}$ and denote by $(e_1, \ldots, e_r)$ the dual basis for $E_{\mathbb{Q}}$. Then the image vectors

$$v_i := P(f_i) \in N_{\mathbb{Q}}, \qquad w_i := Q(e_i) \in K_{\mathbb{Q}}, \qquad 1 \le i \le r,$$

give Gale dual vector configurations $\mathcal{V} = (v_1, \ldots, v_r)$ in $N_{\mathbb{Q}}$ and $\mathcal{W} = (w_1, \ldots, w_r)$ in $K_{\mathbb{Q}}$. Again, the positive orthants in $F_{\mathbb{Q}}$ and $E_{\mathbb{Q}}$ are denoted by

$$\delta := \mathrm{cone}(f_1, \ldots, f_r), \qquad \gamma := \mathrm{cone}(e_1, \ldots, e_r).$$

We split the proofs of the theorems into several propositions. The first two settle Theorem 2.2.2.2 (i) and (ii).

Proposition 2.2.4.1 *The collection $\Sigma(\lambda(w))$ equals the normal quasifan $\mathcal{N}(B_w)$. In particular, $\Sigma(\lambda(w))$ is a normal maximal $\mathcal{V}$-quasifan with support $\mathrm{cone}(\mathcal{V})$.*

Proof The cones of the normal quasifan $\mathcal{N}(B_w)$ correspond to the faces of the polyhedron of $B_w = (Q^{-1}(w) \cap \gamma) - e$, where $e \in E_{\mathbb{Q}}$ is any element with $Q(e) = w$ as follows: Given $B_{w,0} \preceq B_w$, the corresponding cone of $\mathcal{N}(B_w)$ is

$$\sigma_0 := \{v \in N_{\mathbb{Q}}; \, \langle u - u_0, v \rangle \ge 0, \, u \in B_w, \, u_0 \in B_{w,0}\} \subseteq N_{\mathbb{Q}}.$$

Now, let $\gamma_0 \preceq \gamma$ denote the minimal face with $B_{w,0} + e \subseteq \gamma_0$. Then we have $\gamma_0^\circ \cap Q^{-1}(w) \ne \emptyset$. Thus, γ_0 belongs to the Q-lift of $\Theta(\lambda(w))$. For the corresponding face $\delta_0 = \gamma_0^\perp \cap \delta$ of $\delta = \gamma^\vee$, one directly verifies

$$P(\delta_0) = \sigma_0.$$

Thus, the first assertion follows from the observation that the assignment $B_{w,0} \mapsto \gamma_0$ defines a bijection from the faces of B_w to the Q-lift of $\Theta(\lambda(w))$, and the fact that $\Sigma(\lambda)$ equals $P_\downarrow((Q^\dagger \Theta(\lambda(w)))^*)$. The rest follows from

$$\mathrm{Supp}(\mathcal{N}(B_w)) = (P^*(M_{\mathbb{Q}}) \cap \gamma)^\vee = \mathrm{cone}(\mathcal{V}). \qquad \square$$

Proposition 2.2.4.2 *For every chamber $\lambda \in \Lambda(\mathcal{W})$, the associated $\mathcal{W}$-bunch $\Theta(\lambda)$ is maximal.*

Proof Suppose that $\Theta(\lambda)$ is a proper subset of a maximal $\mathcal{W}$-bunch Θ. For the associated saturated separated δ-collections this means that $(Q^\dagger \Theta(\lambda))^*$ is a proper subset of $(Q^\dagger \Theta)^*$. Because P induces a bijection $(Q^\dagger \Theta)^* \to P_\downarrow((Q^\dagger \Theta)^*)$ we obtain that $\Theta(\lambda)^\sharp$ is a proper subset of $\Theta^\sharp$. This contradicts the fact that, due to Proposition 2.2.4.1, the quasifan $\Theta(\lambda)^\sharp$ has $\mathrm{cone}(\mathcal{V})$ as its support. $\qquad \square$

Now we will see that the GKZ decomposition $\Lambda(\mathcal{W})$ is in fact a fan. We prove this for the GKZ decomposition $\Lambda(\mathcal{V})$ of the Gale dual vector configuration and moreover verify the assertions made in Theorem 2.2.2.3.

Proposition 2.2.4.3 *The GKZ decomposition $\Lambda(\mathcal{V})$ is a fan. Moreover, the following statements hold.*

  (i) *The GKZ decomposition $\Lambda(\mathcal{V})$ is the coarsest common refinement of all quasifans $\Sigma(\lambda)$, where $\lambda \in \Lambda(\mathcal{W})$.*
 (ii) *The GKZ decomposition $\Lambda(\mathcal{V})$ is the coarsest common refinement of all $\mathcal{V}$-quasifans having $\mathrm{cone}(\mathcal{V})$ as its support.*
(iii) *Fix for each chamber $\lambda \in \Lambda(\mathcal{W})$ an element $w(\lambda) \in \lambda^\circ$. Then the Minkowski sum over the $B_{w(\lambda)}$ has $\Lambda(\mathcal{V})$ as its normal fan.*

Lemma 2.2.4.4 *For every $\mathcal{V}$-cone $\sigma \subseteq N_{\mathbb{Q}}$, there exists a chamber $\lambda \in \Lambda(\mathcal{W})$ with $\sigma \in \Sigma(\lambda)$.*

Proof Write $\sigma = \mathrm{cone}(v_i \, ; \, i \in I)$, and choose $w \in \mathrm{cone}(w_j \, ; \, j \notin I)^\circ$. Then the associated quasifan $\Sigma(\lambda(w)) = \Theta(\lambda(w))^\sharp$ contains σ. $\qquad \square$

Proof of Proposition 2.2.4.3 Lemma 2.2.4.4 tells us that every $\mathcal{V}$-cone is contained in a normal $\mathcal{V}$-quasifan $\Sigma(\lambda) = \mathcal{N}(B_{w(\lambda)})$, where $\lambda \in \Lambda(\mathcal{W})$ and $w \in \lambda^\circ$ is fixed. It follows that $\Lambda(\mathcal{V})$ is the coarsest common refinement of the $\Sigma(\lambda)$. Because every ray $\mathrm{cone}(v_j) \subseteq N_{\mathbb{Q}}$ belongs to some $\Sigma(\lambda)$, we obtain that $\Lambda(\mathcal{V})$

is even a fan. Moreover, also the second assertion follows from the fact that every $\mathcal{V}$-cone is contained in some $\Sigma(\lambda)$. Finally, we obtained that $\Sigma(\lambda)$ is the normal fan of the Minkowski sum over the $B_{w(\lambda)}$. □

Proposition 2.2.4.5 *For any two* $\lambda_1, \lambda_2 \in \Lambda(\mathcal{W})$ *we have* $\lambda_1 \preceq \lambda_2$ *if and only if* $\Theta(\lambda_1) \leq \Theta(\lambda_2)$ *holds. In particular,* $\lambda_1 \preceq \lambda_2$ *implies* $\Sigma(\lambda_2) \leq \Sigma(\lambda_1)$.

Proof Suppose that $\lambda_1 \preceq \lambda_2$ holds. Then, given any $\tau_2 \in \Theta(\lambda_2)$, we have $\lambda_1 \subseteq \tau_2$ and thus find a face $\tau_1 \preceq \tau_2$ with $\lambda_1^\circ \subseteq \tau_1^\circ$. Consequently $\tau_1 \in \Theta(\lambda_1)$ and $\tau_1 \subseteq \tau_2$ holds. Next suppose that $\Theta(\lambda_1) \leq \Theta(\lambda_2)$ holds. For every $\tau_2 \in \Theta(\lambda_2)$, fix a $\tau_1 \in \Theta(\lambda_1)$ with $\tau_1 \subseteq \tau_2$. This yields $\lambda_1 \subseteq \tau_2$ for all $\tau_2 \in \Theta(\lambda_2)$. We conclude $\lambda_1 \subseteq \lambda_2$ and, because $\Lambda(\mathcal{W})$ is a fan, $\lambda_1 \preceq \lambda_2$. The last assertion is then clear because we have $\Sigma(\lambda_i) = \Theta(\lambda_i)^\sharp$ and $\Theta \to \Theta^\sharp$ is order reversing. □

Proposition 2.2.4.6 *Every normal* $\mathcal{V}$-*quasifan is of the form* $\Sigma(\lambda)$ *with a chamber* $\lambda \in \Lambda(\mathcal{W})$.

Proof Suppose that a $\mathcal{V}$-quasifan Σ is the normal quasifan of a polyhedron $B \subseteq M_{\mathbb{Q}}$. Then the polyhedron B is given by inequalities $\langle u, v_i \rangle \geq c_i$, where $v_i \in \mathcal{V}$ and $c_i \in \mathbb{Q}$. Note that each v_i is the restriction of a coordinate function on $E_{\mathbb{Q}}$ to the subspace $M_{\mathbb{Q}}$. This shows that B may be obtained as the intersection of γ with the parallel translate $e + M_{\mathbb{Q}}$ of the subspace $M_{\mathbb{Q}}$, where $e = -c_1 e_1 - \ldots - c_r e_r \in E_{\mathbb{Q}}$. Thus $B = B_w$, where $w = Q(e)$, and Proposition 2.2.4.1 completes the proof. □

We obtained all the assertions of Theorems 2.2.2.2 and 2.2.2.3 and now turn to the proof of Theorem 2.2.2.6. Recall that we set $\vartheta_i = \mathrm{cone}(w_j; \ j \neq i)$ for $1 \leq i \leq r$.

Proposition 2.2.4.7 *Assume that* $v_1, \ldots, v_r$ *generate pairwise different one-dimensional cones. Then, for* $\lambda \in \Lambda(\mathcal{W})$, *the quasifan* $\Sigma(\lambda)$ *is a fan if and only if* $\lambda^\circ \subseteq \mathrm{cone}(\mathcal{W})^\circ$ *holds.*

Proof Take $w \in \lambda^\circ$. Then the normal quasifan $\mathcal{N}(B_w)$ is a fan if and only if the polyhedron $B_w \subseteq M_{\mathbb{Q}}$ is of full dimension. The latter holds if and only if $Q^{-1}(w)$ intersects γ_0° for a face $\gamma_0 \preceq \gamma$ with $M_{\mathbb{Q}} \subseteq \mathrm{lin}(\gamma_0)$. Because all v_i are nonzero, the latter is equivalent to $Q^{-1}(w) \cap \gamma^\circ \neq \emptyset$. This in turn means precisely $w \in \mathrm{cone}(\mathcal{W})^\circ$. □

Proposition 2.2.4.8 *Assume that* $v_1, \ldots, v_r$ *generate pairwise different one-dimensional cones. Then there exists a true normal* $\mathcal{V}$-*fan* Σ.

Proof Rescale each v_i to $v_i' = c_i v_i$, where $c_i \in \mathbb{Q}_{>0}$, such that every v_i' is a vertex of the convex hull over $0, v_1', \ldots, v_r'$. Then $\Sigma := \mathcal{N}(B)$ is as wanted for

$$B := \{u \in M_{\mathbb{Q}}; \ \langle u, v_i \rangle \geq -1\} \subseteq M_{\mathbb{Q}}.$$ □

Proposition 2.2.4.9 *The cone* Mov($\mathcal{W}$) *is of full dimension in* $K_{\mathbb{Q}}$ *if and only if* $v_1, \ldots, v_r$ *generate pairwise different one-dimensional cones. If one of these statements holds, then we have*

$$\mathrm{Mov}(\mathcal{W})^{\circ} = \vartheta_1^{\circ} \cap \ldots \cap \vartheta_r^{\circ}.$$

Proof First suppose that $v_1, \ldots, v_r$ generate pairwise different one-dimensional cones. Then Proposition 2.2.4.8 provides us with a true normal $\mathcal{V}$-fan Σ and Proposition 2.2.4.6 guarantees $\Sigma = \Sigma(\lambda)$ with a chamber $\lambda \in \Lambda(\mathcal{W})$. Take any $w \in \lambda^{\circ}$. Because each cone(v_i) belongs to $\Sigma = \Theta(\lambda)^{\sharp}$, we obtain that each ϑ_i° contains w. Because every v_i is nonzero, the ϑ_i are of full dimension in $K_{\mathbb{Q}}$ and thus Mov($\mathcal{W}$) is as well.

Conversely, if Mov($\mathcal{W}$) is of full dimension in $K_{\mathbb{Q}}$, then $\vartheta_1, \ldots, \vartheta_r$ together with $Q(\gamma)^{\circ}$ form a true $\mathcal{W}$-bunch. Consequently, cone(f_1), ..., cone(f_r) together with the zero cone, form a true separated δ-collection $\mathfrak{A}$. The P-drop $P_{!}\mathfrak{A}$ is the $\mathcal{V}$-fan Σ consisting of the zero cone and cone(v_1), ..., cone(v_r). Because P induces an injection $\mathfrak{A} \to \Sigma$, we conclude that the cone(v_i) are pairwise different and one-dimensional. $\qquad\square$

Proposition 2.2.4.10 *Assume that the moving cone* Mov($\mathcal{W}$) *is of full dimension in* $K_{\mathbb{Q}}$. *Then we have mutually inverse bijections*

$$\{\lambda \in \Lambda;\ \lambda^{\circ} \subseteq \mathrm{Mov}(\mathcal{W})^{\circ}\} \longleftrightarrow \{\textit{true normal } \mathcal{V}\textit{-fans}\}$$

$$\lambda \mapsto \Sigma(\lambda),$$

$$\bigcap_{\tau \in \Sigma^{\sharp}} \tau \,\leftarrow\!\shortmid\, \Sigma.$$

Proof First we remark that, by Proposition 2.2.4.9, the vectors $v_1, \ldots, v_r$ generate pairwise different one-dimensional cones. Thus, given $\lambda \in \Lambda(\mathcal{W})$, the associated $\mathcal{V}$-fan $\Sigma(\lambda) = \Theta(\lambda)^{\sharp}$ is true if and only if $\Theta(\lambda)$ comprises $\vartheta_1, \ldots, \vartheta_r$, where the latter is equivalent to $\lambda^{\circ} \subseteq \mathrm{Mov}(\mathcal{W})^{\circ}$. Now the assertion follows directly from Propositions 2.2.4.1, 2.2.4.2, and 2.2.4.6 and Theorem 2.2.1.14. $\qquad\square$

2.3 Good toric quotients

2.3.1 Characterization of good toric quotients

We consider the induced action of a closed subgroup of the acting torus on a given toric variety defined over an algebraically closed field $\mathbb{K}$ of characteristic zero. The aim is to present the combinatorial characterization [288, Thm. 4.1] of the existence of a good quotient in the following sense.

Definition 2.3.1.1 Let a reductive group G act on a variety X. We say that a good quotient $p \colon X \to Y$ for the G-action is *separated* if Y is separated.

A first general observation enables us to treat the problem of existence of a separated good quotient entirely in terms of toric geometry.

Proposition 2.3.1.2 *Let X be a toric variety with acting torus T and $p \colon X \to Y$ a separated good quotient for the action of a closed subgroup $H \subseteq T$. Then Y admits a unique structure of a toric variety turning p into a toric morphism.*

Proof First observe that Y inherits normality from X. Next consider the product $T \times X$. Then H acts on the second factor as a subgroup of T and we have a commutative diagram

$$
\begin{array}{ccc}
T \times X & \xrightarrow{\;\mu_X\;} & X \\
{\scriptstyle \mathrm{id} \times p} \downarrow & & \downarrow {\scriptstyle p} \\
T \times Y & \xrightarrow[\;\mu_Y\;]{} & Y
\end{array}
$$

where μ_X describes the T-action and the downwards arrows are good quotients for the respective H-actions. One verifies directly that the induced morphism μ_Y is a T-action with an open dense orbit. The assertion follows. $\qquad\square$

We fix the setup for the rest of this section. Let Δ be a fan in a lattice F and denote by $X := X_\Delta$ the associated toric variety. Moreover, let $Q \colon E \to K$ be an epimorphism from the dual lattice $E = \operatorname{Hom}(F, \mathbb{Z})$ onto an abelian group K; this specifies an embedding of $H := \operatorname{Spec} \mathbb{K}[K]$ into the acting torus $T = \operatorname{Spec} \mathbb{K}[E]$ and thus an action of H on X. Denoting by $M \subseteq E$ the kernel of $Q \colon E \to K$, we obtain mutually dual exact sequences

$$
0 \longrightarrow L \xrightarrow{\;Q^*\;} F \xrightarrow{\;P\;} N
$$

$$
0 \longleftarrow K \xleftarrow[\;Q\;]{} E \xleftarrow[\;P^*\;]{} M \longleftarrow 0
$$

Definition 2.3.1.3 We say that Δ is $L_{\mathbb{Q}}$-*projectable* if any two $\delta_1, \delta_2 \in \Delta^{\max}$ admit an $L_{\mathbb{Q}}$-invariant separating linear form, that is, an element $e \in E_{\mathbb{Q}}$ with

$$
e|_{L_{\mathbb{Q}}} = 0, \qquad e|_{\delta_1} \geq 0, \qquad e|_{\delta_2} \leq 0, \qquad \delta_1 \cap e^{\perp} = \delta_1 \cap \delta_2 = e^{\perp} \cap \delta_2.
$$

Remark 2.3.1.4 In the above setting, the fan Δ is $L_{\mathbb{Q}}$-projectable if and only if every cone $\delta \in \Delta^{\max}$ satisfies

$$P^{-1}(P(\delta)) \cap \mathrm{Supp}(\Delta) = \delta.$$

Construction 2.3.1.5 Suppose that the fan Δ is $L_{\mathbb{Q}}$-projectable. Then we have the collection of those cones of Δ that can be separated by an $L_{\mathbb{Q}}$-invariant linear form from a maximal cone:

$$\mathfrak{A}(\Delta) := \{\delta \in \Delta;\ \delta = e^{\perp} \cap \delta_0 \text{ for some } \delta_0 \in \Delta^{\max} \text{ and } e \in \delta_0^{\vee} \cap L_{\mathbb{Q}}^{\perp}\}.$$

The images $P(\delta)$, where $\delta \in \mathfrak{A}(\Delta)$, form a quasifan Σ in $N_{\mathbb{Q}}$ and all share the same minimal face $\tau \subseteq N_{\mathbb{Q}}$. The map $P \colon F \to N$ is a map of the quasifans Δ and Σ, and we have a bijection

$$\mathfrak{A}(\Delta) \ \to\ \Sigma, \qquad \delta \mapsto P(\delta).$$

Moreover, with $N_1 := N/(\tau \cap N)$ and the canonical map $P_1 \colon F \to N_1$, the cones $P_1(\delta)$, where $\delta \in \mathfrak{A}(\Delta)$, form a fan Σ_1 in N_1. The map $P_1 \colon F \to N_1$ is a map of the fans Δ and Σ_1, and we have a bijection

$$\mathfrak{A}(\Delta) \ \to\ \Sigma_1, \qquad \delta \mapsto P_1(\delta).$$

Proof We claim that any two cones $\delta_i \in \mathfrak{A}(\Delta)$ admit an $L_{\mathbb{Q}}$-invariant separating linear form. Indeed, consider maximal cones $\delta_i' \in \Delta$ with $\delta_i \preceq \delta_i'$. Then δ_1' and δ_2' admit an $L_{\mathbb{Q}}$-invariant separating linear form e, the faces δ_i are separated from δ_i' by $L_{\mathbb{Q}}$-invariant linear forms e_i and any linear combination $e_2 - e_1 + ae$ with a big enough provides the wanted $L_{\mathbb{Q}}$-invariant separating linear form for δ_1 and δ_2. This claim directly implies that the cones $P(\delta)$, where $\delta \in \mathfrak{A}(\Delta)$, form a quasifan Σ and the canonical map $\mathfrak{A}(\Delta) \to \Sigma$ is bijective. By construction, Σ_1 is a fan and the canonical map $\Sigma \to \Sigma_1$ is a bijection. $\square$

Proposition 2.3.1.6 *As in the setup above, let X be the toric variety arising from a fan Δ in a lattice F, and consider the action of a subgroup $H \subseteq T$ given by an epimorphism $Q \colon E \to K$. Then the following statements are equivalent.*

(i) *The action of H on X admits a separated good quotient.*
(ii) *The fan Δ is $L_{\mathbb{Q}}$-projectable.*
(iii) *Every $\delta \in \Delta^{\max}$ satisfies $P^{-1}(P(\delta)) \cap \mathrm{Supp}(\Delta) = \delta$.*

Moreover, if one of these statements holds, then the toric morphism $p_1 \colon X \to Y_1$ arising from the map $P_1 \colon F \to N_1$ of the fans Δ and Σ_1 as in Construction 2.3.1.5 is a separated good quotient for the action of H on X.

Specializing this characterization to the case of geometric quotients with separated quotient space gives the following characterization.

Corollary 2.3.1.7   *As in the setup above, let X be the toric variety arising from a fan Δ in a lattice F, and consider the action of a subgroup $H \subseteq T$ given by an epimorphism $Q: E \to K$. Then the following statements are equivalent.*

(i) *The action of H on X admits a separated geometric quotient.*
(ii) *The restriction $P: \mathrm{Supp}(\Delta) \to N_\mathbb{Q}$ is injective.*

If one of these statements holds, then $\Sigma := \{P(\delta); \ \delta \in \Delta\}$ is a fan in $N_\mathbb{Q}$ and the toric morphism $p: X \to Y$ associated with the map $P: F \to N$ of the fans Δ and Σ is a separated geometric quotient for the action of H on X.

We come to the proof of Proposition 2.3.1.6. The following elementary observation will also be used later.

Lemma 2.3.1.8   *Consider the situation of Construction 2.3.1.5. Then for every $\delta \in \mathfrak{A}(\Delta)$, we have $P_1^{-1}(P_1(\delta)) \cap \mathrm{Supp}(\Delta) = \delta$.*

Proof   Consider $\delta_0 \in \Delta$ with $P_1(\delta_0) \subseteq P_1(\delta)$. Then $P(\delta_0) \subseteq P(\delta)$ holds and thus any $L_\mathbb{Q}$-invariant linear form on F that is nonnegative on δ is necessarily nonnegative on δ_0. It follows that δ_0 is a face of δ. $\square$

Proof of Proposition 2.3.1.6   The equivalence of (ii) and (iii) is elementary. We only show that (i) and (ii) are equivalent. First suppose that the action of H has a good quotient $\pi: X \to Y$. By Proposition 2.3.1.2, the quotient variety Y is toric and π is a toric morphism. So we may assume that π arises from a map $\Pi: F \to N'$ from Δ to a fan Σ' in a lattice N'. Note that the sublattice $L \subseteq F$ is contained in the kernel of Π. We claim that there are bijections of the sets $\Delta^{\max}$ and $(\Sigma')^{\max}$ of maximal cones:

$$\Delta^{\max} \to (\Sigma')^{\max}, \qquad \delta \mapsto \Pi(\delta), \tag{2.3.1}$$

$$(\Sigma')^{\max} \to \Delta^{\max}, \qquad \sigma' \mapsto \Pi^{-1}(\sigma') \cap \mathrm{Supp}(\Delta) \tag{2.3.2}$$

To check that the first map is well-defined, let $\delta \in \Delta^{\max}$. Then the image $\Pi(\delta)$ is contained in some maximal cone $\sigma' \in \Sigma'$. In particular, $\pi(X_\delta) \subseteq X'_{\sigma'}$ holds. Because π is affine, the inverse image $\pi^{-1}(X'_{\sigma'})$ is an invariant affine chart of X, and hence equals X_δ. Because π is surjective, we have $\pi(X_\delta) = X'_{\sigma'}$. This means $\Pi(\delta) = \sigma'$. To see that the second map is well defined, let $\sigma' \in (\Sigma')^{\max}$. The inverse image of the associated affine chart $X'_{\sigma'} \subseteq X'$ is given by

$$\pi^{-1}(X'_{\sigma'}) = \bigcup_{\substack{\tau \in \Delta \\ \Pi(\tau) \subseteq \sigma'}} X_\tau.$$

Because π is affine, this inverse image is an affine invariant chart X_δ given by some cone $\delta \in \Delta$. In follows that

$$\delta = \mathrm{cone}(\tau \in \Delta; \ \Pi(\tau) \subseteq \sigma') = \Pi^{-1}(\sigma') \cap \mathrm{Supp}(\Delta).$$

By surjectivity of π, we have $\Pi(\delta) = \sigma'$. Assume that $\delta \subseteq \vartheta$ for some $\vartheta \in \Delta^{\max}$. As seen earlier, $\Pi(\vartheta)$ is a maximal cone of Σ'. Because $\Pi(\vartheta)$ contains the maximal cone σ', we get $\Pi(\vartheta) = \sigma'$. This implies $\delta = \vartheta$, thus $\delta \in \Delta^{\max}$, and the map (2.3.2) is well defined. Let $\delta_1, \delta_2 \in \Delta^{\max}$ be two different cones. The maximal cones $\sigma_i' := \Pi(\delta_i)$ can be separated by a linear form u' on N'. Then $u := u' \circ \Pi$ is an $L_{\mathbb{Q}}$-invariant separating linear form for the cones δ_1 and δ_2.

Now suppose that the fan Δ is $L_{\mathbb{Q}}$-projectable. We show that the toric morphism $p_1 \colon X \to Y_1$ is affine. Consider an affine chart $Y_1' \subseteq Y_1$ corresponding to a maximal cone $\sigma_1 \subseteq \Sigma_1$. Then there are unique maximal cones $\sigma \in \Sigma$ and $\delta \in \Delta$ projecting onto σ_1. Lemma 2.3.1.8 and the Fiber Formula tell us that $p_1^{-1}(Y_1')$ equals the affine toric chart of X corresponding to δ and thus $p_1 \colon X \to Y_1$ is affine. Proposition 2.1.4.3 then implies that it is a good quotient. $\qquad\square$

Example 2.3.1.9 Let Σ be a fan in a lattice F and consider a cone $\sigma \in \Sigma$. Denote by $\mathrm{star}(\sigma)$ the set of all cones $\tau \in \Sigma$ that contain σ as a face. Then the closure of the toric orbit corresponding to σ is given by

$$\overline{T \cdot x_\sigma} = \bigcup_{\tau \in \mathrm{star}(\sigma)} T \cdot x_\tau.$$

The union $U(\sigma)$ of the affine charts X_τ, where $\tau \in \mathrm{star}(\sigma)$, is an open T-invariant neighborhood of the orbit closure $\overline{T \cdot x_\sigma}$. The set of maximal cones of the fan $\Sigma(\sigma)$ corresponding to $U(\sigma)$ coincides with $\Sigma^{\max} \cap \mathrm{star}(\sigma)$.

Let L be the intersection of the linear span $\mathrm{lin}(\sigma)$ of σ in $F_{\mathbb{Q}}$ with the lattice F, and let $P \colon F \to N := F/L$ denote the projection. Then the cones $P(\tau)$, where τ runs through $\Sigma(\sigma)^{\max}$, are the maximal cones of the quotient fan $\widetilde{\Sigma}(\sigma)$ of $\Sigma(\sigma)$ by L. Moreover, $\widetilde{\Sigma}(\sigma)$ is the fan of $\overline{T \cdot x_\sigma}$ viewed as a toric variety with acting torus T/T_{x_σ}. In other words, the projection defines a good quotient $p \colon U(\sigma) \to \overline{T \cdot x_\sigma}$ by the isotropy group T_{x_σ}.

2.3.2 Combinatorics of good toric quotients

We consider a $\mathbb{Q}$-factorial affine toric variety X with acting torus T and the action of a closed subgroup $H \subseteq T$ on X, all defined over an algebraically closed field $\mathbb{K}$ of characteristic zero. Our aim is a combinatorial description of the "maximal" open subsets $U \subseteq X$ admitting a separated good quotient by the action of H; the main result reformulates [48] in terms of bunches of cones. First we fix the notion of maximality.

Definition 2.3.2.1 Let a reductive affine algebraic group G act on a variety X.

(i) By a *good G-set* in X, we mean an open subset $U \subseteq X$ with a separated good quotient $U \to U /\!\!/ G$.

(ii) We say that a subset $U' \subseteq U$ of a good G-set $U \subseteq X$ is G-*saturated* if it satisfies $U' = p^{-1}(p(U'))$, where $p\colon U \to U /\!\!/ G$ is the quotient.

(iii) We say that a subset $U \subseteq X$ is G-*maximal* if it is a good G-set and maximal w.r.t. G-saturated inclusion.

The key to a combinatorial description of H-maximal subsets for subgroup actions on toric varieties is the following, see [288, Cor. 2.5].

Theorem 2.3.2.2 *Let X be a toric variety with the acting torus T and H be a closed subgroup of T. Then every H-maximal subset of X is T-invariant.*

We fix the setup for the rest of the section. Let F be a lattice, $\delta \subseteq F_{\mathbb{Q}}$ a simplicial cone of full dimension with primitive generators $f_1, \ldots, f_r \in F$, and $Q\colon E \to K$ an epimorphism from the dual lattice $E = \mathrm{Hom}(F, \mathbb{Z})$ onto an abelian group K. Denoting by $M \subseteq E$ the kernel of $Q\colon E \to K$, we obtain mutually dual exact sequences of rational vector spaces

$$0 \longrightarrow L_{\mathbb{Q}} \longrightarrow F_{\mathbb{Q}} \overset{P}{\longrightarrow} N_{\mathbb{Q}} \longrightarrow 0$$

$$0 \longleftarrow K_{\mathbb{Q}} \underset{Q}{\longleftarrow} E_{\mathbb{Q}} \longleftarrow M_{\mathbb{Q}} \longleftarrow 0$$

Let $(e_1, \ldots, e_r)$ be the basis for $E_{\mathbb{Q}}$ dual to the basis $(f_1, \ldots, f_r)$ for $F_{\mathbb{Q}}$. With $v_i := P(f_i)$ and $w_i := Q(e_i)$, we obtain Gale dual vector configurations $\mathcal{V} := (v_1, \ldots, v_r)$ and $\mathcal{W} := (w_1, \ldots, w_r)$. Moreover, let $\gamma \subseteq E_{\mathbb{Q}}$ be the dual cone of $\delta \subseteq F_{\mathbb{Q}}$. We denote by $X := X_\delta$ the affine toric variety associated with the cone δ in the lattice F and consider the action of the subgroup $H := \mathrm{Spec}\, \mathbb{K}[K]$ of the acting torus $T = \mathrm{Spec}\, \mathbb{K}[E]$ on X.

Construction 2.3.2.3 With every saturated connected γ-collection $\mathfrak{B}$ and also to every $\mathcal{W}$-bunch Θ, we associate an $L_{\mathbb{Q}}$-projectable subfan of the fan of faces of δ and the corresponding open toric subsets of X:

$$\Delta(\mathfrak{B}) := \{\delta_0 \preceq \delta;\ \delta_0 \preceq \gamma_0^* \text{ for some } \gamma_0 \in \mathfrak{B}\}, \qquad U(\mathfrak{B}) := X_{\Delta(\mathfrak{B})},$$

$$\Delta(\Theta) := \{\delta_0 \preceq \delta;\ \delta_0 \preceq \gamma_0^* \text{ for some } \gamma_0 \in Q^\dagger \Theta\}, \qquad U(\Theta) := X_{\Delta(\Theta)}.$$

Conversely, any toric good H-set $U \subseteq X$ arises from an $L_{\mathbb{Q}}$-projectable subfan $\Delta(U)$ of the fan of faces of δ and we define an associated saturated connected γ-collection and an associated $\mathcal{W}$-bunch

$$\mathfrak{B}(U) := \mathfrak{A}(\Delta(U))^*, \qquad \Theta(U) := Q_\downarrow(\mathfrak{A}(\Delta(U))^*).$$

Theorem 2.3.2.4 *As in the setup above, let X be the affine toric variety arising from a simplicial cone δ of full dimension in a lattice F, and consider* *the action of a subgroup $H \subseteq T$ given by an epimorphism $Q \colon E \to K$. Then one has order reversing mutually inverse bijections*

$$\{maximal\ W\text{-}bunches\} \longleftrightarrow \{H\text{-}maximal\ subsets\ of\ X\},$$

$$\Theta \mapsto U(\Theta),$$

$$\Theta(U) \leftarrowtail U.$$

Under these bijections, the bunches arising from GKZ chambers correspond to the subsets with a quasiprojective quotient space.

Lemma 2.3.2.5 *Consider the situation of Construction 2.3.1.5. Let $\Delta' \preceq \Delta$ be an $L_{\mathbb{Q}}$-projectable subfan and let X', X denote the toric varieties associated with Δ', Δ respectively. Then X' is H-saturated in X if and only if $\mathfrak{A}(\Delta') \subseteq \mathfrak{A}(\Delta)$ holds.*

Proof We work with the good quotient $p_1 \colon X \to Y_1$ arising from the map $P_1 \colon F \to N_1$ of the fans Δ and Σ_1 provided by Construction 2.3.1.5. First assume that X' is H-saturated in X. Given $\delta' \in \mathfrak{A}(\Delta')$, consider the associate affine toric chart $U' \subseteq X'$. Because Y_1 carries the quotient topology, $p_1(U) \subseteq Y_1$ is open and we obtain $P_1(\delta') \in \Sigma_1$. Lemma 2.3.1.8 yields $\delta' \in \mathfrak{A}(\Delta)$. Conversely, if $\mathfrak{A}(\Delta') \subseteq \mathfrak{A}(\Delta)$ holds, combine Lemma 2.3.1.8 and the Fiber Formula to see that $p^{-1}(p(U'))$ equals U' for every toric affine chart $U' \subseteq X'$. $\qquad\square$

Lemma 2.3.2.6 *Consider the situation before Construction 2.3.2.3. Then we have mutually inverse order reversing bijections*

$$\{saturated\ connected\ \gamma\text{-}collections\} \longleftrightarrow \{L_{\mathbb{Q}}\text{-}projectable\ fans\},$$

$$\mathfrak{B} \mapsto \Delta(\mathfrak{B}),$$

$$\mathfrak{A}(\Delta)^* \leftarrowtail \Delta.$$

$$\{maximal\ W\text{-}bunches\} \longleftrightarrow \{maximal\ L_{\mathbb{Q}}\text{-}projectable\ fans\},$$

$$\Theta \mapsto \Delta(Q^\dagger \Theta),$$

$$Q_\downarrow(\mathfrak{A}(\Delta)^*) \leftarrowtail \Delta.$$

Proof Obviously, the assignment $\Delta \mapsto \mathfrak{A}(\Delta)$ defines a bijection from $L_{\mathbb{Q}}$-projectable fans to saturated separated δ-collections; its inverse is given by

$$\mathfrak{A} \mapsto \{\delta_0 \preceq \delta; \ \delta_0 \preceq \delta_1 \text{ for some } \delta_1 \in \mathfrak{A}\}.$$

Thus, Proposition 2.2.3.5 establishes the first pair of bijections. Using also Proposition 2.2.3.8 gives the second one. $\qquad\square$

Proof of Construction 2.3.2.3 and Theorem 2.3.2.4 Construction 2.3.2.3 is clear by Lemma 2.3.2.6. Theorem 2.3.2.2 ensures that the H-maximal subsets $U \subseteq X$ are toric. According to Proposition 2.3.1.6 and Lemma 2.3.2.5, they correspond to maximal $L_{\mathbb{Q}}$-projectable subfans Δ of the fan of faces of $\delta \subseteq F_{\mathbb{Q}}$. Thus, Lemma 2.3.2.6 provides the desired bijections. The last assertion follows from the characterization of normal fans given in Theorem 2.2.2.2. □

Proposition 2.3.2.7 *Consider the situation of Theorem 2.3.2.4. Let Θ be a maximal $\mathcal{W}$-bunch and $p\colon U(\Theta) \to Y$ the associated good quotient. Then the following statements are equivalent.*

  (i) *The $\mathcal{W}$-bunch Θ consists of full-dimensional cones.*
 (ii) *We have $(Q^{\dagger}\Theta)^* = \Delta(\Theta)$.*
(iii) *The good quotient $p\colon U(\Theta) \to Y$ is geometric.*

Proof The equivalence of (i) and (ii) is elementary and the equivalence of (ii) and (iii) is a direct consequence of the Fiber Formula and the fact that P induces a bijection $(Q^{\dagger}\Theta)^* \to P_{\downarrow}((Q^{\dagger}\Theta)^*)$. □

2.4 Toric varieties and bunches of cones

2.4.1 Toric varieties and lattice bunches

We use the concept of a bunch of cones to describe toric varieties; the ground field $\mathbb{K}$ is algebraically closed of characteristic zero. As with fans, we have to enhance bunches of cones with a lattice structure. We obtain a functor from maximal lattice bunches to "maximal" toric varieties that induces a bijection on isomorphy classes. In fact, for later use, we formulate the assignment first in a more general context.

As earlier, given an abelian group K, we write $K_{\mathbb{Q}} := K \otimes_{\mathbb{Z}} \mathbb{Q}$ for the associated rational vector space and for any $w \in K$, we denote the associated element $w \otimes 1 \in K_{\mathbb{Q}}$ again by w. Moreover, for a homomorphism $Q\colon E \to K$ of abelian groups we denote the associated linear map $E_{\mathbb{Q}} \to K_{\mathbb{Q}}$ again by Q.

Definition 2.4.1.1 A *(true) lattice collection* is a triple $(E \xrightarrow{\varrho} K, \gamma, \mathfrak{B})$, where $Q\colon E \to K$ is an epimorphism from a lattice E with basis $e_1, \ldots, e_r$ onto an abelian group K generated by any $r - 1$ of the $w_i := Q(e_i)$, the cone $\gamma \subseteq E_{\mathbb{Q}}$ is generated by $e_1, \ldots, e_r$ and $\mathfrak{B}$ is a (true) saturated connected γ-collection.

Construction 2.4.1.2 Let $(E \xrightarrow{\varrho} K, \gamma, \mathfrak{B})$ be a true lattice collection. In particular, E is a lattice with basis $e_1, \ldots, e_r$, we have $\gamma := \mathrm{cone}(e_1, \ldots, e_r)$ and $Q\colon E \to K$ an epimorphism onto an abelian group such that any $r - 1$ of the $w_i := Q(e_i)$ generate K as an abelian group. With $M := \ker(Q)$, we have the mutually dual exact sequences

$$0 \longrightarrow L \longrightarrow F \xrightarrow{\ P\ } N$$

$$0 \longleftarrow K \xleftarrow{\ Q\ } E \longleftarrow M \longleftarrow 0$$

Let $f_1, \ldots, f_r$ be the dual basis of $e_1, \ldots, e_r$. Then each $v_i := P(f_i)$ is a primitive lattice vector in N. Let $\delta = \mathrm{cone}(f_1, \ldots, f_r)$ denote the dual cone of $\gamma \subseteq E_{\mathbb{Q}}$ and for $\gamma_0 \preceq \gamma$ let $\gamma_0^* = \gamma_0^\perp \cap \delta$ be the corresponding face. Then one has fans in the lattices F and N:

$$\widehat{\Sigma} := \{\delta_0 \preceq \delta;\ \delta_0 \preceq \gamma_0^* \text{ for some } \gamma_0 \in \mathfrak{B}\}, \qquad \Sigma := \{P(\gamma_0^*);\ \gamma_0 \in \mathfrak{B}\}.$$

The canonical map $\mathfrak{B} \to \Sigma$, $\gamma_0 \mapsto P(\gamma_0^*)$ is an order-reversing bijection. In particular, the fan Σ has exactly r rays, namely $\mathrm{cone}(v_1), \ldots, \mathrm{cone}(v_r)$. The *toric variety associated with* $\mathfrak{B}$ is the toric variety X defined by the fan Σ. The projection $P \colon F \to N$ is a map of the fans $\widehat{\Sigma}$ and Σ and hence defines a toric morphism $\widehat{X} \to X$.

Proof Because any $r - 1$ of the weights w_i generate K as an abelian group, Lemma 2.1.4.1 shows that each $v_i = P(f_i)$ is a primitive lattice vector. Moreover, Propositions 2.2.3.5 and 2.2.3.11 show that Σ is a fan and the canonical map $\mathfrak{B} \to \Sigma$ is an order reversing bijection. $\qquad\square$

Note that every nondegenerate lattice fan can be obtained via this construction. Specializing to bunches of cones, we lose a bit of this generality, but obtain a more concise presentation.

Definition 2.4.1.3 A *(true, maximal) lattice bunch* a triple $(K, \mathcal{W}, \Theta)$, where K is a finitely generated abelian group, $\mathcal{W} = (w_1, \ldots, w_r)$ is a family in K such that any $r - 1$ of the w_i generate K as an abelian group and Θ is a (true, maximal) $\mathcal{W}$-bunch.

Construction 2.4.1.4 Let $(K, \mathcal{W}, \Theta)$ be a true lattice bunch. The associated *projected cone* is $(E \xrightarrow{Q} K, \gamma)$, where $E = \mathbb{Z}^r$, the homomorphism $Q \colon E \to K$ sends the ith canonical basis vector $e_i \in E$ to $w_i \in K$, and $\gamma \subseteq E_{\mathbb{Q}}$ is the cone generated by $e_1, \ldots, e_r$. We have the following collections of cones:

$$Q^\dagger\Theta = \{\gamma_0 \preceq \gamma;\ Q(\gamma_0) \in \Theta\},$$

$$\mathrm{cov}(\Theta) := \{\gamma_0 \in Q^\dagger\Theta;\ \gamma_0 \text{ minimal}\}.$$

The first one is the Q-lift of Θ and we call the second one the *covering collection* of Θ. In the notation of Theorem 2.2.1.14 and Construction 2.4.1.2, it defines fans in F and N:

$$\widehat{\Sigma} := \{\delta_0 \preceq \delta;\ \delta_0 \preceq \gamma_0^* \text{ for some } \gamma_0 \in Q^\dagger\Theta\},$$

$$\Sigma = \Theta^\sharp = \{P(\gamma_0^*);\ \gamma_0 \in Q^\dagger\Theta\}.$$

The *toric variety associated with* $(K, \mathcal{W}, \Theta)$ is the toric variety $X = X_\Theta$ defined by the maximal $\mathcal{V}$-fan $\Sigma = \Theta^\sharp$ corresponding to Θ. The projection $P: F \to N$ defines a map of the fans $\widehat{\Sigma}$ and Σ and thus a toric morphism $\widehat{X} \to X$.

Definition 2.4.1.5 A *map of lattice bunches* $(K_i, \mathcal{W}_i, \Theta_i)$ with associated projected cones $(E_i \xrightarrow{\varrho_i} K_i, \gamma_i)$ is a homomorphism $\varphi: E_1 \to E_2$ such that there is a commutative diagram

$$
\begin{array}{ccc}
E_1 & \xrightarrow{\varphi} & E_2 \\
\scriptstyle Q_1 \downarrow & & \downarrow \scriptstyle Q_2 \\
K_1 & \xrightarrow[\overline{\varphi}]{} & K_2,
\end{array}
$$

where $\varphi(\gamma_1) \subseteq \gamma_2$ holds and for every $\alpha_2 \in \operatorname{cov}(\Theta_2)$ there is $\alpha_1 \in \operatorname{cov}(\Theta_1)$ with $\varphi(\alpha_1) \subseteq \alpha_2$.

Remark 2.4.1.6 Let $(K_i, \mathcal{W}_i, \Theta_i)$ be two true lattice bunches and φ a map between them. Then φ induces a map of the corresponding lattice fans $\Sigma_2 = \Theta_2^\sharp$ and $\Sigma_1 = \Theta_1^\sharp$, and thus a toric morphism $\psi(\varphi): X_{\Theta_2} \to X_{\Theta_1}$.

We say that a toric variety X is *maximal* if it admits no open toric embedding $X \subsetneq X'$ with $X' \setminus X$ of codimension at least 2 in X'. Note that in terms of a defining fan Σ of X, maximality means that Σ does not occur as a proper subfan of some fan Σ' having the same rays as Σ. For example, complete as well as affine toric varieties are maximal.

Theorem 2.4.1.7 *We have a contravariant faithful essentially surjective functor inducing a bijection on the sets of isomorphism classes:*

$$\{\textit{true maximal lattice bunches}\} \longrightarrow \{\textit{maximal toric varieties}\},$$

$$(K, \mathcal{W}, \Theta) \mapsto X_\Theta,$$

$$\varphi \mapsto \psi(\varphi).$$

Proof The assertion follows directly from Construction 2.4.1.4, Remark 2.4.1.6, and Theorem 2.2.1.14. □

Example 2.4.1.8 Consider the true maximal lattice bunch $(K, \mathcal{W}, \Theta)$, where $K := \mathbb{Z}^3$, the family $\mathcal{W} = (w_1, \ldots, w_6)$ is given by

$$w_1 = (1, 0, 0), \quad w_2 = (0, 1, 0), \quad w_3 = (0, 0, 1),$$

$$w_4 = (1, 1, 0), \quad w_5 = (1, 0, 1), \quad w_6 = (0, 1, 1),$$

and, finally, the maximal $\mathcal{W}$-bunch Θ in $K_{\mathbb{Q}}$ has the following four three-dimensional cones as its minimal cones:

$\mathrm{cone}(w_3, w_4, w_5), \quad \mathrm{cone}(w_1, w_4, w_6), \quad \mathrm{cone}(w_2, w_5, w_6), \quad \mathrm{cone}(w_4, w_5, w_6).$

The fan $\Sigma = \Theta^{\sharp}$ is one of the simplest nonpolytopal complete simplicial fans in $N = \mathbb{Z}^3$. It looks as follows. Consider the polytope $B \subset K_{\mathbb{Q}}$ with the vertices

$$(-1, 0, 0), \quad (0, -1, 0), \quad (0, 0, -1), \quad (0, 1, 1), \quad (1, 0, 1), \quad (1, 1, 0)$$

and subdivide the facets of B according to the figure below. Then Σ is the fan generated by the cones over the simplices of this subdivision.

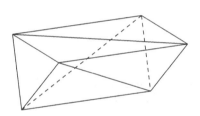 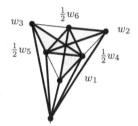

Defining polytope subdivision of Σ Corresponding bunch Θ

2.4.2 Toric geometry via bunches

We describe basic geometric properties of a toric variety in terms of a defining lattice collection or lattice bunch. We consider orbit decomposition, divisor class group, Cox ring, local class groups, Picard group, smoothness, $\mathbb{Q}$-factoriality, cones of movable, semiample, and ample divisors, and intersection numbers. As before, we work over an algebraically closed field $\mathbb{K}$ of characteristic zero.

Proposition 2.4.2.1 (Orbit decomposition II) *Consider the situation of Construction 2.4.1.2. Then we have a bijection*

$$\mathfrak{B} \to \{T\text{-orbits of } X\}, \qquad \gamma_0 \mapsto T \cdot x_{\gamma_0}, \quad \text{where} \quad x_{\gamma_0} := x_{P(\gamma_0^*)}.$$

Moreover, for any two $\gamma_0, \gamma_1 \in \mathfrak{B}$, one has $\gamma_0 \preceq \gamma_1$ if and only if $\overline{T \cdot x_{\gamma_0}} \subseteq \overline{T \cdot x_{\gamma_1}}$ holds.

Proof　The collection $\mathfrak{B}$ is in order-reversing bijection with the defining fan Σ of X via $\gamma_0 \mapsto P(\gamma_0^*)$. Thus, the usual description in Proposition 2.1.2.2 of the orbit decomposition in terms of Σ gives the assertion.　　　　　$\square$

Proposition 2.4.2.2 (Divisor class group and Cox ring)　*Consider the situation of Construction 2.4.1.2. Then there is a commutative diagram of abelian groups*

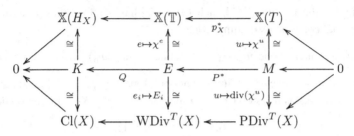

Moreover, $\widehat{X} \to X$ is a characteristic space and for the Cox ring $\mathcal{R}(X)$, we have the following isomorphism of graded algebras

$$\mathbb{K}[E \cap \gamma] = \bigoplus_{w \in K} \mathbb{K}[E \cap \gamma]_w \cong \bigoplus_{[D] \in \mathrm{Cl}(X)} \mathcal{R}(X)_{[D]} = \mathcal{R}(X).$$

Proof　The assertion follows directly from Theorem 2.1.3.2 and the discussion given before this theorem.　　　　　$\square$

Proposition 2.4.2.3　*Consider the situation of Construction 2.4.1.2. For $x \in X$, let $\gamma_0 \in \mathfrak{B}$ be the face with $x \in T \cdot x_{\gamma_0}$. Then the local divisor class group of X at x is given by*

$$\begin{array}{ccc}
\mathrm{Cl}(X) & \longrightarrow & \mathrm{Cl}(X, x) \\
{\scriptstyle\cong}\Big\updownarrow & & \Big\updownarrow{\scriptstyle\cong} \\
K & \longrightarrow & K/Q(\mathrm{lin}(\gamma_0) \cap E)
\end{array}$$

Proof　We may assume that $x = x_\sigma$ with $\sigma = P(\gamma_0^*)$ holds. Set $\widehat{\sigma} := \gamma_0^*$. Using the Fiber Formula 2.1.2.4, we see that the distinguished point $x_{\widehat{\sigma}} \in \widehat{\Sigma}$ has a closed H-orbit in the fiber $p^{-1}(x_\sigma)$, where $p \colon \widehat{X} \to X$ denotes the characteristic space. By Proposition 2.1.4.2, the isotropy group $H_{x_{\widehat{\sigma}}} \subseteq H$ is given by $K \to K/Q(\mathrm{lin}(\gamma_0) \cap E)$. Thus, we may apply Proposition 1.6.2.5 to obtain the assertion.　　　　　$\square$

Corollary 2.4.2.4　*Consider the situation of Construction 2.4.1.2. Inside the divisor class group $\mathrm{Cl}(X) = K$, the Picard group of X is given by*

$$\mathrm{Pic}(X) = \bigcap_{\gamma_0 \in \mathfrak{B}} Q(\mathrm{lin}(\gamma_0) \cap E).$$

Corollary 2.4.2.5 *Consider the situation of Construction 2.4.1.2. For a point $x \in X$, let $\gamma_0 \in \mathfrak{B}$ be the face with $x \in T \cdot x_{\gamma_0}$.*

(i) *The point x is $\mathbb{Q}$-factorial if and only if $\dim Q(\gamma_0)$ equals $\dim K_{\mathbb{Q}}$.*
(ii) *The point x is smooth if and only if $Q(\mathrm{lin}(\gamma_0) \cap E)$ equals K.*

We describe cones of divisors in the rational divisor class group. Recall that a Weil divisor on a variety is called *movable* if it has a positive multiple with base locus of codimension at least 2 and it is called *semiample* if it has a base point free multiple.

Proposition 2.4.2.6 *Consider the situation of Construction 2.4.1.2. The cones of effective, movable, semiample, and ample divisor classes of X in $\mathrm{Cl}_{\mathbb{Q}}(X) = K_{\mathbb{Q}}$ are given as*

$$\mathrm{Eff}(X) = Q(\gamma), \qquad \mathrm{Mov}(X) = \bigcap_{\gamma_0 \text{ facet of } \gamma} Q(\gamma_0),$$

$$\mathrm{SAmple}(X) = \bigcap_{\gamma_0 \in \mathfrak{B}} Q(\gamma_0), \qquad \mathrm{Ample}(X) = \bigcap_{\gamma_0 \in \mathfrak{B}} Q(\gamma_0)^{\circ}.$$

Moreover, if X arises from a lattice bunch $(K, \mathcal{W}, \Theta)$ as in Construction 2.4.1.4, then we have

$$\mathrm{SAmple}(X) = \bigcap_{\tau \in \Theta} \tau, \qquad \mathrm{Ample}(X) = \bigcap_{\tau \in \Theta} \tau^{\circ}.$$

Proof For the descriptions of $\mathrm{SAmple}(X)$ and $\mathrm{Ample}(X)$, let D be an invariant $\mathbb{Q}$-Cartier divisor on X and let $\widehat{w} \in E_{\mathbb{Q}}$ be the corresponding element. Recall that D is semiample (ample) if and only if it is described by a support function (u_{σ}), which is convex (strictly convex) in the sense that $u_{\sigma} - u_{\sigma'}$ is nonnegative (positive) on $\sigma \setminus \sigma'$ for any two $\sigma, \sigma' \in \Sigma$. For $\sigma \in \Sigma$, we denote by $\widehat{\sigma} \in \widehat{\Sigma}$ the cone with $P(\widehat{\sigma}) = \widehat{\sigma}$.

Suppose that D is semiample (ample) with convex (strictly convex) support function (u_{σ}). In terms of $\ell_{\sigma} := \widehat{w} - P^*(u_{\sigma})$ this means that each $\ell_{\sigma'} - \ell_{\sigma}$ is nonnegative (positive) on $\widehat{\sigma} \setminus \widehat{\sigma}'$. Because $\ell_{\sigma} \in \widehat{\sigma}^{\perp}$ holds, this is equivalent to nonnegativity (positivity) of $\ell_{\sigma'}$ on every $\widehat{\sigma} \setminus \widehat{\sigma}'$.

Because all rays of the cone δ occur in the fan $\widehat{\Sigma}$, the latter is valid if and only if $\ell_{\sigma} \in \widehat{\sigma}^*$ (resp. $\ell_{\sigma} \in (\widehat{\sigma}^*)^{\circ}$) holds for all σ. This in turn implies that for every $\sigma \in \Sigma$ we have

$$w = Q(\widehat{w}) = Q(\ell_{\sigma}) \in Q(\widehat{\sigma}^*) \quad (\text{resp. } w \in Q((\widehat{\sigma}^*)^{\circ})). \tag{2.4.1}$$

Now, the $\widehat{\sigma}^*$, where $\sigma \in \Sigma$, are precisely the cones of $\mathfrak{B}$. Because any interior $Q(\widehat{\sigma}^*)^{\circ}$ contains the interior of a cone of Θ, we can conclude that w lies in the respective intersections of the assertion.

Conversely, if w belongs to one of the right-hand side intersections, then we surely arrive at (2.4.1). Thus, for every $\sigma \in \Sigma$, we find an $\ell_\sigma \in \widehat{\sigma}^*$ (an $\ell_\sigma \in (\widehat{\sigma}^*)^\circ$) mapping to w. Reversing the above arguments, we see that $u_\sigma := \widehat{w} - \ell_\sigma$ is a convex (strictly convex) support function describing D. $\qquad\square$

Corollary 2.4.2.7 *Consider the situation of Construction 2.4.1.4. Assume that Θ arises from a chamber $\lambda \subseteq K_{\mathbb{Q}}$ of the GKZ decomposition. Then the toric variety X associated with the true maximal bunch Θ has λ as its semiample cone.*

Proposition 2.4.2.8 *Consider the situation of Construction 2.4.1.2. Let $e_1, \ldots, e_r$ be the primitive generators of γ. Then the canonical divisor class of X is given as*

$$\mathcal{K}_X = -Q(e_1 + \cdots + e_r) \in K.$$

Proof The assertion follows from [141, Sec. 4.3] and Proposition 2.4.2.2.
$\qquad\square$

Example 2.4.2.9 The toric variety $X := X_\Theta$ given in Example 2.4.1.8 is $\mathbb{Q}$-factorial and nonprojective. Moreover, the cone $\mathrm{SAmple}(X)$ of semiample divisors is spanned by the class of the anticanonical divisor.

Example 2.4.2.10 (Kleinschmidt's classification [186]) The smooth maximal toric varieties X with $\mathrm{Cl}(X) \cong \mathbb{Z}^2$ correspond to bunches $(\mathbb{Z}^2, \mathcal{W}, \Theta)$, where

- $\mathcal{W} = (w_{ij};\ 1 \le i \le n,\ 1 \le j \le \mu_i)$ satisfies
  - $w_{1j} := (1, 0)$, and $w_{ij} := (b_i, 1)$ with $0 = b_n < b_{n-1} < \cdots < b_2$,
  - $\mu_1 > 1$, $\mu_n > 0$ and $\mu_2 + \cdots + \mu_n > 1$,
- Θ is the $\mathcal{W}$-bunch arising from the chamber $\lambda = \mathrm{cone}(w_{11}, w_{21})$.

Moreover, the toric variety X defined by such a bunch Θ is always projective, and it is Fano if and only if we have

$$b_2(\mu_2 + \cdots + \mu_n) < \mu_1 + b_2\mu_2 + \cdots + b_{n-1}\mu_{n-1}.$$

Finally, we investigate intersection numbers of a complete $\mathbb{Q}$-factorial toric variety X arising from a lattice bunch $(K, \mathcal{W}, \Theta)$ with $\mathcal{W} = (w_1, \ldots, w_r)$. For $w_{i_1}, \ldots, w_{i_n}$, where $1 < i_1 < \ldots < i_n < r$, denote by $w_{j_1}, \ldots, w_{j_{r-n}}$, where $1 < j_1 < \ldots < j_{r-n} < r$, the complementary weights and set

$$\tau(w_{i_1}, \ldots, w_{i_n}) := \mathrm{cone}(w_{j_1}, \ldots, w_{j_{r-n}}),$$

$$\mu(w_{i_1}, \ldots, w_{i_n}) = [K : \langle w_{j_1}, \ldots, w_{j_{r-n}} \rangle].$$

Proposition 2.4.2.11 *Consider the situation of Construction 2.4.1.4 and assume that X is $\mathbb{Q}$-factorial and complete. The intersection number of classes $w_{i_1}, \ldots, w_{i_n}$, where $n = \dim(X)$ and $1 < i_1 < \ldots < i_n < r$, is given by*

$$w_{i_1} \cdots w_{i_n} = \begin{cases} \mu(w_{i_1}, \ldots, w_{i_n})^{-1}, & \tau(w_{i_1}, \ldots, w_{i_n}) \in \Theta, \\ 0, & \tau(w_{i_1}, \ldots, w_{i_n}) \notin \Theta. \end{cases}$$

Proof Combining Proposition 2.1.2.8 with Lemma 2.1.4.1 gives the assertion. $\square$

Exercises to Chapter 2

Exercise 2.1 (Quasifans and almost homogeneous torus actions)   A variety with almost homogeneous torus action is a variety X with base point $x_0 \in X$ and a (not necessarily effective) torus action $T \times X \to X$ such that the orbit $T \cdot x_0 \subseteq X$ is open and dense in X. A morphism between two such varieties (X, T, x_0) and (X', T', x_0') is a pair $(\varphi, \widetilde{\varphi})$, where $\varphi \colon X \to X'$ is a morphism of varieties and $\widetilde{\varphi} \colon T \to T'$ is a homomorphism of tori such that $\varphi(x_0) = x_0'$ and $\varphi(t \cdot x) = \widetilde{\varphi}(t) \cdot \varphi(x)$ holds for all $x \in X$ and $t \in T$. Establish covariant functors being essentially inverse to each other, between the categories of lattice quasifans and normal varieties with almost homogeneous torus action.

Exercise 2.2 (Affine semigroups and nonnormal affine almost homogeneous torus actions)   An *affine semigroup* is a pair (S, M), where M is a lattice and $S \subseteq M$ a finitely generated submonoid. A morphism of affine semigroups (S, M) and (S', M') is a lattice homomorphism $M \to M'$ sending S to S'. Establish contravariant functors being essentially inverse to each other between the categories of affine semigroups and affine varieties with almost homogeneous torus action.

Exercise 2.3 (Global regular functions)   Consider the toric variety X_Σ arising from a fan Σ in a lattice N. Let M be the dual lattice of N and write $\boldsymbol{\Sigma}$ for the support of Σ, that is, $\boldsymbol{\Sigma} \subseteq N_{\mathbb{Q}}$ is the union of all cones of Σ. Then one has

$$\Gamma(X_\Sigma, \mathcal{O}) \cong \mathbb{K}[\operatorname{cone}(\boldsymbol{\Sigma})^\vee \cap M], \qquad \Gamma(X_\Sigma, \mathcal{O}^*)/\mathbb{K}^* \cong \operatorname{lin}(\boldsymbol{\Sigma})^\perp \cap M.$$

The toric variety X_Σ splits as $X_\Sigma \cong X_{\Sigma'} \times T'$, where Σ' refers to the fan Σ in the lattice $N' := \operatorname{lin}(\boldsymbol{\Sigma}) \cap N$ and $T' = \operatorname{Spec} \mathbb{K}[\operatorname{lin}(\boldsymbol{\Sigma})^\perp \cap M]$. In particular, every toric variety is the product of a nondegenerate one with a torus.

Exercise 2.4   Give an example of a toric characteristic space $p_X \colon \widehat{X} \to X$ such that p_X has a reducible fiber with infinitely many H_X-orbits. Hint: Consider the cone generated by the vectors $e_1, e_2, e_1 + e_3$, and $e_2 + e_3$ in $\mathbb{Q}^3$.

Exercise 2.5 (The face ring) A *simplicial complex* on a finite set S is a collection C of subsets of S such that for each $\vartheta \in C$ also all subsets of ϑ (including $\emptyset$) belong to C. The *face ring*, or *Stanley–Reisner ring*, of a simplicial complex C on $S_r = \{1, \ldots, r\}$ is the factor ring

$$\mathbb{K}[T_1, \ldots, T_r]/I_C, \qquad I_C = \langle T_{j_1} \cdots T_{j_s}; \ j_1, \ldots, j_s \in S_r, \ \{j_1 \ldots, j_s\} \notin C \rangle.$$

The ideal I_C is the *Stanley–Reisner* ideal of the simplicial complex C. Let Σ be a lattice fan with rays $\varrho_1, \ldots, \varrho_r$ and let v_i be the primitive generator of ϱ_i. One obtains a simplicial complex $C(\Sigma)$ on S_r by putting $\vartheta \subseteq S_r$ to $C(\Sigma)$ if and only if $\{v_i; \ i \in \vartheta\}$ is contained in a cone of Σ. Let X_Σ denote the toric variety associated with Σ.

(1) A set $\vartheta \subseteq S_r$ is contained in $C(\Sigma)$ if and only if the intersection of the prime divisors D_{ϱ_i}, where $i \in \vartheta$, on the toric variety X_Σ is nonempty.
(2) If X_Σ is nondegenerate, then the irrelevant ideal coincides with the Stanley–Reisner ideal of the simplicial complex $C(\widehat{\Sigma})$.
(3) The Stanley–Reisner ring of $C(\widehat{\Sigma})$ is isomorphic to the algebra of functions $\Gamma(\overline{X}_\Sigma \setminus \widehat{X}_\Sigma, \mathcal{O})$.
(4) Every simplicial complex is isomorphic to a complex $C(\Sigma)$ for some fan Σ.

Exercise 2.6 (Toric Kempf–Ness sets) The ground field is $\mathbb{C}$. Let X be a complete toric variety arising from a fan Σ in a lattice N. Assume that $\mathrm{Cl}(X)$ is torsion free. Then with the primitive generators $v_1, \ldots, v_r$ of the rays of Σ, one obtains an exact sequence of lattices

$$0 \longrightarrow L \longrightarrow \mathbb{Z}^r \overset{e_i \mapsto v_i}{\longrightarrow} N \longrightarrow 0$$

Regarding the middle three lattices as the lattices of one-parameter subgroups, this leads to associated exact sequences of tori and their maximal compact subgroups:

$$1 \longrightarrow H \longrightarrow \mathbb{T}^r \longrightarrow T \longrightarrow 1 \ ,$$

$$1 \longrightarrow H_c \longrightarrow \mathbb{T}_c^r \longrightarrow T_c \longrightarrow 1 \ .$$

Note that $\mathbb{T}_c^r$ is the r-fold direct product of the unit circle, that is, the boundary of the unit disk $\mathbb{D} := \{z \in \mathbb{C}; \ |z| \leq 1\}$. Define a subset

$$\mathcal{Z}(\Sigma) := \bigcup_{\sigma \in \Sigma} \mathcal{Z}(\sigma) \subseteq \mathbb{D}^r, \quad \text{where} \quad \mathcal{Z}(\sigma) := \{z \in \mathbb{D}^r; \ |z_j| = 1 \text{ if } v_j \notin \sigma\}.$$

Then $\mathcal{Z}(\Sigma)$ is invariant under the action of $\mathbb{T}_c$, it is contained in the characteristic space $\widehat{X} \subseteq \overline{X} = \mathbb{C}^r$, and one has a commutative diagram

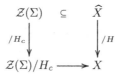

If the fan Σ is simplicial, then $\mathcal{Z}(\Sigma)$ is a compact $(n + r)$-dimensional smooth real manifold, where $n = \dim(X)$, and the induced map $\mathcal{Z}(\Sigma)/H_c \to X$ is a homeomorphism. See [241] for more.

Exercise 2.7 (Cox rings of toric prevarieties)   Generalize Construction 2.1.3.1 and the first statement of Theorem 2.1.3.2 to toric prevarieties. Generalize the second statement of Theorem 2.1.3.2 to toric prevarieties of affine intersection. If you need some background on toric prevarieties, consult [3].

Exercise 2.8 (Properties of the Gale transform)   Let $V = (v_1, \ldots, v_r)$ be a vector configuration and $W = (w_1, \ldots, w_r)$ its Gale transform.

(1) The vector configuration V is the Gale transform of W.
(2) We have $\dim \mathrm{lin}(V) + \dim \mathrm{lin}(W) = r$.
(3) The vector v_i is nonzero if and only if $w_i \in \mathrm{lin}(w_j; \ j \notin i)$ holds.
(4) Let $I \subseteq \{1, \ldots, r\}$. Then $(v_i; \ i \in I)$ is a basis for $\mathrm{lin}(V)$ if and only if $(w_j; \ j \notin I)$ is one for $\mathrm{lin}(W)$.
(5) If $\mathrm{cone}(V) = \mathrm{lin}(V)$ holds, then $\mathrm{cone}(W)$ is pointed.
(6) If $\mathrm{cone}(V)$ is pointed and all v_i are nonzero, then $\mathrm{cone}(W) = \mathrm{lin}(W)$ holds.

Exercise 2.9   Let U be a finite-dimensional vector space, generated by $w_1, \ldots, w_r \in U$. For every $i = 1, \ldots, r$ set $\tau_i := \mathrm{cone}(w_k; \ k \neq i)$. If $\tau_i^\circ \cap \tau_j^\circ \neq \emptyset$ holds for all $1 \leq i, j \leq r$, then $\tau_1^\circ \cap \ldots \cap \tau_r^\circ \neq \emptyset$ holds.

Exercise 2.10 (Combinatorics of fans and bunches)   We use notation of Sections 2.2.3 and 2.2.4.

(1) Give an example of a saturated V-collection whose P-lift is not P-saturated.
(2) For any $\lambda \in \Lambda$, the connected W-collection $\Phi(\lambda)$ is maximal.
(3) Let τ be a W-cone and λ a chamber. Then $\tau \cap \lambda$ is a face of λ.
(4) The condition $\Sigma(\lambda_1) = \Sigma(\lambda_2)$ does not imply $\lambda_1 = \lambda_2$.
(5) Give an example of a nonsimplicial GKZ fan.
(6) Give an example of a simplicial normal fan in $K_\mathbb{Q}$ with convex support that is not the GKZ fan of any vector configuration V in $N_\mathbb{Q}$.
(7) Give an example of a chamber λ of nonmaximal dimension with $\lambda^\circ \subseteq \Lambda^\circ$ such that the V fan $\Sigma(\lambda)$ is simplicial.

(8) If $\lambda \in \Lambda(\mathcal{W})$ is a chamber of full dimension, then the quasifan $\Sigma(\lambda)$ is simplicial.

(9) Let $\Sigma(\lambda)$ and $\Sigma(\lambda')$ be normal proper $\mathcal{V}$-fans. Then $\Sigma(\lambda)$ is a refinement of $\Sigma(\lambda')$ if and only if the chamber λ' is a face of λ.

(10) Give an example of a maximal $\mathcal{V}$-fan whose support does not coincide with cone($\mathcal{V}$).

Exercise 2.11 (Multigraded polynomial rings as Cox rings)   Let K be a finitely generated abelian group and define a K-grading on $\mathbb{K}[T_1, \ldots, T_r]$ by $\deg(T_i) := w_i$ for $1 \leq i \leq r$. Then the K-graded ring $\mathbb{K}[T_1, \ldots, T_r]$ is the Cox ring of a toric variety if and only if the following two conditions hold:

(1) For   every $1 \leq i \leq r$,   the   group K is   generated   by $w_1, \ldots, w_{i-1}, w_{i+1}, \ldots, w_r$.

(2) For   every $1 \leq i < j \leq r$ one   has $\mathrm{cone}(w_k; \, k \neq i)^\circ \cap \mathrm{cone}(w_k; \, k \neq j)^\circ \neq \emptyset$.

Moreover, the K-graded ring $\mathbb{K}[T_1, \ldots, T_r]$ is the Cox ring of a complete toric variety if and only if in addition to (1) and (2) the following condition holds:

(3) the elements $w_1, \ldots, w_r$ generate a pointed cone in $K_\mathbb{Q}$.

If the conditions (1), (2), and (3) hold, then the K-graded ring $\mathbb{K}[T_1, \ldots, T_r]$ is the Cox ring of a $\mathbb{Q}$-factorial projective toric variety.

Exercise 2.12 (Toric quotient presentations)   A *toric quotient presentation* is a toric morphism $\pi : \widetilde{X} \to X$ of nondegenerate toric varieties, where the kernel of the accompanying homomorphism acts strongly stable on $\widetilde{X}$ and π is a good quotient for this action. Each such toric quotient presentation is dominated by Cox's quotient presentation in the sense that there is a commutative diagram of toric morphisms

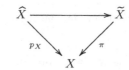

Translate this statement into the category of lattice fans and prove it there. Show that the isomorphy classes of quotient presentations are in one-to-one correspondence with the subgroups of the divisor class group $\mathrm{Cl}(X)$. See Section 4.2.1 for the analogous statement in the nontoric case.

Exercise 2.13 (Good   quotients   and   coordinate   subspace   arrangements) Assume that a quasitorus H acts on $\mathbb{K}^r$ via weights $\chi_1, \ldots, \chi_r \in \mathbb{X}(H)$; that means that we have

$$h \cdot z = (\chi_1(h)z_1, \ldots, \chi_r(h)z_r).$$

Let $U \subseteq \mathbb{K}^r$ such that $\mathbb{K}^r \setminus U$ is a union of coordinate subspaces. We say that a subset $I \subseteq \{1, \ldots, r\}$ is minimal for U if $\text{lin}(e_i; \ i \in I)$ is a minimal coordinate subspace intersecting U. Let $\mathcal{I}$ be the collection of all minimal subsets for U.

(1) The action of H on U admits a good quotient if and only if for any two $I, J \in \mathcal{I}$ one has

$$\text{cone}(w_i; \ i \in I)^\circ \cap \text{cone}(w_j; \ j \in J)^\circ \neq \emptyset.$$

(2) If (1) holds, then the good quotient $U \to U /\!/ H$ is a characteristic space over $U /\!/ H$ if and only if the following conditions hold:

(a) $\cap_{I \in \mathcal{I}} I = \emptyset$,
(b) $w_i \in \text{lin}_{\mathbb{Z}}(w_1, \ldots, w_{i-1}, w_{i+1}, \ldots, w_r)$ for any $i = 1, \ldots, r$,
(c) $\text{cone}(\pm w_i, w_k; \ i \in I, k \in \{1, \ldots, r\} \setminus (I \cup \{j\})) = \text{lin}(w_1, \ldots, w_r)$ for all minimal I and all $j \notin I$.

(3) Consider the $\mathbb{K}^*$-action on $\mathbb{K}^n \setminus \{0\}$ given by $t \cdot z = (t^{m_1} z_1, \ldots, t^{m_r} z_r)$. This action admits a good quotient if and only if either $m_1 = \ldots = m_r = 0$ or all m_i are nonzero of the same sign. In the second case, the good quotient is a characteristic space if and only if $GCD(m_1, \ldots, m_{i-1}, m_{i+1}, \ldots, m_r) = 1$ for any $i = 1, \ldots, r$, and we come to the characteristic space of the weighted projective space $\mathbb{P}_{m_1, \ldots, m_r}$.

Exercise 2.14* (Categorical quotients for subtorus actions)   Let X be a toric variety with acting torus T and consider the action of subtorus $H \subseteq X$; we refer to this setting as a "subtorus action." A *categorical quotient*, for example, in the category of varieties, is an H-invariant morphism $\pi : X \to Y$ of varieties such that every H-invariant morphism of varieties factors uniquely through p. This concept depends essentially on the category.

(1) Give an example of a subtorus action having a categorical quotient in the category of varieties which is not a good quotient.
(2) Give an example of a subtorus action admitting categorical quotients in the categories of varieties and prevarieties such that the respective quotient spaces are not isomorphic to each other.
(3) Give an example of a subtorus action admitting a categorical quotient in the category of varieties but not in the category of prevarieties.
(4) Give an example of a subtorus action admitting a categorical quotient in the category of prevarieties but not in the category of varieties.
(5) Give an example of a subtorus action admitting no categorical quotient in the category of varieties and no categorical quotient in the category of prevarieties.
(6) Characterize in terms of fans when a subtorus action admits a categorical quotient in the category of varieties (in the category of prevarieties).

We refer to [2, 3] for solutions to the first five items. The sixth one is an open problem, even for toric open subsets $X \subseteq \mathbb{K}^r$ with diagonal torus action. Another interesting category in this context is built on constructible sets, see [15, 87].

Exercise 2.15 Describe explicitly fan, Cox ring, characteristic space, irrelevant ideal, and true lattice bunch for products of projective spaces and for the blow-up of a projective space at a toric fixed point.

Exercise 2.16 (Fake weighted projective planes) Let $v_1, v_2, v_3 \in \mathbb{Z}^2$ be three primitive vectors generating $\mathbb{Q}^2$ as a cone. By applying a suitable isomorphism $\mathbb{Z}^2 \to \mathbb{Z}^2$, one can achieve

$$v_1 = (1, 0), \quad v_2 = (a, b), \quad v_3 = (c, d),$$
$$a, b, c, d \in \mathbb{Z}, \quad 0 \le a < b \le -d \le ad - bc.$$

Let Σ be the complete fan in $\mathbb{Q}^2$ having v_1, v_2, v_3 as the primitive generators of its rays and let X_Σ be the associated toric surface. Show

$$\mathrm{Cl}(X_\Sigma) \cong \mathbb{Z} \oplus \mathbb{Z}/\gcd(b, d)\mathbb{Z}, \qquad \mathrm{Pic}(X_\Sigma) \cong \mathbb{Z}.$$

Determine the lattice bunch corresponding to X_Σ. Compute the local divisor class groups of X_Σ and the factor group $\mathrm{Cl}(X_\Sigma)/\mathrm{Pic}(X_\Sigma)$. Generalize your observations to higher dimensional $\mathbb{Q}$-factorial toric varieties of Picard number 1.

Exercise 2.17 Show that every complete toric surface is determined up to isomorphy by its Cox ring and show that it is projective. Construct examples of nonisomorphic projective three-dimensional toric varieties having the same Cox ring.

Exercise 2.18 (A_2-property for toric varieties) Every toric variety whose fan contains at most two maximal cones is quasiprojective. Any two points on a toric variety admit a common affine open neighborhood.

Exercise 2.19 (Toric Kleiman–Chevalley criterion) Let X be a $\mathbb{Q}$-factorial maximal toric variety with $\Gamma(X, \mathcal{O}) = \mathbb{K}$ and set $r := \mathrm{rank}(\mathrm{Cl}(X))$. If any r points of X admit a common affine neighborhood, then X is projective.

Exercise 2.20 Construct the fans of the toric varieties occuring in Kleinschmidt's classification for small n.

3

Cox rings and combinatorics

We present a combinatorial approach to the geometry of varieties with a finitely generated Cox ring. It relies on the observation that basically all varieties X sharing a given K-graded algebra R as their Cox ring arise as good quotients of open sets of $\operatorname{Spec} R$ by the action of the quasitorus $\operatorname{Spec} \mathbb{K}[K]$. The door to combinatorics is opened by geometric invariant theory (GIT), which provides a description of the possible quotients in terms of combinatorial data, certain collections of polyhedral cones living in the rational vector space $K_{\mathbb{Q}}$. In Section 3.1, we develop the geometric invariant theory of quasitorus actions on affine varieties. With every point we associate a convex polyhedral "orbit cone" and based on these data we build up the combinatorial structures describing the variation of quotients. The variation of (semi-)projective quotients is described in terms of the "GIT fan" and for the more general case of torically embeddable quotients the description is in terms of "bunches of orbit cones." In the case of a subtorus action on an affine toric variety, these descriptions coincide with the ones obtained via Gale duality in Chapter 2. In Section 3.2, the concept of a bunched ring is presented; this is basically a factorially graded algebra R together with a bunch of cones living in the grading group K. With any such data we associate an irreducible normal variety X: we consider the action of $\operatorname{Spec} \mathbb{K}[K]$ on $\operatorname{Spec} R$ and take the quotient determined by the bunch of cones. The basic feature of this construction is that the resulting variety X has K as its divisor class group and R as its Cox ring. Moreover, it turns out that X comes with a canonical closed embedding into a toric variety. As first examples we discuss flag varieties and quotients of quadrics. The main task then is to describe the geometry of the variety X in terms of its defining data, the bunched ring. Section 3.3 provides basic results on local divisor class groups, the Picard group, and singularities. Moreover, we determine base loci of divisors as well as the cones of movable, semiample, and ample divisor classes. In the case of a complete intersection, there is a simple formula for the canonical divisor, and intersection numbers can be easily computed. As a rich class of examples, we consider in Section 3.4 complete rational varieties X with an effective torus action $T \times X \to X$ of complexity 1, that is, one has $\dim T = \dim X - 1$. This class naturally generalizes toric varieties and comprises, for example, all rational normal complete $\mathbb{K}^*$-surfaces. The describing bunched rings are explicitly constructed, which leads to a concrete approach to the geometry of these varieties.

3.1 GIT for affine quasitorus actions

3.1.1 Orbit cones

Here we discuss local properties of quasitorus actions and also touch on computational aspects. We work over an algebraically closed field $\mathbb{K}$ of characteristic zero. By K we denote a finitely generated abelian group and we consider an affine K-graded $\mathbb{K}$-algebra

$$A = \bigoplus_{w \in K} A_w.$$

Then the quasitorus $H = \operatorname{Spec} \mathbb{K}[K]$ acts on the affine variety $X := \operatorname{Spec} A$. For a point $x \in X$, we introduced in Definition 1.2.2.7 the *orbit monoid* S_x and the *orbit group* K_x as

$$S_x = \{w \in K; \ f(x) \neq 0 \text{ for some } f \in A_w\} \subseteq K, \quad K_x = S_x - S_x \subseteq K.$$

Let $K_{\mathbb{Q}} := K \otimes_{\mathbb{Z}} \mathbb{Q}$ denote the rational vector space associated with K. Given $w \in K$, we write again w for the element $w \otimes 1 \in K_{\mathbb{Q}}$. The *weight cone* of X is the convex polyhedral cone

$$\omega_X := \omega(A) = \operatorname{cone}(w \in K; \ A_w \neq \{0\}) \subseteq K_{\mathbb{Q}}.$$

Definition 3.1.1.1 The *orbit cone* of a point $x \in X$ is the convex cone $\omega_x \subseteq K_{\mathbb{Q}}$ generated by the weight monoid $S_x \subseteq K$. For the set of all orbit cones we write

$$\Omega_X := \{\omega_x; \ x \in X\}.$$

If we want to specify the acting group H or even the variety X in the orbit data, then we write $K_{H,x}$ or $K_{H,X,x}$ etc.

Remark 3.1.1.2 Let $Y \subseteq X$ be an H-invariant closed subvariety and let $x \in Y$. According to Proposition 1.2.2.5, one has

$$S_{H,Y,x} = S_{H,X,x}, \qquad K_{H,Y,x} = K_{H,X,x}, \qquad \omega_{H,Y,x} = \omega_{H,X,x}.$$

Considering for $x \in X$ its orbit closure $Y := \overline{H \cdot x}$, we infer that $\omega_{H,X,x} = \omega_{H,Y,x}$ equals the weight cone ω_Y and thus is polyhedral.

The following observation reduces the study of orbit cones of a quasitorus action to the case of a torus action.

Remark 3.1.1.3 Let $K^t \subseteq K$ denote the torsion part, set $K^0 := K/K^t$ and let $\alpha \colon K \to K^0$ be the projection. Then we have the coarsened grading

$$A = \bigoplus_{u \in K^0} A_u, \qquad A_u = \bigoplus_{w \in \alpha^{-1}(u)} A_w.$$

This coarsened grading describes the action of the unit component $H^0 \subseteq H$ on X, and we have a commutative diagram

$$
\begin{array}{ccc}
K & \xrightarrow{\ \alpha\ } & K^0 \\
\downarrow & & \downarrow \\
K_{\mathbb{Q}} & \xrightarrow[\alpha]{\ \cong\ } & K_{\mathbb{Q}}^0
\end{array}
$$

For every $x \in X$, the isomorphism $\alpha \colon K_{\mathbb{Q}} \to K_{\mathbb{Q}}^0$ maps the H-orbit cone $\omega_{H,x}$ onto the H^0-orbit cone $\omega_{H^0,x}$.

Proposition 3.1.1.4 *Let $\nu \colon X' \to X$ be the H-equivariant normalization. Then, for every $x' \in X'$, we have $\omega_{x'} = \omega_{\nu(x')}$.*

Proof The inclusion $\omega_{\nu(x')} \subseteq \omega_{x'}$ is clear by equivariance. The reverse inclusion follows from considering equations of integral dependence for the homogeneous elements $f \in \Gamma(X', \mathcal{O})$ with $f(x') \neq 0$. $\qquad\square$

We shall use the orbit cones to describe properties of orbit closures. The basic statement in this regard is the following one.

Proposition 3.1.1.5 *Assume that H is a torus and let $x \in X$. The factor group H/H_x acts with a dense free orbit on the orbit closure $\overline{H \cdot x} \subseteq X$. Moreover, the orbit closure $\overline{H \cdot x}$ has the affine toric variety $\mathrm{Spec}(\mathbb{K}[\omega_x \cap K_x])$ as its (H/H_x)-equivariant normalization.*

Proof The first assertion is obvious, and the second one follows immediately from Proposition 3.1.1.4 and the fact that the algebra of global functions of $\overline{H \cdot x}$ is the semigroup algebra $\mathbb{K}[S_x]$ of the orbit monoid; see Proposition 1.2.2.9. $\quad\square$

The collection of orbits in a given orbit closure $\overline{H \cdot x}$ comes with a partial ordering: we write $H \cdot x_1 \leq H \cdot x_2$ if $H \cdot x_1 \subseteq \overline{H \cdot x_2}$ holds.

Proposition 3.1.1.6 *Let $x \in X$. Then we have a commutative diagram of order preserving bijections*

$$
\begin{array}{ccc}
H\text{-orbits}(\overline{H \cdot x}) & \xrightarrow{\ H \cdot y \,\mapsto\, \omega_{H,y}\ } & \text{faces}(\omega_{H,x}) \\
\cong \Big\updownarrow & & \Big\updownarrow \cong \\
H^0\text{-orbits}(\overline{H^0 \cdot x}) & \xrightarrow[\ H^0 \cdot y \,\mapsto\, \omega_{H^0,y}\ ]{} & \text{faces}(\omega_{H^0,x})
\end{array}
$$

Moreover, for any homogeneous function $f \in A_w$ with $f(x) \neq 0$ and any point $y \in \overline{H \cdot x}$, we have $f(y) \neq 0 \Leftrightarrow w \in \omega_{H,y}$.

Proof The torus $T := H^0/H_x^0$ acts on $Z := \overline{H^0 \cdot x}$, and the T-orbits of Z coincide with its H^0-orbits. According to Proposition 3.1.1.5, the T-equivariant normalization of Z is the affine toric variety $Z' = \operatorname{Spec}(\mathbb{K}[\omega_x \cap K_x])$. The T-orbits of Z' are in order-preserving bijection with the faces of ω_x via $T \cdot z \mapsto \omega_z$; see Proposition 2.1.2.2. The assertion follows from Proposition 3.1.1.4 and the fact that the normalization map $\nu \colon Z' \to Z$ induces an order-preserving bijection between the sets of T-orbits of Z' and Z. $\qquad\square$

We collect some basic observations on the explicit computation of orbit data, when the algebra A is given in terms of homogeneous generators and relations. The following notions will be crucial.

Definition 3.1.1.7 Let K be a finitely generated abelian group, A a K-graded affine $\mathbb{K}$-algebra and $\mathfrak{F} = (f_1, \dots, f_r)$ a system of homogeneous generators for A.

(i) The *projected cone associated with* $\mathfrak{F}$ is $(E \xrightarrow{\ Q\ } K, \gamma)$, where $E := \mathbb{Z}^r$, the homomorphism $Q \colon E \to K$ sends the ith canonical basis vector $e_i \in E$ to $w_i := \deg(f_i) \in K$ and $\gamma \subseteq E_{\mathbb{Q}}$ is the convex cone generated by $e_1, \dots, e_r$.
(ii) A face $\gamma_0 \preceq \gamma$ is called an $\mathfrak{F}$-*face* if the product over all f_i with $e_i \in \gamma_0$ does not belong to the ideal $\sqrt{\langle f_j; \, e_j \notin \gamma_0 \rangle} \subseteq A$.

Construction 3.1.1.8 Fix a system of homogeneous generators $\mathfrak{F} = (f_1, \dots, f_r)$ for our K-graded affine $\mathbb{K}$-algebra A. Setting $\deg(T_i) := w_i := \deg(f_i)$ defines a K-grading on $\mathbb{K}[T_1, \dots, T_r]$ and we have a graded epimorphism

$$
\mathbb{K}[T_1, \dots, T_r] \to A, \qquad T_i \mapsto f_i.
$$

On the geometric side, this gives a diagonal H-action on $\mathbb{K}^r$ via the characters $\chi^{w_1}, \dots, \chi^{w_r}$ and an H-equivariant closed embedding

$$
X \to \mathbb{K}^r, \qquad x \mapsto (f_1(x), \dots, f_r(x)).
$$

With $E = \mathbb{Z}^r$ and $\gamma = \mathrm{cone}(e_1, \ldots, e_r)$, we have $\mathbb{K}[T_1, \ldots, T_r] = \mathbb{K}[E \cap \gamma]$, where T_i is identified with χ^{e_i}. Thus, we may also regard $\mathbb{K}^r$ as the toric variety associated with the cone $\delta := \gamma^\vee \subseteq F_\mathbb{Q}$, where $F := \mathrm{Hom}(E, \mathbb{Z})$.

Proposition 3.1.1.9 *Consider the situation of Construction 3.1.1.8. Consider a face $\gamma_0 \preceq \gamma$ and its corresponding face $\delta_0 \preceq \delta$, that is, we have $\delta_0 = \gamma_0^\perp \cap \delta$. Then the following statements are equivalent.*

(i) *The face $\gamma_0 \preceq \gamma$ is an $\mathfrak{F}$-face.*
(ii) *There is a point $z \in X$ with $z_i \neq 0 \Leftrightarrow e_i \in \gamma_0$ for all $1 \leq i \leq r$.*
(iii) *The toric orbit $\mathbb{T}^r \cdot z_{\delta_0} \subseteq \mathbb{K}^r$ corresponding to $\delta_0 \preceq \delta$ meets X.*

Proof   The equivalence of (i) and (ii) is an immediate consequence of the following equivalence

$$\prod_{e_i \in \gamma_0} f_i \notin \sqrt{\langle f_j; \ e_j \notin \gamma_0 \rangle} \quad\Longleftrightarrow\quad \bigcup_{e_i \in \gamma_0} V(X, f_i) \not\supseteq \bigcap_{e_j \notin \gamma_0} V(X; f_j).$$

The equivalence of (ii) and (iii) is clear by the fact that $\mathbb{T}^r \cdot z_{\delta_0}$ consists exactly of the points $z \in \mathbb{K}^r$ with $z_i \neq 0 \Leftrightarrow e_i \in \gamma_0$ for all $1 \leq i \leq r$. $\square$

Proposition 3.1.1.10 *Consider the situation of Construction 3.1.1.8. Let $\gamma_0 \preceq \gamma$ be an $\mathfrak{F}$-face and let $\delta_0 \preceq \delta$ denote the corresponding face, that is, we have $\delta_0 = \gamma_0^\perp \cap \delta$. Then, for every $x \in X \cap \mathbb{T}^r \cdot z_{\delta_0}$, the orbit data are given by*

$$S_{H,x} = Q(\gamma_0 \cap E), \qquad K_{H,x} = Q(\mathrm{lin}(\gamma_0) \cap E), \qquad \omega_{H,x} = Q(\gamma_0).$$

In particular, the orbit cones of the H-action on X are precisely the projected $\mathfrak{F}$-faces, that is, we have

$$\{\omega_{H,x}; \ x \in X\} = \{Q(\gamma_0); \ \gamma_0 \preceq \gamma \text{ is an } \mathfrak{F}\text{-face}\}.$$

Proof   To obtain the first assertion, observe that the orbit monoid of $x \in X$ is the monoid in K generated by all w_i with $f_i(x) \neq 0$. The other two assertions follow directly. $\square$

Remark 3.1.1.11   Consider the situation of Construction 3.1.1.8. Let $g_1, \ldots, g_s$ generate the kernel of $\mathbb{K}[T_1, \ldots, T_r] \to A$ as an ideal. By an $\mathfrak{F}$-*set*, we mean a subset $I \subseteq \{1, \ldots, r\}$ with the property

$$\prod_{i \in I} T_i \notin \sqrt{\langle g_1^I, \ldots, g_s^I \rangle}, \quad \text{with} \quad g_j^I := g_j(S_1, \ldots, S_r), \quad S_l := \begin{cases} T_l & l \in I, \\ 0 & l \notin I. \end{cases}$$

Then the $\mathfrak{F}$-faces are precisely the $\mathrm{cone}(e_i; \ i \in I)$, where I is an $\mathfrak{F}$-set, and the orbit cones are precisely the $\mathrm{cone}(w_i; \ i \in I)$, where $w_i = \deg(f_i)$ and I is an $\mathfrak{F}$-set.

For the case that $X \subseteq \mathbb{K}^r$ is defined by a single equation, we can easily figure out the $\mathfrak{F}$-faces in a purely combinatorial manner. Recall that the *Newton polytope* of a polynomial $g \in \mathbb{K}[T_1, \ldots, T_r]$ is defined as

$$N(g) := \operatorname{conv}(\nu; \ a_\nu \neq 0) \subseteq \mathbb{Q}^r, \qquad \text{where } g = \sum_{\nu \in \mathbb{Z}_{\geq 0}^r} a_\nu T^\nu.$$

Proposition 3.1.1.12 *Consider the situation of Construction 3.1.1.8. Suppose that $X \subseteq \mathbb{K}^r$ is the zero set of $g \in \mathbb{K}[T_1, \ldots, T_r]$. Then, for every face $\gamma_0 \preceq \gamma$, the following statements are equivalent.*

(i) *The face $\gamma_0 \preceq \gamma$ is an $\mathfrak{F}$-face.*
(ii) *$N(g) \cap \gamma_0$ is empty or contains at least two points.*
(iii) *The number of vertices of $N(g)$ contained in γ_0 differs from 1.*

Proof   The equivalence of (ii) and (iii) is clear. For that of (i) and (ii) write first $g = \sum_{\nu \in \mathbb{Z}_{\geq 0}^r} a_\nu g_\nu$ with $g_\nu := T_1^{\nu_1} \cdots T_r^{\nu_r}$. Then, for any point $z \in \mathbb{K}^r$, we have

$$g_\nu(z) \neq 0 \quad \Longleftrightarrow \quad \nu_i \neq 0 \Rightarrow z_i \neq 0 \text{ holds for } 1 \leq i \leq r.$$

Moreover, if we have $g(z) = 0$, then the number of ν with $a_\nu g_\nu(z) \neq 0$ is different from 1.

Now suppose that (i) holds and let $z \in X$ be as in Proposition 3.1.1.9 (ii). Then we have $g(z) = 0$. If all monomials g_ν with $a_\nu \neq 0$ vanish on z, then $N(g) \cap \gamma_0$ is empty. If not all monomials g_ν with $a_\nu \neq 0$ vanish on z, then there are at least two multi-indices ν, μ with $a_\nu \neq 0 \neq a_\mu$ and $g_\nu(z) \neq 0 \neq g_\mu(z)$. This implies

$$\nu_1 e_1 + \cdots + \nu_r e_r \in N(g) \cap \gamma_0, \qquad \mu_1 e_1 + \cdots + \mu_r e_r \in N(g) \cap \gamma_0.$$

Conversely, suppose that (ii) holds. Define $z \in \mathbb{K}^r$ by $z_i = 1$ if $e_i \in \gamma_0$ and $z_i = 0$ otherwise. If $N(g) \cap \gamma_0$ is empty, then we have $z \in X$, which gives (i). If $N(g) \cap \gamma_0$ contains at least two points, then γ_0 contains at least two vertices ν, μ of $N(g)$, and we find a point $z' \in \mathbb{T}^r \cdot z$ with $g(z') = 0$. $\square$

3.1.2 Semistable quotients

Again, we work over an algebraically closed field $\mathbb{K}$ of characteristic zero. Let K be a finitely generated abelian group and consider an affine K-graded $\mathbb{K}$-algebra

$$A = \bigoplus_{w \in K} A_w.$$

Then the quasitorus $H := \operatorname{Spec} \mathbb{K}[K]$ acts on the affine variety $X := \operatorname{Spec} A$. The following is Mumford's definition [225] of semistability specialized to the case of a "linearization of the trivial line bundle."

Definition 3.1.2.1 The *set of semistable points* associated with an element $w \in K_{\mathbb{Q}}$ is the H-invariant open subset

$$X^{ss}(w) := \{x \in X; \ f(x) \neq 0 \text{ for some } f \in A_{nw}, \ n > 0\} \subseteq X.$$

Note that the set of semistable points $X^{ss}(w)$ is nonempty if and only if w belongs to the weight cone ω_X. The following two statements subsume the basic features of semistable sets.

Proposition 3.1.2.2 *For every $w \in \omega_X$, the H-action on $X^{ss}(w)$ admits a good quotient $\pi \colon X^{ss}(w) \to Y(w)$ with $Y(w)$ projective over $Y(0) = \operatorname{Spec} A_0 = X /\!\!/ H$.*

Proposition 3.1.2.3 *For any two $w_1, w_2 \in \omega_X$ with $X^{ss}(w_1) \subseteq X^{ss}(w_2)$ we have a commutative diagram*

$$
\begin{array}{ccc}
X^{ss}(w_1) & \subseteq & X^{ss}(w_2) \\
\downarrow {\scriptstyle /\!\!/ H} & & \downarrow {\scriptstyle /\!\!/ H} \\
Y(w_1) & \xrightarrow{\ \varphi^{w_1}_{w_2}\ } & Y(w_2)
\end{array}
$$

with a projective surjection $\varphi^{w_1}_{w_2}$. Moreover, we have $\varphi^{w_1}_{w_3} = \varphi^{w_2}_{w_3} \circ \varphi^{w_1}_{w_2}$ whenever composition is possible.

Both propositions are direct consequences of the following two constructions, which we list separately for later use. The first one ensures the existence of the quotient.

Construction 3.1.2.4 For a homogeneous $f \in A$, let $A_{(f)} \subseteq A_f$ denote the degree zero part. Given any two $f \in A_{nw}$ and $g \in A_{mw}$ with $m, n > 0$, we have a commutative diagram of K-graded affine algebras and the associated commutative diagram of affine H-varieties

$$
\begin{array}{ccc}
A_f \longrightarrow A_{fg} \longleftarrow A_g \\
\uparrow \qquad \uparrow \qquad \uparrow \\
A_{(f)} \longrightarrow A_{(fg)} \longleftarrow A_{(g)}
\end{array}
\qquad
\begin{array}{ccc}
X_f \longleftarrow X_{fg} \longrightarrow X_g \\
{\scriptstyle \pi_f}\downarrow \qquad \downarrow \qquad \downarrow {\scriptstyle \pi_g} \\
V_f \longleftarrow V_{fg} \longrightarrow V_g
\end{array}
$$

The morphism $V_{fg} \to V_f$ is an open embedding, and we have $X_{fg} = \pi_f^{-1}(V_{fg})$. Gluing the V_f gives a variety $Y(w)$ and the maps π_f glue together to a good quotient $\pi \colon X^{ss}(w) \to Y(w)$ for the action of H on $X^{ss}(w)$.

The second construction relates the first one to the Proj-construction; see Example 1.2.3.10. This yields projectivity of the quotient space.

Construction 3.1.2.5 In the situation of Construction 3.1.2.4, the weight $w \in K$ defines a Veronese subalgebra

$$A(w) := \bigoplus_{n \in \mathbb{Z}_{\geq 0}} A_{nw} \subseteq \bigoplus_{w' \in K} A_{w'} = A.$$

Proposition 1.1.2.4 ensures that $A(w)$ is finitely generated and thus defines a $\mathbb{K}^*$-variety $X(w) := \operatorname{Spec} A(w)$. For every homogeneous $f \in A(w)$, we have commutative diagrams, where the second one is obtained by applying the Spec functor:

$$
\begin{array}{ccc}
A(w)_f \longrightarrow A_f \\
\uparrow \qquad\quad \uparrow \\
A(w)_{(f)} \underset{\cong}{\longrightarrow} A_{(f)}
\end{array}
\qquad\qquad
\begin{array}{ccc}
X(w)_f \longleftarrow X_f \\
\downarrow \qquad\quad \downarrow \\
U(w)_f \underset{\cong}{\longleftarrow} V_f
\end{array}
$$

Gluing the $U(w)_f$ gives $\operatorname{Proj}(A(w))$ and the isomorphisms $V_f \to U(w)_f$ glue together to an isomorphism $Y(w) \to \operatorname{Proj}(A(w))$. In particular, $Y(w)$ is projective over $Y(0) = \operatorname{Spec} A_0$.

Recall that we defined the orbit cone of a point $x \in X$ to be the convex cone $\omega_x \subseteq K_\mathbb{Q}$ generated by all $w \in K$ that admit an $f \in A_w$ with $f(x) \neq 0$.

Definition 3.1.2.6 The *GIT cone*, or *GIT chamber*, of an element $w \in \omega_X$ is the (nonempty) intersection of all orbit cones containing it:

$$\lambda(w) := \bigcap_{\substack{x \in X, \\ w \in \omega_x}} \omega_x.$$

From finiteness of the number of orbit cones, see Proposition 3.1.1.10, we infer that the GIT cones are polyhedral, and moreover that there are only finitely many of them.

Lemma 3.1.2.7 *For every $w \in \omega_X$, the associated set $X^{ss}(w) \subseteq X$ of semistable points is given by*

$$X^{ss}(w) = \{x \in X;\ w \in \omega_x\} = \{x \in X;\ \lambda(w) \subseteq \omega_x\}.$$

This description shows, in particular, that there are only finitely many sets $X^{ss}(w)$ of semistable points. The following statement describes the collection of all sets $X^{ss}(w)$ of semistable points.

Theorem 3.1.2.8 *Let* $X = \operatorname{Spec} A$ *and* $H = \operatorname{Spec} \mathbb{K}[K]$ *be as before. The collection* $\Lambda(X) := \{\lambda(w); \ w \in \omega_X\}$ *of all GIT cones is a quasifan in* $K_{\mathbb{Q}}$ *having the weight cone* ω_X *as its support. Moreover, for any two* $w_1, w_2 \in \omega_X$, *we have*

$$\lambda(w_1) \subseteq \lambda(w_2) \iff X^{ss}(w_1) \supseteq X^{ss}(w_2),$$

$$\lambda(w_1) = \lambda(w_2) \iff X^{ss}(w_1) = X^{ss}(w_2).$$

Definition 3.1.2.9 The collection $\Lambda(X)$ of Theorem 3.1.2.8, also denoted as $\Lambda(X, H)$ if we want to specify the quasitorus H, is called the *GIT (quasi-)fan* of the H-variety X.

Lemma 3.1.2.10 *Let* $w \in \omega_X \cap K$. *Then the associated GIT cone* $\lambda := \lambda(w) \in \Lambda(X)$ *satisfies*

$$\lambda = \bigcap_{w \in \omega_x^\circ} \omega_x = \bigcap_{\lambda^\circ \subseteq \omega_x^\circ} \omega_x,$$

$$w \in \lambda^\circ = \bigcap_{w \in \omega_x^\circ} \omega_x^\circ = \bigcap_{\lambda^\circ \subseteq \omega_x^\circ} \omega_x^\circ.$$

Proof For any orbit cone ω_x with $w \in \omega_x$, there is a unique minimal face $\omega \preceq \omega_x$ with $w \in \omega$. This face satisfies $w \in \omega^\circ$. According to Proposition 3.1.1.6, the face $\omega \preceq \omega_x$ is again an orbit cone. This gives the first formula. The second one follows from an elementary observation: If the intersection of the relative interiors of a finite number of convex polyhedral cones is nonempty, then it equals the relative interior of the intersection of the cones. $\square$

Proof of Theorem 3.1.2.8 As mentioned, finiteness of the number of orbit cones ensures that $\Lambda(X)$ is a finite collection of convex polyhedral cones. The displayed equivalences are clear by Lemma 3.1.2.7. They allow us in particular to write

$$X^{ss}(\lambda) := X^{ss}(w), \quad \text{where } \lambda = \lambda(w) \text{ for } w \in \omega_X.$$

The only thing we have to show is that $\Lambda(X)$ is a quasifan. This is done below by verifying several claims. For the sake of short notation, we set for the moment $\omega := \omega_x$ and $\Lambda := \Lambda(X)$.

Claim 1. Let $\lambda_1, \lambda_2 \in \Lambda$ with $\lambda_1 \subseteq \lambda_2$. Then, for every $x_1 \in X^{ss}(\lambda_1)$ with $\lambda_1^\circ \subseteq \omega_{x_1}^\circ$, there exists an $x_2 \in X^{ss}(\lambda_2)$ with $\omega_{x_1} \preceq \omega_{x_2}$.

Let us verify the claim. We have $X^{ss}(\lambda_2) \subseteq X^{ss}(\lambda_1)$ and Proposition 3.1.2.3 provides us with a dominant, proper, hence surjective morphism $\varphi\colon Y(\lambda_2) \to Y(\lambda_1)$ of the quotient spaces fitting into the commutative diagram

$$
\begin{array}{ccc}
X^{ss}(\lambda_2) & \subseteq & X^{ss}(\lambda_1) \\
{\scriptstyle /\!\!/ H}\Big\downarrow {\scriptstyle p_2} & & {\scriptstyle p_1}\Big\downarrow {\scriptstyle /\!\!/ H} \\
Y(\lambda_2) & \xrightarrow{\quad\varphi\quad} & Y(\lambda_1)
\end{array}
$$

If a point $x_1 \in X^{ss}(\lambda_1)$ satisfies $\lambda_1^\circ \subseteq \omega_{x_1}^\circ$, then, by Lemma 3.1.2.7 and Proposition 3.1.1.6, its H-orbit is closed in $X^{ss}(\lambda_1)$. Corollary 1.2.3.7 thus tells us that $x_1 \in \overline{H \cdot x_2}$ holds for any point x_2 belonging to the (nonempty) intersection $X^{ss}(\lambda_2) \cap p_1^{-1}(p_1(x_1))$. Using once more Proposition 3.1.1.6 gives Claim 1.

Claim 2. Let $\lambda_1, \lambda_2 \in \Lambda$. Then $\lambda_1 \subseteq \lambda_2$ implies $\lambda_1 \preceq \lambda_2$.

For the verification, let $\tau_2 \preceq \lambda_2$ be the (unique) face with $\lambda_1^\circ \subseteq \tau_2^\circ$, and let $\omega_{1,1}, \ldots, \omega_{1,r}$ be the orbit cones with $\lambda_1^\circ \subseteq \omega_{1,i}^\circ$. Then we obtain, using Lemma 3.1.2.10 for the second observation,

$$
\tau_2^\circ \cap \omega_{1,i}^\circ \neq \emptyset, \qquad \lambda_1 = \omega_{1,1} \cap \ldots \cap \omega_{1,r}.
$$

By Claim 1, we have $\omega_{1,i} \preceq \omega_{2,i}$ with orbit cones $\omega_{2,i}$ satisfying $\lambda_2 \subseteq \omega_{2,i}$, and hence $\tau_2 \subseteq \omega_{2,i}$. The first of the displayed formulas implies $\tau_2 \subseteq \omega_{1,i}$, and the second one thus gives $\tau_2 = \lambda_1$. So, Claim 2 is verified.

Claim 3. Let $\lambda \in \Lambda$. Then every face $\lambda_0 \preceq \lambda$ belongs to Λ.

To see this, consider any $w \in \lambda_0^\circ$. Lemma 3.1.2.10 yields $w \in \lambda(w)^\circ$. By the definition of GIT cones, we have $\lambda(w) \subseteq \lambda$. Claim 2 gives even $\lambda(w) \preceq \lambda$. Thus, we have two faces, λ_0 and $\lambda(w)$ of λ having a common point w in their relative interiors. This means $\lambda_0 = \lambda(w)$, and Claim 3 is verified.

Claim 4. Let $\lambda_1, \lambda_2 \in \Lambda$. Then $\lambda_1 \cap \lambda_2$ is a face of both, λ_1 and λ_2.

Let $\tau_i \preceq \lambda_i$ be the minimal face containing $\lambda_1 \cap \lambda_2$. Choose w in the relative interior of $\lambda_1 \cap \lambda_2$, and consider the GIT cone $\lambda(w)$. By Lemma 3.1.2.10 and the definition of GIT cones, we see

$$
w \in \lambda(w)^\circ \cap \tau_i^\circ, \qquad \lambda(w) \subseteq \lambda_1 \cap \lambda_2 \subseteq \tau_i.
$$

By Claim 2, the second relation implies in particular $\lambda(w) \preceq \lambda_i$. Hence, we can conclude $\lambda(w) = \tau_i$, and hence $\lambda_1 \cap \lambda_2$ is a face of both λ_i. Thus, Claim 4 is verified, and the properties of a quasifan are established for Λ. $\qquad\square$

Remark 3.1.2.11 Theorem 3.1.2.8 provides another proof for the fact that the GKZ decomposition of a vector configuration $W = (w_1, \ldots, w_r)$ in a rational vector space $K_{\mathbb{Q}}$ is a fan; see Theorem 2.2.2.2. Indeed, let $K \subseteq K_{\mathbb{Q}}$ be the lattice generated by $w_1, \ldots, w_r$ and consider the action of the torus

$T = \mathrm{Spec}\ \mathbb{K}[K]$ on $\mathbb{K}^r$ given by $t \cdot z = (\chi^{w_1}(t)z_1, \ldots, \chi^{w_r}(t)z_r)$. Then the orbit cones of this action are precisely the $\mathcal{W}$-cones and thus Theorem 3.1.2.8 gives the result.

Example 3.1.2.12 (A GIT quasifan, which is not a fan) Consider the affine variety

$$X := V(\mathbb{K}^4;\ Z_1 Z_2 + Z_3 Z_4 - 1).$$

Then X is isomorphic to SL(2) and thus it is smooth, factorial, and $\Gamma(X, \mathcal{O})^* = \mathbb{K}^*$ holds. Moreover $\mathbb{K}^*$ acts on X via

$$t \cdot (z_1, z_2, z_3, z_4) := (t z_1, t^{-1} z_2, t z_3, t^{-1} z_4).$$

This action has only one orbit cone, which is the whole line $\mathbb{Q}$. In particular, the resulting GIT quasifan does not consist of pointed cones.

Example 3.1.2.13 Consider the affine space $X := \mathbb{K}^4$ with the action of the torus $H := \mathbb{K}^*$ given by

$$t \cdot (z_1, z_2, z_3 z_4) := (t z_1, t z_2, t^{-1} z_3, t^{-1} z_4).$$

This setting stems from the $\mathbb{Z}$-grading of $\mathbb{K}[Z_1, \ldots, Z_4]$ given by $\deg(Z_1) = \deg(Z_2) = 1$ and $\deg(Z_3) = \deg(Z_4) = -1$. The GIT quasifan lives in $\mathbb{Q}$ and consists of the three cones

$$\lambda(-1) = \mathbb{Q}_{\leq 0}, \qquad \lambda(0) = \{0\}, \qquad \lambda(1) = \mathbb{Q}_{\geq 0}.$$

As always, we have $X^{ss}(0) = X$. The two further sets of semistable points are $X^{ss}(-1) = \{z \in \mathbb{K}^4;\ z_3 \neq 0 \neq z_4\}$ and $X^{ss}(1) = \{z \in \mathbb{K}^4;\ z_1 \neq 0 \neq z_2\}$. The whole GIT system looks as follows:

$$
\begin{array}{ccccc}
X^{ss}(-1) & \subseteq & X^{ss}(0) & \supseteq & X^{ss}(1) \\
\downarrow & & \downarrow & & \downarrow \\
Y(-1) & \longrightarrow & Y(0) & \longleftarrow & Y(1)
\end{array}
$$

Note that $Y(0)$ is the affine cone $V(\mathbb{K}^4;\ T_1 T_2 - T_3 T_4)$ with the apex $y_0 = 0$, the projections $\varphi_0^i \colon Y(i) \to Y(0)$ are isomorphisms over $Y(0) \setminus \{y_0\}$, and the fibers over the apex y_0 are isomorphic to $\mathbb{P}^1$.

3.1.3 A_2-quotients

In the preceding section, we constructed semiprojective quotients via semistable points. Here we look more generally for quotient spaces with the following property. The ground field $\mathbb{K}$ is algebraically closed of characteristic zero.

Definition 3.1.3.1 We say that a prevariety X has the A_2-*property* if any two points $x, x' \in X$ admit a common affine open neighborhood in X.

A prevariety with the A_2-property is necessarily separated, see [224, Prop. I.5.6], and thus we will just speak of A_2-varieties. Examples of A_2-varieties are the quasiprojective ones. By the Kleiman–Chevalley criterion [185], a smooth variety is quasiprojective if and only if every finite subset admits a common affine neighborhood; see also [41, 309] for a generalization to the singular case. By [308, Thm. A], an irreducible normal variety X has the A_2-property if and only if it admits a closed embedding into a toric variety. In particular, toric varieties are A_2-varieties.

We fix the setup for the rest of this section. By K we denote a finitely generated abelian group and we consider an affine K-graded $\mathbb{K}$-algebra

$$A = \bigoplus_{w \in K} A_w.$$

Then the quasitorus $H := \operatorname{Spec} \mathbb{K}[K]$ acts on the affine variety $X := \operatorname{Spec} A$. The necessary data for our quotient construction are again given in terms of orbit cones.

Definition 3.1.3.2 Let Ω_X denote the collection of all orbit cones ω_x, where $x \in X$. A *bunch of orbit cones* is a nonempty collection $\Phi \subseteq \Omega_X$ such that

(i) Given $\omega_1, \omega_2 \in \Phi$, one has $\omega_1^\circ \cap \omega_2^\circ \neq \emptyset$.
(ii) Given $\omega \in \Phi$, every orbit cone $\omega_0 \in \Omega_X$ with $\omega^\circ \subseteq \omega_0^\circ$ belongs to Φ.

A *maximal bunch of orbit cones* is a bunch of orbit cones $\Phi \subseteq \Omega_X$ that cannot be enlarged by adding further orbit cones.

Definition 3.1.3.3 Let $\Phi, \Phi' \subseteq \Omega_X$ be bunches of orbit cones. We say that Φ *refines* Φ' (written $\Phi \leq \Phi'$), if for any $\omega' \in \Phi'$ there is an $\omega \in \Phi$ with $\omega \subseteq \omega'$.

The following example shows that the above notions generalize the combinatorial concepts treated in Section 2.2.

Example 3.1.3.4 Let F be a lattice and $\delta \subseteq F_\mathbb{Q}$ a pointed simplicial cone of full dimension. Let $E := \operatorname{Hom}(F, \mathbb{Z})$ be the dual lattice, and $\gamma := \delta^\vee$ the dual cone. Given a homomorphism $Q \colon E \to K$ to a finitely generated abelian group K, we obtain a K-grading of $A := \mathbb{K}[E]$ via $\deg(\chi^e) = Q(e)$. Consider

$$H = \operatorname{Spec} \mathbb{K}[K], \qquad X_\delta = \operatorname{Spec} \mathbb{K}[\gamma \cap E].$$

The orbit cones of the H-action on X_δ are just the cones $Q(\gamma_0)$, where $\gamma_0 \preceq \gamma$. Moreover, for the primitive generators $e_1, \ldots, e_r$ of γ, set $w_i := Q(e_i) \in K_\mathbb{Q}$.

Then the (maximal) bunches of orbit cones are precisely the (maximal) $\mathcal{W}$-bunches of the vector configuration $\mathcal{W} = (w_1, \ldots, w_r)$ in $K_{\mathbb{Q}}$.

Definition 3.1.3.5 With any collection of orbit cones $\Phi \subseteq \Omega_X$, we associate the following subset:

$$U(\Phi) := \{x \in X; \; \omega_0 \preceq \omega_x \text{ for some } \omega_0 \in \Phi\} \subseteq X.$$

Example 3.1.3.6 Consider the GIT fan $\Lambda(X) = \{\lambda(w); \; w \in \omega_X\}$. Every GIT cone $\lambda = \lambda(w)$ defines a bunch of orbit cones

$$\Phi(w) := \{\omega_x \in \Omega_X; \; w \in \omega_x^\circ\} = \{\omega_x \in \Omega_X; \; \lambda^\circ \subseteq \omega_x^\circ\} =: \Phi(\lambda).$$

For any two $\lambda, \lambda' \in \omega_X$, we have $\Phi(\lambda) \leq \Phi(\lambda')$ if and only if $\lambda \preceq \lambda'$ holds. Moreover, for the set associated with $\Phi(w)$ we have

$$U(\Phi(w)) = X^{ss}(w).$$

Construction 3.1.3.7 Let $\Phi \subseteq \Omega_X$ be a bunch of orbit cones and consider $x \in U(\Phi)$. Fix homogeneous $h_1, \ldots, h_r \in A$ such that $h_i(x) \neq 0$ holds and the orbit cone ω_x is generated by $\deg(h_i)$, where $1 \leq i \leq r$. For $u \in \mathbb{Z}_{>0}^r$, consider the H-invariant open set

$$U(x) := X_{f^u} \subseteq X, \qquad f^u := h_1^{u_1} \cdots h_r^{u_r}.$$

Then the sets $U(x)$ do not depend on the particular choice of $u \in \mathbb{Z}_{>0}^r$. Moreover we have $U(x) \subseteq U(\Phi)$ and for any $w \in \omega_x^\circ$, we find some u with

$$\deg(f^u) = u_1 \deg(h_1) + \cdots + u_r \deg(h_r) \in \mathbb{Q}_{>0}w.$$

Proof We have to verify that every $x' \in X_{f^u}$ belongs to $U(\Phi)$. By construction, we have $\omega_x \subseteq \omega_{x'}$. Consider $\omega_0 \in \Phi$ with $\omega_0 \preceq \omega_x$. Then ω_0° is contained in the relative interior of some face $\omega_0' \preceq \omega_{x'}$. By saturatedness of Φ, we have $\omega_0' \in \Phi$, and hence $x' \in U(\Phi)$. $\qquad\square$

We are ready to list the basic properties of the assignment $\Phi \mapsto U(\Phi)$; The statements are analogous to Propositions 3.1.2.2 and 3.1.2.3. We say that a subset $U \subseteq X$ is *saturated* w.r.t. a map $p \colon X \to Y$ if $U = p^{-1}(p(U))$ holds.

Proposition 3.1.3.8 *Let $\Phi \subseteq \Omega_X$ be a bunch of orbit cones. Then $U(\Phi) \subseteq X$ is H-invariant, open and admits a good quotient $U(\Phi) \to Y(\Phi)$, where the quotient space $Y(\Phi)$ is an A_2-variety. Moreover, the sets $U(x) \subseteq U(\Phi)$ with $H \cdot x$ closed in $U(\Phi)$ are saturated w.r.t. $U(\Phi) \to Y(\Phi)$.*

Proposition 3.1.3.9   *For any two bunches of orbit cones Φ_1, $\Phi_2 \subseteq \Omega_X$ with $\Phi_1 \geq \Phi_2$, we have $U(\Phi_1) \subseteq U(\Phi_2)$, and there is a commutative diagram*

$$
\begin{array}{ccc}
U(\Phi_1) & \subseteq & U(\Phi_2) \\
\big\Vert H & & \big\downarrow \! {\scriptstyle /\!/ H} \\
Y(\Phi_1) & \xrightarrow[\;\varphi^{\Phi_1}_{\Phi_2}\;]{} & Y(\Phi_2)
\end{array}
$$

with a dominant morphism $\varphi^{\Phi_1}_{\Phi_2}$. Moreover, we have $\varphi^{\Phi_1}_{\Phi_3} = \varphi^{\Phi_2}_{\Phi_3} \circ \varphi^{\Phi_1}_{\Phi_2}$ whenever composition is possible.

Lemma 3.1.3.10   *Let $\Phi \subseteq \Omega_X$ satisfy 3.1.3.2 (i) and let $x \in U(\Phi)$. Then the orbit $H \cdot x$ is closed in $U(\Phi)$ if and only if $\omega_x \in \Phi$ holds.*

Proof   First let $H \cdot x$ be closed in $U(\Phi)$. By the definition of $U(\Phi)$, we have $\omega_0 \preceq \omega_x$ for some $\omega_0 \in \Phi$. Consider the closure $C_X(H \cdot x)$ of $H \cdot x$ taken in X, and choose $x_0 \in C_X(H \cdot x)$ with $\omega_{x_0} = \omega_0$. Again by the definition of $U(\Phi)$, we have $x_0 \in U(\Phi)$. Because $H \cdot x$ is closed in $U(\Phi)$, we obtain $x_0 \in H \cdot x$, and hence $\omega = \omega_0 \in \Phi$.

Now, let $\omega_x \in \Phi$. We have to show that any $x_0 \in C_X(H \cdot x) \cap U(\Phi)$ lies in $H \cdot x$. Clearly, $x_0 \in C_X(H \cdot x)$ implies $\omega_{x_0} \preceq \omega_x$. By the definition of $U(\Phi)$, we have $\omega_0 \preceq \omega_{x_0}$ for some $\omega_0 \in \Phi$. Because Φ satisfies 3.1.3.2 (i), we have $\omega_0^\circ \cap \omega_x^\circ \neq \emptyset$. Together with $\omega_0 \preceq \omega_x$ this implies $\omega_0 = \omega_{x_0} = \omega_x$, and hence $x_0 \in H \cdot x$. $\square$

Proof of Proposition 3.1.3.8   We regard $U(\Phi)$ as a union of sets $U(x)$ as provided in Construction 3.1.3.7, where $x \in U(\Phi)$ runs through those points that have a closed H-orbit in $U(\Phi)$; according to Lemma 3.1.3.10 these are precisely the points $x \in U(\Phi)$ with $\omega_x \in \Phi$.

First consider two such $x_1, x_2 \in U(\Phi)$. Then we have $\omega_{x_i} \in \Phi$, and we can choose homogeneous $f_1, f_2 \in A$ such that $\deg(f_1) = \deg(f_2)$ lies in $\omega_{x_1}^\circ \cap \omega_{x_2}^\circ$ and $U(x_i) = X_{f_i}$ holds. Thus, we obtain a commutative diagram

$$
\begin{array}{ccccc}
X_{f_1} & \longleftarrow & X_{f_1 f_2} & \longrightarrow & X_{f_2} \\
\big\downarrow {\scriptstyle /\!/ H} & & \big\downarrow {\scriptstyle /\!/ H} & & \big\downarrow {\scriptstyle /\!/ H} \\
Y_{f_1} & \longleftarrow & Y_{f_1 f_2} & \longrightarrow & Y_{f_2}
\end{array}
$$

where the upper horizontal maps are open embeddings, the downwards maps are good quotients for the respective affine H-varieties, and the lower horizontal arrows indicate the induced morphisms of the affine quotient spaces.

By the choice of f_1 and f_2, the quotient f_2/f_1 is an invariant function on X_{f_1}, and the inclusion $X_{f_1 f_2} \subseteq X_{f_1}$ is just the localization by f_2/f_1. Because

f_2/f_1 is invariant, the latter holds as well for the quotient spaces; that means that the map $Y_{f_1 f_2} \to Y_{f_1}$ is the localization by f_2/f_1.

Now, cover $U(\Phi)$ by sets $U(x_i)$ with $H \cdot x_i$ closed in $U(\Phi)$. The preceding consideration allows gluing of the maps $U(x_i) \to U(x_i)/\!\!/H$ along $U_{ij} \to U_{ij}/\!\!/H$, where $U_{ij} := U(x_i) \cap U(x_j)$. This gives a good quotient $U(\Phi) \to U(\Phi)/\!\!/H$. By construction, the open sets $U(x_i) \subseteq U(\Phi)$ are saturated with respect to the quotient map.

To see that $Y = U(\Phi)/\!\!/H$ is an A_2-variety (and thus in particular separated), consider $y_1, y_2 \in Y$. Then there are f_i as above with $y_i \in Y_{f_i}$. The union $Y_{f_1} \cup Y_{f_2}$ is quasiprojective, because, for example, the set $Y_{f_1} \setminus Y_{f_2}$ defines an ample divisor. It follows that there is a common affine neighborhood of y_1, y_2 in $Y_{f_1} \cup Y_{f_2}$ and hence in Y. $\qquad\square$

Proof of Proposition 3.1.3.9   We only have to show that $\Phi_1 \geq \Phi_2$ implies $U(\Phi_1) \subseteq U(\Phi_2)$. Consider $x \in U(\Phi_1)$. Then we have $\omega_1 \preceq \omega_x$ for some $\omega_1 \in \Phi_1$. Because $\Phi_1 \geq \Phi_2$ holds, there is an $\omega_2 \in \Phi_2$ with $\omega_2 \subseteq \omega_1$. Let $\omega_0 \preceq \omega_1$ be the face with $\omega_2^\circ \subseteq \omega_0^\circ$. By Property 3.1.3.2 (ii), the face ω_0 belongs to Φ_2. Because of $\omega_0 \preceq \omega_x$, we have $x \in U(\Phi_2)$. $\qquad\square$

3.1.4 Quotients of H-factorial affine varieties

We consider a quasitorus action on an irreducible, normal affine variety, where we assume that for every invariant divisor some multiple is principal. The aim is to show that the constructions presented in the preceding two sections provide basically all open subsets admitting quasiprojective or torically embeddable quotients. The results generalize Theorem 2.3.2.4, which settled linear representations of quasitori. The ground field $\mathbb{K}$ is algebraically closed and of characteristic zero.

Definition 3.1.4.1   Let a reductive affine algebraic group G act on a variety X.

 (i) By a *good G-set* in X we mean an open subset $U \subseteq X$ with a good quotient $U \to U/\!\!/G$.
 (ii) We say that a subset $U' \subseteq U$ of a good G-set $U \subseteq X$ is *G-saturated* in U if it satisfies $U' = \pi^{-1}(\pi(U'))$, where $\pi : U \to U/\!\!/G$ is the good quotient.

Note that an open subset $U' \subseteq U$ of a good G-set U is G-saturated in U if and only if any orbit $G \cdot x \subseteq U'$ which is closed in U' is also closed in U. Moreover, for every G-saturated open subset $U' \subseteq U$ of a good G-set $U \subseteq X$ with quotient $\pi : U \to U/\!\!/G$, the image $\pi(U') \subseteq U/\!\!/G$ is open and $\pi : U' \to \pi(U')$ is a good quotient. The latter reduces the problem of describing all good G-sets to the description of the "maximal" ones. Here are the precise concepts.

Definition 3.1.4.2 Let a reductive affine algebraic group G act on a variety X.

(i) By a *qp-maximal subset* of X we mean a good G-set $U \subseteq X$ with $U /\!/ G$ quasiprojective such that U is maximal w.r.t. G-saturated inclusion among all good G-sets $W \subseteq X$ with $W /\!/ G$ quasiprojective.
(ii) By a $(G, 2)$-*maximal subset* of X we mean a good G-set $U \subseteq X$ with $U /\!/ G$ an A_2-variety such that U is maximal w.r.t. G-saturated inclusion among all good G-sets $W \subseteq X$ with $W /\!/ G$ an A_2-variety.

We are ready to formulate the first result. Let K be a finitely generated abelian group and A an integral normal K-graded affine algebra. Then the quasitorus $H = \operatorname{Spec} \mathbb{K}[K]$ acts on the irreducible normal affine variety $X = \operatorname{Spec} A$. For every GIT cone $\lambda \subseteq K_\mathbb{Q}$, we define its set of semistable points to be

$$X^{ss}(\lambda) := X^{ss}(w), \quad \text{where } w \in \lambda^\circ.$$

This does not depend on the particular choice of $w \in \lambda^\circ$. Moreover, with any H-invariant open subset $U \subseteq X$, we associate the cone

$$\lambda(U) := \bigcap_{x \in U} \omega_x \subseteq K_\mathbb{Q}.$$

Theorem 3.1.4.3 *Let $X = \operatorname{Spec} A$ and $H = \operatorname{Spec} \mathbb{K}[K]$ be as above. Assume that for every H-invariant divisor on X some positive multiple is principal. Then, with the GIT fan $\Lambda(X, H)$ of the H-action on X, we have mutually inverse bijections*

$$\Lambda(X, H) \longleftrightarrow \{qp\text{-}maximal \ subsets \ of \ X\}$$

$$\lambda \mapsto X^{ss}(\lambda),$$

$$\lambda(U) \leftarrow\!\shortmid U.$$

These bijections are order-reversing maps of partially ordered sets in the sense that we always have

$$\lambda \preceq \lambda' \iff X^{ss}(\lambda) \supseteq X^{ss}(\lambda').$$

Proof First recall from Proposition 3.1.2.2 that all sets $X^{ss}(w)$ are good H-sets with a quasiprojective quotient space $X^{ss}(w) /\!/ H$. The assertion thus is a direct consequence of the following two claims.

Claim 1. If $U \subseteq X$ is a good H-set such that $U /\!/ H$ is quasiprojective, then U is H-saturated in some set $X^{ss}(w)$ of semistable points.

Set $Y := U /\!/ H$, and let $p \colon U \to Y$ be the quotient map. For a global section f of a divisor D, set for short

$$Z(f) := \operatorname{Supp}(\operatorname{div}_D(f)) = \operatorname{Supp}(D + \operatorname{div}(f)).$$

Choose an (effective) ample divisor E on Y allowing global sections $h_1, \ldots, h_r$ such that the sets $Y \setminus Z(h_i)$ form an affine cover of Y. Consider the pullback data $D' := p^*E$ and $f'_i := p^*(h_i)$. Let $D_1, \ldots, D_s$ be the prime divisors contained in $X \setminus U$. Because the complement $U \setminus Z(f'_i)$ in X is of pure codimension 1, we have

$$U \setminus Z(f'_i) = X \setminus (D_1 \cup \ldots \cup D_s \cup \overline{Z(f'_i)}).$$

Consequently, by closing the components of D' in X and adding a suitably big multiple of $D_1 + \cdots + D_s$, we obtain an H-invariant Weil divisor D on X allowing global sections $f_1, \ldots, f_r$ such that

$$D|_U = D', \quad f_i|_U = f'_i, \quad X \setminus Z(f_i) = U \setminus Z(f'_i) = p^{-1}(Y \setminus Z(h_i)).$$

Replacing D with a suitable positive multiple, we may assume that it is principal, say $D = \mathrm{div}(f)$. Because D is H^0-invariant, f must be H^0-homogeneous. Fix a splitting $H = H^0 \times H^1$ with the unit component $H^0 \subseteq H$ and a finite group $H^1 \subseteq H$ and consider

$$f' := \prod_{h \in H^1} h \cdot f, \quad \text{where } (h \cdot f)(x) := f(h \cdot x).$$

Then f' is even H-homogeneous and its divisor is a multiple of D. So, we may even assume that f is H-homogeneous say of weight $w \in K$. Because the functions f_i are rational H-invariants, also all the $f f_i$ are homogeneous of weight $w \in K$. We infer saturatedness of U in $X^{ss}(w)$ from

$$U = \bigcup_{i=1}^r (U \setminus Z(f'_i)) = \bigcup_{i=1}^r X_{ff_i}.$$

Claim 2. For every $w \in \omega_X$, the associated set of semistable points $X^{ss}(w)$ is qp-maximal.

By Claim 1, it suffices to prove that $X^{ss}(w)$ is not contained as a proper H-saturated subset in some set $X^{ss}(w')$. Any H-saturated inclusion $X^{ss}(w) \subseteq X^{ss}(w')$ gives a commutative diagram

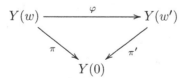

where φ is an open embedding from the quotient space $Y(w)$ into $Y(w')$. Moreover, we know that $\varphi \colon Y(w) \to Y(w')$ is projective and thus φ is an isomorphism. This implies $X^{ss}(w) = X^{ss}(w')$. $\qquad\square$

We prepare the second result. Still K is a finitely generated abelian group, A an integral normal K-graded affine algebra, and we consider the action of the quasitorus $H = \operatorname{Spec} \mathbb{K}[K]$ on the affine variety $X = \operatorname{Spec} A$. With any collection of orbit cones $\Phi \subseteq \Omega_X$, we associated in Definition 3.1.3.5 the subset

$$U(\Phi) = \{x \in X;\ \omega_0 \preceq \omega_x \text{ for some } \omega_0 \in \Phi\} \subseteq X.$$

Conversely, with any H-invariant subset $U \subseteq X$, we associate the following collection of orbit cones

$$\Phi(U) := \{\omega_x;\ x \in U \text{ with } H \cdot x \text{ closed in } U\} \subseteq \Omega_X.$$

Theorem 3.1.4.4 *Let $X = \operatorname{Spec} A$ and $H = \operatorname{Spec} \mathbb{K}[K]$ be as above. Assume that for every H-invariant divisor on X some positive multiple is principal. Then we have mutually inverse bijections*

{maximal bunches of orbit cones in Ω_X} $\longleftrightarrow$ *{$(H, 2)$-maximal subsets of X}*

$$\Phi \mapsto U(\Phi),$$

$$\Phi(U) \leftarrow\!\shortmid U.$$

These bijections are order-reversing maps of partially ordered sets in the sense that we always have

$$\Phi \leq \Phi' \iff U(\Phi) \supseteq U(\Phi').$$

The collections $\Phi(\lambda)$, where $\lambda \in \Lambda(X, H)$, are maximal bunches of orbit cones and they correspond to the qp-maximal subsets of X; in particular, the latter ones are $(H, 2)$-maximal.

Proof According to Proposition 3.1.3.8, for every maximal bunch of orbit cones Φ, the subset $U(\Phi)$ admits a good quotient with an A_2-variety as the quotient space. The following claim gives the converse.

Claim 1. If the H-invariant open set $U \subseteq X$ admits a good quotient $U \to U /\!\!/ H$ with $U /\!\!/ H$ an A_2-variety, then the collection $\Phi(U)$ of orbit cones satisfies 3.1.3.2 (i).

By definition, the elements of $\Phi(U)$ are precisely the orbit cones ω_x, where $H \cdot x$ is a closed subset of U. We have to show that for any two cones $\omega_{x_i} \in \Phi(U)$, their relative interiors intersect nontrivially. Consider the quotient $\pi: U \to U /\!\!/ H$, and let $V \subseteq U /\!\!/ H$ be a common affine neighborhood of $\pi(x_1)$ and $\pi(x_2)$. Then $\pi^{-1}(V)$ is again affine. Thus $X \setminus \pi^{-1}(V)$ is of pure codimension 1 and, by the assumption on X, it is the zero set of a homogeneous function $f \in A$. It follows that the degree of f lies in the relative interior of both cones, ω_{x_1} and ω_{x_2}.

Claim 2. For every collection $\Phi \subseteq \Omega_X$ satisfying 3.1.3.2 (i), we have $\Phi(U(\Phi)) = \Phi$.

Consider any $\omega \in \Phi(U(\Phi))$. By the definition of $\Phi(U(\Phi))$, we have $\omega = \omega_x$ for some $x \in U(\Phi)$ such that $H \cdot x$ is closed in $U(\Phi)$. According to Lemma 3.1.3.10, the latter implies $\omega \in \Phi$. Conversely, let $\omega \in \Phi$. Then we have $x \in U(\Phi)$, for any $x \in X$ with $\omega_x = \omega$. Moreover, Lemma 3.1.3.10 tells us that $H \cdot x$ is closed in $U(\Phi)$. This implies $\omega \in \Phi(U(\Phi))$.

Claim 3. Let $U \subseteq X$ admit a good quotient $U \to U /\!\!/ H$ with an A_2-variety $U /\!\!/ H$, and let $\Phi \subseteq \Omega_X$ be any bunch of orbit cones with $\Phi(U) \subseteq \Phi$. Then we have an H-saturated inclusion $U \subseteq U(\Phi)$.

First let us check that U is in fact a subset of $U(\Phi)$. Given $x \in U$, we may choose $x_0 \in C_X(H \cdot x)$ such that $H \cdot x_0$ is closed in U. By definition of $\Phi(U)$, we have $\omega_{x_0} \in \Phi(U)$, and hence $\omega_{x_0} \in \Phi$. Thus, $\omega_{x_0} \preceq \omega_x$ implies $x \in U(\Phi)$.

To see that the inclusion $U \subseteq U(\Phi)$ is H-saturated, let $x \in U$ with $H \cdot x$ closed in U. We have to show that any $x_0 \in C_X(H \cdot x)$ with $H \cdot x_0$ closed in $U(\Phi)$ belongs to $H \cdot x$. On the one hand, given such x_0, Claim 2 gives us

$$\omega_{x_0} \in \Phi(U(\Phi)) = \Phi.$$

On the other hand, the definition of $\Phi(U)$ yields $\omega_x \in \Phi$, and $x_0 \in C_X(H \cdot x)$ implies $\omega_{x_0} \preceq \omega_x$. Because Φ is a bunch of orbit cones, $\omega_{x_0}^\circ$ and ω_x° intersect nontrivially, and we obtain $\omega_{x_0} = \omega_x$. This gives $x_0 \in H \cdot x$ and Claim 3 is proved.

Now we turn to the assertions of the theorem. First we show that the assignment $\Phi \mapsto U(\Phi)$ is well defined, that is, that $U(\Phi)$ is $(H, 2)$-maximal. Consider any H-saturated inclusion $U(\Phi) \subseteq U$ with an $(H, 2)$-set $U \subseteq X$. Using Claim 2, we obtain

$$\Phi = \Phi(U(\Phi)) \subseteq \Phi(U).$$

By maximality of Φ, this implies $\Phi = \Phi(U)$. Thus, we obtain $U(\Phi) = U(\Phi(U))$. By Claim 3, the latter set contains U as an H-saturated subset. This gives $U(\Phi) = U$ and, consequently, $U(\Phi)$ is $(H, 2)$-maximal.

Thus, we have a well-defined map $\Phi \to U(\Phi)$ from the maximal connected collections in Ω_X to the $(H, 2)$-maximal subsets of X. According to Claim 2, this map is injective. To see surjectivity, consider any $(H, 2)$-maximal $U \subseteq X$. Choose a maximal conneceted collection Φ with $\Phi(U) \subseteq \Phi$. Claim 3 then shows $U = U(\Phi)$. The fact that $\Phi \mapsto U(\Phi)$ and $U \mapsto \Phi(U)$ are inverse to each other is then obvious.

Let us turn to the second statement of the assertion. The subset $U(\Phi')$ is contained in $U(\Phi)$ if and only if any closed H-orbit in $U(\Phi')$ is contained in $U(\Phi)$. By Lemma 3.1.3.10, the points with closed H-orbit in $U(\Phi')$ are precisely the points $x \in X$ with $\omega_x \in \Phi'$. By the definition of $U(\Phi)$, such a point x belongs to $U(\Phi)$ if and only if ω_x has a face contained in Φ.

Finally, for the third statement, we have to show that every set $X^{ss}(w)$ of semistable points is $(H, 2)$-maximal. By what we proved so far, $X^{ss}(w)$ is an H-saturated subset of a set $U(\Phi)$ for some maximal connected collection $\Phi \subseteq \Omega_X$. Thus, we have a diagram of the associated quotient spaces

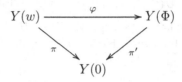

with an open embedding $\varphi \colon Y(w) \to Y(\Phi)$. Because $Y(w) \to Y(0)$ is projective, the morphism $\varphi \colon Y(w) \to Y(\Phi)$ is projective as well. Consequently it is an isomorphism. The claim follows. $\qquad\square$

Corollary 3.1.4.5 *Let $\Phi \subseteq \Omega_X$ be a saturated connected collection. Then the quotient space $U(\Phi)/\!\!/H$ is quasiprojective if and only if we have*

$$\bigcap_{\omega \in \Phi} \omega^{\circ} \neq \emptyset.$$

As a further application of Theorem 3.1.4.4, we obtain a statement in the spirit of [289, Cor. 2.3].

Corollary 3.1.4.6 *Let a quasitorus H act on an irreducible normal affine variety X such that for every H-invariant divisor on X some positive multiple is principal. Moreover, let G be any algebraic group acting on X such that the actions of G and H commute. Then every $(H, 2)$-maximal open subset of X is G-invariant.*

Proof If $U \subseteq X$ is an $(H, 2)$-maximal open subset, then Theorem 3.1.4.4 states that we have $U = U(\Phi)$ for some maximal connected collection Φ of H-orbit cones. Because the actions of H and G commute, the H-orbit cone is constant along G-orbits. Thus, $G \cdot U = U$ holds. $\qquad\square$

3.2 Bunched rings

3.2.1 Bunched rings and their varieties

We present an explicit construction of varieties with prescribed Cox ring. The input is a factorially graded affine algebra R and a collection Φ of pairwise overlapping cones in the grading group K of R. The output variety X has R as its Cox ring and Φ fixes the isomorphy type of X among all varieties sharing R as their Cox ring. We formulate the construction in an elementary manner that turns out to be suitable for explicit applications and requires only minimal background knowledge; the proofs of the basic properties rely on the interpretation in terms of geometric invariant theory (GIT) and are given in Section 3.2.2. We work over an algebraically closed field $\mathbb{K}$ of characteristic zero.

Let K be a finitely generated abelian group and R an integral affine K-graded $\mathbb{K}$-algebra. Recall from Definition 1.5.3.1 that a K-*prime* element in R is a homogeneous nonzero nonunit $f \in R$ such that $f \mid gh$ with homogeneous $g, h \in R$ always implies $f \mid g$ or $f \mid h$. Moreover, we say that R is K-*factorial* or *factorially (K-)graded* if every nonzero homogeneous nonunit $f \in R$ is a product of K-primes. If R is K-factorial, then it admits a system $\mathfrak{F} = (f_1, \ldots, f_r)$ of pairwise nonassociated K-prime generators. By Definition 3.1.1.7, the *projected cone* associated with $\mathfrak{F}$ is $(E \xrightarrow{\varrho} K, \gamma)$, where $E := \mathbb{Z}^r$, the homomorphism $Q \colon E \to K$ sends the ith canonical basis vector $e_i \in E$ to $w_i := \deg(f_i) \in K$ and $\gamma \subseteq E_{\mathbb{Q}}$ is the convex cone generated by $e_1, \ldots, e_r$. Moreover, we introduced in Definition 3.1.1.7 the notion of an $\mathfrak{F}$-*face*: This is a face $\gamma_0 \preceq \gamma$ such that the product over all f_i with $e_i \in \gamma_0$ does not belong to the radical of the ideal $\langle f_j;\ e_j \notin \gamma_0 \rangle \subseteq R$.

Definition 3.2.1.1 Let K be a finitely generated abelian group and R an integral K-graded affine algebra. Moreover, let $\mathfrak{F} = (f_1, \ldots, f_r)$ be a system of pairwise nonassociated K-prime generators for R and $(E \xrightarrow{\varrho} K, \gamma)$ the associated projected cone.

(i) We say that the K-grading of R is *almost free* if for every facet $\gamma_0 \preceq \gamma$ the image $Q(\gamma_0 \cap E)$ generates the abelian group K.
(ii) Let $\Omega_{\mathfrak{F}} = \{Q(\gamma_0);\ \gamma_0 \preceq \gamma\ \mathfrak{F}\text{-face}\}$ denote the collection of projected $\mathfrak{F}$-faces. An $\mathfrak{F}$-*bunch* is a nonempty subset $\Phi \subseteq \Omega_{\mathfrak{F}}$ such that
   a. For any two $\tau_1, \tau_2 \in \Phi$, we have $\tau_1^\circ \cap \tau_2^\circ \neq \emptyset$.
   b. If $\tau_1^\circ \subseteq \tau^\circ$ holds for $\tau_1 \in \Phi$ and $\tau \in \Omega_{\mathfrak{F}}$, then $\tau \in \Phi$ holds.
(iii) We say that an $\mathfrak{F}$-bunch Φ is *true* if for every facet $\gamma_0 \prec \gamma$ the image $Q(\gamma_0)$ belongs to Φ.

We are ready for the central concept of this chapter. We denote by $R_\times^* \subseteq R^*$ the multiplicative group of homogeneous units of R.

Definition 3.2.1.2 A *bunched ring* is a triple $(R, \mathfrak{F}, \Phi)$, where R is an integral, normal, almost freely factorially K-graded affine $\mathbb{K}$-algebra such that $R_\times^* = \mathbb{K}^*$ holds, $\mathfrak{F}$ is a system of pairwise nonassociated K-prime generators for R and Φ is a true $\mathfrak{F}$-bunch.

The following multigraded version of the Proj-construction associates with any bunched ring $(R, \mathfrak{F}, \Phi)$ a variety having R as its Cox ring.

Construction 3.2.1.3 Let $(R, \mathfrak{F}, \Phi)$ be a bunched ring and $(E \xrightarrow{Q} K, \gamma)$ its projected cone. The *collection of relevant faces* and the *covering collection* are

$$\mathrm{rlv}(\Phi) := \{\gamma_0 \preceq \gamma; \ \gamma_0 \text{ an } \mathfrak{F}\text{-face with } Q(\gamma_0) \in \Phi\},$$

$$\mathrm{cov}(\Phi) := \{\gamma_0 \in \mathrm{rlv}(\Phi); \ \gamma_0 \text{ minimal}\}.$$

Consider the action of the quasitorus $H := \mathrm{Spec}\,\mathbb{K}[K]$ on $\overline{X} := \mathrm{Spec}\,R$. We define the localization of $\overline{X}$ with respect to an $\mathfrak{F}$-face $\gamma_0 \preceq \gamma$ to be

$$\overline{X}_{\gamma_0} := \overline{X}_{f_1^{u_1} \dots f_r^{u_r}} \text{ for some } (u_1, \dots, u_r) \in \gamma_0^{\circ}.$$

This does not depend on the particular choice of $(u_1, \dots, u_r) \in \gamma_0^{\circ}$. Moreover, we define an open H-invariant subset of $\overline{X}$ by

$$\widehat{X} := \widehat{X}(R, \mathfrak{F}, \Phi) := \bigcup_{\gamma_0 \in \mathrm{rlv}(\Phi)} \overline{X}_{\gamma_0} = \bigcup_{\gamma_0 \in \mathrm{cov}(\Phi)} \overline{X}_{\gamma_0} = \widehat{X}(\Phi),$$

where $\widehat{X}(\Phi) \subseteq \overline{X}$ is the set associated with the bunch $\Phi \subseteq \Omega_{\mathfrak{F}} = \Omega_{\overline{X}}$ of orbit cones, see Definition 3.1.3.5. Thus, the H-action on $\widehat{X}$ admits a good quotient; we set

$$X := X(R, \mathfrak{F}, \Phi) := \widehat{X}(R, \mathfrak{F}, \Phi) /\!\!/ H$$

and denote the quotient map by $p \colon \widehat{X} \to X$. The affine open subsets $\overline{X}_{\gamma_0} \subseteq \widehat{X}$, where $\gamma_0 \in \mathrm{rlv}(\Phi)$, are H-saturated and their images

$$X_{\gamma_0} := p(\overline{X}_{\gamma_0}) \subseteq X$$

form an affine cover of X. Moreover, every member f_i of $\mathfrak{F}$ defines a prime divisor $D_X^i := p(V(\widehat{X}, f_i))$ on X.

Theorem 3.2.1.4 Let $\widehat{X} := \widehat{X}(R, \mathfrak{F}, \Phi)$ and $X := X(R, \mathfrak{F}, \Phi)$ arise from a bunched ring $(R, \mathfrak{F}, \Phi)$. Then X is an irreducible normal A_2-variety with

$$\dim(X) = \dim(R) - \dim(K_{\mathbb{Q}}), \qquad \Gamma(X, \mathcal{O}^*) = \mathbb{K}^*,$$

there is an isomorphism $\mathrm{Cl}(X) \to K$ sending $[D_X^i]$ to $\deg(f_i)$, the map $p \colon \widehat{X} \to X$ is a characteristic space, and the Cox ring $\mathcal{R}(X)$ is isomorphic to R.

Let us illustrate this with a couple of examples. The first one shows how toric varieties fit into the picture of bunched rings.

Example 3.2.1.5 (Bunched polynomial rings)   Set $R := \mathbb{K}[T_1, \dots, T_r]$, let K be a finitely generated abelian group and assume that R is almost freely K-graded via $\deg(T_i) = w_i$ with $w_1, \dots, w_r \in K$. Then R is factorial and has $\mathfrak{F} := (T_1, \dots, T_r)$ as a system of pairwise nonassociated homogeneous K-prime generators. Every face $\gamma_0 \preceq \gamma$ is an $\mathfrak{F}$-face and thus a true $\mathfrak{F}$-bunch is nothing but a true $\mathcal{W}$-bunch for the vector configuration $\mathcal{W} = (w_1, \dots, w_r)$

in $K_{\mathbb{Q}}$ in the sense of Definition 2.2.1.10. The variety associated with such a bunched ring $(R, \mathfrak{F}, \Phi)$ is the toric variety X associated with the $\mathcal{W}$-bunch Φ and its fan $\Sigma = \Phi^\sharp$ is obtained from Φ via Gale duality as described in Theorem 2.2.1.14. Note that the sets $X_{\gamma_0} \subseteq X$ are precisely the affine toric charts, where the cones $\gamma_0 \in \mathrm{cov}(\Phi)$ provide the maximal ones. Moreover, the divisors D_X^i are the toric prime divisors.

Example 3.2.1.6 (A singular del Pezzo surface) Set $K := \mathbb{Z}^2$ and consider the K-grading of $\mathbb{K}[T_1, \ldots, T_5]$ defined by $\deg(T_i) := w_i$, where w_i is the ith column of the matrix

$$Q := \begin{bmatrix} 1 & -1 & 0 & -1 & 1 \\ 1 & 1 & 1 & 0 & 2 \end{bmatrix}.$$

Then this K-grading descends to a K-grading of the following residue algebra, which is factorial due to Proposition 3.2.4.1:

$$R := \mathbb{K}[T_1, \ldots, T_5] / \langle T_1 T_2 + T_3^2 + T_4 T_5 \rangle.$$

The classes $f_i \in R$ of $T_i \in \mathbb{K}[T_1, \ldots, T_5]$, where $1 \le i \le 5$, form a system $\mathfrak{F}$ of pairwise nonassociated K-prime generators of R. We have

$$E = \mathbb{Z}^5, \qquad \gamma = \mathrm{cone}(e_1, \ldots, e_5)$$

and the K-grading is almost free. The $\mathfrak{F}$-faces can be directly computed using the definition; see also Remark 3.2.4.6; writing $\gamma_{i_1,\ldots,i_k} := \mathrm{cone}(e_{i_1}, \ldots, e_{i_k})$, they are given as

$$\{0\}, \; \gamma_1, \; \gamma_2, \; \gamma_4, \; \gamma_5, \; \gamma_{1,4}, \; \gamma_{1,5}, \; \gamma_{2,4}, \; \gamma_{2,5}, \; \gamma_{1,2,3}, \; \gamma_{3,4,5},$$

$$\gamma_{1,2,3,4}, \; \gamma_{1,2,3,5}, \; \gamma_{1,2,4,5}, \; \gamma_{1,3,4,5}, \; \gamma_{2,3,4,5}, \; \gamma_{1,2,3,4,5}.$$

In particular, we see that there is precisely one true maximal $\mathfrak{F}$-bunch Φ; it has $\tau := \mathrm{cone}(w_2, w_5)$ as its unique minimal cone.

Note that $\Phi = \Phi(w_3)$ is the bunch arising from $w_3 \in \tau^\circ$ as in Example 3.1.3.6. The collection of relevant faces and the covering collection are

$$\mathrm{rlv}(\Phi) =$$

$$\{\gamma_{1,4}, \gamma_{2,5}, \gamma_{1,2,3}, \gamma_{3,4,5}, \gamma_{1,2,3,4}, \gamma_{1,2,3,5}, \gamma_{1,2,4,5}, \gamma_{1,3,4,5}, \gamma_{2,3,4,5}, \gamma_{1,2,3,4,5}\},$$

$$\mathrm{cov}(\Phi) = \{\gamma_{1,4}, \gamma_{2,5}, \gamma_{1,2,3}, \gamma_{3,4,5}\}.$$

The open set $\widehat{X}(R, \mathfrak{F}, \Phi)$ in $\overline{X} = V(\mathbb{K}^5; T_1T_2 + T_3^2 + T_4T_5)$ equals $\overline{X}^{ss}(w_3)$ and is the union of four affine charts:

$$\widehat{X}(R, \mathfrak{F}, \Phi) = \overline{X}_{f_1f_4} \cup \overline{X}_{f_2f_5} \cup \overline{X}_{f_1f_2f_3} \cup \overline{X}_{f_3f_4f_5}.$$

Because $\widehat{X}(R, \mathfrak{F}, \Phi)$ is a set of semistable points, the resulting variety $X = X(R, \mathfrak{F}, \Phi)$ is projective. Moreover, we have

$$\dim(X) = 2, \qquad \mathrm{Cl}(X) = K, \qquad \mathcal{R}(X) = R.$$

In fact, the methods presented later show that X is a $\mathbb{Q}$-factorial Gorenstein del Pezzo $\mathbb{K}^*$-surface with one singularity, of type A_2, see 3.2.5.8, 3.3.1.14, 3.3.2.10, 3.3.3.5 and 5.4.3.4.

Example 3.2.1.7 (The smooth del Pezzo surface of degree 5) Consider the polynomial ring $A(2, 5) := \mathbb{K}[T_{ij}; 1 \le i < j \le 5]$ and the ideal $I(2, 5) \subseteq A(2, 5)$ generated by the Plücker relations:

$$T_{12}T_{34} - T_{13}T_{24} + T_{14}T_{23}, \qquad T_{12}T_{35} - T_{13}T_{25} + T_{15}T_{23},$$

$$T_{12}T_{45} - T_{14}T_{25} + T_{15}T_{24}, \qquad T_{13}T_{45} - T_{14}T_{35} + T_{15}T_{34},$$

$$T_{23}T_{45} - T_{24}T_{35} + T_{25}T_{34}.$$

The ring $R := A(2, 5)/I(2, 5)$ is factorial [264, Prop. 8.5], and the classes $f_{ij} \in R$ of T_{ij} define a system $\mathfrak{F} = (f_{ij})$ of pairwise nonassociated prime generators. The Plücker relations are homogeneous w.r.t. the grading by $K = \mathbb{Z}^5$, which associates with T_{ij} the ijth column of the matrix

$$Q = \begin{bmatrix} 0 & 0 & 0 & 0 & 1 & 1 & 1 & 1 & 1 & 1 \\ 1 & 0 & 0 & 0 & -1 & -1 & -1 & 0 & 0 & 0 \\ 0 & 1 & 0 & 0 & -1 & 0 & 0 & -1 & -1 & 0 \\ 0 & 0 & 1 & 0 & 0 & -1 & 0 & -1 & 0 & -1 \\ 0 & 0 & 0 & 1 & 0 & 0 & -1 & 0 & -1 & -1 \end{bmatrix}$$

$$\quad\; 12 \quad 13 \quad 14 \quad 15 \quad 23 \quad 24 \quad 25 \quad 34 \quad 35 \quad 45$$

In particular, R is K-graded. These data describe the cone over the Grassmannian $G(2, 5)$ with an effective action of the five-dimensional torus. The half sum $w = (3, -1, -1, -1, -1)$ of the columns of Q defines a true $\mathfrak{F}$-bunch $\Phi = \Phi(w)$ and hence we have a bunched ring $(R, \mathfrak{F}, \Phi)$. As we will see in Example 4.1.4.1, the associated variety $X(R, \mathfrak{F}, \Phi)$ is the smooth del Pezzo surface of degree 5, that is, the blow-up of $\mathbb{P}^2$ in four points in general position; compare also [277, Prop. 3.2].

We say that a variety X is A_2-*maximal* if it has the A_2-property and admits no big open embedding $X \subsetneq X'$ into an A_2-variety X', where *big* means that

$X' \setminus X$ is of codimension at least 2 in X'. Moreover, we call an $\mathfrak{F}$-bunch *maximal* if it cannot be enlarged by adding further projected $\mathfrak{F}$-faces.

Proposition 3.2.1.8 *Let X arise from a bunched ring $(R, \mathfrak{F}, \Phi)$. Then X is A_2-maximal if and only if Φ is maximal.*

The following statement shows in particular that every irreducible normal A_2-variety with finitely generated Cox ring can be realized as a big open subset in some A_2-maximal one and that the latter ones are obtained by Construction 3.2.1.3.

Theorem 3.2.1.9 *Let X be an irreducible normal A_2-variety with $\Gamma(X, \mathcal{O}^*) = \mathbb{K}^*$, finitely generated divisor class group $K := \mathrm{Cl}(X)$ and finitely generated Cox ring $R := \mathcal{R}(X)$. Let $\mathfrak{F}$ be any finite system of pairwise nonassociated K-prime generators for R.*

(i) *There exist a maximal $\mathfrak{F}$-bunch Φ and a big open embedding $X \to X(R, \mathfrak{F}, \Phi)$.*
(ii) *If X is A_2-maximal, then $X \cong X(R, \mathfrak{F}, \Phi)$ holds with some maximal $\mathfrak{F}$-bunch Φ.*

Corollary 3.2.1.10 *Let X be an irreducible normal complete A_2-variety with finitely generated divisor class group and finitely generated Cox ring. Then $X \cong X(R, \mathfrak{F}, \Phi)$ holds with some bunched ring $(R, \mathfrak{F}, \Phi)$.*

Corollary 3.2.1.11 *Let X be an irreducible normal projective variety with finitely generated divisor class group and finitely generated Cox ring. Then $X \cong X(R, \mathfrak{F}, \Phi)$ holds with some bunched ring $(R, \mathfrak{F}, \Phi)$.*

3.2.2 Proofs to Section 3.2.1

We enter the detailed discussion of Construction 3.2.1.3. First we consider almost free gradings and show that this notion is indeed a property of the grading and does not depend on the choice of a system $\mathfrak{F}$ of generators. The ground field $\mathbb{K}$ is algebraically closed and of characteristic zero.

Construction 3.2.2.1 *Let K be a finitely generated abelian group, R an integral normal factorially K-graded affine algebra with $R_\times^* = \mathbb{K}^*$, and $\mathfrak{F} = (f_1, \ldots, f_r)$ any system of pairwise nonassociated K-prime generators for R. Consider the action of $H = \mathrm{Spec}\, \mathbb{K}[K]$ on $\overline{X} = \mathrm{Spec}\, R$ and set*

$$\widehat{X}(\mathfrak{F}) := \overline{X} \setminus \bigcup_{i \neq j} V(\overline{X}; f_i, f_j) = \bigcup_{i=1}^{r} \overline{X}_{f_1 \cdots f_{i-1} f_{i+1} \cdots f_r} \subseteq \overline{X}.$$

Then the subset $\widehat{X}(\mathfrak{F}) \subseteq \overline{X}$ is open, H-invariant and its complement $\overline{X} \setminus \widehat{X}(\mathfrak{F})$ is of codimension at least two in $\overline{X}$. In terms of Construction 3.2.1.3, the set $\widehat{X}(\mathfrak{F})$ is the union of the localizations $\overline{X}_{\gamma_0}$, where $\gamma_0 \preceq \gamma$ is a facet.

Proposition 3.2.2.2 *Let K be a finitely generated abelian group and R an integral normal factorially K-graded affine $\mathbb{K}$-algebra with $R_\times^* = \mathbb{K}^*$. Moreover, let $\mathfrak{F} = (f_1, \ldots, f_r)$ be a system of pairwise nonassociated K-prime generators for R with associated projected cone $(E \xrightarrow{\varrho} K, \gamma)$. Then the following statements are equivalent.*

(i) *For every facet $\gamma_0 \preceq \gamma$, the image $Q(\gamma_0 \cap E)$ generates the abelian group K.*
(ii) *H acts freely on $\widehat{X}(\mathfrak{F})$.*
(iii) *H acts freely on an invariant open subset $W \subseteq \overline{X}$ with $\operatorname{codim}(\overline{X} \setminus W) \geq 2$.*

Proof Suppose that (i) holds. By construction of $\widehat{X}(\mathfrak{F})$, we find for every $x \in \widehat{X}(\mathfrak{F})$ some $1 \leq i \leq r$ such that $f_j(x) \neq 0$ holds for all $j \neq i$. By (i) the weights $w_j = \deg(f_j)$, where $j \neq i$, generate the abelian group K. Thus, Proposition 1.2.2.8 tells us that the isotropy group H_x is trivial, which gives (ii). The implication "(ii)⇒(iii)" is obvious. Finally, suppose that (iii) holds. Because $W \subseteq \overline{X}$ is big and the f_i are pairwise nonassociated K-primes, we find for every $1 \leq i \leq r$ a point $x \in W$ with $f_i(x) = 0$ but $f_j(x) \neq 0$ whenever $j \neq i$. Clearly, the weights w_j with $j \neq i$ generate the orbit group K_x. Because H_x is trivial, we infer $K_x = K$ from Proposition 1.2.2.8. □

We turn to $\mathfrak{F}$-bunches. The following observation enables us to apply the geometric invariant theory for quasitorus actions on affine varieties developed in the preceding section.

Remark 3.2.2.3 Let K be a finitely generated abelian group and R an integral normal factorially K-graded affine algebra with $R_\times^* = \mathbb{K}^*$. Moreover, let $\mathfrak{F} = (f_1, \ldots, f_r)$ be a system of pairwise nonassociated K-prime generators for R and $(E \xrightarrow{\varrho} K, \gamma)$ the associated projected cone. By Proposition 3.1.1.10, the projected $\mathfrak{F}$-faces are precisely the orbit cones of the action $H = \operatorname{Spec} \mathbb{K}[K]$ on $\overline{X} = \operatorname{Spec} R$. Thus, the (maximal) $\mathfrak{F}$-bunches are precisely the (maximal) bunches of orbit cones. According to Proposition 3.1.3.8, the $\mathfrak{F}$-bunch Φ determines a good H-set in $\overline{X}$; concretely this set is given as

$$\widehat{X}(\Phi) := \{x \in \overline{X}; \ \omega_0 \in \Phi \text{ for some } \omega_0 \preceq \omega_x\} \subseteq \overline{X}.$$

According to Theorem 3.1.4.4, the set $\widehat{X}(\Phi)$ is $(H, 2)$-maximal if and only if Φ is maximal. Moreover, the closed H-orbits of $\widehat{X}(\Phi)$ are precisely the orbits $H \cdot x \subseteq \overline{X}$ with $\omega_x \in \Phi$.

Note that the interpretation in terms of orbit cones shows that an $\mathfrak{F}$-bunch does not depend on the particular choice of the system of generators $\mathfrak{F}$. The following two statements comprise in particular the assertions made in Construction 3.2.1.3.

Proposition 3.2.2.4 *Consider the situation of Construction 3.2.1.3. The good H-subset $\widehat{X}(\Phi) \subseteq \overline{X}$ associated with Φ satisfies*

$$\widehat{X}(\Phi) = \bigcup_{\gamma_0 \in \mathrm{rlv}(\Phi)} \overline{X}_{\gamma_0} = \widehat{X}(R, \mathfrak{F}, \Phi).$$

Moreover, the localizations $\overline{X}_{\gamma_0} \subseteq \overline{X}$, where $\gamma_0 \in \mathrm{rlv}(\Phi)$, are H-saturated subsets of $\widehat{X}(\Phi)$.

Proof Consider $z \in \widehat{X}(\Phi)$ with $H \cdot z$ closed in $\widehat{X}(\Phi)$. By Lemma 3.1.3.10, the orbit cone ω_z belongs to Φ. In terms of $\mathfrak{F} = (f_1, \ldots, f_r)$ and its projected cone $(E \xrightarrow{\varrho} K, \gamma)$ we have

$$\omega_z = \mathrm{cone}(\deg(f_i); \ f_i(z) \neq 0) = Q(\gamma_0), \quad \text{where } \gamma_0 := \mathrm{cone}(e_i; \ f_i(z) \neq 0).$$

This shows that $\overline{X}_{\gamma_0}$ is a neighborhood $U(z)$ as considered in Construction 3.1.3.7. In particular, we have $\overline{X}_{\gamma_0} \subseteq \widehat{X}(\Phi)$, and Proposition 3.1.3.8 ensures that this inclusion is H-saturated. Going through the points with closed orbit in $\widehat{X}$, we see that $\widehat{X}(\Phi)$ is a union of certain H-saturated subsets $\overline{X}_{\gamma_0}$ with $\gamma_0 \in \mathrm{rlv}(\Phi)$.

To conclude the proof we have to show that in fact every $\overline{X}_{\gamma_0}$ with $\gamma_0 \in \mathrm{rlv}(\Phi)$ is an H-saturated subset of $\widehat{X}(\Phi)$. Given $\gamma_0 \in \mathrm{rlv}(\Phi)$, choose a point $z \in \overline{X}$ satisfying $f_i(z) \neq 0$ if and only if $e_i \in \gamma_0$. Then we have $\omega_z = Q(\gamma_0)$. This implies $z \in \widehat{X}(\Phi)$ and Lemma 3.1.3.10 says that $H \cdot z$ is closed in $\widehat{X}(\Phi)$. Thus, the preceding consideration shows that $\overline{X}(\gamma_0)$ is an H-saturated subset of $\widehat{X}(\Phi)$. □

Proposition 3.2.2.5 *Let K be a finitely generated abelian group, R an integral normal almost freely and factorially K-graded affine $\mathbb{K}$-algebra with $R^*_\times = \mathbb{K}^*$, and $\mathfrak{F} = (f_1, \ldots, f_r)$ a system of pairwise nonassociated K-prime generators for R. Then, for any $\mathfrak{F}$-bunch Φ, the following statements are equivalent.*

(i) *The $\mathfrak{F}$-bunch Φ is true.*
(ii) *We have an H-saturated inclusion $\widehat{X}(\mathfrak{F}) \subseteq \widehat{X}(\Phi)$.*

Moreover, if $(R, \mathfrak{F}, \Phi)$ is a bunched ring, then, in the notation of Construction 3.2.1.3, every $D^i_X = p(V(\widehat{X}, f_i))$ is a prime divisor on X.

Proof If Φ is a proper $\mathfrak{F}$-bunch, then Proposition 3.2.2.4 ensures that we have H-saturated inclusions $\overline{X}_{\gamma_0} \subseteq \widehat{X}(\Phi)$, where $\gamma_0 \subseteq \gamma$ is a facet. Thus,

$\widehat{X}(\mathfrak{F}) \subseteq \widehat{X}(\Phi)$ is H-saturated. Conversely, if $\widehat{X}(\mathfrak{F}) \subseteq \widehat{X}(\Phi)$ is H-saturated, then we look at points $z_i \in \widehat{X}(\mathfrak{F})$ with $f_i(z_i) = 0$ and $f_j(z_i) \neq 0$ for all $j \neq i$. The orbits $H \cdot z_i$ are closed in $\widehat{X}(\Phi)$ and thus the corresponding orbit cones ω_i belong to Φ; see Theorem 3.1.4.4. In the setting of Definition 3.2.1.1, the orbit cones ω_i are exactly the Q-images of the facets of γ. It follows that the $\mathfrak{F}$-bunch Φ is proper. The last statement is then clear because the restriction $p \colon \widehat{X}(\mathfrak{F}) \to p(\widehat{X}(\mathfrak{F}))$ is a geometric quotient for a free H-action. $\qquad\square$

Proof of Theorem 3.2.1.4 According to Proposition 3.2.2.4, the good quotient $p \colon \widehat{X} \to X$ exists and X is an A_2-variety. Proposition 3.2.2.2 tells us that H acts freely on $\widehat{X}(\mathfrak{F})$ and by Proposition 3.2.2.5 we have an H-saturated inclusion $\widehat{X}(\mathfrak{F}) \subseteq \widehat{X}$. Thus, the action of H on $\widehat{X}$ is strongly stable. We conclude that $X = \widehat{X} /\!\!/ H$ is of dimension $\dim(R) - \dim(H)$. Moreover, $R_\times^* = \mathbb{K}^*$ implies $\Gamma(X, \mathcal{O}^*) = \mathbb{K}^*$. Proposition 1.6.4.5 provides the desired isomorphism $\mathrm{Cl}(X) \cong K$ and Theorem 1.6.4.3 shows that $p \colon \widehat{X} \to X$ is a characteristic space. The latter implies $\mathcal{R}(X) \cong R$. $\qquad\square$

Proof of Proposition 3.2.1.8 Let X be A_2-maximal. If Φ were not maximal, then we would have $\Phi \subsetneq \Phi'$ with some $\mathfrak{F}$-bunch Φ'. This gives an H-saturated inclusion $\widehat{X}(\Phi) \subsetneq \widehat{X}(\Phi')$ and thus an open embedding $X \subsetneq X'$ of the quotient varieties. Because $\widehat{X}(\Phi)$ is big in $\widehat{X}(\Phi')$, also X is big in X', which is contradiction.

Now let Φ be maximal. Consider a big open embedding $X \subseteq X'$ into an A_2-variety. Replacing, if necessary, X' with its normalization, we may assume that X' is normal. Then X and X' share the same Cox ring R and thus occur as good quotients of open subsets $\widehat{X} \subseteq \widehat{X}'$ of their common total coordinate space $\overline{X}$. By $(H, 2)$-maximality of $\widehat{X} \subseteq \overline{X}$, we obtain $\widehat{X} = \widehat{X}'$ and thus $X = X'$. $\qquad\square$

Proof of Theorem 3.2.1.9 Theorem 1.5.1.1 says that the Cox ring R is integral and normal. Moreover, Proposition 1.5.2.5 (i) ensures $R_\times^* = \mathbb{K}^*$. By Theorem 1.5.3.7, the Cox ring R is factorially K-graded. Consider the corresponding total coordinate space $\overline{X} = \mathrm{Spec}\, R$ with its action of $H = \mathrm{Spec}\, \mathbb{K}[K]$ and the characteristic space $p_X \colon \widehat{X} \to X$, which is a good quotient for the action of H. By Proposition 1.6.1.6, we have a small complement $\overline{X} \setminus p_X^{-1}(X')$, where $X' \subseteq X$ denotes the set of smooth points, and H acts freely on $p_X^{-1}(X')$. Proposition 3.2.2.2 thus tells us that the K-grading is almost free. Next observe that $\widehat{X}$ is an H-saturated subset of some $(H, 2)$-maximal subset of $\overline{X}$. According to Theorem 3.1.4.4, the latter is of the form $\widehat{X}(\Phi)$ with a maximal $\mathfrak{F}$-bunch $\Phi \subseteq \Omega_{\mathfrak{F}} = \Omega_{\overline{X}}$. Propositions 3.2.2.2 and 3.2.2.5 show that Φ is true. Assertions (i) and (ii) of the theorem follow. $\qquad\square$

3.2.3 Example: Flag varieties

We show how flag varieties fit into the language of bunched rings. As before, we work over an algebraically closed field $\mathbb{K}$ of characteristic zero. Let G be a connected affine algebraic group. Recall that a *Borel subgroup* $B \subseteq G$ is a maximal connected solvable subgroup of G and, more generally, a *parabolic subgroup* $P \subseteq G$ is a subgroup containing some Borel subgroup. The homogeneous space G/P is a smooth projective variety, called a *flag variety*. The following example explains this name.

Example 3.2.3.1 Consider the special linear group SL_n. The subgroup $B_n \subseteq \mathrm{SL}_n$ of upper triangular matrices is a Borel subgroup; it is the stabilizer of the standard complete flag

$$\mathfrak{f}_n = \mathbb{K}^1 \subset \ldots \subset \mathbb{K}^{n-1} \in \mathrm{Gr}(1, n) \times \ldots \times \mathrm{Gr}(n-1, n)$$

under the diagonal SL_n-action on the product of Grassmannians. Thus, $\mathrm{SL}_n/B_n \cong \mathrm{SL}_n \cdot \mathfrak{f}_n$ is the set of all complete flags. Similarly, any sequence $0 < d_1 < d_2 < \ldots < d_s < n$ defines a parabolic $P_{d_1,\ldots,d_s} \subseteq \mathrm{SL}_n$, namely the stabilizer of the partial flag

$$\mathfrak{f}_{d_1,\ldots,d_s} = \mathbb{K}^{d_1} \subset \ldots \subset \mathbb{K}^{d_s} \in \mathrm{Gr}(d_1, n) \times \ldots \times \mathrm{Gr}(d_s, n)$$

and the possible flag varieties SL_n/P are precisely the SL_n-orbits $\mathrm{SL}_n \cdot \mathfrak{f}_{d_1,\ldots,d_s}$. If $s = 1$ and $d_1 = k$ holds, then $P_k \subseteq \mathrm{SL}_n$ is a maximal parabolic subgroup, and G/P_k is nothing but the Grassmannian $\mathrm{Gr}(k, n)$.

Example 3.2.3.2 In SL_3 we have two maximal parabolic subgroups P_1 and P_2 with $B_3 = P_1 \cap P_2$, namely

$$P_1 = \left\{ \begin{pmatrix} * & * & * \\ 0 & * & * \\ 0 & * & * \end{pmatrix} \right\}, \qquad P_2 = \left\{ \begin{pmatrix} * & * & * \\ * & * & * \\ 0 & 0 & * \end{pmatrix} \right\}.$$

The flag varieties SL_3/P_1 and SL_3/P_2 both are isomorphic to $\mathbb{P}^2$, while the variety of complete flags SL_3/B_3 is given in $\mathbb{P}^2 \times \mathbb{P}^2$ as

$$\{([x_0, x_1, x_2], [y_0, y_1, y_2]) ; \ x_0 y_0 + x_1 y_1 + x_2 y_2 = 0\}.$$

We determine the characteristic space of a flag variety G/P for simply connected semisimple G. The following observation is an important ingredient.

Proposition 3.2.3.3 *Let a connected affine algebraic group G with $\mathbb{X}(G) = \{1\}$ act rationally by means of algebra automorphisms on an affine $\mathbb{K}$-algebra A with $A^* = \mathbb{K}^*$. If A is factorial, then the algebra of invariants A^G is factorial as well.*

Proof Given a nonzero nonunit $a \in A^G$, consider its decomposition into primes $a = a_1^{\nu_1} \cdots a_r^{\nu_r}$ in A. Then, for every $g \in G$, one has

$$a = g \cdot a = (g \cdot a_1)^{\nu_1} \cdots (g \cdot a_r)^{\nu_r}.$$

Using $A^* = \mathbb{K}^*$, we see that g permutes the a_i up to multiplication by a constant. Because G is connected and $\mathbb{X}(G) = \{1\}$ holds, we can conclude $a_1, \ldots, a_r \in A^G$. $\square$

Now consider a parabolic subgroup $P \subseteq G$ and let $P' \subseteq P$ denote its commutator group. Then $H := P/P'$ is a torus and it acts freely via multiplication from the right on the homogeneous space G/P'. The canonical map $G/P' \to G/P$ is a geometric quotient for this H-action.

Proposition 3.2.3.4 *Let G be a simply connected semisimple affine algebraic group and $P \subseteq G$ a parabolic subgroup. Then we have* $\mathrm{Cl}(G/P) \cong \mathbb{X}(H)$ *and* $G/P' \overset{/H}{\longrightarrow} G/P$ *is a characteristic space.*

Proof By the assumptions on G, we have $\Gamma(G, \mathcal{O}^*) = \mathbb{K}^*$ and the algebra $\Gamma(G, \mathcal{O})$ is factorial; see [190, Prop. 1.2,] and [189, Prop. 4.6,]. Moreover, P' is connected and has a trivial character group. Consider the action of P' on G by multiplication from the right. Using Chevalley's theorem [178, Thm. 11.2], we see that G/P' is a quasiaffine variety. Proposition 3.2.3.3 applied to the induced representation of P' on $\Gamma(G, \mathcal{O})$ shows that $\Gamma(G/P', \mathcal{O})$ is factorial. This gives $\mathrm{Cl}(G/P') = 0$. The assertion now follows from Proposition 1.6.4.5 and Theorem 1.6.4.3. $\square$

Let us express this in terms of bunched rings. We first consider the case of a Borel subgroup $B \subseteq G$. Then the commutator $B' \subseteq B$ is a maximal unipotent subgroup $U \subseteq G$ and some maximal torus $T \subseteq G$ projects isomorphically onto $H = B/B'$. In particular, we have $B = TU$.

To proceed, we recall some basic representation theory; we refer to [143] for details. For every simple (finite-dimensional, rational) G-module V, the subspace V^U of U-invariant vectors is one-dimensional, and T acts on V^U by a character $\mu_V \in \mathbb{X}(T)$, called the *highest weight* of V. The set of the highest weights of the simple G-modules is a submonoid $X_+(G) \subseteq \mathbb{X}(T)$ and $V \mapsto \mu_V$ induces a bijection

$$\{\text{isomorphism classes of simple } G\text{-modules}\} \longrightarrow X_+(G).$$

The elements of $X_+(G)$ are called *dominant weights* of the group G (with respect to the pair (B, T)). The cone $C_+ \subseteq \mathbb{X}(T)_{\mathbb{Q}}$ generated by $X_+(G) \subseteq \mathbb{X}(T)$ is called the *positive Weyl chamber*. The intersection $C_+ \cap \mathbb{X}(T)$ coincides with $X_+(G)$. The monoid $X_+(G)$ has a unique system of free generators ϖ_i, $1 \leq i \leq s$, where s is the rank of the lattice $\mathbb{X}(T)$; the ϖ_i are called the *fundamental weights* of G.

The algebra $\Gamma(G/U, \mathcal{O})$ comes with the structure of a rational G-module induced from the left G-action on G/U. For every $\mu \in X_+(G)$, there is a unique simple G-submodule $V(\mu) \subseteq \Gamma(G/U, \mathcal{O})$ having μ as highest weight, see [301, Thm. 3.12]; in other words, we have the isotypic decomposition

$$\Gamma(G/U, \mathcal{O}) = \bigoplus_{\mu \in X_+(G)} V(\mu).$$

The T-action on $\Gamma(G/U, \mathcal{O})$ coming from right T-multiplication on G/U induces scalar multiplication on every submodule $V(\mu)$ defined by the highest weight μ^* of the dual G-module $V(\mu)^*$. Thus, the above isotypic decomposition is a G-equivariant $X_+(G)$-grading which we consider as a $\mathbb{X}(T)$-grading; note that $X_+(G)$ generates $\mathbb{X}(T)$ as a lattice.

We are ready to turn $R := \Gamma(G/U, \mathcal{O})$ into a bunched ring. It is factorial, graded by $K := \mathbb{X}(T)$, and we have $R^* = \mathbb{K}^*$. For every fundamental weight $\varpi_i \in X_+(G)$ fix a basis $\mathfrak{F}_i = (f_{ij}, 1 \le j \le s_i)$ of $V(\varpi_i)$ and put all these bases together to a family $\mathfrak{F} = (f_{ij})$. Finally, set $\Phi := \{C_+\}$.

Proposition 3.2.3.5 *The triple $(R, \mathfrak{F}, \Phi)$ is a bunched ring having the flag variety G/B as its associated variety.*

In the proof and also later, we make use of the following simple criterion for K-primality.

Lemma 3.2.3.6 *Let K be finitely generated abelian group and A be a factorially K-graded $\mathbb{K}$-algebra. If $w \in S(A)$ is indecomposable in $S(A)$, then every $0 \ne f \in A_w$ is K-prime.*

Proof of Proposition 3.2.3.5 First recall that the fundamental weights ϖ_i generate the weight monoid $X_+(G)$. Next note that for any two dominant weights $\mu, \mu' \in X_+(G)$ the G-module $V(\mu + \mu')$ is simple and thus we have a surjective multiplication map

$$V(\mu) \times V(\mu') \rightarrow V(\mu + \mu').$$

Consequently, the f_{ij} generate R. By Lemma 3.2.3.6, they are K-prime elements and clearly they are pairwise nonassociated. Because each $V(\varpi_i)$ is of dimension at least 2, Φ is a true $\mathfrak{F}$-bunch. Moreover, Φ is the only possible true $\mathfrak{F}$-bunch and thus the associated open subset of Spec R must be G/B'. □

Remark 3.2.3.7 The total coordinate space Spec R of G/B admits an explicit realization as a G-orbit closure in a representation space; see [150, Sec. 5]:

$$\mathrm{Spec}\, \Gamma(G/U, \mathcal{O}) \cong \overline{G \cdot (v_{\varpi_1}, \ldots, v_{\varpi_r})} \subseteq V(\varpi_1) \oplus \ldots \oplus V(\varpi_r).$$

We turn to arbitrary flag varieties. Fix a Borel subgroup $B \subseteq G$ and a maximal torus $T \subseteq B$. Let $C \preceq C_+$ be a face of the positive Weyl chamber. Set

$K_C := K \cap \mathrm{lin}(C)$ and consider the associated Veronese subalgebra

$$R_C = \bigoplus_{u \in K_C} R_u.$$

Let $\mathfrak{F}_C$ denote the system of generators obtained by putting together the bases $\mathfrak{F}_i$ with $\varpi_i \in C$. Moreover, set $\Phi_C := \{C\}$.

Proposition 3.2.3.8 *The triple $(R_C, \mathfrak{F}_C, \Phi_C)$ is a bunched ring with associated variety G/P_C, where $P_C \subseteq G$ is the parabolic subgroup defined by the index set $\{i; \varpi_i \in C\}$. Moreover, there is a commutative diagram*

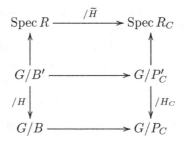

Finally, given any parabolic subgroup $P \subseteq G$, the associated flag variety G/P is isomorphic to some G/P_C.

Proof of Proposition 3.2.3.8 Using [301, Thm. 3.12], one verifies that R_C is the ring of functions of G/P'_C. Then the statement follows from Propositions 3.2.3.4 and 3.2.3.5. The last assertion is due to the fact that any parabolic $P \subseteq G$ is conjugate to some P_C, see [178, Thm. 30.1]. $\square$

Example 3.2.3.9 Consider $G = \mathrm{SL}_n$ and an extremal ray $C = \mathrm{cone}(\varpi_k)$ of the positive Weyl chamber C_+. Then $P_C = P_k$ holds, G/P_C is the Grassmannian $\mathrm{Gr}(k, n)$, and the total coordinate space $\mathrm{Spec}\, R_C$ is the affine cone over the Plücker embedding of $\mathrm{Gr}(k, n)$ with the standard $\mathbb{Z}_{\geq 0}$-grading.

Example 3.2.3.10 Consider again the special linear group SL_3 and the Borel subgroup $B_3 \subseteq \mathrm{SL}_3$. Then the commutator $U = B'_3$ and a maximal torus with $B = TU$ are

$$U = \left\{ \begin{pmatrix} 1 & * & * \\ 0 & 1 & * \\ 0 & 0 & 1 \end{pmatrix} \right\}, \qquad T = \left\{ \begin{pmatrix} t_1 & 0 & 0 \\ 0 & t_2 & 0 \\ 0 & 0 & t_3 \end{pmatrix}; t_1 t_2 t_3 = 1 \right\}.$$

To determine SL_3/U explicitly, consider the SL_3-module $\mathbb{K}^3 \times (\mathbb{K}^3)^*$, where SL_3 acts canonically on $\mathbb{K}^3$ and $(\mathbb{K}^3)^*$ is the dual module. The point (e_1, e_3^*) has U as its stabilizer, and its orbit closure is the affine quadric

$$Z := \overline{\mathrm{SL}_3 \cdot (e_1, e_3^*)} = V(\mathbb{K}^3 \times (\mathbb{K}^3)^*; X_1 Y_1 + X_2 Y_2 + X_3 Y_3).$$

In fact, $SL_3/U \cong SL_3 \cdot (e_1, e_3^*)$ is obtained from Z by removing the coordinate subspaces $V(Z; X_1, X_2, X_3)$ and $V(Z; Y_1, Y_2, Y_3)$. The T-action on SL_3/U via multiplication from the right appears in coordinates of $\mathbb{K}^3 \times (\mathbb{K}^3)^*$ as

$$t \cdot ((x_1, x_2, x_3), (y_1, y_2, y_3)) = ((t_1 x_1, t_1 x_2, t_1 x_3), (t_1 t_2 y_1, t_1 t_2 y_2, t_1 t_2 y_3)),$$

and the orbit $SL_3 \cdot (e_1, e_3^*)$ coincides with the set of semistable points $Z^{ss}(\chi)$, where the weight $\chi \in \mathbb{X}(T)$ may be taken as $\chi(t) = t_1^2 t_2$. Thus, we recover the characteristic space via the identifications

$$
\begin{array}{ccc}
Z^{ss}(\chi) & \xrightarrow{\ /T\ } & Z^{ss}(\chi)/T \\
\cong \big\uparrow & & \big\uparrow \cong \\
G/U & \xrightarrow{\ /T\ } & G/B
\end{array}
$$

Let $\chi_1, \chi_2 \in \mathbb{X}(T)$ denote the characters with $\chi_i(t) = t_i$. Then the fundamental weights are $\varpi_1 = \chi_1$ and $\varpi_2 = \chi_1 + \chi_2$. Thus, a weight $c_1 \chi_1 + c_2 \chi_2$ is dominant if and only if $c_1 \geq c_2$ holds.

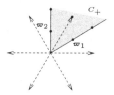

The fundamental modules are $V(\varpi_1) \cong \mathbb{K}^3$ and $V(\varpi_2) \cong \Lambda^2 \mathbb{K}^3 \cong (\mathbb{K}^3)^*$. Thus, we may take $\mathfrak{F}_1 = (X_1, X_2, X_3)$ as a basis for $V(\varpi_2)$ and $\mathfrak{F}_2 = (Y_1, Y_2, Y_3)$ as a basis for $V(\varpi_1)$. The ring $R = \Gamma(G/U, \mathcal{O})$ is then given as

$$R = \mathbb{K}[X_1, X_2, X_3, Y_1, Y_2, Y_3]/\langle X_1 Y_1 + X_2 Y_2 + X_3 Y_3 \rangle.$$

with $\deg(X_i) = \varpi_1$ and $\deg(Y_i) = \varpi_2$. Finally, the Veronese subalgebras of R producing the bunched rings of the flag varieties G/P_1 and G/P_2 are $\mathbb{K}[X_1, X_2, X_3]$ and $\mathbb{K}[Y_1, Y_2, Y_3]$, respectively.

3.2.4 Example: Quotients of quadrics

Here we consider bunched rings arising from a nondegenerate affine quadric. The resulting varieties are quotients of suitable quasitorus actions on the quadric; we call them *full intrinsic quadrics*. We work over an algebraically closed field $\mathbb{K}$ of characteristic zero.

We first give a guide to concrete examples of full intrinsic quadrics, and later see that the general ones are isomorphic to these. For $m \in \mathbb{Z}_{\geq 1}$ consider the quadratic forms

$$g_{2m} := T_1 T_2 + \cdots + T_{2m-1} T_{2m},$$

$$g_{2m+1} := T_1 T_2 + \cdots + T_{2m-1} T_{2m} + T_{2m+1}^2.$$

Write $R(r) := \mathbb{K}[T_1, \ldots, T_r]/\langle g_r \rangle$ for the factor ring, let $f_i \in R(r)$ denote the class of the variable T_i, and set $\mathfrak{F}(r) = (f_1, \ldots, f_r)$. The following holds even for non algebraically closed fields $\mathbb{K}$, see [264, Thm. 8.2]; in our case the proof is simple.

Proposition 3.2.4.1 (Klein–Nagata)   *For $r \geq 5$, the ring $R(r)$ is factorial and $\mathfrak{F}(r)$ is a system of pairwise nonassociated prime generators.*

Proof   For $r \geq 5$, the polynomial g_{r-2} is irreducible, and thus we have an integral factor ring

$$\mathbb{K}[T_2, \ldots, T_r]\langle g_{r-2}(T_3, \ldots, T_r)\rangle \cong \mathbb{K}[T_1, \ldots, T_r]\langle T_1, g_r\rangle \cong R(r)/\langle f_1\rangle.$$

In other words, f_1 is prime in $R(r)$. Moreover, localizing $R(r)$ by f_1 gives a factorial ring isomorphic to $\mathbb{K}[T_1^{\pm 1}, T_3, \ldots, T_r]$. Thus, $R(r)$ is factorial. $\square$

To find suitable gradings of $R(r)$, we first observe that there is a unique maximal grading keeping the variables homogeneous, and any other grading keeping the variables homogeneous is a coarsening of this maximal one.

Construction 3.2.4.2 (Maximal diagonal grading)   Consider any polynomial $g \in \mathbb{K}[T_1, \ldots, T_r]$ of the form

$$g = a_0 T_1^{l_{01}} \cdots T_r^{l_{0r}} + \cdots + a_k T_1^{l_{k1}} \cdots T_r^{l_{kr}}.$$

First, we build a $k \times r$ matrix P_g from the exponents l_{ij} of g. Define row vectors $l_i := (l_{i1}, \ldots, l_{ir})$ and set

$$P_g = \begin{bmatrix} l_1 - l_0 \\ \vdots \\ l_k - l_0 \end{bmatrix}$$

With the row lattice $M_g \subseteq \mathbb{Z}^r$ of P_g, we define the *gradiator* of g to be the projection $Q_g \colon \mathbb{Z}^r \to K_g := \mathbb{Z}^r / M_g$. It gives rise to a K_g-grading on $\mathbb{K}[T_1, \ldots, T_r]$ via

$$\deg(T_1) := Q_g(e_1), \qquad \ldots, \qquad \deg(T_r) := Q_g(e_r).$$

This grading is effective and $T_1, \ldots, T_r, g$ are homogeneous. Moreover, given any other such grading, say by an abelian group K, there is a commutative diagram

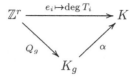

If, instead of one g, we have several $g_1, \ldots, g_s$, then, replacing P_g with the stack matrix of $P_{g_1}, \ldots, P_{g_s}$, we obtain a gradiator $Q_{g_1, \ldots, g_s}$ with the analogous property.

For our polynomials g_{2m} and g_{2m+1} the gradiators are easily obtained by a direct computation.

Proposition 3.2.4.3 *For the polynomials g_{2m} and g_{2m+1}, the associated maximal grading groups are $K_{2m} = K_{2m+1} = \mathbb{Z}^{m+1}$ and gradiators are given by*

$$
Q_{2m} = \begin{bmatrix} 1 & -1 & 0 & 0 & 0 & 0 \\ 0 & 0 & \ddots & \ddots & 0 & 0 \\ 0 & 0 & 0 & 0 & 1 & -1 \\ 0 & 1 & \cdots & \cdots & 0 & 1 \end{bmatrix},
$$

$$
Q_{2m+1} = \begin{bmatrix} 1 & -1 & 0 & 0 & 0 & 0 & 0 \\ 0 & 0 & \ddots & \ddots & 0 & 0 & 0 \\ 0 & 0 & 0 & 0 & 1 & -1 & 0 \\ 1 & 1 & \cdots & \cdots & 1 & 1 & 1 \end{bmatrix}.
$$

From Definition 3.2.1.1, we directly extract the following (necessary and sufficient) conditions that a grading gives rise to a bunched ring.

Remark 3.2.4.4 Let K be a finitely generated abelian group, R an integral normal factorially K-graded affine $\mathbb{K}$-algebra with $R_\times^* = \mathbb{K}^*$, and $\mathfrak{F} = (f_1, \ldots, f_r)$ a system of pairwise nonassociated K-prime generators with projected cone $(E \xrightarrow{Q} K, \gamma)$.

(i) The K-grading of R is almost free if and only if $Q(\gamma_0 \cap \mathbb{Z}^r)$ generates K as an abelian group for every facet $\gamma_0 \preceq \gamma$.

(ii) $\mathfrak{F} = (f_1, \ldots, f_r)$ admits a true $\mathfrak{F}$-bunch if and only if $Q(\gamma_1)^\circ \cap Q(\gamma_2)^\circ \neq \emptyset$ holds in $K_{\mathbb{Q}}$ for any two facets $\gamma_1, \gamma_2 \preceq \gamma$.

Let us illustrate this construction of bunched rings with an example in the case of six variables; the resulting variety has torsion in its divisor class group.

Example 3.2.4.5 For the polynomial $g = T_1 T_2 + T_3 T_4 + T_5 T_6$ in $\mathbb{K}[T_1, \ldots, T_6]$, we have $K_g = \mathbb{Z}^4$. Consider $K := \mathbb{Z} \oplus \mathbb{Z}/3\mathbb{Z}$ and the coarsening

$$\alpha: K_g \to K, \qquad e_1, e_2, e_3 \mapsto (1, \bar{1}), \quad e_4 \mapsto (2, \bar{0}).$$

Then $Q := \alpha \circ Q_g: \mathbb{Z}^6 \to K$ fullfills the conditions of Remark 3.2.4.4; more explicitly, the K-grading of the factor ring $R = \mathbb{K}[T_1, \ldots, T_6]/\langle g \rangle$ is given by

$$\deg(f_1) = (1, \bar{1}), \quad \deg(f_2) = (1, \bar{2}), \quad \deg(f_3) = (1, \bar{1}),$$

$$\deg(f_4) = (1, \bar{2}), \quad \deg(f_5) = (1, \bar{1}), \quad \deg(f_6) = (1, \bar{2}),$$

where f_i is the class of T_i. For $\mathfrak{F} = (f_1, \ldots, f_6)$, there is one true $\mathfrak{F}$-bunch, namely $\Phi = \{\mathbb{Q}_{\geq 0}\}$. The variety $X = X(R, \mathfrak{F}, \Phi)$ is a $\mathbb{Q}$-factorial projective 4-fold.

In particular for the treatment of more advanced examples, explicit knowledge of the $\mathfrak{F}(r)$-faces may be useful. A simple recipe to determine them now follows.

Remark 3.2.4.6 First let $r = 2m$ and arrange the set $\mathfrak{R} := \{1, \ldots, r\}$ according to the following scheme

$$2 \qquad 4 \qquad \ldots \qquad 2m - 2 \qquad 2m$$

$$1 \qquad 3 \qquad \ldots \qquad 2m - 3 \qquad 2m - 1$$

The *column sets* are $\{1, 2\}, \ldots, \{2m - 1, 2m\}$. Consider subsets $U \cup V \subseteq \mathfrak{R}$, where U is located in the upper row and V in the lower one. We look for the types:

(i) The union $U \cup V$ contains no column set, for example

(ii) The union $U \cup V$ contains at least two column sets, for example

By Proposition 3.1.1.9, the possible $\mathfrak{F}$-faces for $R(2m)$ and $\mathfrak{F} = \mathfrak{F}(2m)$ are $\gamma_0 = \text{cone}(e_i;\ i \in U \cup V) \preceq \gamma$ with $U \cup V$ of type (i) or (ii).

Now, let $r = 2m + 1$. Then we arrange the set $\mathfrak{R} := \{1, \ldots, r\}$ according to the following scheme

2	4	...	$2m - 2$	$2m$	$2m + 1$
1	3	...	$2m - 3$	$2m - 1$	$2m + 1$

This time the column sets are $\{1, 2\}, \ldots, \{2m - 1, 2m\}$ and $\{2m + 1\}$. Again, we consider two types of $U \cup V \subseteq \mathfrak{R}$ with U in the upper row and V in the lower one:

 (i) The union $U \cup V$ contains no column set.
(ii) The union $U \cup V$ contains at least two column sets.

Then, as before, the possible $\mathfrak{F}$-faces for $R(2m + 1)$ and $\mathfrak{F} = \mathfrak{F}(2m + 1)$ are $\gamma_0 = \mathrm{cone}(e_i; \ i \in U \cup V) \preceq \gamma$ with a constellation $U \cup V$ of type (i) or (ii).

By a *full intrinsic quadric* we mean a variety X with Cox ring of the form $\mathcal{R}(X) \cong \mathbb{K}[T_1, \ldots, T_r]/\langle g \rangle$, where g is a quadratic form of rank r and the classes $f_i \in \mathcal{R}(X)$ of the variables T_i are $\mathrm{Cl}(X)$-homogeneous. The following statement shows that any full intrinsic quadric with torsion free divisor class group is isomorphic to the variety arising from a bunched ring $(R(r), \mathfrak{F}(r), \Phi)$, as previously discussed.

Proposition 3.2.4.7 *Let $g \in \mathbb{K}[T_1, \ldots, T_r]$ be a quadratic form of rank r. Consider an effective grading of $\mathbb{K}[T_1, \ldots, T_r]$ by a torsion-free abelian group K such that $T_1, \ldots, T_r$ and g are K-homogeneous. Then there exist linearly independent K-homogeneous linear forms $S_1, \ldots, S_r$ in $T_1, \ldots, T_r$ with $g(T_1, \ldots, T_r) = g_r(S_1, \ldots, S_r)$.*

Proof We may assume that the group K is finitely generated. The K-grading on $\mathbb{K}[T_1, \ldots, T_r]$ is given by a linear action of a torus $H := \mathrm{Spec}\,\mathbb{K}[K]$ on $\mathbb{K}^r$. Let $O(g)$ be the subgroup of $\mathrm{GL}_r(\mathbb{K})$ consisting of linear transformations preserving the quadratic form g and $EO(g)$ be the extension of $O(g)$ by the subgroup of scalar matrices.

By assumption, H is a subgroup of $EO(g)$ consisting of diagonal matrices. Because g is of rank r, the subgroup $EO(g)$ is conjugate to $EO(g_r)$, and this conjugation sends H to a subgroup H' of $EO(g_r)$. As any torus, H' is contained in a maximal torus of $EO(g_r)$, see [178, Sec. 22.3, Cor. B].

On the other hand, the intersection of $EO(g_r)$ with the subgroup of diagonal matrices is a maximal torus of $EO(g_r)$. Any two maximal tori of an affine algebraic group are conjugate, and we may assume that H' is a subgroup of $EO(g_r)$ consisting of diagonal matrices. So the conjugation sends H to H'.

This means that every new basis vector is a linear combination of the old ones having the same degree, and the assertion follows. □

We conclude the section with two classification results on full intrinsic quadrics with small divisor class groups from [44].

Proposition 3.2.4.8 *Let X be a full intrinsic quadric with $\mathrm{Cl}(X) \cong \mathbb{Z}$. Then X arises from a bunched ring $(R, \mathfrak{F}, \Phi)$ with Φ given by*

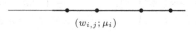

with positive integers $w_{i,j}$, where $0 \leq i \leq n$ and $1 \leq j \leq \mu_i$, such that $w_{i,1} < w_{i+1,1}$, $w_{ij} = w_{ik}$ and $w_{i,1} + w_{n-i,1} = w$ holds for some fixed $w \in \mathbb{N}$, and the μ_i satisfy

$$\mu_i \geq 1, \qquad \mu_i = \mu_{n-i}, \qquad \mu_0 + \cdots + \mu_n \geq 5.$$

The variety X is always $\mathbb{Q}$-factorial, projective, and $\mathbb{Q}$-Fano, and for its dimension, we have

$$\dim(X) = \mu_1 + \cdots + \mu_n - 2.$$

Moreover, X is smooth if and only if $n = 0$ and $w_{01} = 1$ hold, and in this case it is a smooth projective quadric.

The second result is the "intrinsic quadrics version" of Kleinschmidt's classification [186] of smooth complete toric varieties with Picard number 2; compare also Example 2.4.2.10.

Theorem 3.2.4.9 *Let X be a smooth full intrinsic quadric with $\mathrm{Cl}(X) \cong \mathbb{Z}^2$. Then X arises from a bunched ring $(R, \mathfrak{F}, \Phi)$ with an $\mathfrak{F}$-bunch Φ given by one of the following figures:*

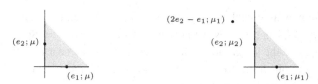

where in the left-hand-side case, $\mu \geq 3$ holds, and in the right-hand-side case, one has $\mu_i \geq 1$ and $2\mu_1 + \mu_2 \geq 5$. Any such X is projective, and its dimension is given by

$$\dim(X) = 2\mu - 3, \quad \text{or} \quad \dim(X) = 2\mu_1 + \mu_2 - 3,$$

where the first equation corresponds to the left-hand-side case, and the second one to the right-hand-side case. The variety X is Fano if and only if Φ belongs *to the left-hand-side case. Moreover, different figures define nonisomorphic varieties.*

Example 3.2.4.10 In the setting of Theorem 3.2.4.9, consider the bunch arising from the left-hand-side picture with $\mu = 3$. Then R is defined by a quadratic form of rank six and X equals the flag variety SL_3/B_3 discussed in Example 3.2.3.10

3.2.5 The canonical toric embedding

As we will see here, every variety defined by a bunched ring allows a closed embedding into a toric variety with nice properties. For a fixed bunched ring, we give a canonical construction in 3.2.5.3. The toric ambient variety arising from this construction may be (noncanonically) completed, which is discussed in Constructions 3.2.5.6 and 3.2.5.7. The ground field $\mathbb{K}$ is algebraically closed and of characteristic zero.

Remark 3.2.5.1 Consider a morphism $\varphi \colon X \to Z$ from an irreducible normal variety X to a toric variety Z with acting torus T and base point $z_0 \in Z$. Let $D_Z^i = \overline{T \cdot z_i}$, where $1 \le i \le r$, be the T-invariant prime divisors and set

$$Z' := T \cdot z_0 \cup T \cdot z_1 \cup \ldots \cup T \cdot z_r.$$

Suppose that $\varphi^{-1}(D_Z^i)$, where $1 \le i \le r$, are pairwise different irreducible hypersurfaces in X. Then the complement $X \setminus \varphi^{-1}(Z')$ is of codimension at least two in X, and we have a canonical pullback homomorphism

$$\mathrm{WDiv}^T(Z) = \mathrm{CDiv}^T(Z') \xrightarrow{\varphi^*} \mathrm{CDiv}(\varphi^{-1}(Z')) \subseteq \mathrm{WDiv}(X)$$

It sends principal divisors to principal divisors and consequently induces a pullback homomorphism $\varphi^* \colon \mathrm{Cl}(Z) \to \mathrm{Cl}(X)$ on the level of divisor class groups.

Definition 3.2.5.2 Let X be an irreducible, normal variety and Z a toric variety with acting torus T and invariant prime divisors $D_Z^i = \overline{T \cdot z_i}$, where $1 \le i \le r$. A *neat embedding* of X into Z is a closed embedding $\iota \colon X \to Z$ such that $\iota^{-1}(D_Z^i)$, where $1 \le i \le r$, are pairwise different irreducible hypersurfaces in X and the pullback homomorphism $\iota^* \colon \mathrm{Cl}(Z) \to \mathrm{Cl}(X)$ is an isomorphism.

Construction 3.2.5.3 Let $(R, \mathfrak{F}, \Phi)$ be a bunched ring and $(E \xrightarrow{Q} K, \gamma)$ the associated projected cone. Then, with $M := \ker(Q)$, we have the mutually dual exact sequences

$$0 \longrightarrow L \longrightarrow F \overset{P}{\longrightarrow} N$$

$$0 \longleftarrow K \underset{Q}{\longleftarrow} E \longleftarrow M \longleftarrow 0$$

The *envelope* of the collection $\mathrm{rlv}(\Phi)$ of relevant $\mathfrak{F}$-faces is the saturated Q-connected γ-collection

$$\mathrm{Env}(\Phi) := \{\gamma_0 \preceq \gamma; \ \gamma_1 \preceq \gamma_0 \text{ and } Q(\gamma_1)^\circ \subseteq Q(\gamma_0)^\circ \text{ for some } \gamma_1 \in \mathrm{rlv}(\Phi)\}.$$

Let $\delta \subseteq F_{\mathbb{Q}}$ denote the dual cone of $\gamma \subseteq E_{\mathbb{Q}}$ and for $\gamma_0 \preceq \gamma$ let $\gamma_0^* = \gamma_0^\perp \cap \delta$ be the corresponding face. Then one has fans in the lattices F and N:

$$\widehat{\Sigma} := \{\delta_0 \preceq \delta; \ \delta_0 \preceq \gamma_0^* \text{ for some } \gamma_0 \in \mathrm{Env}(\Phi)\},$$

$$\Sigma := \{P(\gamma_0^*); \ \gamma_0 \in \mathrm{Env}(\Phi)\}.$$

Consider the action of $H := \mathrm{Spec}\,\mathbb{K}[K]$ on $\overline{X} := \mathrm{Spec}\,R$. Set $\widehat{X} := \widehat{X}(R, \mathfrak{F}, \Phi)$ and $X := X(R, \mathfrak{F}, \Phi)$. Let $\overline{Z} = \mathbb{K}^r$ denote the toric variety associated with the cone δ in F. The system $\mathfrak{F} = (f_1, \ldots, f_r)$ of generators of R defines a closed embedding

$$\bar{\imath} \colon \overline{X} \to \overline{Z}, \qquad z \mapsto (f_1(z), \ldots, f_r(z)),$$

which becomes H-equivariant if we endow $\overline{Z}$ with the diagonal H-action given by the characters $\chi^{w_1}, \ldots, \chi^{w_r}$, where $w_i = \deg(f_i) \in K$. Denoting by $\widehat{Z}$ and Z the toric varieties associated with the fans $\widehat{\Sigma}$ and Σ, we obtain a commutative diagram

$$
\begin{array}{ccc}
\overline{X} & \overset{\bar{\imath}}{\longrightarrow} & \overline{Z} \\
\uparrow & & \uparrow \\
\widehat{X} & \overset{\widehat{\imath}}{\longrightarrow} & \widehat{Z} \\
{\scriptstyle /\!/ H}\downarrow{\scriptstyle p_X} & & {\scriptstyle p_Z}\downarrow{\scriptstyle /\!/ H} \\
X & \underset{\imath}{\longrightarrow} & Z
\end{array}
$$

where the map $\widehat{\imath}$ is the restriction of $\bar{\imath}$ and the toric morphism $p_Z \colon \widehat{Z} \to Z$ arises from $P \colon F \to N$. We call the induced map of quotients $\imath \colon X \to Z$ the *canonical toric embedding* associated with the bunched ring $(R, \mathfrak{F}, \Phi)$.

Proposition 3.2.5.4 *In the setting of Construction 3.2.5.3, the following statements hold.*

(i) *The quotient morphism $p_Z \colon \widehat{Z} \to Z$ is a toric characteristic space, we have $\widehat{X} = \bar{\iota}^{-1}(\widehat{Z})$ and $\iota \colon X \to Z$ is a closed embedding.*

(ii) *For any $\gamma_0 \in \mathrm{rlv}(\Phi)$ and the associated toric affine chart $Z_{P(\gamma_0^*)} \subseteq Z$ we have $X_{\gamma_0} = \iota^{-1}(Z_{P(\gamma_0^*)})$.*

(iii) *The prime divisors $D_X^i = p_X(V(\widehat{X}, f_i))$ and $D_Z^i = p_Z(V(\widehat{Z}, T_i))$ satisfy $D_X^i = \iota^*(D_Z^i)$ and we have a commutative diagram*

$$
\begin{array}{ccc}
\mathrm{Cl}(X) & \xleftarrow{\ \ \iota^* \ \ }_{\cong} & \mathrm{Cl}(Z) \\
{\scriptstyle [D_X^i] \mapsto \deg(f_i)} \Big\downarrow {\scriptstyle \cong} & & {\scriptstyle \cong} \Big\downarrow {\scriptstyle [D_Z^i] \mapsto \deg(T_i)} \\
K & =\!\!=\!\!= & K
\end{array}
$$

In particular, the embedding $\iota \colon X \to Z$ of the quotient varieties is a neat embedding.

(iv) *The maximal cones of the fan Σ are precisely the cones $P(\gamma_0^*) \in \Sigma$, where $\gamma_0 \in \mathrm{cov}(\Phi)$.*

(v) *The image $\iota(X) \subseteq Z$ intersects every closed toric orbit of Z nontrivially.*

Generalizing the classical homogeneous coordinates on the projective space, the canonical toric embedding leads to global coordinates in the following sense.

Remark 3.2.5.5 Let $(R, \mathfrak{F}, \Phi)$ be a bunched ring and $X = X(R, \mathfrak{F}, \Phi)$ the associated variety. The commutative diagram of Construction 3.2.5.3 allows us to represent every point $x \in X$ by its *Cox coordinates* $[z_1, \ldots, z_r]$, where

$$
z = (z_1, \ldots, z_r) \in p_X^{-1}(x) \subseteq \widehat{X} \subseteq \overline{Z} = \mathbb{K}^r
$$

is any point of the fiber p_X^{-1} having a closed H-orbit in $\widehat{X}$. Note that, for the Cox coordinates $[z]$ of a given point $x \in X$, the element $z \in \mathbb{K}^r$ is unique up to applying elements of H.

Proof of Construction 3.2.5.3 and Proposition 3.2.5.4 First note that by Proposition 2.2.3.5, the collection of cones Σ is indeed a fan. Moreover, by Theorem 2.1.3.2, the toric morphism $p_Z \colon \widehat{Z} \to Z$ is a characteristic space. For each $\gamma_0 \in \mathrm{rlv}(\Phi)$, the affine toric chart of $\widehat{Z}$ corresponding to $\gamma_0^* \in \widehat{\Sigma}$ is the localization $\overline{Z}_{\gamma_0} = \overline{Z}_{T^u}$, where $u \in \gamma_0^\circ$ and $T^u = T_1^{u_1} \cdots T_r^{u_r}$. Thus, we obtain

$$
\bar{\iota}^{-1}(\overline{Z}_{\gamma_0}) = \bar{\iota}^{-1}(\overline{Z}_{T_1^{u_1} \cdots T_r^{u_r}}) = \overline{X}_{f_1^{u_1} \cdots f_r^{u_r}} = \overline{X}_{\gamma_0}.
$$

Because $\widehat{Z}$ and $\widehat{X}$ are the union of the localizations of $\overline{Z}$ and $\overline{X}$ by faces $\gamma_0 \in \mathrm{rlv}(\Phi)$, we conclude $\widehat{X} = \bar{\iota}^{-1}(\widehat{Z})$. By Theorem 1.2.3.6, the induced map $\iota \colon X \to Z$ is a closed embedding. This establishes the construction and the first two items of the proposition.

We turn to the third assertion of the proposition. By the commutative diagram given in the construction, we have

$$\imath^{-1}(D_Z^i) \;=\; p_X(\bar{\imath}^{-1}(p_Z^{-1}(D_Z^i))) \;=\; p_X(V(\widehat{X}; f_i)) \;=\; D_X^i.$$

Thus, denoting by T the acting torus of Z, we have a well-defined pull-back homomorphism $\imath^*\colon \mathrm{WDiv}^T(Z) \to \mathrm{WDiv}(X)$. It satisfies $\imath^*(D_z^i) = D_X^i$ because of

$$\bar{\imath}^*(p_Z^*(D_Z^i)) \;=\; \bar{\imath}^*(\mathrm{div}(T_i)) \;=\; \mathrm{div}(f_i) \;=\; p_X^*(D_X^i).$$

As a consequence, we obtain that the diagram of (iii) is commutative and thus the embedding $\imath\colon X \to Z$ is neat.

We show (iv) and (v) of the proposition. By definition, the envelope $\mathrm{Env}(\Phi)$ has the covering collection $\mathrm{cov}(\Phi)$ as its collection of minimal cones. Consequently, the maximal cones of $\widehat{\Sigma}$ and Σ are given by

$$\widehat{\Sigma}^{\mathrm{max}} \;=\; \{\gamma_0^*;\; \gamma_0 \in \mathrm{cov}(\Phi)\}, \qquad\qquad \Sigma^{\mathrm{max}} \;=\; \{P(\gamma_0^*);\; \gamma_0 \in \mathrm{cov}(\Phi)\}.$$

This verifies in particular the fourth assertion. The last one is then a simple consequence. $\qquad\qquad\qquad\qquad\qquad\qquad\qquad\qquad\qquad\qquad\qquad\square$

Note that, if the variety X associated with the bunched ring $(R, \mathfrak{F}, \Phi)$ is complete (projective), then the canonical ambient toric variety Z need not be complete (projective). Passing to completions of Z, means to give up the last two properties of the above proposition. However, the following construction preserves the first three properties.

Construction 3.2.5.6 Situation as in Construction 3.2.5.3. Let $\mathfrak{B} \subseteq \mathrm{faces}(\gamma)$ be any saturated Q-connected collection comprising $\mathrm{Env}(\Phi)$. Then $\mathfrak{B}$ defines fans in F in N:

$$\widehat{\Sigma}_1 := \{\delta_0 \preceq \delta;\; \delta_0 \preceq \gamma_0^* \text{ for some } \gamma_0 \in \mathfrak{B}\},$$

$$\Sigma_1 = \{P(\gamma_0^*);\; \gamma_0 \in \Theta\}.$$

The fans $\widehat{\Sigma}$ and Σ defined by the envelope $\mathrm{Env}(\Phi)$ are subfans of $\widehat{\Sigma}_1$ and Σ_1 respectively. With the toric varieties $\widehat{Z}_1$ and Z_1, associated with $\widehat{\Sigma}_1$ and Σ_1,

we obtain a commutative diagram

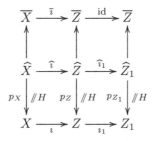

where $\widehat{\imath}_1$ and $\imath_1$ are open embeddings, and the compositions $\overline{\imath}_1 \circ \overline{\imath}$, $\widehat{\imath}_1 \circ \widehat{\imath}$ and $\imath_1 \circ \imath$ are closed embeddings satisfying the assertions (i), (ii), and (iii) of Proposition 3.2.5.4. In particular, $\imath_1 \circ \imath \colon X \to Z_1$ is a neat embedding.

Proof Because $\widehat{X} \subseteq \overline{X}$ is $(H, 2)$-maximal, we obtain $\widehat{X} = (\overline{\imath}_1 \circ \overline{\imath})^{-1}(\widehat{Z}_1)$. The remaining assertions are then obvious. □

The following special case of the preceding construction provides projective toric ambient varieties for the case that our variety arising from the bunched ring is projective.

Construction 3.2.5.7 Consider the situation of Construction 3.2.5.3. Let $\Lambda(\overline{X}, H)$ and $\Lambda(\overline{Z}, H)$ denote the GIT fans of the actions of H on $\overline{X}$ and $\overline{Z}$, respectively. Suppose that the $\mathfrak{F}$-bunch Φ arises from a GIT cone $\lambda \in \Lambda(\overline{X}, H)$, which means that we have

$$\Phi = \{Q(\gamma_0); \ \gamma_0 \preceq \gamma \ \mathfrak{F}\text{-face with } \lambda^\circ \subseteq Q(\gamma_0)\}.$$

Then $X = \overline{X}^{ss}(\lambda)$ holds. Moreover, for any GIT cone $\eta_1 \in \Lambda(\overline{Z}, H)$ with $\eta_1^\circ \subseteq \lambda^\circ$, Construction 3.2.5.6 provides a neat embedding $X \to Z_1$ into the projective toric variety $Z_1 = \overline{Z}^{ss}(\eta_1) /\!\!/ H$ associated with the bunch of cones arising from η_1.

Example 3.2.5.8 Consider again the bunched ring $(R, \mathfrak{F}, \Phi)$ with $K = \mathbb{Z}^2$ from 3.2.1.6. The K-graded ring R is given by

$$R = \mathbb{K}[T_1, \ldots, T_5] / \langle T_1 T_2 + T_3^2 + T_4 T_5 \rangle,$$

$$Q = \begin{bmatrix} 1 & -1 & 0 & -1 & 1 \\ 1 & 1 & 1 & 0 & 2 \end{bmatrix},$$

where $\deg(T_i) = w_i \in K$ with the ith column w_i of Q. The classes f_i of T_i give $\mathfrak{F} = (f_1, \ldots, f_5)$ and the $\mathfrak{F}$-bunch is $\Phi = \Phi(w_3)$. In the associated projected cone $(E \xrightarrow{\varrho} K, \gamma)$ we have $E = \mathbb{Z}^5$ and $\gamma = \text{cone}(e_1, \dots, e_5)$. The covering collection is

$$\text{cov}(\Phi) = \{\gamma_{1,4}, \gamma_{2,5}, \gamma_{1,2,3}, \gamma_{3,4,5}\},$$

where $\gamma_{i_1,\dots,i_k} := \text{cone}(e_{i_1}, \dots, e_{i_k})$. A Gale dual map $P \colon F \to N$ for $Q \colon E \to K$ is given by the matrix

$$P = \begin{bmatrix} -1 & -1 & 2 & 0 & 0 \\ -1 & -1 & 0 & 1 & 1 \\ -1 & 0 & 1 & -1 & 0 \end{bmatrix}.$$

The maximal cones of the fan Σ constructed via $\text{Env}(\Phi)$ in 3.2.5.3 correspond to the members of the covering collection; in terms of the columns $v_1, \dots, v_r$ of P they are given as

$$\text{cone}(v_2, v_3, v_5), \quad \text{cone}(v_1, v_3, v_4), \quad \text{cone}(v_4, v_5), \quad \text{cone}(v_1, v_2).$$

In particular, we see that the canonical toric ambient variety Z of X determined by $(R, \mathfrak{F}, \Phi)$ is not complete. Completions of Z are obtained as in Construction 3.2.5.7. Recall that

$$X = X(R, \mathfrak{F}, \Phi) = \overline{X}^{ss}(w_3) /\!\!/ H$$

holds with the action of $H = \text{Spec}\,\mathbb{K}[K]$ on $\overline{X} = \text{Spec}\,R$. The two GIT fans $\Lambda(\overline{X}, H)$ and $\Lambda(\overline{Z}, H)$ are

$$\Lambda(\overline{X}, H) \qquad\qquad\qquad \Lambda(\overline{Z}, H)$$

In $\Lambda(\overline{X}, H)$, the weight w_3 belongs to the GIT cone $\lambda = \text{cone}(w_2, w_5)$. In $\Lambda(\overline{Z}, H)$, we have three choices:

$$\eta_1 := \text{cone}(w_2, w_3), \qquad \eta_{12} := \text{cone}(w_3), \qquad \eta_2 := \text{cone}(w_3, w_5).$$

The GIT cones η_1, η_2 provide $\mathbb{Q}$-factorial projective toric completions Z_1, Z_2 of Z, whereas η_{12} gives a projective toric completion with non-$\mathbb{Q}$-factorial singularities.

3.3 Geometry via defining data

3.3.1 Stratification and local properties

We observe that the variety arising from a bunched ring comes with a decomposition into locally closed subvarieties; these turn out to be the intersections with the toric orbits of the canonical toric ambient variety. We show that the local divisor class groups are constant along the pieces and conclude some local properties from this. The ground field $\mathbb{K}$ is algebraically closed and of characteristic zero.

Construction 3.3.1.1 Let $(R, \mathfrak{F}, \Phi)$ be a bunched ring; then consider the action of $H := \operatorname{Spec} \mathbb{K}[K]$ on $\overline{X} := \operatorname{Spec} R$ and set

$$\widehat{X} := \widehat{X}(R, \mathfrak{F}, \Phi), \qquad X := X(R, \mathfrak{F}, \Phi).$$

Let $(E \xrightarrow{Q} K, \gamma)$ be the projected cone associated with $\mathfrak{F} = (f_1, \ldots, f_r)$. With any $\mathfrak{F}$-face $\gamma_0 \preceq \gamma$, we associate a locally closed subset

$$\overline{X}(\gamma_0) := \{z \in \overline{X}; \; f_i(z) \neq 0 \Leftrightarrow e_i \in \gamma_0 \text{ for } 1 \leq i \leq r\} \subseteq \overline{X}.$$

These sets are pairwise disjoint and cover the whole $\overline{X}$. Taking the pieces defined by relevant $\mathfrak{F}$-faces, one obtains a constructible subset

$$\widetilde{X} := \bigcup_{\gamma_0 \in \mathrm{rlv}(\Phi)} \overline{X}(\gamma_0) \subseteq \widehat{X}.$$

Note that $\widetilde{X}$ is the union of all closed H-orbits of $\widehat{X}$. The images of the pieces inside $\widetilde{X}$ form a decomposition of X into pairwise disjoint locally closed pieces:

$$X = \bigcup_{\gamma_0 \in \mathrm{rlv}(\Phi)} X(\gamma_0), \qquad \text{where } X(\gamma_0) := p_X(\overline{X}(\gamma_0)).$$

Example 3.3.1.2 If we have $R = \mathbb{K}[T_1, \ldots, T_r]$ and $\mathfrak{F} = \{T_1, \ldots, T_r\}$, then X is the toric variety arising from the image fan Σ associated with $\mathrm{rlv}(\Phi)$, and for any $\gamma_0 \in \mathrm{rlv}(\Phi)$, the piece $X(\gamma_0) \subseteq X$ is precisely the toric orbit corresponding to the cone $P(\gamma_0^*) \in \Sigma$.

Proposition 3.3.1.3 *Consider the situation of Construction 3.3.1.1. For every* $\gamma_0 \in \mathrm{rlv}(\Phi)$, *the associated piece* $X(\gamma_0) \subseteq X$ *has the following descriptions.*

(i) *In terms of the embedding* $X \subseteq Z$ *constructed in 3.2.5.3, the piece* $X(\gamma_0)$ *is the intersection of* X *with the toric orbit of* Z *given by* $P(\gamma_0^*) \in \Sigma$.

(ii) *In terms of the prime divisors* $D_X^i \subseteq X$ *defined in Construction 3.2.1.3 via the generators* $f_i \in \mathfrak{F}$, *the piece* $X(\gamma_0)$ *is given as*

$$X(\gamma_0) = \bigcap_{e_i \notin \gamma_0} D_X^i \setminus \bigcup_{e_j \in \gamma_0} D_X^j$$

(iii) *In terms of the open subsets $X_{\gamma_1} \subseteq X$ and $\overline{X}_{\gamma_1} \subseteq \overline{X}$ defined in Construction 3.2.1.3 via relevant faces $\gamma_1 \in \mathrm{rlv}(\Phi)$, we have*

$$X(\gamma_0) = X_{\gamma_0} \setminus \bigcup_{\gamma_0 \prec \gamma_1 \in \mathrm{rlv}(\Phi)} X_{\gamma_1},$$

$$p_X^{-1}(X(\gamma_0)) = \overline{X}_{\gamma_0} \setminus \bigcup_{\gamma_0 \prec \gamma_1 \in \mathrm{rlv}(\Phi)} \overline{X}_{\gamma_1}.$$

Proof of Construction 3.3.1.1 and Proposition 3.3.1.3   Obviously, $\overline{X}$ is the union of the locally closed pieces $\overline{X}(\gamma_0)$, where $\gamma_0 \preceq \gamma$ runs through the $\mathfrak{F}$-faces. Proposition 3.2.2.4 tells us that $\widehat{X} \subseteq \overline{X}$ is the $(H, 2)$-maximal subset given by the bunch of orbit cones

$$\Phi = \{\omega_z; \ H \cdot z \text{ closed in } \widehat{X}\}.$$

Moreover, it says that the closed orbits of $\widehat{X}$ are precisely the orbits $H \cdot z \subseteq \overline{X}$ with $\omega_z \in \Phi$. Given $z \in \widetilde{X}$, we have $\omega_z = Q(\gamma_0)$ for some $\gamma_0 \in \mathrm{rlv}(\Phi)$ and thus $H \cdot z$ is closed in $\widehat{X}$. Conversely, if $H \cdot z$ is closed in $\widehat{X}$, consider the $\mathfrak{F}$-face

$$\gamma_0 := \mathrm{cone}(e_i; \ f_i(z) \neq 0) \preceq \gamma.$$

Then we have $z \in \overline{X}(\gamma_0)$ and $Q(\gamma_0) = \omega_z \in \Phi$. The latter shows that γ_0 is a relevant face. This implies $z \in \widetilde{X}$.

All further statements are most easily seen by means of a neat embedding $X \subseteq Z$ as constructed in 3.2.5.3. Let $\overline{X} \subseteq \overline{Z} = \mathbb{K}^r$ denote the closed H-equivariant embedding arising from $\mathfrak{F}$. Then, $\overline{X}$ intersects precisely the $\overline{Z}(\gamma_0)$, where γ_0 is an $\mathfrak{F}$-face, and in these cases we have

$$\overline{X}(\gamma_0) = \overline{Z}(\gamma_0) \cap \overline{X}.$$

As mentioned in Example 3.3.1.2, the images $Z(\gamma_0) = p_Z(\overline{Z}(\gamma_0))$, where $\gamma_0 \in \mathrm{rlv}(\Phi)$, are precisely the toric orbits of Z. Moreover, we have

$$\widehat{X} = \widehat{Z} \cap \overline{X}, \qquad \widetilde{X} = \widetilde{Z} \cap \overline{X}.$$

Because p_Z separates H-orbits along $\widetilde{Z}$, we obtain $X(\gamma_0) = Z(\gamma_0) \cap X$ for every $\gamma_0 \in \mathrm{rlv}(\Phi)$. Consequently, the $X(\gamma_0)$, where $\gamma_0 \in \mathrm{rlv}(\Phi)$, are pairwise disjoint and form a decomposition of X into locally closed pieces. Finally, using $X(\gamma_0) = Z(\gamma_0) \cap X$ and $D_X^i = \iota^*(D_Z^i)$, we obtain assertions 3.3.1.3 (ii) and (iii) directly from the corresponding representations of the toric orbit $Z(\gamma_0)$. $\square$

We use the decomposition into pieces to study local properties of the variety associated with a bunched ring. First recall from Proposition 1.6.2.4 that we associated with any point $x \in X$ of a variety X with characteristic space

$p_X \colon \widehat{X} \to X$ a submonoid of the divisor class group as follows. Let $\widehat{x} \in p_X^{-1}(x)$ be a point with closed H_X-orbit and set

$$S_x := \{[D] \in \mathrm{Cl}(X); \ f(\widehat{x}) \neq 0 \text{ for some } f \in \Gamma(X, \mathcal{R}_{[D]})\} \subseteq \mathrm{Cl}(X).$$

A first task is to express this for a variety X arising from a bunched ring $(R, \mathfrak{F}, \Phi)$ in terms of the defining data. For this we use the isomorphism $\mathrm{Cl}(X) \to K$ of Theorem 3.2.1.4 sending the class of the prime divisor $D_X^i \subseteq X$ defined by $f_i \in \mathfrak{F}$ to the degree $\deg(f_i) \in K$.

Proposition 3.3.1.4 *Consider the situation of Construction 3.3.1.1. Let $\gamma_0 \in \mathrm{rlv}(\Phi)$ and $x \in X(\gamma_0)$. Then, under the isomorphism $\mathrm{Cl}(X) \to K$ of Theorem 3.2.1.4, the monoid S_x corresponds to $Q(\gamma_0 \cap E)$.*

Proof According to Theorem 1.6.4.3 and Remark 1.6.4.7, we may identify the quasitorus $H = \mathrm{Spec}\,\mathbb{K}[K]$ with $H_X = \mathrm{Spec}\,\mathbb{K}[\mathrm{Cl}(X)]$ and the characteristic space $p \colon \widehat{X}(R, \mathfrak{F}, \Phi) \to X$ with $p_X \colon \widehat{X} \to X$ constructed from a Cox sheaf. Let $z \in \overline{X}(\gamma_0)$. Then z is a point with closed H-orbit in $\widehat{X}$ and thus, for $x = p(z)$, we see that S_x equals $Q(\gamma_0 \cap E)$. $\qquad\square$

As an application, we compute the local class group $\mathrm{Cl}(X, x)$; recall that this is the group of Weil divisors $\mathrm{WDiv}(X)$ modulo the subgroup of all divisors being principal on a neighborhood of $x \in X$.

Proposition 3.3.1.5 *Consider the situation of Construction 3.3.1.1. Let $\gamma_0 \in \mathrm{rlv}(\Phi)$ and $x \in X(\gamma_0)$. Then we have a commutative diagram*

$$
\begin{array}{ccc}
\mathrm{Cl}(X) & \longrightarrow & \mathrm{Cl}(X,x) \\
{\scriptstyle\cong}\big\uparrow & & \big\uparrow{\scriptstyle\cong} \\
K & \longrightarrow & K/Q(\mathrm{lin}(\gamma_0) \cap E))
\end{array}
$$

In particular, the local divisor class groups are constant along the pieces $X(\gamma_0)$, where $\gamma_0 \in \mathrm{rlv}(\Phi)$.

Proof By Proposition 1.6.2.4, the kernel of $\mathrm{Cl}(X) \to \mathrm{Cl}(X, x)$ is the subgroup of $\mathrm{Cl}(X)$ generated by S_x. Thus, Proposition 3.3.1.4 gives the assertion. $\qquad\square$

Corollary 3.3.1.6 *Consider the situation of Construction 3.3.1.1. Inside the divisor class group $\mathrm{Cl}(X) \cong K$, the Picard group of X is given by*

$$\mathrm{Pic}(X) \cong \bigcap_{\gamma_0 \in \mathrm{cov}(\Phi)} Q(\mathrm{lin}(\gamma_0) \cap E).$$

Proof The divisor class given by $w \in K$ stems from a Cartier divisor if and only if it defines the zero class in $\mathrm{Cl}(X, x)$ for any $x \in X$. According to Proposition 3.3.1.5, the latter is equivalent to $w \in Q(\mathrm{lin}(\gamma_0) \cap E)$ for all $\gamma_0 \in \mathrm{rlv}(\Phi)$. Because we have $\mathrm{cov}(\Phi) \subseteq \mathrm{rlv}(\Phi)$, and for any $\gamma_0 \in \mathrm{rlv}(\Phi)$, there is a $\gamma_1 \in \mathrm{cov}(\Phi)$ with $\gamma_1 \preceq \gamma_0$, it suffices to take the $\gamma_0 \in \mathrm{cov}(\Phi)$.   □

Corollary 3.3.1.7 *Let X be the variety associated with a bunched ring and $X \subseteq Z$ the associated canonical toric embedding as provided in Construction 3.2.5.3.*

(i) *For every $x \in X$ we have an isomorphism $\mathrm{Cl}(X, x) \cong \mathrm{Cl}(Z, x)$.*
(ii) *We have an isomorphism $\mathrm{Pic}(X) \cong \mathrm{Pic}(Z)$.*

In particular, if the ambient variety Z has a toric fixed point, then the Picard group $\mathrm{Pic}(X)$ is torsion free.

Proof The two items are clear. For the last statement, recall that a toric variety with toric fixed point has a free Picard group; see, for example, [141, Sec. 3.4].   □

A point $x \in X$ is factorial ($\mathbb{Q}$-factorial) if and only if every Weil divisor is Cartier ($\mathbb{Q}$-Cartier) at x. Thus, Proposition 3.3.1.5 has the following application to singularities.

Corollary 3.3.1.8 *Consider the situation of Construction 3.3.1.1. Consider a relevant face $\gamma_0 \in \mathrm{rlv}(\Phi)$ and point $x \in X(\gamma_0)$.*

(i) *The point x is factorial if and only if Q maps $\mathrm{lin}(\gamma_0) \cap E$ onto K.*
(ii) *The point x is $\mathbb{Q}$-factorial if and only if $Q(\gamma_0)$ is of full dimension.*

Corollary 3.3.1.9 *The variety X arising from a bunched ring $(R, \mathfrak{F}, \Phi)$ is $\mathbb{Q}$-factorial if and only if Φ consists of cones of full dimension.*

Whereas local factoriality admits a simple combinatorial characterization, smoothness involves algebraic properties of the Cox ring, as the following statement shows.

Proposition 3.3.1.10 *Consider the situation of Construction 3.3.1.1. Consider $\gamma_0 \in \mathrm{rlv}(\Phi)$, a point $x \in X(\gamma_0)$, and $\widehat{x} \in p^{-1}(x)$. Then the following statements are equivalent.*

(i) *The point $x \in X$ is smooth*
(ii) *Q maps $\mathrm{lin}(\gamma_0) \cap E$ onto K and $\widehat{x} \in \widehat{X}$ is smooth.*

Proof If (i) holds, then x is factorial and, by Corollary 3.3.1.8, the map Q sends $\mathrm{lin}(\gamma_0) \cap E$ onto K. Moreover, Proposition 1.6.1.6 shows that $\widehat{x} \in \widehat{X}$ is a smooth point.

Suppose that (ii) holds. Because Q maps $\text{lin}(\gamma_0) \cap E$ onto K, Propositions 1.2.2.8 and 3.1.1.10 yield that the fiber $p^{-1}(x)$ consists of a single free H-orbit. Consequently, H acts freely over an open neighborhood $U \subseteq X$ of x and thus p is an étale H-principal bundle over U. Because $\widehat{x} \in \widehat{X}$ is smooth, we can achieve by appropriate shrinking that the whole $p^{-1}(U)$ is smooth. Because étale morphisms preserve smoothness, we conclude that U and hence x are smooth. $\qquad \square$

Corollary 3.3.1.11 *Consider the situation of Construction 3.3.1.1. Suppose that $\widehat{X}$ is smooth; let $\gamma_0 \in \text{rlv}(\Phi)$ and $x \in X(\gamma_0)$. Then x is a smooth point of X if and only if Q maps $\text{lin}(\gamma_0) \cap E$ onto K.*

Corollary 3.3.1.12 *Let X be the variety associated with a bunched ring and $X \subseteq Z$ the embedding into the toric variety Z constructed in 3.2.5.3. Moreover, let $x \in X$.*

(i) *The point x is a factorial ($\mathbb{Q}$-factorial) point of X if and only if it is a smooth ($\mathbb{Q}$-factorial) point of Z.*
(ii) *The point x is a smooth point of X if and only if it is a smooth point of Z and some $\widehat{x} \in p^{-1}(x)$ is a smooth point of $\widehat{X}$.*

As an immediate consequence of general results on quotient singularities (see [64] and [174]), one obtains the following.

Proposition 3.3.1.13 *Let $(R, \mathfrak{F}, \Phi)$ be a bunched ring. Set $X = X(R, \mathfrak{F}, \Phi)$ and $\widehat{X} = \widehat{X}(R, \mathfrak{F}, \Phi)$. Suppose that $\widehat{X}$ is smooth. Then X has at most rational singularities. In particular, X is Cohen–Macaulay.*

Example 3.3.1.14 Consider the surface $X = X(R, \mathfrak{F}, \Phi)$ from Example 3.2.1.6. Recall that R is graded by $K = \mathbb{Z}^2$ and is explicitly given as

$$R = \mathbb{K}[T_1, \ldots, T_5] / \langle T_1 T_2 + T_3^2 + T_4 T_5 \rangle,$$

$$Q = \begin{bmatrix} 1 & -1 & 0 & -1 & 1 \\ 1 & 1 & 1 & 0 & 2 \end{bmatrix},$$

where $\deg(T_i) = w_i$ with the ith column w_i of Q. The system of generators $\mathfrak{F}$ consists of the classes of $T_1, \ldots, T_5$ and the $\mathfrak{F}$-bunch is $\Phi = \Phi(w_3)$. In the projected cone $(E \xrightarrow{\varrho} K, \gamma)$, we have $E = \mathbb{Z}^5$ and $\gamma = \text{cone}(e_1, \ldots, e_5)$. With $\gamma_{i_1,\ldots,i_k} := \text{cone}(e_{i_1}, \ldots, e_{i_k})$, the collection of relevant faces is

$$\text{rlv}(\Phi) =$$

$$\{\gamma_{1,4}, \gamma_{2,5}, \gamma_{1,2,3}, \gamma_{3,4,5}, \gamma_{1,2,3,4}, \gamma_{1,2,3,5}, \gamma_{1,2,4,5}, \gamma_{1,3,4,5}, \gamma_{2,3,4,5}, \gamma_{1,2,3,4,5}\}.$$

The stratum $X(\gamma_{1,2,3,4,5}) \subseteq X$ is open. The facets of γ define open sets of the prime divisors D_X^i divisors and we have four other strata, each consisting of one point:

$$X(\gamma_{1,2,3,4}) = D_X^5, \quad \ldots, \quad X(\gamma_{2,3,4,5}) = D_X^1,$$

$$X(\gamma_{1,4}), \quad X(\gamma_{2,5}), \quad X(\gamma_{1,2,3}), \quad X(\gamma_{3,4,5}).$$

To determine the local class groups, note that for the relevant faces $\gamma_0 \preceq \gamma$ we obtain

$$Q(\gamma_0 \cap E) = \begin{cases} \mathbb{Z} \cdot (-1, 1) + \mathbb{Z} \cdot (0, 3) & \gamma_0 = \gamma_{2,5}, \\ \mathbb{Z}^2 & \text{else.} \end{cases}$$

Thus, all points different from the point $x_0 = [0, 1, 0, 0, 1] \in X(\gamma_{2,5})$ have a trivial local class group. Moreover,

$$\mathrm{Cl}(X, x_0) = \mathbb{Z}/3\mathbb{Z}, \qquad \mathrm{Pic}(X) = \mathbb{Z} \cdot (-1, 1) + \mathbb{Z} \cdot (0, 3) \subseteq \mathbb{Z}^2 = \mathrm{Cl}(X).$$

Clearly, X is $\mathbb{Q}$-factorial. Because $\widehat{X}(R, \mathfrak{F}, \Phi)$ is smooth, the singular locus is determined by the combinatorial part of the data; it is $X(\gamma_{2,5}) = \{x_0\}$.

3.3.2 Base loci and cones of divisors

We first provide general descriptions of (stable) base loci as well as the cones of effective, movable, semiample, and ample divisor classes on a variety in terms of its Cox ring. Then we interpret the results in the language of bunched rings. The ground field $\mathbb{K}$ is algebraically closed and of characteristic zero.

Recall that a Weil divisor $D \in \mathrm{WDiv}(X)$ on an irreducible, normal prevariety X is effective if its multiplicities are all nonnegative. The *effective cone* is the cone $\mathrm{Eff}(X) \subseteq \mathrm{Cl}_{\mathbb{Q}}(X)$ generated by the classes of effective divisors. Note that $\mathrm{Eff}(X)$ is convex, and, given $D \in \mathrm{WDiv}(X)$, we have $[D] \in \mathrm{Eff}(X)$ if and only if there is a nonzero $f \in \Gamma(X, \mathcal{O}_X(nD))$ for some $n > 0$.

Proposition 3.3.2.1 *Let X be an irreducible, normal prevariety with $\Gamma(X, \mathcal{O}^*) = \mathbb{K}^*$ and finitely generated divisor class group $\mathrm{Cl}(X)$. Let f_i, $i \in I$, be any system of nonzero homogeneous generators of the Cox ring $\mathcal{R}(X)$. Then the cone of effective divisor classes of X is given by*

$$\mathrm{Eff}(X) = \mathrm{cone}(\deg(f_i); \ i \in I).$$

Proof Clearly each $\deg(f_i)$ is the class of an effective divisor and thus the cone on the right-hand side is contained in $\mathrm{Eff}(X)$. Conversely, if $[D]$ belongs to $\mathrm{Eff}(X)$, then some multiple $n[D]$ is represented by an effective $nD \in \mathrm{WDiv}(X)$ and the canonical section 1_{nD} defines a nonzero element in $\Gamma(X, \mathcal{R}_{[nD]})$ that is a polynomial in the f_i. Consequently, $[D]$ is a nonnegative linear combination of the classes $\deg(f_i)$. □

For a Weil divisor D on an irreducible, normal prevariety X and a section $f \in \Gamma(X, \mathcal{O}_X(D))$ we introduced the D-divisor as $\operatorname{div}_D(f) = \operatorname{div}(f) + D$. The *base locus* and the *stable base locus* of D, or of the linear system associated with D if X is projective, are defined as

$$\operatorname{Bs}|D| := \bigcap_{f \in \Gamma(X, \mathcal{O}_X(D))} \operatorname{Supp}(\operatorname{div}_D(f)), \qquad \mathbf{B}|D| := \bigcap_{n \in \mathbb{Z}_{\geq 1}} \operatorname{Bs}|nD|.$$

Remark 3.3.2.2 The base locus and the stable base locus of a Weil divisor D on an irreducible, normal prevariety X only depend on the class $[D] \in \operatorname{Cl}(X)$. Thus, we can speak of the (stable) base locus of a class $[D] \in \operatorname{Cl}(X)$. Moreover, if we have $\Gamma(X, \mathcal{O}^*) = \mathbb{K}^*$ and $\operatorname{Cl}(X)$ is finitely generated, then we may write the base locus of a class $w := [D] \in \operatorname{Cl}(X)$ in terms of the Cox ring as

$$\operatorname{Bs}(w) := \operatorname{Bs}|D| = \bigcap_{f \in \Gamma(X, \mathcal{R}_{[D]})} \operatorname{Supp}(\operatorname{div}_{[D]}(f)), \qquad \mathbf{B}(w) := \mathbf{B}|D|.$$

A Weil divisor $D \in \operatorname{WDiv}(X)$ is called *movable* if its stable base locus is of codimension at least 2 in X. The *moving cone* $\operatorname{Mov}(X) \subseteq \operatorname{Cl}_{\mathbb{Q}}(X)$ is the cone consisting of the classes of movable divisors. Note that $\operatorname{Mov}(X)$ is convex.

Proposition 3.3.2.3 *Let X be an irreducible, normal complete variety with finitely generated divisor class group $\operatorname{Cl}(X)$ and let f_i, $i \in I$, be any system of pairwise nonassociated $\operatorname{Cl}(X)$-prime generators for the Cox ring $\mathcal{R}(X)$. Then the moving cone is given as*

$$\operatorname{Mov}(X) = \bigcap_{i \in I} \operatorname{cone}(\deg(f_j); \ j \in I \setminus \{i\}).$$

Lemma 3.3.2.4 *Let X be an irreducible, normal complete variety with $\operatorname{Cl}(X)$ finitely generated and let $w \in \operatorname{Cl}(X)$ be effective. Then the following two statements are equivalent.*

(i) *The stable base locus of the class $w \in \operatorname{Cl}(X)$ contains a divisor.*
(ii) *There exist an $w_0 \in \operatorname{Cl}(X)$ with $\dim \Gamma(X, \mathcal{R}_{nw_0}) = 1$ for any $n \in \mathbb{Z}_{\geq 0}$ and an $f_0 \in \Gamma(X, \mathcal{R}_{w_0})$ such that for any $m \in \mathbb{Z}_{\geq 1}$ and $f \in \Gamma(X, \mathcal{R}_{mw})$ one has $f = f' f_0$ with some $f' \in \Gamma(X, \mathcal{R}_{mw-w_0})$.*

Proof The implication "(ii)⇒(i)" is obvious. So, assume that (i) holds. The class $w \in \operatorname{Cl}(X)$ is represented by some nonnegative divisor D. Let D_0 be a prime component of D that occurs in the base locus of any positive multiple of D, and let $w_0 \in \operatorname{Cl}(X)$ be the class of D_0. Then the canonical section of D_0 defines an element $f_0 \in \Gamma(X, \mathcal{R}_{w_0})$ that by Proposition 1.5.3.5 divides any $f \in \Gamma(X, \mathcal{R}_{mw})$, where $m \in \mathbb{Z}_{\geq 1}$. Note that $\Gamma(X, \mathcal{R}_{nw_0})$ is of dimension 1 for every $n \in \mathbb{Z}_{\geq 1}$, because otherwise $\Gamma(X, \mathcal{R}_{na_0w_0})$ where $a_0 > 0$ is the multiplicity of D_0 in D, would provide enough sections in $\Gamma(X, \mathcal{R}_{nw})$ to move $na_0 D_0$.   □

Proof of Proposition 3.3.2.3 Set $w_i := \deg(f_i)$ and denote by $I_0 \subseteq I$ the set of indices with $\dim \Gamma(X, \mathcal{R}_{nw_0}) \leq 1$ for all $n \in \mathbb{N}$. Let $w \in \mathrm{Mov}(X)$. Then Lemma 3.3.2.4 tells us that for any $i \in I_0$, there must be a monomial of the form $\prod_{j \neq i} f_j^{n_j}$ in some $\Gamma(X, \mathcal{R}_{nw})$. Consequently, w lies in the cone of the right-hand side. Conversely, consider an element w of the cone of the right-hand side. Then, for every $i \in I_0$, a product $\prod_{j \neq i} f_j^{n_j}$ belongs to some $\Gamma(X, \mathcal{R}_{nw})$. Hence none of the f_i, $i \in I_0$, divides all elements of $\Gamma(X, \mathcal{R}_{nw})$. Again by Lemma 3.3.2.4, we conclude $w \in \mathrm{Mov}(X)$.   □

Now suppose that we are in the situation of Construction 1.6.1.3. That means that X is of affine intersection, $\Gamma(X, \mathcal{O}^*) = \mathbb{K}^*$ holds, $\mathrm{Cl}(X)$ is finitely generated, and the Cox sheaf $\mathcal{R}$ is locally of finite type. Then we have a characteristic space $p_X \colon \widehat{X} \to X$, where $\widehat{X} = \mathrm{Spec}_X \mathcal{R}$, which is a good quotient for the action of $H_X = \mathrm{Spec}\, \mathbb{K}[\mathrm{Cl}(X)]$. For $x \in X$, we already considered the submonoid

$$S_x = \{[D] \in \mathrm{Cl}(X);\ f(\widehat{x}) \neq 0 \text{ for some } f \in \Gamma(X, \mathcal{R}_{[D]})\} \subseteq \mathrm{Cl}(X),$$

where $\widehat{x} \in p_X^{-1}(x)$ is a point with closed H_X-orbit. Let $\omega_x \subseteq \mathrm{Cl}_{\mathbb{Q}}(X)$ denote the convex cone generated by S_x.

Proposition 3.3.2.5 *Consider the situation of Construction 1.6.1.3. Then the base locus and the stable base locus of a Weil divisor D on X are given as*

$$\mathrm{Bs}\,|D| = \{x \in X;\ [D] \notin S_x\}, \qquad \mathbf{B}\,|D| = \{x \in X;\ [D] \notin \omega_x\}.$$

Proof The assertions immediately follow from the description of the monoid S_x in terms of the $[D]$-divisors provided by Corollary 1.6.2.2:

$$S_x = \{[D] \in \mathrm{Cl}(X);\ x \notin \mathrm{div}_{[D]}(f) \text{ for some } f \in \Gamma(X, \mathcal{R}_{[D]})\}$$

$$= \{[D] \in \mathrm{Cl}(X);\ D \geq 0,\ x \notin \mathrm{Supp}(D)\}. \qquad\qquad □$$

Recall that a Weil divisor D in an irreducible, normal prevariety X is said to be *semiample* if its stable base locus is empty. The cone $\mathrm{SAmple}(X)$ in $\mathrm{Cl}_{\mathbb{Q}}(X)$ generated by the classes of semiample divisors is convex. Moreover, D is *ample* if X is covered by affine sets $X_{nD,f}$ for some $n \in \mathbb{Z}_{\geq 1}$. The classes of ample divisors generate a cone $\mathrm{Ample}(X)$ in $\mathrm{Cl}_{\mathbb{Q}}(X)$ that is again convex.

Proposition 3.3.2.6 *Consider the situation of Construction 1.6.1.3. The cones of semiample and ample divisor classes in $\mathrm{Cl}_{\mathbb{Q}}(X)$ are given as*

$$\mathrm{SAmple}(X) = \bigcap_{x \in X} \omega_x, \qquad \mathrm{Ample}(X) = \bigcap_{x \in X} \omega_x^\circ.$$

Proof The statement on the semiample cone follows directly from the description of the stable base locus given in Proposition 3.3.2.5.

We turn to the ample cone. Surely, $\mathrm{Ample}(X)$ is the intersection of all cones $\omega_x^a \subseteq \mathrm{Cl}_{\mathbb{Q}}(X)$ generated by the submonoid

$$S_x^a := \{[D] \in \mathrm{Cl}(X);\ x \in X_{[D],f} \text{ for an } f \in \Gamma(X, \mathcal{R}_{[D]}) \text{ with } X_{[D],f} \text{ affine}\}$$
$$\subseteq \mathrm{Cl}(X).$$

Thus, we have to verify $\omega_x^a = \omega_x^\circ$ for any $x \in X$. We first show $\omega_x^a \supseteq \omega_x^\circ$. Let $[E] \in \omega_x^\circ$. Then $[E]$ admits a neighborhood in ω_x° of the form

$$[E] \in \mathrm{cone}([E_1], \ldots, [E_k]) \subseteq \omega_x^\circ.$$

Here, we may assume that there are $g_i \in \Gamma(X, \mathcal{R}_{[E_i]})$ with $x \in X_{[E_i],g_i}$. Take any $[D] \in \omega_x^a$ and $f \in \Gamma(X, \mathcal{R}_{[D]})$ such that $X_{[D],f}$ is affine. Then we have

$$p_X^{-1}(x) \subseteq \widehat{X}_{f g_1 \cdots g_r} \subseteq \widehat{X}_f = p_X^{-1}(X_{[D],f}).$$

Because the restricted quotient map $p_X \colon \widehat{X}_f \to X_{[D],f}$ separates the disjoint closed H_X-invariant subsets $\widehat{X}_f \setminus \widehat{X}_{f g_1 \cdots g_r}$, we find a function $h \in \Gamma(X_{[D],f}, \mathcal{O}_X)$ with

$$h(x) \neq 0, \qquad h|_{p_X(\widehat{X}_f \setminus \widehat{X}_{f g_1 \cdots g_r})} = 0.$$

Fix $n \in \mathbb{Z}_{\geq 1}$ with $f' := h f^n \in \Gamma(X, \mathcal{R}_{[nD]})$. For any choice $a_1, \ldots, a_k$ of positive integers, we obtain a p_X-saturated affine open neighborhood of the fiber over x:

$$p_X^{-1}(x) \subseteq \widehat{X}_{f' g_1^{a_1} \cdots g_r^{a_r}} \subseteq \widehat{X}_f = p_X^{-1}(X_{[D],f}).$$

Thus, $X_{n[D]+a_1[E_1]+\cdots+a_k[E_k], f' g_1^{a_1} \cdots g_r^{a_r}}$ is an affine neighborhood of x. Consequently, all $n[D] + a_1[E_1] + \cdots + a_k[E_k]$ belong to ω_x^a. We conclude $[E] \in \omega_x^a$.

To show $\omega_x^a \subseteq \omega_x^\circ$, let $[D] \in \omega_x^a$. Replacing D with a suitable positive multiple, we may assume that $\widehat{X}_f$ is an affine neighborhood of $p_X^{-1}(x)$ for some $f \in \Gamma(X, \mathcal{R}_{[D]})$. This enables us to choose an H-equivariant affine closure $\widehat{X} \subseteq Z$ with $\widehat{X}_f = Z_f$. Let $H_X \cdot \widehat{x}$ be the closed orbit of the fiber $p_X^{-1}(x)$. Then we have

$$[D] \in \omega_{Z,\widehat{x}} \subseteq \omega_x.$$

Let $\omega \preceq \omega_{Z,\widehat{x}}$ be the face with $[D] \in \omega^\circ$. Then $\omega = \omega_{Z,z}$ holds for some $z \in \overline{H_X \cdot \widehat{x}}$, where the closure is taken with respect to Z. Proposition 3.1.1.6 tells us $f(z) \neq 0$. Because $q^{-1}(x)$ and hence $H_X \cdot \widehat{x}$ are closed in $\widehat{X}_f$ and f vanishes along $Z \setminus \widehat{X}_f$, we obtain $z \in H_X \cdot \widehat{x}$, which implies $\omega = \omega_{Z,\widehat{x}}$ and $[D] \in \omega_{Z,\widehat{x}}^\circ \subseteq \omega_x^\circ$. $\qquad\square$

Now we consider the variety X arising from a bunched ring $(R, \mathfrak{F}, \Phi)$. We first take a close look at the canonical isomorphism $K \cong \mathrm{Cl}(X)$ provided by Theorem 3.2.1.4. Set

$$E(R) := \bigcup_{w \in K} E(R)_w,$$

$$E(R)_w := \left\{ \frac{g}{h}; \ g, h \in R \text{ homog., } \deg(g) - \deg(h) = w \right\}.$$

Then the vector space $E(R)_w$ contains precisely the homogeneous rational functions of weight w on $\overline{X}$; use H-factoriality of $\overline{X}$ to see this. In the actual context, Proposition 1.6.4.5 tells us the following.

Proposition 3.3.2.7 *Consider the situation of Construction 3.2.1.3. Then there is an epimorphism of abelian groups*

$$\delta: E(R) \to \mathrm{WDiv}(X), \qquad f \mapsto p_*(\mathrm{div}(f)).$$

We have $\mathrm{div}(f) = p^*(p_*(\mathrm{div}(f)))$ *for every* $f \in E(R)$. *The epimorphism* δ *induces a well-defined isomorphism*

$$K \to \mathrm{Cl}(X), \qquad w \mapsto [\delta(f)], \quad \text{with any } f \in E(R)_w.$$

Fix $f \in E(R)_w$ *and set* $D := \delta(f)$. *Then, for every open set* $U \subseteq X$, *we have an isomorphism of* $\Gamma(U, \mathcal{O})$-*modules*

$$\Gamma(U, \mathcal{O}_X(D)) \to \Gamma(p^{-1}(U), \mathcal{O}_{\widehat{X}})_w, \qquad g \mapsto fp^*(g).$$

Moreover, for any section $g = h/f \in \Gamma(X, \mathcal{O}_X(D))$, *the corresponding* D-*divisor satisfies*

$$\mathrm{div}_D(g) = p_*(\mathrm{div}(h)), \qquad p^*(\mathrm{div}_D(g)) = \mathrm{div}(h).$$

If in this situation the open subset $X_{D,g} \subseteq X$ *is affine, then its inverse image is given as* $p^{-1}(X_{D,g}) = \overline{X}_h$.

Proposition 3.3.2.8 *Consider the situation of Construction 3.2.1.3. Then, for every* $w \in K \cong \mathrm{Cl}(X)$, *the base locus and the stable base locus are given as*

$$\mathrm{Bs}(w) = \bigcup_{\substack{\gamma_0 \in \mathrm{rlv}(\Phi) \\ w \notin Q(\gamma_0 \cap E)}} X(\gamma_0), \qquad \mathbf{B}(w) = \bigcup_{\substack{\gamma_0 \in \mathrm{rlv}(\Phi) \\ w \notin Q(\gamma_0)}} X(\gamma_0).$$

Proof Proposition 3.3.1.4 tells us that, for every $\gamma_0 \in \mathrm{rlv}(\Phi)$ and $x \in X(\gamma_0)$, the monoid $S_x \subseteq \mathrm{Cl}(X)$ corresponds to $Q(\gamma_0 \cap E)$. Thus, Proposition 3.3.2.5 gives the desired statements. $\square$

Proposition 3.3.2.9 *Consider the situation of Construction 3.2.1.3. The cones of effective, movable, semiample, and ample divisor classes of X in* $\mathrm{Cl}_{\mathbb{Q}}(X) = K_{\mathbb{Q}}$ *are given as*

$$\mathrm{Eff}(X) \;=\; Q(\gamma), \qquad\qquad \mathrm{Mov}(X) \;=\; \bigcap_{\gamma_0 \ \text{facet of} \ \gamma} Q(\gamma_0),$$

$$\mathrm{SAmple}(X) \;=\; \bigcap_{\tau \in \Phi} \tau, \qquad\qquad \mathrm{Ample}(X) \;=\; \bigcap_{\tau \in \Phi} \tau^{\circ}.$$

Moreover, if $X \subseteq Z$ *is the canonical toric embedding constructed in 3.2.5.3, then the cones of effective, movable, semiample, and ample divisor classes in* $\mathrm{Cl}_{\mathbb{Q}}(X) = K_{\mathbb{Q}} = \mathrm{Cl}_{\mathbb{Q}}(Z)$ *coincide for X and Z.*

Proof The descriptions of the effective and the moving cones are clear by Propositions 3.3.2.1 and 3.3.2.3. For the semiample and the ample cones, note as before that $S_x \subseteq \mathrm{Cl}(X)$ corresponds to $Q(\gamma_0 \cap E) \subseteq K$ and apply Proposition 3.3.2.6. $\square$

Example 3.3.2.10 We continue the study of the surface $X = X(R, \mathfrak{F}, \Phi)$ from 3.2.1.6. The ring R and its grading by $K = \mathbb{Z}^2$ are given as

$$R = \mathbb{K}[T_1, \ldots, T_5] \, / \, \langle T_1 T_2 + T_3^2 + T_4 T_5 \rangle,$$

$$Q = \begin{bmatrix} 1 & -1 & 0 & -1 & 1 \\ 1 & 1 & 1 & 0 & 2 \end{bmatrix},$$

where $\deg(T_i) = w_i$ with the ith column w_i of Q. The system $\mathfrak{F}$ consists of the classes of $T_1, \ldots, T_5$ and the $\mathfrak{F}$-bunch is $\Phi = \Phi(w_3)$. The cones of effective, movable, and semiample divisor classes in $\mathrm{Cl}_{\mathbb{Q}}(X) = \mathbb{Q}^2$ are given by

$$\mathrm{Eff}(X) \;=\; \mathrm{cone}(w_1, w_4), \qquad \mathrm{Mov}(X) \;=\; \mathrm{SAmple}(X) \;=\; \mathrm{cone}(w_2, w_5).$$

In the projected cone $(E \xrightarrow{\;\varrho\;} K, \gamma)$, we have $E = \mathbb{Z}^5$ and $\gamma = \mathrm{cone}(e_1, \ldots, e_5)$. With $\gamma_{i_1,\ldots,i_k} := \mathrm{cone}(e_{i_1}, \ldots, e_{i_k})$, the collection of relevant faces is

$$\mathrm{rlv}(\Phi) =$$

$$\{\gamma_{1,4}, \gamma_{2,5}, \gamma_{1,2,3}, \gamma_{3,4,5}, \gamma_{1,2,3,4}, \gamma_{1,2,3,5}, \gamma_{1,2,4,5}, \gamma_{1,3,4,5}, \gamma_{2,3,4,5}, \gamma_{1,2,3,4,5}\}.$$

Recall that $X(\gamma_{2,5})$ consists of the singular point of X. For the (stable) base loci of low multiples of $w := (0, 1)$, we obtain

$$\mathrm{Bs}(w) \;=\; \mathrm{Bs}(2w) \;=\; X(\gamma_{2,5}), \qquad \mathrm{Bs}(3w) \;=\; \mathbf{B}(w) \;=\; \emptyset.$$

3.3.3 Complete intersections

We consider a bunched ring $(R, \mathfrak{F}, \Phi)$, where R is a homogeneous complete intersection in the sense defined below. In this situation, there is a simple description of the canonical divisor and intersection numbers can easily be computed. The ground field $\mathbb{K}$ is algebraically closed and of characteristic zero.

Definition 3.3.3.1 Let $(R, \mathfrak{F}, \Phi)$ be a bunched ring with grading group K and $\mathfrak{F} = (f_1, \ldots, f_r)$. We say that $(R, \mathfrak{F}, \Phi)$ is a *complete intersection* if the kernel $I(\mathfrak{F})$ of the epimorphism

$$\mathbb{K}[T_1, \ldots, T_r] \;\to\; R, \qquad T_i \mapsto f_i.$$

is generated by K-homogeneous polynomials $g_1, \ldots, g_d \in \mathbb{K}[T_1, \ldots, T_r]$, where $d = r - \dim R$. In this situation, we call $(w_1, \ldots, w_r)$, where $w_i := \deg(f_i) \in K$ and $(u_1, \ldots, u_d)$, where $u_i := \deg(g_i) \in K$, *degree vectors* for $(R, \mathfrak{F}, \Phi)$.

Proposition 3.3.3.2 *Let the bunched ring $(R, \mathfrak{F}, \Phi)$ be a complete intersection with degree vectors $(w_1, \ldots, w_r)$ and $(u_1, \ldots, u_d)$. Then the canonical divisor class of $X = X(R, \mathfrak{F}, \Phi)$ is given in $\mathrm{Cl}(X) = K$ by*

$$w_X^{\mathrm{can}} = \sum_{j=1}^{d} u_j - \sum_{i=1}^{r} w_i.$$

Proof Consider the embedding $X \to Z$ into a toric variety Z of X as constructed in 3.2.5.3. Then Proposition 3.3.1.12 tells us that the respective embeddings of the smooth loci fit into a commutative diagram

$$
\begin{array}{ccc}
X_{\mathrm{reg}} & \longrightarrow & Z_{\mathrm{reg}} \\
\downarrow & & \downarrow \\
X & \longrightarrow & Z
\end{array}
$$

Note that all maps induce isomorphisms on the respective divisor class groups. Moreover, the restrictions of the canonical divisors K_X and K_Z of X and Z give the canonical divisors of X_{reg} and Z_{reg}, respectively; see, for example, [215, p. 164]. Thus, we may assume that X and its ambient toric variety Z are smooth.

Let $\mathcal{I} \subset \mathcal{O}_Z$ be the ideal sheaf of X. Then the normal sheaf of X in Z is the rank d locally free sheaf $\mathcal{N}_X := (\mathcal{I}/\mathcal{I}^2)^*$, and a canonical bundle on X can be obtained as follows; see, for example, [160, Prop. II.8.20]:

$$\mathcal{K}_X = \mathcal{K}_Z|_X \otimes \left(\bigwedge^{d} \mathcal{N}_X \right).$$

Choose a cover of Z by open subsets U_i such that $\mathcal{I}/\mathcal{I}^2$ is free over U_i. Then the g_l generate the relations of the f_j over $\widehat{U}_i := p_Z^{-1}(U_i)$. Thus, after suitably refining the cover, we find functions $h_{il} \in \mathcal{O}^*(\widehat{U}_i)$ of degree $\deg(g_l)$ such that $\mathcal{I}/\mathcal{I}^2(U_i)$ is generated by $g_1/h_{i1}, \ldots, g_d/h_{id}$. Therefore over U_i, the dth exterior power of $\mathcal{I}/\mathcal{I}^2$ is generated by the function

$$\frac{g_1}{h_{i1}} \wedge \cdots \wedge \frac{g_d}{h_{id}}.$$

Proposition 3.3.2.7 tells us that the class of $\bigwedge^d \mathcal{I}/\mathcal{I}^2$ in K is minus the sum of the degrees of the g_j. As $\bigwedge^d \mathcal{N}_X$ is the dual sheaf, its class is $\deg(g_1) + \cdots + \deg(g_l)$. Furthermore, from Proposition 2.4.2.8 we know that the class of the canonical divisor of Z in K is given by $-(w_1 + \cdots + w_r)$. Putting all together we arrive at the assertion. $\qquad\square$

A variety is called *(Q-)Gorenstein* if (some multiple of) its anticanonical divisor is Cartier. Moreover, a *Fano variety* is an irreducible, normal projective variety with an ample anticanonical divisor.

Corollary 3.3.3.3 *Let the bunched ring* $(R, \mathfrak{F}, \Phi)$ *be a complete intersection with degree vectors* $(w_1, \ldots, w_r)$ *and* $(u_1, \ldots, u_d)$ *and let* $X = X(R, \mathfrak{F}, \Phi)$ *be the associated variety.*

(i) *X is Q-Gorenstein if and only if*

$$\sum_{i=1}^r w_i - \sum_{j=1}^d u_j \in \bigcap_{\tau \in \Phi} \mathrm{lin}(\tau),$$

(ii) *X is Gorenstein if and only if*

$$\sum_{i=1}^r w_i - \sum_{j=1}^d u_j \in \bigcap_{\gamma_0 \in \mathrm{cov}(\Phi)} Q(\mathrm{lin}(\gamma_0) \cap E).$$

(iii) *X is Fano if and only if we have*

$$\sum_{i=1}^r w_i - \sum_{j=1}^d u_j \in \bigcap_{\tau \in \Phi} \tau^\circ,$$

Construction 3.3.3.4 (Computing intersection numbers) Let the bunched ring $(R, \mathfrak{F}, \Phi)$ be a complete intersection with degree vectors $(w_1, \ldots, w_r)$ and $(u_1, \ldots, u_d)$ and suppose that $\Phi = \Phi(\lambda)$ holds with a full-dimensional GIT cone $\lambda \in \Lambda(\overline{X}, H)$. Fix a full-dimensional $\eta \in \Lambda(\overline{Z}, H)$ with $\eta^\circ \subseteq \lambda^\circ$. For $w_{i_1}, \ldots, w_{i_{n+d}}$ let $w_{j_1}, \ldots, w_{j_{r-n-d}}$ denote the complementary weights and set

$$\tau(w_{i_1}, \ldots, w_{i_{n+d}}) := \mathrm{cone}(w_{j_1}, \ldots, w_{j_{r-n-d}}),$$

$$\mu(w_{i_1}, \ldots, w_{i_{n+d}}) := [K : \langle w_{j_1}, \ldots, w_{j_{r-n-d}} \rangle].$$

Then the intersection product $K_{\mathbb{Q}}^{n+d} \to \mathbb{Q}$ of the ($\mathbb{Q}$-factorial) toric variety Z_1 associated with $\Phi(\eta)$ is determined by the values

$$
w_{i_1} \cdots w_{i_{n+d}} = \begin{cases} \mu(w_{i_1}, \ldots, w_{i_{n+d}})^{-1}, & \eta \subseteq \tau(w_{i_1}, \ldots, w_{i_{n+d}}), \\ 0, & \eta \not\subseteq \tau(w_{i_1}, \ldots, w_{i_{n+d}}). \end{cases}
$$

As a complete intersection, $X \subseteq Z_1$ inherits intersection theory. For a tuple $D_X^{i_1}, \ldots, D_X^{i_n}$ on X, its intersection number can be computed by

$$
D_X^{i_1} \cdots D_X^{i_n} = w_{i_1} \cdots w_{i_n} \cdot u_1 \cdots u_d.
$$

Note that the intersection number $D_X^{i_1} \cdots D_X^{i_n}$ vanishes if cone($e_{i_1}, \ldots, e_{i_n}$) does not belong to rlv(Φ).

Example 3.3.3.5 We continue the study of the surface $X = X(R, \mathfrak{F}, \Phi)$ from Example 3.2.1.6. The ring R and its grading by $K = \mathbb{Z}^2$ are given as

$$
R = \mathbb{K}[T_1, \ldots, T_5] / \langle T_1 T_2 + T_3^2 + T_4 T_5 \rangle,
$$

$$
Q = \begin{bmatrix} 1 & -1 & 0 & -1 & 1 \\ 1 & 1 & 1 & 0 & 2 \end{bmatrix},
$$

where $\deg(T_i) = w_i$ with the ith column w_i of Q. The system $\mathfrak{F}$ consists of the classes of $T_1, \ldots, T_5$ and the $\mathfrak{F}$-bunch $\Phi = \Phi(w_3)$ has $\tau = \mathrm{cone}(w_2, w_5)$ as the unique minimal cone. The degree of the defining relation is $\deg(T_1 T_2 + T_3^2 + T_4 T_5) = 2w_3$ and thus the canonical class of X is given as

$$
w_X^c = 2w_3 - (w_1 + w_2 + w_3 + w_4 + w_5) = -3w_3.
$$

In particular, the anticanonical class is ample and, by Example 3.3.1.14, it lies in Pic(X). Thus, X is a Gorenstein singular del Pezzo surface. The self-intersection number of the canonical class is

$$
(-3w_3)^2 = \frac{9(w_1 + w_2)(w_4 + w_5)}{4} = \frac{9}{4}(w_1 \cdot w_4 + w_2 \cdot w_4 + w_1 \cdot w_5 + w_2 \cdot w_5)
$$

The $w_i \cdot w_j$ equal the toric intersection numbers $2w_i \cdot w_j \cdot w_3$. To compute these numbers, let w_{ij}^1, w_{ij}^2 denote the weights in $\{w_1, \ldots, w_5\} \setminus \{w_i, w_j, w_3\}$. Then we have

$$
w_i \cdot w_j \cdot w_3 = \begin{cases} \mu(w_i, w_j, w_3)^{-1} & \tau \subseteq \mathrm{cone}(w_{ij}^1, w_{ij}^2), \\ 0 & \tau \not\subseteq \mathrm{cone}(w_{ij}^1, w_{ij}^2), \end{cases}
$$

where the multiplicity $\mu(w_i, w_j, w_3)^{-1}$ is the absolute value of $\det(w_{ij}^1, w_{ij}^2)$. Thus, we can proceed in the computation:

$$
(-3w_3)^2 = \frac{9 \cdot 2}{4} \left(|\det(w_2, w_5)|^{-1} + |\det(w_1, w_4)|^{-1} \right) = \frac{9}{2} \cdot \frac{4}{3} = 6.
$$

Finally, we apply the techniques for complete intersections to complete d-dimensional varieties X with divisor class group $\mathrm{Cl}(X) \cong \mathbb{Z}$. The Cox ring $\mathcal{R}(X)$ is finitely generated and the total coordinate space $\overline{X} := \mathrm{Spec}\,\mathcal{R}(X)$ is a factorial affine variety coming with an action of $\mathbb{K}^*$ defined by the $\mathrm{Cl}(X)$-grading of $\mathcal{R}(X)$. Choose a system $f_1, \ldots, f_\nu$ of homogeneous pairwise nonassociated prime generators for $\mathcal{R}(X)$. This provides an $\mathbb{K}^*$-equivariant embedding

$$\overline{X} \to \mathbb{K}^\nu, \qquad \overline{x} \mapsto (f_1(\overline{x}), \ldots, f_\nu(\overline{x})).$$

where $\mathbb{K}^*$ acts diagonally with the weights $w_i = \deg(f_i) \in \mathrm{Cl}(X) \cong \mathbb{Z}$ on $\mathbb{K}^\nu$. Moreover, X is the geometric $\mathbb{K}^*$-quotient of $\widehat{X} := \overline{X} \setminus \{0\}$, and the quotient map $p \colon \widehat{X} \to X$ is a characteristic space.

Proposition 3.3.3.6 *For any $\overline{x} = (\overline{x}_1, \ldots, \overline{x}_\nu) \in \widehat{X}$ the local divisor class group $\mathrm{Cl}(X, x)$ of $x := p(\overline{x})$ is finite of order $\gcd(w_i; \ \overline{x}_i \neq 0)$. The index of the Picard group $\mathrm{Pic}(X)$ in $\mathrm{Cl}(X)$ is given by*

$$[\mathrm{Cl}(X) : \mathrm{Pic}(X)] = \mathrm{lcm}_{x \in X}(|\,\mathrm{Cl}(X, x)|).$$

Suppose that the ideal of $\overline{X} \subseteq \mathbb{K}^\nu$ is generated by $\mathrm{Cl}(X)$-homogeneous polynomials $g_1, \ldots, g_{\nu-d-1}$ of degree $\gamma_j := \deg(g_j)$. Then one obtains

$$-\mathcal{K}_X = \sum_{i=1}^{\nu} w_i - \sum_{j=1}^{\nu-d-1} \gamma_j, \qquad (-\mathcal{K}_X)^d = \left(\sum_{i=1}^{\nu} w_i - \sum_{j=1}^{\nu-d-1} \gamma_j \right)^d \frac{\gamma_1 \cdots \gamma_{\nu-d-1}}{w_1 \cdots w_\nu}$$

for the anticanonical class $-\mathcal{K}_X \in \mathrm{Cl}(X) \cong \mathbb{Z}$. In particular, X is a $\mathbb{Q}$-Fano variety if and only if the following inequality holds:

$$\sum_{j=1}^{\nu-d-1} \gamma_j < \sum_{i=1}^{\nu} w_i.$$

3.3.4 Mori dream spaces

We take a closer look at projective varieties with a finitely generated Cox ring, as before, defined over an algebraically closed field $\mathbb{K}$ of characteristic zero. Hu and Keel investigated the birational geometry of these varieties in [176], and also gave them a name.

Definition 3.3.4.1 A *Mori dream space* is an irreducible, normal projective variety X with finitely generated divisor class group $\mathrm{Cl}(X)$ and finitely generated Cox ring $\mathcal{R}(X)$.

We first discuss the relations between Mori dream spaces sharing a given Cox ring in terms of the GIT fan of the characteristic quasitorus action. By the *moving cone* of the K-graded algebra R we mean here the intersection $\mathrm{Mov}(R)$ over all $\mathrm{cone}(w_1, \ldots, w_{i-1}, w_{i+1}, \ldots, w_r)$, where the w_i are the degrees of any system of pairwise nonassociated homogeneous K-prime generators for R.

Remark 3.3.4.2 Let $R = \bigoplus_K R_w$ be an almost freely factorially graded affine algebra with $R_0 = \mathbb{K}$ and consider the GIT fan $\Lambda(\overline{X}, H)$ of the action of $H = \mathrm{Spec}\,\mathbb{K}[K]$ on $\overline{X} = \mathrm{Spec}\,R$.

Then every GIT cone $\lambda \in \Lambda(\overline{X}, H)$ defines a projective variety $X(\lambda) := \overline{X}^{ss}(\lambda) /\!\!/ H$. If $\lambda^{\circ} \subseteq \mathrm{Mov}(R)^{\circ}$ holds, then $X(\lambda)$ is the variety associated with the bunched ring $(R, \mathfrak{F}, \Phi(\lambda))$ with $\Phi(\lambda)$ defined as in 3.1.3.6. In particular, in this case we have

$$\mathrm{Cl}(X(\lambda)) \;=\; K, \qquad \mathcal{R}(X(\lambda)) \;=\; R,$$

$$\mathrm{Mov}(X(\lambda)) \;=\; \mathrm{Mov}(R), \qquad \mathrm{SAmple}(X(\lambda)) \;=\; \lambda.$$

All projective varieties with Cox ring R are isomorphic to some $X(\lambda)$ with $\lambda^{\circ} \subseteq \mathrm{Mov}(R)^{\circ}$, and the Mori dream spaces among them are precisely those arising from a full-dimensional λ.

Let X be the variety arising from a bunched ring $(R, \mathfrak{F}, \Phi)$. Every Weil divisor D on X defines a positively graded sheaf

$$\mathcal{S}^{+}(D) \;:=\; \bigoplus_{n \in \mathbb{Z}_{\geq 0}} \mathcal{S}_n^{+}(D), \qquad \mathcal{S}_n^{+}(D) \;:=\; \mathcal{O}_X(nD).$$

Using Propositions 1.1.2.4, 1.6.1.1, and 1.6.1.4, we see that this sheaf is locally of finite type, and thus we obtain a rational map

$$\varphi(D)\colon X \dashrightarrow X(D), \qquad X(D) \;:=\; \mathrm{Proj}(\Gamma(X, \mathcal{S}^{+}(D))).$$

Note that $X(D)$ is explicitly given as the closure of the image of the rational map $X \to \mathbb{P}^m$ determined by the linear system of a sufficiently big multiple nD.

Remark 3.3.4.3 Consider the GIT fan $\Lambda(\overline{X}, H)$ of the action of $H = \mathrm{Spec}\,\mathbb{K}[K]$ on $\overline{X} = \mathrm{Spec}\,R$. Let $\lambda \in \Lambda(\overline{X}, H)$ be the cone with $[D] \in \lambda^{\circ}$ and

$W \subseteq \overline{X}$ the open subset obtained by removing the zero sets of the generators $f_1, \ldots, f_r \in R$. Then we obtain a commutative diagram

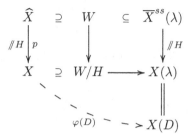

The maximal area of definition of the rational map $\varphi \colon X \dashrightarrow X(D)$ is precisely $X \setminus \mathbf{B}\,|D|$, the complement of the stable base locus $\mathbf{B}\,|D|$. Note that $\mathbf{B}\,|D|$ equals the image $p(\widehat{X} \setminus \overline{X}^{ss}(\lambda))$.

We list basic properties of the maps $\varphi(D)$; by a *small quasimodification* between two varieties we mean a rational map defining an isomorphism of open subsets with complement of codimension 2.

Remark 3.3.4.4 Let $D \in \mathrm{WDiv}(X)$ be any Weil divisor, and denote by $[D] \in \mathrm{Cl}(X)$ its class. Then the assocciated rational map $\varphi(D) \colon X \to X(D)$ is

   (i) birational if and only if $[D] \in \mathrm{Eff}(X)^\circ$ holds.
  (ii) a small quasimodification if and only if $[D] \in \mathrm{Mov}(X)^\circ$ holds.
 (iii) a morphism if and only if $[D] \in \mathrm{SAmple}(X)$ holds.
 (iv) an isomorphism if and only if $[D] \in \mathrm{Ample}(X)$ holds.

Recall that two Weil divisors $D, D' \in \mathrm{WDiv}(X)$ are said to be *Mori equivalent* if $\mathbf{B}\,|D| = \mathbf{B}|D'|$ holds and there is a commutative diagram of rational maps

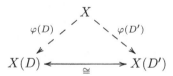

where the horizontal arrow stands for an isomorphism of varieties. The Mori equivalence is described by the GIT fan of the characteristic quasitorus action on the total coordinate space.

Proposition 3.3.4.5 *Let* $X = X(R, \mathfrak{F}, \Phi)$ *be the variety arising from a bunched ring and let* $\Lambda(\overline{X}, H)$ *be the GIT fan of the action of* $H = \mathrm{Spec}\,\mathbb{K}[K]$

on $\overline{X} = \operatorname{Spec} R$. Then for any two D, $D' \in \operatorname{WDiv}(X)$, the following statements are equivalent.

(i) *The divisors D and D' are Mori equivalent.*
(ii) *One has $[D], [D'] \in \lambda^{\circ}$ for some GIT chamber $\lambda \in \Lambda(\overline{X}, H)$.*

Proof The assertion follows immediately from the observation that $X(D)$ is the GIT quotient associated with the chamber $\lambda \in \Lambda(\overline{X}, H)$ with $[D] \in \lambda^{\circ}$. □

We now provide Zariski decomposition for Weil divisors on Mori dream spaces X. Recall that a *rational divisor*, or *$\mathbb{Q}$-divisor*, on X is a $\mathbb{Q}$-linear combination of irreducible hypersurfaces that means an element of

$$\operatorname{WDiv}_{\mathbb{Q}}(X) := \operatorname{WDiv}(X) \otimes_{\mathbb{Z}} \mathbb{Q}.$$

Note that we have canonical epimorphism $\operatorname{WDiv}_{\mathbb{Q}}(X) \to \operatorname{Cl}_{\mathbb{Q}}(X)$; we write $[D]$ for the image of D.

Definition 3.3.4.6 Let X be an irreducible $\mathbb{Q}$-factorial normal variety and $D \in \operatorname{WDiv}(X)$ an effective Weil divisor on X. A *Zariski decomposition* of D is a presentation $D = P + N$ with P, $N \in \operatorname{WDiv}_{\mathbb{Q}}(X)$ such that $[N] \in \operatorname{Eff}(X)$ and $[P] \in \operatorname{Mov}(X)$ hold and for some $n \in \mathbb{Z}_{\geq 1}$ such that nP is integral, the canonical map $\Gamma(X, \mathcal{S}^{+}(nP)) \to \Gamma(X, \mathcal{S}^{+}(nD))$ is an isomorphism of graded algebras.

Proposition 3.3.4.7 *An effective Weil divisor D on an irreducible, normal variety X admits at most one Zariski decomposition.*

Proof Consider Zariski decompositions $D = P + N = P' + N'$. Fix $n > 0$ such that nD, nP, nP' are integral and the latter two have base locus of codimension at least 2. Then we have

$$\Gamma(X, \mathcal{O}(nP)) \cong \Gamma(X, \mathcal{O}(nD)) \cong \Gamma(X, \mathcal{O}(nP')).$$

It follows that nN and nN' are the unique effective divisors with $\operatorname{div}_{nD}(f) \geq nN$ and $\operatorname{div}_{nD}(f) = nN'$ for all $f \in \Gamma(X, \mathcal{O}(nD))$. This gives $N = N'$ and $P = P'$. □

Theorem 3.3.4.8 *Let X be the variety arising from a bunched ring $(R, \mathfrak{F}, \Phi)$ as in Construction 3.2.1.3. Asumme X is $\mathbb{Q}$-factorial with $\Gamma(X, \mathcal{O}) = \mathbb{K}$ and let $D \in \operatorname{WDiv}(X)$ be a positive combination of the divisors D_X^i. Consider the cone*

$$\tau_i := \operatorname{cone}(w_D, -w_i) \cap \operatorname{cone}(w_1, \ldots, w_{i-1}, w_{i+1}, \ldots, w_r) \subseteq K_{\mathbb{Q}} \cong \operatorname{Cl}_{\mathbb{Q}}(X),$$

where w_D, w_i denote the class of D, D_X^i in $\mathbb{K}_\mathbb{Q}$. Let μ_i be the unique nonnegative rational number such that $w_D - \mu_i w_i$ is an extremal ray of τ_i. Then D admits a Zariski decomposition

$$D = P + N, \quad \text{where} \quad N := \mu_1 D_X^1 + \cdots + \mu_r D_X^r, \quad P := D - N,$$

with unique rational Weil divisors P and N on X. Moreover, for any $n \in \mathbb{Z}_{\geq 0}$ such that nN is an integral Weil divisor, $\Gamma(X, \mathcal{O}(nN))$ is of dimension at most 1.

Proof First of all observe that $\mu_i \in \mathbb{Q}_{\geq 0}$ is the smallest number such that the class w_D admits a presentation

$$w_D = \mu_i w_i + \sum_{j \neq i} \alpha_j w_j, \quad \text{where } \alpha_j \in \mathbb{Q}_{\geq 0}.$$

In particular, we see that the class $w_N = \sum \mu_i w_i$ of N is effective. Moreover, the class $w_P = w_D - w_N$ of $P = D - N$ belongs to $\text{Mov}(X)$, because by construction it admits for every i a presentation

$$w_P = \sum_{j \neq i} \beta_j w_j, \quad \text{where } \beta_j \in \mathbb{Q}_{\geq 0},$$

using Proposition 3.3.2.9. Now, fix $n \in \mathbb{Z}_{\geq 1}$ such that nP is integral. Consider any monomial $f_1^{a_1} \cdots f_r^{a_r} \in R_{knw_D}$. By the definition of μ_i, we have $a_i \geq kn\mu_i$. This gives a presentation

$$f_1^{a_1} \cdots f_r^{a_r} = \left(f_1^{a_1 - kn\mu_1} \cdots f_r^{a_r - kn\mu_r} \right) \cdot \left(f_1^{kn\mu_1} \cdots f_r^{kn\mu_r} \right),$$

where the first monomial of the right-hand side belongs to R_{knw_P}. We conclude that the canonical map

$$R(nw_P) \to R(nw_D), \quad R_{knw_P} \ni g \mapsto \left(f_1^{n\mu_1} \cdots f_r^{n\mu_r} \right)^k g \in R_{knw_D}$$

of Veronese subalgebras of the Cox ring $R = \mathcal{R}(X)$ is surjective. Consequently, also $\Gamma(X, \mathcal{S}^+(nP)) \to \Gamma(X, \mathcal{S}^+(nD))$ is an isomorphism of graded algebras.

For the last statement, we show that the representation $w_N = \mu_1 w_1 + \cdots + \mu_r w_r$ as a nonnegative linear combination in $K_\mathbb{Q}$ is unique. Let $w_N = \nu_1 w_1 + \cdots + \nu_r w_r$ be another one. Because w_i generate a pointed cone, we must have $\nu_i < \mu_i$ for some i. This provides a presentation $w_D = w_P + \nu_1 w_1 + \cdots + \nu_r w_r$, which contradicts the definition of μ_i. $\square$

Corollary 3.3.4.9 *Let X be a Mori dream space. Then any divisor D of X admits a unique Zariski decomposition.*

Proof It is enough to observe that X comes from a bunched ring and that D is linearly equivalent to a divisor satisfying the hypothesis of Theorem 3.3.4.8. $\square$

Remark 3.3.4.10 If X is a Mori dream space and D is an effective divisor of X then, after applying a suitable small quasimodification $X \dashrightarrow X'$, the divisor P in the Zariski decomposition of D becomes semiample.

3.4 Varieties with a torus action of complexity 1

3.4.1 Detecting factorial gradings

In this section, $\mathbb{K}$ is an arbitrary field. We consider algebras R graded by an abelian group K and ask when this grading is factorial. The main result shows that it suffices to check this property for a suitable Veronese subalgebra of some localization of R. The original reference for this section is [40].

Definition 3.4.1.1 Let R be a K-graded $\mathbb{K}$-algebra. We say that R is 1-*noetherian* if for every nonempty set of K-homogeneous elements $f \in R$ the associated set of ideals $\langle f \rangle$ has maximal elements.

Example 3.4.1.2 Let X be an irreducible, normal variety with an action of a quasitorus H. This gives an algebra $R = \Gamma(X, \mathcal{O})$ graded by $K = \mathbb{X}(H)$, which is always 1-noetherian but not necessarily noetherian. For example, according to Proposition 4.3.4.5, the blow-up of the projective plane at nine points in very general position has a nonfinitely generated Cox ring. The associated characteristic space has a nonfinitely generated, hence nonnoetherian algebra of functions, using [65, Lemma 2.4].

Construction 3.4.1.3 Let R be a K-graded $\mathbb{K}$-algebra and $S \subseteq R$ a *homogeneous multiplicative monoid*, that is, a multiplicative monoid consisting of homogeneous elements. Then the localization $S^{-1}R$ becomes K-graded by defining the component of degree $w \in K$ as

$$(S^{-1}R)_w := \left\{ \frac{r}{s} \in S^{-1}R; \ r \in R_\times, \ s \in S, \ \deg(r) - \deg(s) = w \right\}.$$

Definition 3.4.1.4 Let K be an abelian group and R a K-graded integral $\mathbb{K}$-algebra. We call (K', S) a *reducing pair* if $K' \subseteq K$ is a subgroup, $S \subseteq R$ is a homogeneous multiplicative monoid generated by K-prime elements and in the K-graded localization $S^{-1}R$ every homogeneous element is associated with a homogeneous element of the Veronese subalgebra $(S^{-1}R)(K')$.

Theorem 3.4.1.5 *Let K be an abelian group and R a K-graded integral 1-noetherian $\mathbb{K}$-algebra. Then the following statements are equivalent.*

(i) *The algebra R is factorially K-graded.*

(ii) *There is a reducing pair* (K', S) *such that* $(S^{-1}R)(K')$ *is factorially* K'*-graded.*

(iii) *For every reducing pair* (K', S) *the algebra* $(S^{-1}R)(K')$ *is factorially* K'*-graded.*

The following special case of the theorem relates the concept of a factorially graded algebra to factoriality in the usual sense.

Corollary 3.4.1.6 *Let* K *be an abelian group,* R *a* K*-graded integral* $\mathbb{K}$*-algebra with* K*-prime generators* $f_1, \ldots, f_r \in R$, *and* $S \subseteq R$ *the multiplicative monoid generated by the* $f_1, \ldots, f_r$. *Then the following statements are equivalent.*

(i) *The algebra* R *is factorially* K*-graded.*
(ii) *The algebra* $(S^{-1}R)_0$ *is factorial.*

For the proof of the theorem, we study how the property of being factorially graded behaves with respect to (suitable) homogeneous localization and with respect to passing to (suitable) Veronese subalgebras. The first step means establishing a homogeneous version of Nagata's characterization of factorial rings [216, Thm. 20.2].

Lemma 3.4.1.7 *Let* K *be an abelian group,* R *a* K*-graded integral 1-noetherian* $\mathbb{K}$*-algebra, and* $S \subseteq R$ *a homogeneous multiplicative monoid generated by* K*-primes.*

(i) *Every homogeneous* $f \in R$ *admits a presentation* $f = s'f'$ *with* $s' \in S$ *and a homogeneous* $f' \in R$ *coprime to all members of* S.
(ii) *Let* $f \in R$ *be a homogeneous element coprime to all members of* S. *Then* f *is* K*-prime in* R *if and only if* f *is* K*-prime in* $S^{-1}R$.

Proof We verify (i). If f is coprime to all members of S, we may take $s' = 1$ and $f' = f$. If f has factors in S, then we consider the (nonempty) set M of all principal ideals $\langle h \rangle \subseteq R$, where $h \in R$ is K-homogeneous such that $f = sh$ holds with $s \in S$. Because R is 1-noetherian, there is a maximal member $\langle f' \rangle \in M$. Then we have $f = s'f'$ with $s' \in S$ and f', as wanted.

We prove (ii). First assume that f is K-prime in R. Then $f \in S^{-1}R$ is nonzero and it is a nonunit, because it is coprime to all members of S. Let f divide a product $(g/s)(h/r)$ of homogeneous elements in $S^{-1}R$, that is, we have $(g/s)(h/r) = (f/1)(a/b)$. Then $fasr = ghb$ holds in R. Because f is coprime to $b \in S$, we see that f divides gh in R. Thus it divides one of the factors, say g. Then f divides g/s in $S^{-1}R$.

Now assume that f is K-prime in $S^{-1}R$. Then f is a nonzero nonunit in R. Suppose that f divides a product gh of homogeneous elements g and h in R. Then f divides gh as well in $S^{-1}R$. Thus, f divides one of the factors g, h, say

g, in $S^{-1}R$. This means $gs = g'f$ with some $s \in S$ and a homogeneous $g' \in R$. Because s is a product of K-primes and f is coprime to s, we obtain that s divides g' and hence a presentation $g = g''f$ with a homogeneous $g'' \in R$.   □

Proposition 3.4.1.8 *Let K be an abelian group, R a K-graded integral 1-noetherian $\mathbb{K}$-algebra, and $S \subseteq R$ a homogeneous multiplicative monoid generated by K-prime elements. Then the following statements are equivalent.*

(i) *The algebra R is factorially K-graded.*
(ii) *The algebra $S^{-1}R$ is factorially K-graded.*

Proof Assume that (i) holds. Let $f/s \in S^{-1}R$ be a homogeneous nonzero nonunit. Then $f \in R$ is a homogeneous nonzero nonunit and thus admits a decomposition $f = f_1 \cdots f_r$ into K-primes $f_i \in R$. Because S is generated by K-primes, a given f_i is either associated with some element of S or coprime to all elements of S. Suitably renumbering, we ensure that the first k of the f_i are precisely the ones associated with an element of S. Then $g := f_1 \cdots f_k$ is a unit in $S^{-1}R$ and, according to Lemma 3.4.1.7 (ii), the elements $f_{k+1}, \ldots, f_r$ are K-prime in $S^{-1}R$. Thus, the desired decomposition of f/s is $(g/s)f_{k+1} \cdots f_r$.

Now assume that (ii) holds. Let $f \in R$ be a homogeneous nonzero nonunit. Then, we have a decomposition $f = (f_1/s_1) \cdots (f_r/s_r)$ into K-primes $f_i/s_i \in S^{-1}R$. Write $f_i = s_i'f_i'$ as in Lemma 3.4.1.7 (i). Then each f_i' is K-prime in $S^{-1}R$ and thus by Lemma 3.4.1.7 (ii) also K-prime in R. Moreover, in R, we have

$$s_1 \cdots s_r \cdot f = s_1' \cdots s_r' \cdot f_1' \cdots f_r'.$$

Because all $s_i, s_i' \in S$ are products of K-primes from S, we obtain from this a presentation $f = s_1'' \cdots s_r'' \cdot f_1' \cdots f_r'$ with $s_i'' \in S$. Writing each s_i'' as a product of K-primes, we finally arrive at the desired decomposition of f into K-primes of R.   □

Lemma 3.4.1.9 *Let K be an abelian group, R a K-graded integral $\mathbb{K}$-algebra, and $K' \subseteq K$ a subgroup such that for every homogeneous $f \in R$, there is a homogeneous unit $c_f \in R$ with $c_f f \in R(K')$. Then the following statements are equivalent.*

(i) *The element $f \in R$ is K-prime.*
(ii) *The element $c_f f \in R(K')$ is K'-prime.*

Proof Assume that (i) holds. Clearly $c_f f \in R(K')$ is a homogeneous nonzero nonunit. Let $c_f f$ divide a product gh of homogeneous elements $g, h \in R(K')$. Then f divides gh in R, and we may assume that $g = fg'$ holds with a homogeneous $g' \in R$. This implies $g = c_f f c_f^{-1} g'$. Because $\deg(c_f^{-1} g')$ equals $\deg(g) - \deg(c_f f) \in K'$, we have $c_f^{-1} g' \in R(K')$.

Now assume that (ii) holds. Then $f = c_f^{-1}(c_f f) \in R$ is a homogeneous nonzero nonunit. Let f divide a product gh of homogeneous elements $g, h \in R$. Then $c_f f$ divides $(c_g g)(c_h h)$ in $R(K')$ and we may assume that $c_g g = c_f f g'$ with a homogeneous $g' \in R(K')$ holds. This gives $g = f c_f c_g^{-1} g'$. $\qquad\square$

Proposition 3.4.1.10 *Let K be an abelian group, R a K-graded integral $\mathbb{K}$-algebra, and $K' \subseteq K$ a subgroup such that every homogeneous element of R is associated with a homogeneous element of $R(K')$. Then the following statements are equivalent.*

(i) *The algebra R is factorially K-graded.*
(ii) *The algebra $R(K')$ is factorially K'-graded.*

Proof Assume that (i) holds. Given a homogeneous nonzero nonunit $f \in R(K')$, we can write it as a product $f = f_1 \cdots f_r$ with K-primes $f_i \in R$. Choose homogeneous units c_i with $c_i f_i \in R(K')$. Then Lemma 3.4.1.9 tells us that each $c_i f_i$ is K'-prime. Moreover, with $c := c_1 \cdots c_r$, we have

$$cf = c_1 f_1 \cdots c_r f_r, \qquad \deg(c) + \deg(f) = \deg(c_1 f_1) + \cdots + \deg(c_r f_r).$$

We conclude $\deg(c) \in K'$ and thus $c \in R(K')$. Consequently, $f = c^{-1}(c_1 f_1) \cdots (c_r f_r)$ is the desired decomposition into K'-primes in $R(K')$.

Assume that (ii) holds. Let $f \in R$ be a homogeneous nonzero nonunit. Choose a homogeneous unit $c_f \in R$ with $c_f f \in R(K')$. Then we can write $c_f f = f_1 \cdots f_r$ with K'-primes $f_i \in R(K')$. According to Lemma 3.4.1.9, the f_i are K-primes in R. Thus $f = c_f^{-1} f_1 \cdots f_r$ is the desired decomposition into K-primes. $\qquad\square$

Proof of Theorem 3.4.1.5 To prove "(i)$\Rightarrow$(iii)", let (K', S) be a reducing pair. Then Proposition 3.4.1.8 yields that $S^{-1} R$ is factorially K-graded and Proposition 3.4.1.10 says that $(S^{-1} R)(K')$ is factorially K'-graded. The implication "(iii)$\Rightarrow$(ii)" is clear; one may, for example, take the trivial reducing pair $(K, \{1\})$. To obtain "(ii)$\Rightarrow$(i)", we apply first Proposition 3.4.1.10 and see that $(S^{-1} R)$ factorially K-graded. Then Proposition 3.4.1.8 shows that R is factorially K-graded. $\qquad\square$

As an application, we show that the concepts "factorially graded" and "factorial" coincide for a finitely generated integral algebra graded by a finitely generated torsion free abelian group, provided that $\mathbb{K}$ is algebraically closed of characteristic zero; see also [6].

Theorem 3.4.1.11 *Assume that $\mathbb{K}$ is algebraically closed of characteristic zero. Let K be a torsion-free abelian group and $R = \oplus_K R_w$ a K-graded finitely generated integral $\mathbb{K}$-algebra. Then the following statements hold.*

(i) *A K-homogeneous element $f \in R$ is K-prime if and only if it is prime.*
(ii) *The ring R is factorially K-graded if and only if it is factorial.*

Proof Because R is finitely generated, we may assume that the grading group K is finitely generated. Moreover, we may assume that and the grading is effective.

We verify (i). If a K-homogeneous f is prime, then it is clearly K-prime. For the converse, we may assume $K = \mathbb{Z}^k$ and thus have the lexicographic order on K. Let f be K-prime and assume that f divides a product gh with two $g, h \in R$. Write

$$g = \widehat{g} + g_1, \qquad h = \widehat{h} + h_1,$$

where $\widehat{g}$ and $\widehat{h}$ denote the leading terms, that is, the K-homogeneous components of g and h of maximal degree with respect to the lexicographic order on K. Then we see that f divides $\widehat{gh} = \widehat{g}\widehat{h}$. Because f is K-prime, it divides $\widehat{g}$ or $\widehat{h}$.

If f divides both leading terms, then it divides $g_1 h_1$ and hence $\widehat{g_1}$ or $\widehat{h_1}$. Iterating leads either to the situation that f divides g and h, or, for some i, it does not divide one of the leading terms, say $\widehat{h_i}$. Then f divides $\widehat{g_i}$ and hence $g_{i+1} h_i$, where $g_{i+1} = g_i - \widehat{g_i}$. Iterating further shows that f divides all homogeneous components of g and thus g itself.

We turn to (ii). First assume R is factorially K-graded. We consider the affine variety $Y := \operatorname{Spec} R$ with the action of the torus $H := \operatorname{Spec} \mathbb{K}[K]$. Take nonzero homogeneous elements $f_1, \ldots, f_r \in R$ such that their degrees generate $K \cong \mathbb{Z}^k$ and set $f := f_1 \cdots f_r$. Then H acts freely on Y_f and we obtain an isomorphism

$$Y_f \;\to\; Y_0 \times \mathbb{T}^k, \qquad y \mapsto (\pi(y), g_1(y), \ldots, g_k(y)),$$

where $\pi \colon Y_f \to Y_0 := Y_f/H$ is the quotient map and $g_j \in R_f$ are any homogeneous elements such that their degrees form a basis of K. Write every f_i as a product of K-primes f_{ij} and denote by $S \subseteq R$ the multiplicative monoid generated by all the f_{ij}. Corollary 3.4.1.6 shows that

$$(S^{-1}R)_0 = (R_f)_0 = \Gamma(Y_0, \mathcal{O})$$

is a factorial ring. Consequently, Y_0 has trivial divisor class group $\operatorname{Cl}(Y_0)$. Then also $Y_f = Y_0 \times \mathbb{T}^k$ has trivial divisor class group. In other words, $R_f = S^{-1}R$ is a factorial ring. By (i), the f_{ij} are even prime. Thus, Nagata's criterion [216, Thm. 20.2] says that R is a factorial ring.

Now, if R is factorial, then every homogeneous f is a product of primes $p_1, \ldots, p_r$. Because K is torsion free, we can conclude that the p_i are K-homogeneous and thus K-prime. $\qquad\square$

3.4.2 Factorially graded rings of complexity 1

We present a construction associating with a pair A, P_0 of certain matrices a finitely generated normal algebra $R(A, P_0)$ with an effective pointed factorial grading of complexity 1. As we will see later in 4.4.2.3, this gives in fact all algebras of the latter type. The algebras $R(A, P_0)$ turn out to be complete intersections defined by specific trinomial relations. The original references for this section are [180, 221] for factorial rings in dimensions 2 and 3 and [166, 167] for factorially graded rings in arbitrary dimension. We work over an algebraically closed field $\mathbb{K}$ of characteristic zero.

Construction 3.4.2.1 Fix an integer $r \in \mathbb{Z}_{\geq 1}$, a sequence $n_0, \ldots, n_r \in \mathbb{Z}_{\geq 1}$, set $n := n_0 + \cdots + n_r$ and let $m \in \mathbb{Z}_{\geq 0}$. The input data are

- A matrix $A := [a_0, \ldots, a_r]$ with pairwise linearly independent column vectors $a_0, \ldots, a_r \in \mathbb{K}^2$
- An integral $r \times (n + m)$ block matrix $P_0 = (L, 0)$, where L is a $r \times n$ matrix build from tuples $l_i := (l_{i1}, \ldots, l_{in_i}) \in \mathbb{Z}_{\geq 1}^{n_i}$ as follows:

$$
L = \begin{bmatrix} -l_0 & l_1 & \cdots & 0 \\ \vdots & \vdots & \ddots & \vdots \\ -l_0 & 0 & \cdots & l_r \end{bmatrix}.
$$

Consider the polynomial ring $\mathbb{K}[T_{ij}, S_k]$ in the variables T_{ij}, where $0 \leq i \leq r$, $1 \leq j \leq n_i$, and S_k, where $1 \leq k \leq m$. For every $0 \leq i \leq r$, define a monomial

$$
T_i^{l_i} := T_{i1}^{l_{i1}} \cdots T_{in_i}^{l_{in_i}}.
$$

Denote by $\mathfrak{J}$ the set of all triples $I = (i_1, i_2, i_3)$ with $0 \leq i_1 < i_2 < i_3 \leq r$ and define for any $I \in \mathfrak{J}$ a trinomial

$$
g_I := g_{i_1, i_2, i_3} := \det \begin{bmatrix} T_{i_1}^{l_{i_1}} & T_{i_2}^{l_{i_2}} & T_{i_3}^{l_{i_3}} \\ a_{i_1} & a_{i_2} & a_{i_3} \end{bmatrix}.
$$

Let P_0^* denote the transpose of P_0. We introduce a grading on $\mathbb{K}[T_{ij}, S_k]$ by the factor group $K_0 := \mathbb{Z}^{n+m}/\mathrm{im}(P_0^*)$. Let $Q_0 \colon \mathbb{Z}^{n+m} \to K_0$ be the projection and set

$$
\deg(T_{ij}) := w_{ij} := Q_0(e_{ij}), \qquad \deg(S_k) := w_k := Q_0(e_k),
$$

where $e_{ij} \in \mathbb{Z}^{n+m}$, for $0 \leq i \leq r$, $1 \leq j \leq n_i$, and $e_k \in \mathbb{Z}^{n+m}$, for $1 \leq k \leq m$, are the canonical basis vectors. Note that all the g_I are K_0-homogeneous of degree

$$\mu := l_{01} w_{01} + \cdots + l_{0n_0} w_{0n_0} = \cdots = l_{r1} w_{r1} + \cdots + l_{rn_r} w_{rn_r} \in K_0.$$

In particular, the trinomials g_I generate a K_0-homogeneous ideal and thus we obtain a K_0-graded factor algebra

$$R(A, P_0) := \mathbb{K}[T_{ij}, S_k] / \langle g_I; \ I \in \mathfrak{I} \rangle.$$

Example 3.4.2.2 Let $r = 2$, take the sequence $(n_0, n_1, n_2) = (1, 1, 1)$, let $m = 0$, and consider the data

$$A = \begin{bmatrix} 0 & -1 & 1 \\ 1 & -1 & 0 \end{bmatrix}, \qquad P_0 = \begin{bmatrix} -2 & 2 & 0 \\ -2 & 0 & 2 \end{bmatrix}.$$

Then we have exactly one triple in $\mathfrak{I}$, namely $I = (0, 1, 2)$, and, as a ring, $R(A, P_0)$ is given by

$$R(A, P_0) = \mathbb{K}[T_{01}, T_{11}, T_{21}]/\langle T_{01}^2 + T_{11}^2 + T_{21}^2 \rangle.$$

The grading group $K_0 = \mathbb{Z}^3 / \mathrm{im}(P_0)$ is isomorphic to $\mathbb{Z} \oplus \mathbb{Z}/2\mathbb{Z} \oplus \mathbb{Z}/2\mathbb{Z}$. Concretely this grading can be realized as

$$\deg(T_{01}) = (1, \overline{0}, \overline{0}), \qquad \deg(T_{11}) = (1, \overline{1}, \overline{0}), \qquad \deg(T_{21}) = (1, \overline{0}, \overline{1}).$$

The algebra $R(A, P_0)$ is factorially K_0-graded but it is not factorial, because $X := V(T_{01}^2 + T_{11}^2 + T_{21}^2)$ has divisor class group $\mathbb{Z}/2\mathbb{Z}$.

We recall the necessary terminology on gradings. Given an integral affine $\mathbb{K}$-algebra $A = \oplus_K A_w$ graded by a finitely generated abelian group K, we say that the grading is *effective* if the weight monoid $S(A)$ generates K as a group and we call it *pointed* if $A_0 = \mathbb{K}$ holds and the weight cone $\omega(A)$ contains no line. Moreover, we say that the grading is of *complexity 1* if $\dim(K_\mathbb{Q})$ equals $\dim(A) - 1$.

Theorem 3.4.2.3 *With the notation of Construction 3.4.2.1, the following statements hold.*

(i) *The ring $R(A, P_0)$ is an integral normal complete intersection of dimension $n + m - r + 1$ with $R(A, P_0)^* = \mathbb{K}^*$.*

(ii) *The K_0-grading of ring $R(A, P_0)$ is effective, pointed, factorial, and of complexity 1.*

(iii) *The variables T_{ij} and S_k define a system of pairwise nonassociated K_0-prime generators of $R(A, P_0)$.*

(iv) *Suppose $r \geq 2$ and $n_i l_{ij} > 1$ for all i, j. Then the following statements are equivalent.*
  - *The ring $R(A, P_0)$ is factorial.*
  - *The numbers $l_i := \gcd(l_{i1}, \ldots, l_{in_i})$ are pairwise coprime.*
  - *The group K_0 is torsion free.*

As we will see in 4.4.2.3, every integral normal affine $\mathbb{K}$-algebra with an effective pointed factorial grading of complexity 1 is isomorphic to some $R(A, P_0)$. First, however, we prove Theorem 3.4.2.3. We begin with an observation concerning the structure of the ideal $\langle g_I; \ I \in \mathfrak{J} \rangle$.

Lemma 3.4.2.4 *In the setting of Construction 3.4.2.1, set $\alpha_{ij} := \det(a_i, a_j)$. Then, for any $0 \leq i < j < k < l \leq r$, one has the identities*

$$g_{i,k,l} \ = \ \alpha_{kl} \cdot g_{i,j,k} + \alpha_{ik} \cdot g_{j,k,l}, \qquad\qquad g_{i,j,l} \ = \ \alpha_{jl} \cdot g_{i,j,k} + \alpha_{ij} \cdot g_{j,k,l}.$$

In particular, the trinomials $g_i := g_{i,i+1,i+2}$, where $0 \leq i \leq r-2$, generate the whole ideal $\langle g_I; \ I \in \mathfrak{J} \rangle$.

Proof The identities are easily obtained by direct computation; note that for this one may assume $a_j = (1, 0)$ and $a_k = (0, 1)$. The last statement then follows by repeated application of the identities. $\square$

Denote the points of $\mathbb{K}^{n+m}$ as tuples $z = (z_{ij}, z_k)$ according to the variables T_{ij} and S_k. We will consider the zero set of the relations

$$\overline{X} \ := \ V(g_I; \ I \in \mathfrak{J}) \subseteq \mathbb{K}^{n+m}.$$

Once we know that $R(A, P_0)$ is reduced, we have $\overline{X} = \operatorname{Spec} R(A, P_0)$. Moreover, let H_0 be the kernel of the homomorphism

$$\mathbb{T}^{n+m} \ \to \ \mathbb{T}^r, \qquad (t_{ij}, t_k) \ \mapsto \ \left(\frac{t_1^{l_1}}{t_0^{l_0}}, \ldots, \frac{t_r^{l_r}}{t_0^{l_0}} \right), \qquad \text{where } t_i^{l_i} := t_{i1}^{l_{i1}} \cdots t_{in_i}^{l_{in_i}}.$$

Then H_0 is a quasitorus, isomorphic to $\operatorname{Spec} \mathbb{K}[K_0]$. It acts as a subgroup of $\mathbb{T}^{n+m}$ on $\mathbb{K}^{n+m}$ and $\overline{X}$ is invariant under this action.

Lemma 3.4.2.5 *In the setting of Construction 3.4.2.1 and the above notation, let $z \in \overline{X}$. If we have $T_i^{l_i}(z) = T_j^{l_j}(z) = 0$ for two $0 \leq i < j \leq r$, then $T_k^{l_k}(z) = 0$ holds for all $0 \leq k \leq r$.*

Proof Take a trinomial g_I, where I consists of the indices i, j, k suitably ordered. Then $g_I(z) = 0$ implies $T_k^{l_k}(z) = 0$. $\square$

Proposition 3.4.2.6 *The ring $R(A, P_0)$ constructed in 3.4.2.1 is an integral normal complete intersection of dimension $n + m - r + 1$ with $R(A, P_0)^* = \mathbb{K}^*$. Moreover, the K_0-grading of $R(A, P_0)$ is effective, pointed, and of complexity 1.*

Proof As a first step we show that $\overline{X} \subseteq \mathbb{K}^{n+m}$ is connected. For this, consider the one-parameter subgroup of $H_0 \subseteq \mathbb{T}^{n+m}$ given by

$$\lambda \colon \mathbb{K}^* \to H_0, \quad t \mapsto (t^{\zeta_{01}}, \ldots, t^{\zeta_{rn_r}}, t, \ldots, t), \quad \zeta_{ij} := n_i^{-1} l_{ij}^{-1} \prod_k n_k \prod_{k,m} l_{km}.$$

Because all ζ_{ij} are positive, any orbit of this one-parameter subgroup in $\mathbb{K}^{n+m}$ has the origin in its closure. As λ leaves $\overline{X}$ invariant, we conclude that $\overline{X}$ is connected.

We show now that $R(A, P_0)$ is a normal complete intersection of dimension $n + m - r + 1$. By Serre's criterion [132, Thm. 18.15] and Lemma 3.4.2.4, it suffices to show that the set of points of $z \in \overline{X}$ where the Jacobian of $g = (g_0, \ldots, g_{r-2})$ is not of full rank is of codimension at least 2 in $\overline{X}$. The Jacobian of g is of the shape $(J_g, 0)$ with a zero block arising from the variables S_k and a block

$$J_g = \begin{bmatrix} \delta_{00} & \delta_{01} & \delta_{02} & 0 & & & & 0 \\ & 0 & \delta_{11} & \delta_{12} & \delta_{13} & 0 & & \\ & & & & \vdots & & & \\ & & & & 0 & \delta_{r-3\,r-3} & \delta_{r-3\,r-2} & \delta_{r-3\,r-1} & 0 \\ 0 & & & & & 0 & \delta_{r-2\,r-2} & \delta_{r-2\,r-1} & \delta_{r-2\,r} \end{bmatrix}$$

given by the variables T_{ij}, where each δ_{ti} is a nonzero multiple of the gradient $\delta_i := \operatorname{grad} T_i^{l_i}$. Now consider $z \in \overline{X}$ with $J_g(z)$ not of full rank. Then $\delta_i(z) = 0 = \delta_k(z)$ holds with some $0 \le i < k \le r$. This implies $z_{ij} = 0 = z_{kl}$ for some $1 \le j \le n_i$ and $1 \le l \le n_k$. It follows $T_i^{l_i}(z) = 0 = T_k^{l_k}(z)$. Lemma 3.4.2.5 gives $T_s^{l_s}(z) = 0$, for all $0 \le s \le r$. Thus, some coordinate z_{st} must vanish for every $0 \le s \le r$. This shows that z belongs to a closed subset of $\overline{X}$ having codimension at least 2 in $\overline{X}$.

We consider the K_0-grading. Effectivity is given by construction, because the degrees $\deg(T_{ij}) = Q_0(e_{ij})$ and $\deg(S_k) = Q_0(e_k)$ generate K_0. Because the columns of the matrix P_0 generate $\mathbb{Q}^r$ as a cone, we obtain that the weight cone of the K_0-grading contains no lines and every variable T_{ij}, S_k defines a nonzero degree in $(K_0)_{\mathbb{Q}}$. We conclude $R(A, P_0)_0 = \mathbb{K}$ and $R(A, P_0)^* = \mathbb{K}^*$. Moreover, as the dimension of $R(A, P_0)$ is one larger than that of $(K_0)_{\mathbb{Q}}$, the K_0-grading is of complexity 1. $\square$

Lemma 3.4.2.7 *The notation is as in Construction 3.4.2.1. Then the variable T_{ij} defines a prime ideal in $R(A, P_0)$ if and only if the numbers $\gcd(l_{k1}, \ldots, l_{kn_k})$, where $k \neq i$, are pairwise coprime.*

Proof   We treat exemplarily T_{01}. Using Lemma 3.4.2.4, we see that the ideal of relations of $R(A, \mathfrak{n}, L)$ can be presented as follows

$$\langle g_{s,s+1,s+2};\ 0 \le s \le r - 2 \rangle = \langle g_{0,s,s+1};\ 1 \le s \le r - 1 \rangle.$$

Thus, the ideal $\langle T_{01} \rangle \subseteq R(A, \mathfrak{n}, L)$ is prime if and only if the following binomial ideal is prime

$$\mathfrak{a} := \langle \alpha_{s+10} f_s + \alpha_{0s} f_{s+1};\ 1 \le s \le r - 1 \rangle \subseteq \mathbb{K}[T_{ij};\ (i, j) \ne (0, 1)].$$

Set $l_i := (l_{i1}, \dots, l_{in_i})$. Then the ideal $\mathfrak{a}$ is prime if and only if the following family can be complemented to a lattice basis

$$(l_1, -l_2, 0, \dots, 0),\ \dots,\ (0, \dots, 0, l_{r-1}, -l_r).$$

This in turn is equivalent to the statement that the numbers $\gcd(l_{k1}, \dots, l_{kn_k})$, where $1 \le k \le r$, are pairwise coprime. $\square$

Proposition 3.4.2.8   *The notation is Construction 3.4.2.1. Regard the variables T_{ij} as regular functions on the affine variety $\overline{X} = \operatorname{Spec} R(A, P_0)$.*

(i) *The divisors of the T_{ij} on $\overline{X}$ are H_0-prime and pairwise different. In particular, the T_{ij} define pairwise nonassociated K_0-prime elements in $R(A, P_0)$.*
(ii) *If the ring $R(A, P_0)$ is factorial and $n_i l_{ij} > 1$ holds, then the divisor of T_{ij} on $\overline{X}$ is even prime.*

Proof   For (i), we exemplarily show that the divisor of T_{01} is H_0-prime. First note that by Lemma 3.4.2.4, the zero set $V(\overline{X}; T_{01})$ is described in $\mathbb{K}^{n+m}$ by the equations

$$T_{01} = 0, \qquad \alpha_{s+10} T_s^{l_s} + \alpha_{0s} T_{s+1}^{l_{s+1}} = 0, \quad 1 \le s \le r - 1. \qquad (3.4.1)$$

Let h denote the product of all T_{ij} with $(i, j) \ne (0, 1)$. Then, in $\mathbb{K}_h^{n+m}$, equations (3.4.1) are equivalent to

$$T_{01} = 0, \qquad -\frac{\alpha_{s+10} T_s^{l_s}}{\alpha_{0s} T_{s+1}^{l_{s+1}}} = 1, \quad 1 \le s \le r - 1.$$

Now, choose a point $z \in \mathbb{K}_h^{n+m}$ satisfying these equations. Then z_{01} is the only vanishing coordinate of z. Any other such point z' is of the form

$$z' = (0, t_{02} z_{02}, \dots, t_{rn_r} z_{rn_r}), \qquad t_{ij} \in \mathbb{K}^*,\ t_s^{l_s} = t_{s+1}^{l_{s+1}},\ 1 \le s \le r - 1.$$

Setting $t_{01} := t_{02}^{-l_{02}} \cdots t_{0n_0}^{-l_{0n_0}} t_1^{l_1}$, we obtain an element $t = (t_{ij}) \in H_0$ such that the above point z' equals $t \cdot z$. This consideration shows

$$V(\overline{X}_h; T_{01}) = H_0 \cdot z.$$

Using Lemma 3.4.2.5, we see that $V(\overline{X}; T_{01}, T_{ij})$ is of codimension at least 2 in $\overline{X}$ whenever $(i, j) \neq (0, 1)$. This allows us to conclude

$$V(\overline{X}; T_{01}) = \overline{H \cdot z}.$$

Thus, to obtain that T_{01} defines an H_0-prime divisor on $\overline{X}$, we need only that the equations (3.4.1) define a radical ideal. This in turn follows from the fact that their Jacobian at the point $z \in V(\overline{X}; T_{01})$ is of full rank.

To verify (ii), let $R(A, P_0)$ be factorial. Assume that the divisor of T_{ij} is not prime. Then we have $T_{ij} = h_1 \cdots h_s$ with prime elements $h_l \in R(A, P_0)$. Consider their decomposition into homogeneous parts

$$h_l = \sum_{w \in K} h_{l,w}.$$

Plugging this into the product $h_1 \cdots h_s$ shows that $\deg(T_{ij})$ is a positive combination of some $\deg(T_{kl})$ with $(k, l) \neq (i, j)$. Thus, there is a vector $(c_{kl}) \in \ker(Q) \subseteq E$ with $c_{ij} = 1$ and $c_{kl} \leq 0$ whenever $(k, l) \neq (i, j)$. Because, $\ker(Q_0)$ is spanned by the rows of P_0, we must have $n_i = 1$ and $l_{ij} = 1$, a contradiction to our assumptions. □

Proof of Theorem 3.4.2.3 Assertions (i), (ii), and (iii) are almost settled by Propositions 3.4.2.6 and 3.4.2.8. The only thing we still have to verify is factoriality of the K_0-grading. For this, consider once more the homomorphism of tori

$$\varphi \colon \mathbb{T}^{n+m} \to \mathbb{T}^r, \qquad (t_{ij}, t_k) \mapsto \left(\frac{t_1^{l_1}}{t_0^{l_0}}, \ldots, \frac{t_r^{l_r}}{t_0^{l_0}} \right), \qquad \text{where } t_i^{l_i} := t_{i1}^{l_{i1}} \cdots t_{in_i}^{l_{in_i}}.$$

Denoting by $U_1, \ldots, U_r$ the coordinates on $\mathbb{T}^r$ and setting $U_0 := 1$, we see the relations g_I are the pullbacks of certain affine-linear forms

$$g_{i_1,i_2,i_3} = T_0^{l_0} \varphi^*(h_{i_1,i_2,i_3}), \qquad h_{i_1,i_2,i_3} := \det \begin{bmatrix} U_{i_1} & U_{i_2} & U_{i_3} \\ a_{i_1} & a_{i_2} & a_{i_3} \end{bmatrix}.$$

Note that the affine linear forms h_{i_1,i_2,i_3} generate the vanishing ideal of an $r + 1$ times punctured projective line Y. In particular, $\mathbb{K}[U_i]/\langle h_I; I \in \mathfrak{I} \rangle$ is a factorial ring. By construction it is isomorphic to $(R(A, P_0)_t)_0$, the K_0-degree zero part of the localization by the product t over all T_{ij} and S_k. Thus, Corollary 3.4.1.6 tells us that $R(A, P_0)$ is factorially K_0-graded.

We prove (iv). If K_0 is torsion free, then K_0-factoriality of $R(A, P_0)$ implies factoriality; see Theorem 3.4.1.11. If $R(A, P_0)$ is factorial, then the generators T_{ij} are prime by Proposition 3.4.2.8. From Lemma 3.4.2.7, we then infer that the numbers $\gcd(l_{i1}, \ldots, l_{in_i})$ are pairwise coprime. The latter in turn implies that the rows of P_0 generate a primitive sublattice of $\mathbb{Z}^{n+m}$ and thus K_0 is torsion free. □

3.4.3 T-varieties of complexity 1 via bunched rings

Suitably downgrading the rings $R(A, P_0)$ from the preceding section, we produce bunched rings. The resulting varieties are rational varieties coming with a torus action of complexity 1, that is, the maximal orbit dimension is one less than the dimension of the variety. As we show later in Theorem 4.4.1.6, this construction produces all A_2-maximal varieties with only constant global functions coming with an effective torus action of complexity 1. Everything takes place over an algebraically closed fild $\mathbb{K}$ of characteristic zero. The original references for this section are [166, 167]. For an example class, we also refer to Section 5.4 treating $\mathbb{K}^*$-surfaces explicitly.

Construction 3.4.3.1 Fix $r \in \mathbb{Z}_{\geq 1}$, a sequence $n_0, \ldots, n_r \in \mathbb{Z}_{\geq 1}$, set $n :=$ $n_0 + \cdots + n_r$, and fix integers $m \in \mathbb{Z}_{\geq 0}$ and $0 < s < n + m - r$. The input data are

- A matrix $A := [a_0, \ldots, a_r]$ with pairwise linearly independent column vectors $a_0, \ldots, a_r \in \mathbb{K}^2$,
- An integral block matrix P of size $(r + s) \times (n + m)$ the columns of which are pairwise different primitive vectors generating $\mathbb{Q}^{r+s}$ as a cone.

$$P = \begin{bmatrix} L & 0 \\ d & d' \end{bmatrix},$$

where d is an $(s \times n)$-matrix, d' an $(s \times m)$-matrix, and L an $(r \times n)$-matrix build from tuples $l_i := (l_{i1}, \ldots, l_{in_i}) \in \mathbb{Z}_{\geq 1}^{n_i}$ as in Construction 3.4.2.1.

Let P^* denote the transpose of P, and consider the factor group $K :=$ $\mathbb{Z}^{n+m}/\mathrm{im}(P^*)$ and the projection $Q: \mathbb{Z}^{n+m} \to K$. We define a K-grading on $\mathbb{K}[T_{ij}, S_k]$ by setting

$$\deg(T_{ij}) := Q(e_{ij}), \qquad \deg(S_k) := Q(e_k).$$

The trinomials g_I of Construction 3.4.2.1 are K-homogeneous, all of the same degree. In particular, we obtain a K-graded factor ring

$$R(A, P) := \mathbb{K}[T_{ij}, S_k; \ 0 \leq i \leq r, \ 1 \leq j \leq n_i, 1 \leq k \leq m] / \langle g_I; \ I \in \mathfrak{J} \rangle.$$

Remark 3.4.3.2 As rings $R(A, P_0)$ and $R(A, P)$ coincide but the K_0-grading is finer than the K-grading. The downgrading map $K_0 \to K$ fits into the following commutative diagram built from exact sequences:

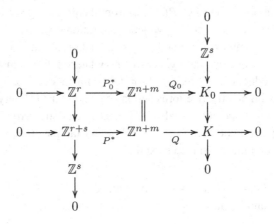

The snake lemma allows us to identify the direct factor $\mathbb{Z}^s$ of $\mathbb{Z}^{r+s}$ with the kernel of the downgrading map $K_0 \to K$. Note that for the quasitori T, H_0, and H associated with the abelian groups $\mathbb{Z}^s$, K_0, and K, we have $T = H_0/H$.

Example 3.4.3.3 (The E_6-singular cubic I)   Let $r = 2$, $n_0 = 2$, $n_1 = n_2 = 1$, $m = 0$, $s = 1$ and consider the data

$$A = \begin{bmatrix} 0 & -1 & 1 \\ 1 & -1 & 0 \end{bmatrix}, \qquad P = \begin{bmatrix} -3 & -1 & 3 & 0 \\ -3 & -1 & 0 & 2 \\ -2 & -1 & 1 & 1 \end{bmatrix}.$$

Then we have exactly one triple $I = (1, 2, 3)$ and, as a ring, $R(A, P)$ is given by

$$R(A, P) = \mathbb{K}[T_{01}, T_{02}, T_{11}, T_{21}] \,/\, \langle T_{01}^3 T_{02} + T_{11}^3 + T_{21}^2 \rangle.$$

The grading group $K = \mathbb{Z}^4/\mathrm{im}(P^*)$ is isomorphic to $\mathbb{Z}$ and the grading can be given explicitly via

$$\deg(T_{01}) = 1, \quad \deg(T_{02}) = 3, \quad \deg(T_{11}) = 2, \quad \deg(T_{21}) = 3.$$

As we will see in Examples 3.4.4.2, 3.4.4.10, and 5.4.3.5, the ring $R(A, P)$ is the Cox ring of the E_6-singular cubic surface X given in the projective space as

$$X = V(z_0^2 z_1 + z_0 z_2^2 + z_3^3) \subseteq \mathbb{P}^3.$$

Generalizing the foretelling of this example, the following statement says in particular that the K-graded algebras $R(A, P)$ are always Cox rings.

Theorem 3.4.3.4 *With the notation of Construction 3.4.3.1, the following statements hold.*

(i) *The K-grading of the ring $R(A, P)$ is factorial, pointed, and almost free, that is, K is generated by any $n + m - 1$ of the $\deg(T_{ij})$, $\deg(S_k)$.*

(ii) *The variables T_{ij} and S_k define a system of pairwise nonassociated K-prime generators of $R(A, P)$.*

Lemma 3.4.3.5 *Let K_0 be a finitely generated abelian group and $R = \bigoplus_K R_w$ a K_0-graded algebra. Moreover, let $\delta : K_0 \to K$ be an epimorphism with torsion-free kernel $K_0' \subseteq K_0$ and let $f_1, \ldots, f_r \in R$ be pairwise nonassociated K_0-primes such that any $r - 1$ of the images $\delta(\deg(f_i))$ generate K. Then $f_1, \ldots, f_r$ are pairwise nonassociated K-primes.*

Proof Consider f_1 and let $S \subseteq R$ be the multiplicative monoid generated by $f_2, \ldots, f_r$. Lemma 3.4.1.7 yields that f_1 is K_0-prime in the localization $S^{-1}R$. Because K is generated by $\delta(\deg(f_i))$, where $2 \leq i \leq r$, Lemma 3.4.1.9 shows that f_1 defines a K_0'-prime in the Veronese subalgebra $S^{-1}R(K_0')$. Theorem 3.4.1.11 (i) says that f_1 is even prime in $S^{-1}R(K_0')$. Using again Lemma 3.4.1.9, we see that f_1 is K-prime in $S^{-1}R$.

To see that f_1 is K-prime in R, assume that f_1 divides the product gh of two K-homogeneous $g, h \in R$. Then f_1 divides g, h as well in $S^{-1}R$ and thus one of the factors, say g. This means $gs = g'f_1$ with some $s \in S$ and $g' \in R$. Because s is a product of K_0-primes and coprime to f_1, we see that s divides all K_0-homogeneous components of g' in R and thus g'. We obtain $g = g''f_1$ with some $g'' \in R$. $\square$

Proof of Theorem 3.4.3.4 Because the columns of the matrix P are pairwise different primitive vectors, Lemma 2.1.4.1 tells us that any $n + m - 1$ of the images $Q(e_{ij})$, $Q(e_k)$ generate the group K. Thus, the K-grading is almost free. According to Lemma 3.4.3.5, variables T_{ij} and S_k define pairwise nonassociated K-primes in $R(A, P)$.

Now, let $S \subseteq R(A, P_0)$ be the multiplicative monoid generated by the variables. Consider the kernel $\mathbb{Z}^s \subseteq K_0$ of the downgrading map $K_0 \to K$. By Theorem 3.4.1.5, the ring $(S^{-1}R(A, P_0))(\mathbb{Z}^s)$ is factorially $\mathbb{Z}^s$-graded and hence, by Theorem 3.4.1.11 (ii), factorial. By construction, $(S^{-1}R(A, P_0))(\mathbb{Z}^s)$ equals $(S^{-1}R(A, P))_0$. Thus, Corollary 3.4.1.6 shows that $R(A, P)$ is factorially K-graded.

We show that the K-grading is pointed. Because the columns of P generate $\mathbb{Q}^{r+s}$ as a cone, we obtain that the weight cone of the K-grading contains no lines and all generators T_{ij}, S_k have a nonzero degree in $K_{\mathbb{Q}}$. This implies $R(A, P)_0 = \mathbb{K}$. $\square$

Theorem 3.4.3.4 allows us to apply the machinery of bunched rings. This leads to varieties with a torus action of complexity 1.

Construction 3.4.3.6 Let A and P be matrices as in Construction 3.4.3.1; consider the associated K-graded ring $R := R(A, P)$ and let $\mathfrak{F} \subseteq R$ denote the system of generators defined by the variables T_{ij} and S_k. The associated projected cone is $(E \xrightarrow{\varrho} K, \gamma)$, where

$$E := \mathbb{Z}^{n+m}, \qquad \gamma := \mathrm{cone}(e_{ij}, e_k) \subseteq E_{\mathbb{Q}}, \qquad K = E/\mathrm{im}(P^*)$$

with the canonical basis vectors $e_{ij}, e_k \in E$ corresponding to T_{ij}, S_k and $Q \colon E \to K$ is the projection. Every true $\mathfrak{F}$-bunch Φ gives rise to a bunched ring $(R, \mathfrak{F}, \Phi)$. As in Construction 3.2.1.3, we consider $\overline{X} := \overline{X}(A, P) := \mathrm{Spec}\, R$ with the action of $H := \mathrm{Spec}\, \mathbb{K}[K]$ and obtain varieties

$$\widehat{X} := \widehat{X}(A, P, \Phi) := \widehat{X}(R, \mathfrak{F}, \Phi), \qquad X := X(A, P, \Phi) := X(R, \mathfrak{F}, \Phi),$$

where X is the quotient $\widehat{X}/\!\!/ H$ by the action of H. The quotient map $p \colon \widehat{X} \to X = \widehat{X}/H$ defines prime divisors

$$D_X^{ij} := p(V(\widehat{X}, T_{ij})) \subseteq X, \qquad D_X^k := p(V(\widehat{X}, S_k)) \subseteq X.$$

The action of $H_0 := \mathrm{Spec}\, \mathbb{K}[K_0]$ on $\overline{X}$ leaves $\widehat{X}$ invariant and thus there is an induced effective action of the torus $T := H_0/H = \mathrm{Spec}\, \mathbb{K}[\mathbb{Z}^s]$ on X.

Theorem 3.4.3.7 *In the setting of Construction 3.4.3.6, consider $R := R(A, P)$ and a variety $X := X(A, P, \Phi)$ with its action of the torus $T = \mathrm{Spec}\, \mathbb{K}[\mathbb{Z}^s]$. Then X is an irreducible, normal A_2-variety with*

$$\dim(X) = \dim(R) - \dim(K_{\mathbb{Q}}) = s + 1, \qquad \Gamma(X, \mathcal{O}) = \mathbb{K}.$$

There is an isomorphism $K \to \mathrm{Cl}(X)$ sending the degrees of T_{ij}, S_k to the classes of D_X^{ij}, D_X^k, the map $p \colon \widehat{X} \to X$ is a characteristic space over X, the K-graded ring R is the Cox ring of X, and T acts effectively with maximal orbit dimension $\dim(X) - 1$ on X.

As we will prove in Theorem 4.4.1.6, all A_2-maximal varieties with only constant global functions coming with an effective torus action of complexity 1 arise from Construction 3.4.3.6.

Construction 3.4.3.8 In the setting of Construction 3.4.3.6, consider $R := R(A, P)$ and the varieties $\widehat{X} := \widehat{X}(A, P, \Phi)$ and $X := X(A, P, \Phi)$. The canonical toric embedding of X presented in Construction 3.2.5.3 was constructed via a commutative diagram of characteristic spaces

$$\begin{array}{ccc} \widehat{X} & \longrightarrow & \widehat{Z} \\ {\scriptstyle /\!/ H} \downarrow & & \downarrow {\scriptstyle /\!/ H} \\ X & \longrightarrow & Z \end{array}$$

where the fan Σ of the toric variety Z has the columns of the matrix P as its rays. The inclusion $(0 \times \mathbb{Z}^s) \subseteq \mathbb{Z}^{r+s}$ defines an inclusion $T \subseteq T_Z$ into the acting torus of Z that turns $X \to Z$ into a T-equivariant embedding.

Example 3.4.3.9 (The E_6-singular cubic II)   As in Example 3.4.3.3, let $r = 2$, $n_0 = 2, n_1 = n_2 = 1, m = 0, s = 1$ and consider the data

$$A = \begin{bmatrix} 0 & -1 & 1 \\ 1 & -1 & 0 \end{bmatrix}, \qquad P = \begin{bmatrix} -3 & -1 & 3 & 0 \\ -3 & -1 & 0 & 2 \\ -2 & -1 & 1 & 1 \end{bmatrix}.$$

Then $R(A, P) = \mathbb{K}[T_{01}, T_{02}, T_{11}, T_{21}]/\langle T_{01}^3 T_{02} + T_{11}^3 + T_{21}^2 \rangle$ has $\mathfrak{F} = (T_{01}, T_{02}, T_{11}, T_{21})$ as a system of pairwise nonassociated generators. The associated projected cone $(E \xrightarrow{\varrho} K, \gamma)$ is given by $E = \mathbb{Z}^4$, $K = \mathbb{Z}$ and

$$Q = \begin{bmatrix} 1 & 3 & 2 & 3 \end{bmatrix}.$$

There is a unique $\mathfrak{F}$-bunch $\Phi = \{\mathbb{Q}_{\geq 0}\}$. The resulting variety $X := X(A, P, \Phi)$ is a $\mathbb{K}^*$-surface. The canonical toric ambient variety Z of X is constructed via the commutative diagram

$$\begin{array}{ccc} \widehat{X} \longrightarrow \widehat{Z} \longrightarrow \mathbb{K}^4 \setminus \{0\} \\ \downarrow \qquad \downarrow \qquad \downarrow \\ X \longrightarrow Z \longrightarrow \mathbb{P}_{1,3,2,3} \end{array}$$

Thus, Z is an open toric subvariety of the weighted projective space $\mathbb{P}_{1,3,2,3}$. Let us look at the fan of Z. In terms of the columns $v_{01}, v_{02}, v_{11}, v_{21}$ of P, its maximal cones are

$$\sigma^+ := \mathrm{cone}(v_{01}, v_{11}, v_{21}), \quad \sigma^- := \mathrm{cone}(v_{02}, v_{11}, v_{21}), \quad \tau_{01} := \mathrm{cone}(v_{01}, v_{02}).$$

Proposition 3.4.3.10   *In the setting of Construction 3.4.3.6, consider the variety $X := X(A, P, \Phi)$ with its action of the torus $T = \mathrm{Spec}\,\mathbb{K}[\mathbb{Z}^s]$.*

(i) *The divisors $D_X^k \subseteq X$, where $1 \leq k \leq m$, have generic isotropy group G_k isomorphic to the one-torus $\mathbb{K}^*$.*

(ii) *The divisors $D_X^{ij} \subseteq X$, where $0 \leq i \leq r$ and $1 \leq j \leq n_i$ have generic isotropy group G_{ij} isomorphic to $\mathbb{Z}/l_{ij}\mathbb{Z}$ and the greatest common divisor of the entries of d_{ij} defines the nonzero weight of the cotangent representation of G_{ij} at any general $x \in D_X^{ij}$.*

(iii) *Let $X \subseteq Z$ be the canonical toric embedding and $Z_0 \subseteq Z$ the open toric subvariety defined by the subfan $\Sigma_0 \preceq \Sigma$ having the first n rays as its maximal cones. Then we have a commutative diagram*

$$
\begin{array}{ccc}
X_0 & \subseteq & Z_0 \\
\downarrow & & \downarrow \\
C & \longrightarrow & \mathbb{P}^r
\end{array}
$$

where $X_0 = Z_0 \cap X$ is the union of all T-orbits with at most finite isotropy group, the downwards map is the toric morphism corresponding to the projection $\mathbb{Z}^{r+s} \to \mathbb{Z}^r$, and the image $C \subseteq \mathbb{P}^r$ of $X_0 \subseteq Z_0$ is the projective line given as $V(h_I; \ I \in \mathfrak{I})$, where for $I = (i_1, i_2, i_3)$ we set

$$
h_I(z_0, \dots, z_r) := \det \begin{bmatrix} z_{i_1} & z_{i_2} & z_{i_3} \\ a_{i_1} & a_{i_2} & a_{i_3} \end{bmatrix}.
$$

Proof The canonical toric embedding of Construction 3.4.3.8 is equivariant with respect to the T-action. This allows us to deduce (i) and (ii) from Proposition 2.1.4.2. For (iii), consult the proof of Theorem 3.4.2.3. $\square$

3.4.4 Geometry of T-varieties of complexity 1

The description of rational T-varieties of complexity 1 in terms of bunched rings $R(A, P, \Phi)$ allows us to study explicitly their geometry. We mainly focus here on two items relying on the special shape of the occurring complete intersection Cox rings: the canonical divisor and the resolution of singularities. Again, we also refer to Section 5.4 treating $\mathbb{K}^*$-surfaces as a simple example class. The ground field $\mathbb{K}$ is algebraically closed of characteristic zero. The original reference for this section is [177].

Proposition 3.4.4.1 *The notation is Construction 3.4.3.6. Let $w_{ij}, w_k \in K$ denote the degrees of T_{ij}, S_k. Then, for any $0 \leq i_0 \leq r$, the canonical divisor class of $X := X(A, P, \Phi)$ is given in $\mathrm{Cl}(X) = K$ by*

$$
w_X^{\mathrm{can}} = (r-1) \sum_{j=1}^{n_{i_0}} l_{i_0 j} w_{i_0 j} - \sum_{i,j} w_{ij} - \sum_k w_k.
$$

Proof The Cox ring $R(A, P)$ of $X(A, P)$ is a complete intersection. Thus, Proposition 3.3.3.2 tells us that the canonical divisor class equals the sum of the degrees of the relations minus the sum of the degrees of the generators. $\square$

Example 3.4.4.2 (The E_6-singular cubic III) As in Examples 3.4.3.3 and 3.4.3.9, let $r = 2$, $n_0 = 2$, $n_1 = n_2 = 1$, $m = 0$, $s = 1$ and consider the data

$$A = \begin{bmatrix} 0 & -1 & 1 \\ 1 & -1 & 0 \end{bmatrix}, \qquad P = \begin{bmatrix} -3 & -1 & 3 & 0 \\ -3 & -1 & 0 & 2 \\ -2 & -1 & 1 & 1 \end{bmatrix}.$$

and the (unique) $\mathfrak{F}$-bunch $\Phi = \{\mathbb{Q}_{\geq 0}\}$. The resulting variety $X := X(A, P, \Phi)$ is a $\mathbb{K}^*$-surface embedded into the weighted projective space $\mathbb{P}_{3,1,2,3}$. The canonical class in $K = \mathbb{Z}$ is given as

$$w_X^{\mathrm{can}} = l_{01} w_{01} + l_{02} w_{02} - w_{01} - w_{02} - w_{11} - w_{21} = -3.$$

In particular, using Proposition 3.3.2.6, we see that the anticanonical divisor is ample. The homogeneous part $R(A, P)_3$ of degree $-w_X^{\mathrm{can}} = 3$ in the Cox ring $\mathcal{R}(X) = R(A, P)$ is generated by

$$z_0 := T_{01}^3, \qquad z_1 := T_{02}, \qquad z_2 := T_{21}, \qquad z_3 := T_{01} T_{11}.$$

Note that in degree 9, we have the relation $z_0^2 z_1 + z_0 z_2^2 + z_3^3$ between these sections. One directly checks that $\widehat{X} = \widehat{X}(A, P, \Phi)$ is covered by the affine sets $\overline{X}_{z_i}$ and we have an induced embedding $\imath : X \to \mathbb{P}^3$ where the image is given by

$$\imath(X) = V(z_0^2 z_1 + z_0 z_2^2 + z_3^3) \subseteq \mathbb{P}^3.$$

Construction 3.4.4.3 Let $X = X(A, P, \Phi)$ be a variety as provided by Construction 3.4.3.6. Consider the fan $\Delta(r) := \{0, \varrho_0, \ldots, \varrho_r\}$ in $\mathbb{Z}^r$, where the ϱ_i are rays defined in terms of the canonical basis vectors $e_1, \ldots, e_r \in \mathbb{Z}^r$ as follows:

$$\varrho_0 := \mathrm{cone}(-e_1 - \cdots - e_r), \qquad \varrho_i := \mathrm{cone}(e_i), \quad 1 \leq i \leq r.$$

Then, with the matrices P_0 as in Construction 3.4.2.1 and P as in Construction 3.4.3.1 and the projection matrix $P_1 = [E_r, 0]$, where E_r is the $r \times r$ unit matrix, we have a commutative diagram

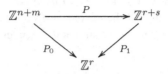

Note that P_1 sends an ijth column v_{ij} of P into the ray ϱ_i and the columns v_k to zero. Now consider the canonical toric embedding $X \subseteq Z$ from 3.4.3.8 and let Σ denote the fan of Z. Define a new fan in $\mathbb{Z}^{r+s}$ by

$$\Sigma' := \Sigma'(A, P) := \{\sigma \cap P_1^{-1}(\varrho_i); \ \sigma \in \Sigma, \ 0 \le i \le r\}.$$

Then Σ' refines Σ and the birational toric morphism $Z' \to Z$ arising from the map of fans $\Sigma' \to \Sigma$ fits into a commutative diagram

$$\begin{array}{ccc} X' & \longrightarrow & Z' \\ \downarrow & & \downarrow \\ X & \longrightarrow & Z \end{array}$$

where $X' \subseteq Z'$ is the proper transform, that is, the closure of $X \cap \mathbb{T}^{r+s}$ in Z'. We call $X' \to X$ the *weak tropical resolution* of X. We say that X is *weakly tropical* if $\Sigma' = \Sigma$ holds.

The word "tropical" emphasizes the relation to tropical geometry. For a general introduction, we refer to [213]; here we restrict ourselves to the following.

Remark 3.4.4.4 In the notation of Construction 3.4.4.3, the inverse image $P_1^{-1}(\mathrm{Supp}(\Delta(r)))$ is the tropical variety of $X \cap \mathbb{T}^{r+s}$. Recall that to a closed subvariety $Y \subseteq \mathbb{T}^k$ of a torus one associates the *tropical variety*

$$\mathrm{trop}(Y) := \bigcap_{f \in I(Y)} \mathrm{Supp}(\Sigma(f)) \subseteq \mathbb{Q}^k$$

where $\Sigma(f)$ denotes the fan of cones of codimension at least 1 in the normal quasifan of the Newton polytope of the Laurent polynomial $f \in I(Y)$. In fact, finitely many f are enough to define $\mathrm{trop}(Y)$; any such finite collection is called a tropical basis for the vanishing ideal $I(Y)$.

Remark 3.4.4.5 Let $X = X(A, P, \Phi)$ be a variety as provided by Construction 3.4.3.6 and let $X' \to X$ be the weak tropical resolution. Then X' is of the form $X' = X'(A, P', \Phi')$, where A is the same as for X, columns of P' are the primitive generators of the fan Σ' and Φ' is the Gale dual of Σ'.

Proposition 3.4.4.6 *Let $X = X(A, P, \Phi)$ be weakly tropical, let $X \subseteq Z$ be the canonical toric embedding, denote by $Y(r) \subseteq \mathbb{P}^r$ the open toric subvariety* *corresponding to the fan $\Delta(r)$, and let $\pi_1 \colon Z \to Y(r)$ be the toric morphism defined by the projection $P_1 \colon \mathbb{Z}^{r+s} \to \mathbb{Z}^r$. Then there is a commutative diagram*

$$
\begin{array}{ccc}
X & \longrightarrow & Z \\
\pi_1 \downarrow & & \downarrow \pi_1 \\
C & \longrightarrow & Y(r)
\end{array}
$$

where the set $C := \pi_1(X) \subseteq Y(r)$ is a projective line and we have $X = \pi_1^{-1}(C)$. For $0 \le i \le r$, let $Y_i \subseteq Y(r)$ be the affine toric variety defined by the ray $\varrho_i \in \Delta(r)$ and $F_i \subseteq Y_i$ the closure of the image of the one-parameter group associated with ϱ_i. Then there is a commutative diagram

$$
\begin{array}{ccccc}
X \cap \pi_1^{-1}(Y_i) & \overset{\cong}{\longleftrightarrow} & X_i' & \subseteq & X_i \\
\pi_1 \downarrow & & \downarrow & & \downarrow \pi_1 \\
C \cap Y_i & \underset{\cong}{\longleftrightarrow} & F_i' & \subseteq & F_i
\end{array}
$$

where the subset $F_i' \subseteq F_i$ is obtained by removing r points from $F_i \cap \mathbb{T}^r$, we have $X_i' := \pi_1^{-1}(F_i')$, and $X_i = \pi_1^{-1}(F_i)$ is the $(s+1)$-dimensional toric variety with defining fan $\Sigma_i := \{\sigma \cap P_1^{-1}(\varrho_i); \ \sigma \in \Sigma\}$ in the lattice $N_i := \mathbb{Z}^{r+s} \cap \mathrm{lin}_{\mathbb{Q}}(P_1^{-1}(\varrho_i))$.

Proof Because X is weakly tropical, the fan Σ maps via the projection P_1 to the fan $\Delta(r)$. This gives the commutative diagram. Note that the image $C = \pi_1(X)$ is given by the equations

$$
C \ = \ \pi_1(X) \ = \ V(h_I; \ I \in \mathfrak{J}), \qquad h_{i_1, i_2, i_3} := \det \begin{bmatrix} U_{i_1} & U_{i_2} & U_{i_3} \\ a_{i_1} & a_{i_2} & a_{i_3} \end{bmatrix},
$$

where $U_0, \ldots, U_r$ denote the homogeneous coordinates on $\mathbb{P}^r$. In particular, we see that $\pi_1(X)$ is a projective line intersecting each toric divisor $V(U_j)$ in exactly one point.

We consider the situation exemplarily for $i = r$. Note that Y_r is contained in the affine toric chart $\mathbb{P}^r \setminus V(T_0) \cong \mathbb{K}^r$ and there the image $\pi_1(X)$ is parameterized with respect to inhomogeneous coordinates by

$$
\eta \colon \mathbb{K} \ \to \ \mathbb{K}^r, \qquad t \mapsto (\eta_1(t), \ldots, \eta_r(t))
$$

with affine linear maps η_j, where we choose η_r to be linear. For $x \in X$ mapping to Y_r, let $t(x) := \eta^{-1}(\pi_1(x))$ denote its parameter. Moreover, set $W_r := X \cap \pi_1^{-1}(Y_r)$. Then we have a T-equivariant morphism

$$
\varphi \colon W_r \to \pi_1^{-1}(Y_r), \qquad x \mapsto \left(\frac{1}{\eta_1(t(x))}, \ldots, \frac{1}{\eta_{r-1}(t(x))}, \frac{t(x)}{\eta_r(t(x))}, 1, \ldots, 1 \right) \cdot x
$$

If $\lambda \colon \mathbb{K}^* \to \mathbb{T}^r$ denotes the one-parameter subgroup defined by the primitive vector in ϱ_r, then $\pi_1(\varphi(x))$ equals $\lambda(t(x))$. We conclude that φ maps W_r isomorphically onto its image

$$\varphi(W_r) \ = \ X'_r \ = \ \pi_1^{-1}(F'_r),$$

where $F'_r \subseteq F_r = \overline{\lambda(\mathbb{K}^*)}$ is the set obtained by removing all points $\lambda(t) \in V(U_j)$, for $j = 0, \dots, r-1$. Here, X'_r is an open π_1-saturated subset of the toric variety $X_r = \pi_1^{-1}(F_r)$, which in turn is described by the fan Σ_r.   □

Corollary 3.4.4.7 *Let $X = X(A, P, \Phi)$ be weakly tropical and let $X \subseteq Z$ be the canonical toric embedding. Then, for any $\gamma_0 \in \mathrm{rlv}(\Phi)$, the dimension of the associated stratum $X(\gamma_0)$ is given by*

$$\dim(X(\gamma_0)) \ = \ \dim(Z(\gamma_0)) - r + 1 \ = \ s + 1 - \dim(P(\gamma_0^*)).$$

Proposition 3.4.4.8 *The variety $X(A, P, \Phi)$ is complete if and only if the fan $\Sigma'(A, P)$ has $P_1^{-1}(\mathrm{Supp}(\Delta(r)))$ as its support.*

Proof If the fan $\Sigma'(A, P)$ has $P_1^{-1}(\mathrm{Supp}(\Delta(r)))$ as its support, then the morphism $X' \to C$ of Proposition 3.4.4.6 is proper. Thus, X' is complete. Because we have a dominant morphism $X' \to X$, also X is complete. For the converse, apply Tevelev's criterion [294].   □

Theorem 3.4.4.9 *Let $X = X(A, P, \Phi)$ be a variety as provided by Construction 3.4.3.6, let $X \subseteq Z$ be the canonical toric embedding and Σ the fan of Z. Then a T-equivariant desingularization $X'' \to X$ of X is obtained as follows:*

- *Determine the fan $\Sigma' := \Sigma'(A, P, \Phi)$ and compute a regular subdivision Σ'' of Σ'. This gives a map of fans $\Sigma'' \to \Sigma$.*
- *Consider the toric morphism $Z'' \to Z$ defined by $\Sigma'' \to \Sigma$, let X'' be the closure of $X \cap \mathbb{T}^{r+s}$ in Z'' and $X'' \to X$ the induced morphism.*

The resulting variety X'' is smooth and of the form $X'' = X(A, P'', \Phi'')$, where A is the same as for X, the columns of P'' are the primitive generators of the fan Σ'', and Φ'' is the Gale dual of Σ''.

Proof The first step provides the weak tropical resolution $X' \to X$. As observed in Proposition 3.4.4.6, the variety X' is covered by open subsets of toric varieties X'_i with fans Σ'_i. Any regular subdivision of Σ' provides a regular subdivisions of the Σ'_i. Applying Proposition 3.4.4.6 once more, we see that the resulting variety X'' is smooth.   □

Example 3.4.4.10 (The E_6-singular cubic IV) As in Examples 3.4.3.3 and 3.4.3.9, let $r = 2$, $n_0 = 2$, $n_1 = n_2 = 1$, $m = 0$, $s = 1$ and consider the data

$$A = \begin{bmatrix} 0 & -1 & 1 \\ 1 & -1 & 0 \end{bmatrix}, \qquad P = \begin{bmatrix} -3 & -1 & 3 & 0 \\ -3 & -1 & 0 & 2 \\ -2 & -1 & 1 & 1 \end{bmatrix}$$

and the (unique) $\mathfrak{F}$-bunch $\Phi = \{\mathbb{Q}_{\geq 0}\}$. As observed in Example 3.4.4.2, the fan of the canonical toric ambient variety has the following maximal cones given in terms of the columns v_{01}, v_{02}, v_{11}, v_{21} of P:

$$\sigma^+ := \mathrm{cone}(v_{01}, v_{11}, v_{21}), \quad \sigma^- := \mathrm{cone}(v_{02}, v_{11}, v_{21}), \quad \tau_{01} := \mathrm{cone}(v_{01}, v_{02}).$$

The toric fixed point corresponding to σ^+ is the singularity $x_0 = [0, 1, 0, 0] \in X$. It is singular for two reasons; compare Corollary 3.3.1.12: first, σ^+ is singular, and second, the total coordinate space

$$\overline{X} = V(T_{01}^3 T_{02} + T_{11}^3 + T_{21}^2)$$

is singular along the fiber $p^{-1}(x_0) = 0 \times \mathbb{K}^* \times 0 \times 0$ over x_0. Now we resolve according to Theorem 3.4.4.9. The first step means inserting the rays through $v_1 := (0, 0, 1)$ and $v_2 := (0, 0, -1)$. The new matrix then is

$$P' = \begin{bmatrix} -1 & -3 & 3 & 0 & 0 & 0 \\ -1 & -3 & 0 & 2 & 0 & 0 \\ -1 & -2 & 0 & 1 & 1 & -1 \end{bmatrix}.$$

In the resulting fan, there are now five singular cones left, namely $\mathrm{cone}(v_{i1}, v_1)$, where $i = 0, 1, 2$ and $\mathrm{cone}(v_{in_i}, v_2)$, where $i = 1, 2$. Suitable subdividing gives

$$P'' = \begin{bmatrix} -1 & -2 & -3 & -1 & 1 & 2 & 3 & 1 & 0 & 0 & 0 & 0 & 0 \\ -1 & -2 & -3 & -1 & 0 & 0 & 0 & 0 & 1 & 2 & 1 & 0 & 0 \\ 0 & -1 & -2 & -1 & 1 & 1 & 1 & 0 & 1 & 1 & 0 & 1 & -1 \end{bmatrix}.$$

Because we consider a surface, there is only one possible bunch Φ''. Consequently, the resolution $X(A, P'', \Phi'')$ is found. We take up again this example in 5.4.3.5, where we determine a minimal resolution and identify the E_6 resolution graph.

Exercises to Chapter 3

Exercise 3.1 Let a quasitorus H act on an irreducible normal affine variety X. Then the GIT quasifan $\Lambda(X, H)$ is the normal quasifan of some convex polyhedron.

Exercise 3.2 (King's theorem)   Characterize semistability on factorial affine varieties in the framework of quiver representations following A.D. King [184]. A *quiver* is a directed graph Q consisting of finite sets Q_0 (vertices) and Q_1 (arrows) together with maps

$$h: Q_1 \rightarrow Q_0, \qquad t: Q_1 \rightarrow Q_0,$$

where $h(a) \in Q_0$ is the head and $t(a) \in Q_0$ the tail of the arrow $a \in Q_1$. A *representation* of a quiver Q is a pair (W, φ), where $W = (W_v)_{v \in Q_0}$ is a family of finite-dimensional $\mathbb{K}$-vector spaces W_v indexed by the vertices and $\varphi = (\varphi_a)_{a \in Q_1}$ is a family of linear maps $\varphi_a: W_{t(a)} \rightarrow W_{h(a)}$ indexed by the arrows. The dimension vector of (W, φ) is

$$\alpha := \dim(W) := (\dim(W_v); \ v \in Q_0) \in \mathbb{Z}^{Q_0}.$$

The isomorphy classes of representations of Q with dimension vector α correspond to the orbits of the group $\mathrm{GL}(\alpha)$ in the representation space $Rep(Q, \alpha)$, where

$$\mathrm{GL}(\alpha) := \prod_{v \in Q_0} \mathrm{GL}(W_v), \qquad Rep(Q, \alpha) := \bigoplus_{a \in Q_1} \mathrm{Hom}(W_{t(a)}, W_{h(a)})$$

and an $g = (g_v)_{v \in Q_0} \in \mathrm{GL}(\alpha)$ acts via $(g \cdot \varphi)_a = g_{h(a)} \circ \varphi_a \circ g_{t(a)}^{-1}$, where $g = (g_v)_{v \in Q_0} \in \mathrm{GL}(\alpha)$. Consider the action of the subgroup

$$\mathrm{SL}(\alpha) := \prod_{v \in Q_0} \mathrm{SL}(W_v) \subseteq \prod_{v \in Q_0} \mathrm{GL}(W_v) = \mathrm{GL}(\alpha)$$

on the representation space and the good quotient

$$\pi: Rep(Q, \alpha) \rightarrow X, \qquad X := Rep(Q, \alpha) /\!/ \mathrm{SL}(\alpha).$$

The affine variety X is factorial and carries a natural action of the factor torus $T(\alpha) := \mathrm{GL}(\alpha)/\mathrm{SL}(\alpha)$. The kernel T' of the $T(\alpha)$-action on X is a one-parameter subgroup, and we have an effective action of the factor $T := T(\alpha)/T'$ on X. Every character of the group $\mathrm{GL}(\alpha)$ has the form

$$\omega: \mathrm{GL}(\alpha) \rightarrow \mathbb{K}^*, \qquad (g_v)_{v \in Q_0} \mapsto \prod_{v \in Q_0} (\det g_v)^{\omega_v}$$

with numbers $\omega_v \in \mathbb{Z}$ uniquely determined by ω. This gives a concrete identification $\mathbb{X}(T(\alpha)) = \mathbb{Z}^{Q_0}$. For every $\beta \in \mathbb{Z}^{Q_0}$ we set

$$\langle \omega, \beta \rangle := \sum_{v \in Q_0} \omega_v \beta_v \in \mathbb{Z}.$$

Then a character ω descends to T if and only if $\langle \omega, \alpha \rangle = 0$ holds. A *subrepresentation* of a representation (W, φ) of the quiver Q is a family $U = (U_v)_{v \in Q_0}$

of vector subspaces $U_v \subseteq W_v$ such that $\varphi_a(U_{t(a)}) \subseteq U_{h(a)}$ holds for all $a \in Q_1$. For a character $\omega \in \mathbb{X}(T)$ and a representation (W, φ), the following conditions are equivalent.

(1) The image of (W, φ) under π is contained in the set of semistable points $X(\omega)$.
(2) For any subrepresentation $(U_v)_{v \in Q_0}$ of (W, φ), we have $\langle \omega, \dim(U) \rangle \leq 0$.

Exercise 3.3 (GIT fan for reductive groups)   Let a connected reductive group G act on an irreducible affine variety $X = \operatorname{Spec} A$. Write $K := \mathbb{X}(G)$ for the character group and let $A_w \subseteq A$ denote the vector space of w-homogeneous functions. Every $w \in K$ defines a set of semistable points:

$$X^{ss}(w) := \{x \in X; w \in \omega_x\}, \qquad \omega_x := \operatorname{cone}(u \in K; f(x) \neq 0 \text{ for an } f \in A_u).$$

There is a good quotient $X^{ss}(w) \to Y(w)$ and the quotient space $Y(w)$ is projective over $\operatorname{Spec} A_0$. The *GIT fan* of the G-action on X is

$$\Lambda(X, G) := \{\lambda(w); w \in K\}, \qquad \lambda(w) := \bigcap_{w \in \omega_x} \omega_x.$$

Indeed, $\Lambda(X, G)$ is a (polyhedral) quasifan in $K_{\mathbb{Q}}$ that is supported on the weight cone $\omega_X = \operatorname{cone}(w \in K; A_w \neq 0)$. Moreover, $\Lambda(X, G)$ reflects the GIT equivalence in the sense that we have

$$X^{ss}(w) \subseteq X^{ss}(w') \iff \lambda(w) \succeq \lambda(w').$$

If X is factorial, then $\Lambda(X, G)$ is in bijection with the qp-maximal good G-sets of X via $\lambda \mapsto X^{ss}(w)$, where $w \in \lambda^\circ$. Hint: Reduce the problem to the torus case by passing to $Y := X /\!\!/ G^s$, where $G^s \subseteq G$ is the maximal connected semisimple subgroup, and the induced action of $T := G/G^s$ on Y. See [20] for a solution.

Exercise 3.4*   Let Z be a factorial affine variety with an action of a quasitorus H. Find a combinatorial description for

(1) The good H-subsets $U \subseteq Z$ that are maximal w.r.t. saturated inclusion
(2) The good H-subsets $U \subseteq Z$ that admit a complete quotient space

Exercise 3.5*   A variety X is called an A_k-*variety* if any k points $x_1, \ldots, x_k \in X$ admit a common neighborhood in X.

(1) If X is an A_k-variety, then also every locally closed subset $Y \subseteq X$ is an A_k-variety.
(2) Any variety contains only finitely many maximal open subsets that are A_k-varieties.

(3) Prove or disprove that every irreducible normal A_k-variety admits a closed embedding into a toric A_k-variety.
(4) Give a combinatorial description of all (H, k)-maximal (defined analogously to $(H, 2)$-maximal) subsets of an H-factorial affine variety.

The solution of (2) can be found in [289]. An affirmative answer to (3) is conjectured in [308]. A related concept is k-divisoriality [163].

Exercise 3.6 (Total spaces of line bundles)   Let X be a variety and $L \to X$ be a line bundle. If X is (quasi-)affine, then L is (quasi-)affine as well. Give an example with X quasiaffine, $\Gamma(X, \mathcal{O})$ finitely generated, but $\Gamma(L, \mathcal{O})$ not.

Exercise 3.7   An irreducible normal prevariety X is a quasiprojective variety if and only if there are a Weil divisor D on X and $f_1, \ldots, f_r \in \Gamma(X, \mathcal{O}_X(D))$ such that the subsets $X \setminus \mathrm{Supp}(\mathrm{div}_D(f_i))$ form an affine cover of X.

Exercise 3.8   (A smooth complete three-dimensional variety without A_2-property)   Compare [289, Example 6.4]. Let $X := V(T_1 T_6 + T_2 T_5 + T_3 T_4) \subseteq \mathbb{K}^6$ be the affine cone over the Grassmannian $G(2, 4)$ and consider the action of the 2-torus $H := \mathbb{T}^2$ given by

$$t \cdot z = (t_1 t_2^6 z_1, \; t_1 t_2^5 z_2, \; t_1 t_2^4 z_3, \; t_1 t_2^3 z_4, \; t_1 t_2^2 z_5, \; t_1 t_2 z_6).$$

Let $U \subseteq X$ be the H-invariant open subset obtained by removing from X the sets $V(T_1, T_2, T_4)$ and $V(T_3, T_5, T_6)$. Then U admits a geometric good quotient $\pi \colon U \to U/H$ with a smooth complete three-dimensional quotient space U/H that does not have the A_2-property. Hint: Use Theorem 3.1.4.3 and consider the points $\pi(x_i)$ for

$$x_1 := (1, 0, 1, 0, 0, 0), \qquad\qquad x_2 := (0, 0, 0, 1, 0, 1).$$

Exercise 3.9*   Let X be an irreducible normal variety with only constant invertible global functions and finitely generated divisor class group such that the Cox ring of X is a polynomial ring.

(1) If X is complete, then X is a toric variety.
(2) If X is affine, is it true that X is a toric variety?

Note that a positive solution of the linearization problem implies a positive answer to the question of (2).

Exercise 3.10 There exist irreducible normal A_2-maximal varieties X with $\Gamma(X, \mathcal{O}) = \mathbb{K}$ that are not complete. Hint: Look at [136].

Exercise 3.11 A bunched ring $(R, \mathfrak{F}, \Phi)$ defines projective variety if and only if the cone $Q(\gamma) \subseteq K_{\mathbb{Q}}$ is pointed and no generator degree $\deg(T_i)$ maps to the apex.

Exercise 3.12 Let G be a connected semisimple group. Then a parabolic subgroup $P \subseteq G$ is maximal if and only if $\mathrm{Cl}(G/P) \cong \mathbb{Z}$ holds.

Exercise 3.13 Consider the factor algebra $R(r) = \mathbb{K}[T_1, \ldots, T_r]/\langle g_r \rangle$, where the quadratic polynomials g_r are defined by

$$g_{2m} := T_1 T_2 + \cdots + T_{2m-1} T_{2m}, \qquad g_{2m+1} := T_1 T_2 + \cdots + T_{2m-1} T_{2m} + T_{2m+1}^2.$$

Set $\mathfrak{F}(r) := (f_1, \ldots, f_r)$, where $f_i \in R(r)$ denotes the class of the variable T_i. Then the number of $\mathfrak{F}(r)$-faces equals $2^{2m} - 3^{m-1}m$ if $r = 2m$ and $2^{2m+1} - 3^{m-1}(m + 3)$ if $r = 2m + 1$.

Exercise 3.14 Let $g = T_1^2 + \cdots + T_r^2$. Consider the effective $(\mathbb{Z}/2\mathbb{Z})^r$-grading on $\mathbb{K}[T_1, \ldots, T_r]$ given by $\deg(T_i) = (\bar{0}, \ldots, \bar{1}, \ldots, \bar{0})$. Then $T_1, \ldots, T_r$ and g are homogeneous, and there exist no linearly independent homogeneous linear forms $S_1, \ldots, S_r$ in $T_1, \ldots, T_r$ with $g(T_1, \ldots, T_r) = g_r(S_1, \ldots, S_r)$.

Exercise 3.15 (Kempf–Ness sets) The ground field is $\mathbb{C}$. Let $X = X(R, \mathfrak{F}, \Phi)$ be the variety arising from a bunched ring, where we assume that $K = \mathrm{Cl}(X)$ is free. Consider a toric embedding $X \to Z$ obtained by completing the canonical toric ambient variety according to Construction 3.2.5.6 and let Σ denote the describing fan of Z. Then we have a commutative diagram

$$
\begin{array}{ccccccc}
\mathcal{Z}(\Phi) & \subseteq & \mathcal{Z}(\Sigma) & \subseteq & \widehat{Z} & \supseteq & \widehat{X} \\
\downarrow {\scriptstyle /H_c} & & \downarrow {\scriptstyle /H_c} & & \downarrow {\scriptstyle /H} & & \downarrow {\scriptstyle /H} \\
\mathcal{Z}(\Phi)/H_c & \subseteq & \mathcal{Z}(\Sigma)/H_c & \longrightarrow & Z & \supseteq & X
\end{array}
$$

where $\mathcal{Z}(\Sigma) \subseteq Z$ is the toric Kempf–Ness set, $H_c \subseteq H$ is a maximal compact subgroup and we set $\mathcal{Z}(\Phi) := \mathcal{Z}(\Sigma) \cap \widehat{X}$; see Exercise 2.6 for the necessary definitions. If the fan Σ is simplicial, then the induced map $\mathcal{Z}(\Phi)/H_c \to X$ is a homeomorphism.

Exercise 3.16 (Kleiman–Chevalley criterion for varieties with finitely generated Cox ring)   Let X be a $\mathbb{Q}$-factorial A_2-maximal variety with $\Gamma(X, \mathcal{O}) = \mathbb{K}$ and finitely generated Cox ring. Set $r := \operatorname{rank}(\operatorname{Cl}(X))$. If any r points of X admit a common affine neighborhood, then X is projective; see [44, Thm. 8.1] for the case that $\operatorname{Cl}(X)$ is torsion free.

Exercise 3.17 (Strongly stable torus quotients of $G(2, 4)$)   Consider the Grassmannian $G(2, 4) = V(T_1 T_6 - T_2 T_5 + T_3 T_4) \subseteq \mathbb{P}^5$ together with the action of the three-dimensional torus $\mathbb{T}^3$ given by

$$t \cdot [z] = [t_1 z_1, t_2 z_2, t_3 z_3, z_4, t_3 t_2^{-1} z_5, t_3 t_1^{-1} z_6].$$

Note that $\mathbb{T}^3$ defines this way a maximal torus in the automorphism group of $G(2, 4)$. Let $H \subseteq \mathbb{T}^3$ be a nontrivial subtorus and $U \subseteq G(2, 4)$ an H-invariant open subset such that $G(2, 4) \setminus U$ is of codimension at least 2 in $G(2, 4)$, there is a good quotient

$$p \colon U \ \to \ X, \qquad\qquad X := U /\!\!/ H,$$

the action of H on U is strongly stable, and X is a complete variety with the A_2-property; compare Exercise 3.8. Then $2 \le \dim(X) \le 3$ holds, X comes with a torus action $T \times X \to X$ of complexity 1, and the defining matrix P for X can be chosen as

$$P = \begin{bmatrix} -1 & -1 & 1 & 1 & 0 & 0 \\ -1 & -1 & 0 & 0 & 1 & 1 \\ a & b & 0 & c & 0 & d \end{bmatrix}, \quad P = \begin{bmatrix} -1 & -1 & 1 & 1 & 0 & 0 \\ -1 & -1 & 0 & 0 & 1 & 1 \\ a_1 & b_1 & 0 & c_1 & 0 & d_1 \\ a_2 & b_2 & 0 & c_2 & 0 & d_2 \end{bmatrix},$$

where the first matrix corresponds to a surface and the second one to a three-dimensional variety. In both cases, show that X is projective and describe in terms of P the smooth varieties X, the Fano-varieties X, and those with a smooth (normalized) general T-orbit closure.

Exercise 3.18* Consider full intrinsic quadrics X in the sense of Section 3.2.4. Prove (or, unlikely, disprove) that for any pair (d, l) of positive integers, there are, up to isomorphy, at most finitely many X of dimension d such that $-K_X$ is ample and $l K_X$ is Cartier. Give effective bounds for the number and classifications for small values. Generalize to other classes of hypersurface Cox rings.

Exercise 3.19 (A total coordinate space with non finitely generated divisor class group) Consider the projective $\mathbb{K}^*$-surface $X := X(A, P, \Phi)$ defined by the data

$$
A := \begin{bmatrix} -1 & 1 & 0 & 1 \\ -1 & 0 & 1 & -1 \end{bmatrix}, \quad P := \begin{bmatrix} -1 & -1 & 3 & 0 & 0 \\ -1 & -1 & 0 & 3 & 0 \\ -1 & -1 & 0 & 0 & 3 \\ -1 & -2 & 1 & 1 & 2 \end{bmatrix}, \quad \Phi := \{\mathbb{Q}_{\geq 0}\}.
$$

The divisor class group of X is isomorphic to $K = \mathbb{Z} \oplus \mathbb{Z}/3\mathbb{Z} \oplus \mathbb{Z}/3\mathbb{Z}$ and the Cox ring of X is given as

$$
R(A, P) = \mathbb{K}[T_{01}, T_{02}, T_{11}, T_{21}, T_{31}]/\langle T_{01} T_{02} + T_{11}^3 + T_{21}^3, \ -T_{11}^3 + T_{21}^3 + T_{31}^3 \rangle,
$$

$$
Q = \begin{bmatrix} 2 & 1 & 1 & 1 & 1 \\ \bar{1} & \bar{2} & \bar{1} & \bar{1} & \bar{0} \\ \bar{1} & \bar{2} & \bar{2} & \bar{0} & \bar{0} \end{bmatrix},
$$

where the columns of Q are the K-degrees of the variables T_{ij}. With $H = \operatorname{Spec} \mathbb{K}[K]$, the unit component $H^0 \subseteq H$ and $\Gamma := H/H^0$ we obtain a commutative diagram

where the dashed downwards arrows indicate the separation of the quotient by the $\mathbb{K}^*$-action which is induced by the one on $\mathbb{K}^5$ with weights $1, -1, 0, 0, 0$; see Proposition 3.4.4.6. The smooth curves C' and C are given as

$$
C' = V(T_0 + T_1^3 + T_2^3, \ -T_1^3 + T_2^3 + T_3^3) \subseteq \mathbb{P}_{3,1,1,1},
$$

$$
C = V(T_0 + T_1 + T_2, \ -T_1 + T_2 + T_3) \subseteq \mathbb{P}^3.
$$

We have $C \cong \mathbb{P}^1$ and $C' \to C$ is a branched covering of degree 9. There are precisely nine branch points, each of order 2; they lie in the intersections with the coordinate axes

$$
C' \cap V(T_i) \subseteq \mathbb{P}_{3,1,1,1}, \qquad i = 1, 2, 3.
$$

Thus, the Riemann–Hurwitz formula says that C' is of genus 1; alternatively, observe that C' is isomorphic to the Fermat cubic. As a consequence, the divisor class groups of X', $\widehat{X}$ and the total coordinate space $\overline{X}$ are not finitely generated.

Exercise 3.20 Generalize the tropical completeness criterion 3.4.4.8 to varieties $X(R, \mathfrak{F}, \Phi)$ arising from an arbitrary bunched ring $(R, \mathfrak{F}, \Phi)$.

Exercise 3.21* Generalize the desingularization algorithm 3.4.4.9 to varieties $X(R, \mathfrak{F}, \Phi)$ arising from larger classes of bunched rings $(R, \mathfrak{F}, \Phi)$.

4

Selected topics

We discuss various topics around Cox rings. The first section is devoted to birational maps, for example, blow-ups. We figure out the class of modifications that preserve finite generation and we show how to compute the Cox ring of the modified variety in terms of the Cox ring of the initial one. In Section 4.2, we first introduce a class of quotient presentations dominated by the characteristic space and comprising, for example, the classical affine cones. Then we provide a lifting result for connected subgroups of the automorphism group. In the case of a finitely generated Cox ring, this gives an approach to the whole automorphism group, showing in particular that it is affine algebraic. The topic of Section 4.3 is finite generation of the Cox ring. We provide general criteria and characterizations; for example, the finiteness characterization of Hu and Keel in terms of polyhedral subdivisions of the moving cone is proven and we discuss finite generation of the Cox ring for Fano varieties. In Section 4.4 we consider varieties coming with a torus action. We express their Cox ring in terms of data of the action. In the case of a rational variety with an action of complexity 1, we see that the Cox ring is given by trinomial relations as in Section 3.4. In particular, the constructions given there produce indeed all rational normal complete A_2-varieties with a torus action of complexity 1. Section 4.5 is about almost homogeneous varieties, that is, equivariant open embeddings of homogeneous spaces. We first describe the Cox ring of a homogeneous space G/H. Embeddings X of homogeneous spaces G/H with finitely generated Cox ring and small boundary $X \setminus G/H$ allow an immediate description via bunched rings and provide a rich example class. Finally, we survey in this section results on the case of wonderful and more general spherical varieties.

4.1 Toric ambient modifications

4.1.1 The Cox ring of an embedded variety

We consider a prevariety embedded into another one and relate its characteristic space and Cox ring to that of its ambient space. As an application, we obtain Lefschetz theorems for the Cox rings of hypersurfaces in the projective space and more generally in toric varieties. The ground field $\mathbb{K}$ is algebraically closed of characteristic zero. The original reference for this section is [165].

We fix the setting for this section. Let Z be an irreducible normal prevariety with $\Gamma(Z, \mathcal{O}^*) = \mathbb{K}^*$, $K_Z \subseteq \mathrm{WDiv}(Z)$ a subgroup generating the divisor class group $\mathrm{Cl}(Z)$, and $\mathcal{R}_Z$ a Cox sheaf arising from K_Z. Assume that $\mathcal{R}_Z$ is locally of finite type, so that we have the associated characteristic space $p_Z \colon \widehat{Z} \to Z$. Given a closed irreducible normal $X \subseteq Z$ with $\Gamma(X, \mathcal{O}^*) = \mathbb{K}^*$, we have a commutative diagram

$$
\begin{array}{ccc}
\widehat{X} & \xrightarrow{\ \widehat{\imath}\ } & \widehat{Z} \\
{\scriptstyle p_X}\downarrow & & \downarrow{\scriptstyle p_Z} \\
X & \xrightarrow{\ \imath\ } & Z
\end{array}
$$

where $\widehat{X} \subseteq \widehat{Z}$ is the closure of the preimage $p_Z^{-1}(X')$ of the intersection $X' := X \cap Z'$ with the set of smooth points $Z' \subseteq Z$; the maps $\imath, \widehat{\imath}$ denote the respective inclusions and p_X is the restriction of p_Z. We assume that X is not contained in the support of any $D \in K_Z$ and that X' has a complement of codimension at least 2 in X. Then there is a canonical pullback homomorphism

$$
K_Z = K_Z \cap \mathrm{CDiv}(Z') \xrightarrow{\ \imath^*\ } \mathrm{CDiv}(X') \subseteq \mathrm{WDiv}(X).
$$

It sends principal divisors to principal divisors and consequently induces a pullback homomorphism $\imath^* \colon \mathrm{Cl}(Z) \to \mathrm{Cl}(X)$ on the level of divisor class groups.

Theorem 4.1.1.1 *In the above setting, assume that Z is smooth and the pullback $\imath^* \colon \mathrm{Cl}(X) \to \mathrm{Cl}(Z)$ is an isomorphism. Then, for any Cox sheaf $\mathcal{R}_X$ on X, there is an isomorphism of $\mathrm{Cl}(X)$-graded $\mathcal{O}_X$-algebras $\mathcal{R}_X \cong (p_X)_* \mathcal{O}_{\widehat{X}}$. In particular, $\widehat{X} = p_Z^{-1}(X)$ is irreducible and normal, and $p_X \colon \widehat{X} \to X$ is a characteristic space for X.*

Proof Consider the subgroup $K_Z \subseteq \mathrm{WDiv}(Z)$ fixed before and the associated sheaf $\mathcal{S}_Z$ of divisorial algebras on Z. Recall from Construction 1.4.2.1 that the Cox sheaf is obtained as $\mathcal{R}_Z = \mathcal{S}_Z / \mathcal{I}_Z$, where the ideal $\mathcal{I}_Z$ is defined via a character

$$
\chi \colon K_Z \cap \mathrm{PDiv}(Z) \ \to\ \mathbb{K}(Z)^*, \qquad E \mapsto \chi(E).
$$

With the pullback group $K_X := \iota^* K_Z$ and the pullback character, we obtain a sheaf of divisorial algebras $\mathcal{S}_X$, an ideal $\mathcal{I}_X$, and a Cox sheaf $\mathcal{R}_X$ on X. Moreover, we have a commutative diagram

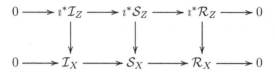

where the rows are exact and the downwards arrows are isomorphisms; here we use smoothness of Z to see that $\iota^* \mathcal{S}_Z$ maps isomorphically onto $\mathcal{S}_X$. Recall that we have $\mathcal{R}_Z \cong (p_Z)_* \mathcal{O}_{\widehat{Z}}$. The first assertion follows from the graded isomorphisms

$$\mathcal{R}_X \cong \iota^* \mathcal{R}_Z \cong \iota^* (p_Z)_* \mathcal{O}_{\widehat{Z}} \cong (p_X)_* \mathcal{O}_{\widehat{X}}.$$

For the last statement, observe that $\widehat{X}$ is the relative spectrum of $(p_X)_* \mathcal{O}_{\widehat{X}}$ and use Theorem 1.5.1.1 to obtain irreducibility and normality of $\widehat{X}$. $\qquad\square$

Corollary 4.1.1.2 *In the above setting, assume that $\iota^* \colon \mathrm{Cl}(X) \to \mathrm{Cl}(Z)$ is an isomorphism, $\widehat{X}$ is normal, and the complement $\widehat{X} \setminus p_X^{-1}(X')$ is of codimension at least 2 in $\widehat{X}$. Then, for any Cox sheaf $\mathcal{R}_X$ on X, there is an isomorphism of $\mathrm{Cl}(X)$-graded $\mathcal{O}_X$-algebras $\mathcal{R}_X \cong (p_X)_* \mathcal{O}_{\widehat{X}}$. In particular, $p_X \colon \widehat{X} \to X$ is a characteristic space for X.*

Proof Recall that X' is the intersection of X with the smooth locus Z' of Z. Thus, Theorem 4.1.1.1 tells us that we have an isomorphism $\mathcal{R}_X \cong (p_X)_* \mathcal{O}_{\widehat{X}}$ over X'. By the assumptions, this isomorphism extends over the whole X. $\qquad\square$

If the Cox ring $\mathcal{R}_Z$ of the ambient prevariety Z is finitely generated, then we have a total coordinate space $\overline{Z}$ containing $\widehat{Z}$ as an open H_Z-invariant subset and we define $\overline{X}$ to be the closure of $p_X^{-1}(X')$.

Corollary 4.1.1.3 *In the above setting, assume that $\iota^* \colon \mathrm{Cl}(X) \to \mathrm{Cl}(Z)$ is an isomorphism and $\mathcal{R}(Z)$ is finitely generated. If $\overline{X} \setminus p_X^{-1}(X')$ is of codimension at least 2 in $\overline{X}$, then the following hold.*

(i) *The H_X-equivariant normalization of $\overline{X}$ is a total coordinate space of X. In particular, the Cox ring $\mathcal{R}(X)$ of X is finitely generated.*

(ii) *If $\overline{X}$ is normal, then it is the total coordinate space of X. In particular, we then have $\mathcal{R}(X) \cong \Gamma(\overline{X}, \mathcal{O})$.*

Proof Only for (i) is there something to show. By Theorem 4.1.1.1, the $\mathrm{Cl}(X')$-graded algebra $\Gamma(p_X^{-1}(X'), \mathcal{O})$ equals the Cox ring of X'. Moreover, $p_X^{-1}(X')$ is normal and thus is an open subset of the equivariant normalization $\widetilde{X}$ of $\overline{X}$.

The assumption ensures that the complement $\widetilde{X} \setminus p_X^{-1}(X')$ is of codimension at least 2 in $\widetilde{X}$. Thus, $\Gamma(\widetilde{X}, \mathcal{O})$ equals the Cox ring of X' and hence that of X. $\qquad\square$

Remark 4.1.1.4 If, in the situation of Corollary 4.1.1.3, the subvariety $X \subseteq Z$ is a normal hypersurface, then $\overline{X}$ is normal and the Cox ring of X is $\mathcal{R}(Z)/\langle f \rangle$, where $f \in \mathcal{R}(Z)$ is the homogeneous element with $\mathrm{div}(f) = \overline{X}$; see [181, Thm. 6] for the smooth case and and [11, Thm. 2.1] for the general case.

We specialize to the case that the ambient space Z is a toric prevariety of affine intersection; see [3]. We take $K_Z \subseteq \mathrm{WDiv}(Z)$ to be the group generated by the invariant prime divisors $D_1, \ldots, D_r$ of Z and denote by $T_1, \ldots, T_r$ the corresponding variables of the Cox ring $\mathcal{R}(X) = \mathbb{K}[T_1, \ldots, T_r]$. Recall that $\deg(T_i)$ is the class of D_i in $\mathrm{Cl}(X)$.

Corollary 4.1.1.5 *In the above setting, assume that Z is a toric prevariety of affine intersection with invariant prime divisors $D_1, \ldots, D_r$ and that $\iota^*\colon \mathrm{Cl}(X) \to \mathrm{Cl}(Z)$ is an isomorphism. If $\overline{X}$ is normal and $\overline{X} \setminus \widehat{X}$ is of codimension at least 2 in $\overline{X}$, then the Cox ring of X is given by*

$$\mathcal{R}(X) = \mathbb{K}[T_1, \ldots, T_r]/I(\overline{X}), \qquad \deg(T_i) = [D_i] \in \mathrm{Cl}(X) = \mathrm{Cl}(Z).$$

Moreover, if the pullbacks $\iota^*(D_1), \ldots, \iota^*(D_r)$ are pairwise different prime divisors on X, then the corresponding variables $T_1, \ldots, T_r$ define pairwise nonassociated $\mathrm{Cl}(X)$-primes in $\mathcal{R}(X)$.

Example 4.1.1.6 Let $\mathfrak{n} = (n_0, \ldots, n_r)$ be a tuple of positive integers and consider the smooth toric prevariety $Z(\mathfrak{n})$ obtained from the projective space $\mathbb{P}^r$ by n_i-plying the ith toric divisor $V(\mathbb{P}^r, S_i)$, where $\mathbb{K}[S_0, \ldots, S_r]$ is the homogeneous coordinate ring. Then the toric characteristic space is $\widehat{Z}(\mathfrak{n}) \to Z(\mathfrak{n})$ with an open toric subset $\widehat{Z}(\mathfrak{n}) \subseteq \mathbb{K}^n$, where $n = n_0 + \cdots + n_r$. On the respective acting tori the morphism is given by

$$(t_{ij}) \mapsto \left(\frac{t_{11} \cdots t_{1n_1}}{t_{01} \cdots t_{0n_0}}, \ldots, \frac{t_{r1} \cdots t_{rn_r}}{t_{01} \cdots t_{0n_0}} \right).$$

Now, let $A = (a_0, \ldots, a_r)$ be a sequence of pairwise noncollinear points in $a_i = (b_i, c_i) \in \mathbb{K}^2 \setminus \{0\}$. For every $0 \le i \le r - 2$ set $k = j + 1 = i + 2$ and consider the linear forms

$$h_i := (b_j c_k - b_k c_j)S_i + (b_k c_i - b_i c_k)S_j + (b_i c_j - b_j c_i)S_k.$$

The set of common zeroes $X(A, \mathfrak{n})$ of $h_0, \ldots, h_r$ in $Z(\mathfrak{n})$ is a projective line with multiplied points isomorphic to the space $\mathbb{P}^1(A, \mathfrak{n})$ constructed in Example 1.4.1.6. Its divisor class group is

$$\mathrm{Cl}(X(A, \mathfrak{n})) = \bigoplus_{j=1}^{n_0} \mathbb{Z} \cdot [x_{0j}] \oplus \bigoplus_{i=1}^{r} \left(\bigoplus_{j=1}^{n_i - 1} \mathbb{Z} \cdot [x_{ij}] \right) \cong \mathrm{Cl}(Z(\mathfrak{n})),$$

where $x_{ij} \in X(A, \mathfrak{n})$ denotes the intersection of $X(A, \mathfrak{n})$ with the jth divisor D_{ij} lying over $V(S_i)$. Set $T_i := T_{i1} \cdots T_{in_i}$. Then pulling back the relations h_i to the toric characteristic space gives us the following equations for $\overline{X(A, \mathfrak{n})} \subseteq \mathbb{K}^n$:

$$g_i := (b_j c_k - b_k c_j) T_i + (b_k c_i - b_i c_k) T_j + (b_i c_j - b_j c_i) T_k.$$

for $0 \le i \le r - 2$ and $k = j + 1 = i + 2$. One directly checks that these relations even generate the vanishing ideal and we are in the situation of Corollary 4.1.1.5. Thus, for $r \le 1$ the Cox ring of $X(A, \mathfrak{n})$ is isomorphic to the polynomial ring $\mathbb{K}[T_{ij}]$, and for $r \ge 2$ it has a presentation

$$\mathcal{R}(X(A, \mathfrak{n})) \cong \mathbb{K}[T_{ij}; \ 0 \le i \le r, \ 1 \le j \le n_i] / \langle g_i; \ 0 \le i \le r - 2 \rangle.$$

We specialize further to the classical case that the ambient variety Z is the projective space.

Corollary 4.1.1.7 *Let $n \ge 3$ and consider an n-dimensional smooth complete intersection $X = V(g_1, \dots, g_k) \subseteq \mathbb{P}^{n+k}$ not contained in any coordinate hyperplane. Then the Cox ring of X is given by*

$$\mathcal{R}(X) = \mathbb{K}[T_0, \dots, T_{n+k}]/\langle g_1, \dots, g_k \rangle, \qquad \deg(T_i) = 1 \in \mathrm{Cl}(X) = \mathbb{Z}.$$

Moreover, if the pullbacks $\imath^*(D_0), \dots, \imath^*(D_{n+k})$ are pairwise different prime divisors on X, then the corresponding variables $T_0, \dots, T_{n+k}$ define pairwise nonassociated $\mathrm{Cl}(X)$-primes in $\mathcal{R}(X)$.

Proof The Lefschetz theorem for Picard groups [159] guarantees that the pullback $\imath^*: \mathrm{Cl}(\mathbb{P}^n) \to \mathrm{Cl}(X)$ is an isomorphism. Because X is a smooth complete intersection, we can use Serre's criterion to see that $\overline{X} \subseteq \mathbb{K}^{n+k+1}$ is normal. Moreover, we have $\overline{X} \setminus \widehat{X} = \{0\}$. Thus, Corollary 4.1.1.5 applies. $\square$

Generalized Lefschetz hyperplane theorems for the divisor class group as, for example, provided in [256, 257] can be used to weaken the assumptions. Combining this with Remark 4.1.1.4, one obtains the following, see [11, Thm. 3.1].

Example 4.1.1.8 Let Z be a smooth toric Fano variety of dimension 4 and let X be a smooth element of $|-K_Z|$, so that X is a Calabi–Yau threefold. Then the Cox ring of X admits a single defining relation if and only if Z is $\mathbb{P}^4$ or the projectivization of the vector bundle $\mathcal{O}_{\mathbb{P}^2} \oplus \mathcal{O}_{\mathbb{P}^2}(a) \oplus \mathcal{O}_{\mathbb{P}^2}(b)$, where $a, b \ge 0$ and $a + b \le 2$.

Recall that for a Laurent polynomial $f = \sum a_\nu T^\nu \in \mathbb{K}[T_1^{\pm 1}, \dots, T_n^{\pm 1}]$ one defines its Newton polytope to be the convex hull of its exponent vectors:

$$N(f) := \mathrm{conv}(\nu \in \mathbb{Z}^n; \ a_\nu \ne 0) \subseteq \mathbb{Q}^n.$$

For a face $\Delta \preceq N(f)$, we define the face polynomial to be $f_\Delta := \sum_{\nu \in \Delta} a_\nu T^\nu$. We say that the Laurent polynomial f is *nondegenerate* if none of its face polynomials $\mathrm{grad}(f_\Delta)$ has zeroes along $\mathbb{T}^n$ and every one-dimensional face $\Delta \preceq N(f)$ defines a regular cone

$$\mathrm{cone}(u - u_0;\ u \in N(f),\ u_0 \in \Delta)^\vee \subseteq \mathbb{Q}^n.$$

Moreover, given a homomorphism $\pi\colon \mathbb{T}^r \to \mathbb{T}^n$, let $f_\pi \in \mathbb{K}[T_1, \ldots, T_r]$ be the unique generator of the radical of $\langle \pi^* f \rangle \cap \mathbb{K}[T_1, \ldots, T_r]$. We say that f is π-*admissible* if the variables T_i define pairwise nonassociated $\mathbb{X}(\mathrm{ker}(\pi))$-prime elements in $\mathbb{K}[T_1, \ldots, T_r]/\langle f_\pi \rangle$.

Proposition 4.1.1.9 *Let Z be a toric variety and let the characteristic space $\widehat{Z} \to Z$ have $\pi\colon \mathbb{T}^r \to \mathbb{T}^n$ as the associated homomorphism of acting tori. Let $f \in \mathbb{K}[T_1^{\pm 1}, \ldots, T_n^{\pm 1}]$ be an irreducible nondegenerate π-admissible polynomial with Newton polytope of dimension at least 4. Then the closure $X \subseteq Z$ of $V(f) \subseteq \mathbb{T}^n$ has Cox ring*

$$\mathcal{R}(X) = \mathbb{K}[T_1, \ldots, T_r]/\langle f_\pi \rangle, \qquad \mathrm{deg}(T_i) = [D_i] \in \mathrm{Cl}(Z) = \mathrm{Cl}(X).$$

Proof The fact that f is nondegenerate ensures that $V(f) \subseteq \mathbb{T}^n$ has trivial divisor class group; see [125]. Thus, Corollary 3.4.1.6 tells us that the ring $\mathbb{K}[T_1, \ldots, T_r]/\langle f_\pi \rangle$ is factorially $\mathrm{Cl}(Z)$-graded via $\mathrm{deg}(T_i) = [D_i]$. The assertion follows. $\qquad\square$

4.1.2 Algebraic modification

We provide a machinery for constructing new factorially graded rings out of given ones. The method will in particular apply to the computation of the Cox ring of certain blow-ups. In this section, $\mathbb{K}$ is any field. The original reference for this section is [28].

Consider a grading of the polynomial ring $\mathbb{K}[T_1, \ldots, T_{r_1}]$ by a finitely generated abelian group K_1 such that the variables T_i are homogeneous. Then we have a pair of exact sequences

$$0 \longrightarrow \mathbb{Z}^{k_1} \xrightarrow{\ Q_1^* \ } \mathbb{Z}^{r_1} \xrightarrow{\ P_1 \ } \mathbb{Z}^n$$

$$0 \longleftarrow K_1 \xleftarrow[\ Q_1 \ ]{} \mathbb{Z}^{r_1} \xleftarrow[\ P_1^* \ ]{} \mathbb{Z}^n \longleftarrow 0$$

where $Q_1\colon \mathbb{Z}^{r_1} \to K_1$ is the degree map sending the ith canonical basis vector e_i to $\mathrm{deg}(T_i) \in K_1$. We enlarge P_1 to a $n \times r_2$ matrix P_2 by concatenating further $r_2 - r_1$ columns. This gives a new pair of exact sequences

$$0 \longrightarrow \mathbb{Z}^{k_2} \xrightarrow{\ Q_2^* \ } \mathbb{Z}^{r_2} \xrightarrow{\ P_2 \ } \mathbb{Z}^n$$

$$0 \longleftarrow K_2 \xleftarrow[\ Q_2 \ ]{} \mathbb{Z}^{r_2} \xleftarrow[\ P_2^* \ ]{} \mathbb{Z}^n \longleftarrow 0$$

Construction 4.1.2.1 Given a K_1-homogeneous ideal $I_1 \subseteq \mathbb{K}[T_1, \ldots, T_{r_1}]$, we transfer it to a K_2-homogeneous ideal $I_2 \subseteq \mathbb{K}[T_1, \ldots, T_{r_2}]$ by taking extensions and contractions according to the scheme

$$
\begin{array}{ccc}
\mathbb{K}[T_1, \ldots, T_{r_2}] & & \mathbb{K}[T_1, \ldots, T_{r_1}] \\
\Big\downarrow{\imath_2} & & \Big\downarrow{\imath_1} \\
\mathbb{K}[T_1^{\pm 1}, \ldots, T_{r_2}^{\pm 1}] \xleftarrow{\pi_2^*} \mathbb{K}[S_1^{\pm 1}, \ldots, S_n^{\pm 1}] \xrightarrow{\pi_1^*} \mathbb{K}[T_1^{\pm 1}, \ldots, T_{r_1}^{\pm 1}]
\end{array}
$$

where $\imath_1, \imath_2$ are the canonical embeddings and π_i^* are the homomorphisms of group algebras defined by $P_i^* \colon \mathbb{Z}^n \to \mathbb{Z}^{r_i}$.

Now let $I_1 \subseteq \mathbb{K}[T_1, \ldots, T_{r_1}]$ be a K_1-homogeneous ideal and $I_2 \subseteq \mathbb{K}[T_1, \ldots, T_{r_2}]$ the transferred K_2-homogeneous ideal. Our result relates factoriality properties of the algebras $R_1 := \mathbb{K}[T_1, \ldots, T_{r_1}]/I_1$ and $R_2 := \mathbb{K}[T_1, \ldots, T_{r_2}]/I_2$ to each other.

Theorem 4.1.2.2 *Assume that R_1, R_2 are integral, $T_1, \ldots, T_{r_1}$ define K_1-primes in R_1, and $T_1, \ldots, T_{r_2}$ define K_2-primes in R_2. Then the following statements are equivalent.*

(i) *The algebra R_1 is factorially K_1-graded.*
(ii) *The algebra R_2 is factorially K_2-graded.*

Proof First observe that the homomorphisms π_j^* embed $\mathbb{K}[S_1^{\pm 1}, \ldots, S_n^{\pm 1}]$ as the degree zero part of the respective K_j-grading and fit into a commutative diagram

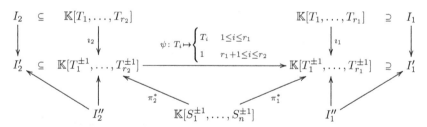

The factor ring R_1' of the extension $I_1' := \langle \imath_1(I_1) \rangle$ is obtained from R_1 by localization with respect to K_1-primes $T_1, \ldots, T_{r_1}$:

$$
R_1' := \mathbb{K}[T_1^{\pm 1}, \ldots, T_{r_1}^{\pm 1}]/I_1' \cong (R_1)_{T_{r_1} \cdots T_{r_1}}.
$$

The ideal I_1'' is the degree zero part of I_1'. Thus, its factor algebra is the degree zero part of R_1':

$$
R_1'' := \mathbb{K}[T_1^{\pm 1}, \ldots, T_{r_1}^{\pm 1}]_0/I_1'' \cong (R_1')_0.
$$

Note that $\mathbb{K}[T_1^{\pm 1}, \ldots, T_{r_1}^{\pm 1}]$ and hence R_1' admit units in every degree. Thus, Corollary 3.4.1.6 yields that R_1 is factorially K_1-graded if and only if R_1'' is a unique factorization domain.

The homomorphism ψ restricts to an isomorphism ψ_0 of the respective degree zero parts. Thus, the shifted ideal $I_2'' := \psi_0^{-1}(I_1'')$ defines an algebra R_2'' isomorphic to R_1'':

$$R_2'' := \mathbb{K}[T_1^{\pm 1}, \ldots, T_{r_2}^{\pm 1}]_0/I_2'' \cong R_1''.$$

The ideal $I_2' := \langle \pi_2^*((\pi_1^*)^{-1}(I_1')) \rangle$ has I_2'' as its degree zero part and $\mathbb{K}[T_1^{\pm 1}, \ldots, T_{r_2}^{\pm 1}]$ admits units in every degree. The associated K_2-graded algebra

$$R_2' := \mathbb{K}[T_1^{\pm 1}, \ldots, T_{r_2}^{\pm 1}]/I_2'$$

is the localization of R_2 by the K_2-primes $T_1, \ldots, T_{r_2}$. Again by Corollary 3.4.1.6, we obtain that R_2'' is a unique factorization domain if and only if R_2 is factorially K_2-graded. □

The rest of the section is devoted to practical aspects around Theorem 4.1.2.2. The following construction is an alternative to the transferring scheme 4.1.2.1. We use it in Proposition 4.1.2.5 to weaken the assumptions in Theorem 4.1.2.2.

Definition 4.1.2.3 Consider an $n \times r$ matrix P and an $n \times l$ matrix B, both integral. A *weak B-lifting (with respect to P)* is an integral $r \times l$ matrix A allowing a commutative diagram

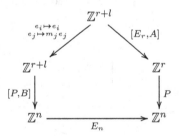

where the e_i are the first r; the e_j the last l canonical basis vectors of $\mathbb{Z}^{r+l}$; the m_j are positive integers; and E_n, E_r denote the unit matrices of size n, r respectively.

Note that weak B-liftings A always exist. Given such A, consider the following homomorphism of Laurent polynomial rings:

$$\psi_A \colon \mathbb{K}[T_1^{\pm 1}, \ldots, T_r^{\pm 1}] \to \mathbb{K}[T_1^{\pm 1}, \ldots, T_r^{\pm 1}, S_1^{\pm 1}, \ldots, S_l^{\pm 1}],$$

$$\sum \alpha_\nu T^\nu \mapsto \sum \alpha_\nu T^\nu S^{A^t \cdot \nu}.$$

Set $K_1 := \mathbb{Z}^r / P^*(\mathbb{Z}^n)$. Then the left-hand side algebra is K_1-graded by assigning to the ith variable the class of e_i in K_1.

Proposition 4.1.2.4 *In the above notation, let $g_1 \in \mathbb{K}[T_1^{\pm 1}, \ldots, T_r^{\pm 1}]$ be a K_1-homogeneous Laurent polynomial.*

(i) *We have $T^\nu S^\mu \psi_A(g_1) = g_2'$ with $\nu \in \mathbb{Z}^r$, $\mu \in \mathbb{Z}^l$ and a unique monomial free $g_2' \in \mathbb{K}[T_1, \ldots, T_r, S_1, \ldots, S_l]$.*

(ii) *The polynomial g_2' is of the form $g_2' = g_2(T_1, \ldots, T_r, S_1^{m_1}, \ldots, S_l^{m_1})$ with a $g_2 \in \mathbb{K}[T_1, \ldots, T_r, S_1, \ldots, S_l]$ not depending on the choice of A.*

(iii) *If, in the setting of Construction 4.1.2.1, we have $I_1 = \langle g_1 \rangle$, then the transferred ideal is given by $I_2 = \langle g_2 \rangle$.*

(iv) *The variable T_i defines a prime element in $\mathbb{K}[T_0, \ldots, T_{r+l}]/\langle g_2 \rangle$ if and only if the polynomial $g_2(T_1, \ldots, T_{i-1}, 0, T_{i+1}, \ldots, T_{r+l})$ is irreducible.*

Proof Consider the commutative diagram of group algebras corresponding to the dualized diagram 4.1.2.3. There, ψ_A occurs as the homomorphism of group algebras defined by the transpose $[E_r, A]^*$. Let T^κ be any monomial of g_1. Then $g_1' := T^{-\kappa} g_1$ gives rise to the same g_2, but g_1' is of K_1-degree zero and hence a pullback $g_1' = \psi_{P^*}(h)$. The latter allows us to use commutativity of the diagram which gives (i) and (ii). Assertions (iii) and (iv) are clear. $\square$

Proposition 4.1.2.5 *In the notation of Theorem 4.1.2.2 assume that R_1 is integral and $T_1, \ldots, T_{r_1}$ define nonzero elements in R_1. Then R_2 is integral.*

Proof Write $P_2 = [P_1, B]$ with a suitable matrix B and consider a weak B-lifting with respect to P_1:

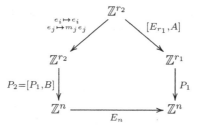

The assumptions on the T_i and the fact that I_1 is prime ensure that $\langle \iota_1(I_1) \rangle$ is prime. Because the lattice homomorphism $F := [E_{r_1}, A]$ is surjective, the image of the dual homomorphism $F^*: \mathbb{Z}^{r_1} \to \mathbb{Z}^{r_2}$ is a direct summand. Thus, the homomorphism of Laurent polynomial rings corresponding to F^* extends $\langle \iota_1(I_1) \rangle$ to a prime ideal. We conclude that I_2 is prime. $\square$

The following observation provides a normality criterion and reduces the number of necessary primality tests for Theorem 4.1.2.2 in many cases.

Proposition 4.1.2.6 *In the notation of Theorem 4.1.2.2 assume that the canonical map $K_2 \to K_1$ admits a section (for example, K_1 is free).*

 (i) *If R_1 is integral (normal) and $T_{r_1+1}, \ldots, T_{r_2}$ define primes in R_2 (for example, they are K_2-prime and K_2 is free), then R_2 is integral (normal).*
 (ii) *Assume that R_1 is integral, $T_1, \ldots, T_{r_1}$ define K_1-primes in R_1, and $T_{r_1+1}, \ldots, T_{r_2}$ define K_2-primes in R_2. If no T_j with $j \geq r_1 + 1$ divides a T_i with $i \leq r_1$ in R_2, then also $T_1, \ldots, T_{r_1}$ define K_2-primes in R_2.*
 (iii) *If R_2 is integral (normal) then R_1 is integral (normal). Moreover, if $T_1, \ldots, T_{r_2}$ define K_2-primes in R_2 and no T_j with $j \geq r_1 + 1$ divides a T_i with $i \leq r_1$ in R_2, then $T_1, \ldots, T_{r_1}$ define K_1-primes in R_1.*

Lemma 4.1.2.7 *Let R be an integral ring and $f \in R$ a prime element. If R_f is normal, then R is normal.*

Proof Let k be the quotient field of R and $h \in k$ be integral over R. Then there are $a_0, \ldots, a_{n-1} \in R$ such that

$$h^n + a_{n-1}h^{n-1} + \cdots + a_1 h + a_0 = 0 \in k.$$

Then h is integral over R_f as well and thus we have $h \in R_f$. Thus $h = g/f^l$ holds with $g \in R$ and $l \in \mathbb{Z}_{\geq 0}$. We choose g in this presentation such that l is minimal. We claim that $l = 0$ holds. Otherwise, plugging $h = g/h^l$ into the equation of integral dependence shows $f \mid g^n$. Because f is prime, we obtain $f \mid g$, a contradiction to the minimality of l. $\qquad\square$

Proof of Proposition 4.1.2.6 The exact sequences involving the grading groups K_1 and K_2 fit into a commutative diagram where the upwards sequences are exact and $\mathbb{Z}^{r_2 - r_1} \to K_2'$ is an isomorphism:

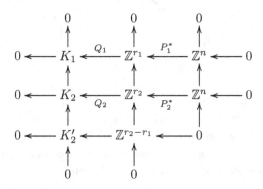

Moreover, denoting by $K_1' \subseteq K_2$ the image of the section $K_1 \to K_2$, there is a splitting $K_2 = K_2' \oplus K_1'$. As $K_2' \subseteq K_2$ is the subgroup generated by the degrees of $T_{r_1+1}, \ldots, T_{r_2}$, we obtain a commutative diagram

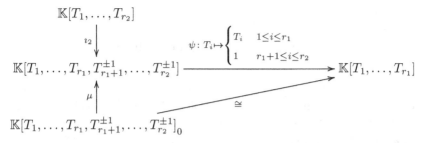

where the map μ denotes the embedding of the degree zero part with respect to the K_2'-grading. By the splitting $K_2 = K_2' \oplus K_1'$, the image of μ is precisely the Veronese subalgebra associated with the subgroup $K_1' \subseteq K_2$. For the factor rings R_2 and R_1 by the ideals I_2 and I_1, the above diagram leads to the following situation

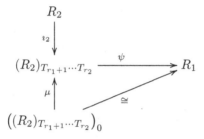

We prove (i). As just observed, the degree zero part $((R_2)_{T_{r_1+1} \cdots T_{r_2}})_0$ of the K_2'-grading is isomorphic to R_1 and thus integral (normal) if R_1 is so. Moreover, we infer that $(R_2)_{T_{r_1+1} \cdots T_{r_2}}$ is a Laurent polynomial algebra over R_1. Thus, $(R_2)_{T_{r_1+1} \cdots T_{r_2}}$ is integral (normal) if R_1 is so. Construction 4.1.2.1 gives that R_2 is integral. Moreover, if $T_{r_1+1}, \ldots, T_{r_2}$ define primes in R_2, then an iterated application of Lemma 4.1.2.7 shows that R_2 is normal.

To prove (ii), consider a variable T_i with $1 \leq i \leq r_1$. We have to show that T_i defines a K_2-prime element in R_2. By the above diagram, T_i defines a K_1'-prime element in $((R_2)_{T_{r_1+1} \cdots T_{r_2}})_0$, the Veronese subalgebra of R_2 defined by $K_1' \subseteq K_2$. Because every K_2-homogeneous element of $(R_2)_{T_{r_1+1} \cdots T_{r_2}}$ can be shifted by a homogeneous unit into $((R_2)_{T_{r_1+1} \cdots T_{r_2}})_0$, Lemma 3.4.1.9 yields that T_i defines a K_2-prime in $(R_2)_{T_{r_1+1} \cdots T_{r_2}}$. By assumption, $T_{r_1+1}, \ldots, T_{r_2}$ define K_2-primes in R_2 and are all coprime to T_i. It follows from Lemma 3.4.1.7 that T_i defines a K_2-prime in R_2.

We turn to assertion (iii). As seen in the above diagram, R_1 occurs as the K_2'-degree zero Veronese subalgebra of $(R_2)_{T_{r_1+1} \cdots T_{r_2}}$. In particular, R_1 is integral and inherits normality. Now assume that $T_1, \ldots, T_{r_2}$ define K_2-primes in R_2 and no T_j with $j \geq r_1+1$ divides a T_i with $i \leq r_1$ in R_2. Then, using as before Lemmas 3.4.1.7 and 3.4.1.9, we see that $T_1, \ldots, T_{r_1}$ define K_1-primes in R_1. □

4.1.3 Toric ambient modifications

We use the results of the preceding section to study the effect of toric ambient modifications on the Cox ring of an embedded variety. This applies in particular to the setting of a canonical toric embedding of a variety associated with a bunched ring. We demonstrate this technique by computing the Cox ring of the blow-up of the projective plane at four points in general position. The original references are [28, 165, 168]. The ground field $\mathbb{K}$ is algebraically closed of characteristic zero.

We begin with the contraction problem, that is, given a modification $X_2 \to X_1$ of complete varieties we aim to describe the Cox ring of X_1 in terms of the Cox ring of X_2. A first general statement now follows.

Proposition 4.1.3.1 *Let* $\pi \colon X_2 \to X_1$ *be a birational morphism of irreducible normal complete varieties with center* $C \subseteq X_1$. *Set* $K_i := \mathrm{Cl}(X_i)$ *and* $R_i := \mathcal{R}(X_i)$ *and identify* $U := X_2 \setminus \pi^{-1}(C)$ *with* $X_1 \setminus C$. *Then we have canonical surjective pushforward maps*

$$\pi_* \colon K_2 \to K_1, \ [D] \mapsto [\pi_* D], \quad \pi_* \colon R_2 \to R_1, (R_2)_{[D]} \ni f \mapsto f|_U \in (R_1)_{[\pi_* D]}.$$

Now suppose that $\mathcal{R}(X_2)$ *is finitely generated, let* $E_1, \ldots, E_l \subseteq X_2$ *denote the exceptional prime divisors and* $f_1, \ldots, f_l \in \mathcal{R}(X_2)$ *the corresponding canonical sections. Then we have a commutative diagram*

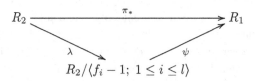

of morphisms of graded algebras, where λ *is the canonical projection and the induced map* ψ *is an isomorphism.*

Lemma 4.1.3.2 *Let* R *be a* K_2-*graded domain,* $f \in R_w$ *with* w *of infinite order in* K_2, *and consider the downgrading of* R *given by* $K_2 \to K_1 := K_2/\langle w \rangle$. *Then* $f - 1$ *is* K_1-*prime.*

Proof Let $(f - 1)g = ab$, where $g, a, b \in R$ are K_1-homogeneous elements. Because w has infinite order, any K_1-homogeneous element $u \in R$ can be uniquely written as a sum $u = u_0 + \cdots + u_n$, where each u_i is K_2-homogeneous, both u_0 and u_n are nonzero, and $\deg_{K_2}(u_i) = \deg_{K_2}(u_0) + i w$

holds for each i such that u_i is nonzero. According to this observation we write

$$(f-1)\sum_{i=0}^{n} g_i = \Big(\sum_{i=0}^{n_1} a_i\Big)\Big(\sum_{i=0}^{n_2} b_i\Big).$$

We have $-g_0 = a_0 b_0$ because otherwise, by equating the K_2-homogeneous elements of the same degree, we would obtain either $g_0 = 0$ or $a_0 b_0 = 0$. Similarly, we see $f g_n = a_{n_1} b_{n_2}$. Thus

$$f g_n = a_{n_1} b_{n_1}, \quad f g_{n-1} - g_n = a_{n_1} b_{n_2-1} + a_{n_1-1} b_{n_2}, \quad \dots \quad -g_0 = a_0 b_0.$$

Eliminating the g_i gives

$$(a_0 f^{n_1} + a_1 f^{n_1-1} + \cdots + a_{n_1})(b_0 f^{n_2} + b_1 f^{n_2-1} + \cdots + b_{n_2}) = 0.$$

Because R is integral, one of the two factors must be zero, say the first one. Then $f-1$ divides

$$a = a_0 + \cdots + a_{n_1} = a_0(1 - f^{n_1}) + \cdots + a_{n_1-1}(1 - f). \qquad \square$$

Proof of Proposition 4.1.3.1 Let $x_i \in X_i$ be smooth points with $\pi(x_2) = x_1$ such that x_2 is not contained in any of the exceptional divisors. Consider the divisorial sheaf $\mathcal{S}_i$ on X_i associated with the subgroup of divisors avoiding the point x_i as in Construction 1.4.2.3. Then we have canonical morphisms of graded rings

$$\Gamma(X_2, \mathcal{S}_2) \to \Gamma(U_2, \mathcal{S}_2) \to \Gamma(X_1, \mathcal{S}_1),$$

where $U_2 \subseteq X_2$ is the open subset obtained by removing the exceptional divisors of $\pi\colon X_2 \to X_1$ and the accompanying homomorphisms of the grading groups are the respective pushforwards of Weil divisors. The homomorphisms are compatible with the relations of the Cox sheaves $\mathcal{R}_i$; see again Construction 1.4.2.3, and thus induce canonical morphisms of graded rings

$$\Gamma(X_2, \mathcal{R}_2) \to \Gamma(U_2, \mathcal{R}_2) \to \Gamma(X_1, \mathcal{R}_1).$$

This establishes the surjection $\pi_*\colon R_2 \to R_1$ with the canonical pushforward of divisor class groups as the accompanying homomorphism. Clearly, the canonical sections f_i of the exceptional divisors are sent to $1 \in R_1$.

We show that the induced map ψ is an isomorphism. As we may proceed by induction on l, it suffices to treat the case $l = 1$. Lemma 4.1.3.2 tells us that $f_1 - 1$ is K_1-prime. From Proposition 1.5.3.3 we infer that $\langle f_1 - 1 \rangle$ is a radical ideal in R_2. Because $\mathrm{Spec}(\psi)$ is a closed embedding of varieties of the same dimension and equivariant with respect to the action of the quasitorus $\mathrm{Spec}\,\mathbb{K}[K_1]$, the assertion follows. $\qquad \square$

Let us look at the contraction problem in the setting of ambient modifications. Consider a proper birational toric morphism $Z_2 \to Z_1$ arising from a refinement $\Sigma_2 \to \Sigma_1$ of nondegenerate fans. Then, with the toric Cox constructions $p_i \colon \widehat{Z}_i \to Z_i$ and the open embeddings into the toric total coordinate spaces $\widehat{Z}_i \subseteq \mathbb{K}^{r_i}$, the situation is the following:

$$
\begin{array}{ccccc}
\mathbb{K}^{r_2} & \supseteq & \widehat{Z}_2 & \widehat{Z}_1 & \subseteq & \mathbb{K}^{r_1} \\
& & \downarrow{\scriptstyle p_2} & \downarrow{\scriptstyle p_1} & & \\
& & Z_2 & \longrightarrow & Z_1 &
\end{array}
$$

The toric morphisms p_i arise from lattice homomorphisms $P_i \colon \mathbb{Z}^{r_i} \to \mathbb{Z}^n$ and the polynomial rings $\mathbb{K}[T_1, \ldots, T_{r_i}]$ are graded by $K_i := \mathbb{Z}^{r_i}/P_i^*(\mathbb{Z}^n)$. Now, let a K_2-homogeneous ideal $I_2 \subseteq \mathbb{K}[T_1, \ldots, T_{r_2}]$ be given and consider the geometric data

$$
\overline{X}_2 := V(I_2) \subseteq \mathbb{K}^{r_2}, \qquad \widehat{X}_2 := \overline{X}_2 \cap \widehat{Z}_2, \qquad X_2 := p_2(\widehat{X}_2) \subseteq Z_2.
$$

Via the transferring scheme 4.1.2.1, the ideal I_2 defines a K_1-homogeneous ideal I_1 in $\mathbb{K}[T_1, \ldots, T_{r_i}]$ and we have the K_i-graded rings $R_i := \mathbb{K}[T_1, \ldots, T_{r_i}]/I_i$. Now assume that X_2 is complete, R_2 is factorially K_2-graded with $R_2^* = \mathbb{K}^*$, and $T_1, \ldots, T_{r_2}$ define pairwise nonassociated K_2-prime elements in R_2. Then R_2 is the Cox ring of X_2. Our statement concerns the Cox ring of the image $X_1 \subseteq Z_1$ of $X_2 \subseteq Z_2$ under $Z_2 \to Z_1$.

Theorem 4.1.3.3 *In the above setting, the K_1-graded ring R_1 is the Cox ring of X_1 and the variables $T_1, \ldots, T_{r_1}$ define pairwise nonassociated K_1-prime elements in R_1.*

Proof Viewing the P_i as matrices, we may assume that $P_2 = [P_1, B]$ with a matrix B of size $n \times (r_2 - r_1)$. Take a weak B-lifting with respect to P_1 and look at the corresponding diagram of homomorphisms of tori:

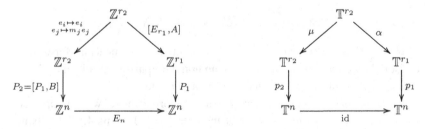

Observe that for every element $f \in I_2$, the Laurent polynomials $\mu^*(f)$ and $\alpha^*(f(t_1, \ldots, t_{r_1}, 1, \ldots, 1))$ differ by a monomial factor. We conclude

$$
\mathbb{K}[T_1^{\pm 1}, \ldots, T_{r_2}^{\pm 1}] \cdot I_2 = \langle f(t_1, \ldots, t_{r_1}, 1, \ldots, 1); f \in I_2 \rangle \subseteq \mathbb{K}[T_1^{\pm 1}, \ldots, T_{r_2}^{\pm 1}].
$$

Thus, Proposition 4.1.3.1 tells us that R_1 is the Cox ring of X_1. Because $T_1, \ldots, T_{r_1}$ define pairwise different prime divisors in X_1, they are pairwise nonassociated K_1-primes in R_1. □

For practical puposes, the following criterion is helpful to decide contractibility of a divisor; see [165, Prop. 6.7]. Given a factorially K-graded $\mathbb{K}$-algebra $R = \oplus_K R_w$ with pairwise nonassociated K-prime generators $f_1, \ldots, f_r$, we say that the weight $w_i \in K$ of $f_i \in R$ is *exceptional* if it generates an extremal ray $\varrho_i \preceq \omega(R)$ of the weight cone $\omega(R) \subseteq K_\mathbb{Q}$ and $w_i \notin \varrho_i$ holds for the weights w_j of f_j with $j \neq i$.

Remark 4.1.3.4 Consider the canonical toric embedding $X \subseteq Z$ of a $\mathbb{Q}$-factorial projective variety X arising from a bunched ring $(R, \mathfrak{F}, \Phi)$, where $\Phi = \Phi(\lambda)$ is given by a GIT cone $\lambda \in \Lambda(\overline{X}, H)$. Then, for a toric prime divisor $D_Z^i \subseteq Z$ and the prime divisor $D_X^i = D_Z^i \cap X$ the following statements are equivalent.

(i) The class $[D_X^i]$ is exceptional for $R = \mathcal{R}(X)$ and λ shares a common facet with a full-dimensional $\lambda' \in \Lambda(\overline{X}, H)$ satisfying $[D_X^i] \in \lambda$.
(ii) There exists a morphism $X \to X'$ onto a $\mathbb{Q}$-factorial projective variety X' contracting precisely $D_X^i \subseteq X$.
(iii) The class $[D_Z^i]$ is exceptional for $\mathcal{R}(Z)$ and λ shares a common facet with a full-dimensional $\lambda' \in \Lambda(\overline{X}, H)$ satisfying $[D_Z^i] \in \lambda$.
(iv) There exists a toric morphism $Z \to Z'$ onto a $\mathbb{Q}$-factorial quasiprojective toric variety Z' contracting precisely D_Z^i.

If one of these properties holds, then the restriction of $Z \to Z'$ of (iv) gives a morphism $X \to X'$ as in (ii). Recall from Construction 3.2.5.3 that the ambient varieties Z and Z' are not projective in general.

We turn to the problem of determining the Cox ring of the modified, for example, blown up, variety. The setting is the same as before: we consider a toric modifcation $Z_2 \to Z_1$, let P_i denote the matrices defining the toric Cox constructions $p_i \colon \widehat{Z}_i \to Z_i$, etc. This time, we start with a variety $X_1 \subseteq Z_1$, concretely given by a K_1-homogeneous ideal $I_1 \subseteq \mathbb{K}[T_1, \ldots, T_r]$ and

$$\overline{X}_1 := V(I_1) \subseteq \mathbb{K}^{r_1}, \qquad \widehat{X}_1 := \overline{X}_1 \cap \widehat{Z}_1, \qquad X_1 := p_1(\widehat{X}_1) \subseteq Z_1.$$

Assume that R_1 is factorially K_1-graded with $R_1^* = \mathbb{K}^*$ and $T_1, \ldots, T_{r_1}$ define pairwise nonassociated K_1-prime elements in R_1. Then R_1 is the Cox ring of X_1. Our statement concerns the Cox ring of the proper transform $X_2 \subseteq Z_2$ of $X_1 \subseteq Z_1$ with respect to $Z_2 \to Z_1$.

Theorem 4.1.3.5 *In the above setting, assume that R_2 is normal and the variables $T_1, \ldots, T_{r_2}$ define pairwise nonassociated K_2-prime elements in R_2. Then the K_2-graded ring R_2 is the Cox ring of X_2.*

Proof By Proposition 4.1.2.5 the ring R_2 is integral and by Theorem 4.1.2.2, it is factorially K_2-graded. Moreover, with the toric Cox construction $\pi_2 \colon \widehat{Z}_2 \to Z_2$, we obtain that R_2 is the algebra of functions of the closure $\widehat{X}_2 \subseteq \widehat{Z}_2$ of $\pi_2^{-1}(X_2 \cap \mathbb{T}^{r_2})$. Thus, R_2 is the Cox ring of X_2.   □

As a consequence of Theorems 4.1.3.3 and 4.1.3.5, we obtain that the modifications preserving finite generation are exactly those arising from toric modifications as discussed. More precisely, let $Z_2 \to Z_1$ be a toric modification mapping $X_2 \subseteq Z_2$ onto $X_1 \subseteq Z_1$. We call $Z_2 \to Z_1$ a *good* toric ambient modification if it is as in Theorem 4.1.3.5 (i).

Corollary 4.1.3.6   *Let $X_2 \to X_1$ be a birational morphism of $\mathbb{Q}$-factorial projective varieties such that the Cox ring $\mathcal{R}(X_1)$ is finitely generated. Then the following statements are equivalent.*

(i) *The Cox ring $\mathcal{R}(X_2)$ is finitely generated.*
(ii) *The morphism $X_2 \to X_1$ arises from a good toric ambient modification.*

Proof The implication "(ii)⇒(i)" is Theorem 4.1.3.5. For the reverse direction, observe first that the prime components of the exceptional divisor have exceptional classes. Thus, the desired toric ambient modifications can be obtained by applying successively Remark 4.1.3.4.   □

We conclude with a general statement describing the Cox ring of X_2 for a blow-up $\pi \colon X_2 \to X_1$ of a Mori dream space X_1 at an irreducible subvariety $C \subseteq X_1$ contained in the smooth locus. Write $K_i := \mathrm{Cl}(X_i)$ for the divisor class groups and $R_i := \mathcal{R}(X_i)$ for the Cox rings. Then we have the canonical pullback maps

$$\pi^* \colon K_1 \to K_2, \; [D] \mapsto [\pi^* D], \quad \pi^* \colon R_1 \to R_2, \; (R_1)_{[D]} \ni f \mapsto \pi^* f \in (R_2)_{[\pi^* D]}.$$

Moreover, identifying $U := X_2 \setminus \pi^{-1}(C)$ with $X_1 \setminus C$, we obtain canonical pushforward maps

$$\pi_* \colon K_2 \to K_1, \; [D] \mapsto [\pi_* D], \quad \pi_* \colon R_2 \to R_1, \; (R_2)_{[D]} \ni f \mapsto f|_U \in (R_1)_{[\pi_* D]}.$$

Let $J \subseteq R_1$ be the irrelevant ideal, that is, the vanishing ideal of $\overline{X}_1 \setminus \widehat{X}_1$, and $I \subseteq R_1$ the vanishing ideal of $p_1^{-1}(C) \subseteq \overline{X}_1$. We define the *saturated Rees algebra* to be the subalgebra

$$R_1[I]^{\mathrm{sat}} := \bigoplus_{d \in \mathbb{Z}} (I^{-d} : J^\infty) t^d \subseteq R_1[t^{\pm 1}], \qquad \text{where } I^k := R_1 \text{ for } k \leq 0.$$

Remark 4.1.3.7   The usual Rees algebra $R_1[I] = \oplus_{\mathbb{Z}} I^{-d} t^d$ is a subalgebra of the saturated Rees algebra $R_1[I]^{\mathrm{sat}}$. In the above situation, $I \subseteq R_1$ is a

K_1-prime ideal and thus we have $I : J^\infty = I$. Consequently, $R_1[I]^{\text{sat}}$ equals $R_1[I]$ if and only if $R_1[I]^{\text{sat}}$ is generated in the $\mathbb{Z}$-degrees 0 and ± 1. In this case, $R_1[I]^{\text{sat}}$ is finitely generated because $R_1[I]$ is so.

Note that the saturated Rees algebra $R_1[I]^{\text{sat}}$ is naturally graded by $K_1 \times \mathbb{Z}$ as R_1 is K_1-graded and the ideals I, J are homogeneous. Let $E = \pi^{-1}(C)$ denote the exceptional divisor. Then we have a splitting $K_2 = \pi^* K_1 \times \mathbb{Z} \cdot [E] \cong K_1 \times \mathbb{Z}$.

Proposition 4.1.3.8 *In the above situation, we have the following mutually inverse isomorphisms of graded algebras:*

$$R_2 \longleftrightarrow R_1[I]^{\text{sat}},$$

$$(R_2)_{[\pi^* D] + d[E]} \ni f \;\mapsto\; \pi_* f \cdot t^d \in (R_1[I]^{\text{sat}})_{[D]+d},$$

$$(R_2)_{[\pi^* D] + d[E]} \ni \pi^* f \cdot 1_E^d \;\longleftarrow\; f \cdot t^d \in (R_1[I]^{\text{sat}})_{[D]+d}.$$

Lemma 4.1.3.9 *In the above situation, consider the characteristic spaces* $p_i \colon \widehat{X}_i \to X_i$ *and let* $\mathcal{I}_C$, $\mathcal{I}_{\widehat{C}}$, $\mathcal{I}_E$, $\mathcal{I}_{\widehat{E}}$ *be the ideal sheaves of* C, $\widehat{C} = p_1^{-1}(C)$, E, $\widehat{E} = p_2^{-1}(E)$ *on* X_1, $\widehat{X}_1$, X_2, $\widehat{X}_2$ *respectively. Then, for any* $m > 0$, *we have*

$$p_1^*(\mathcal{I}_C^m) = \mathcal{I}_{\widehat{C}}^m, \qquad p_2^*(\mathcal{I}_E^m) = \mathcal{I}_{\widehat{E}}^m, \qquad \pi^*(\mathcal{I}_C^m) = \mathcal{I}_E^m.$$

Moreover, with the vanishing ideals $I \subseteq R_1$ *of the closure of* $\widehat{C}$ *and* $J \subseteq R_1$ *of* $\operatorname{Spec} R_1 \setminus \widehat{X}_1$, *we have* $\Gamma(\widehat{X}_1, \mathcal{I}_{\widehat{C}}^m) = I^m : J^\infty$ *for any* $m > 0$.

Proof For the first equality, we use that C is contained in the smooth locus of X_1. This implies that p_1 has no multiple fibers near C and the claim follows. The second equality is obtained by the same reasoning. The third one is a standard fact on blowing up [151, Prop. 8.1.7 and Cor. 8.1.8]. The last statement follows from the fact that $\widehat{X}_1$ is quasiaffine. $\qquad \square$

Proof of Proposition 4.1.3.8 We only have to prove that the maps are well defined. For the map from R_2 to $R_1[I]^{\text{sat}}$ consider $f \in (R_2)_{[\pi^* D]+d}$. We have to show that for $d < 0$, the pushforward $\pi_* f$ belongs to $I^{-d} : J^\infty$. Note that $\pi_* f = \pi_* f'$ holds with $f' := f \cdot 1_E^{-d} \in (R_2)_{[\pi^* D]}$. Pushing f' locally to X_2, then to X_1, and finally lifting it to $\widehat{X}_1$, we see using Lemma 4.1.3.9 that $\pi_* f'$ is a global section of the $-d$-th power of the ideal sheaf of $p_1^{-1}(C) \subseteq \widehat{X}_1$. This gives $\pi_* f' \in I^{-d} : J^\infty \subseteq R_1$.

For the map from $R_1[I]^{\text{sat}}$ to R_2, consider $f \cdot t^d \in (R_1[I]^{\text{sat}})_{[D]+d}$, where $d < 0$. We need that 1_E^{-d} divides $\pi^* f$ in R_2. By definition, there exist an $h \in J$ and a $k \geq 0$ such that $h \notin I$ and $f h^k \in I^{-d}$. Then we have $(\pi^* f)(\pi^* h)^k \in$

$(\pi^*I)^{-d}$. Using $\langle \pi^*I \rangle = \langle 1_E \rangle$ and the fact that 1_E is a K_2-prime not dividing any power of π^*h we see that 1_E^{-d} divides π^*f. $\square$

4.1.4 Computing examples

Discussing the basic example of blowing up for points in general position on the projective plane, we indicate how to use the technique of toric ambient modifications to determine Cox rings. As a result of computational application [168] we then present the Cox rings of the Gorenstein log del Pezzo surfaces of Picard number 1 that do not admit a $\mathbb{K}^*$-action. The ground field $\mathbb{K}$ is algebraically closed of characteristic zero.

Example 4.1.4.1 We consider the blow-up of the projective plane $\mathbb{P}^2$ in four points x_1, x_2, x_3, x_4 in general position, that is, no three of them lying on a projective line. After applying a suitable linear transformation, we may assume that the points are given by

$$x_0 = [1, 1, 1], \qquad x_1 = [1, 0, 0], \qquad x_2 = [0, 1, 0], \qquad x_3 = [0, 0, 1].$$

The points x_1, x_2, x_3 are toric fixed points. So, blowing up these points leads to the toric surface X_0 given by the fan Σ_0 in $\mathbb{Z}^2$ having the columns of the following matrix as the primitive generators of its rays:

$$P_0 := \begin{bmatrix} 1 & 0 & -1 & 1 & 0 & -1 \\ 0 & 1 & -1 & 1 & -1 & 0 \end{bmatrix}.$$

The toric total coordinate space of X_0 is $\overline{X}_0 = \mathbb{K}^6$, and the $\mathbb{Z}^4$-grading of the Cox ring $\mathbb{K}[T_1, \ldots, T_6]$ of X_0 is given by a Gale dual Q_0 of P_0; we may take

$$Q_0 := \begin{bmatrix} 1 & 1 & 1 & 0 & 0 & 0 \\ -1 & -1 & 0 & 1 & 0 & 0 \\ -1 & 0 & -1 & 0 & 1 & 0 \\ 0 & -1 & -1 & 0 & 0 & 1 \end{bmatrix}.$$

Moreover, the Cox coordinates of $x_0 \in X_0$ are $[1, \ldots, 1] \in \mathbb{K}^6$ and the toric blow-up $\pi_{123} \colon X_0 \to \mathbb{P}^2$ lifts to a map of the total coordinate spaces

$$\overline{\pi}_{123} \colon \mathbb{K}^6 \to \mathbb{K}^3, \qquad z \mapsto (z_1 z_4 z_5, z_2 z_4 z_6, z_3 z_5 z_6).$$

To blow-up $x_0 \in X_0$ by a toric ambient modification, we have to embed X_0 suitably into some toric variety Z_1. For this, consider the lines through x_0, x_i:

$$L_i := V(z_j - z_k) \subseteq \mathbb{P}^2, \qquad i = 1, 2, 3, \quad j, k \neq i.$$

Their proper transforms $L'_i \subseteq X_1$ under $\pi_{123} \colon X_1 \to \mathbb{P}^2$ are given in Cox coordinates by $L'_i = V(g_i)$, where the polynomials $g_i \in \mathbb{K}[T_1, \dots, T_6]$ are

$$g_1 := T_2T_4 - T_3T_5, \qquad g_2 := T_1T_4 - T_3T_6, \qquad g_3 := T_1T_5 - T_2T_6.$$

To get $x_0 \in X_0$ cut out by toric divisors, we realize X_0 as a neat subvariety of a toric variety Z_1 with total coordinate space $\overline{Z}_1 = \mathbb{K}^9$ by

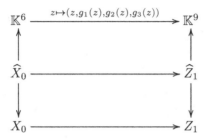

We write X_1 for the image of X_0. The Cox ring R_1 of X_1 is then presented as $\mathbb{K}[T_1, \dots, T_9]/I_1$ with the ideal I_1 generated by $f_i = T_{6+i} - g_i$, where $i = 1, 2, 3$. The $\mathbb{Z}^4$-grading of R_1 is given by the matrix Q_1 obtained by concatenating the degrees of g_1, g_2, g_3 to Q_0:

$$Q_1 := \begin{bmatrix} 1 & 1 & 1 & 0 & 0 & 0 & 1 & 1 & 1 \\ -1 & -1 & 0 & 1 & 0 & 0 & 0 & 0 & -1 \\ -1 & 0 & -1 & 0 & 1 & 0 & 0 & -1 & 0 \\ 0 & -1 & -1 & 0 & 0 & 1 & -1 & 0 & 0 \end{bmatrix}$$

Because X_1 is a surface, $\widehat{X}_1$ is the set of semistable points for any weight from the moving cone of Z_1. The matrix P_1 describing the toric morphism $\widehat{Z}_1 \to Z_1$ is Gale dual to Q_1 and may be taken as

$$P_1 := \begin{bmatrix} 1 & 0 & -1 & 1 & 0 & -1 & 0 & 0 & 0 \\ 0 & 1 & -1 & 1 & -1 & 0 & 0 & 0 & 0 \\ 0 & -1 & 0 & -1 & 0 & 0 & 1 & 0 & 0 \\ -1 & 0 & 0 & -1 & 0 & 0 & 0 & 1 & 0 \\ -1 & 0 & 0 & 0 & -1 & 0 & 0 & 0 & 1 \end{bmatrix}$$

In this setting, the point x_0 has Cox coordinates $[1, 1, 1, 1, 1, 1, 0, 0, 0]$. Blowing up the toric orbit through x_0 means to perform a barycentric subdivision of the cone generated by the last three columns of P_1. This gives the matrix

$$P_2 := \begin{bmatrix} 1 & 0 & -1 & 1 & 0 & -1 & 0 & 0 & 0 & 0 \\ 0 & 1 & -1 & 1 & -1 & 0 & 0 & 0 & 0 & 0 \\ 0 & -1 & 0 & -1 & 0 & 0 & 1 & 0 & 0 & 1 \\ -1 & 0 & 0 & -1 & 0 & 0 & 0 & 1 & 0 & 1 \\ -1 & 0 & 0 & 0 & -1 & 0 & 0 & 0 & 1 & 1 \end{bmatrix}$$

Under the corresponding toric blow-up $Z_2 \to Z_1$, the proper transform of $X_2 \subseteq Z_2$ is the blow-up of $X_1 \subseteq Z_1$ at x_0. Transferring the defining the ideal $I_1 \subseteq \mathbb{K}[T_1, \ldots, T_9]$ according to the scheme 4.1.2.1 gives an ideal $I_2 \subseteq \mathbb{K}[T_1, \ldots, T_{10}]$ generated by

$$T_7 T_{10} - T_2 T_4 + T_3 T_5, \qquad T_8 T_{10} - T_1 T_4 + T_3 T_6, \qquad T_9 T_{10} - T_1 T_5 + T_2 T_6,$$

$$T_6 T_7 - T_5 T_8 + T_4 T_9, \qquad T_1 T_7 - T_2 T_8 + T_3 T_9.$$

Note that the last two relations show up when passing from the Laurent polynomial ring in $T_1, \ldots, T_{10}$ to the polynomial ring in $T_1, \ldots, T_{10}$. Moreover, one directly checks the conditions of Proposition 4.1.2.6.

Now we apply Corollary 4.1.3.5 and obtain that the Cox ring of X_2 is $R_2 = \mathbb{K}[T_1, \ldots, T_{10}]/I_2$ together with the $\mathbb{Z}^5$-grading given by the Gale dual Q_2 of P_2, which one may choose as

$$Q_2 := \begin{bmatrix} 1 & 1 & 1 & 0 & 0 & 0 & 1 & 1 & 1 & 0 \\ -1 & -1 & 0 & 1 & 0 & 0 & 0 & 0 & -1 & 0 \\ -1 & 0 & -1 & 0 & 1 & 0 & 0 & -1 & 0 & 0 \\ 0 & -1 & -1 & 0 & 0 & 1 & -1 & 0 & 0 & 0 \\ 0 & 0 & 0 & 0 & 0 & 0 & -1 & -1 & -1 & 1 \end{bmatrix}$$

Thus, suitably renaming the variables, we see that the Cox ring of the projective plane blown up at four points in general position equals the ring of functions of the affine cone over $G(2, 5)$. In particular, we see that the latter is factorial.

We turn to the Gorenstein log terminal del Pezzo surfaces X. These are $\mathbb{Q}$-factorial projective surfaces with an ample anticanonical divisor and at most rational double points as singularities. A classification according to the singularity type, that is, the configuration $S(X)$ of singularities, in the case of Picard number 1 can be found in [5, Thm. 4.3]. We concentrate here on the X that do not admit a nontrivial $\mathbb{K}^*$-action. For the remaining cases and some more background, we refer to Sections 5.4.4 and 5.4.3

Below, we will write a Cox ring as a quotient $\mathbb{K}[T_1, \ldots, T_r]/I$ and specify generators for the ideal I. The $\mathrm{Cl}(X)$-grading is encoded by a degree matrix, that is, a matrix with $\deg(T_1), \ldots, \deg(T_r) \in \mathrm{Cl}(X)$ as columns.

Theorem 4.1.4.2 *The following table lists the Cox rings of Gorenstein log terminal del Pezzo surfaces X of Picard number 1 that do not allow a nontrivial $\mathbb{K}^*$-action.*

S(X)	Cox ring $\mathcal{R}(X)$	Cl(X) *and degree matrix*
$A_5 A_2 A_1$	$\mathbb{K}[T_1, \dots, T_7]/I$ *with I generated by* $T_5^2 + T_6^2 - T_7 T_1,\ T_4 T_5 + T_6 T_1 - T_2 T_6 - T_7^2,$ $T_1 T_5 - T_2 T_5 - T_4 T_6 + T_7 T_3,\ T_3 T_4 - T_6^2 + T_7 T_1 - T_2 T_7,$ $-T_3 T_6 - T_5 T_7 + T_1 T_4,\ T_3^2 - T_6 T_1 + T_7^2,$ $T_1 T_3 - T_2 T_3 + T_6 T_5 - T_7 T_4,\ T_1 T_2 - T_2^2 - T_4^2 + T_3 T_5,$ $T_1^2 - T_2^2 - T_4^2 + 2T_3 T_5 - T_7 T_6$	$\mathbb{Z} \oplus \mathbb{Z}/6\mathbb{Z},$ $\begin{bmatrix} 1 & 1 & 1 & 1 & 1 & 1 & 1 \\ \bar{2} & \bar{2} & \bar{3} & \bar{5} & \bar{1} & \bar{4} & \bar{0} \end{bmatrix}$
$2A_4$	$\mathbb{K}[T_1, \dots, T_6]/I$ *with I generated by* $-T_2 T_5 + T_3 T_4 + T_6^2,\ -T_2 T_4 + T_3^2 + T_5 T_6,$ $T_1 T_6 - T_3 T_5 + T_4^2,\ T_1 T_3 - T_4 T_6 + T_5^2,$ $T_1 T_2 - T_3 T_6 + T_4 T_5$	$\mathbb{Z} \oplus \mathbb{Z}/5\mathbb{Z}$ $\begin{bmatrix} 1 & 1 & 1 & 1 & 1 & 1 \\ \bar{2} & \bar{2} & \bar{3} & \bar{4} & \bar{0} & \bar{1} \end{bmatrix}$
D_8	$\mathbb{K}[T_1, \dots, T_4]/I$ *with I generated by* $T_1^2 - T_4^2 T_2 T_3 + T_4^4 + T_3^4$	$\mathbb{Z} \oplus \mathbb{Z}/2\mathbb{Z}$ $\begin{bmatrix} 2 & 1 & 1 & 1 \\ \bar{1} & \bar{1} & \bar{1} & \bar{0} \end{bmatrix}$
$D_5 A_3$	$\mathbb{K}[T_1, \dots, T_5]/I$ *with I generated by* $T_1 T_3 - T_4^2 - T_5^2,\ T_1 T_2 - T_3^2 + T_4 T_5$	$\mathbb{Z} \oplus \mathbb{Z}/4\mathbb{Z}$ $\begin{bmatrix} 1 & 1 & 1 & 1 & 1 \\ \bar{2} & \bar{2} & \bar{0} & \bar{3} & \bar{1} \end{bmatrix}$
$D_6 2A_1$	$\mathbb{K}[T_1, \dots, T_5]/I$ *with I generated by* $T_5 T_2 - T_5^2 + T_3^2 + T_4^2,\ -T_2^2 + T_5 T_2 + T_1^2 - T_4^2$	$\mathbb{Z} \oplus \mathbb{Z}/2\mathbb{Z} \oplus \mathbb{Z}/2\mathbb{Z}$ $\begin{bmatrix} 1 & 1 & 1 & 1 & 1 \\ \bar{1} & \bar{0} & \bar{0} & \bar{1} & \bar{0} \\ \bar{0} & \bar{1} & \bar{0} & \bar{1} & \bar{1} \end{bmatrix}$
$E_6 A_2$	$\mathbb{K}[T_1, \dots, T_4]/I$ *with I generated by* $-T_1 T_4^2 + T_2^3 + T_2 T_3 T_4 + T_3^3$	$\mathbb{Z} \oplus \mathbb{Z}/3\mathbb{Z}$ $\begin{bmatrix} 1 & 1 & 1 & 1 \\ \bar{1} & \bar{2} & \bar{0} & \bar{1} \end{bmatrix}$
$E_7 A_1$	$\mathbb{K}[T_1, \dots, T_4]/I$ *with I generated by* $-T_1 T_3^3 - T_2^2 + T_2 T_3 T_4 + T_4^4$	$\mathbb{Z} \oplus \mathbb{Z}/2\mathbb{Z}$ $\begin{bmatrix} 1 & 2 & 1 & 1 \\ \bar{1} & \bar{1} & \bar{1} & \bar{0} \end{bmatrix}$
E_8	$\mathbb{K}[T_1, \dots, T_4]/I$ *with I generated by* $T_1^3 + T_1^2 T_4^2 + T_2^2 - T_3 T_4^5$	$\mathbb{Z}$ $\begin{bmatrix} 2 & 3 & 1 & 1 \end{bmatrix}$
A_7	$\mathbb{K}[T_1, \dots, T_4]/I$ *with I generated by* $T_1^2 - T_4 T_2 T_3 + T_4^4 + T_3^4$	$\mathbb{Z} \oplus \mathbb{Z}/2\mathbb{Z}$ $\begin{bmatrix} 2 & 2 & 1 & 1 \\ \bar{1} & \bar{1} & \bar{1} & \bar{0} \end{bmatrix}$
A_8	$\mathbb{K}[T_1, \dots, T_4]/I$ *with I generated by* $-T_1 T_2 T_3 + T_2^3 + T_3^3 + T_4^3$	$\mathbb{Z} \oplus \mathbb{Z}/3\mathbb{Z}$ $\begin{bmatrix} 1 & 1 & 1 & 1 \\ \bar{1} & \bar{2} & \bar{0} & \bar{1} \end{bmatrix}$
$A_7 A_1$	$\mathbb{K}[T_1, \dots, T_5]/I$ *with I generated by* $-T_2 T_3 + T_4^2 - T_5^2,\ T_1^2 - T_3^2 + T_4 T_5$	$\mathbb{Z} \oplus \mathbb{Z}/4\mathbb{Z}$ $\begin{bmatrix} 1 & 1 & 1 & 1 & 1 \\ \bar{2} & \bar{2} & \bar{0} & \bar{3} & \bar{1} \end{bmatrix}$

$2A_3A_1$

$\mathbb{K}[T_1, \ldots, T_9]/I$ with I *generated by*

$-\frac{1}{2}T_4^2 + T_5^2 + \frac{1}{2}T_7T_9$, $\quad -\frac{1}{2}T_3T_8 - \frac{1}{2}T_4T_5 + T_6^2$,

$-\frac{1}{2}T_3T_4 + T_5T_8 + \frac{1}{2}T_6T_9$, $\quad T_2T_6 - T_7T_9 - 4T_8^2$,

$T_5T_2 - 2T_3T_7 + T_8T_9$, $\quad -\frac{1}{4}T_2T_4 + \frac{1}{4}T_3T_9 + T_7T_8$,

$T_1T_7 + T_2T_7 - 4T_3T_4 + 2T_6T_9$, $\quad T_1T_6 - 2T_4^2 + T_7T_9$,

$\frac{1}{2}T_1T_6 - \frac{1}{2}T_2T_6 + T_3^2 - T_4^2$, $\quad T_1T_8 - 2T_4T_7 + T_5T_9$,

$T_1^2 - 16T_4T_5 + 8T_7^2 - T_9^2$, $\quad T_2^2 - 16T_3T_8 + 8T_7^2 - T_9^2$,

$T_2T_3 - T_4T_9 + 4T_5T_7 - 8T_6T_8$, $\quad T_1T_2 - 8T_7^2 - T_9^2$,

$T_1T_5 + 2T_3T_7 - 4T_4T_6 + T_8T_9$, $\quad T_1T_3 - T_4T_9 - 4T_5T_7$,

$-\frac{1}{8}T_4T_1 + \frac{1}{8}T_2T_4 + T_5T_6 - T_7T_8$,

$-\frac{1}{16}T_9T_1 + \frac{1}{16}T_2T_9 - T_4T_8 + T_6T_7$,

$-\frac{1}{8}T_9T_1 + \frac{1}{8}T_2T_9 + T_3T_5 - T_4T_8$,

$\frac{1}{4}T_1T_8 - \frac{1}{4}T_2T_8 + T_6T_3 - T_4T_7$,

$\mathbb{Z} \oplus \mathbb{Z}/2\mathbb{Z} \oplus \mathbb{Z}/4\mathbb{Z}$

$$\begin{bmatrix} 1 & 1 & 1 & 1 & 1 & 1 & 1 & 1 & 1 \\ \bar{1} & \bar{1} & \bar{0} & \bar{1} & \bar{1} & \bar{1} & \bar{0} & \bar{0} & \bar{0} \\ \bar{3} & \bar{3} & \bar{2} & \bar{0} & \bar{2} & \bar{1} & \bar{3} & \bar{0} & \bar{1} \end{bmatrix}$$

$4A_2$

$\mathbb{K}[T_1, \ldots, T_{10}]/I$ with I *generated by*

$3T_3T_6 + 3T_4T_7\zeta + (-3\zeta - 3)T_5T_8$,

$(\zeta - 1)T_2T_8 + 3T_3^2 + (-\zeta - 2)T_6T_9$,

$3T_2T_7\zeta + 3T_6T_{10} + (-3\zeta - 3)T_8T_9$,

$(-\zeta + 1)T_2T_5 + (\zeta - 1)T_4T_9 + 3T_6T_8$,

$-\zeta T_1T_{10} + T_2T_{10}\zeta + 3T_4T_7 - 3T_5T_8$,

$(\zeta + 1)T_1T_{10} - T_2T_{10}\zeta + 3T_5T_8 - T_9^2$,

$(-\zeta - 1)T_1T_9 + (\zeta + 1)T_2T_9 - 3T_3T_4 + 3T_5T_6$,

$-T_1T_9\zeta - T_2T_9 + 3T_3T_7 + (\zeta + 1)T_{10}^2$,

$-T_1T_9\zeta + T_2T_9\zeta + 3T_3T_7 - 3T_8T_4$,

$(-2\zeta - 1)T_1T_8 + (\zeta - 1)T_2T_8 + 3T_3^2 + (\zeta - 1)T_7T_{10}$,

$(-\zeta + 1)T_1T_8 + (\zeta - 1)T_2T_8 + 3T_3^2 - 3T_4T_5$,

$(-\zeta - 2)T_1T_7 + (2\zeta + 1)T_2T_7 + 3T_5^2 + (-\zeta - 2)T_8T_9$,

$(\zeta + 2)T_1T_7 + (-\zeta - 2)T_2T_7 + 3T_4T_3 - 3T_5^2$,

$-3T_1T_6 + (3\zeta + 3)T_2T_6 + (-3\zeta - 3)T_7T_9 + 3T_8T_{10}$,

$(-\zeta - 2)T_1T_6 + (2\zeta + 1)T_2T_6 + 3T_3T_5 + (-\zeta - 2)T_7T_9$,

$(2\zeta + 1)T_1T_6 + (-2\zeta - 1)T_2T_6 - 3T_3T_5 + 3T_4^2$,

$(-\zeta + 1)T_1T_5 + (-\zeta - 2)T_2T_5 + (2\zeta + 1)T_3T_{10} + 3T_7^2$,

$(-\zeta + 1)T_1T_5 + (\zeta - 1)T_2T_5 - 3T_6T_8 + 3T_7^2$,

$-3T_1T_4 + (3\zeta + 3)T_2T_4 + (-3\zeta - 3)T_3T_9 + 3T_5T_{10}$,

$(-2\zeta - 1)T_1T_4 + (2\zeta + 1)T_5T_{10} + 3T_6^2$,

$(\zeta + 2)T_1T_4 + (-2\zeta - 1)T_2T_4 + (\zeta - 1)T_3T_9 + 3T_7T_8$,

$(-\zeta + 1)T_1T_3 + (\zeta - 1)T_5T_9 + 3T_6T_7$,

$3\zeta T_1T_3 + 3T_4T_{10} + (-3\zeta - 3)T_5T_9$,

$(\zeta + 2)T_1T_3 + (-2\zeta - 1)T_2T_3 + (\zeta - 1)T_5T_9 + 3T_8^2$,

$T_2T_1 + (-\zeta - 1)T_1^2T_2 + 3T_8T_3 + T_9T_{10}\zeta$,

$-\zeta T_1T_2 + 3T_4T_6 + T_9T_{10}\zeta$,

$T_1^2 + (-\zeta - 1)T_1T_2 + 3T_5T_7 + T_9T_{10}\zeta$

where ζ is a primitive third root of unity.

The class group and degree matrix are

$\mathbb{Z} \oplus \mathbb{Z}/3\mathbb{Z} \oplus \mathbb{Z}/3\mathbb{Z}$

$$\begin{bmatrix} 1 & 1 & 1 & 1 & 1 & 1 & 1 & 1 & 1 & 1 \\ \bar{2} & \bar{2} & \bar{1} & \bar{0} & \bar{2} & \bar{1} & \bar{2} & \bar{0} & \bar{1} & \bar{0} \\ \bar{1} & \bar{1} & \bar{2} & \bar{2} & \bar{2} & \bar{0} & \bar{0} & \bar{0} & \bar{1} & \bar{1} \end{bmatrix}.$$

4.2 Lifting automorphisms

4.2.1 Quotient presentations

We consider irreducible normal prevarieties X with only constant invertible global functions and finitely generated divisor class group $\mathrm{Cl}(X)$. Generalizing the construction of the characteristic space to arbitrary subgroups of $\mathrm{Cl}(X)$, we obtain various quotient presentations of X. For example, this construction gives all classical affine cones over projective X. The ground field $\mathbb{K}$ is algebraically closed of characteristic zero.

Recall that a *morphism* of two prevarieties Y_1, Y_2, coming with actions of algebraic groups G_1, G_2 respectively, is a pair $(\varphi, \widetilde{\varphi})$, where $\varphi: Y_1 \to Y_2$ is a morphism of prevarieties and $\widetilde{\varphi}: G_1 \to G_2$ is a homomorphism of algebraic groups such that $\varphi(g \cdot y) = \widetilde{\varphi}(g) \cdot \varphi(y)$ holds for all $g \in G_1$ and $y \in Y_1$.

Definition 4.2.1.1 Let X be an irreducible normal prevariety with only constant invertible global functions and finitely generated divisor class group.

(i) A *quotient presentation* of X is a good quotient $\pi \colon \widetilde{X} \xrightarrow{/\!/H} X$, where $\widetilde{X}$ is an irreducible normal prevariety with a strongly stable action of a quasitorus H such that every H-homogeneous invertible function on $\widetilde{X}$ is constant.

(ii) A *morphism* of two quotient presentations $\pi_i \colon \widetilde{X}_i \xrightarrow{/\!/H_i} X$ is a morphism $(\varphi, \widetilde{\varphi})$ from the H_1-variety $\widetilde{X}_1$ to the H_2-variety $\widetilde{X}_2$ such that the following diagram is commutative

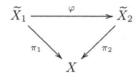

Construction 4.2.1.2 Let X be an irreducible normal prevariety with only constant invertible global functions, finitely generated divisor class group $\mathrm{Cl}(X)$, and Cox sheaf $\mathcal{R}$ locally of finite type. With any subgroup $L \subseteq \mathrm{Cl}(X)$ we associate the sheaf of Veronese subalgebras

$$\mathcal{R}(L) := \bigoplus_{[D] \in L} \mathcal{R}_{[D]}.$$

This sheaf is locally of finite type and we have the relative spectrum $\widetilde{X}(L) := \mathrm{Spec}_X \mathcal{R}(L)$ with the canonical morphism $\pi_L \colon \widetilde{X}(L) \to X$. Altogether, we arrive at a commutative diagram of quotient presentations

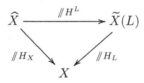

where H^L and H_L are defined by the exact sequence $1 \leftarrow H_L \leftarrow H_X \leftarrow H^L \leftarrow 1$ of quasitori associated with the exact sequence $0 \to L \to \mathrm{Cl}(X) \to \mathrm{Cl}(X)/L \to 0$ of abelian groups and $\widehat{X} \to X$ is the characteristic space defined by $\mathcal{R}$.

Lemma 4.2.1.3 *Let G be a connected reductive group, $H \subseteq G$ a normal, reductive subgroup, and Z an irreducible G-prevariety.*

(i) *If the good quotient $Z \to Z /\!/ H$ exists, then there is a unique G/H-action on $Z /\!/ H$ making $Z \to Z /\!/ H$ and $G \to G/H$ to a morphism of prevarieties with group action.*

(ii) *The good quotient $Z \to Z/\!\!/G$ exists if and only if the good quotients $Z \to Z/\!\!/H$ and $Z/\!\!/H \to (Z/\!\!/H)/\!\!/(G/H)$ exist. In this case, one has a commmutative diagram*

$$
\begin{array}{ccc}
Z & \xrightarrow{\ /\!\!/H\ } & Z/\!\!/H \\
{\scriptstyle /\!\!/G} \downarrow & & \downarrow {\scriptstyle /\!\!/(G/H)} \\
Z/\!\!/G & \xleftarrow[\ \cong\ ]{} & (Z/\!\!/H)/\!\!/(G/H)
\end{array}
$$

Proof In the setting of (i), universality of the good quotient allows us to push down the G-action to $Z/\!\!/H$; see [46, Thm. 7.1.4]. In the setting of (ii), if $Z \to Z/\!\!/G$ exists, then also $Z \to Z/\!\!/H$ exists; see [47, Cor. 10], and one directly verifies that the induced morphism $Z/\!\!/H \to Z/\!\!/G$ is a good quotient for the action of G/H. Conversely, if the stepwise good quotients exist, then one directly verifies that their composition is a good quotient for the G-prevariety Z. $\square$

Proof of Construction 4.2.1.2 Existence of the above diagram with arrows being good quotients follows from Lemma 4.2.1.3. The properties of a quotient presentation for the involved maps follow from the corresponding properties of the characteristic space. $\square$

To formulate the main statement on quotient presentations, let us turn the set of all subgroups of the divisor class group $\mathrm{Cl}(X)$ into a category by defining the inclusions $L \subseteq L' \subseteq \mathrm{Cl}(X)$ as morphisms.

Theorem 4.2.1.4 *Let X be an irreducible normal prevariety with only constant invertible global functions, finitely generated divisor class group $\mathrm{Cl}(X)$ and Cox sheaf $\mathcal{R}$ locally of finite type. Then there is a contravariant essentially surjective functor*

$$\{\textit{subgroups of } \mathrm{Cl}(X)\} \longrightarrow \{\textit{quotient presentations of } X\},$$

$$L \mapsto \pi_L \colon \widetilde{X}(L) \to X.$$

Proof Functoriality of the assignment $L \mapsto \pi_L$ can be directly verified. To show that the functor is essentially surjective, let $\pi \colon \widetilde{X} \xrightarrow{/\!\!H} X$ be a quotient presentation. Set $L := \mathbb{X}(H)$. Then Proposition 1.6.4.5 provides a monomorphism

$$L \to \mathrm{Cl}(X), \qquad w \mapsto [\pi_*(\mathrm{div}(f))], \quad \text{with any } f \in E(\widetilde{X})_w,$$

where $E(\widetilde{X})$ denotes the multiplicative group of nonzero homogeneous rational functions. We regard L as a subgroup of $\mathrm{Cl}(X)$ via this identification. The idea is to construct an isomorphism $\mathcal{R}(L) \to \pi_*(\mathcal{O}_{\widetilde{X}})$, where $\mathcal{R}(L)$ is the Veronese subalgebra of the Cox sheaf $\mathcal{R}$ given by $L \subseteq \mathrm{Cl}(X)$.

Choose a subgroup $K \subseteq \mathrm{WDiv}(X)$ mapping onto $\mathrm{Cl}(X)$ and let $K' \subseteq K$ be the preimage of $L \subseteq \mathrm{Cl}(X)$. Then we have $\mathcal{R} = \mathcal{S}/\mathcal{I}$ with the sheaf of divisorial algebras $\mathcal{S}$ associated with K and the sheaf of ideals $\mathcal{I}$ defined as in Construction 1.4.2.1 via a character $\chi \colon K^0 \to \mathbb{K}(X)^*$, where $K^0 \subseteq K$ is the kernel of $K \to \mathrm{Cl}(X)$. Note that $\mathcal{R}(L)$ is the factor ring $\mathcal{S}(L)/\mathcal{I}(L)$ of the corresponding Veronese subalgebra of $\mathcal{S}$ by the L-degrees of $\mathcal{I}$.

Choose a basis $D_1, \ldots, D_s$ for K' such that $K' \cap K^0$ is freely generated by suitable multiples $c_1 D_1, \ldots, c_k D_k$ of the first k divisors. Then $D_i = \delta(h_i)$ holds with homogeneous rational functions $h_i \in E(\widetilde{X})_{w_i}$, where $w_i \in L = \mathbb{X}(H)$ denotes the character corresponding to the class $[D_i] \in \mathrm{Cl}(X)$. Indeed, each D_i is of the form $\delta(h_i') + \mathrm{div}(f_i)$ with some $h_i' \in E(\widetilde{X})_{w_i}$ and $f_i \in \mathbb{K}(X)$. Thus, $h_i = h_i' \pi^*(f_i)$ is as wanted.

For $D = a_1 D_1 + \cdots + a_s D_s$, set $h_D := h_1^{a_1} \cdots h_s^{a_s}$. Suitably rescaling $h_1, \ldots, h_k$, we achieve $h_E = \pi^*(\chi(E))$ for all $E \in K' \cap K^0$; here we use the fact that the invertible functions $h_i/\pi^*(\chi(c_i D_i))$ on $\widetilde{X}$ are constant. Proposition 1.6.4.5 provides an isomorphism of $\mathbb{K}$-vector spaces

$$\Phi_{U,D} \colon \Gamma(U, \mathcal{S}_D) \ \to \ \Gamma(\pi^{-1}(U), \mathcal{O}_{\widetilde{X}})_w, \qquad g \mapsto \pi^*(g) h_D.$$

for any open set $U \subseteq X$ and any $D \in K'$, where $w \in L = \mathbb{X}(H)$ denotes the character corresponding to the class of D. The $\Phi_{U,D}$ fit together to an epimorphism of graded sheaves $\Phi \colon \mathcal{S}(L) \to \pi_*(\mathcal{O}_{\widetilde{X}})$. As in the proof of Theorem 1.6.4.3, we see that Φ has the sheaf of ideals $\mathcal{I}(L)$ as its kernel. This gives an isomorphism $\mathcal{R}(L) \to \pi_*(\mathcal{O}_{\widetilde{X}})$, which in turn yields the desired isomorphism $\widetilde{X} \to \widetilde{X}(L)$. $\qquad\square$

Definition 4.2.1.5 Consider the situation of Construction 4.2.1.2. We call a subgroup $L \subseteq \mathrm{Cl}(X)$ *ample* if X is covered by affine $[D]$-localizations $X_{[D],f}$, where $[D] \in L$ and $f \in \Gamma(X, \mathcal{R}_{[D]})$.

Note that we do not require that an ample group of divisors consists of classes of ample divisors. The latter ones fit as follows in the concept.

Example 4.2.1.6 Consider the situation of Construction 4.2.1.2. If D is an ample divisor on X, then the subgroup $L \subseteq \mathrm{Cl}(X)$ generated by $[D] \in \mathrm{Cl}(X)$ is ample and the quotient presentation $\pi_L \colon \widetilde{X}(L) \xrightarrow{/\!\!/ H_L} X$ is the $\mathbb{K}^*$-bundle obtained by removing the zero section from the line bundle associated with $\mathcal{O}_X(D)$.

Proposition 4.2.1.7 *Let $\pi_L \colon \widetilde{X}(L) \to X$ be a quotient presentation as provided by Construction 4.2.1.2 and assume that X is of affine intersection. Then the following statements are equivalent.*

(i) *The group L is ample.*
(ii) *The prevariety $\widetilde{X}(L)$ is quasiaffine.*

Proof If L is ample, then X is covered by affine $[D]$-localizations $X_{[D],f}$ with $[D] \in L$ and $f \in \Gamma(X, \mathcal{R}_{[D]})$. Proposition 1.6.2.1 yields that $\widehat{X}_f = p_X^{-1}(X_{[D],f})$ is affine. It follows that

$$\widetilde{X}(L)_f = \widehat{X}_f /\!\!/ H^L = \pi_L^{-1}(X_{[D],f})$$

is affine and $\widetilde{X}(L)$ is covered by the sets $\widetilde{X}(L)_f$. Thus, $\widetilde{X}(L)$ is quasiaffine. Conversely, if $\widetilde{X}(L)$ is quasiaffine, then it is covered by affine sets $\widetilde{X}(L)_f$ with L-homogeneous functions f. Then $\widehat{X}$ is covered by affine sets $\widehat{X}(L)_f$ with L-homogeneous functions f and Proposition 1.6.2.1 shows that X is covered by affine $[D]$-localizations $X_{[D],f}$ with L-homogeneous sections f. $\square$

Proposition 4.2.1.8 *Let $\pi_L \colon \widetilde{X}(L) \to X$ be a quotient presentation as provided by Construction 4.2.1.2.*

(i) *The group $L \subseteq \mathrm{Cl}(X)$ consists of classes of $\mathbb{Q}$-Cartier divisors if and only if $\pi_L \colon \widetilde{X}(L) \to X$ is a geometric quotient.*

(ii) *The group $L \subseteq \mathrm{Cl}(X)$ consists of classes of Cartier divisors if and only if $\pi_L \colon \widetilde{X}(L) \to X$ is an étale H_L-principal bundle.*

Proof Consider a point $x \in X$ and a class $[D] \in L$. Then $[D]$ is Cartier near x if and only if there is a section $f \in \Gamma(U, \mathcal{R}_{[D]})$ defined on an open neighborhood $U \subseteq X$ of x such that $x \notin \mathrm{Supp}(\mathrm{div}_{[D]}(f))$ holds. The latter merely means that there is a $[D]$-homogeneous invertible section on $\pi_L^{-1}(x)$. From this we conclude that $\pi_L^{-1}(x)$ consists of a single (a single free) orbit if and only if almost all (all) classes $[D] \in L$ are Cartier near x. The assertions follow. $\square$

Example 4.2.1.9 (The Picard quotient presentation) Consider the subgroup $\mathrm{Pic}(X) \subseteq \mathrm{Cl}(X)$ of all Cartier divisor classes. Then the associated quotient presentation $\widetilde{X} \to X$ is an étale $H_{\mathrm{Pic}(X)}$-principal bundle. Moreover, $\mathrm{Pic}(X)$ is ample if and only if X is *divisorial* in the sense that X is covered by affine open sets of the form $X \setminus \mathrm{Supp}(D)$ with Cartier divisors D. See also Exercise 4.8.

Construction 4.2.1.10 Let $\pi_L \colon \widetilde{X}(L) \to X$ be a quotient presentation provided by an ample subgroup $L \subseteq \mathrm{Cl}(X)$ on a variety X. Write $\widetilde{X} := \widetilde{X}(L)$ and $\pi := \pi_L$ for short. Assume that $S := \Gamma(\widetilde{X}, \mathcal{O})$ is finitely generated as a $\mathbb{K}$-algebra; this holds, for example, if the Cox ring $\mathcal{R}(X)$ is finitely generated. Given an L-graded S-module M, let $\widetilde{M}$ be the associated quasicoherent sheaf on $Z := \mathrm{Spec}\, S$. Take the restriction $\iota^* \widetilde{M}$ to $\widetilde{X}$, where $\iota \colon \widetilde{X} \to Z$ is the inclusion, and denote by $(\pi_* \iota^* \widetilde{M})_0$ the part of degree zero of the image sheaf.

Proposition 4.2.1.11 *Consider the situation of Construction 4.2.1.10. The assignment $M \mapsto (\pi_* \iota^* \tilde{M})_0$ defines exact essentially surjective functors*

(i) *from the category of L-graded S-modules to the category of quasicoherent sheaves of $\mathcal{O}_X$-modules,*
(ii) *from the category of finitely generated L-graded S-modules to the category of coherent sheaves of $\mathcal{O}_X$-modules.*

Proof Using the fact that $\pi \colon \tilde{X} \to X$ is an affine morphism, one directly verifies exactness. Moreover, finitely generated M are clearly sent to coherent sheaves. To check essential surjectivity, let $\mathcal{N}$ be a quasicoherent sheaf of $\mathcal{O}_X$-modules. Because $\mathcal{S} := \pi_* \mathcal{O}_{\tilde{X}}$ is a sheaf of L-graded $\mathcal{O}_X$-algebras, $\mathcal{N}' := \mathcal{S} \otimes_{\mathcal{O}_X} \mathcal{N}$ is a sheaf of L-graded $\mathcal{S}$-modules on X. Obviously $\mathcal{N}'_0 \cong \mathcal{N}$ holds. Setting $\mathcal{M} := \pi^* \mathcal{N}'$ gives a quasicoherent sheaf of L-graded $\mathcal{O}_{\tilde{X}}$-modules on $\tilde{X}$. Observe that $\pi_* \mathcal{M} \cong \mathcal{N}'$ holds. By $M := \Gamma(\tilde{X}, \mathcal{M})$, we define an L-graded S-module satisfying $\mathcal{M} \cong \iota^* \tilde{M}$ and hence $\mathcal{N} \cong (\pi_* \iota^* \tilde{M})_0$. If $\mathcal{N}$ was coherent, then we can replace M with a suitable finitely generated L-graded submodule $M' \subseteq M$ satisfying $\mathcal{M} \cong \tilde{\iota}^* M'$. $\square$

4.2.2 Linearization of line bundles

The ground field $\mathbb{K}$ is algebraically closed of characteristic zero. We provide the basic statements on linearizations of line bundles over a prevariety with an affine algebraic group action; see Theorem 4.2.2.5. The first step is to study the Picard group of an affine algebraic group; note that in the approach of [189], this occurs as an application of linearization.

Theorem 4.2.2.1 *Let G be a connected affine algebraic group. Then $\mathrm{Pic}(G)$ is finite and there is a finite epimorphism $G' \to G$ with a connected affine algebraic group G' satisfying $\mathrm{Pic}(G') = 0$.*

The crucial point in the proof is to show that the characteristic space of a semisimple group is again a semisimple group; see [92] for a related approach. The following ingredients will also be used later.

Remark 4.2.2.2 For the invertible global regular functions on prevarieties, one has the following satements, see also [190, Sec. 1.1].

(i) For every prevariety X, the factor group $\Gamma(X, \mathcal{O}^*)/\mathbb{K}^*$ is finitely generated.
(ii) For any two irreducible prevarieties X_1, X_2, we have an epimorphism of abelian groups
$$\Gamma(X_1, \mathcal{O}^*) \times \Gamma(X_2, \mathcal{O}^*) \mapsto \Gamma(X_1 \times X_2, \mathcal{O}^*), \quad (f_1, f_2) \mapsto f_1 \otimes f_2.$$
(iii) On a connected affine algebraic group G, every element $f \in \Gamma(G, \mathcal{O}^*)$ satisfying $f(e_G) = 1$ is a character.

Lemma 4.2.2.3 *Let X_1, X_2 be irreducible normal prevarieties with only constant invertible global functions, finitely generated divisor class groups $\mathrm{Cl}(X_i)$, and Cox sheaves $\mathcal{R}_{X_i}$. Suppose that one of the X_i is rational. Set $X := X_1 \times X_2$.*

(i) *The divisor class group of X is generated by divisors $D_1 \times X_2$ and $X_1 \times D_2$, where $D_i \subseteq X_i$ is prime. In particular, we have*

$$\mathrm{Cl}(X) = \mathrm{Cl}(X_1) \times \mathrm{Cl}(X_2).$$

Moreover, the Cox sheaf and the Cox ring of the product $X = X_1 \times X_2$ are given by

$$\mathcal{R}_X = \mathcal{R}_{X_1} \otimes_{\mathbb{K}} \mathcal{R}_{X_2}, \qquad \mathcal{R}(X) = \mathcal{R}(X_1) \otimes_{\mathbb{K}} \mathcal{R}(X_2),$$

where the tensor products are endowed with the canonical grading by $\mathrm{Cl}(X_1) \times \mathrm{Cl}(X_2)$.

(ii) *If each $\mathcal{R}_{X_i}$ is locally of finite type then $\mathcal{R}_X$ is so and if each $\mathcal{R}(X_i)$ is finitely generated, then $\mathcal{R}(X)$ is so.*

(iii) *If $p_i \colon \widehat{X}_i \to X_i$ are characteristic spaces with the characteristic quasitori H_i, then $p := p_1 \times p_2$ and $\widehat{X} := \widehat{X}_1 \times \widehat{X}_2$ with the canonical action of $H := H_1 \times H_2$ and $p \colon \widehat{X} \to X$ define a characteristic space for X.*

(iv) *If X_i admit total coordinate spaces $\overline{X}_i$ with the characteristic quasitori H_i, then $\overline{X} := \overline{X}_1 \times \overline{X}_2$ together with the canonical action of $H := H_1 \times H_2$ is a total coordinate space for X.*

Proof The statement on the divisor class groups in (i) is basic. The second isomorphy of (i) follows from the fact that the Cox sheaves $\mathcal{R}_{X_i}$ are quasi-coherent. Both together imply (iii). Now, regarding the Cox ring $\mathcal{R}(X)$ as the algebra of functions on the characteristic space $\widehat{X}$, the Künneth formula gives the third isomorphism of (i). Assertions (ii) and (iv) follow. □

Proof of Theorem 4.2.2.1 First we consider the case that G is semisimple. We show that the characteristic space $p_G \colon \widehat{G} \to G$ is a finite epimorphism of affine algebraic groups, which gives the assertion. Recall that G has only trivial characters and thus Remark 4.2.2.2 (iii) yields $\Gamma(G, \mathcal{O}^*) = \mathbb{K}^*$. Moreover, G is a rational variety and thus $\mathrm{Cl}(G)$ is finitely generated. In particular, the characteristic space $p_G \colon \widehat{G} \to G$ exists. Note that $\widehat{G}$ is a smooth affine variety with only constant H-homogeneous invertible global functions, where H is the characteristic quasitorus of G.

We claim that $\widehat{G}$ comes with a natural structure of an affine algebraic group. Fix the neutral element $e \in G$ as a base point of G. We work with the canonical Cox sheaves provided by Construction 1.4.2.3: let $\mathcal{S}_G$ be the sheaf of divisorial algebras over G arising from the group of all divisors avoiding e. Similarly, let $\mathcal{S}_{G \times G}$ be the sheaf of divisorial algebras associated with the base point $(e, e) \in G \times G$. Denote by S_G and $S_{G \times G}$ the respective algebras of global sections. Then we have canonical morphisms

$$\mu^* \colon S_G \ \to \ S_{G \times G}, \qquad \iota^* \colon S_G \ \to \ S_G$$

of graded algebras arising from pullback via the multiplication $\mu \colon G \times G \to G$ and the inversion $\iota \colon G \to G$. With the ideals $I_G \subseteq S_G$ and $I_{G \times G} \subseteq S_{G \times G}$ defined by the canonical characters χ^e and $\chi^{(e,e)}$ of Construction 1.4.2.3, we obtain the Cox rings $R_G := S_G/I_G$ of G and $R_{G \times G} := S_{G \times G}/I_{G \times G}$ of $G \times G$. By Lemma 4.2.2.3, we have a canonical isomorphism $R_{G \times G} \cong R_G \otimes R_G$. Because the above homomorphisms respect the defining ideals I_G and $I_{G \times G}$, we obtain induced homomorphisms

$$\mu^* \colon R_G \ \to \ R_G \otimes R_G, \qquad \iota^* \colon R_G \ \to \ R_G, \qquad \eta^* \colon \mathbb{K} \ \to \ R_G$$

where the additional one, η^*, comes from the evaluation $\eta \colon R_G \to \mathbb{K}$ at a point $\widehat{e}$ chosen in the fiber $p_G^{-1}(e)$ as a base point for $\widehat{G}$. By construction, these data turn $R_G = \Gamma(\widehat{G}, \mathcal{O})$ into a Hopf algebra and thus $\widehat{G}$ into an affine algebraic group. Note that the inclusion of $\Gamma(G, \mathcal{O})$ into R_G as the component of degree zero is a homomorphism of Hopf algebras. Thus $p_G \colon \widehat{G} \to G$ is a homomorphism of affine algebraic groups. The kernel is isomorphic to H.

We claim that $\widehat{G}$ is semisimple. Indeed, as the multiplication of $\widehat{G}$ respects the grading, H lies in the center, and thus $\widehat{G}$ is reductive. It follows that H is finite, because otherwise $\widehat{G}$ would have nonconstant characters, that is, nonconstant invertible H-homogeneous global functions. In particular we conclude that $\mathrm{Pic}(G) = \mathbb{X}(H)$ is finite. If $\mathrm{Pic}(\widehat{G}) \neq 0$ holds, then we iterate the above process. This leads to a sequence of finite coverings of semisimple groups that necessarily terminates at a semisimple G' with trivial Picard group.

Now let G be arbitrary. Then G is a semidirect product of a maximal reductive subgroup G^{red} and the unipotent radical $R_u(G)$. Because $R_u(G)$ has a trivial Picard group, we obtain $\mathrm{Pic}(G) = \mathrm{Pic}(G^{\mathrm{red}})$. The reductive group G^{red} is of the form $(T \times G^{ss})/\Gamma$ with a torus T, a semisimple group G^{ss}, and a finite central subgroup Γ. We conclude that $\mathrm{Pic}(G^{\mathrm{red}})$ is finite. Now replace G^{ss} with a covering $\widehat{G}^{ss}$ as provided by the first part of the proof. This gives the desired epimorphism $G' \to G$, where the covering group G' has $T \times \widehat{G}^{ss}$ as its reductive part and hence has trivial Picard group. $\qquad \square$

We come to the linearization of line bundles. Recall that a *(geometric) line bundle* over a prevariety X is a locally trivial morphism $\pi \colon L \to X$ with typical fiber $\mathbb{K}$ and structure group $\mathbb{K}^*$. In particular, every fiber $L_x = \pi^{-1}(x)$ comes with the structure of a one-dimensional vector space. Morphisms of line bundles over X are fiberwise linear morphisms over X. The isomorphy classes of line bundles over X correspond to the elements of the Picard group $\mathrm{Pic}(X)$, where the addition reflects the tensor product operation on line bundles.

Definition 4.2.2.4 Let X be a prevariety with an action of an algebraic group G. A *G-linearization* of a line bundle $L \to X$ is a G-action on L such that $L \to X$ becomes G-equivariant and for all $(g, x) \in G \times X$ the map $L_x \to L_{g \cdot x}$ is linear.

Theorem 4.2.2.5 *Let X be an irreducible normal prevariety with an action of a connected affine algebraic group G and denote by m the order of $\mathrm{Pic}(G)$. Then for any line bundle $L \to X$, the mth power $L^{\otimes m} \to X$ admits a G-linearization.*

For practical purposes, the following two consequences are often sufficient; the second one uses the covering $G' \to G$ provided by Theorem 4.2.2.1.

Corollary 4.2.2.6 *Let X be an irreducible normal variety with an action of a connected affine algebraic group G and $L \to X$ a line bundle.*

 (i) *If G has trivial Picard group, then the line bundle $L \to X$ admits a G-linearization.*
 (ii) *There is a finite epimorphism $G' \to G$ such that $L \to X$ admits a G'-linearization.*

Remark 4.2.2.7 Let X be a prevariety with $\Gamma(X, \mathcal{O}^*) = \mathbb{K}^*$ and let a connected affine algebraic group G act on X. If a line bundle $L \to X$ admits two G-linearizations, say $(g, l) \mapsto g * l$ and $(g, l) \mapsto g \star l$, then there is a $\chi \in \mathbb{X}(G)$ with $g^{-1} \star g * l = \chi(g)l$ for all $l \in L$ and $g \in G$; use Remark 4.2.2.2 (ii).

We prepare the proof of Theorem 4.2.2.5. Given a line bundle $\pi : L \to X$ and a morphism $\varphi : X' \to X$, the *pullback bundle* is

$$\varphi^*(L) := X' \times_X L = \{(x', l) \in X' \times L; \ \varphi(x') = \pi(l)\} \subseteq X' \times L$$

with the canonical map $\pi' : \varphi^*(L) \to X'$ sending (x', l) to x'. The following observation relates existence of a G-linearization to comparing certain pullback bundles.

Lemma 4.2.2.8 *Let X be an irreducible prevariety coming with an action of a connected algebraic group G, let $\pi : L \to X$ be a line bundle and consider the two morphisms*

$$\mathrm{pr}_X : G \times X \to X, \quad (g, x) \mapsto x, \qquad \mu : G \times X \to X, \quad (g, x) \mapsto g \cdot x.$$

If $\pi : L \to X$ is G-linearized, then we have mutually inverse isomorphisms of line bundles over $G \times X$:

$$\mathrm{pr}_X^*(L) \longleftrightarrow \mu^*(L)$$

$$((g, x), l) \mapsto ((g, x), g \cdot l),$$

$$((g, x), g^{-1} \cdot l) \leftarrow ((g, x), l).$$

Conversely, if $\mathrm{pr}_X^*(L)$ *and* $\mu^*(L)$ *are isomorphic as line bundles over* $G \times X$, *then* L *admits a* G-*linearization.*

Proof Only the second observation needs a proof. Let $\alpha \colon \mathrm{pr}_X^*(L) \to \mu^*(L)$ be an isomorphism of line bundles. Then we have morphisms

$$G \times L \xrightarrow[\cong]{(g,l) \mapsto ((g,\pi(l)),l)} \mathrm{pr}_X^*(L) \xrightarrow[\cong]{\alpha} \mu^*(L) \xrightarrow{((g,x),l) \mapsto l} L \ .$$

We claim that the composition ν of these morphisms is a G-linearization. So we have to check that ν is a fiberwise linear action. Clearly, ν fits into a commutative diagram

$$
\begin{array}{ccc}
G \times L & \xrightarrow{\ \nu\ } & L \\
{\scriptstyle \mathrm{id} \times \pi} \big\downarrow & & \big\downarrow {\scriptstyle \pi} \\
G \times X & \xrightarrow[\ \mu\ ]{} & X
\end{array}
$$

and by assumption, the induced maps $L_x \to L_{g \cdot x}$, are linear isomorphisms. Write for short $g * l := \nu(g, l)$. Scaling ν with a suitable function of $\Gamma(X, \mathcal{O}^*)$, we achieve $e_G * l = l$ for all $l \in L$. Moreover, for some $\eta \in \mathcal{O}^*(G \times G \times L)$, we have

$$(gh) * l \ = \ \eta(g, h, l)(h * (g * l)), \quad \text{for all } g, h \in G, \ l \in L.$$

Here, existence of such η as function of sets is clear and regularitiy can be checked on local trivializations. Remark 4.2.2.2 (ii) tells us that $\eta(g, h, l) = \eta_1(g)\eta_2(h)\eta_3(l)$ holds with invertible functions η_i. Combining this with $e_G * l = l$ leads to $\eta = 1$. $\square$

Proof of Theorem 4.2.2.5 According to Lemma 4.2.2.8, we only have to show that $\mathrm{pr}_X^*(L^{\otimes m})$ and $\mu^*(L^{\otimes m})$ are isomorphic as line bundles, that is, define the same class in $\mathrm{Pic}(G \times X)$. Because X and G are normal, we may work in terms of Cartier divisors. For a Cartier divisor D on X representing the line bundle L, consider

$$E \ := \ \mathrm{pr}_X^*(D) - \mu^*(D) \ \in \ \mathrm{CDiv}(G \times X).$$

Our task is to show that mE is principal. Because G is rational, Lemma 4.2.2.3 tells us that the divisor E is linearly equivalent to a divisor of the form

$$C \ = \ \sum a_i(D_i^G \times X) + \sum b_j(G \times D_j^X) \ \in \ \mathrm{CDiv}(G \times X).$$

with prime divisors D_i^G on G and prime divisors D_j^X on X. Write C^G for the first and C^X for the second sum of C. Note that both C^G and C^X are Cartier. Moreover, we may assume that no D_i^G passes through the neutral element

$e_G \in G$. Then, with the embedding $\imath : X \to G \times X$ sending x to (e_G, x) we have

$$\imath^*(E) = 0, \qquad \imath^*(C) = \imath^*(C^X) = \sum b_j D_j^X.$$

Because E and C differ by a principal divisor, we conclude that $\imath^*(C^X)$ is principal. But then C^X is also principal. Thus, E is linearly equivalent to C^G. Theorem 4.2.2.1 yields that mC^G and hence mE is principal. $\qquad\square$

Remark 4.2.2.9 Consider an action $\mu : G \times X \to X$ of a connected algebraic group G on an irredicible prevariety X; let $L \to X$ be a G-linearized bundle and denote by $\mathcal{L}$ the sheaf of sections of L. Then, in the setting of Lemma 4.2.2.8, the isomorphism $\mu^*(L) \to \mathrm{pr}_X^*(L)$ of line bundles gives rise to an isomorphism of invertible sheaves $\varphi : \mu^*\mathcal{L} \to \mathrm{pr}_X^*\mathcal{L}$ such that one has a commutative diagram

$$
\begin{array}{ccccc}
(\mathrm{id}_G \times \mu)^*\mu^*\mathcal{L} & \xrightarrow{(\mathrm{id}_G \times \mu)^*\varphi} & (\mathrm{id}_G \times \mu)^*\mathrm{pr}_X^*\mathcal{L} & = & \mathrm{pr}_{G \times X}^*\mu^*\mathcal{L} \\
\Big\| & & & & \Big\downarrow{\mathrm{pr}_{G \times X}^*\varphi} \\
(\mathrm{m} \times \mathrm{id}_X)^*\mu^*\mathcal{L} & \xrightarrow[(\mathrm{m} \times \mathrm{id}_X)^*\varphi]{} & (\mathrm{m} \times \mathrm{id}_X)^*\mathrm{pr}_X^*\mathcal{L} & = & \mathrm{pr}_{G \times X}^*\mathrm{pr}_X^*\mathcal{L}
\end{array}
$$

where $\mathrm{m} : G \times G \to G$ is the multiplication map and $\mathrm{pr}_{G \times X} : G \times G \times X \to G \times X$ sends (g_1, g_2, x) to (g_2, x). This means that any G-linearization of $L \to X$ defines a G-linearization of the invertible sheaf $\mathcal{L}$ in the sense of Mumford [227, Def. 1.6]. In particular, the existence statements Theorem 4.2.2.5 and Corollary 4.2.2.6 on linearizations carry over to invertible sheaves on X.

4.2.3 Lifting group actions

Here we treat the problem of lifting group actions to quotient presentations in the sense of Definition 4.2.1.1. The ground field $\mathbb{K}$ is algebraically closed and of characteristic zero. The precise concept of a lifting we are going to study is the following.

Definition 4.2.3.1 Let X be an irreducible normal prevariety with a quotient presentation $\pi : \widetilde{X} \xrightarrow{/H} X$. Given an algebraic group action $\mu : G \times X \to X$, we say that a finite epimorphism $\varepsilon : G' \to G$ and an action $\mu' : G' \times \widetilde{X} \to \widetilde{X}$ *lift* the G-action to the quotient presentation, or call (ε, μ') a *lifting* of μ if

(i) the actions of G' and H on $\widetilde{X}$ commute, that is, for all $g' \in G'$, $h \in H$ and $\widetilde{x} \in \widetilde{X}$, we have $g' \cdot (h \cdot \widetilde{x}) = h \cdot (g' \cdot \widetilde{x})$,

(ii) the epimorphism $\varepsilon : G' \to G$ satisfies $\pi(g' \cdot \widetilde{x}) = \varepsilon(g') \cdot \pi(\widetilde{x})$ for all $g' \in G'$ and $\widetilde{x} \in \widetilde{X}$.

The main result of the section ensures existence of liftings and gives particular statements on the case that G is semisimple and simply connected or a torus.

Theorem 4.2.3.2 *Let X be an irreducible normal prevariety with only constant invertible global functions and finitely generated divisor class group. Let $\pi : \widetilde{X} \xrightarrow{/H} X$ be a quotient presentation and $G \times X \to X$ be an action of a connected affine algebraic group.*

(i) *There exist a finite epimorphism $\varepsilon : G' \to G$ of connected affine algebraic groups and an action $G' \times \widetilde{X} \to \widetilde{X}$ that lift the G-action to the quotient presentation.*

(ii) *Given a finite epimorphism $\varepsilon : G' \to G$ and two G'-actions $*$ and $\star$ on $\widetilde{X}$ that provide liftings of $G \times X \to X$, there is a homomorphism $\eta : G' \to H$ with $g' * \widetilde{x} = \eta(g') \cdot g' \star \widetilde{x}$ for all $g' \in G'$ and $\widetilde{x} \in \widetilde{X}$.*

(iii) *If G is a simply connected semisimple group then there is a lifting (id_G, μ') of the G-action to the quotient presentation.*

(iv) *If G is a torus, then there is a lifting (ε, μ') of the G-action to the quotient presentation with $\varepsilon : G \to G$, $g \mapsto g^b$ for some $b \in \mathbb{Z}_{\geq 1}$; if in addition $\mathrm{Cl}(X)$ is free, then one may choose $b = 1$.*

The proof uses linearizations of sheaves of graded algebras following [164, 225]. We briefly fix the setting. Let X be an irreducible prevariety, K be a finitely generated abelian group and consider a sheaf of K-graded $\mathbb{K}$-algebras

$$\mathcal{A} = \bigoplus_{w \in K} \mathcal{A}_w$$

on X satisfying $\mathcal{A}_0 = \mathcal{O}_X$. Let an affine algebraic group G act on X via the morphism $\mu : G \times X \to X$ and denote by $\mathrm{m} : G \times G \to G$ the multiplication map. Then we have a commutative diagram

$$
\begin{array}{ccc}
G \times G \times X & \xrightarrow{\;\mathrm{id}_G \times \mu\;} & G \times X \\
{\scriptstyle \mathrm{m} \times \mathrm{id}_X}\big\downarrow & & \big\downarrow{\scriptstyle \mu} \\
G \times X & \xrightarrow{\quad\mu\quad} & X
\end{array}
$$

Similarly to [225, Def. 1.6], the definition of a G-linearization of the sheaf $\mathcal{A}$ is formulated in terms of the above maps and the projection maps

$$\mathrm{pr}_{G \times X} : G \times G \times X \to G \times X, \qquad (g_1, g_2, x) \mapsto (g_2, x),$$

$$\mathrm{pr}_X : G \times X \to X, \qquad (g, x) \mapsto x.$$

Definition 4.2.3.3 Let X, $\mathcal{A}$, and G be as above. A *G-linearization* of $\mathcal{A}$ is an isomorphism $\Phi\colon \mu^*\mathcal{A} \to \mathrm{pr}_X^*\mathcal{A}$ of K-graded $\mathcal{O}_{G\times X}$-algebras such that Φ is the identity in degree zero, and the following diagram is commutative:

$$
\begin{array}{ccccc}
(\mathrm{id}_G \times \mu)^*\mu^*\mathcal{A} & \xrightarrow{\ (\mathrm{id}_G \times \mu)^*\Phi\ } & (\mathrm{id}_G \times \mu)^*\mathrm{pr}_X^*\mathcal{A} & = & \mathrm{pr}_{G\times X}^*\mu^*\mathcal{A} \\
\Big\| & & & & \Big\downarrow \mathrm{pr}_{G\times X}^*\Phi \\
(\mathrm{m} \times \mathrm{id}_X)^*\mu^*\mathcal{A} & \xrightarrow[\ (\mathrm{m}\times\mathrm{id}_X)^*\Phi\ ]{} & (\mathrm{m} \times \mathrm{id}_X)^*\mathrm{pr}_X^*\mathcal{A} & = & \mathrm{pr}_{G\times X}^*\mathrm{pr}_X^*\mathcal{A}
\end{array}
$$

Example 4.2.3.4 Consider the sheaf of divisorial algebras $\mathcal{A} = \oplus_{n\in\mathbb{Z}}\mathcal{O}_X(nD)$ defined by a Cartier divisor D on X. Then the G-linearizations $\Phi\colon \mu^*\mathcal{A} \to \mathrm{pr}_X^*\mathcal{A}$ of $\mathcal{A}$ correspond to the G-linearizations of the invertible sheaf $\mathcal{O}_X(D)$ in the sense of Remark 4.2.2.9 via passing to the degree one part $\Phi_1\colon \mu^*\mathcal{O}_X(D) \to \mathrm{pr}_X^*\mathcal{O}_X(D)$.

We will see now that from the geometric point of view, a G-linearization of the sheaf $\mathcal{A}$ is merely a lifting of the G-action on X to a G-action on the relative spectrum of $\mathcal{A}$. The quasitorus $H := \mathrm{Spec}\,\mathbb{K}[K]$ acts on $\widetilde{X} := \mathrm{Spec}_X\mathcal{A}$ and the canonical map $q\colon \widetilde{X} \to X$ is a good quotient for this action.

Construction 4.2.3.5 Let X, $\mathcal{A}$, $\widetilde{X}$, and G be as above. Any G-linearization $\Phi\colon \mu^*\mathcal{A} \to \mathrm{pr}_X^*\mathcal{A}$ of $\mathcal{A}$ defines a commutative diagram

$$
\begin{array}{ccccc}
\mathrm{Spec}_{G\times X}(\mathrm{pr}_X^*\mathcal{A}) & \xrightarrow{\ \mathrm{Spec}(\Phi)\ } & \mathrm{Spec}_{G\times X}(\mu^*\mathcal{A}) & \longrightarrow & \mathrm{Spec}_X(\mathcal{A}) \\
\Big\| & & & & \Big\| \\
G \times \widetilde{X} & & \xrightarrow[\widetilde{\mu}]{\hspace{8cm}} & & \widetilde{X}
\end{array}
$$

$$(4.2.1)$$

Here, $\mathrm{Spec}_{G\times X}(\mu^*\mathcal{A})$ is the fiber product of $\mu\colon G \times X \to X$ and the canonical map $q\colon \widetilde{X} \to X$ and the right upper arrow is the projection to $\widetilde{X}$.

Proposition 4.2.3.6 *Let X, $\mathcal{A}$, $\widetilde{X}$, and G be as above. Then the following statements hold.*

(i) *If $\Phi\colon \mu^*\mathcal{A} \to \mathrm{pr}_X^*\mathcal{A}$ is a G-linearization, then the map $\widetilde{\mu}\colon G \times \widetilde{X} \to \widetilde{X}$ of 4.2.3.5 is a G-action that commutes with the H-action, and makes $q\colon \widetilde{X} \to X$ equivariant.*

(ii) *For every action $\widetilde{\mu}\colon G \times \widetilde{X} \to \widetilde{X}$ commuting with the H-action and making $q\colon \widetilde{X} \to X$ equivariant, there is a unique G-linearization $\Phi\colon \mu^*\mathcal{A} \to \mathrm{pr}_X^*\mathcal{A}$ such that the diagram (4.2.1) becomes commutative.*

Proof For (i), note that $q \circ \widetilde{\mu}$ equals $\mu \circ (\mathrm{id}_G \times q)$, because Φ is the identity in degree zero. Moreover, the commutative diagram of Definition 4.2.3.3 yields the associativity law of a group action for $\widetilde{\mu}$, and $e_G \in G$ acts trivially because Φ is an isomorphism. Finally, the actions of G and H commute, because $\widetilde{\mu}$ has graded comorphisms.

To verify (ii), we use that $\mathrm{Spec}_{G \times X}(\mu^* \mathcal{A})$ is the fiber product of $\mu \colon G \times X \to X$ and $\widetilde{X} \to X$. By the universal property, $\widetilde{\mu} \colon G \times \widetilde{X} \to \widetilde{X}$ lifts to a unique morphism $\mathrm{Spec}_{G \times X}(\mathrm{pr}_X^* \mathcal{A}) \to \mathrm{Spec}_{G \times X}(\mu^* \mathcal{A})$. It is straightforward to check that this morphism stems from a G-linearization $\Phi \colon \mu^* \mathcal{A} \to \mathrm{pr}_X^* \mathcal{A}$. $\square$

Eventually, via the lifted G-action on $\widetilde{X}$, we associate with any G-linearization of $\mathcal{A}$ a *graded G-sheaf structure* on $\mathcal{A}$. The latter is a collection of graded $\mathcal{O}(U)$-algebra homomorphisms $\mathcal{A}(U) \to \mathcal{A}(g \cdot U)$, $f \mapsto g \cdot f$, being compatible with group operations in G and with restriction and algebra operations in $\mathcal{A}$; thereby G acts as usual on the structure sheaf $\mathcal{O}_X$ via $g \cdot f(x) := f(g^{-1} \cdot x)$.

Proposition 4.2.3.7 *Let X, $\mathcal{A}$, $\widetilde{X}$ and G be as above and let $\Phi \colon \mu^* \mathcal{A} \to \mathrm{pr}_X^* \mathcal{A}$ be a G-linearization. Then there is a unique graded G-sheaf structure on $\mathcal{A}$ satisfying $g \cdot f(\widetilde{x}) := f(g^{-1} \cdot \widetilde{x})$ for all $g \in G$ and $\widetilde{x} \in \widetilde{X}$. For every G-invariant open $U \subseteq X$, the induced representation of G on $\mathcal{A}(U)$ is rational.*

Proof The existence of the G-sheaf structure is clear by $\mathcal{A} = q_* \mathcal{O}_{\widetilde{X}}$. Rationality of the induced representations follows, for example, from [189, Lemma 2.5]. $\square$

Remark 4.2.3.8 Let $\Phi \colon \mu^* \mathcal{A} \to \mathrm{pr}_X^* \mathcal{A}$ be a G-linearization. Then a section $f \in \mathcal{A}(X)$ is invariant with respect to the induced G-representation on $\mathcal{A}(X)$ if and only if $\Phi(\mu^*(f)) = \mathrm{pr}_X^*(f)$ holds.

For the proof of Theorem 4.2.3.2, we need the following observation on affine algebraic groups.

Lemma 4.2.3.9 *Let G be a connected affine algebraic group and b be a positive integer. Then there is a finite covering $\varepsilon \colon G' \to G$ such that for any $\zeta \in \mathbb{X}(G)$ there is an $\eta \in \mathbb{X}(G')$ with $\varepsilon^*(\zeta) = \eta^b$.*

Proof As any connected affine algebraic group, G is the semidirect product of a maximal reductive subgroup G^{red} and the unipotent radical $R_{\mathrm{u}}(G)$. Replacing G^{red} with a finite covering, we may assume that it is a direct product of a torus T and a simply connected semisimple group G^{ss}. Then we are in the situation

$$G = (T \times G^{\mathrm{ss}}) \ltimes R_{\mathrm{u}}(G).$$

Because any character of G is trivial on $R_{\mathrm{u}}(G)$ and G^{ss}, we may canonically identify the character group of G with that of T. Consider the finite covering

$T \to T, t \mapsto t^b$. It gives rise to another semidirect product

$$G' := (T \times G^{ss}) \ltimes R_u(G),$$

Then the canonical homomorphism $\varepsilon \colon G' \to G$ sending (t, g, u) to (t^b, g, u) has the desired property. □

Proof of Theorem 4.2.3.2 We first prove the uniqueness statement (ii). Denote the two lifted G-actions on $\widetilde{X}$ by $g * \widetilde{x}$ and $g \star \widetilde{x}$. Because the action of H is strongly stable, we may assume that H acts freely on $\widetilde{X}$. Then $\pi \colon \widetilde{X} \to X$ is in particular a geometric quotient. Consider the morphism $G \times \widetilde{X} \to \widetilde{X}$ sending $(g, \widetilde{x})$ to $g^{-1} \star g * \widetilde{x}$. Then we have

$$g^{-1} \star g * \widetilde{x} = h(g, \widetilde{x}) \cdot \widetilde{x}$$

with a unique element $h(g, \widetilde{x}) \in H$. In fact, because G is connected, $h(g, \widetilde{x})$ even lies in the unit component $H^0 \subseteq H$. Using the fact that, by freeness of the H-action, $\widetilde{X} \to \widetilde{X}/H^0$ is locally trivial with fiber H^0, we see that the assignment defines a morphism

$$G \times \widetilde{X} \to H^0, \qquad (g, \widetilde{x}) \mapsto h(g, \widetilde{x}).$$

By Remark 4.2.2.2 (ii), we have $h(g, \widetilde{x}) = \eta(g)\beta(\widetilde{x})$ with a homomorphism $\eta \colon G \to H^0$ of algebraic groups and a morphism $\beta \colon \widetilde{X} \to H^0$. Because the actions of G and H commute and $\widetilde{X}$ admits only constant invertible global H-homogeneous functions, we can conclude that β maps everything to the unit element of H.

We prove the existence statement (i). By Construction 4.2.1.2 and Theorem 4.2.1.4, it suffices to prove the statement for the characteristic space $p_X \colon \widehat{X} \to X$. By normality and Proposition 4.2.3.6, it is enough to provide a G'-linearization of the Cox sheaf $\mathcal{R}$ over the smooth locus for a suitable finite covering group G' of G. In particular, we may assume that G has trivial Picard group and X is smooth.

Let $K \subseteq \mathrm{WDiv}(X)$ be a finitely generated group mapping onto $\mathrm{Cl}(X)$ and let S denote the sheaf of divisorial algebras associated with K. Choose a basis $D_1, \ldots, D_r$ of K such that the kernel $K^0 \subseteq K$ of $K \to \mathrm{Cl}(X)$ has a basis of the form $a_i D_i$, where $1 \leq i \leq s$ with some $s \leq r$. Corollary 4.2.2.6 (i) gives us G-linearizations of invertible sheaves in the sense of Remark 4.2.2.9:

$$\Phi_i \colon \mu^* \mathcal{O}_X(D_i) \to \mathrm{pr}_X^* \mathcal{O}_X(D_i).$$

Tensoring the Φ_i, we obtain for each $D \in K$ an isomorphism $\mu^* S_D \to \mathrm{pr}_X^* S_D$. These maps are compatible with the multiplicative structures of $\mu^* S$ and $\mathrm{pr}_X^* S$, and hence fit together to a G-linearization of S.

The Cox sheaf of X is obtained as $\mathcal{R} = \mathcal{S}/\mathcal{I}$ with the ideal sheaf $\mathcal{I}$ generated by the elements $1 - \chi_E$, where $E \in K^0$, as introduced in Construction 1.4.2.1. We now adjust the G-action and the G-linearization of $\mathcal{S}$ such that the sections $\chi_E \in \Gamma(X, \mathcal{S}_{-E})$ and hence the ideal $\mathcal{I}$ are G-invariant.

Because $a_i D_i$ is a principal divisor, the sheaf $\mathcal{S}_{a_i D_i}$ is globally trivial, and the relative spectrum $\widehat{X}_i$ of the sheaf of Veronese subalgebras $\mathcal{S}(a_i \mathbb{Z} D_i)$ admits a global trivialization as a $\mathbb{K}^*$-principal bundle

$$\widehat{X}_i \;\rightarrow\; X \times \mathbb{K}^*, \qquad \widehat{x} \;\mapsto\; (p_i(\widehat{x}), \chi_{a_i D_i}(\widehat{x})),$$

where $p_i \colon \widehat{X}_i \to X$ is the canonical morphism. Proposition 4.2.3.6 provides us with a G-action on $\widehat{X}_i$ and thus on $X \times \mathbb{K}^*$. Using Remark 4.2.2.2, we see that the latter is given by a character $\zeta_i \in \mathbb{X}(G)$ as

$$g \cdot (x, t) \;=\; (g \cdot x, \zeta_i(g)t).$$

Set $b := a_1 \cdots a_s$. Then Lemma 4.2.3.9 yields a finite epimorphism $\varepsilon \colon G' \to G$ such that $\varepsilon^*(\zeta_i) = \eta_i^b$ holds with $\eta_i \in \mathbb{X}(G')$ for all i. Consider the G'-action

$$G' \times X \;\rightarrow\; X, \qquad (g', x) \;\mapsto\; \varepsilon(g') \cdot x.$$

The sheaf of divisorial algebras $\mathcal{S}$ is as well linearized with respect to this action. Twisting each G'-linearization of D_i with η_i^{-b/a_i}, we achieve that each $\chi_{a_i D_i}$ is invariant with respect to the twisted new G-linearization of $\mathcal{S}$. Consequently, $\mathcal{I}$ is invariant and the Cox sheaf $\mathcal{R} = \mathcal{S}/\mathcal{I}$ becomes G-linearized.

Assertions (iii) and (iv) follow directly from the construction of liftings performed in the proof of assertion (i). $\qquad\square$

4.2.4 Automorphisms of Mori dream spaces

Given a variety with finitely generated Cox ring, we study its automorphism group in terms of the Cox ring. Our approach extends ideas of [104, Sec. 4] treating the case of toric varieties. The original reference for this section is [21, Sec. 2]. We work over an algebraically closed field $\mathbb{K}$ of characteristic zero.

Let X be an irreducible normal complete variety with finitely generated divisor class group $\mathrm{Cl}(X)$ and finitely generated Cox ring $\mathcal{R}(X)$, for example, a Mori dream space. Recall that the characteristic quasitorus, the characteristic space, and the total coordinate space fit into the following picture:

$$\mathrm{Spec}_X \mathcal{R} \;=\; \widehat{X} \;\subseteq\; \overline{X} \;=\; \mathrm{Spec}\,\mathcal{R}(X)$$
$$p_X \Big\downarrow \,/\!\!/ H_X$$
$$X$$

We study automorphisms of X in terms of automorphisms of $\overline{X}$ and $\widehat{X}$. Recall that an H_X-equivariant automorphism of $\overline{X}$ is a pair $(\varphi, \widetilde{\varphi})$, where $\varphi \colon \overline{X} \to \overline{X}$ is an automorphism of varieties and $\widetilde{\varphi} \colon H_X \to H_X$ is an automorphism of affine algebraic groups satisfying

$$\varphi(t \cdot x) = \widetilde{\varphi}(t) \cdot \varphi(x) \quad \text{for all } x \in \overline{X}, \ t \in H_X.$$

The group of H_X-equivariant automorphisms of $\overline{X}$ is denoted by $\mathrm{Aut}(\overline{X}, H_X)$. Analogously, one defines the group $\mathrm{Aut}(\widehat{X}, H_X)$ of H_X-equivariant automorphisms of $\widehat{X}$. A *weak automorphism* of X is a birational map $\varphi \colon X \to X$ admitting open subsets $U_1, U_2 \subseteq X$ with complement $X \setminus U_i$ of codimension at least 2 in X such that $\varphi|_{U_1} \colon U_1 \to U_2$ is a regular isomorphism. We write $\mathrm{Bir}_2(X)$ for the group of weak automorphisms of X. The main result brings all these groups together.

Theorem 4.2.4.1 *Let X be an irreducible normal complete variety with finitely generated Cox ring. Then there is a commutative diagram of morphisms of affine algebraic groups where the rows are exact sequences and the upwards inclusions are of finite index:*

$$
\begin{array}{ccccccccc}
1 & \longrightarrow & H_X & \longrightarrow & \mathrm{Aut}(\overline{X}, H_X) & \longrightarrow & \mathrm{Bir}_2(X) & \longrightarrow & 1 \\
 & & \| & & \uparrow & & \uparrow & & \\
1 & \longrightarrow & H_X & \longrightarrow & \mathrm{Aut}(\widehat{X}, H_X) & \longrightarrow & \mathrm{Aut}(X) & \longrightarrow & 1
\end{array}
$$

Moreover, there is a big open subset $U \subseteq X$ with $\mathrm{Aut}(U) = \mathrm{Bir}_2(X)$ and the groups $\mathrm{Aut}(\overline{X}, H_X)$, $\mathrm{Bir}_2(X)$, $\mathrm{Aut}(\widehat{X}, H_X)$, and $\mathrm{Aut}(X)$ act morphically on $\overline{X}$, U, $\widehat{X}$, and X, respectively.

The proof of this theorem is given at the end of the section. We first list some consequences.

Corollary 4.2.4.2 *The automorphism group of an irreducible normal complete variety with finitely generated Cox ring is affine algebraic and acts morphically. Moreover, if two such varieties are isomorphic to each other outside closed subsets of codimension at least 2, then the unit components of their automorphism groups are isomorphic to each other.*

Let $\mathrm{CAut}(\overline{X}, H_X)$ denote the centralizer of H_X in the automorphism group $\mathrm{Aut}(\overline{X})$. Then $\mathrm{CAut}(\overline{X}, H_X)$ consists of all automorphisms $\varphi \colon \overline{X} \to \overline{X}$ satisfying

$$\varphi(t \cdot x) = t \cdot \varphi(x) \quad \text{for all } x \in \overline{X}, \ t \in H_X.$$

In particular, we have $\mathrm{CAut}(\overline{X}, H_X) \subseteq \mathrm{Aut}(\overline{X}, H_X)$. The group $\mathrm{CAut}(\overline{X}, H_X)$ may be used to detect the unit component $\mathrm{Aut}(X)^0$ of the automorphism group of X.

Corollary 4.2.4.3 *Let X be an irreducible normal complete variety with finitely generated Cox ring. Then there is an exact sequence of affine algebraic groups*

$$1 \longrightarrow H_X \longrightarrow \mathrm{CAut}(\overline{X}, H_X)^0 \longrightarrow \mathrm{Aut}(X)^0 \longrightarrow 1.$$

Proof According to [288, Cor. 2.3], the group $\mathrm{CAut}(\overline{X}, H_X)^0$ leaves $\widehat{X}$ invariant. Thus, we have $\mathrm{CAut}(\overline{X}, H_X)^0 \subseteq \mathrm{Aut}(\widehat{X}, H_X)$ and the sequence is well defined. Moreover, for any $\varphi \in \mathrm{Aut}(X)^0$, the pullback $\varphi^* \colon \mathrm{Cl}(X) \to \mathrm{Cl}(X)$ is the identity. Consequently, φ lifts to an element of $\mathrm{CAut}(\overline{X}, H_X)$. Exactness of the sequence thus follows for dimension reasons. $\qquad\square$

Corollary 4.2.4.4 *Let X be an irreducible normal complete variety with finitely generated Cox ring. Then, for any closed subgroup $F \subseteq \mathrm{Aut}(X)^0$, there is a closed subgroup $F' \subseteq \mathrm{CAut}(\overline{X}, H_X)^0$ such that the induced epimorphism $F' \to F$ is finite.*

Note that the above statement provides another proof of Theorem 4.2.3.2 for the case that X has a finitely generated Cox ring.

Corollary 4.2.4.5 *Let X be an irreducible normal complete variety with finitely generated Cox ring such that the group $\mathrm{CAut}(\overline{X}, H_X)$ is connected, for example, a toric variety; then there is an exact sequence of affine algebraic groups*

$$1 \longrightarrow H_X \longrightarrow \mathrm{CAut}(\overline{X}, H_X) \longrightarrow \mathrm{Aut}(X)^0 \longrightarrow 1.$$

Example 4.2.4.6 Consider the nondegenerate quadric X in the projective space $\mathbb{P}^{n+1}$, where $n \geq 4$ is even. Then the Cox ring of X is the $\mathbb{Z}$-graded ring

$$\mathcal{R}(X) = \mathbb{K}[T_0, \ldots, T_{n+1}] / \langle T_0^2 + \cdots + T_{n+1}^2 \rangle,$$

$$\deg(T_0) = \cdots = \deg(T_{n+1}) = 1.$$

The characteristic quasitorus is $H_X = \mathbb{K}^*$. Moreover, for the equivariant automorphisms and the centralizer of H_X we obtain

$$\mathrm{Aut}(\overline{X}, H_X) = \mathrm{CAut}(\overline{X}, H_X) = \mathbb{K}^* E_{n+2} \cdot O_{n+2}.$$

Thus, $\mathrm{CAut}(\overline{X}, H_X)$ has two connected components. Note that for $n = 4$, the quadric X comes with a torus action of complexity 1.

We enter the proof of Theorem 4.2.4.1. In a first step, we study the automorphism group of any finitely generated $\mathbb{K}$-algebra R graded by a finitely generated abelian group K. Recall that an *automorphism* of the K-graded algebra R is a pair (ψ, F), where $\psi \colon R \to R$ is an isomorphism of $\mathbb{K}$-algebras and $F \colon K \to K$ is an isomorphism of abelian groups such that $\psi(R_w) = R_{F(w)}$ holds for all $w \in K$. We denote the group of such automorphisms of R by $\mathrm{Aut}(R, K)$.

Proposition 4.2.4.7 *Let K be a finitely generated abelian group and R a K-graded finitely generated integral $\mathbb{K}$-algebra with $R^* = \mathbb{K}^*$. Suppose that the grading is pointed. Then $\mathrm{Aut}(R, K)$ is an affine algebraic group over $\mathbb{K}$ and R is a rational $\mathrm{Aut}(R, K)$-module.*

Proof The idea is to represent the automorphism group $\mathrm{Aut}(R, K)$ as a closed subgroup of the linear automophism group of a suitable finite-dimensional vector subspace $V^0 \subseteq R$. In the subsequent construction of V^0, we may assume that the weight cone ω generates $K_{\mathbb{Q}}$ as a vector space.

Consider the subgroup $\Gamma \subseteq \mathrm{Aut}(K)$ of $\mathbb{Z}$-module automorphisms $K \to K$ such that the induced linear isomorphism $K_{\mathbb{Q}} \to K_{\mathbb{Q}}$ leaves the weight cone $\omega \subseteq K_{\mathbb{Q}}$ invariant. By finite generation of R, the cone ω is polyhedral and thus Γ is finite. Let $f_1, \ldots, f_r \in R$ be homogeneous generators and denote by $w_i := \deg(f_i) \in K$ their degrees. Define a finite Γ-invariant subset and a vector subspace

$$S^0 := \Gamma \cdot \{w_1, \ldots, w_r\} \subseteq K, \qquad V^0 := \bigoplus_{w \in S^0} R_w \subseteq R.$$

For every automorphism (ψ, F) of the graded algebra R, we have $F(S^0) = S^0$ and thus $\psi(V^0) = V^0$. Moreover, (ψ, F) is uniquely determined by its restriction on V^0. Consequently, we may regard the automorphism group $H := \mathrm{Aut}(R, K)$ as a subgroup of the general linear group $\mathrm{GL}(V^0)$. Note that every $g \in H$

(i) permutes the components R_w of the decomposition $V^0 = \bigoplus_{w \in S^0} R_w$,
(ii) satisfies $\sum_\nu a_\nu g(f_1)^{\nu_1} \cdots g(f_r)^{\nu_r} = 0$ for any relation $\sum_\nu a_\nu f_1^{\nu_1} \cdots f_r^{\nu_r} = 0$.

Obviously, these are algebraic conditions. Moreover, every $g \in \mathrm{GL}(V^0)$ satisfying the above conditions can be extended uniquely to an element of $\mathrm{Aut}(R, K)$ via

$$g\left(\sum_\nu a_\nu f_1^{\nu_1} \cdots f_r^{\nu_r}\right) := \sum_\nu a_\nu g(f_1)^{\nu_1} \cdots g(f_r)^{\nu_r}.$$

Thus, we saw that $H \subseteq \mathrm{GL}(V^0)$ is precisely the closed subgroup defined by conditions (i) and (ii). In particular $H = \mathrm{Aut}(R, K)$ is affine algebraic. Moreover, the symmetric algebra SV^0 is a rational $\mathrm{GL}(V^0)$-module; hence SV^0 is a rational H-module for the algebraic subgroup H of $\mathrm{GL}(V^0)$, and so is its factor module R. $\qquad\square$

Corollary 4.2.4.8 *Let K be a finitely generated abelian group, R a K-graded finitely generated integral $\mathbb{K}$-algebra with $R^* = \mathbb{K}^*$, and consider the corresponding action of $H := \operatorname{Spec} \mathbb{K}[K]$ on $\overline{X} := \operatorname{Spec} R$. Then we have a canonical isomorphism*

$$\operatorname{Aut}(\overline{X}, H) \ \to \ \operatorname{Aut}(R, K), \qquad (\varphi, \widetilde{\varphi}) \ \mapsto \ (\varphi^*, \widetilde{\varphi}^*),$$

where φ^* is the pullback of regular functions and $\widetilde{\varphi}^*$ the pullback of characters. If the K-grading is pointed, then $\operatorname{Aut}(\overline{X}, H)$ is an affine algebraic group acting morphically on $\overline{X}$.

Proof of Theorem 4.2.4.1 We set $G_{\overline{X}} := \operatorname{Aut}(\overline{X}, H_X)$ for short. According to Corollary 4.2.4.8, the group $G_{\overline{X}}$ is affine algebraic and acts morphically on $\overline{X}$. Looking at the representations of H_X and $G_{\overline{X}}$ on $\Gamma(\overline{X}, \mathcal{O}) = \mathcal{R}(X)$ defined by the respective actions, we see that the canonical inclusion $H_X \to G_{\overline{X}}$ is a morphism of affine algebraic groups.

Next we construct the subset $U \subseteq X$ from the last part of the statement. Consider the translates $g \cdot \widehat{X}$, where $g \in G_{\overline{X}}$. Each of them admits a good quotient with a complete quotient space:

$$p_{X,g} \colon g \cdot \widehat{X} \ \to \ (g \cdot \widehat{X}) /\!\!/ H_X.$$

By [45], there are only finitely many open subsets of $\overline{X}$ with such a good quotient. In particular, the number of translates $g \cdot \widehat{X}$ is finite.

Let $W \subseteq X$ denote the maximal open subset such that the restricted quotient $\widehat{W} \to W$, where $\widehat{W} := p_X^{-1}(W)$, is geometric, that is, has the H_X-orbits as its fibers. Then, for any $g \in G_{\overline{X}}$, the translate $g \cdot \widehat{W} \subseteq g \cdot \widehat{X}$ is the (unique) maximal open subset which is saturated with respect to the quotient map $p_{X,g}$ and defines a geometric quotient. Consider

$$\widehat{U} \ := \ \bigcap_{g \in G_{\overline{X}}} g \cdot \widehat{W} \ \subseteq \ \widehat{X}.$$

By the preceding considerations $\widehat{U}$ is open, and by construction it is $G_{\overline{X}}$-invariant and saturated with respect to p_X. Proposition 1.6.1.6 (ii) states that the set $\widehat{W}$ is big in $\overline{X}$. Consequently, $\widehat{U}$ is also big in $\overline{X}$. Thus, the (open) set $U := p_X(\widehat{U})$ is big in X. By the universal property of the geometric quotient, there is a unique morphical action of $G_{\overline{X}}$ on U making $p_X \colon \widehat{U} \to U$ equivariant. Thus, we have a homomorphism of groups

$$G_{\overline{X}} \ \to \ \operatorname{Aut}(U) \ \subseteq \ \operatorname{Bir}_2(X).$$

We show that $G_{\overline{X}} \to \operatorname{Bir}_2(X)$ is surjective. Consider a weak automorphism $\varphi \colon X \to X$. The pullback defines an automorphism of the group of Weil divisors

$$\varphi^* \colon \operatorname{WDiv}(X) \to \operatorname{WDiv}(X), \qquad D \mapsto \varphi^* D.$$

As in the construction of the Cox sheaf, consider the sheaf of divisorial algebras $\mathcal{S} = \oplus \mathcal{S}_D$ associated with $\mathrm{WDiv}(X)$ and fix a character $\chi \colon \mathrm{PDiv}(X) \to \mathbb{K}(X)^*$ with $\mathrm{div}(\chi(E)) = E$ for any $E \in \mathrm{PDiv}(X)$. Then we obtain a homomorphism

$$\alpha \colon \mathrm{PDiv}(X) \;\to\; \mathbb{K}^*, \qquad E \mapsto \frac{\varphi^*(\chi(E))}{\chi(\varphi^*(E))}.$$

We extend this to a homomorphism $\alpha \colon \mathrm{WDiv}(X) \to \mathbb{K}^*$ as follows. Write $\mathrm{Cl}(X)$ as a direct sum of a free part and cyclic groups $\Gamma_1, \ldots, \Gamma_s$ of order n_i. Take $D_1, \ldots, D_r \in \mathrm{WDiv}(X)$ such that the classes of $D_1, \ldots, D_s$ are generators for $\Gamma_1, \ldots, \Gamma_s$ and the remaining ones define a basis of the free part. Set

$$\alpha(D_i) := \sqrt[n_i]{\alpha(n_i D_i)} \text{ for } 1 \le i \le s, \qquad \alpha(D_i) := 1 \text{ for } s+1 \le i \le r.$$

Then one directly checks that this extends α to a homomorphism $\mathrm{WDiv}(X) \to \mathbb{K}^*$. Using $\alpha(E)$ as a "correction term," we define an automorphism of the graded sheaf $\mathcal{S}$ of divisorial algebras: for any open set $V \subseteq X$ we set

$$\varphi^* \colon \Gamma(V, \mathcal{S}_D) \;\to\; \Gamma(\varphi^{-1}(V), \mathcal{S}_{\varphi^*(D)}), \qquad f \mapsto \alpha(D) f \circ \varphi.$$

By construction φ^* sends the ideal $\mathcal{I}$ arising from the character χ to itself. Consequently, φ^* descends to an automorphism (ψ, F) of the (graded) Cox sheaf $\mathcal{R}$; note that F is the pullback of divisor classes via φ. The degree zero part of ψ equals the usual pullback of regular functions on X via φ. Thus, the element in $\mathrm{Aut}(\overline{X}, H_X)$ defined by $\mathrm{Spec}\,\psi \colon \widehat{U} \to \widehat{U}$ maps to φ.

Clearly, H_X lies in the kernel of $\pi \colon G_{\overline{X}} \to \mathrm{Bir}_2(X)$. For the reverse inclusion, consider an element $g \in \ker(\pi)$. Then g is a pair $(\varphi, \widetilde{\varphi})$ and, by the construction of π, we have a commutative diagram

$$
\begin{array}{ccc}
\widehat{U} & \xrightarrow{\ \varphi\ } & \widehat{U} \\
{\scriptstyle p_X}\downarrow & & \downarrow{\scriptstyle p_X} \\
U & \xrightarrow[\ \mathrm{id}\ ]{} & U
\end{array}
$$

In particular, φ stabilizes all H_X-invariant divisors. It follows that the pullback φ^* on $\Gamma(\widehat{U}, \mathcal{O}) = \mathcal{R}(X)$ stabilizes the homogeneous components. Thus, for any homogeneous f of degree w, we have $\varphi^*(f) = \lambda(w) f$ with a homomorphism $\lambda \colon K \to \mathbb{K}^*$. Consequently $\varphi(x) = h \cdot x$ holds with an element $h \in H_X$. The statements concerning the upper sequence are verified.

Now, consider the lower sequence. Because $\widehat{X}$ is big in $\overline{X}$, every automorphism of $\widehat{X}$ extends to an automorphism of $\overline{X}$. We conclude that $\mathrm{Aut}(\widehat{X}, H_X)$ is the (closed) subgroup of $G_{\overline{X}}$ leaving the complement $\overline{X} \setminus \widehat{X}$ invariant. As seen before, the collection of translates $G_{\overline{X}} \cdot \widehat{X}$ is finite and thus the subgroup

$\mathrm{Aut}(\widehat{X}, H_X)$ of $G_{\overline{X}}$ is of finite index. Moreover, lifting $\varphi \in \mathrm{Aut}(X)$ as before gives an element of $\mathrm{Aut}(\overline{X}, H_X)$ leaving $\widehat{X}$ invariant. Thus, $\mathrm{Aut}(\widehat{X}, H_X) \to \mathrm{Aut}(X)$ is surjective with kernel H_X. By the universal property of the qood quotient $\widehat{X} \to X$, the action of $\mathrm{Aut}(X)$ on X is morphical. $\qquad\square$

4.3 Finite generation

4.3.1 General criteria

We gather general statements ensuring finite generation of the Cox ring of a given variety. The original references for this section are [27, 42, 188]. We begin with a basic observation relating finite generation of the Cox ring to finite generation of section rings. The ground field $\mathbb{K}$ is algebraically closed and of characteristic zero.

Proposition 4.3.1.1 *Let X be an irreducible normal prevariety with only constant invertible global functions and finitely generated divisor class group. Then the following statements are equivalent.*

(i) *The Cox ring $\mathcal{R}(X)$ is finitely generated.*
(ii) *For any finitely generated subgroup $K \subseteq \mathrm{WDiv}(X)$ and any open subset $U \subseteq X$, the algebra of sections $\Gamma(U, \mathcal{S})$ of the sheaf of divisorial algebras $\mathcal{S}$ associated with K is finitely generated.*

Proof Only for the implication "(i)$\Rightarrow$(ii)" is there something to show. First assume that K projects onto $\mathrm{Cl}(X)$. The first part of the proof of Proposition 1.1.2.7 shows that finite generation of $\mathcal{R}(X)$ implies finite generation of $\Gamma(X, \mathcal{S})$.

Let $U \subsetneq X$ be an open subset. Then the complement $X \setminus U$ can be written as a union of the support of an effective divisor E and a closed subset of codimension at least 2. Let $D \in K$ be a divisor that is linearly equivalent to E, that is, $E = D + \mathrm{div}(f)$ with a suitable rational function f. Then f is contained in $\Gamma(X, \mathcal{S}_D)$ and Remark 1.3.1.7 shows that $\Gamma(U, \mathcal{S}) = \Gamma(X, \mathcal{S})_f$ is finitely generated.

Finally, if $K \subseteq \mathrm{WDiv}(X)$ does not project onto $\mathrm{Cl}(X)$, then we take any finitely generated group $K' \subseteq \mathrm{WDiv}(X)$ with $K \subseteq K'$ projecting onto $\mathrm{Cl}(X)$ and obtain finite generation of $\Gamma(U, \mathcal{S}')$ for the associated sheaf $\mathcal{S}'$ of divisorial algebras. This gives finite generation for the Veronese subalgebra $\Gamma(U, \mathcal{S}) \subseteq \Gamma(U, \mathcal{S}')$ corresponding to $K \subseteq K'$. $\qquad\square$

Corollary 4.3.1.2 *Let X be an irreducible normal prevariety with finitely generated Cox ring. Then for every open subset $U \subseteq X$ the algebra $\Gamma(U, \mathcal{O})$ is finitely generated.*

We take a brief look at the affine case. There finite generation of the Cox ring turns out to be a local property and we observe a relation to the behavior of singularities.

Proposition 4.3.1.3 *Let X be an irreducible normal affine variety with only constant invertible global functions and finitely generated divisor class group. Then the following statements are equivalent.*

(i) *The Cox ring $\mathcal{R}(X)$ is finitely generated.*
(ii) *The Cox sheaf $\mathcal{R}$ is locally of finite type on X.*

Proof Implication "(i)⇒(ii)" is proven in Proposition 1.6.1.1. For the converse, recall that the characteristic space is an affine morphism. Thus, the variety $\widehat{X}$ is affine and the algebra $\mathcal{R}(X) \cong \Gamma(\widehat{X}, \mathcal{O})$ is finitely generated. □

Corollary 4.3.1.4 *Let X be an irreducible normal affine variety with only constant invertible global functions. Suppose that there is an open subset $U \subseteq X$ with $\mathrm{Cl}(U) = 0$ and $\Gamma(U, \mathcal{O})$ not finitely generated. Then $\mathrm{Cl}(X)$ is finitely generated and the Cox sheaf $\mathcal{R}$ on X is not locally of finite type. In particular, the variety X has non-$\mathbb{Q}$-factorial singularities.*

Proof The group $\mathrm{Cl}(X)$ is generated by the classes of the prime divisors supported in $X \setminus U$. Assume that $\mathcal{R}$ is locally of finite type on X. Then Proposition 4.3.1.3 states that $\mathcal{R}(X)$ is finitely generated and Corollary 4.3.1.2 implies that $\Gamma(U, \mathcal{O})$ is finitely generated, a contradiction. The last statement follows from Proposition 1.6.1.2. □

We turn to irreducible varieties X coming with an action of a connected reductive algebraic group G. Following [212] and [298], we define the *complexity $c(X)$* of the G-action on X to be the codimension of a general orbit $B \cdot x$ of a Borel subgroup $B \subseteq G$. Varieties with an action of complexity zero are called *spherical*. By Rosenlicht's theorem [301, Thm. 2.3], the complexity is equal to the transcendence degree of the field $\mathbb{K}(X)^B$ of rational B-invariants in $\mathbb{K}(X)$.

The first aim is to extend a result of Knop [188] on function rings of irreducible normal unirational G-varieties X of complexity $c(X) \leq 1$ to Cox rings. Recall that an irreducible variety X is *unirational* if there is a rational dominant morphism $\mathbb{K}^m \to X$ for some $m > 0$. This is equivalent to saying that the field of functions $\mathbb{K}(X)$ is isomorphic to a subfield of the field of rational functions $\mathbb{K}(T_1, \ldots, T_m)$.

Theorem 4.3.1.5 *Let G be a connected reductive algebraic group and X be an irreducible normal unirational G-variety of complexity $c(X) \leq 1$ with only constant invertible global functions. Then the divisor class group $\mathrm{Cl}(X)$ and the Cox ring $\mathcal{R}(X)$ are finitely generated.*

Proposition 4.3.1.6 *Let X be an irreducible normal prevariety with an action of a factorial connected affine algebraic group G. Moreover, let $K \subseteq \mathrm{WDiv}(X)$ be a finitely generated subgroup.*

(i) *The sheaf $\mathcal{S}$ of divisorial algebras associated with K admits a G-linearization.*

(ii) *If $\mathcal{S}$ is locally of finite type, then there is a G-action on $\widetilde{X} := \mathrm{Spec}_X \mathcal{S}$ making $\widetilde{X} \to X$ equivariant and commuting with the action of $H = \mathrm{Spec}\, \mathbb{K}[K]$.*

(iii) *In the setting of (ii), if G is reductive and the G-action on X is of complexity c, then the action of $G \times H$ on $\widetilde{X}$ is also of complexity c.*

Proof For (i) we proceed as in the proof of Theorem 4.2.3.2. Choose a basis $D_1, \ldots, D_r$ of K. Corollary 4.2.2.6 (i) gives us G-linearizations of invertible sheaves in the sense of Remark 4.2.2.9:

$$\Phi_i \colon \mu^* \mathcal{O}_X(D_i) \ \to \ \mathrm{pr}_X^* \mathcal{O}_X(D_i).$$

Tensoring the Φ_i, we obtain for each $D \in K$ an isomorphism $\mu^* \mathcal{S}_D \to \mathrm{pr}_X^* \mathcal{S}_D$. These maps are compatible with the multiplicative structures of $\mu^* \mathcal{S}$ and $\mathrm{pr}_X^* \mathcal{S}$, and hence fit together to a G-linearization of $\mathcal{S}$.

The second assertion is a direct application of Proposition 4.2.3.6. For the third assertion note that if B is a Borel subgroup of G then $B \times H$ is a Borel subgroup of $G \times H$. The claim then follows from the facts that $\widetilde{X} \to X$ is G-equivariant and a geometric quotient for a free H-action over the set of smooth points. □

Proposition 4.3.1.7 *Let G be a connected reductive algebraic group and X an irreducible normal unirational G-variety with $c(X) \leq 1$. Then X is rational.*

Proof By the local structure theorem, see [296, Thm. 4.1], there are a parabolic subgroup $P = L P^u$ of G with a Levi subgroup L and the unipotent radical P^u and a locally closed L-invariant affine subvariety $Z \subseteq X$ such that

- $X^0 := P \cdot Z$ is an open affine subset of X.
- There is a geometric quotient $X^0 \to X^0/P^u = \mathrm{Spec}\, \Gamma(X^0, \mathcal{O})^{P^u}$.
- The P-action on X^0 gives rise to an isomorphism $P^u \times Z \cong P \times_L Z \cong X^0$.
- The kernel L_0 of the natural action of $L = P/P^u$ on X^0/P^u contains the commutant L'.
- $X^0/P^u \cong Z \cong L/L_0 \times C$, where the torus L/L_0 acts on C trivially.

All Borel subgroups in G are conjugate, so we may assume that B is a semidirect product of P^u and some Borel subgroup in L. Then X^0 is isomorphic to $Z^0 \times C$, where $Z^0 = P^u \times L/L_0$ is a B-orbit and B acts on C trivially. By assumptions, C is either a point ($c(X) = 0$) or a curve ($c(X) = 1$). In the second case, because $\mathbb{K}(C)$ is contained in $\mathbb{K}(X)$ and X is unirational, it follows from Lüroth's theorem that the curve C is rational. Because $X^0 \cong P^u \times L/L_0 \times C$, where P^u is isomorphic to an affine space and L/L_0 is a torus, the variety X is rational. $\qquad\qquad\Box$

Proof of Theorem 4.3.1.5   We may assume that X is smooth. Because X is rational, the group $\mathrm{Cl}(X)$ is finitely generated. Fix a finitely generated subgroup $K \subseteq \mathrm{WDiv}(X)$ projecting onto $\mathrm{Cl}(X)$. Let $\mathcal{S}$ be the associated sheaf of divisorial algebras and consider the relative spectrum $\widetilde{X} := \mathrm{Spec}_X \mathcal{S}$ over X. Choose a lifting of the G-action to $\widetilde{X}$ as in Propostion 4.3.1.6. Then we have $c(\widetilde{X}) = c(X) \le 1$, and by Remark 1.3.2.7, the variety $\widetilde{X}$ is unirational. Knop's theorem [188] then implies that $\Gamma(\widetilde{X}, \mathcal{O}) \cong \Gamma(X, \mathcal{S})$ is finitely generated. By Lemma 1.5.1.2 and Lemma 1.4.3.5, the Cox ring $\mathcal{R}(X)$ is finitely generated as well. $\qquad\qquad\Box$

The following result from [27] says in particular that GIT quotients of Mori dream space are again Mori dream spaces.

Theorem 4.3.1.8   *Let a reductive affine algebraic group G act on an irreducible normal variety X with finitely generated Cox ring $\mathcal{R}(X)$ and let $U \subseteq X$ be an open invariant subset admitting a good quotient $\pi : U \to U /\!\!/ G$ such that $U /\!\!/ G$ has only constant invertible global functions. Then the Cox ring $\mathcal{R}(U /\!\!/ G)$ is finitely generated as well.*

For the proof we need the *canonical linearization* of a sheaf $\mathcal{S}$ of divisorial algebras associated with a group $K \subseteq \mathrm{WDiv}(X)$ of G-invariant divisors. Recall that for an irreducible normal Z and a dominant morphism $p : Z \to X$ such that the inverse image $p^{-1}(X')$ of the smooth locus $X' \subset X$ has a complement of codimension at least 2 in Z, the pullback $\mathrm{CDiv}(X') \to \mathrm{CDiv}(p^{-1}(X'))$ induces a map $p^* : \mathrm{WDiv}(X) \to \mathrm{WDiv}(Z)$, and we obtain

$$p^*\mathcal{S} = \bigoplus_{D \in K} \mathcal{O}_Z(p^*D).$$

Proposition 4.3.1.9   *Let an affine algebraic group G act on an irreducible normal variety X. Let $K \subseteq \mathrm{WDiv}(X)$ be a subgroup generated by finitely many G-invariant divisors and denote by $\mathcal{S}$ the associated sheaf of divisorial algebras. Then there is a canonical G-linearization*

$$\mu^*\mathcal{S} = \bigoplus_{D \in \Lambda} \mathcal{O}_{G \times X}(\mu^*D) = \bigoplus_{D \in \Lambda} \mathcal{O}_{G \times X}(\mathrm{pr}_X^*D) = \mathrm{pr}_X^*\mathcal{S},$$

where $\mu : G \times X \to X$ is the action and $\mathrm{pr}_X : G \times X \to X$ the projection. The induced G-sheaf structure on $\mathcal{S}$ is given by the usual action of G on the function field $\mathbb{K}(X)$ via $g \cdot f(x) = f(g^{-1} \cdot x)$.

Proof We have to show that any G-invariant Weil divisor $D = \sum n_E E$ satisfies $\mu^* D = \mathrm{pr}_X^* D$. For this, we consider the isomorphism

$$\beta \colon G \times X \to G \times X, \qquad (g, x) \mapsto (g, g^{-1} \cdot x).$$

Then we have $\beta^* \mathrm{pr}_X^* D = \mathrm{pr}_X^* D$, and $\mathrm{pr}_X^* D = \beta^* \mu^* D$. Because β^* has an inverse, the assertion follows. $\qquad\square$

Lemma 4.3.1.10 *Let a reductive group G act on an irreducible normal variety X and $U \subseteq X$ be an open G-invariant subset that admits a good quotient $\pi \colon U \to U /\!\!/ G$. If $\mathrm{Cl}(X)$ is finitely generated then $\mathrm{Cl}(U /\!\!/ G)$ is finitely generated as well.*

Proof Without loss of generality we assume X and $U /\!\!/ G$ to be smooth. From [190, Prop. 4.2] we infer that the pullback homomorphism $\pi^* \colon \mathrm{Pic}(U /\!\!/ G) \to \mathrm{Pic}_G(U)$ into the classes of G-linearized line bundles is injective. It therefore suffices to show that $\mathrm{Pic}_G(U)$ is finitely generated. By [190, Lemma 2.2] the following sequence is exact

$$\mathrm{H}^1_{\mathrm{alg}}(G, \mathcal{O}(U)^*) \longrightarrow \mathrm{Pic}_G(U) \longrightarrow \mathrm{Pic}(U) \ .$$

Note that the group of algebraic cocycles $\mathrm{H}^1_{\mathrm{alg}}(G, \mathcal{O}(U)^*)$ is finitely generated by the exact sequence in [190, Prop. 2.3]

$$\mathbb{X}(G) \longrightarrow \mathrm{H}^1_{\mathrm{alg}}(G, \mathcal{O}(U)^*) \longrightarrow \mathrm{H}^1\left(G/G^0, E(U)\right) \ ,$$

where G/G^0 is finite and $E(U) = \mathcal{O}(U)^* / \mathbb{K}^*$ is finitely generated by [190, Prop. 1.3]. $\qquad\square$

Proof of Theorem 4.3.1.8 We may assume that X and $U /\!\!/ G$ are smooth. By Lemma 4.3.1.10 we can choose a finitely generated group K of Weil divisors on the quotient space $U /\!\!/ G$ projecting surjectively onto the divisor class group $\mathrm{Cl}(U /\!\!/ G)$. With $\mathcal{S}$ denoting the sheaf of divisorial algebras associated with K, the Cox ring $\mathcal{R}(U /\!\!/ G)$ is the quotient of $\Gamma(U /\!\!/ G, \mathcal{S})$ by the ideal $\Gamma(U /\!\!/ G, \mathcal{I})$. Thus it suffices to show that the algebra of global sections $\Gamma(U /\!\!/ G, \mathcal{S})$ is finitely generated.

The pullback group $\pi^* K$ consists of invariant Weil divisors on U. It is therefore canonically G-linearized and we have the corresponding G-representation on the algebra $\Gamma(U, \mathcal{T})$, where $\mathcal{T}$ denotes the sheaf of divisorial algebras associated with the group $\pi^* K$. We claim that we have a pullback homomorphism mapping $\Gamma(U /\!\!/ G, \mathcal{S})$ injectively onto the algebra $\Gamma(U, \mathcal{T})^G$ of invariant sections of $\Gamma(U, \mathcal{T})$:

$$\pi^* \colon \Gamma(U /\!\!/ G, \mathcal{S}) \to \Gamma(U, \mathcal{T})^G, \quad \Gamma(U /\!\!/ G, \mathcal{S}_D) \ni f \mapsto \pi^* f \in \Gamma(U, \mathcal{T}_{\pi^* D}).$$

We first note that every pullback section $\pi^* f \in \Gamma(U, \mathcal{T}_{\pi^* D})$ is indeed G-invariant because $\pi^* K$ is canonically G-linearized and $\pi^* f$ is G-invariant as a rational function on U. On each homogeneous component of $\Gamma(U /\!/ G, \mathcal{S})$ the map π^* is injective because it is the pullback with respect to the surjective morphism $\pi : U \to U /\!/ G$. Because π^* is graded this yields injectivity of π^* as an algebra homomorphism.

For surjectivity it suffices to show that every homogeneous G-invariant section is a pullback section because the actions of G and H commute and, thus, $\Gamma(U, \mathcal{T})^G$ is a graded subalgebra of $\Gamma(U, \mathcal{T})$. Consider a G-invariant homogeneous section $f \in \Gamma(U, \mathcal{T}_{\pi^* D})$. Because f is invariant as a rational function in $\mathbb{K}(U)$ and it is regular on $U' := U \backslash \pi^{-1}(\operatorname{Supp}(D))$, it descends to a regular function $\tilde{f}$ on $\pi(U')$, which is an open subset of $U /\!/ G$. Observe that we have

$$\pi^*(\operatorname{div}(\tilde{f}) + D) = \operatorname{div}(f) + \pi^* D \geq 0.$$

In particular, we obtain that the divisor $\operatorname{div}(\tilde{f}) + D$ is effective and thus $\tilde{f}$ is a section in $\Gamma(U /\!/ G, \mathcal{S}_D)$. By construction f equals the pullback $\pi^* \tilde{f}$; hence our claim follows.

Thus the algebras $\Gamma(U /\!/ G, \mathcal{S})$ and $\Gamma(U, \mathcal{T})^G$ are isomorphic. The algebra $\Gamma(U, \mathcal{T})$ is finitely generated by Proposition 4.3.1.1. Hilbert's finiteness theorem then shows that the algebra of invariants $\Gamma(U, \mathcal{T})^G$ is finitely generated as well. $\square$

4.3.2 Finite generation via multiplication map

We characterize finite generation of an integral $\mathbb{K}$-algebra A graded by a finitely generated abelian group K in terms of surjectivity properties of the multiplication map of A. The original reference for this section is [19].

The coefficient field $\mathbb{K}$ is algebraically closed and of characteristic zero. As earlier, we denote for any $u \in K$ the associated element $u \otimes 1$ in $K_{\mathbb{Q}} = K \otimes_{\mathbb{Z}} \mathbb{Q}$ again by u. Recall that the weight cone of A is the cone $\omega(A) \subseteq K_{\mathbb{Q}}$ generated by all $u \in K$ with $A_u \neq 0$.

Definition 4.3.2.1 Let K be a finitely generated abelian group and A a K-graded integral $\mathbb{K}$-algebra.

(i) We call a pair $u, v \in K$ *generating* if u, v belong to $\omega(A)$ and there is an $m > 0$ such that for any $k > 0$ the multiplication map defines a surjection

$$\mu_{km} : A_{kmu} \otimes_{\mathbb{K}} A_{kmv} \to A_{km(u+v)}, \qquad f \otimes g \mapsto fg.$$

(ii) By a *generating fan* for A we mean a quasifan Λ in $K_{\mathbb{Q}}$ with support $\omega(A)$ such that for all $\lambda \in \Lambda$, $u \in \lambda$, and $v \in \lambda^{\circ}$ the pair (u, v) is generating.

The first statement shows that for finitely generated K-graded algebras A, the GIT quasifan $\Lambda(A)$ of the action of $H := \operatorname{Spec} \mathbb{K}[K]$ on $X := \operatorname{Spec} A$ is a generating quasifan for A and even the coarsest possible one.

Theorem 4.3.2.2 *Let K be a finitely generated abelian group and A an integral finitely generated K-graded $\mathbb{K}$-algebra with GIT quasifan $\Lambda(A)$.*

(i) *If $u, v \in \omega(A) \cap K$ is a generating pair, then the weights u, v lie in a common GIT cone $\lambda \in \Lambda(A)$.*

(ii) *If $u, v \in \omega(A) \cap K$ lie in a common GIT cone $\lambda \in \Lambda(A)$ and u belongs to the relative interior λ°, then u, v is a generating pair.*

The second result shows that in particular for Cox rings of complete varieties the existence of a generating quasifan is also sufficient for finite generation.

Theorem 4.3.2.3 *Let K be a finitely generated abelian group and A an integral K-graded $\mathbb{K}$-algebra. Assume that all homogeneous components A_u are finite-dimensional $\mathbb{K}$-vector spaces and the weight cone $\omega(A)$ is pointed. Then the following conditions are equivalent.*

(i) *The algebra A is finitely generated.*

(ii) *There is a generating quasifan for A.*

We come to proofs of the above theorems. In a first step, we give a more algebraic characterization of the GIT quasifan. For $u, v \in \omega(A) \cap K$, we will work in terms of the following subalgebras:

$$A(u) := \bigoplus_{n \in \mathbb{Z}_{\geq 0}} A_{nu}, \qquad A(u, v) := \bigoplus_{n \in \mathbb{Z}_{\geq 0}} A_{nu} \cdot A_{nv}.$$

Clearly, $A(u, v)$ is contained in $A(u + v)$. We call $A(u, v)$ *large* in $A(u + v)$, if the ideals $A(u, v)_+ \subseteq A(u, v)$, and $A(u + v)_+ \subseteq A(u + v)$ generated by the homogeneous parts of strictly positive degree satisfy

$$\sqrt{\langle A(u, v)_+ \rangle} = A(u + v)_+ \subseteq A(u + v).$$

Proposition 4.3.2.4 *Let K be a lattice and A a K-graded finitely generated integral $\mathbb{K}$-algebra. Then, for any two $u, v \in \omega(A)$, the following statements are equivalent.*

(i) *There is a GIT cone $\lambda \in \Lambda$ satisfying $u, v \in \lambda$.*

(ii) *We have $X^{ss}(u) \cap X^{ss}(v) = X^{ss}(u + v)$.*

(iii) *The subalgebra $A(u, v)$ is large in $A(u + v)$.*

Proof We begin with the equivalence of (i) and (ii). If (i) holds, then every orbit cone ω_x containing $u + v$ must contain u and v as well. This gives

$$x \in X^{ss}(u) \cap X^{ss}(v) \iff u, v \in \omega_x$$
$$\iff u + v \in \omega_x$$
$$\iff x \in X^{ss}(u + v).$$

Conversely, if (ii) holds, then we see that $\lambda(u)$ and $\lambda(v)$ are faces of $\lambda(u + v)$. Thus, we have $u, v \in \lambda(u + v)$.

For the equivalence of (ii) and (iii) note that for any $w \in \omega(A) \cap K$ the complement $X \setminus X^{ss}(w)$ equals the zero set $V(A(w)_+)$. Thus, setting $w := u + v$, we obtain

$$X^{ss}(u) \cap X^{ss}(v) = X^{ss}(w) \iff V(A(u)_+) \cup V(A(v)_+) = V(A(w)_+)$$
$$\iff V(A(u)_+ \cdot A(v)_+) = V(A(w)_+).$$

The latter property holds if and only if the ideals generated by $A(u)_+ \cdot A(v)_+$ and $A(w)_+$ have the same radical in A. This holds if and only if they generate the same radical ideal in $A(w)$, which eventually is equivalent to $A(u, v)$ being a large subalgebra of $A(w)$. $\square$

For $X^{ss}(u) \subseteq X^{ss}(v)$, there is an induced projective morphism $Y(u) \to Y(v)$ of the quotient spaces. In particular, if u, v lie in a common GIT cone, then we obtain a commutative diagram

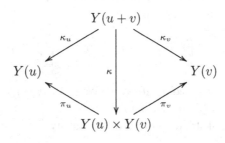

$$(4.3.1)$$

The next step is a geometric characterization of the GIT quasifan. It is given in terms of the map $\kappa \colon Y(u + v) \to Y(u) \times Y(v)$ introduced in diagram 4.3.1.

Proposition 4.3.2.5 *Let $u, v \in \omega(A) \cap K$ belong to a common GIT cone $\lambda \in \Lambda(A)$. Then, in the setting of diagram 4.3.1, the following statements are equivalent.*

 (i) *The pair $u, v \in \omega(A) \cap K$ is generating.*
 (ii) *The map $\kappa \colon Y(u + v) \to Y(u) \times Y(v)$ is a closed embedding.*

Proof Recall that the quotient spaces $Y(w) = \text{Proj}(A(w))$ are projective over $Y_0 = \text{Spec } A_0$. Moreover, denoting by $q : X^{ss}(w) \to Y(w)$ the quotient map, we obtain for $n \in \mathbb{Z}_{\geq 0}$ a sheaf on $Y(w)$, namely

$$\mathcal{L}_{nw} := \left(q_* \mathcal{O}_{X^{ss}(w)}\right)_{nw} = \mathcal{O}_{Y(w)}(n).$$

Replacing u with a large multiple, we may assume that $A(u)$ is generated as an A_0-algebra by the component A_u, and that for any $n \in \mathbb{Z}_{\geq 1}$ the canonical maps

$$\iota_{nu} : A_{nu} \to \Gamma(Y(u), \mathcal{L}_{nu})$$

are surjective; see [160, Exercise 2.5.9]. Note that then $\mathcal{L}_u$ is an ample invertible sheaf on $Y(u)$. Of course, we may arrange the same situation for v and $u + v$.

On $Y(u) \times Y(v)$ we have the ample invertible sheaves $\mathcal{E}_n := \pi_u^* \mathcal{L}_{nu} \otimes \pi_v^* \mathcal{L}_{nv}$. We claim that the natural map

$$\Gamma(Y(u), \mathcal{L}_{nu}) \otimes \Gamma(Y(v), \mathcal{L}_{nv}) \to \Gamma(Y(u) \times Y(v), \mathcal{E}_n)$$

is an isomorphism. Indeed, using the projection formula, we obtain canonical isomorphisms

$$\Gamma(Y(u) \times Y(v), \mathcal{E}_n) \cong \Gamma(Y(u), \pi_{u*} \mathcal{E}_n) \cong \Gamma(Y(u), \mathcal{L}_{nu} \otimes \pi_{u*} \pi_v^* \mathcal{L}_{nv}).$$

We look a bit closer at $\pi_{u*} \pi_v^* \mathcal{L}_{nv}$. Given an open subset $U \subseteq Y(u)$, we denote by $\pi_v^U : U \times Y(v) \to Y(v)$ the restricted projection. Then we have

$$\Gamma(U, \pi_{u*} \pi_v^* \mathcal{L}_{nv}) = \Gamma(U \times Y(v), \pi_v^* \mathcal{L}_{nv}) \cong \Gamma(Y(v), \mathcal{L}_{nv} \otimes \pi_{v\,*}^U \mathcal{O}_{U \times Y(v)}).$$

Likewise, one obtains $\pi_{v\,*}^U \mathcal{O}_{U \times Y(v)} = \Gamma(U, \mathcal{O}_U) \otimes \mathcal{O}_{Y(v)}$ for any affine open set $U \subseteq Y(u)$. Consequently, we have a canonical isomorphism

$$\Gamma(U, \pi_{u*} \pi_v^* \mathcal{L}_{nv}) = \Gamma(U, \mathcal{O}_U) \otimes \Gamma(Y(v), \mathcal{L}_{nv}).$$

This in turn shows $\pi_{u*} \pi_v^* \mathcal{L}_{nv} \cong \mathcal{O}_{Y(u)} \otimes \Gamma(Y(v), \mathcal{L}_{nv})$, and our claim follows. Thus, we arrive at a commutative diagram

$$
\begin{array}{ccc}
A_{nu} \otimes A_{nv} & \xrightarrow{\ \mu_n\ } & A_{nu+nv} \\
\cong \downarrow & & \downarrow \cong \\
\Gamma(Y(u) \times Y(v), \mathcal{E}_n) & \xrightarrow{\ \kappa_n^*\ } & \Gamma(Y(u+v), \mathcal{L}_{nu+nv})
\end{array}
$$

where the upper horizontal arrow is the multiplication map we are interested in, and the lower horizontal arrow is the canonical pullback map

$$\kappa_n^* : \Gamma(Y(u) \times Y(v), \mathcal{E}_n) \to \Gamma(Y(u+v), \mathcal{L}_{nu+nv})$$

$$\pi_u^* f \otimes \pi_v^* g \mapsto \kappa_u^* f \cdot \kappa_v^* g.$$

Now, note that the morphism $\kappa\colon Y(u+v) \to Y(u) \times Y(u)$ is induced from the multiplication map, because we have

$$Y(u) \times Y(v) = \mathrm{Proj}\left(\bigoplus_{n\geq 0} A_{nu} \otimes A_{nv}\right), \quad Y(u+v) = \mathrm{Proj}\left(\bigoplus_{n\geq 0} A_{nu+nv}\right),$$

Thus, the assertion follows from the basic fact that κ is a closed embedding if and only if there is an $l > 1$ such that μ_{ln} are surjective for any $n > 0$. $\qquad\square$

Proof of Theorem 4.3.2.2 If $u, v \in \omega(A) \cap K$ is a generating pair, then the algebra $A(u, v)$ is large in $A(u + v)$. Thus, the first assertion follows from Proposition 4.3.2.4. To see the second one, note that both u and $u + v$ lie in the relative interior λ° of the GIT cone $\lambda \in \Lambda(A)$. Thus, $Y(u+v) \to Y(u)$ is an isomorphism, and the statement follows from Proposition 4.3.2.5. $\qquad\square$

Proposition 4.3.2.6 *Let A be an integral $\mathbb{Z}_{\geq 0}$-graded $\mathbb{K}$-algebra. Assume that the homogeneous component A_1 is a nonzero finite-dimensional $\mathbb{K}$-vector space and that there are infinitely many $d > 0$ with $\mathbb{K}[A_1]_d = A_d$. Then A is finitely generated.*

Proof Consider the subalgebra $B := \mathbb{K}[A_1] \subseteq A$. We first show that the inclusion $\mathrm{Quot}(B) \subseteq \mathrm{Quot}(A)$ is in fact an equality. Because A_1 is of finite dimension and A is integral, we obtain that A_0 is also of finite dimension. Because A_0 is an integral algebra, we conclude $A_0 = \mathbb{K}$ and thus $B_0 = A_0$. Given $k > 1$, choose $d > k$ with $B_d = A_d$ and fix $0 \neq g \in B_{d-k}$. Then we have $A_k g \subseteq B$ and thus $A_k \subseteq \mathrm{Quot}(B)$. This implies $\mathrm{Quot}(B) = \mathrm{Quot}(A)$.

Next we show that A is integral over B. Let B' be the integral closure of B in its quotient field. Then B' is a finitely generated $\mathbb{K}$-algebra and we have the affine $\mathbb{K}^*$-varieties $X = \mathrm{Spec}\, B$ and $X' := \mathrm{Spec}\, B'$ and the equivariant normalization $X' \to X$. If X' is of dimension 1, then $X' = \mathbb{K}$ holds and we obtain $B = A = B'$, because $A_1 \neq 0$. So, let X be of dimension at least 2. Given $f \in A_k$, we regard f as a homogeneous rational function on X'. We verify $f \in B'$ by checking that f has nonnegative order along any invariant prime divisor D' on X'. The image $D \subseteq X$ of $D' \subseteq X'$ is an invariant irreducible hypersurface. Some $l > 0$ admits a $g \in B_l$ that does not vanish along D. Because g is a homogeneous polynomial in elements of B_1, we even find an $h \in B_1$ not vanishing along D. Take $d > k$ with $B_d = A_d$. Then we have $fh^{d-k} \in B$. This implies that f has nonnegative order along $D' \subseteq X'$.

Because B' is finitely generated as a $\mathbb{K}$-algebra, it is a finitely generated B-module. Because B is noetherian, the B-submodule $A \subseteq B'$ is also finitely generated. It follows that A is a finitely generated $\mathbb{K}$-algebra. $\qquad\square$

Proof of Theorem 4.3.2.3 The implication "(i)$\Rightarrow$(ii)" follows from Theorem 4.3.2.2. We prove "(ii)$\Rightarrow$(i)." Passing to a suitable Veronese subalgebra, we may assume that K is torsion free. Subdividing the generating fan Λ, we may assume that its cones are regular. Moreover, it suffices to show that for any maximal cone λ of the generating quasifan the algebra

$$A(\lambda) := \bigoplus_{w \in \lambda \cap K} A_w$$

is finitely generated. Thus, we may assume that Λ is the fan of faces of a single regular cone.

We claim that for any $w \in \lambda$, the Veronese subalgebra $A(w)$ is finitely generated. Because the pair w, w is generating, there exists an $r \in \mathbb{Z}_{>0}$ such that $A_{srw} \cdot A_{srw}$ equals A_{2srw} for all $s \in \mathbb{Z}_{\geq 0}$. This gives $A_{2^p rw} = A_{rw}^{2^p}$ for all $p > 0$, where we may assume that A_{rw} is nonzero. Proposition 4.3.2.6 yields that $A(rw)$ is finitely generated. It follows that $A(w)$ is finitely generated. As a consequence, we remark that there is an $m > 0$ such that for every $s \geq 0$, the Veronese subalgebra $A(smw)$ is generated by the component A_{smw}, see [58, Prop. III.3.3].

Let $u_1, \ldots, u_d$ be the primitive generators of λ. Moreover, for any $I \subseteq \{1, \ldots, d\}$ set $u_I := \sum_{i \in I} u_i$. The preceding observation says that passing once more to a suitable Veronese subalgebra of A, we may assume that every $A(u_I)$ is generated by A_{u_I}. Let $B \subseteq A$ denote the subalgebra generated by all A_{u_I}.

We claim that for every $w \in \lambda \cap K$, there exists an $m \in \mathbb{Z}_{>0}$ such that B_{smw} equals A_{smw} for all $s \in \mathbb{Z}_{\geq 0}$. Passing to a suitable multiple, we may assume that $w = \alpha_1 u_1 + \cdots + \alpha_d u_d$ holds with nonnegative integers α_i. We exemplarily consider the case that $\alpha_1 \leq \cdots \leq \alpha_d$ holds. Set

$$w_1 := \alpha_1 u_{\{1,\ldots,d\}}, \quad w_2 := (\alpha_2 - \alpha_1) u_{\{2,\ldots,d\}}, \quad \ldots \quad w_d := (\alpha_d - \alpha_{d-1}) u_{\{d\}}.$$

Then we have $w = w_1 + \cdots + w_d$. Because w_1, w_2 is a generating pair, we find an $m_1 > 0$ such that $A_{sm_1 w_1} A_{sm_1 w_2}$ equals $A_{sm_1(w_1+w_2)}$ for all $s \in \mathbb{Z}_{\geq 0}$. Similarly, because $w_1 + w_2, w_3$ is generating, there is an m_2 such that $A_{sm_1 m_2(w_1+w_2)} A_{sm_1 m_2 w_3}$ equals $A_{sm_1 m_2(w_1+w_2+w_3)}$ for all $s \in \mathbb{Z}_{\geq 0}$. Proceeding like this, we arrive at $m = m_1 \cdots m_d$ such that for all $s \in \mathbb{Z}_{\geq 0}$ we have

$$B_{smw} = B_{smw_1} B_{smw_2} \cdots B_{smw_d} = A_{smw_1} A_{smw_2} \cdots A_{smw_d} = A_{smw}.$$

Having verified the claim, we use it to see that B and A have the same quotient field. Let $f \in A_w$. Then we have $B_{mw} = A_{mw}$ for some $m > 0$. With any nonzero $g \in B_{(m-1)w}$, we have $fg \in B$ and thus f lies in the quotient field of B. Moreover, we obtain that A is integral over B. Given $f \in A_w$, choose $m > 0$ with $B_{mw} = A_{mw}$. Then $f^m \in B$ give the desired integral dependence. Both facts together imply that A is finitely generated. $\qquad\square$

4.3.3 Finite generation after Hu and Keel

We present the characterization of finite generation of the Cox ring given in [176]. Recall that we denote by $\mathrm{SAmple}(X) \subseteq \mathrm{Cl}_{\mathbb{Q}}(X)$ the cone of semiample divisor classes of a variety X, that is, classes having a basepoint free positive multiple and that a *small quasimodification* $X \dashrightarrow Y$ is a rational map that defines an isomorphism $U \to V$ of open subsets $U \subseteq X$ and $V \subseteq Y$ such that the complements $X \setminus U$ and $Y \setminus V$ are of codimension at least 2.

Theorem 4.3.3.1 *Let X be an irreducible normal complete variety with finitely generated divisor class group. Then the following statements are equivalent.*

(i) *The Cox ring $\mathcal{R}(X)$ is finitely generated.*
(ii) *There are small quasimodifications $\pi_i\colon X \dashrightarrow X_i$, where $i = 1, \ldots, r$, such that each semiample cone $\mathrm{SAmple}(X_i) \subseteq \mathrm{Cl}_{\mathbb{Q}}(X)$ is polyhedral and*

$$\mathrm{Mov}(X) = \pi_1^*(\mathrm{SAmple}(X_1)) \cup \cdots \cup \pi_r^*(\mathrm{SAmple}(X_r)).$$

Moreover, if one of these two statements holds, then the X_i from (ii) can be taken $\mathbb{Q}$-factorial and projective.

Lemma 4.3.3.2 *Let X be an irreducible normal complete variety with $\mathrm{Cl}(X)$ finitely generated. Then $\mathrm{Mov}(X)$ is of full dimension in the rational divisor class group $\mathrm{Cl}_{\mathbb{Q}}(X)$.*

Proof Using Chow's lemma and resolution of singularities, we obtain a birational morphism $\pi\colon X' \to X$ with a smooth projective variety X'. Let $D_1, \ldots, D_r \in \mathrm{WDiv}(X)$ be prime divisors generating $\mathrm{Cl}(X)$, and consider their proper transforms $D_1', \ldots, D_r' \in \mathrm{WDiv}(X')$. Moreover, let $E' \in \mathrm{WDiv}(X')$ be very ample such that all $E' + D_i'$ are also very ample, and denote by $E \in \mathrm{WDiv}(X)$ its pushforward. Then the classes E and $E + D_i$ generate a full-dimensional cone $\tau \subseteq \mathrm{Cl}_{\mathbb{Q}}(X)$ and, because E' and the $E' + D_i'$ are movable, we have $\tau \subseteq \mathrm{Mov}(X)$. $\qquad\square$

Lemma 4.3.3.3 *Let X be an irreducible normal complete variety with finitely generated divisor class group $\mathrm{Cl}(X)$. If $\mathrm{Mov}(X)$ is polyhedral then also $\mathrm{Eff}(X)$ is polyhedral.*

Proof In the situation of Proposition 3.3.2.3, set $w_i := \deg(f_i)$. Because $\mathrm{Mov}(X)$ is polyhedral, it has only finitely many facets and these are cut out by hyperplanes $H_1, \ldots, H_m$. Let H_k^+ denote the closed half space bounded by H_k that comprises $\mathrm{Mov}(X)$. We claim that for every k, there is at most one w_i with $w_i \notin H_k^+$. Otherwise, we have two $w_i, w_j \notin H_k^+$. Let $\sigma_k = \mathrm{Mov}(X) \cap H_k$ be the facet of $\mathrm{Mov}(X)$ cut out by H_k. Then the cones

$$\tau_i := \mathbb{Q}_{\geq 0} \cdot w_i + \sigma_k, \qquad\qquad \tau_j := \mathbb{Q}_{\geq 0} \cdot w_j + \sigma_k,$$

are of full dimension and their relative interiors intersect nontrivially. Consider any $w \in \tau_i^\circ \cap \tau_j^\circ$. Then, by the description of Mov(X) given in Proposition 3.3.2.3, we have $w \in \text{Mov}(X)$. On the other hand, we have $w \notin H_k^+$, a contradiction. This proves our claim, that is, each H_k^+ has at most the generator w_i in its complement. Thus, besides the generators of Mov(X), only finitely many w_i are needed to generate Eff(X). $\qquad\square$

Lemma 4.3.3.4 *Let X be an irreducible normal complete variety and $L \subseteq$ WDiv(X) a submonoid generated by finitely many semiample divisors. Then $\oplus_{D \in L} \Gamma(X, \mathcal{O}(D))$ is a finitely generated $\mathbb{K}$-algebra.*

Proof Take generators $D_1, \ldots, D_r$ for L, let $\mathcal{A}_i$ denote the sheaf of divisorial algebras associated with D_i, and consider the sheaf of symmetric algebras $\mathcal{A} := \text{Sym}(\mathcal{A}_1, \ldots, \mathcal{A}_r)$. Then $A = \Gamma(X, \mathcal{A})$ maps onto $\oplus_{D \in L} \Gamma(X, \mathcal{O}(D))$. Thus, it suffices to show that A is finitely generated.

We begin with the case $r = 1$. First let D_1 be a very ample Cartier divisor. Then A is $\mathbb{Z}$-graded with weight monoid $\mathbb{Z}_{\geq 0}$ and we have $S_0 = \mathbb{K}$. Moreover, $\widetilde{X} := \text{Spec}_X A$ is an irreducible normal quasiaffine variety and with a $\mathbb{K}^*$-action having only constant global functions of degree zero. Let $\widetilde{X} \subseteq \overline{X}$ be a $\mathbb{K}^*$-equivariant affine closure such that $\overline{X}$ has only constant global functions of degree zero. Then $\overline{X} \setminus \widetilde{X}$ consists of the (unique) fixed point. Normalizing $\overline{X}$, we achieve that $\widetilde{X}$ and $\overline{X}$ have the same functions. In particular $A = \Gamma(\widetilde{X}, \mathcal{O})$ is finitely generated. Now suppose that D_1 is semiample. Then some positive multiple D_1' of D_1 is a basepoint free Cartier divisor defining a contraction morphism $X \to Y$ with connected fibers and, after passing to a higher multiple, D_1' is the pullback of a very ample Cartier divisor E_1'. As just observed, the sheaf of divisorial algebras defined by E_1' has a finitely generated algebra B' of global sections. Because B' has the same global sections as the Veronese subalgebra $A' \subseteq A$ defined by D_1', Proposition 1.1.2.5 ensures that A is finitely generated.

Now let r be general. Passing to a suitable Veronese subalgebra, we may assume that the D_i are base point free Cartier divisors; use again Proposition 1.1.2.5. Note for $\widetilde{X} := \text{Spec}_X \mathcal{A}$, the canonical map $\widetilde{X} \to X$ is a rank r-vector bundle over X. For $u \in \mathbb{N}^r$, denote $|u| := u_1 + \cdots + u_r$ and set

$$\mathcal{B} := \bigoplus_{k \in \mathbb{N}} \mathcal{B}_k, \qquad \mathcal{B}_k := \bigoplus_{|u|=k} \mathcal{A}_u.$$

Consider the corresponding projective space bundle $X' := \text{Proj}_X(\mathcal{B})$ with its projection $\varphi : X' \to X$ and $\mathcal{L}' := \mathcal{O}_{X'}(1)$, which means that $\varphi_*(\mathcal{L}'^{\otimes k}) = \mathcal{B}_k$. Then we obtain an $\mathcal{O}_{X'}$-algebra and an associated variety:

$$\mathcal{A}' := \bigoplus_{k \in \mathbb{N}} \mathcal{L}'^{\otimes k}, \qquad \widetilde{X}' := \text{Spec}_{X'}(\mathcal{A}').$$

Note that $\widetilde{X}'$ is obtained from the rank r vector bundle $\widetilde{X}$ over X by blowing up the zero section $s_0 \colon X \hookrightarrow \widetilde{X}$. For the respective rings of global sections, we obtain

$$\bigoplus_{k \in \mathbb{N}} \Gamma(X', \mathcal{L}'^{\otimes k}) = \bigoplus_{k \in \mathbb{N}} \Gamma(X, \varphi_* \mathcal{L}'^{\otimes k}) = \bigoplus_{k \in \mathbb{N}} \Gamma(X, \mathcal{B}_k) = \bigoplus_{u \in \mathbb{N}^r} \Gamma(X, \mathcal{A}_u).$$

To reduce our problem to the case $r = 1$, we have to ascertain that $\mathcal{L}'$ is base-point free. Because $\pi(\widetilde{X}'_f) = X'_f$ holds for all homogeneous $f \in \Gamma(X', \mathcal{A}')$, it suffices to show that any given $x \in \widetilde{X}' \setminus X'$ admits such an f of degree $n \in \mathbb{N}_{>0}$ with $f(x) \neq 0$. But this is easy: Consider the canonical projections

$$\widetilde{X} \to \widetilde{X}_i := \mathrm{Spec}_X(\mathcal{A}_i), \quad \text{where } \mathcal{A}_i := \bigoplus_{m \in \mathbb{N}} \mathcal{O}(m D_i).$$

Because $\widetilde{X}' \setminus X'$ equals $\widetilde{X} \setminus X$, at least one of these maps does not send $x \in \widetilde{X} \setminus X$ to the zero section of $\widetilde{X}_i$. By semiampleness of D_i, there is a homogeneous $f \in \Gamma(X, \mathcal{A}_i)$ of nontrivial degree $m e_i$ with $f(x) \neq 0$. $\square$

Proof of Theorem 4.3.3.1 Set $K := \mathrm{Cl}(X)$ and $R := \mathcal{R}(X)$. Let $\mathfrak{F} = (f_1, \ldots, f_r)$ be a system of pairwise nonassociated homogeneous prime generators of R and set $w_i := \deg(f_i)$. By Proposition 1.6.1.6 and Construction 1.6.3.1, the group $H := \mathrm{Spec}\, \mathbb{K}[K]$ acts freely on an open subset $W \subseteq \overline{X}$ of $\overline{X} = \mathrm{Spec}\, R$ such that $\overline{X} \setminus W$ is of codimension at least 2 in $\overline{X}$. Proposition 3.2.2.2 then tells us that the corresponding K-grading of R is almost free. Moreover, by Lemma 4.3.3.2, the moving cone of X is of full dimension, and by Proposition 3.3.2.3, it is given as

$$\mathrm{Mov}(X) = \bigcap_{i=1}^{r} \mathrm{cone}(w_j; \ j \neq i).$$

Thus, we are in the setting of Remark 3.3.4.2. That means that $\mathrm{Mov}(X)$ is a union of full-dimensional GIT chambers $\lambda_1, \ldots, \lambda_r$, the relative interiors of which are contained in the relative interior of $\mathrm{Mov}(X)$ and the associated projective varieties $X_i := \widehat{X}_i /\!\!/ H$, where $\widehat{X}_i := \overline{X}^{ss}(\lambda_i)$, are $\mathbb{Q}$-factorial, have $\mathcal{R}(X)$ as their Cox ring and λ_i as their semiample cone.

   Moreover, if $q \colon \widehat{X} \to X$ and $q_i \colon \widehat{X}_i \to X_i$ denote the associated characteristic spaces, then the desired small quasimodifications $\pi_i \colon X \dashrightarrow X_i$ are obtained as follows. Let $X' \subseteq X$ and $X'_i \subseteq X_i$ be the respective sets of smooth points. Then, by Proposition 1.6.1.6, the sets $q^{-1}(X')$ and $q_i^{-1}(X'_i)$ have a small complement in $\overline{X}$ and thus we obtain open embeddings with a small complement

$$X \longleftarrow (q^{-1}(X') \cap q_i^{-1}(X'_i)) /\!\!/ H \longrightarrow X_i.$$

Now suppose that (ii) holds. Then $\mathrm{Mov}(X)$ is polyhedral and hence, by Lemma 4.3.3.3, also $\mathrm{Eff}(X)$ is polyhedral. Let $w_1, \ldots, w_d \in \mathrm{Eff}(X)$ be those primitive generators of extremal rays of $\mathrm{Eff}(X)$ that satisfy $\dim \Gamma(X, \mathcal{R}_{nw_i}) \leq 1$ for any $n \in \mathbb{Z}_{\geq 0}$ and fix $0 \neq f_i \in \mathcal{R}(X)_{n_i w_i}$ with n_i minimal. Then we have

$$\bigoplus_{n \in \mathbb{Z}_{\geq 0}} \mathcal{R}(X)_{nw_i} = \mathbb{K}[f_i].$$

Set $\lambda_i := \pi_i^*(\mathrm{SAmple}(X_i))$. Then, by Gordan's lemma and Lemma 4.3.3.4, we have another finitely generated subalgebra of the Cox ring, namely

$$\mathcal{S}(X) := \bigoplus_{w \in \mathrm{Mov}(X)} \mathcal{R}(X)_w = \sum_{i=1}^{r} \left(\bigoplus_{w \in \lambda_i} \mathcal{R}(X)_w \right).$$

We show that $\mathcal{R}(X)$ is generated by $\mathcal{S}(X)$ and the $f_i \in \mathcal{R}(X)_{n_i w_i}$. Consider any $0 \neq f \in \mathcal{R}(X)_w$ with $w \notin \mathrm{Mov}(X)$. Then, by Lemma 3.3.2.4, we have $f = f^{(1)} f_i$ for some $1 \leq i \leq d$ and some $f^{(1)} \in \mathcal{R}(X)$ homogeneous of degree $w(1) := w - n_i w_i$. If $w(1) \notin \mathrm{Mov}(X)$ holds, then we repeat this procedure with $f^{(1)}$ and obtain $f = f^{(2)} f_i f_j$ with $f^{(2)}$ homogeneous of degree $w(2)$. At some point, we must end with $w(n) = \deg(f^{(n)}) \in \mathrm{Mov}(X)$, because otherwise the sequence of the $w(n)$'s would leave the effective cone. $\square$

Theorem 4.3.3.5 *Let X be an irreducible normal complete surface with finitely generated divisor class group $\mathrm{Cl}(X)$. Then the following statements are equivalent.*

(i) *The Cox ring $\mathcal{R}(X)$ is finitely generated.*
(ii) *One has $\mathrm{Mov}(X) = \mathrm{SAmple}(X)$ and this cone is polyhedral.*

Moreover, if one of these two statements holds, then the surface X is $\mathbb{Q}$-factorial and projective.

Proof For "(i)$\Rightarrow$(ii)," we only have to show that the moving cone coincides with the semiample cone. Clearly, we have $\mathrm{SAmple}(X) \subseteq \mathrm{Mov}(X)$. Suppose that $\mathrm{SAmple}(X) \neq \mathrm{Mov}(X)$ holds. Then $\mathrm{Mov}(X)$ is properly subdivided into GIT chambers; see Remark 3.3.4.2. In particular, we find two chambers λ' and λ both intersecting the relative interior of $\mathrm{Mov}(X)$ such that λ' is a proper face of λ. The associated GIT quotients Y' and Y of the total coordinate space $\overline{X}$ have λ' and λ as their respective semiample cones. Moreover, the inclusion $\lambda' \subseteq \lambda$ gives rise to a proper morphism $Y \to Y'$, which is an isomorphism in codimension 1. As Y and Y' are normal surfaces, we obtain $Y \cong Y'$, which contradicts the fact that the semiample cones of Y and Y' are of different dimension.

The verification of "(ii)⇒(i)" runs as in the preceding proof; this time one uses the finitely generated subalgebra

$$S(X) := \bigoplus_{w \in \text{Mov}(X)} \mathcal{R}(X)_w = \bigoplus_{w \in \text{SAmple}(X)} \mathcal{R}(X)_w.$$

Moreover, by Theorem 4.3.3.1, there is a small quasimodification $X \dashrightarrow X'$ with X' projective and $\mathbb{Q}$-factorial. As X and X' are complete surfaces, this map already defines an isomorphism. □

In the case of a $\mathbb{Q}$-factorial surface X, we obtain the following simpler characterization involving the cone $\text{Nef}(X) \subseteq \text{Cl}_\mathbb{Q}(X)$ of *numerically effective* divisor classes, that is, the cone generated by the classes of the Weil divisors D satisfying $D \cdot C \geq 0$ for all irreducible curves $C \subseteq X$. Note that the implication "(ii)⇒(i)" was obtained for smooth surfaces in [146, Cor. 1].

Corollary 4.3.3.6 *Let X be a $\mathbb{Q}$-factorial projective surface with finitely generated divisor class group $\text{Cl}(X)$. Then the following statements are equivalent.*

(i) *The Cox ring $\mathcal{R}(X)$ is finitely generated.*
(ii) *The effective cone $\text{Eff}(X) \subseteq \text{Cl}_\mathbb{Q}(X)$ is polyhedral and $\text{Nef}(X) = \text{SAmple}(X)$ holds.*

Proof If (i) holds, then we infer from [43, Cor. 7.4] that the semiample cone and the nef cone of X coincide. Now suppose that (ii) holds. From

$$\text{SAmple}(X) \subseteq \text{Mov}(X) \subseteq \text{Nef}(X)$$

we then conclude $\text{Mov}(X) = \text{Nef}(X)$. Moreover, because $\text{Eff}(X)$ is polyhedral, $\text{Nef}(X)$ is given by a finite number of inequalities and hence is also polyhedral. Thus, we can apply Theorem 4.3.3.5. □

We conclude with a discussion of results from birational geometry showing in particular that smooth Fano varieties are Mori dream spaces. First we recall the necessary background on the singularities of the minimal model program; see [192, Sec. 2.3] for more.

Consider an irreducible complete variety X coming with a boundary divisor Δ, that is, we have $\Delta = \sum b_i D_i$ with prime divisors D_i and $b_i \in \mathbb{Q}_{\geq 0}$. Let K_X be a canonical divisor on X and assume that some positive multiple of $K_X + \Delta$ is Cartier. Then, for any proper birational morphism $\pi : \tilde{X} \to X$ with an irreducible normal variety $\tilde{X}$, there exists a (unique) canonical divisor $K_{\tilde{X}}$ on $\tilde{X}$ satisfying

$$K_{\tilde{X}} = \pi^*(K_X + \Delta) + \sum_E a(E, X, \Delta)E,$$

where E runs over the prime divisors of $\tilde{X}$ and the pullback of $K_X + \Delta$ is defined as $m^{-1}(\pi^*(mK_X + m\Delta))$ with a positive integer such that $mK_X + m\Delta$

is an (integral) Cartier divisor. Following [192, Def. 2.34] we say that the pair (X, Δ) is

(i) *log canonical* if for all $\pi : \widetilde{X} \to X$ as above and all prime divisors $E \subseteq \widetilde{X}$ one has $a(E, X, \Delta) \geq -1$,
(ii) *purely log terminal (plt)* if for all $\pi : \widetilde{X} \to X$ as above and all prime divisors $E \subseteq \widetilde{X}$ one has $a(E, X, \Delta) > -1$,
(iii) *Kawamata log terminal (klt)* if the coefficients of $\Delta = \sum b_i D_i$ all satisfy $b_i < 1$ and (X, Δ) is purely log terminal.

One says that X has at most *log canonical (log terminal) singularities* if the pair $(X, 0)$, that means X with the trivial boundary divisor, is log canonical (purely log terminal).

We say that X is of *Fano type* if X admits an effective rational boundary divisor Δ such that $-K_X + \Delta$ is ample and the pair (X, Δ) is Kawamata log terminal. Note that every Fano variety with at most log terminal singularities is of Fano type.

Theorem 4.3.3.7 ([148, Thm. 1.1]) *Let X be an irreducible $\mathbb{Q}$-factorial normal projective variety. Then the following statements are equivalent.*

(i) *The variety X is of Fano type.*
(ii) *The variety X is a Mori dream space and the total coordinate space $\overline{X}$ has at most log terminal singularities.*

This result implies in particular that $\mathbb{Q}$-factorial Fano varieties with at most log terminal singularities are Mori dream spaces, as obtained earlier in [53, Cor. 1.3.2]. Observe that a variety X of Fano type has a big anticanonical divisor. As we infer from [162, Example 1.8], the converse does not hold in general:

Example 4.3.3.8 Let $\pi : X \to \mathbb{P}^3$ be the blow-up of $\mathbb{P}^3$ at the nine points that are a complete intersection of two general cubic curves C_1 and C_2 contained in a plane $V \subseteq \mathbb{P}^3$. Then the anticanonical linear system of X consists of quartic surfaces singular at the nine points so that $|-K_X| = |Q| + 2V$, where $2V = \mathrm{Bs} |-K_X|$ and $|Q|$ is the pullback of the linear system of quadrics of $\mathbb{P}^3$. Thus $-K_X$ is big. Using the generality assumption on C_1 and C_2 one can show that V contains infinitely many (-1)-curves and that the class of any such curve spans an extremal ray of the Mori cone of X. Hence, by duality, the nef cone of X is not polyhedral so that X is not a Mori dream space and in particular it is not of Fano type.

It is shown in [217, Lemma 4.10] that a smooth projective variety X with finitely generated divisor class group and a big, movable $-K_X$ is of Fano type if and only if it is a Mori dream space.

We say that X is of *Calabi–Yau type* if X admits an effective boundary $\mathbb{Q}$-divisor Δ such that some positive mutiple of $K_X + \Delta$ is principal and the pair (X, Δ) is log canonical. Note that every variety with at most canonical singularities and trivial canonical divisor class is of Calabi–Yau type.

Theorem 4.3.3.9 ([183, Thm. 1.1]) *Let X be a Mori dream space. Then the following statements are equivalent.*

(i) *The variety X is of Calabi–Yau type.*
(ii) *The total coordinate space $\overline{X}$ has at most log canonical singularities.*

4.3.4 Cox–Nagata rings

The idea behind Nagata's counterexample [228] to Hilbert's fourteenth problem is to relate the Cox ring of a blown up projective space to the ring of invariants of a certain vector group action; this led to the name "Cox–Nagata rings"; see [284]. We first present Nagata's approach and then discuss applications to rings of invariants and further results in this direction.

Consider a sequence $a_1, \ldots, a_r \in \mathbb{K}^{n+1}$ of pairwise linearly independent vectors and let X be the blow-up of $\mathbb{P}^n$ at the corresponding points. Denote by $D_0 \subseteq X$ the proper transform of a hyperplane, by $D_1, \ldots, D_r \subseteq X$ the exceptional divisors, and by $K \subseteq \mathrm{WDiv}(X)$ the subgroup generated by all these divisors. Then K projects isomorphically onto $\mathrm{Cl}(X)$ and thus the Cox ring of X is of the form

$$\mathcal{R}(X) = \bigoplus_{D \in K} \Gamma(X, \mathcal{O}_X(D)).$$

To relate this Cox ring to the ring of invariants of a vector group action, consider the $(n + 1) \times r$ matrix A having the homogeneous coordinate vectors $a_1, \ldots, a_r$ as its columns. We assume that A has rank $n + 1$. Let $U \subseteq \mathbb{K}^r$ be the nullspace of A. Then U is an (additive) unipotent group and, with $u * z := (u_1 z_1, \ldots, u_r z_r)$, we have an action

$$U \times (\mathbb{K}^r \times \mathbb{K}^r) \to (\mathbb{K}^r \times \mathbb{K}^r), \qquad (u, (z, w)) \mapsto (z, w + u * z).$$

Write $\mathbb{K}[T, S]$ for the polynomial ring in the variables $T_1, \ldots, T_r$ and $S_1, \ldots, S_r$ corresponding to the coordinates z_i and w_i. Then $\mathbb{K}[T, S]$ becomes $\mathbb{Z}^r$-graded by setting $\deg(T_i) = \deg(S_i) = e_i$. Because the action of U respects this grading, the subalgebra of invariants $\mathbb{K}[T, S]^U$ is $\mathbb{Z}^r$-graded as well.

To formulate the result, we need to downgrade the Cox ring $\mathcal{R}(X)$ appropriately. Consider the linear map $Q \colon K \to \mathbb{Z}^r$ given by $D_0 \mapsto (1, \ldots, 1)$ and $D_i \mapsto e_i$. This gives us a $\mathbb{Z}^r$-grading of the Cox ring $\mathcal{R}(X)$, namely

$$\mathcal{R}(X) = \bigoplus_{v \in \mathbb{Z}^r} \mathcal{R}(X)_v, \qquad \mathcal{R}(X)_v := \bigoplus_{D \in Q^{-1}(v)} \Gamma(X, \mathcal{O}_X(D)).$$

Theorem 4.3.4.1 *In the above situation, $\mathcal{R}(X)$ and $\mathbb{K}[T, S]^U$ are isomorphic as $\mathbb{Z}^r$-graded $\mathbb{K}$-algebras.*

We follow Mukai's proof [223]. A first step is to represent the algebra of invariants $\mathbb{K}[T, S]^U$ as a kind of Rees algebra; more precisely, given an algebra $R \subseteq \mathbb{K}[T^\pm, S]$ of Laurent polynomials and ideals $I_1, \ldots, I_r \subseteq R$, we define the associated *extended multi-Rees algebra* to be the subalgebra

$$(R : I_1, \ldots, I_r) := \left(R + \sum_{d \in \mathbb{Z}^r_{\geq 0}} (I_1^{d_1} \cap \ldots \cap I_r^{d_r}) T^{-d} \right)[T] \subseteq R[T^{\pm 1}].$$

Let us define the algebra R and the ideals $I_1, \ldots, I_r$ in our concrete setting. First introduce Laurent polynomials $J_1, \ldots, J_{n+1} \in \mathbb{K}[T^\pm, S]$ as the components of the vector

$$J := T_0 \cdot A \cdot \frac{S}{T} \in \mathbb{K}[T^\pm, S]^{n+1}, \quad T_0 := T_1 \cdots T_r, \quad \frac{S}{T} := \left(\frac{S_1}{T_1}, \ldots, \frac{S_r}{T_r} \right),$$

where A is an $(n + 1) \times r$ matrix of rank $n + 1$; the condition on the columns of A posed at the beginning are not needed for the moment. Our algebra R will be

$$\mathbb{K}[J] := \mathbb{K}[J_1, \ldots, J_{n+1}] \subseteq \mathbb{K}[T^\pm, S].$$

To define the ideals, let $V \subseteq \mathbb{K}[J]$ be the linear span of the generators $J_1, \ldots, J_r$, denote by $V_i \subseteq V$ the vector subspace consisting of all Laurent polynomials not owning the monomial $T_0 T_i^{-1} S_i$, and let $I_i \subseteq \mathbb{K}[J]$ be the ideal generated by V_i.

Lemma 4.3.4.2 *In the above notation, the algebra $(\mathbb{K}[J] : I_1, \ldots, I_r)$ is contained in $\mathbb{K}[T, S]$ and we have $\mathbb{K}[T, S]^U = (\mathbb{K}[J] : I_1, \ldots, I_r)$.*

Proof By definition of the polynomials J_i and the ideals I_i, the monomial T^d divides any element of $I_1^{d_1} \cap \ldots \cap I_r^{d_r}$ and thus the algebra $(\mathbb{K}[J] : I_1, \ldots, I_r)$ indeed consists of polynomials in T and S. Moreover, because all $T_i^{\pm 1}$ and all J_i are U-invariants, we see that $(\mathbb{K}[J] : I_1, \ldots, I_r)$ is contained in $\mathbb{K}[T, S]^U$.

Before proceeding, consider briefly the restricted action of U on the open subset obtained by localizing with respect to T_0; it is is given as

$$U \times (\mathbb{T}^r \times \mathbb{K}^r) \to (\mathbb{T}^r \times \mathbb{K}^r), \quad (u, (z, w)) \mapsto (z, w + u * z).$$

Here we obtain right from the definitions that $\mathbb{K}[T^\pm, S]^U$ equals $\mathbb{K}[J][T^\pm]$. As a direct consequence, we obtain

$$\mathbb{K}[T, S]^U = \mathbb{K}[T^{\pm 1}, S]^U \cap \mathbb{K}[T, S] = \mathbb{K}[J][T^{\pm 1}] \cap \mathbb{K}[T, S].$$

We are ready to show that $\mathbb{K}[T, S]^U$ is contained in $(\mathbb{K}[J] : I_1, \ldots, I_r)$. Take a homogeneous element $h(T, S) \in \mathbb{K}[T, S]^U_v$ of degree $v \in \mathbb{Z}^r$. By the preceding consideration, we obtain a presentation

$$h(T, S) = \sum_{d \in \mathbb{Z}^r} f_d T^{-d}, \qquad \text{with } f_d \in \mathbb{K}[J] \text{ homogeneous of degree } v + d.$$

We claim that given any $d \in \mathbb{Z}_{\geq 0}^r$, all monomials $T_i^{d_i}$ divide f_d. Indeed, every generator J_k of the algebra $\mathbb{K}[J]$ has degree $(1, \ldots, 1)$, and thus $v + d = (m, \ldots, m)$, where m is the degree of f_d in S. Thus that different vectors d define different terms of the polynomial $h(T, S)$ in the variables S. So the $f_d T^{-d}$ are polynomials in T and S, and the claim follows.

It remains to show that $T_i^{d_i}$ divides $f(J)$ if and only if $f(J)$ belongs to $I_i^{d_i}$. If $a_i = 0$ then T_i divides all $J_1, \ldots, J_{n+1}$, $V_i = V$, but none of J_k is divisible by T_i^2. So, suppose $a_i \neq 0$. By renumbering we may assume that $a_{i1} \neq 0$. We define

$$Z_1 := J_1/a_{i1}, \quad Z_2 := J_2 - a_{i2}Z_1, \quad \ldots, \quad Z_{n+1} := J_{n+1} - a_{in+1}Z_1.$$

Then $f(J) = f(a_{i1}Z_1, a_{i2}Z_1 + Z_2, \ldots, a_{in+1}Z_1 + Z_{n+1})$ holds and I_i is the ideal generated by all linear forms in the Z_k vanishing at $(1, 0, \ldots, 0)$. Using this, we obtain that $f(J)$ belongs to $I_i^{d_i}$ if and only if

$$f(a_{i1}Z_1, a_{i2}Z_1 + Z_2, \ldots, a_{in+1}Z_1 + Z_{n+1}) \in \langle Z_2, \ldots, Z_{n+1} \rangle^{d_i}.$$

But T_i divides all Z_j with $j \neq 1$ and does not divide Z_1. So the latter condition is equivalent to $T_i^{d_i}$ divides $f(J)$. $\qquad \square$

Proof of Theorem 4.3.4.1 Given a divisor $D := m_0 D_0 + \cdots + m_r D_r \in K$, the corresponding homogeneous component of the Cox ring is

$$\mathcal{R}(X)_{[D]} = \Gamma(X, \mathcal{O}_X(m_0 D_0 + m_1 D_1 + \cdots + m_r D_r)).$$

It may be identified with the space of homogeneous polynomials of degree m_0 in $J_1, \ldots, J_{n+1}$ that have zeroes/poles of order $\geq -m_i$ at the point a_i for all i. The latter space coincides with

$$I_1^{-m_1} \cap \ldots \cap I_r^{-m_r} \subseteq \mathbb{K}[J],$$

where we set $I_i^b = \mathbb{K}[J]$ for all $b \leq 0$. Together with Lemma 4.3.4.2, these observations show that for $v = (m_0 + m_1, \ldots, m_0 + m_r)$ the homogeneous component $\mathbb{K}[T, S]_v^U$ equals $\mathcal{R}(X)_{[D]}$. The assertion follows. $\qquad \square$

Example 4.3.4.3 Assume that $r = n + 1$ and that $a_1, \ldots, a_r$ are linearly independent. Then the blow-up of $\mathbb{P}^n$ at $a_1, \ldots, a_r$ is a toric variety. In terms of Theorem 4.3.4.1 this means $U = \{0\}$ and the algebra of invariants is just $\mathbb{K}[T, S]$.

Example 4.3.4.4 Consider four points in the projective plane $\mathbb{P}^2$ lying in general position, for example, take the matrix

$$A = \begin{bmatrix} 1 & 0 & 0 & -1 \\ 0 & 1 & 0 & -1 \\ 0 & 0 & 1 & -1 \end{bmatrix}.$$

Then we have $U \cong (\mathbb{K}, +)$ and the action is given by $(T, S) \to (T, S - uT)$. The algebra of invariants is generated by

$$T_1, \ldots, T_4, \qquad Q_{12}, \ldots, Q_{34}, \text{ where } \quad Q_{jk} := S_j T_k - S_k T_j.$$

By direct computation, we see that the associated ideal of relations is the Plücker ideal of $G(2, 5)$; it is generated by the five quadrics

$$Q_{12}Q_{34} - Q_{13}Q_{24} + Q_{14}Q_{23}, \qquad T_i Q_{jk} - T_j Q_{ki} + T_k Q_{ij}.$$

Recall that we computed the Cox ring of the blow-up of $\mathbb{P}^2$ in the points a_1, a_2, a_3, a_4 also via toric ambient modifications in Example 4.1.4.1.

For his counterexample to Hilbert's fourteenth problem, Nagata [228] showed that the monoid of effective divisor classes of the blow-up of $\mathbb{P}^2$ at s^2 generic points is not polyhedral for any $s \geq 4$. Steinberg [282] gave a smaller example by taking nine points on a rational cubic curve of $\mathbb{P}^2$. For other approaches to nonfinitely generated rings of invariants see [1, 261].

Let us give a simple example using Nagata's principle. We say that nine points in $\mathbb{P}^2$ are in *very general position* if there is unique smooth cubic Γ through them and for some line $\ell \subseteq \mathbb{P}^2$ $3\ell|_\Gamma - p_1 - \cdots - p_9$ is not a torsion element in $\mathrm{Pic}^0(\Gamma)$.

Proposition 4.3.4.5 *Let X be the blow-up of $\mathbb{P}^2$ at nine points in a very general position. Then the Cox ring of X is not finitely generated.*

Proof Let Γ be the unique smooth cubic curve through the nine points $p_1, \ldots, p_9$ and let C be its strict transform. The blow-up morphism $\pi : X \to \mathbb{P}^2$ induces an isomorphism between C and Γ. Via this isomorphism the degree zero Cartier divisor $\pi_*(C|_C)$ is linearly equivalent to the divisor $3\ell - p_1 - \cdots - p_9$, where ℓ is a general line of $\mathbb{P}^2$. Because the points are in very general position, then $C|_C$ is nontorsion in $\mathrm{Pic}^0(C)$ so that $H^0(C, nC|_C)$ is trivial for any $n > 0$. Hence by the above and the exact sequence

$$0 \longrightarrow H^0(X, (n-1)C) \longrightarrow H^0(X, nC) \longrightarrow H^0(C, nC|_C)$$

we deduce, by induction on n, that $H^0(X, nC)$ is isomorphic to $H^0(X, \mathcal{O}_X)$ and thus it is one-dimensional for any $n > 0$. As a consequence the class of C spans an extremal ray of the effective cone $\mathrm{Eff}(X)$. Because $C^2 = 0$ we conclude that $\mathrm{Eff}(X)$ is not polyhedral; see Lemma 5.1.1.6. $\qquad\qquad\square$

Corollary 4.3.4.6 *Let A be a* 3×9 *matrix the columns of which define a very general point configuration in* $\mathbb{P}^2$. *Then the action of the nullspace* $U \cong \mathbb{K}^6$ *of A on* $\mathbb{K}^{18}$ *has a nonfinitely generated algebra of invariants.*

Mukai [223] showed that the blow-up of $\mathbb{P}^n$ in r general points has finitely generated Cox ring if and only if one has

$$\frac{1}{n+1} + \frac{1}{r-n-1} \geq \frac{1}{2}.$$

We conclude the section with two theorems of Castravet and Tevelev [86], generalizing in particular Mukai's result.

Theorem 4.3.4.7 *Let* $X_{a,b,c}$ *be the variety obtained by blowing up* $(\mathbb{P}^{c-1})^{a-1}$ *at* $b + c$ *points in general position. Then the following statements are equivalent.*

(i) *The Cox ring* $\mathcal{R}(X_{a,b,c})$ *is finitely generated.*
(ii) *The effective cone* $\mathrm{Eff}(X_{a,b,c})$ *is polyhedral.*
(iii) *One has* $\dfrac{1}{a} + \dfrac{1}{b} + \dfrac{1}{c} > 1$.

For $a = 2$ one obtains Mukai's results about the blow-up of the projective space at points in general position. For $a = 2$ and $c = 3$ one recognizes the result of Batyrev and Popov [33] about finite generation of the Cox ring of a smooth del Pezzo surface. Note that the Cox ring of $X_{s+1,1,n+1}$, the blow-up of $(\mathbb{P}^n)^s$ at $n + 2$ points is, as a ring, isomorphic to the homogeneous coordinate ring of the Grassmannian $G(s + 1, n + s + 2)$.

Theorem 4.3.4.8 *Let X be the blow-up of* $\mathbb{P}^n$ *at* $r \geq n + 3$ *distinct points lying on a rational normal curve of degree n. Then the Cox ring of X is generated by global sections of the exceptional divisors* $D_1, \ldots, D_r$ *and of the divisors*

$$D = kD_0 - k \sum_{i \in I} D_i - (k-1) \sum_{i \in I^c} D_i,$$

where D_0 *is the pullback of the hyperplane class, I varies among all the cardinality* $n + 2 - 2k$ *subsets of* $\{1, \ldots, r\}$, *and* $1 \leq k \leq 1 + n/2$.

The cases left out by the theorem are well known. For $r \leq n + 2$ the variety X is toric and its Cox ring is a polynomial ring. The case $r = n + 2$ is $X_{2,1,n+1}$ from before, the blow-up of $\mathbb{P}^n$ in $n + 2$ points with the (suitably graded) homogeneous coordinate ring of the Grassmannian $G(2, n + 3)$ as its Cox ring.

Sturmfels and Velasco determined in [283] the ideal of relations of $\mathcal{R}(X)$ for $r = n + 3$ and $n \leq 8$ and formulated a conjecture for higher n. Moreover in [284] Cox–Nagata rings are used to construct toric degenerations.

4.4 Varieties with torus action

4.4.1 The Cox ring of a variety with torus action

We describe the Cox ring of a variety with torus action following [169]. The ground field $\mathbb{K}$ is algebraically closed and of characteristic zero. First recall from Theorem 2.1.3.2 the case of a complete toric variety X. Its Cox ring is given in terms of the prime divisors $E_1, \ldots, E_r$ in the boundary $X \setminus T \cdot x_0$ of the open orbit:

$$\mathcal{R}(X) = \mathbb{K}[T_1, \ldots, T_r], \qquad \deg(T_k) = [E_k] \in \mathrm{Cl}(X).$$

Now let X be any irreducible normal variety with only constant invertible global functions and finitely generated divisor class group. Consider an effective algebraic torus action $T \times X \to X$, where $\dim(T)$ may be less than $\dim(X)$. For a point $x \in X$, denote by $T_x \subseteq T$ its isotropy group. The points with finite isotropy group form a nonempty T-invariant open subset

$$X_0 := \{x \in X; \ T_x \text{ is finite}\} \subseteq X.$$

This set will replace the open orbit of a toric variety. Let E_k, where $1 \le k \le m$, denote the prime divisors in $X \setminus X_0$; note that each E_k is T-invariant with infinite generic isotropy, that is, the subgroup of T acting trivially on E_k is infinite. According to a result of Sumihiro [285, Cor. 3], there is a geometric quotient $q \colon X_0 \to X_0/T$ with an irreducible normal but possibly nonseparated orbit space X_0/T.

Example 4.4.1.1 Consider $\mathbb{P}^1 \times \mathbb{P}^1$ and the $\mathbb{K}^*$-action given in inhomogeneous coordinates by $t \cdot (z, w) = (z, tw)$. Let X be the $\mathbb{K}^*$-equivariant blow-up of $\mathbb{P}^1 \times \mathbb{P}^1$ at the fixed points $(0, 0)$, $(1, 0)$, and $(\infty, 0)$.

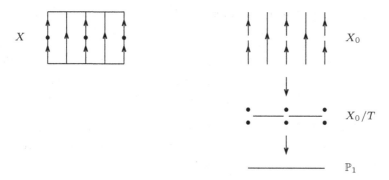

The open set X_0 is obtained by removing the two fixed point curves and the three isolated fixed points. The quotient X_0/T is the nonseparated projective line with the points $0, 1, \infty$ doubled; note that there is a canonical map $X_0/T \to \mathbb{P}^1$.

Besides the divisors E_k, we fix T-invariant prime divisors $D_1, \ldots, D_n \subseteq X$ such that each D_i has finite generic isotropy of order $l_i \geq 1$ and any T-invariant prime divisor $D \subseteq X$ with nontrivial finite isotropy occurs among the D_i. Note that each D_i intersects X_0. In the toric case, there are no D_i with $l_i > 1$. For any prime divisor D, we denote by 1_D the corresponding canonical section in the Cox ring.

Theorem 4.4.1.2 *Let X be an irreducible normal variety with only constant invertible global functions and finitely generated divisor class group. Assume that $T \times X \to X$ is an effective algebraic torus action.*

(i) *The orbit space X_0/T is an irreducible normal prevariety with only constant invertible global functions and finitely generated divisor class group.*

(ii) *There is a graded injection $q^*\colon \mathcal{R}(X_0/T) \to \mathcal{R}(X)$ of Cox rings and the assignments $T_i \mapsto 1_{D_i}$ and $S_k \mapsto 1_{E_k}$ induce an isomorphism*

$$\mathcal{R}(X) \cong \mathcal{R}(X_0/T)[T_i, S_k] \,/\, \langle T_i^{l_i} - 1_{q(D_i)};\ 1 \leq i \leq n \rangle$$

of $\mathrm{Cl}(X)$-graded rings, where $\mathrm{Cl}(X)$-grading on the right-hand side associates with each T_i the class of D_i and to each S_k the class of E_k.

As we will see in Proposition 4.4.3.2, one always finds a *separation* for the orbit space in our setting: there exist a rational map $\pi\colon X_0/T \dashrightarrow Y$ to a variety Y, an open subset $W \subseteq X_0/T$, and prime divisors $C_0, \ldots, C_r$ on Y such that the following hold:

- The complement of W in X_0/T is of codimension at least 2 and the restriction $\pi\colon W \to Y$ is a surjective local isomorphism.
- Each inverse image $\pi^{-1}(C_i)$ is a disjoint union of prime divisors C_{ij}, where $1 \leq j \leq n_i$.
- The map π is an isomorphism over $Y \setminus (C_0 \cup \ldots \cup C_r)$ and all prime divisors of X_0 with nontrivial generic isotropy occur among the $D_{ij} := \overline{q^{-1}(C_{ij})}$.

Let $l_{ij} \in \mathbb{Z}_{\geq 1}$ be the order of the generic isotropy group of the T-action on the prime divisor D_{ij} and define monomials $T_i^{l_i} := T_{i1}^{l_{i1}} \cdots T_{in_i}^{l_{in_i}}$ in the variables T_{ij}. Moreover, let 1_{E_k} and $1_{D_{ij}}$ denote the canonical sections of E_k and D_{ij}.

Theorem 4.4.1.3 *Let X be an irreducible normal variety with only constant invertible global functions and finitely generated divisor class group. Assume that $T \times X \to X$ is an effective algebraic torus action.*

(i) *The separation Y of X_0/T is an irreducible normal variety with only constant invertible global functions and finitely generated divisor class group.*

(ii) *There is a graded injection* $\mathcal{R}(Y) \to \mathcal{R}(X)$ *of Cox rings and* $T_{ij} \mapsto 1_{D_{ij}}$, $S_k \mapsto 1_{E_k}$ *defines an isomorphism of* $\mathrm{Cl}(X)$*-graded rings*

$$\mathcal{R}(X) \cong \mathcal{R}(Y)[T_{ij}, S_k] / \langle T_i^{l_i} - 1_{C_i}; \ 0 \le i \le r \rangle,$$

where the $\mathrm{Cl}(X)$*-grading on the right-hand side is defined by associating to* T_{ij} *the class of* D_{ij} *and to* S_k *the class of* E_k.

As a direct consequence, we obtain that finite generation of the Cox ring of X is equivalent to finite generation of the Cox ring of X_0/T or that of Y.

Corollary 4.4.1.4 *Let X be an irreducible normal variety with only constant invertible global functions, finitely generated divisor class group, and an effective torus action $T \times X \to X$. Then the following statements are equivalent.*

(i) *The Cox ring $\mathcal{R}(X)$ is finitely generated.*
(ii) *The Cox ring $\mathcal{R}(X_0/T)$ of the orbit space X_0/T is finitely generated.*
(iii) *The Cox ring $\mathcal{R}(Y)$ of the separation Y of X_0/T is finitely generated.*

Proof Clearly, if one of the algebras $\mathcal{R}(X_0/T)$ and $\mathcal{R}(Y)$ is finitely generated, then also $\mathcal{R}(X)$ is finitely generated. So, let $\mathcal{R}(X)$ be finitely generated. Because $\mathcal{R}(X)$ is a finite module over $\mathcal{R}(X_0/T)[S_k]$, we conclude that $\mathcal{R}(X_0/T)$ is finitely generated. To see that $\mathcal{R}(Y)$ is finitely generated, consider the abelian group

$$K := \mathbb{Z}^m \oplus \bigoplus_{i=0}^{r} \mathbb{Z}^{n_i}/\mathbb{Z} \cdot (l_{i1}, \ldots, l_{in_i})$$

and the canonical projection $Q \colon \mathbb{Z}^m \oplus \mathbb{Z}^n \to K$, where $n = n_0 + \cdots + n_r$. Then we obtain a K-grading on $\mathcal{R}(X)$ given by setting $\deg(g) = 0$ for all $g \in \mathcal{R}(Y)$ and

$$\deg(T_{ij}) := Q(e_{ij}), \qquad \deg(S_k) := Q(e_k),$$

where $e_{ij} \in \mathbb{Z}^n$ and $e_k \in \mathbb{Z}^m$ are the canonical basis vectors. Then $\mathcal{R}(Y) \subseteq \mathcal{R}(X)$ consists precisely of the K-homogeneous elements of degree zero and thus is finitely generated by Corollary 1.1.2.6. □

We specialize to the case that the T-action on X is of complexity 1, that is, its general orbits are of codimension 1 in X. Then the orbit space X_0/T is of dimension 1 and smooth.

Remark 4.4.1.5 Let X be an irreducible normal variety with $\Gamma(X, \mathcal{O}) = \mathbb{K}$ and $T \times X \to X$ a torus action of complexity 1. Then the following statements are equivalent.

(i) $\mathrm{Cl}(X)$ is finitely generated.
(ii) X is rational.

Moreover, if one of these statements holds, then the separation of the orbit space is a morphism $\pi : X_0/T \to \mathbb{P}^1$.

The former prime divisors $C_i \subseteq Y$ are now points $a_0, \ldots, a_r \in \mathbb{P}^1$ and $\pi^{-1}(a_i)$ consists of points $x_{i1}, \ldots, x_{in_i} \in X_0/T$. As before, we may assume that all prime divisors of X with nontrivial generic isotropy occur among the prime divisors $D_{ij} = \overline{q^{-1}(x_{ij})}$.

Theorem 4.4.1.6 *Let X be an irreducible normal variety with $\Gamma(X, \mathcal{O}) = \mathbb{K}$, finitely generated divisor class group, and an effective algebraic torus action $T \times X \to X$ of complexity 1.*

(i) *The $\mathrm{Cl}(X)$-graded ring $\mathcal{R}(X)$ is isomorphic to a K-graded algebra $R(A, P)$ as constructed in Construction 3.4.3.1, where $A = [a_0, \ldots, a_r]$ and the L_0-block of P is built from the l_{ij}.*

(ii) *If X is A_2-maximal, for example, projective, then X is isomorphic as a T-variety to one of the varieties $X(A, P, \Phi)$ as constructed in 3.4.3.6.*

Let us discuss some applications of Theorem 4.4.1.6. In a first application we compute the Cox ring of a surface obtained by repeatedly blowing up points of the projective plane that lie on a given line; the case $n_0 = \cdots = n_r = 1$ was done by other methods in [239].

Example 4.4.1.7 (Blowing up points on a line)   Consider a line $Y \subseteq \mathbb{P}^2$ and points $p_0, \ldots, p_r \in Y$. Let X be the surface obtained by blowing up n_i times the point p_i, where $0 \le i \le r$; in every step, we identify Y with its proper transform and p_i with the point in the intersection of Y with the exceptional curve. Set

$$g_i := (b_j c_k - c_j b_k)T_i + (b_k c_i - c_k b_i)T_j + (b_i c_j - c_i b_j)T_k,$$

with $p_i = [b_i, c_i]$ in $Y = \mathbb{P}^1$, the monomials $T_i = T_{i0} \cdots T_{in_i}$, and the indices $k = i + 2$, $j = i + 1$. Then the Cox ring of the surface X is given as

$$\mathcal{R}(X) = \mathbb{K}[T_{ij}, S; \ 0 \le i \le r, \ 0 \le j \le n_i] \, / \, \langle g_i; \ 0 \le i \le r - 2 \rangle.$$

We verify this using a $\mathbb{K}^*$-action; note that blowing up $\mathbb{K}^*$-fixed points always can be made equivariant. With respect to suitable homogeneous coordinates z_0, z_1, z_2, we have $Y = V(z_0)$. Let $\mathbb{K}^*$ act via

$$t \cdot [z] := [z_0, tz_1, tz_2].$$

Then $E_1 := Y$ is a fixed point curve, the jth (equivariant) blow-up of p_i produces an invariant exceptional divisor D_{ij} with a free $\mathbb{K}^*$-orbit inside and Theorem 4.4.1.6 gives the claim.

As a further application we indicate a general recipe for computing the Cox ring of a rational hypersurface given by a trinomial equation in the projective space. We perform this for the E_6 cubic surface in $\mathbb{P}^3$; the Cox ring of the resolution of this surface has been computed in [162]; compare also the example series started in Example 3.4.3.3.

Example 4.4.1.8 (The E_6 cubic surface)   There is a cubic surface X in the projective space having singular locus $X^{\text{sing}} = \{x_0\}$ and x_0 of type E_6. The surface is unique up to projective linear transformation and can be realized as follows:

$$X = V(z_1 z_2^2 + z_2 z_0^2 + z_3^3) \subseteq \mathbb{P}^3.$$

Note that the defining equation is a trinomial but not of the shape of those occurring in Theorem 4.4.1.6. However, any trinomial hypersurface in a projective space comes with a complexity 1 torus action. Here, we have the $\mathbb{K}^*$-action

$$t \cdot [z_0, \ldots, z_3] = [z_0, t^{-3} z_1, t^3 z_2, t z_3].$$

This allows us to use Theorem 4.4.1.6 for computing the Cox ring. The task is to find the divisors E_k, D_{ij} and the orders l_{ij} of the isotropy groups. Note that $\mathbb{K}^*$ acts freely on the open toric orbit of $\mathbb{P}^3$. The intersections of X with the toric prime divisors $V(z_i) \subseteq \mathbb{P}^3$ are given as

$$X \cap V(z_0) = V(z_0, z_1 z_2^2 + z_3^3), \qquad X \cap V(z_1) = V(z_1, z_2 z_0^2 + z_3^3),$$

$$X \cap V(z_3) = V(z_3, z_2(z_1 z_2 + z_0^2)) = (X \cap V(z_2)) \cup (X \cap V(z_1 z_2 + z_0^2)).$$

The first two sets are irreducible and both of them intersect the big torus orbit of the respective toric prime divisors $V(z_0)$ and $V(z_1)$. To achieve this also for $V(z_2)$, $V(z_3)$, we use a suitable weighted blow-up of $\mathbb{P}^3$ at $V(z_2) \cap V(z_3)$. In terms of fans, this means to perform a certain stellar subdivision. Consider the matrices

$$P = \begin{bmatrix} -1 & 1 & 0 & 0 \\ -1 & 0 & 1 & 0 \\ -1 & 0 & 0 & 1 \end{bmatrix}, \qquad P' = \begin{bmatrix} -1 & 1 & 0 & 0 & 0 \\ -1 & 0 & 1 & 0 & 3 \\ -1 & 0 & 0 & 1 & 1 \end{bmatrix}.$$

The columns $v_0, \ldots, v_3$ of P are the primitive generators of the fan Σ of $Z := \mathbb{P}^3$ and we obtain a fan Σ' subdividing Σ at the last column v_4 of P'; note that v_4 is located on the tropical variety $\text{trop}(X)$. Consider the associated toric morphism and the proper transform

$$\pi : Z' \to Z, \qquad X' := \overline{\pi^{-1}(X \cap \mathbb{T}^3)} \subseteq Z'.$$

A simple computation shows that the intersection of X' with the toric prime divisors of Z' is irreducible and intersects their big orbits. Moreover, $\pi : X' \to X$ is an isomorphism, because along X' nothing gets contracted. To proceed, note that we have no divisors of type E_k and that there is a commutative diagram

$$\begin{array}{ccc} X_0' & \longrightarrow & Z_0' \\ {\scriptstyle /\mathbb{K}^*}\downarrow & & \downarrow{\scriptstyle /\mathbb{K}^*} \\ X_0'/\mathbb{K}^* & \longrightarrow & Z_0'/\mathbb{K}^* \end{array}$$

We determine the quotient $Z_0' \to Z_0'/\mathbb{K}^*$. The group $\mathbb{K}^*$ acts on Z' via homomorphism $\lambda_v \colon \mathbb{K}^* \to \mathbb{T}^3$ corresponding to $v = (-3, 3, 1) \in \mathbb{Z}^3$. The quotient by this action is the toric morphism given by any map $S \colon \mathbb{Z}^3 \to \mathbb{Z}^2$ having $\mathbb{Z} \cdot v$ as its kernel. We take S as follows and compute the images of the columns $v_0, \ldots, v_4$ of P':

$$S := \begin{bmatrix} 1 & 0 & 3 \\ 0 & 1 & -3 \end{bmatrix}, \qquad S \cdot P' = \begin{bmatrix} -4 & 1 & 0 & 3 & 3 \\ 2 & 0 & 1 & -3 & 0 \end{bmatrix}.$$

This shows that the toric divisors D_Z^1 and D_Z^4 corresponding to v_1 and v_4 are mapped to a doubled divisor in the nonseparated quotient; see also [3]. The generic isotropy group of the $\mathbb{K}^*$-action along the toric divisor D_Z^i is given as the gcd l_i of the entries of the ith column of $S \cdot P'$. We obtain

$$l_0 = 2, \qquad l_1 = 1, \qquad l_2 = 1, \qquad l_3 = 3, \qquad l_4 = 3.$$

By construction, the divisors $D_X^i := D_Z^i \cap X_0$ of the embedded variety $X_0' \subseteq Z_0'$ inherit the orders l_i of the isotropy groups and the behavior with respect to the quotient map $X_0' \to X_0'/\mathbb{K}^*$. Renaming these divisors by

$$D_{01} := D_X^1, \qquad D_{02} := D_X^4, \qquad D_{11} := D_X^3, \qquad D_{21} := D_X^0,$$

we arrive in the setting of Theorem 4.4.1.6; because D_X^2 has trivial isotropy group and does not get multiplied by the quotient map, it does not occur here. The Cox ring of $X \cong X'$ is then given as

$$\mathcal{R}(X) \cong \mathbb{K}[T_{01}, T_{02}, T_{11}, T_{21}]/\langle T_{01} T_{02}^3 + T_{11}^3 + T_{21}^2 \rangle.$$

4.4.2 *H*-factorial quasiaffine varieties

We consider an irreducible normal quasiaffine variety X with an effective action of a quasitorus H. Provided that every invertible homogeneous function on X is constant and X is H-factorial, we relate the algebra $\Gamma(X, \mathcal{O})$ to the Cox ring of a maximal geometric quotient. We work over an algebraically closed field $\mathbb{K}$ of characteristic zero. The ground field $\mathbb{K}$ is algebraically closed and of characteristic zero. The original reference for this section is [169].

Denote by $E_1, \ldots, E_m$ the H-prime divisors in X having infinite generic isotropy, that is, the subgroup of H leaving E_i pointwise fixed is infinite, and let $D_1, \ldots, D_n \subseteq X$ be a collection of H-prime divisors with a finite generic isotropy group of order $l_i \geq 1$ such that all H-prime divisors with nontrivial finite generic isotropy occur among the D_i. Finally, define an H-invariant open set

$$X_0 := \{x \in X; \ H_x \text{ is finite}\} \subseteq X.$$

Theorem 4.4.2.1 *In the above setting assume that X is H-factorial and every invertible homogeneous function on X is constant. Let f_i, g_k be H-homogeneous functions on X with divisors D_i, E_k respectively. Then there is a unique subgroup $H' \subseteq H$ such that in the commutative diagram*

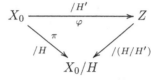

the map $Z \to X_0/H$ is a characteristic space for X_0/H. Moreover, for every regular function $h \in \Gamma(\mathcal{O}, Z)$ the pullback $\varphi^*(h) \in \Gamma(X_0, \mathcal{O})$ extends regularly to X and we have an isomophism of $\mathbb{X}(H)$-graded $\mathbb{K}$-algebras

$$\mathcal{R}(X_0/H)[T_1, \ldots, T_n, S_1, \ldots, S_m]/\langle T_i^{l_i} - f_i^{l_i}; \ 1 \leq i \leq n \rangle \ \to \ \Gamma(X, \mathcal{O}),$$

where $\mathcal{R}(X_0/H) = \Gamma(Z, \mathcal{O})$ is graded by $\mathbb{X}(H/H') \subseteq \mathbb{X}(H)$; with the variables T_i, S_k we associate the degrees of f_i, g_k in $\mathbb{X}(H)$, any $h \in \mathcal{R}(X_0/H)$ is sent to $\varphi^*(h)$, and T_i, S_k are sent to f_i, g_k.

If the H-action is of complexity 1 and X admits only constant global H-invariant functions, then X_0/H turns out to be a possibly nonseparated rational curve lying over $\mathbb{P}^1$. The Cox ring of X_0/T was computed earlier, which finally leads to the following.

Theorem 4.4.2.2 *Let $H \times X \to X$ be an effective action of a quasitorus H on an irreducible normal quasiaffine variety X such that $\dim(H) = \dim(X) - 1$ holds. Assume that all invariant as well as all invertible homogeneous functions on X are constant and that X is H-factorial.*

(i) *There is a separation $X_0/H \to \mathbb{P}^1$ and points $a_0, \ldots, a_r \in \mathbb{P}^1$ such that for any $a \in \mathbb{P}^1$ different from the a_i, the inverse image of a in X_0 is a free H-orbit.*

(ii) *The $\mathbb{X}(H)$-graded algebra $\Gamma(X, \mathcal{O})$ is isomorphic to a K_0-graded algebra $R(A, P_0)$ as constructed in Theorem 3.4.2.1, where $A = [a_0, \ldots, a_r]$ and the entries l_{ij} of P_0 are the orders of the generic isotropy groups of the H-prime divisors D_{ij} lying over a_i.*

Note that this result implies in particular finite generation of $\Gamma(X, \mathcal{O})$ in the case of complexity 1.

Corollary 4.4.2.3 *Let K_0 be a finitely generated abelian group and R a finitely generated integral $\mathbb{K}$-algebra with an effective pointed factorial K_0-grading. Then R is isomorphic to a K_0-graded algebra $R(A, P_0)$ as constructed in Theorem 3.4.2.1.*

We come to the proofs of the theorems. In the first lemma, X might even be any irreducible normal prevariety, but starting from the second one, X is quasiaffine.

Lemma 4.4.2.4 *If there is an H-fixed point $x \in X$, then every H-homogeneous function $f \in \Gamma(X, \mathcal{O})$ with $f(x) \neq 0$ is H-invariant.*

Proof Let $\chi \in \mathbb{X}(H)$ be the weight of $f \in \Gamma(X, \mathcal{O})$. Then, for every $h \in H$, we have $f(x) = f(h \cdot x) = \chi(h)f(x)$. This implies $\chi(h) = 1$ for all $h \in H$. $\square$

Lemma 4.4.2.5 *Let $B_1, \ldots, B_m \subseteq X$ be H-prime divisors; suppose that $B_i = \operatorname{div}(f_i)$ holds with H-homogeneous $f_i \in \Gamma(X, \mathcal{O})$ of degree $\chi_i \in \mathbb{X}(H)$, let $H_i \subseteq H$ be the generic isotropy group of B_i, and set $H_0 := H_1 \cdots H_m \subseteq H$.*

(i) *The restriction of χ_i to H_i generates the character group $\mathbb{X}(H_i)$.*
(ii) *For any two i, j with $j \neq i$, the function f_i is H_j-invariant.*
(iii) *The group H_0 is isomorphic to the direct product of the $H_i \subseteq H$.*
(iv) *$\Gamma(X, \mathcal{O})$ is generated by $f_1, \ldots, f_m$ and the H_0-invariant functions of X.*

Proof Choose H-invariant affine open subsets $U_i \subseteq X$ such that $A_i := U_i \cap B_i$ is nonempty and $U_i \cap B_j$ is empty for every $j \neq i$.

To prove (i), let $\xi_i \in \mathbb{X}(H_i)$ be given. Then ξ_i is the restriction of some $\eta_i \in \mathbb{X}(H)$. Let $V_i \subseteq U_i$ be an H-invariant affine open subset on which H acts freely, and choose a nontrivial H-homogeneous function g_i of weight η_i on V_i. Suitably shrinking U_i, we achieve that g_i is regular without zeroes on $U_i \setminus A_i$. Then, on U_i, the divisor $\operatorname{div}(h_i)$ is a multiple of the H-prime divisor $A_i = \operatorname{div}(f_i)$ and hence $g_i = a_i f_i^k$ holds with an H-homogeneous invertible function a_i on U_i and some $k \in \mathbb{Z}$. By Lemma 4.4.2.4, the function a_i is H_i-invariant. We conclude $\eta_i = k\chi_i$ on H_i.

Assertion (ii) is clear by Lemma 4.4.2.4. To obtain (iii), it suffices to show that χ_i is trivial on H_j for any two i, j with $j \neq i$. But, according to (ii), we have $f_i = \chi_i(h)f_i$ for every $h \in H_j$, which gives the claim. Finally, we verify (iv). Given an H-homogeneous function $f \in \Gamma(X, \mathcal{O})$, we may write $f = f' f_1^{\nu_1} \cdots f_m^{\nu_m}$ with $\nu_i \in \mathbb{Z}_{\geq 0}$ and a regular function f' on X, which is homogeneous with respect to some weight $\chi' \in \mathbb{X}(H)$ and has order zero along each H-prime divisor B_i. By Lemma 4.4.2.4, the function f' is invariant under every H_i and thus under H_0. $\square$

As before, let $E_1, \ldots, E_m \subseteq X$ denote the H-prime divisors of X with infinite generic isotropy group. Note that the E_k are precisely the H-prime divisors contained in the open H-invariant set

$$X_0 = \{x \in X;\ H_x \text{ is finite}\} \subseteq X.$$

Proposition 4.4.2.6 *In the above setting, assume that $E_k = \mathrm{div}(g_k)$ holds with H-homogeneous functions $g_k \in \Gamma(X, \mathcal{O})$, let $H_k \subseteq H$ be the generic isotropy group of E_k, and set $H_0 := H_1 \cdots H_m \subseteq H$.*

(i) *Each H_k is a one-dimensional torus and each E_k is a prime divisor. Moreover, there is a nonempty open subset $X' \subseteq X_0$ such that for any $x \in X'$ and any k, the orbit closure $\overline{H_k \cdot x} \subseteq X$ intersects E_k.*

(ii) *The H_0-action on X_0 is free, admits a geometric quotient $\lambda \colon X_0 \to Y$, and the isotropy groups of the induced action of $G := H/H_0$ on Y satisfy $G_{\lambda(x)} \cong H_x$ for every $x \in X_0$.*

(iii) *Y is irreducible normal, quasiaffine, and moreover, if X is H-factorial (admits only constant invertible (H-homogeneous) global functions), then Y is G-factorial (admits only constant invertible (G-homogeneous) global functions).*

(iv) *Every H_0-invariant rational function of X has neither poles nor zeroes outside X_0. Moreover, there is an isomorphism*

$$\Gamma(X_0, \mathcal{O})^{H_0}[S_1, \ldots, S_m] \to \Gamma(X, \mathcal{O}), \qquad S_k \mapsto g_k.$$

Proof We prove (i). By Lemma 4.4.2.5 (i), every H_k is a one-dimensional torus. To proceed, take any H_0-equivariant affine closure $X \subseteq \overline{X}$ and consider the quotient $\lambda_k \colon \overline{X} \to \overline{X} /\!\!/ H_k$. It maps the fixed point set of the H_k-action isomorphically onto its image in the quotient space $\overline{X} /\!\!/ H_k$. Because $\overline{X} /\!\!/ H_k$ is irreducible and of dimension at most $\dim(\overline{X}) - 1$, we obtain $\lambda_k(\overline{E}_k) = \overline{X} /\!\!/ H_k$ for the closure $\overline{E}_k$ of E_k in $\overline{X}$. It follows that $\overline{E}_k$ is irreducible, equals the whole fixed point set of H_k in $\overline{X}$, and any H_k-orbit of $\overline{X}$ has a point of $\overline{E}_k$ in its closure.

We turn to (ii). Because none of the g_k has a zero inside X_0, we infer from Lemma 4.4.2.5 that H_0 acts freely on X_0. In particular, the action of H_0 on X_0 admits a geometric quotient $\lambda \colon X_0 \to Y$ with an irreducible normal prevariety Y. The statement on the isotropy groups of the G-action on Y is obvious.

We prove the statements made in (iii) and (iv). Denoting by $\mathbb{T}^m$ the standard m-torus $(\mathbb{K}^*)^m$, we have a well-defined morphism of irreducible normal prevarieties

$$\varphi \colon X_0 \to Y \times \mathbb{T}^m, \qquad x \mapsto (\lambda(x), f_1(x), \ldots, f_m(x)).$$

According to Lemma 4.4.2.5, the weight χ_k of g_k generates the character group of H_k for $k = 1, \ldots, m$. Using this and the fact that H_0 is the direct product of $H_1, \ldots, H_m$, we conclude that φ is bijective and thus an isomorphism. In particular, we conclude that Y is quasiaffine, because U and hence X_0 is so.

Now, endow Y with the induced action of $G = H/H_0$ and $\mathbb{T}^m$ with the diagonal H_0-action given by the weights $\chi_1, \ldots, \chi_m$ of $g_1, \ldots, g_m$. Then φ becomes H-equivariant, where H acts via the splitting $H = G \times H_0$ on $Y \times \mathbb{T}^m$. Using this, we see that H-factoriality of X_0 implies G-factoriality of Y.

We show now that every H_0-invariant rational function $f \in \mathbb{K}(X_0)$ has neither zeroes nor poles outside X_0. The prime divisors inside $X \setminus X_0$ are precisely the E_k which in turn are the fixed point sets of the H_k-actions on X. Because the general orbit $H_0 \cdot x \subseteq X$ has a point $x_k \in B_k$ in its closure, we see that f has neither poles nor zeroes along the B_k. In particular, if f is regular on X_0 then it is so on the whole X. As a consequence, we see that every invertible global (G-homogeneous) function on Y is constant provided that every invertible global (H-homogeneous) function on X is constant.

Finally, according to Lemma 4.4.2.5 (iv), the homomorphism of (iv) is surjective. Moreover, because the weights $\chi_1, \ldots, \chi_m$ of the $g_1, \ldots, g_m$ are a basis of the character group of H_0, there are no relations among the g_k. $\qquad\square$

Let a quasitorus G act effectively with at most finite isotropy groups on a quasiaffine variety Y. Suppose that every invertible G-homogeneous function on Y constant and that Y is G-factorial. Let $C_1, \ldots, C_n \subseteq Y$ be G-prime divisors of Y such that all G-prime divisors with nontrivial generic isotropy group occur among the C_i. Moreover, let l_i be the order of the generic isotropy group $G_i \subseteq G$ of C_i, let l_i be the order of G_i, and let $h_1, \ldots, h_n$ be homogeneous functions on Y with $\mathrm{div}(h_i) = C_i$.

Proposition 4.4.2.7 *Consider the action of $G_0 := G_1 \cdots G_n \subseteq G$ on Y, and let $\kappa : Y \to Z$ be the associated quotient.*

(i) *Z is a quasiaffine variety with induced action of $F := G/G_0$ and there is a big open F-invariant subset $Z_0 \subseteq Z$ such that F acts freely on Z_0.*

(ii) *The assignments $h \mapsto \kappa^*(h)$ and $T_i \mapsto h_i$ define an isomorphism of $\mathbb{X}(G)$-graded algebras*

$$\Gamma(Z, \mathcal{O})[T_1, \ldots, T_n]/\langle T_i^{l_i} - h_i^{l_i}; \ i = 1, \ldots, n \rangle \ \to \ \Gamma(Y, \mathcal{O}).$$

(iii) *If Y is G-factorial with only constant invertible global G-homogeneous functions, then Z is F-factorial with only constant invertible global F-homogeneous functions.*

Proof We prove (i). The quotient Z is quasiaffine, because Y is so and G_0 is finite. Let $Y_0 \subseteq Y$ denote the subset consisting of all points $y \in Y$ that have either trivial isotropy group $H_{0,y}$ or belong to some D_i and have isotropy group $H_{0,y} = H_i$. Note that $Y_0 \subseteq Y$ is big, H-invariant, and open. Set $Z_0 := \kappa(Y_0)$. Then $Z_0 \subseteq Z$ is big and the restriction $\kappa : Y_0 \to Z_0$ is a quotient for the action of G_0. By construction, $F = G/G_0$ acts freely on Z_0.

We turn to (ii). By Lemma 4.4.2.5 (i), every G_i is cyclic. Moreover, Lemma 4.4.2.5 (iv) says that there is an epimorphism

$$\Gamma(Z, \mathcal{O})[T_1, \ldots, T_n] \;\to\; \Gamma(Y, \mathcal{O}), \qquad h \mapsto \kappa^*(h), \quad T_i \mapsto h_i.$$

From Lemma 4.4.2.5 (iii) we infer that $G_0 \subseteq G$ is the direct product of the cyclic groups $G_1, \ldots, G_n \subseteq G$. Thus, the quotient $\kappa \colon Y \to Z$ can also be obtained via dividing stepwise by effective actions of the G_i. Using this, one directly checks that the kernel of this epimorphism is the ideal generated by $T_i^{l_i} - h_i^{l_i}$, where $1 \leq i \leq n$.

We prove (iii). Clearly, Z has only constant invertible global (F-homogeneous) functions if Y has only constant invertible global (G-homogeneous) functions. Let us show that Z is F-factorial. The projection $G \to F$ corresponds to an inclusion $K_0 \subseteq K$ of character groups. Moreover, the action of G on Y turns $A := \Gamma(Y, \mathcal{O})$ into a K-graded algebra and the subalgebra of G_0-invariants is the Veronese subalgebra $A(K_0) \subseteq A$. Assertion (ii) yields

$$\left(A_{h_1 \cdots h_n} \right)(K_0) = A(K_0)_{h_1^{l_1} \cdots h_n^{l_n}}.$$

Because Y_0 is G-factorial, the algebra A is factorially K-graded. From Theorem 3.4.1.5 we infer that the left-hand side algebra is factorially K_0-graded. We claim that the $g_i^{l_i} \in A(K_0)$ are K_0-prime. Indeed, we have $\kappa^* \mathrm{div}(g_i^{l_i}) = l_i \mathrm{div}(g_i)$ and the quotient map κ is ramified of order l_i along $C_i = \mathrm{div}(g_i)$. We see that $\mathrm{div}(g_i^{l_i}) = \kappa(C_i)$ is F-prime and the claim is verified. Proposition 3.4.1.10 says that $A(K_0)$ is factorially K_0-graded and thus Z is F-factorial. $\qquad\Box$

Proof of Theorem 4.4.2.1 We use the notation of Proposition 4.4.2.6 and Proposition 4.4.2.7. Then the geometric quotient $X_0 \to X_0/H$ splits into a series of geometric quotients

$$
\begin{array}{ccc}
X_0 \xrightarrow{\;/H_0\;} Y \xrightarrow{\;/G_0\;} Z \\[4pt]
{\scriptstyle /H}\Big\downarrow \qquad\qquad \Big\downarrow {\scriptstyle /F} \\[4pt]
X_0/H \xrightarrow[\;\cong\;]{} Z/F
\end{array}
$$

Define $H' \subseteq H$ to be the subgroup generated by the generic isotropy groups of all the divisors D_i and E_k. Note that we have $H_0 \subseteq H'$ and $G = H'/H_0$. In particular, $X_0 \to Z$ is the geometric quotient by the H'-action and we have $F = H/H'$.

Proposition 4.4.2.6 (iii) ensures that Y is irreducible, normal, and G-factorial with only constant invertible global homogeneous functions for $G = H/H_0$. Parts (i) and (iii) of Proposition 4.4.2.7 then tell us that Z is

F-factorial with only constant invertible global homogeneous functions for $F = G/G_0$ and, moreover, the action of F is strongly stable. According to Theorem 1.6.4.3, the quotient $Z \to Z/F$ is a characteristic space. This implies $\mathcal{R}(X) \cong \Gamma(Z, \mathcal{O})$. The assertion now follows from Proposition 4.4.2.7 (ii) and Proposition 4.4.2.6 (iv).   □

Proof of Theorem 4.4.2.2   Because we have $\dim(H) = \dim(X) - 1$, the quotient X_0/H is a possibly nonseparated curve. By Theorem 4.4.2.1 it has finitely generated divisor class group and there is a graded injection $\mathcal{R}(X/H_0) \to \Gamma(X, \mathcal{O})$. We conclude that X/H_0 is rational and has only constant global functions. It follows that there is a separation $\sigma : X/H_0 \to \mathbb{P}^1$.

Take a collection $a_0, \ldots, a_r \in \mathbb{P}^1$ of points comprising all $a \in \mathbb{P}^1$ such that $\pi^{-1}(\sigma^{-1}(a))$ contains an H-prime divisor with nontrivial generic isotropy group or at least two H-prime divisors. Denote by $D_{i1}, \ldots, D_{in_i}$ the H-prime divisors mapping to a_i. As seen in Example 1.4.1.6, the Cox ring of X/H is given as

$$\mathcal{R}(X/H) = \mathbb{K}[T'_{ij}; \ 0 \le i \le r, \ 0 \le j \le n_i] \, / \, \langle g_i; \ 0 \le i \le r - 2 \rangle.$$

where

$$g_i := (b_j c_k - c_j b_k) T'_i + (b_k c_i - c_k b_i) T'_j + (b_i c_j - c_i b_j) T'_k,$$

with the homogeneous coordinates $a_i = [b_i, c_i] \in \mathbb{P}^1$, the monomials $T'_i = T'_{i1} \cdots T'_{in_i}$, and the indices $k = i + 2$, $j = i + 1$. Let l_{ij} be the generic isotropy group of D_{ij}. Then Theorem 4.4.2.1 gives us the relations of the algebra of functions of X:

$$\Gamma(X, \mathcal{O}) = \mathbb{K}[T_{ij}, S_k] \, / \, \langle g_i; \ 0 \le i \le r - 2 \rangle,$$

where

$$g_i := (b_j c_k - c_j b_k) T_i^{l_i} + (b_k c_i - c_k b_i) T_j^{l_j} + (b_i c_j - c_i b_j) T_k^{l_k}$$

with $T_i^{l_i} := T_{i1}^{l_{i1}} \cdots T_{in_i}^{l_{in_i}}$ and $k = i + 2$, $j = i + 1$. Thus, as a ring $\mathcal{R}(X)$ equals $R(A, P_0)$, where P_0 is built up with the l_{ij}. Moreover, the variables T_{ij} and S_k are homogeneous with respect to H. Thus, the $\mathbb{X}(H)$-grading coarsens the K_0-grading of $R(A, P_0)$.

It remains to show that the downgrading map $K_0 \to \mathbb{X}(H)$ is an isomorphism. Consider the inclusion $H \subseteq H_0$ of the associated quasitori. Take one of the H-prime divisors D_{ij}. Its generic isotropy H_{ij} is of order l_{ij}. Clearly, D_{ij} is H_0-invariant. Moreover, the generic H_0-isotropy of D_{ij} is also of order l_{ij} und thus equals H_{ij}. For a general $x \in D_{ij}$, we have

$$H/H_{ij} \cong H \cdot x = H_0 \cdot x \cong H_0/H_{ij},$$

where $H \cdot x = H_0 \cdot x$ is the dense open (H_0-invariant) subset of D_{ij} consisting of all points with H-isotropy H_{ij}. We conclude that the canonical map $H/H_{ij} \to H_0/H_{ij}$ is an isorphism. This gives $H = H_0$ and thus $K_0 \to \mathbb{X}(H)$ is an isomorphism. □

4.4.3 Proof of Theorems 4.4.1.2, 4.4.1.3, and 4.4.1.6

Besides the announced proofs, we provide here an existence statement for the separation and relate the Cox ring of a prevariety to that of its separation. The ground field $\mathbb{K}$ is algebraically closed and of characteristic zero. Our approach is based on [166, 169].

Lemma 4.4.3.1 *Consider a quotient presentation* $p \colon \widehat{X} \xrightarrow{/\!\!/ H_X} X$. *Assume that a torus T acts on $\widehat{X}$, X and for some epimorphism $\varepsilon \colon T \to T$ one has*

$$t \cdot h \cdot z = h \cdot t \cdot z, \qquad p(t \cdot z) = \varepsilon(t) \cdot p(z)$$

for all $h \in H_X$, $t \in T$ and $z \in \widehat{X}$. Let $z \in \widehat{X}$ such that $p(z) \in X$ is smooth. Then the trivially acting subgroup $H' \subseteq T \times H_X$ is given by

$$H' = \{(t, h) \in (T \times H_X)_z; \ \varepsilon(t) = 1\}.$$

Moreover, for the induced effective action of $H := (T \times H_X)/H'$ on $\widehat{X}$, there is an isomorphism of isotropy groups $H_z \cong T_{p(z)}$.

Proof Let $(t, h) \in H'$ be given. Then $(t, h) \cdot z' = z'$ holds for every point $z' \in \widehat{X}$. In particular, (t, h) belongs to $(T \times H_X)_z$. Moreover, we obtain $\varepsilon(t) \cdot p(z') = p(z')$ for every $z' \in \widehat{X}$. Because $p \colon \widehat{X} \to X$ is surjective and T acts effectively on X, this implies $\varepsilon(t) = 1$.

Now consider $(t, h) \in (T \times H_X)_z$ with $\varepsilon(t) = 1$. Then, for every $z' \in \widehat{X}$, we have $p((t, h) \cdot z') = p(z')$. Consequently $t \cdot z' = h(t, z') \cdot z'$ holds with a uniquely determined $h(t, z') \in H_X$. Consider the assignment

$$\eta \colon \widehat{X} \to H_X, \qquad z' \mapsto h(t, z').$$

Because H_X acts freely we may choose for any z' homogeneous functions $f_1, \ldots, f_r$, defined near z' with $f_i(z') = 1$ such that their weights $\chi_1, \ldots, \chi_r$ form a basis of the character group of H_X. Then, near z', we have a commutative diagram

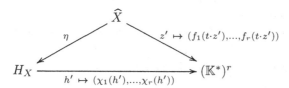

Consequently, the map η is is a morphism. Moreover, pulling back characters of H_X via η gives invertible H_X-homogeneous functions on $\widehat{X}$ that are all constant. Thus, η is constant. This means that $h(t) := h(t, z')$ does not depend on z'. By construction, $(t, h(t)^{-1})$ belongs to H'. Moreover, $t \cdot z = h^{-1} \cdot z$ and freeness of the H_X-action give $h(t) = h^{-1}$. This implies $(t, h) \in H'$.

We come to the second claim. Note that $(t, h) \mapsto \varepsilon(t)$ defines a homomorphism $\beta \colon (T \times H_X)_z \to T_{p(z)}$. We claim that β is surjective. Given $t \in T_{p(z)}$, choose $t' \in T$ with $\varepsilon(t') = t$. Then we have

$$p(t' \cdot z) = \varepsilon(t') \cdot p(z) = t \cdot p(z) = p(z).$$

Consequently, $t' \cdot z = h \cdot z$ holds for some $h \in H_X$. Thus, $(t', h^{-1}) \in (T \times H_X)_z$ is mapped by β to $t \in T_{p(z)}$. By the first step, the kernel of β is just H'. This gives a commutative diagram

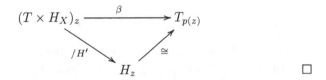

Proof of Theorem 4.4.1.2   Consider the characteristic space $p \colon \widehat{X} \to X$ with its action of the characteristic quasitorus H_X. According to Theorem 4.2.3.2, there is a T-action on $\widehat{X}$ and an epimorphism $\varepsilon \colon T \to T$ such that for all $h \in H_X, t \in T$ and $z \in \widehat{X}$ one has

$$t \cdot h \cdot z = h \cdot t \cdot z, \qquad p(t \cdot z) = \varepsilon(t) \cdot p(z).$$

Thus, we have an action of $T \times H_X$ on $\widehat{X}$. Let $H' \subseteq T \times H_X$ be the trivially acting subgroup and consider the induced effective action of $H := (T \times H_X)/H'$ on $\widehat{X}$. Then the quasiaffine variety $\widehat{X}$ is H-factorial and has only constant invertible global H-homogeneous functions. The idea is to apply Theorem 4.4.2.1.

Consider the T-invariant prime divisors $E_1, \ldots, E_m \subseteq X$ supported in $X \setminus X_0$. Their inverse images $\widehat{E}_k := p^{-1}(E_k)$ are H-prime divisors and, because $\widehat{X}$ is H-factorial, we have $\widehat{E}_k = \mathrm{div}(g_k)$ with some H-homogeneous $g_k \in \mathcal{R}(X)$. By Lemma 4.4.3.1, the $\widehat{E}_k$ are precisely the H-prime divisors supported in $\widehat{X} \setminus \widehat{X}_0$. Similarly, for the T-invariant prime divisors $D_i \subseteq X$ with finite generic isotropy group of order $l_i > 1$, Lemma 4.4.3.1 ensures that their (H-prime) inverse images $\widehat{D}_i := p^{-1}(D_i)$ have finite generic isotropy group of order l_i as well. Again by H-factoriality, we have $\widehat{D}_i = \mathrm{div}(f_i)$ holds with some H-homogeneous $f_i \in \mathcal{R}(X)$. Note that the divisors $\widehat{E}_k$ and $\widehat{D}_i$ are pairwise different. Moreover, we may view the functions g_k and h_i as the canonical sections of the divisors E_k and D_j. Because we have $X_0/T \cong \widehat{X}_0/H$, Theorem 4.4.2.1 shows that $\mathcal{R}(X) = \Gamma(\widehat{X}, \mathcal{O})$ looks as wanted. $\square$

Proposition 4.4.3.2 *Let $\mathcal{X}$ be an irreducible normal quasiaffine variety with a free action of a diagonalizable group H. Suppose that every invertible homogeneous function on $\mathcal{X}$ is constant and that $\mathcal{X}$ is H-factorial. Then $X := \mathcal{X}/H$ admits a separation.*

Proof We first treat the case of a certain toric variety. Consider the standard action of $\mathbb{T}^r = (\mathbb{K}^*)^r$ on $\mathbb{K}^r$, let $\mathcal{Z} \subseteq \mathbb{K}^r$ be the union of all orbits of the big torus $\mathbb{T}^r \subseteq \mathbb{K}^r$ of dimension at least $r - 1$, and let $H \subseteq \mathbb{T}^r$ be a closed subgroup acting freely on $\mathcal{Z}$. The fan Σ of $\mathcal{Z}$ has the extremal rays of the positive orthant $\mathbb{Q}^r_{\geq 0}$ as its maximal cones and $Z := \mathcal{Z}/H$ is the toric prevariety obtained by gluing the orbit spaces $\mathcal{Z}_\varrho/H$ along their common big torus T/H, where $\mathcal{Z}_\varrho \subseteq \mathcal{Z}$ denotes the affine toric chart corresponding to $\varrho \in \Sigma$. The embedding $H \to \mathbb{T}^r$ corresponds to a surjection $\mathbb{Z}^r \to K$ of the respective character groups. Let $P \colon \mathbb{Z}^r \to N$ be a map having $\mathrm{Hom}(K, \mathbb{Z})$ as its kernel. Then we obtain a canonical separation $Z \to Z'$ onto a toric variety Z', the fan of which lives in N and consists of the cones $P(\varrho)$, where $\varrho \in \Sigma$.

In the general case, choose a finitely generated graded subalgebra $A \subseteq \Gamma(\mathcal{X}, \mathcal{O})$ such that we obtain an open embedding $\mathcal{X} \subseteq \overline{X}$, where $\overline{X} := \mathrm{Spec}\, A$. Properly enlarging A, we may assume that it admits a system $f_1, \ldots, f_r$ of homogeneous generators such that each $\mathrm{div}(f_i)$ is H-prime in $\widehat{X}$. Consider the H-equivariant closed embedding $\overline{X} \to \mathbb{K}^r$ defined by $f_1, \ldots, f_r$ and let $\mathcal{Z} \subseteq \mathbb{K}^r$ be as above. By construction, $\mathcal{U} := \mathcal{Z} \cap \mathcal{X}$ is a big H-invariant open subset of $\mathcal{X}$, and we obtain a commutative diagram

$$
\begin{array}{ccc}
\mathcal{U} & \longrightarrow & \mathcal{Z} \\
{\scriptstyle /H}\downarrow & & \downarrow{\scriptstyle /H} \\
U & \longrightarrow & Z
\end{array}
$$

where the induced map $U \to Z$ of quotients is a locally closed embedding and Z is a toric prevariety. Again by construction, the intersection of the invariant prime divisors of Z with U are prime divisors on U. Consequently, the restriction of $Z \to Z'$ defines the desired separation $U \to U'$. $\qquad\square$

Remark 4.4.3.3 Let $\varphi \colon X \dashrightarrow Y$ be a separation. Then there are big open subsets $U \subseteq X$ and $V \subseteq Y$ such that $\varphi \colon U \to V$ is a local isomorphism and moreover there are prime divisors $C_0, \ldots, C_r$ on V such that

(i) φ maps $U \setminus \varphi^{-1}(C_0 \cup \ldots \cup C_r)$ isomorphically onto $V \setminus (C_0 \cup \ldots \cup C_r)$.
(ii) Each $\varphi^{-1}(C_i)$ is a disjoint union of prime divisors C_{ij} of U.

Proposition 4.4.3.4 *Let $\varphi \colon X \dashrightarrow Y$ be a separation, $C_0, \ldots, C_r$ prime divisors on Y as in Remark 4.4.3.3, and $\varphi^{-1}(C_i) = C_{i1} \cup \ldots \cup C_{in_i}$ with pairwise* *disjoint prime divisors C_{ij} on X. Then $\varphi^* \colon \mathrm{Cl}(Y) \to \mathrm{Cl}(X)$ is injective, and we have*

$$\mathrm{Cl}(X) = \varphi^* \, \mathrm{Cl}(Y) \; \oplus \; \bigoplus_{\substack{0 \le i \le r, \\ 1 \le j \le n_i - 1}} \mathbb{Z}[C_{ij}].$$

If $\Gamma(X, \mathcal{O}^*) = \mathbb{K}^*$ holds and $\mathrm{Cl}(X)$ is finitely generated, then there is a canonical injective pullback homomorphism $\varphi^* \colon \mathcal{R}(Y) \to \mathcal{R}(X)$ of Cox rings. Moreover, with $\deg(T_{ij}) := [C_{ij}]$ and $T_i := T_{i1} \cdots T_{in_i}$, the assignment $T_{ij} \mapsto 1_{C_{ij}}$ defines a $\mathrm{Cl}(X)$-graded isomorphism

$$\mathcal{R}(Y)[T_{ij}; \; 0 \le i \le r, \; 1 \le j \le n_i] \, / \, \langle T_i - 1_{C_i}; \; 0 \le i \le r \rangle \to \mathcal{R}(X).$$

Proof Because the divisor class group and Cox ring do not change when passing to big open subsets, we may assume $U = X$ and $V = Y$ in the setting of Remark 4.4.3.3. The assertion on the divisor class group follows immediately from the facts that the principal divisors of X are precisely the pullbacks of principal divisors on Y and that the divisor class group of X is generated by all pullback divisors and the classes $[C_{ij}]$, where $0 \le i \le r$ and $1 \le j \le n_i - 1$.

We turn to the Cox rings. Let $K_Y \subseteq \mathrm{WDiv}(Y)$ be a finitely generated subgroup containing $C_0, \ldots, C_r$ and mapping onto $\mathrm{Cl}(Y)$. Moreover, let $K_X \subseteq \mathrm{WDiv}(X)$ be the subgroup generated by $\varphi^*(K_Y)$ and the divisors C_{ij}, where $0 \le i \le r$ and $1 \le j \le n_i - 1$; note that $C_{in_i} \in K_X$ holds. Consider the associated graded sheaves

$$\mathcal{S}_Y := \bigoplus_{E \in K_Y} \mathcal{O}_Y(E), \qquad \mathcal{S}_X := \bigoplus_{D \in K_X} \mathcal{O}_X(D).$$

Then we have a graded injective pullback homomorphism $\varphi^* \colon \mathcal{S}_Y \to \mathcal{S}_X$, which in turn extends to a homomorphism

$$\psi \colon \mathcal{S}_Y[T_{ij}; \; 0 \le i \le r, \; 1 \le j \le n_i] \to \mathcal{S}_X, \qquad T_{ij} \mapsto 1_{C_{ij}}.$$

We show that ψ is surjective. Given a section h of $\mathcal{S}_X$ of degree $D \in K_X$, consider its divisor $D(h) = \mathrm{div}(h) + D$. If there occurs a $C_{ij} \in K_X$ in $D(h)$, then we may divide h in $\mathcal{S}_X$ by the corresponding $1_{C_{ij}}$. Doing this as often as possible, we arrive at some section h' of $\mathcal{S}_X$, homogeneous of some degree $D' \in K_X$, such that $D(h') = \mathrm{div}(h') + D'$ has no components C_{ij}. But then D' is a pullback divisor and h' is a pullback section. This in turn means that h' is a polynomial over $\varphi^* \mathcal{S}_Y$ and the $1_{C_{ij}}$.

Next, we determine the kernel of ψ, which amounts to determining the relations among the sections $s_{ij} := 1_{C_{ij}}$. Consider two coprime monomials F, F' in the s_{ij} and two homogeneous pullback sections h, h' of $\varphi^*(\mathcal{S}_Y)$. If $\deg(hF) = \deg(h'F')$ holds in K_X, then the difference $\deg(F') - \deg(F)$ must be a linear combination of some $\varphi^*(C_i) \in K_X$ and hence F and F' are products of some $\varphi^* 1_{C_i}$. As a consequence, we obtain that any homogeneous (and hence any) relation among the s_{ij} is generated by the relations $T_i - 1_{C_i}$.

Finally, fix a character χ_Y for $\mathcal{S}_Y$ as in Construction 1.4.2.1. Because $K_X^0 = \varphi^*(K_Y^0)$ holds, the pullback character $\varphi^* \chi_Y$ extends uniquely to a character χ_X for $\mathcal{S}_X$. We conclude $\mathcal{I}_X = \varphi^*(\mathcal{I}_Y)$ and hence obtain a well-defined graded pullback homomorphism $\varphi^* \colon \mathcal{R}(Y) \to \mathcal{R}(X)$, which is injective, because $\varphi^* \colon K_Y/K_Y^0 \to K_X/K_X^0$ is so and $\varphi^* \colon \mathcal{S}_Y \to \mathcal{S}_X$ is an isomorphism when restricted to homogeneous components. Now one directly verifies that the above epimorphism ψ induces the desired isomorphism. $\square$

Proof of Theorem 4.4.1.3 By Proposition 4.4.3.2, the orbit space X_0/T admits a separation $\pi \colon X_0/T \to Y$. According to Remark 4.4.3.3, we may assume that there are prime divisors $C_0, \ldots, C_r$ on Y such that each $\pi^{-1}(C_i)$ is a disjoint union of prime divisors $C_{ij} \subseteq X_0/T$, where $1 \le j \le n_i$, the map π is an isomorphism over $Y \setminus (C_0 \cup \ldots \cup C_r)$, and all the D_j occur among the divisors $D_{ij} := q^{-1}(C_{ij})$. Then, according to Proposition 4.4.3.4, we have

$$\mathcal{R}(X_0/T) \cong \mathcal{R}(Y)[\widetilde{T}_{ij};\ 0 \le i \le r,\ 1 \le j \le n_i] \,/\, \langle \widetilde{T}_i - 1_{C_i};\ 0 \le i \le r \rangle,$$

where the variables $\widetilde{T}_{ij}$ correspond to the canonical sections $1_{C_{ij}}$ and we define $\widetilde{T}_i := \widetilde{T}_{i1} \cdots \widetilde{T}_{in_i}$. Let $l_{ij} \in \mathbb{Z}_{\ge 1}$ denote the order of the generic isotropy group of $D_{ij} = q^{-1}(C_{ij})$. Then, by Theorem 4.4.1.2, we have

$$\mathcal{R}(X) \cong \mathcal{R}(X_0/T)[S_1, \ldots, S_m;\ T_{ij}] \,/\, \langle T_{ij}^{l_{ij}} - 1_{C_{ij}} \rangle,$$

where the variables T_{ij} correspond to the canonical sections $1_{C_{ij}}$; note that $1_{C_{ij}}$ and $1_{D_{ij}}$ are identified for $l_{ij} = 1$. Putting these two presentations of Cox rings together, we arrive at the assertion. $\square$

Proof of Theorem 4.4.1.6 Consider the characteristic space $p_X \colon \widehat{X} \to X$. Note that every H_X-invariant function on $\widehat{X}$ is constant, because we required $\Gamma(X, \mathcal{O}) = \mathbb{K}$. By Theorem 4.2.3.2, the T-action on X lifts $\widehat{X}$. Dividing $H_X \times T$ by the kernel H' of ineffectivity, we obtain an effective quasitorus action $H \times \widehat{X} \to \widehat{X}$ of complexity 1. According to Lemma 4.4.3.1, for any $\widehat{x} \in \widehat{X}$ the isotropy group $H_{\widehat{x}}$ is isomorphic to $T_{p_X(\widehat{x})}$.

Theorem 4.4.2.2 tells us that the $\mathbb{X}(H)$-graded algebra $\Gamma(\widehat{X}, \mathcal{O})$ is isomorphic to a K_0-graded algebra $R(A, P_0)$. In particular, $\Gamma(\widehat{X}, \mathcal{O})$ is finitely generated and we have the total coordinate space given by trinomial equations

$$\overline{X} = V(g_i;\ 0 \le i \le r - 2) \subseteq \mathbb{K}^{n+m}.$$

Recall that the K_0-grading of $R(A, P_0)$ is the finest possible, making the variables T_{ij} and S_k corresponding to D_{ij} and S_k homogeneous. Because T_{ij} and S_k are also homogeneous with respect to the grading by $K := \mathrm{Cl}(X)$, we have a downgrading homomomorphism $K_0 \to K$. It fits into the commutative diagram

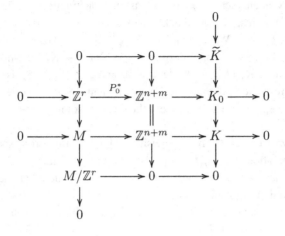

$$(4.4.1)$$

In particular we extract from this the following two commutative triangles, where the second one is obtained by dualizing the first one

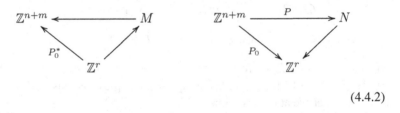

$$(4.4.2)$$

We claim that the kernel $\widetilde{K}$ is free. Consider $\widetilde{H} := \operatorname{Spec} \mathbb{K}[\widetilde{K}] = H/H_X$. Using the description of the kernel of ineffectivity $H' \subseteq T \times H_X$ provided in Lemma 4.4.3.1, we obtain a commutative diagram

$$
\begin{array}{ccc}
H & \xrightarrow{\ /H_X\ } & \widetilde{H} \\
{\scriptstyle /H'}\Big\uparrow & & \Big\uparrow{\scriptstyle \cong} \\
T \times H_X & \xrightarrow[(t,h)\mapsto\varepsilon(t)]{} & T
\end{array}
$$

$$(4.4.3)$$

Consequently $\widetilde{H}$ is a torus and thus $\widetilde{K}$ is free. Now the snake lemma tells us that $M/\mathbb{Z}^r$ is free as well. In particular, the first vertical sequence of (4.4.1) splits. Thus, we obtain the desired matrix presentation of P from rewriting the second commutative triangle of (4.4.2) as

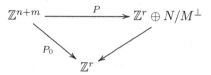

This proves the first assertion of the theorem. To verify the second one, we only have to show that the induced action of $\widetilde{H} = H/H'$ coincides with the original T-action up to an isomorphism $T \to \widetilde{H}$. This is directly seen via the commutative diagram (4.4.3). $\square$

4.5 Almost homogeneous varieties

4.5.1 Homogeneous spaces

We consider homogeneous spaces G/F, where G is a connected affine algebraic group and $F \subseteq G$ a closed subgroup defined over an algebraically closed field $\mathbb{K}$ of characteristic zero. By Chevalley's theorem, G/F is a smooth quasiprojective variety. Following [18] we describe the characteristic space and the Cox ring of G/F in the case that $\operatorname{Pic}(G)$ and $\mathbb{X}(G)$ are trivial; recall that the latter two properties hold if and only if G is a semidirect product of a semisimple simply connected group G^s and the unipotent radical G^u.

First we have to care about the Picard group $\operatorname{Pic}(G/F)$ of the homogeneous space G/F; the reference for details is [254, Sec. 2]. The key is the following construction associating to every character $\chi \in \mathbb{X}(F)$ a G-linearized line bundle $E_\chi \to G/F$.

Construction 4.5.1.1 Let G be a connected affine algebraic group, $F \subseteq G$ a closed subgroup, and $\chi \in \mathbb{X}(F)$ a character. Consider the action of F on $G \times \mathbb{K}$ given by

$$h \cdot (g, a) := (gh^{-1}, \chi(h)a).$$

This action admits a categorical quotient $G \times \mathbb{K} \to E_\chi$ having the F-orbits as its fibers; we denote by $[g, a] \in E_\chi$ the image of $(g, a) \in G \times \mathbb{K}$. The map $G \times \mathbb{K} \to G/F, (g, a) \mapsto gF$, induces a morphism

$$E_\chi \to G/F, \qquad [g, a] \mapsto gF.$$

It turns out that $E_\chi \to G/F$ is a line bundle over the homogeneous space G/F. It comes with a G-linearization given by the following natural G-action on the total space

$$G \times E_\chi \to E_\chi, \qquad g' \cdot [g, a] := [g'g, a].$$

Moreover, the assignment $\chi \mapsto E_\chi$ induces a homomorphism to the group $\mathrm{Pic}_G(G/F)$ of equivariant isomorphism classes of G-linearized line bundles on G/F equipped with the G-action by multiplication from the left:

$$\nu\colon \mathbb{X}(F) \;\to\; \mathrm{Pic}_G(G/F), \qquad \chi \mapsto [E_\chi].$$

As we will see now, the homomorphism ν relates the character group $\mathbb{X}(F)$ to the Picard group $\mathrm{Pic}(G/F)$. We denote by $\mathbb{X}_G(F) \subseteq \mathbb{X}(F)$ the subgroup of characters extending to a character of G.

Theorem 4.5.1.2 *Let G be a connected affine algebraic group and $F \subseteq G$ a closed subgroup. Then $\nu\colon \mathbb{X}(F) \to \mathrm{Pic}_G(G/F)$ is an isomorphism. Moreover, if $\mathrm{Pic}(G) = 0$ holds, then there is an exact sequence*

$$0 \longrightarrow \mathbb{X}_G(F) \longrightarrow \mathbb{X}(F) \xrightarrow[\cong]{\;\nu\;} \mathrm{Pic}_G(G/F) \longrightarrow \mathrm{Pic}(G/F) \longrightarrow 0.$$

If $\mathrm{Pic}(G) = 0$ and $\mathbb{X}(G) = 0$ hold, for example, the group G is semisimple simply connected, then one has $\mathbb{X}(F) \cong \mathrm{Pic}(G/F)$.

Proof Given a G-linearized line bundle $L \to G/F$, the group F acts on the fiber $L_{eF} \cong \mathbb{K}$ by a character χ and the canonical map

$$G \times L_{eF} \;\to\; L, \qquad\qquad (g,l) \mapsto g \cdot l$$

induces a G-eqivariant isomorphism $E_\chi \cong L$. It follows that ν is surjective. Injectivity is obvious.

If $\mathrm{Pic}(G) = 0$ holds, then every line bundle on G/F is G-linearizable and thus the forgetting map $\mathrm{Pic}_G(G/F) \to \mathrm{Pic}(G/F)$ is surjective. To see exactness at $\mathbb{X}(F)$, assume that $[E_\chi]$ is trivial. This means that there is a function $f \in \Gamma(G, \mathcal{O})^*$ such that $f(gh) = f(g)\chi(h)$ for all $g \in G, h \in F$, and $f(e) = 1$. Then f is a character of G and because of $f(eh) = \chi(h)$ it extends χ.

If $\mathrm{Pic}(G) = 0$ and $\mathbb{X}(G) = 0$ hold, then every line bundle on G/F has a unique G-linearization and thus the forgetting map $\mathrm{Pic}_G(G/F) \to \mathrm{Pic}(G/F)$ is an isomorphism. This gives the last claim. $\square$

Example 4.5.1.3 Let G be a semisimple simply connected affine algebraic group and $P \subseteq G$ a parabolic subgroup. Then $\mathrm{Pic}(G/P)$ is the lattice $\mathbb{X}(P)$ whose rank may vary from 1, when P is maximal, up to the rank of G, when P is a Borel subgroup.

Corollary 4.5.1.4 *Let G be a connected affine algebraic group and $F \subseteq G$ a closed subgroup.*

(i) *The divisor class group $\mathrm{Cl}(G/F)$ equals the Picard group $\mathrm{Pic}(G/F)$ and is finitely generated.*

(ii) *If $G/F \subseteq X$ is an open embedding into an irreducible normal variety, then $\mathrm{Cl}(X)$ is finitely generated.*

Proof Replacing G with a finite covering group, we may assume that $\mathrm{Pic}(G) = 0$ holds. By Remark 4.2.2.2, the group $\mathbb{X}(F)$ is finitely generated. Theorem 4.5.1.2 shows that $\mathrm{Pic}(G/F)$ as the image of $\mathrm{Pic}_G(G/F) \cong \mathbb{X}(F)$ is finitely generated. This proves the first assertion. For the second one, observe that $\mathrm{Cl}(X)$ is generated by generators of $\mathrm{Cl}(G/F)$ and the prime divisors in $X \setminus G/F$. $\qquad\square$

Corollary 4.5.1.5 *Let G be a semisimple simply connected group and $F \subseteq G$ a semisimple closed subgroup. Then G/F is smooth affine with $\mathrm{Cl}(G/F) = 0$.*

Proof For a reductive, for example, semisimple, subgroup F the quotient space G/F is given as $\mathrm{Spec}\,\Gamma(G, \mathcal{O})^F$ and thus is affine. According to Theorem 4.5.1.2, we have $\mathrm{Cl}(G/F) \cong \mathbb{X}(F) = 0$. $\qquad\square$

Example 4.5.1.6 Consider the smooth affine quadric $Z_{n-1} = V(T_1^2 + \cdots + T_n^2 - 1)$ in $\mathbb{K}^n$. The group SO_n acts transitively on Z_{n-1} and the isotropy group of a point on Z_{n-1} is SO_{n-1}. For $n \geq 3$, the universal 2-covering is $\mathrm{Spin}_n \to \mathrm{SO}_n$ and for $n \geq 4$, we have

$$Z_{n-1} \cong \mathrm{SO}_n/\mathrm{SO}_{n-1} \cong \mathrm{Spin}_n/\mathrm{Spin}_{n-1}.$$

Because the group Spin_n is semisimple simply connected and $\mathbb{X}(\mathrm{Spin}_{n-1})$ is trivial, the variety Z_{n-1} is factorial; compare with Proposition 3.2.4.1. In particular, $Z_3 \cong \mathrm{SL}_2 \times SL_2/SL_2$.

We enter the description of the characteristic space over G/F; it will be realized as a homogeneous space G/F_1 with a subgroup $F_1 \subseteq F$ constructed below.

Construction 4.5.1.7 Let G be a connected affine algebraic group and let $F \subseteq G$ be a closed subgroup. Set

$$K := \mathbb{X}(F), \qquad H := \mathrm{Spec}\,\mathbb{K}[K].$$

The inclusion $\mathbb{K}[K] \subseteq \Gamma(F, \mathcal{O})$ defines an epimorphism $\pi : F \to H$ onto the quasitorus H. The kernel is given as

$$F_1 := \mathrm{Ker}\,(\pi) = \bigcap_{\chi \in K} \mathrm{Ker}\,(\chi) \subseteq F.$$

Because F_1 is normal in F, the group $H \cong F/F_1$ acts on G/F_1 by right multiplication. Moreover, we have a commutative diagram

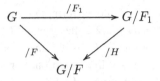

We are ready for the main result of this section. Note that it generalizes Proposition 3.2.3.4, which settles the special case of flag varieties.

Theorem 4.5.1.8 *Let G be a connected affine algebraic group with* $\mathrm{Pic}(G) = 0$ *and* $\mathbb{X}(G) = 0$ *and let* $F \subseteq G$ *be a closed subgroup. Then* $H := \mathrm{Spec}\, \mathbb{K}[\mathbb{X}(F)]$ *is the characteristic quasitorus of G/F and the geometric quotient* $q \colon G/F_1 \to G/F$ *by the action of H is a characteristic space over G/F. In particular, the Cox ring of G/F is given by*

$$\mathcal{R}(G/F) \cong \Gamma(G/F_1, \mathcal{O}) \cong \Gamma(G, \mathcal{O})^{F_1} \cong \bigoplus_{\chi \in \mathbb{X}(F)} \Gamma(G, \mathcal{O})_\chi^{F_1}.$$

The property of the characteristic space of being quasiaffine and the finite generation of the Cox ring are reflected in general concepts on affine algebraic groups; compare [18, 150].

Definition 4.5.1.9 Let G be a connected affine algebraic group.

(i) A closed subgroup $F \subseteq G$ is called *observable* if the homogeneous space G/F is quasiaffine.
(ii) A closed subgroup $F \subseteq G$ is called a *Grosshans subgroup* if it is observable and $\Gamma(G/F, \mathcal{O})$ is finitely generated.
(iii) A closed subgroup $F \subseteq G$ is called a *Grosshans extension* if the subgroup $F_1 \subseteq G$ of 4.5.1.7 is a Grosshans subgroup of G.

Lemma 4.5.1.10 *Let G be a connected affine algebraic group, $F \subseteq G$ a closed subgroup, and $F_1 \subseteq F$ as in Construction 4.5.1.7. Then F_1 is an observable subgroup of G.*

Proof By Chevalley's Theorem [178, Thm. 11.2], there exist a rational finite-dimensional G-module V and a nonzero vector $v \in V$ such that F is the isotropy group of the line $\mathbb{K}v$. Let $\chi_0 \in K$ be a character such that $g \cdot v = \chi_0(g)v$ for any $g \in F$. Then $F_0 := \mathrm{Ker}(\chi_0)$ equals the isotropy group G_v, and thus is observable in G, see [150, Thm. 2.1]. We have

$$F_1 = \bigcap_{\chi \in K} \mathrm{Ker}(\chi|_{F_0}) \subseteq F_0.$$

Because the intersection of observable subgroups is again observable, it suffices to show that each $F_\chi := \mathrm{Ker}\,(\chi|_{F_0})$ is observable. For this, we use again [150, Thm. 2.1] to realize the one-dimensional F_0-module given by $\chi|_{F_0}$ as an F_0-submodule $\mathbb{K}v_\chi$ of a G-module V_χ. Then F_χ is the isotropy group of (v, v_χ) in the G-module $V \oplus V_\chi$, and hence it is observable. $\qquad\square$

Proof of Theorem 4.5.1.8 We verify the conditions of Theorem 1.6.4.3 for $q: G/F_1 \to G/F$. By Lemma 4.5.1.10, the homogeneous space G/F_1 is quasi-affine. Clearly, the H-action on G/F_1 by right multiplication is free and commutes with the G-action. Because $\Gamma(G, \mathcal{O})^{F_1} \cong \Gamma(G/F_1, \mathcal{O})$ is a subalgebra of $\Gamma(G, \mathcal{O})$, we have $\Gamma(G/F_1, \mathcal{O})^* = \mathbb{K}^*$.

It remains to verify that the variety G/F_1 is H-factorial, that is, any H-invariant divisor on G/F_1 is principal. Note that H-invariant divisors on G/F_1 are precisely the divisors $q^*(D)$ for some $D \in \mathrm{WDiv}(G/F)$. By Theorem 4.5.1.2, any divisor D on the homogeneous space G/F is the divisor of a section of the line bundle $E_\chi \to G/F$ with some $\chi \in K$. We have to prove that the pullback of E_χ to G/F_1 is the trivial bundle. Consider the pullback diagram

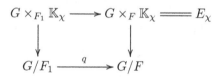

Because the restriction of χ to F_1 is trivial, the line bundle $q^*(E_\chi) \cong G \times_{F_1} \mathbb{K}_\chi$ is trivial as well. Thus, all conditions of Theorem 1.6.4.3 holds for $q: G/F_1 \to g/F$ and the assertion follows. $\qquad\square$

As a consequence we obtain a lifting result in the spirit of Theorem 4.2.3.2 with the additional statement that transitivity is respected.

Corollary 4.5.1.11 *Let a connected affine algebraic group G with $\mathrm{Pic}(G) = 0$ and $\mathbb{X}(G) = 0$ act transitively on a variety X coming with a quotient presentation $q: \widehat{X} \to X$. Then there exists a transitive G-action on $\widehat{X}$ commuting with the characteristic quasitorus action on $\widehat{X}$ and turning q into a G-equivariant morphism.*

Remark 4.5.1.12 Let G be a connected affine algebraic group with $\mathrm{Pic}(G) = 0$ and $\mathbb{X}(G) = 0$ and $F \subseteq G$ a closed subgroup. If F is connected, then $H \cong F/F_1$ is a torus and thus, by Proposition 1.4.1.5, the Cox ring $\mathcal{R}(G/F)$ is a unique factorization domain.

The following examples indicate how Theorem 4.5.1.8 can be used to construct nonfactorial Cox rings.

Example 4.5.1.13 Consider the special linear group $G = \mathrm{SL}_2$ and let $F \subseteq G$ be the normalizer of the standard maximal torus, that is,

$$F = T \cup nT, \qquad T = \left\{ \begin{bmatrix} t & 0 \\ 0 & t^{-1} \end{bmatrix} \right\}, \qquad n = \begin{bmatrix} 0 & 1 \\ -1 & 0 \end{bmatrix}.$$

Then F has only one nontrivial character defined by $\chi(n) = -1$. Thus we have $F_1 = T$ and $H \cong \mathbb{Z}/2\mathbb{Z}$. The variety $G/F_1 = \mathrm{SL}_2/T$ is H-factorial and we have

$$\mathrm{Cl}(\mathrm{SL}_2/T) \cong \mathbb{X}(T) \cong \mathbb{Z}.$$

In fact, the homogeneous space $G/F_1 = \mathrm{SL}_2/T$ is the smooth quadric surface in $\mathbb{K}^3$ treated in Example 1.4.4.1.

Example 4.5.1.14 [13, Example 4.3]. The example given above generalizes as follows: Given any nonfree finitely generated abelian group

$$A = \mathbb{Z}/d_1\mathbb{Z} \oplus \cdots \oplus \mathbb{Z}/d_s\mathbb{Z} \oplus \mathbb{Z}^n,$$

we construct a smooth affine variety X with divisor class group $\mathrm{Cl}(X) \cong A$ and a nonfactorial Cox ring $\mathcal{R}(X)$. Set $X = G/F$, where

$$G = \mathrm{SL}_{d_1} \times \cdots \times \mathrm{SL}_{d_s} \times \mathrm{SL}_2 \times \cdots \times \mathrm{SL}_2,$$

$$F = F(1) \times \cdots \times F(s) \times T \times \cdots \times T$$

with $F(i)$ being an extension of a maximal torus of the group SL_{d_i} by elements of its normalizer that act as powers of one d_i-cycle. Then we have

$$\mathrm{Cl}(G/F) \cong \mathbb{X}(F) \cong A, \qquad \mathrm{Cl}(G/F_1) \cong \mathbb{X}(F_1) \cong \mathbb{Z}^{d_1 + \cdots + d_s - s}.$$

In particular, the Cox ring $\mathcal{R}(G/F)$ of the homogeneous space $X = G/F$ is not a unique factorization domain.

Definition 4.5.1.15 Let G be a connected affine algebraic group and $F \subseteq G$ a Grosshans subgroup. Then the *canonical embedding* of G/F is the affine variety $\mathrm{CE}(G/F) := \mathrm{Spec}\,\Gamma(G/F, \mathcal{O})$ together with the G-action extending the multiplication from the left.

Note that the canonical embedding $\mathrm{CE}(G/F)$ of a homogeneous space G/F is an irreducible normal affine variety and the complement of the open G-orbit G/F is of codimension at least 2.

Remark 4.5.1.16 Let G be a connected affine algebraic group with $\mathrm{Pic}(G) = 0$ and $\mathbb{X}(G) = 0$ and let $F \subseteq G$ be a closed subgroup. Then the following statements are equivalent:

(i) The subgroup $F \subseteq G$ is a Grosshans extension.
(ii) The Cox ring $\mathcal{R}(G/F)$ is finitely generated.

If one of these statements holds, then the total coordinate space of G/F equals the canonical embedding of G/F_1 and the irrelevant ideal of G/F is the smallest nonzero G-invariant ideal in $\Gamma(G/F_1, \mathcal{O})$.

We conclude this section with some remarks on homogeneous spaces with a nonfinitely generated Cox ring. Examples are produced as follows.

Remark 4.5.1.17 Let F_1 be a unipotent subgroup of a connected simply connected semisimple algebraic group G such that the algebra $\Gamma(G, \mathcal{O})^{F_1}$ is not finitely generated. Taking F to be an extension of F_1 by some quasitorus H from the normalizer $N_G(F_1)$, we obtain other examples of homogeneous spaces G/F with a nonfinitely generated Cox ring $\Gamma(G/F_1, \mathcal{O})$. Moreover, for any open embedding $G/F_1 \subseteq X$ into a normal affine variety X the Cox ring $\mathcal{R}(X)$ is not finitely generated and X has non-$\mathbb{Q}$-factorial singularities; use Corollary 4.3.1.4.

Remark 4.5.1.18 Nagata's principle discussed in Section 4.3.4 provides unipotent subgroups $F_1 \subseteq \mathrm{SL}_n$ such that $\mathbb{K}[T_1, \ldots, T_r]^{F_1}$ is not finitely generated. By Grosshans' theorem [150, Thm. 9.3], the algebra $\Gamma(\mathrm{SL}_n/F_1, \mathcal{O})$ is not finitely generated as well.

4.5.2 Small embeddings

We consider equivariant embeddings of homogeneous spaces having a small boundary and finitely generated Cox ring. The main result gives a complete description and opens a way to treat such spaces via bunched rings. The original reference to this section is [18]. The ground field $\mathbb{K}$ is algebraically closed and of characteristic zero.

We consider homogeneous spaces G/F arising from Grosshans extensions $F \subseteq G$. Thus, the map $F \to H$, where $H = \mathrm{Spec}\, \mathbb{K}[\mathbb{X}(F)]$, has a Grosshans subgroup F_1 of G as its kernel, which means that the homogeneous space G/F_1 is quasiaffine with a finitely generated algebra of regular functions; see Definition 4.5.1.9.

Remark 4.5.2.1 Let G be a connected affine algebraic group. In each of the following cases, the subgroup $F_1 \subset G$ is a Grosshans subgroup:

- G/F_1 is quasiaffine and spherical or of complexity 1; see [188].
- F_1 is the unipotent radical of a parabolic subgroup of G; see [150].
- F_1 is the generic stabilizer of a factorial affine G-variety; see [150].

If a Grosshans subgroup $F_1 \subseteq G$ has only trivial characters, then any quasitorus $H \subseteq N_G(F_1)$ in the normalizer of F_1 in G gives Grosshans extension $F := H \cdot F_1 \subseteq G$ with $H = \mathrm{Spec}\, \mathbb{K}[\mathbb{X}(F)]$ and F_1 being the kernel of $F \to H$.

Example 4.5.2.2 Consider the special linear group $G := \mathrm{SL}_3$ and the two subgroups

$$F_1 := \left\{ \begin{bmatrix} 1 & 0 & a \\ 0 & 1 & b \\ 0 & 0 & 1 \end{bmatrix}; \; a, b \in \mathbb{K} \right\} \subseteq \{A \in G; \; a_{31} = a_{32} = 0\} =: P.$$

Then F_1 is the unipotent radical of the parabolic subgroup $P \subseteq G$ and thus $F_1 \subseteq G$ is a Grosshans subgroup. Set

$$F := \left\{ \begin{bmatrix} t & 0 & a \\ 0 & t & b \\ 0 & 0 & t^{-2} \end{bmatrix}; \; t \in \mathbb{K}^*, \; a, b \in \mathbb{K} \right\},$$

$$F' := \left\{ \begin{bmatrix} t_1 & 0 & a \\ 0 & t_2 & b \\ 0 & 0 & t_1^{-1}t_2^{-1} \end{bmatrix}; \; t_1, t_2 \in \mathbb{K}^*, \; a, b \in \mathbb{K} \right\}.$$

Then $F \subseteq G$ as well as $F' \subseteq G$ are Grosshans extensions with $\mathbb{X}(F) = \mathbb{Z}$ and $\mathbb{X}(F') = \mathbb{Z}^2$. For the tori $H = \mathrm{Spec}\,\mathbb{K}[\mathbb{Z}]$ and $H' = \mathrm{Spec}\,\mathbb{K}[\mathbb{Z}^2]$, we have

$$H \cong F/F_1, \qquad H' \cong F'/F_1.$$

We come to embeddings. Let G be a connected affine algebraic group and $F \subset G$ a closed subgroup. An *equivariant embedding* of the homogeneous space G/F is an irreducible, normal G-variety X together with a base point $x_0 \in X$ such that F equals the isotropy group G_{x_0} of $x_0 \in X$ and the morphism

$$G/F \to X, \qquad gF \mapsto g \cdot x_0$$

is a G-equivariant open embedding. A *morphism* of two equivariant embeddings X and X' of G/F is a G-equivariant morphism $X \to X'$ sending the base point $x_0 \in X$ to the base point $x_0' \in X'$. Note that if a morphism of G/F-embeddings exists, then it is unique.

Definition 4.5.2.3 Let G be a connected affine algebraic group with $\mathrm{Pic}(G) = 0$ and $\mathbb{X}(G) = 0$ and let $F \subseteq G$ be a Grosshans extension.

(i) A *small G/F-embedding* is an equivariant G/F-embedding X such that $X \setminus G \cdot x_0$ is of codimension at least 2 in X.

(ii) A *morphism* of two small G/F-embeddings is a morphism of equivariant G/F-embeddings.

Construction 4.5.2.4 Let G be a connected affine algebraic group with $\mathrm{Pic}(G) = 0$ and $\mathbb{X}(G) = 0$, let $F \subseteq G$ be a Grosshans extension and $F_1 \subseteq F$ the subgroup constructed in 4.5.1.7. Set

$$W := G/F_1 \subseteq \mathrm{CE}(G/F_1) = \mathrm{Spec}\,\Gamma(G/F_1, \mathcal{O}) =: Z.$$

The canonical embedding Z is the total coordinate space of G/F. The group G and the quasitorus $H = \operatorname{Spec} \mathbb{K}[\mathbb{X}(F)] \cong F/F_1$ of G/F act on Z via

$$g \cdot (g'F_1) := gg'F_1, \qquad (hF_1) \cdot g'F_1 := g'h^{-1}F_1.$$

Let $\widehat{X} \subseteq Z$ be a G-invariant good H-set containing W as H-saturated subset. Then $X := \widehat{X} /\!/ H$ is an irreducible normal variety with $\Gamma(X, \mathcal{O}^*) = \mathbb{K}^*$ and its divisor class group and Cox ring are given by

$$\operatorname{Cl}(X) = \mathbb{X}(F), \qquad \mathcal{R}(X) = \Gamma(Z, \mathcal{O}) = \bigoplus_{\chi \in \mathbb{X}(F)} \Gamma(Z, \mathcal{O})_\chi.$$

Moreover, $p_X \colon \widehat{X} \to \widehat{X} /\!/ H$ is a characteristic space over X and the G-action on $\widehat{X}$ induces a G-action on X turning X into a small G/F-embedding.

Proof Because $W \subseteq \widehat{X}$ is an H-saturated inclusion, we obtain a commutative diagram of morphisms

$$
\begin{array}{ccc}
W & \subseteq & \widehat{X} \\
\Big\downarrow{\scriptstyle /H} & & {\scriptstyle \pi}\Big\downarrow{\scriptstyle /\!/H} \\
G/F & \longrightarrow & X
\end{array}
$$

where the induced map of quotient spaces is an open embedding and the G-action on $\widehat{X}$ descends to a G-action on X. Note that X is a small G-embedding, because $Z \setminus W$ is of codimension at least 2 in Z. Due to the assumption $\mathbb{X}(G) = 0$, we have $\Gamma(Z, \mathcal{O}^*) = \mathbb{K}^*$ and by Theorem 4.5.1.8, the affine variety Z is H-factorial. Thus, the remaining statements follow from Theorem 1.6.4.3. $\qquad\square$

Example 4.5.2.5 Consider the group $G := \operatorname{SL}_3$ and the Grosshans subgroup $F_1 \subseteq G$ of Example 4.5.2.2. Then we have

$$Z := \operatorname{CE}(G/F_1) = \operatorname{Spec}(\Gamma(G, \mathcal{O})^{H_1}) \cong \mathbb{K}^3 \oplus \mathbb{K}^3.$$

For the Grosshans extension $F \subseteq G$ of Example 4.5.2.2 with $H = F/F_1 \cong \mathbb{K}^*$, we obtain a small G/F-embedding

$$X = \widehat{X} /\!/ H \cong \mathbb{P}(\mathbb{K}^3) \times \mathbb{P}(\mathbb{K}^3), \quad \widehat{X} = \{(v_1, v_2) \in Z;\ v_1 \neq 0 \neq v_2\} \subseteq Z.$$

For the Grosshans extension $F' \subseteq G$ of 4.5.2.2 with $H' = F'/F_1 \cong (\mathbb{K}^*)^2$, we obtain a small G/F'-embedding

$$X' = \widehat{X}' /\!/ H' \cong \mathbb{P}(\mathbb{K}^3 \oplus \mathbb{K}^3), \quad \widehat{X}' = \{(v_1, v_2) \in Z;\ v_1 \neq 0 \text{ or } v_2 \neq 0\} \subseteq Z.$$

To obtain our description of small G/F-embeddings, we use the combinatorial construction of GIT quotients via orbit cones presented in Section 3.1. Given an affine variety $Z = \operatorname{Spec} R$ with an action of a quasitorus $H = \operatorname{Spec} \mathbb{K}[K]$ arising from a K-grading of the affine algebra R, we introduced the assignments

$$\Phi \mapsto U(\Phi), \qquad \lambda \mapsto U(\lambda)$$

from bunches of orbit cones Φ to the good H-sets of Z or from GIT cones λ to the good H-sets of Z. These assignments are order reversing with respect to the realtions "$\leq$" on bunches and "$\preceq$" on the GIT cones.

If R is factorially K-graded with only constant invertible global homogeneous functions, then the bunches of orbit cones are precisely the $\mathfrak{F}$-bunches for any system $\mathfrak{F}$ of pairwise nonassociated K-prime generators of R; see Remark 3.2.2.3. This allows us to transfer the concepts "maximal" and "true" from $\mathfrak{F}$-bunches to bunches of orbit cones. By an *interior* GIT cone we mean one whose relative interior intersects the relative interior of the moving cone, that is, the associated bunch of orbit cones is true.

Theorem 4.5.2.6 *Let G be a connected affine algebraic group with $\operatorname{Pic}(G) = 0$ and $\mathbb{X}(G) = 0$, let $F \subseteq G$ be a Grosshans extension and consider the action of $H := \operatorname{Spec} \mathbb{K}[\mathbb{X}(F)]$ on $Z := \Gamma(G/F_1, \mathcal{O})$. Then we have a contravariant equivalence of categories:*

$$\{\textit{true maximal bunches of } H\textit{-orbit cones}\} \to \left\{ \begin{array}{l} A_2\textit{-maximal small} \\ G/F\textit{-embeddings} \end{array} \right\}$$

$$\Phi \mapsto U(\Phi) /\!\!/ H.$$

If $\Gamma(G/F, \mathcal{O}) = \mathbb{K}$ holds, then we have in addition a contravariant equivalence of categories:

$$\{\textit{interior GIT cones}\} \to \left\{ \begin{array}{l} \textit{projective small} \\ G/F\textit{-embeddings} \end{array} \right\}$$

$$\lambda \mapsto U(\lambda) /\!\!/ H.$$

In both cases, the affine H-variety is the total coordinate space and the respective quotient maps are the characteristic spaces.

Proof According to Corollary 3.1.4.6, the sets $U(\Phi) \subseteq Z$ defined by maximal bunches $\Phi \subseteq \Omega_Z$ are G-invariant. Moreover, any true maximal bunch $\Phi \subseteq \Omega_Z$ contains the generic orbit cone ω_{eF_1}. Consequently, $G/F_1 \subseteq U(\Phi)$ holds, and this is an H-saturated inclusion. Thus, Construction 4.5.2.4 shows that $U(\Phi) /\!\!/ H$ is a small G/F-embedding. Moreover, Theorem 3.1.4.4 yields that $\Phi \mapsto U(\Phi) /\!\!/ H$ is a contravariant functor.

Now consider any small G/F-embedding X. Then Z is also a total coordinate space for X. This gives rise to an H-saturated open embedding $G/F_1 \to \widehat{X}$

fitting into a commutative diagram

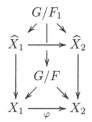

For any morphism $\varphi \colon X_1 \to X_2$ of two small G/F-embeddings X_1 and X_2, the associated commutative diagrams fit together to commutative diagram

$$
\begin{array}{ccc}
& G/F_1 & \\
\widehat{X}_1 & \longrightarrow & \widehat{X}_2 \\
& G/F & \\
X_1 & \underset{\varphi}{\longrightarrow} & X_2
\end{array}
$$

where $\widehat{X}_1 \to \widehat{X}_2$ is an open inclusion. Moreover, $\varphi \colon X_1 \to X_2$ is an open embedding if and only if $\widehat{X}_1 \subseteq \widehat{X}_2$ is H-saturated. From this we conclude that a maximal Φ defines an A_2-maximal X and that φ is an essentially surjective fully faithful functor.

The statements concerning the projective case are immediate consequences of the last part of Theorem 3.1.4.4 and the A_2-maximal case just settled. $\quad\square$

Remark 4.5.2.7 If we drop the assumption $\Gamma(G/F, \mathcal{O}) = \mathbb{K}$ in the second part of Theorem 4.5.2.6, then the interior GIT cones correspond to those A_2-maximal small G/F-embeddings that are projective over $\operatorname{Spec} \Gamma(G/F, \mathcal{O})$.

In the case of a small character group $\mathbb{X}(F)$ and $\Gamma(G/F, \mathcal{O}) = \mathbb{K}$, there are only projective A_2-maximal small G/F-embeddings. More precisely, we obtain the following.

Corollary 4.5.2.8 *Let G be a connected affine algebraic group with $\operatorname{Pic}(G) = 0$ and $\mathbb{X}(G) = 0$ and let $F \subseteq G$ be Grosshans extension with $\Gamma(G/F, \mathcal{O}) = \mathbb{K}$.*

(i) *If $\mathbb{X}(F)$ is of rank at most 2, then every A_2-maximal small G/F-embedding is projective.*

(ii) *If $\mathbb{X}(F) \cong \mathbb{Z}$ holds, then there is up to isomorphy exactly one projective small G/F-embedding.*

We conclude with some words on the condition $\Gamma(G/F, \mathcal{O}) = \mathbb{K}$. By definition, this means that the subgroup $F \subseteq G$ is *epimorphic*. Any parabolic subgroup is epimorphic, but the class of epimorphic subgroups is much wider;

various characterizations of epimorphic subgroups and concrete examples can be found in [49, 50], and [150, Sec. 23 B]. If $G/F \subseteq X$ is a small completion, then $\Gamma(X, \mathcal{O}) = \Gamma(G/F, \mathcal{O}) = \mathbb{K}$, and thus F is epimorphic.

Using counterexamples to Hilbert's fourteenth problem, see Remark 4.5.1.17, one can construct epimorphic subgroups $F \subseteq G$ such that the G/F has no small completion, see [14, 24, 50, 51]. Moreover, a homogeneous space G/F admits a small projective embedding if and only if F is epimorphic and there is a character χ of F such that $\text{Ker}(\chi)$ is a Grosshans subgroup of G; see [14, Thm. 1.1]. More precisely, projective small embeddings of G/F are defined by such characters χ, that is,

$$X = \text{Proj}(S), \quad \text{where} \quad S = \Gamma(G/\text{Ker}(\chi), \mathcal{O}) = \bigoplus_{k \geq 0} \Gamma(G/\text{Ker}(\chi), \mathcal{O})_{k\chi}.$$

By Theorem 4.5.2.6, two characters, χ_1 and χ_2, define isomorphic projective small embeddings if and only if their GIT cones coincide.

4.5.3 Examples of small embeddings

We provide a series of examples for small embbedings. Combining the description of the preceding section with the approach via bunched rings developed in Chapter 3 allows one to describe geometric properties. The original reference to this section is [18]. The ground field $\mathbb{K}$ is algebraically closed of characteristic zero.

Proposition 4.5.3.1 *Let $G := \text{SL}_m$ act diagonally on $(\mathbb{K}^m)^s$, where $s \leq m - 1$, and consider the isotropy subgroup*

$$S := G_{(e_1,\dots,e_s)} = \left\{ \begin{bmatrix} E_s & A \\ 0 & B \end{bmatrix}; \ B \in \text{SL}_{m-s}, \ A \in \text{Mat}(s \times (m - s)) \right\}.$$

Then S is a connected Grosshans subgroup of G. A maximal torus $T_S \subseteq N_G(S)/S$ is given as the isomorphic image under $\pi : N_G(S) \to N_G(S)/S$ of

$$T_S' := \{\text{diag}(t_1,\dots,t_s, t^{-1}, 1,\dots, 1); \ t_i \in \mathbb{K}^*, \ t = t_1 \cdots t_s\} \subseteq N_G(S).$$

Moreover, we have $Z = \text{Spec}\,\Gamma(G, \mathcal{O})^S = (\mathbb{K}^m)^s$, the torus T_S acts on the variety Z via

$$t \cdot (v_1,\dots, v_s) = (t_1^{-1}v_1,\dots, t_s^{-1}v_s),$$

and the T_S-orbit cones in $\mathbb{X}_{\mathbb{Q}}(T) = \mathbb{Q}^s$ are precisely the cones generated by weights of the coordinate functions.

Proof The complement of the open G-orbit in $(\mathbb{K}^m)^s$ is the variety of collections of linearly dependent vectors; thus it has codimension ≥ 2. This implies

$$Z = \text{Spec}\,\Gamma(G/S, \mathcal{O}) \cong (\mathbb{K}^m)^s.$$

In particular, S is a Grosshans subgroup of G. The normalizer $N_G(S)$ comprises, and hence coincides with the maximal parabolic subgroup

$$P = \left\{ \begin{bmatrix} C & A \\ 0 & B \end{bmatrix} \right\} \subseteq \mathrm{SL}_m$$

and we have $N_G(S)/S \cong \mathrm{GL}_s$. Clearly, the projection $\pi : N_G(S) \to N_G(S)/S$ maps T_S' isomorphically onto a maximal torus of GL_s. The further statements are obvious. $\qquad\square$

Remark 4.5.3.2 In the notation of Proposition 4.5.3.1, consider a surjection of abelain groups $Q: \mathbb{X}(T_S) \to K$ and the corresponding closed subgroup $H \subseteq T_S$. Then $F_H := \pi^{-1}(H)$ is a Grosshans extension in G with $(F_H)_1 = S$.

Example 4.5.3.3 In the notation of Proposition 4.5.3.1, take $m = 3$ and $s = 2$. So, we have $G = \mathrm{SL}_3$ acting diagonally on $(\mathbb{K}^3)^2$. Consider the subtorus

$$H' := \mathrm{diag}(t, t^{-1}, 1) \subseteq T_S' \subseteq N_G(S),$$

and set $H = \pi(H')$. Then the corresponding Grosshans extension is

$$F = F_H = \left\{ \begin{bmatrix} t & 0 & a \\ 0 & t^{-1} & b \\ 0 & 0 & 1 \end{bmatrix}; t \in \mathbb{K}^*,\, a, b \in \mathbb{K} \right\},$$

The algebra $\Gamma(G, \mathcal{O})^F = \Gamma(Z, \mathcal{O})$ is generated by the coordinate functions of $Z = (\mathbb{K}^3)^2$, and hence the weights of the generators in $\mathbb{Z} = \mathbb{X}(H)$ are $u_1 = 1$ and $u_2 = -1$. The collection of orbit cones and the possible maximal bunches are given by

$$\Omega_Z = \{\mathbb{Q}, \mathbb{Q}_{\geq 0}, \mathbb{Q}_{\leq 0}, 0\}, \quad \Phi_0 = \{\mathbb{Q}, 0\}, \quad \Phi_1 = \{\mathbb{Q}, \mathbb{Q}_{\geq 0}\}, \quad \Phi_2 = \{\mathbb{Q}, \mathbb{Q}_{\leq 0}\}.$$

Thus, we see that for the homogeneous space G/F there are, up to isomorphy, precisely three A_2-maximal small G/F-embeddings. We discuss them below a little more in detail.

The set $U(\Phi_0)$ associated with Φ_0 is the whole $Z = (\mathbb{K}^3)^2$. The resulting small G/F-embedding $X_0 = U(\Phi_0)/\!/H$ is an affine cone with apex $x_1 \in X_0$; it may be realized in the G-module $\mathbb{K}^3 \otimes \mathbb{K}^3$ as the closure of the G-orbit through $(e_1 \otimes e_2)$ with the quotient map

$$U(\Phi_0) \to X_0, \qquad (v_1, v_2) \mapsto v_1 \otimes v_2.$$

For the collection Φ_1, one has $U(\Phi_1) = \{(v_1, v_2);\ v_1 \neq 0\}$. The resulting small G/F-embedding $X_1 = U(\Phi_1)/\!/H$ is quasiprojective but not affine. Indeed, the quotient map may be realized via

$$U(\Phi_1) \to (\mathbb{K}^3 \otimes \mathbb{K}^3) \oplus \mathbb{P}^2, \qquad (v_1, v_2) \mapsto (v_1 \otimes v_2, \langle v_1 \rangle).$$

From Theorem 4.5.2.6 we know that there is a morphism $X_1 \to X_0$ of G/F-embeddings. In fact, this is the projection to $\mathbb{K}^3 \otimes \mathbb{K}^3$; this map is an isomorphism over $X_0 \setminus \{x_1\}$, and the fiber over the apex x_1 is isomorphic to $\mathbb{P}^2$.

The variety $X_2 = U(\Phi_2)/\!/H$ is isomorphic to X_1 as a G-variety, but not as a G/F-embedding (there is no base point preserving equivariant morphism). We may realize X_2 by the same construction as X_1 but twisted by the automorphism θ of $Z = \mathbb{K}^3 \oplus \mathbb{K}^3$, given by $\theta(v_1, v_2) = (v_2, v_1)$.

The next example gives smooth projective small G/F-embeddings; recall from [51, Sec. 5] that the existence of a smooth projective small G/F-embedding implies that G/F is generically rationally connected.

Example 4.5.3.4   In the setting of Proposition 4.5.3.1, let $m := 4$ and $s := 3$. Then the torus $T_S \subset N_G(S)/S$ is of dimension 3, and it may be identified with

$$\{\mathrm{diag}(t_1, t_2, t_3, t_1^{-1}t_2^{-1}t_3^{-1}); \ t_i \in \mathbb{K}^*\} \subset N_G(S).$$

We consider the projection $\mathbb{X}(T_S) \to \mathbb{Z}^2$ sending the canonical generators of the character group $\mathbb{X}(T_S)$ to the following lattice vectors:

$$u_1 := (1, 0), \qquad u_0 := (1, 1), \qquad u_2 := (0, 1).$$

Thus, speaking more concretely, we deal with $G = \mathrm{SL}_4$, the Grosshans subgroup $S \subset G$ as in 4.5.3.1, a two-dimensional torus $H \subset N_G(S)/S$, and the Grosshans extension $F = F_H \subset G$ given by

$$F = \left\{ \begin{bmatrix} t_1 & 0 & 0 & a_1 \\ 0 & t_1 t_2 & 0 & a_2 \\ 0 & 0 & t_2 & a_3 \\ 0 & 0 & 0 & t_1^{-2} t_2^{-2} \end{bmatrix}; \ t_1, t_2 \in \mathbb{K}^*, \ a_i \in \mathbb{K} \right\},$$

Let us determine the bunched rings $(R, \mathfrak{F}, \Phi)$ describing the possible A_2-maximal small G/F-embeddings. First of all, $R = \Gamma(G, \mathcal{O})^S$ is the ring of functions of $G/S = (\mathbb{K}^4)^3$. So, R is a polynomial ring, and as a system of generators $\{f_1, \ldots, f_{12}\} \subset R$, one may take the collection of indeterminates.

The remaining task is to determine the possible $\mathfrak{F}$-bunches. As we know from Corollary 4.5.2.8, these bunches correspond to projective varieties, and hence we only need to know the GIT fan in $\mathbb{Z}^2 \cong \mathbb{X}(H)$ of the action of H on G/S. This in turn is easy to determine; it looks as follows:

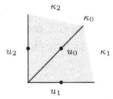

So, the possible $\mathfrak{F}$-bunches are those arising from the interior GIT cones κ_1, κ_0, and κ_2 as indicated above, and they are explicitly given by

$$\Phi_1 = \{\kappa_1\}, \qquad \Phi_0 = \{\kappa_0\}, \qquad \Phi_2 = \{\kappa_2\}.$$

Let X_i denote the small G/F-embedding corresponding to the $\mathfrak{F}$-bunch Φ_i. Then, applying the results on the geometry of X_i provided in Section 3.3, we see, for example, that X_1 and X_2 are smooth, whereas X_0 has non-$\mathbb{Q}$-factorial singularities.

Moreover, all varieties X_i have a divisor class group of rank 2, and X_0 has a Picard group of rank 1. Finally, Theorem 4.5.2.6 tells us that the possible morphisms of G/F-embeddings are

$$X_1 \longrightarrow X_0 \longleftarrow X_2.$$

By determining explicitly the varieties $U(\kappa_i)$ over X_i one may describe these morphisms. It turns out that for $i = 1, 2$ the exceptional locus of $X_i \to X_0$ is isomorphic to $\mathbb{P}^3 \times \mathbb{P}^3$ and is contracted to a $\mathbb{P}^3$ lying in X_0.

Moreover, one obtains that, as G-varieties, X_1 and X_2 are isomorphic, but, of course, not as G/F-embeddings.

By slight modification of the preceding example, we present a homogeneous space SL_4/F that does not admit any smooth small completion. Existence of such examples is due to M. Brion, as mentioned in [51].

Example 4.5.3.5 In the setting of Proposition 4.5.3.1, let $m := 4$ and $s := 3$. As before, T_S is of dimension 3, but now we consider the projection $\mathbb{X}(T_S) \to \mathbb{Z}^2$ sending the canonical generators to

$$u_1 := (1, 0), \qquad u_0 := (2, 3), \qquad u_2 := (0, 1).$$

Concretely this means that we have again the Grosshans subgroup $S \subset G = SL_4$, but another two-dimensional torus $H \subset N_G(S)/S$. The Grosshans extension $F = F_H \subset G$ this time is given by

$$F = \left\{ \begin{bmatrix} t_1 & 0 & 0 & a_1 \\ 0 & t_1^2 t_2^3 & 0 & a_2 \\ 0 & 0 & t_2 & a_3 \\ 0 & 0 & 0 & t_1^{-3} t_2^{-4} \end{bmatrix} ; \ t_1, t_2 \in \mathbb{K}^*, \ a_i \in \mathbb{K} \right\},$$

The possible small G/F-completions arise from H-maximal open subsets of $Z = (\mathbb{K}^4)^3$. All of them are toric, hence A_2-maximal, hence projective; use, for example, Corollary 4.5.2.8. Thus the GIT fan for the H-action on $Z = (\mathbb{K}^4)^3$ gives the full information; it looks as follows:

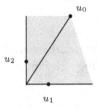

Using Proposition 3.3.1.11 (i), we see that each of the three possible projective small G/F-embeddings is singular; in two cases, we have $\mathbb{Q}$-factorial singularities, in the remaining one, the singularities are even worse.

Example 4.5.3.6   In the setting of Proposition 4.5.3.1, let $m = 7$ and $s = 6$. So, we have $G = \mathrm{SL}_7$ acting diagonally on $(\mathbb{K}^7)^6$, and the torus T_S is of dimension 6. Set $K := \mathbb{Z}^3$ and consider the map $\mathbb{X}(T_S) \to K$ sending the canonical generators to

$$e_1, \quad e_2, \quad e_3, \quad w_1 := e_1 + e_2, \quad w_2 := e_1 + e_3, \quad w_3 := e_2 + e_3.$$

Let $\mathfrak{F} = \{f_1, \ldots, f_{42}\}$ be the indeterminates of the polynomial ring $\Gamma(G, \mathcal{O})^S$. Then the following cones in $K_\mathbb{Q}$ define an $\mathfrak{F}$-bunch:

$$\mathrm{cone}(e_3, w_1, w_2), \quad \mathrm{cone}(e_1, w_1, w_3), \quad \mathrm{cone}(e_2, w_2, w_3), \quad \mathrm{cone}(w_1, w_1, w_2),$$

Combining [43, Example 11.2] and [43, Constr. 11.4] shows that the corresponding small G/F-embedding $X(R, \mathfrak{F}, \Phi)$ is complete and $\mathbb{Q}$-factorial but not projective.

All small embeddings considered up to now are toric varieties. A classification of 2-complete small embeddings $G/F \subset X$, where G is a simple group and X admits the structure of a toric variety, is given in [18, Prop. 4.7]. Let us provide some examples of nontoric small embeddings.

Proposition 4.5.3.7   *Let $\mathbb{K}^{2m}$ be the symplectic vector space with the skew-symmetric bilinear form $\langle . , . \rangle$, given as*

$$\begin{bmatrix} 0 & E_m \\ -E_m & 0 \end{bmatrix},$$

and $G = \mathrm{Sp}_{2m}$ be the symplectic group. Consider the diagonal G-action on $(\mathbb{K}^{2m})^s$, where $s \leq m$, and the isotropy group

$$S := G_{(e_1, \ldots, e_s)}.$$

Then S is a connected Grosshans subgroup of G, and a maximal torus $T_S \subset N_G(S)/S$ is the isomorphic image of

$$T'_S := \{\operatorname{diag}(t_1, \ldots, t_s, 1, \ldots, 1, t_1^{-1}, \ldots, t_s^{-1}, 1, \ldots, 1,); t_i \in \mathbb{K}^*\} \subseteq N_G(S).$$

The affine variety $Z = \operatorname{Spec} \Gamma(G, \mathcal{O})^S$ can be realized as the G-orbit closure of $(e_1, \ldots, e_s)$ and is given by

$$Z = \{(v_1, \ldots, v_s); \langle v_i, v_j \rangle = 0 \; \forall \, i, j\}.$$

The action of T_S on the variety Z is given as

$$t \cdot (v_1, \ldots, v_s) = (t_1^{-1} v_1, \ldots, t_s^{-1} v_s),$$

and every cone generated by weights of the restricted coordinate functions is a T_S-orbit cone.

Proof First, note that we have $Z = G \cdot (L)^s$, where $L = \langle e_1, \ldots, e_m \rangle$ is the Lagrangian subspace. This shows that the complement of the open orbit $G \cdot (e_1, \ldots, e_s)$ has codimension at least 2 in Z. Moreover, Serre's normality criterion implies that Z is normal. We conclude $\Gamma(Z, \mathcal{O}) = \Gamma(G/S, \mathcal{O})$. Second, the normalizer $N_G(S)$ is again a maximal parabolic subgroup of G, we have $N_G(S)/S \cong \operatorname{GL}_s$, and the assertion follows. $\qquad\square$

Example 4.5.3.8 In the setting of Proposition 4.5.3.7, let $m = 3$, and $s = 3$. Then the maximal torus $T_S \subset N_G(S)/S$ is of dimension 3. Consider a surjection $\mathbb{X}(T_S) \to \mathbb{Z}^2$, sending the canonical generators to $u_1, u_0, u_2 \in \mathbb{Z}^2$. Then our Grosshans extension $F = F_H \subset \operatorname{Sp}_6$ consists of the matrices

$$\begin{bmatrix} \chi^{u_1}(t) & 0 & 0 & a_{11} & a_{12} & a_{13} \\ 0 & \chi^{u_2}(t) & 0 & a_{12} & a_{22} & a_{23} \\ 0 & 0 & \chi^{u_3}(t) & a_{13} & a_{23} & a_{33} \\ 0 & 0 & 0 & \chi^{-u_1}(t) & 0 & 0 \\ 0 & 0 & 0 & 0 & \chi^{-u_2}(t) & 0 \\ 0 & 0 & 0 & 0 & 0 & \chi^{-u_3}(t) \end{bmatrix}, \quad t \in T_S; \; a_{ij} \in \mathbb{K}.$$

Taking u_i as in Example 4.5.3.4, we obtain three projective small G/F-embeddings X_1, X_0, and X_2. By Corollary 3.3.1.8, the varieties X_1 and X_2 are locally factorial and X_3 is not $\mathbb{Q}$-factorial. In fact, an analysis of the singular locus of the cone $Z = \operatorname{Spec} \Gamma(G, \mathcal{O})^S$ shows that open subsets $U(\Phi_1)$ and $U(\Phi_2)$ lying over X_1 and X_2 respectively are smooth. Thus, Proposition 3.3.1.11 yields that the varieties X_1 and X_2 are even smooth.

Taking u_i as in Example 4.5.3.5, one obtains another torus H, and other projective small G/F-embeddings X_1, X_0, and X_2. Then Corollary 3.3.1.8 tells us that X_1 and X_2 are $\mathbb{Q}$-factorial, but not locally factorial.

4.5.4 Spherical varieties

Spherical varieties form a special class of almost homogeneous varieties naturally extending the class of toric varieties; see [187, 296] for general background. We provide the basic notions, describe the divisor class group, provide generators for the Cox ring and briefly discuss the ideal of relations in the Cox ring. In this section, we follow the ideas of [76]. The ground field $\mathbb{K}$ is algebraically closed and of characteristic zero.

Definition 4.5.4.1 A *spherical variety* is an irreducible normal variety X together with an action $G \times X \to X$ of a connected reductive affine algebraic group G, a Borel subgroup $B \subseteq G$, and a base point $x_0 \subseteq X$ such that the orbit map $B \to X$, $g \mapsto g \cdot x_0$ is an open embedding.

Observe that this directly extends the definition of a toric variety given in 2.1.1.1: there, the acting group G is a torus and the Borel group B equals G. If we want to specify the acting group, the Borel subgroup, and the base point for spherical variety X, we write (X, G, B, x_0).

Definition 4.5.4.2 Let (X, G, B, x_0) be a spherical variety. One distinguishes between two types of B-invariant prime divisors:

(i) A *boundary divisor* of X is a G-invariant prime divisor on X,
(ii) A *color* of X is a B-invariant prime divisor on X that is not G-invariant.

In the case of a toric variety, there are no colors due to $B = G$. Another example is the projective space $\mathbb{P}^n$ with the natural action $A \cdot [z] = [Az]$ of SL_{n+1}: together with the Borel subgroup B consisting of all upper triangular matrices of SL_{n+1} and the base point $[0, \ldots, 0, 1]$ it is a spherical variety without boundary divisors and precisely one color, namely $V(z_n) \subseteq \mathbb{P}^n$.

Example 4.5.4.3 Consider the special linear group $G := SL_3$ and the Borel subgroup $B \subseteq G$ consisting of the upper triangular matrices. Then G acts diagonally on $Y := \mathbb{P}^2 \times \mathbb{P}^2$:

$$A \cdot (z, w) := (Az, Aw), \qquad z = [z_0, z_1, z_2], \quad w = [w_0, w_1, w_2].$$

With the base point $y_0 := ([0, 0, 1], [0, 1, 1]) \in Y$, we have a spherical variety (Y, G, B, y_0); in fact, the orbit of Borel subgroup $B \subseteq G$ through the point y_0 is the open subset

$$B \cdot y_0 = \{(z, w) \in Y;\ z_2 \neq 0,\ w_2 \neq 0,\ z_1 w_2 - z_2 w_1 \neq 0\}.$$

The divisors $V(z_2)$, $V(w_2)$, and $V(z_1 w_2 - z_2 w_1)$ are colors in Y and there are no boundary divisors. Consider the G-invariant two-dimensional subset of Y given by the diagonal

$$\Delta := \{(z, z);\ z \in \mathbb{P}^2\} \subseteq Y.$$

The blow-up X of Y at Δ comes with a unique G-action making the contraction $\kappa \colon X \to Y$ equivariant. With the point $x_0 \in X$ mapped to $y_0 \in Y$, we have a spherical variety (X, G, B, x_0). Denoting by $\kappa^\sharp(C)$ the proper transform of a prime divisor $C \subseteq Y$, the boundary divisors and colors of X are given as

$$E_1 := \kappa^\sharp(\Delta), \qquad D_1 := \kappa^\sharp(V(z_2)),$$

$$D_2 := \kappa^\sharp(V(w_2)), \qquad D_3 := \kappa^\sharp(V(z_1 w_2 - z_2 w_1)).$$

For a toric variety, the monoid of effective divisor classes and thus the divisors class group are generated by the boundary divisors; see Proposition 2.1.2.7. For a general spherical variety, we also have to take the colors into account.

Proposition 4.5.4.4 *Let (X, G, B, x_0) be a spherical variety.*

(i) *There are only finitely many boundary divisors $E_1, \ldots, E_r$ and only finitely many colors $D_1, \ldots, D_s$ on X and we have*

$$X \setminus B \cdot x_0 = E_1 \cup \ldots \cup E_r \cup D_1 \cup \ldots \cup D_s.$$

(ii) *The classes of the E_k and D_i generate the monoid of effective divisor classes. In particular they generate $\mathrm{Cl}(X)$ as a group and $\mathrm{Eff}(X) \subseteq \mathrm{Cl}_{\mathbb{Q}}(X)$ as a cone.*

(iii) *Let $\eta_1, \ldots, \eta_l$ be a basis for the sublattice $M(X) \subseteq \mathbb{X}(B)$ of weights of nonzero B-homogeneous rational functions on X. Then the kernel of the epimorphism $\mathbb{Z}^{r+s} \to \mathrm{Cl}(X)$ given by $e_i \mapsto [E_i]$ for $i = 1, \ldots, r$ and $e_i \mapsto [D_i]$ for $i = r + 1, \ldots, r + s$ is generated by the vectors*

$$(\mathrm{ord}_{E_1}(\eta_i), \ldots, \mathrm{ord}_{E_r}(\eta_i), \mathrm{ord}_{D_1}(\eta_i), \ldots, \mathrm{ord}_{D_s}(\eta_i)), \qquad 1 \le i \le l.$$

Lemma 4.5.4.5 *Let X be an irreducible normal variety with an action $B \times X \to X$ of a connected solvable affine algebraic group B. Then every (effective) divisor on X is linearly equivalent to an (effective) B-invariant one.*

Proof We may assume that X is smooth and it suffices to consider effective divisors D on X. Linearizing the line bundle associated with D, we obtain a representation of B on $\Gamma(X, \mathcal{O}(D))$. Because B is connected solvable, this representation has a B-eigenvector f. The divisor D is linearly equivalent to $\mathrm{div}_D(f)$ and the latter is B-invariant. $\qquad\square$

Proof of Proposition 4.5.4.4 Because $G \cdot x_0$ and $B \cdot x_0$ are open in X, there exist only finitely many boundary divisors and colors on X. The displayed formula follows from the fact that $B \cdot x_0$ is affine and hence its complement is the union of the B-invariant prime divisors. This proves (i). Assertion (ii) follows directly from Lemma 4.5.4.5. To obtain (iii) note that (ii) provides us with an exact sequence

$$M(X) \longrightarrow \mathrm{WDiv}^B(X) \longrightarrow \mathrm{Cl}(X) \longrightarrow 0.$$

Indeed, the B-invariant principal divisors stem from B-homogeneous rational functions $f \in \mathbb{K}(X)$; see [301, Thm. 3.1]. Any such f is determined up to a scalar by its weight $\chi \in \mathbb{X}(B)$, because B acts with an open orbit on X. Thus, $\chi \mapsto \mathrm{div}(f)$ gives the desired homomorphism $M(X) \to \mathrm{WDiv}^B(X)$. Passing to coordinates with respect to the η_i and the E_k, D_i gives the assertion. $\qquad\square$

We turn to the Cox ring of a spherical variety (X, G, B, x_0). Theorem 4.2.3.2 provides us with a lifting over the set of smooth points $X_{\mathrm{reg}} \subseteq X$. That means that we have a finite covering $\varepsilon \colon G' \to G$ of connected affine algebraic groups and a G'-action on $\widehat{X}_{\mathrm{reg}}$ commuting with the H_X-action such that $q \colon \widehat{X}_{\mathrm{reg}} \to X_{\mathrm{reg}}$ satisfies $q(g \cdot z) = \varepsilon(g) \cdot q(z)$. In particular, $(g \cdot f)(z) := f(g^{-1} \cdot z)$ defines a G'-representation on the Cox ring

$$\mathcal{R}(X) = \mathcal{R}(X_{\mathrm{reg}}) = \Gamma(\widehat{X}_{\mathrm{reg}}, \mathcal{O})$$

respecting the $\mathrm{Cl}(X)$-grading. Fix a Borel subgroup $B' \subseteq G'$ with $\varepsilon(B') = B$ and a maximal unipotent subgroup $U' \subseteq B'$ such that $\varepsilon(U') \subseteq B$ is a maximal unipotent subgroup of G. Note that if G is semisimple and simply connected; then Theorem 4.2.3.2 (iii) ensures that we may choose $G' = G$ with $\varepsilon = \mathrm{id}$; in particular, we then have $U = U' \subseteq B' = B$.

Theorem 4.5.4.6 *Let (X, G, B, x_0) be a spherical variety with $\Gamma(X, \mathcal{O}^*) = \mathbb{K}^*$. Denote by $E_1, \ldots, E_r$ the boundary divisors and by $D_1, \ldots, D_s$ the colors of X. Then, with the representation of G' on $\mathcal{R}(X)$ and a maximal unipotent subgroup $U' \subseteq B'$ as introduced, the following hold.*

(i) *The subalgebra of U'-invariants $\mathcal{R}(X)^{U'} \subseteq \mathcal{R}(X)$ is freely generated by the canonical sections 1_{E_k} of the boundary divisors and 1_{D_i} of the colors.*

(ii) *The subalgebra of G'-invariants $\mathcal{R}(X)^{G'} \subseteq \mathcal{R}(X)$ is freely generated by the canonical sections 1_{E_k} of the boundary divisors.*

(iii) *The Cox ring $\mathcal{R}(X)$ is a free $\mathcal{R}(X)^{G'}$-module. As a $\mathbb{K}$-algebra, $\mathcal{R}(X)$ is generated by $1_{E_1}, \ldots, 1_{E_r}$ and the finite-dimensional vector subspaces $\mathrm{lin}_{\mathbb{K}}(G' \cdot 1_{D_i}) \subseteq \mathcal{R}(X)$, where $1 \le i \le s$.*

Proof We may assume that X is smooth and thus the action of G lifts to a G'-action on $\widehat{X}$ commuting with the action of the characteristic quasitorus H_X.

To prove (i), observe first that the action of $B' \times H_X$ on $\widehat{X}$ defines a grading of the algebra of U'-invariants:

$$\mathcal{R}(X)^{U'} = \bigoplus_{\mathbb{X}(B') \times \mathrm{Cl}(X)} \mathcal{R}(X)_\chi^{U'}.$$

The canonical sections 1_{D_i} and 1_{E_k} have $B' \times H_X$-invariant divisors. Consequently, they are $B' \times H_X$-homogeneous and thus belong to $\mathcal{R}(X)^{U'}$.

Because $B' \times H_X$ acts with an open orbit on $\widehat{X}$, all homogeneous components of $\mathcal{R}(X)^{U'}$ are one-dimensional $\mathbb{K}$-vector spaces. Thus any collection of pairwise nonproportional homogeneous elements in $\mathcal{R}(X)^{U'}$ is linearly independent. We conclude that any set of pairwise different monomials in the 1_{D_i} and 1_{E_k} is linearly independent, because two different monomials of this type have different divisors on $\widehat{X}$ and thus are nonproportional. In other words, the 1_{D_i} and 1_{E_k} generate a polynomial ring over $\mathbb{K}$.

Given $f \in \mathcal{R}(X)^{U'}$, write it as a sum $f = \sum f_\chi$ of nonzero homogeneous elements. The degrees are pairs $\chi = (\eta, [D])$, where $[D]$ is the usual $\mathrm{Cl}(X)$-degree of f_χ in $\mathcal{R}(X)$ and f_χ is B'-homogeneous with respect to the character $\eta \in \mathbb{X}(B')$. In particular, the divisor of f_χ on $\widehat{X}$ is B'-invariant. Then the $[D]$-divisor of f_χ on X is B-invariant and Proposition 4.5.4.4 gives

$$\mathrm{div}_{[D]}(f_\chi) = \sum a_i D_i + \sum b_k E_k$$

with nonnegative integers a_i and b_k. Thus, Proposition 1.5.3.5 tells us that, up to a constant, f_χ is the product of the powers $1_{D_i}^{a_i}$ and $1_{E_k}^{b_k}$. We conclude that f is a polynomial in the elements 1_{E_k} and 1_{D_i}. This proves (i).

Assertion (ii) follows from (i) and the observation that the G'-invariant elements of $\mathcal{R}(X)$ are precisely the polynomials in $1_{E_1}, \ldots, 1_{E_r}$.

To prove (iii), consider the G'-submodule $M \subseteq \mathcal{R}(X)$ generated by all monomials in the canonical sections $1_{D_1}, \ldots, 1_{D_s}$. Note that $M^{U'}$ equals the polynomial ring $\mathbb{K}[1_{D_1}, \ldots, 1_{D_s}]$. The multiplication map defines a homomorphism of G'-modules

$$\mu \colon \mathcal{R}(X)^{G'} \otimes M \rightarrow \mathcal{R}(X).$$

By (i), the restriction to the U'-invariants $(\mathcal{R}(X)^{G'} \otimes M)^{U'} = \mathcal{R}(X)^{G'} \otimes M^{U'}$ defines an isomorphism onto $\mathcal{R}(X)^{U'}$. Using basic theory of G-modules, see [301, Sec. 3.14], we can conclude that μ is an isomorphism. Moreover, as a G'-module, $\mathcal{R}(X)$ is generated by $\mathcal{R}(X)^{U'}$. Consequently, $\mathcal{R}(X)$ is generated as a $\mathbb{K}$-algebra by the 1_{E_k} and the elements of the $\mathrm{lin}_{\mathbb{K}}(G' \cdot 1_{D_i})$. $\qquad\square$

An explicit description of the ideal of relations between the above generators is involved in general, see Brion's result on the wonderful case stated in Proposition 4.5.5.5. The following remark shows that the main difficulties are located in the embedded homogeneous space.

Remark 4.5.4.7 In the setting of Theorem 4.5.4.6, consider the canonical sections 1_{E_k} of the boundary divisors and 1_{D_i} of the colors. Choose a basis $f_{i1}, \ldots, f_{in_i}$ for each $\mathrm{lin}_{\mathbb{K}}(G' \cdot 1_{D_i})$ with $f_{i1} = 1_{D_i}$. Then Theorem 4.5.4.6 provides us with an epimorphism

$$\Phi \colon \mathbb{K}[S_k, T_{ij}] \ \to \ \mathcal{R}(X), \qquad S_k \mapsto 1_{E_k}, \quad T_{ij} \mapsto f_{ij}.$$

and an associated $G' \times H_X$-equivariant closed embedding $\overline{X} \subseteq \mathbb{K}^{r+n}$, where $n = n_1 + \cdots + n_s$. Set $\vartheta := S_1 \cdots S_r \cdot T_{11} \cdots T_{s1}$. Then we obtain a commutative diagram

$$
\begin{array}{ccc}
\overline{X}_\vartheta & \subseteq & \mathbb{K}^{r+n}_\vartheta \\
{\scriptstyle /H_X}\downarrow & & \downarrow{\scriptstyle /H_X} \\
B \cdot x_0 & \subseteq & Z
\end{array}
$$

with the affine toric variety $Z = \mathbb{K}^{r+n}_\vartheta / H_X$. The homomorphism $\pi \colon \mathbb{T}^{r+n} \to \mathbb{T}^m$ of acting tori, where $m = \dim(Z)$, is given by the relation matrix P of the generator degrees in $\mathrm{Cl}(X)$ which in turn can be read off from Proposition 4.5.4.4, as indicated below. Then, with the ideal I of $B \cdot x_0$ in Z, the ideal $\ker(\Phi)$ of relations of the Cox ring $\mathcal{R}(X)$ is the saturation

$$\ker(\Phi) \ = \ \pi^* I : \vartheta^\infty \ \subseteq \ \mathbb{K}[S_k, T_{ij}].$$

A closer look to the ideal I gives a more intrinsic description. We may assume that the first l rows of P describe the relations in $\mathrm{Cl}(X)$ as in 4.5.4.4 (iii); in particular, they are linear combinations in the $e_k, e_{i1} \in \mathbb{Z}^{r+n}$. Each of the remaining rows can be chosen as a linear combination in the e_k, e_{ij} and precisely one e_{ij} with $j \neq 1$. Write $U_1, \ldots, U_m$ for the coordinates of the acting torus $\mathbb{T}^m$ of Z and consider

$$h_t := U_t|_{Bx_0} \in \Gamma(B \cdot x_0, \mathcal{O}), \qquad 1 \leq t \leq m.$$

Then the weights of $h_1, \ldots, h_l$ form a basis for the lattice $M(X) \subseteq \mathbb{X}(B)$ of weights stemming from B-homogeneous rational functions and the $h_{l+1}, \ldots, h_m$ correspond to the f_{ij} with $j \neq 1$. By construction, I is the ideal of relations between the generators $h_1, \ldots, h_m$ of $\Gamma(B \cdot x_0, \mathcal{O})$. Note that in [144], the saturation $\pi^* I : \vartheta^\infty$ is basically expressed in terms of I and data of the spherical variety (X, G, B, x_0).

Example 4.5.4.8 Consider again the blow-up X of the diagonal $\Delta \subseteq \mathbb{P}^2 \times \mathbb{P}^2$ with the action of $G := \mathrm{SL}_3$ as introduced in Example 4.5.4.3. According to Theorem 4.5.4.6, the Cox ring of X is generated by the canonical sections 1_{E_1}

of the boundary divisor and the canonical sections $1_{D_{ij}}$ of the translated colors

$$D_{11} := \kappa^{\sharp}(V(z_2)), \quad D_{12} := \kappa^{\sharp}(V(z_1)), \quad D_{13} := \kappa^{\sharp}(V(z_0)),$$

$$D_{21} := \kappa^{\sharp}(V(w_2)), \quad D_{22} := \kappa^{\sharp}(V(w_1)), \quad D_{23} := \kappa^{\sharp}(V(w_0)),$$

$$D_{31} := \kappa^{\sharp}(V(z_1 w_2 - z_2 w_1)), \qquad D_{32} := \kappa^{\sharp}(V(z_0 w_2 - z_2 w_0)),$$

$$D_{33} := \kappa^{\sharp}(V(z_0 w_1 - z_1 w_0)),$$

where $\kappa^{\sharp}$ denotes the strict transform with respect to the contraction map $\kappa \colon X \to \mathbb{P}^2 \times \mathbb{P}^2$. The divisor class group of X is $\mathrm{Cl}(X) = \mathbb{Z}^3$ and the degrees of the generators $1_{D_{1j}}, 1_{D_{2j}}, 1_{D_{3j}}, 1_{E_1}$ are the columns of the matrix

$$Q := \begin{bmatrix} 1 & 1 & 1 & 0 & 0 & 0 & 1 & 1 & 1 & 0 \\ 0 & 0 & 0 & 1 & 1 & 1 & 1 & 1 & 1 & 0 \\ 0 & 0 & 0 & 0 & 0 & 0 & -1 & -1 & -1 & 1 \end{bmatrix}$$

Finally, set $T_{ij} = 1_{D_{ij}}$ and $S_1 = 1_{E_k}$. Then the ideal of relations between the generators T_{ij} and S_k of the Cox ring $\mathcal{R}(X)$ is generated by the quadrics

$$T_{11}T_{22} - T_{12}T_{21} + T_{31}S_1, \quad T_{11}T_{23} - T_{13}T_{21} + T_{32}S_1,$$

$$T_{12}T_{23} - T_{13}T_{22} + T_{33}S_1, \quad T_{13}T_{31} - T_{12}T_{32} + T_{11}T_{33},$$

$$T_{23}T_{31} - T_{22}T_{32} + T_{21}T_{33}.$$

In particular we see that, as a ring, $\mathcal{R}(X)$ is the homogeneous coordinate ring of the Grassmannian $G(2, 5)$. The Gale dual matrix P of the grading matrix Q is, as indicated in Remark 4.5.4.7, of the form

$$P := \begin{bmatrix} 1 & 0 & 0 & 1 & 0 & 0 & -1 & 0 & 0 & -1 \\ 0 & 1 & 0 & 1 & 0 & 0 & -1 & 0 & 0 & -1 \\ 0 & 0 & 1 & 1 & 0 & 0 & -1 & 0 & 0 & -1 \\ 1 & 0 & 0 & 0 & 1 & 0 & -1 & 0 & 0 & -1 \\ 1 & 0 & 0 & 0 & 0 & 1 & -1 & 0 & 0 & -1 \\ 1 & 0 & 0 & 1 & 0 & 0 & 0 & -1 & 0 & -1 \\ 1 & 0 & 0 & 1 & 0 & 0 & 0 & 0 & -1 & -1 \end{bmatrix}$$

4.5.5 Wonderful varieties and algebraic monoids

We take a closer look to wonderful varieties, which form a special case of spherical varieties. We present Brion's description of their Cox ring following [76], and, moreover, discuss the relations to algebraic monoids due to Vinberg [299]. The ground field $\mathbb{K}$ is algebraically closed and of characteristic zero.

Definition 4.5.5.1 A *wonderful variety* is a smooth complete variety X together with an action $G \times X \to X$ of semisimple simply connected group G such that

(i) There is a point $x_0 \in X$ with open G-orbit and the complement $X \setminus G \cdot x_0$ is a union of prime divisors $E_1, \ldots, E_r$ having normal crossings.
(ii) The G-orbit closures in X are precisely the intersections $\cap_{k \in I} E_k$, where I runs through the subsets of $\{1, \ldots, r\}$.

Cheap examples of wonderful varieties are the flag varieties G/P, where G is semisimple simply connected and $P \subseteq G$ is parabolic. In Example 4.5.4.3, the variety X is wonderful with presicely two G-orbits, whereas the variety Y is spherical with two orbits but not wonderful. Many important examples occur among the complete symmetric varieties arising from De Concini and Procesi's completion of homogeneous spaces [106]; see Example 4.5.5.7 for the case of an adjoint semisimple group.

Remark 4.5.5.2 Any wonderful variety X with acting group G admits a unique closed G-orbit $G \cdot x_\infty$, namely the intersection of all G-orbit closures. Clearly, X is the only G-invariant open set containing $G \cdot x_\infty$. On the other hand, Sumihiro's Theorem [285, 189] ensures existence of a quasiprojective G-invariant open neighborhood of $G \cdot x_\infty$. Thus, X is projective.

Remark 4.5.5.3 Every wonderful variety is spherical, as proven by Luna [211]. For a wonderful variety (X, G, B, x_0) with colors $D_1, \ldots, D_s$, the *big cell*

$$X_B := X \setminus (D_1 \cup \ldots \cup D_s)$$

is an affine space, see [296, Sec. 30.1]. In particular, X_B admits only constant invertible global functions. As a consequence, we obtain

$$\mathrm{Cl}(X) = \mathbb{Z}[D_1] \oplus \ldots \oplus \mathbb{Z}[D_s].$$

Let (X, G, B, x_0) be a wonderful variety with boundary divisors $E_1, \ldots, E_r$ and colors $D_1, \ldots, D_s$. Remark 4.5.5.3 gives us identifications $\mathrm{Cl}(X) = \mathbb{Z}^s$ for the divisor class group and $H_X = \mathbb{T}^s$ for the characteristic quasitorus. Moreover, the G-action on X lifts uniquely to a G-action on the characteristic space $\widehat{X} \to X$.

We fix a maximal torus $T \subseteq B$ such that $T \cap B_0$ is a maximal quasitorus in the isotropy group $B_0 \subseteq B$ of $x_0 \in X$. Denoting by $\omega_i \in \Lambda := \mathbb{X}(B) = \mathbb{X}(T)$ the B-weight of the canonical section $1_{D_i} \in \mathcal{R}(X)$, we obtain a homomorphism

$$T \to H_X = \mathbb{T}^s, \qquad t \mapsto (\omega_1(t), \ldots, \omega_s(t))$$

and thus an action $T \times \widehat{X} \to \widehat{X}$ commuting with the action of G. This turns the Cox ring $\mathcal{R}(X)$ into a $(G \times T)$-algebra. Brion's description is given in terms of representation theory and the spherical roots of G. We briefly recall the necessary notions.

We denote by $\Lambda^+ \subseteq \Lambda$ the set of dominant weights, that is, the set of highest weights of simple G-modules. As usual, we denote for a G-module V and $\lambda \in \Lambda^+$ by $V_{(\lambda)}$ the corresponding isotypic component, that is, the sum of all simple G-submodules in V having λ as highest weight; see also Section 3.2.3. The *spherical roots* are the negatives $\gamma_1, \ldots, \gamma_r \in \Lambda$ of the weights of the canonical sections $1_{E_k} \in \mathcal{R}(X)$ of the boundary divisors. We denote by $\Lambda(X) \subseteq \Lambda$ the sublattice generated by the spherical roots. For $\lambda, \mu \in \Lambda$ write $\lambda \leq_X \mu$ if the difference $\mu - \lambda$ is a nonnegative linear combination of spherical roots.

We are ready to formulate Brion's structure theorem [76, Thm. 3.2.3] on the Cox ring. Let $G_1 \subseteq G_0$ be the intersection of the kernels of all characters of the isotropy group $G_0 \subseteq G$ of $x_0 \in X$; see also Construction 4.5.1.7. Below we will understand the invariants $\Gamma(G, \mathcal{O})^{G_1}$ with respect to the action of G_1 by multiplication from the right. For each color D_i, we denote by $f_i \in \Gamma(G, \mathcal{O})$ the function with $f_i(e_G) = 1$ and $\mathrm{div}(f_i) = \alpha^{-1}(D_i)$, where $\alpha \colon G \to X, g \mapsto g \cdot x_0$ is the orbit map. Observe that f_i is homogeneous with respect to $B \times G_0$, where B acts from the left and G_0 from the right. The B-weight of $f_i \in \Gamma(G, \mathcal{O})$ equals the B-weight ω_i of $1_{D_i} \in \mathcal{R}(X)$ and f_i is invariant under the right G_1-action.

Theorem 4.5.5.4 *Let (X, G, B, x_0) be a wonderful variety and fix a Cox sheaf $\mathcal{R}$ on X. Then, with the above notation, the Cox ring $\mathcal{R}(X) = \Gamma(X, \mathcal{R})$ fits into a commutative diagram of $G \times T$-algebras:*

$$
\begin{array}{ccccc}
\Gamma(X, \mathcal{R}) & \subseteq & \Gamma(G \cdot x_0, \mathcal{R}) & \subseteq & \Gamma(G \times T, \mathcal{O}) \\
\cong \big\uparrow\big\downarrow & & \cong \big\uparrow\big\downarrow & & \cong \big\uparrow\big\downarrow \\
\displaystyle\bigoplus_{\mu \in \Lambda} \bigoplus_{\substack{\lambda \in \Lambda^+ \\ \lambda \leq_X \mu}} \Gamma(G, \mathcal{O})^{G_1}_{(\lambda)}\mu & \subseteq & \displaystyle\bigoplus_{\mu \in \Lambda} \bigoplus_{\substack{\lambda \in \Lambda^+ \\ \mu - \lambda \in \Lambda(X)}} \Gamma(G, \mathcal{O})^{G_1}_{(\lambda)}\mu & \subseteq & \displaystyle\bigoplus_{\mu \in \Lambda} \Gamma(G, \mathcal{O})\mu
\end{array}
$$

Under the downwards isomorphism, the canonical section $1_{E_k} \in \mathcal{R}(X)$ of the boundary divisor E_k corresponds to the spherical root $\gamma_k \in \mathbb{X}(T)$ and the canonical section $1_{D_i} \in \mathcal{R}(X)$ of the color D_i corresponds to $f_i \omega_i \in \Gamma(G, \mathcal{O})^{G_1} \mathbb{X}(\mathbb{T})$.

Using the terminology just introduced, we now formulate Brion's description of the ideal of relations between the generators of the Cox ring of a wonderful variety (X, G, B, x_0). According to Theorem 4.5.4.6 we can write

$$
\mathcal{R}(X) = S/I, \qquad S := \mathbb{K}[S_1, \ldots, S_r] \otimes \mathrm{Sym}\left(\bigoplus_{i=1}^{s} V_i\right),
$$

where the variables S_k represent the canonical section 1_{E_k} of the boundary divisors and $V_i \subseteq \mathcal{R}(X)$ is the G-submodule generated by the canonical section 1_{D_i} of the color $D_i \subseteq X$.

Via Theorem 4.5.5.4, we may regard V_i as the G-submodule of $\Gamma(G, \mathcal{O})^{G_1}$ generated by the function $f_i \in \Gamma(G, \mathcal{O})$ defined before, where G acts from the left and the G_1-invariants are understood with respect to multiplication from the right. Moreover, there is a decomposition

$$\Gamma(G, \mathcal{O})^{G_1} = \bigoplus_{a \in \mathbb{Z}_{\geq 0}^s} V_a, \qquad V_a := \lim_{\mathbb{K}}(G \cdot f_1^{a_1} \cdots f_s^{a_s}).$$

In terms of the B-weight $\omega_i \in \Lambda$ and the G_0-weight $\chi_i \in \mathbb{X}(G_0)$ of f_i one can decompose any product $V_i V_j \subseteq \Gamma(G/G_0, \mathcal{O})$ into a sum over certain V_a's, where

$$\sum_{t=1}^s a_t \omega_t \leq_X \omega_i + \omega_j, \qquad \sum_{t=1}^s a_t \chi_t = \chi_i + \chi_j.$$

In particular, the components V_a of the product $V_i V_j$ may be indexed by their highest weight $\lambda = \sum_t a_t \omega_t$ and each component $V_{(\lambda)}$ embeds into $\otimes_t V_t^{\otimes a_t}$ as its Cartan component. This leads to morphisms of G-modules

$$V_i \otimes V_j \ \to \ V_i V_j \ \to \ V_{(\lambda)} \ \to \ \bigotimes_{t=1}^s V_t^{\otimes a_t} \ \to \ \mathrm{Sym}\left(\bigoplus_{i=1}^s V_i\right).$$

Call their composition p_{ij}^λ. Note that the morphism p_{ij}^λ is a priori defined only up to multiplication with a nonzero scalar. Brion's description of the ideal of relations reads as follows; see [76, Prop. 3.3.1].

Proposition 4.5.5.5 *The notation is as above. There exist unique scalings of the p_{ij}^λ such that the ideal I of relations of the Cox ring $\mathcal{R}(X)$ is generated by*

$$v_i \otimes v_j \ - \ \sum_{\substack{\lambda \in \Lambda^+ \\ \lambda \leq_X \omega_i + \omega_j}} p_{ij}^\lambda(v_i \otimes v_j) \prod_{k=1}^r S_k^{b_k},$$

where $v_i \in V_i$, $v_j \in V_j$ and $b_1, \ldots, b_r \in \mathbb{Z}_{\geq 0}$ are the coefficients of $\omega_1 + \omega_2 - \lambda$ with respect to the spherical roots, that is, $\omega_1 + \omega_2 - \lambda = b_1 \gamma_1 + \cdots + b_r \gamma_r$.

For a certain class of complete symmetric varieties, the ideal of relations admits a more direct description, as carried out by Chiviri, Littelmann, and Maffei in [89]. Moreover, Chiviri and Maffei [90] developed a "standard monomial theory" to study the Cox ring of a complete symmetric variety.

Remark 4.5.5.6 Let G be a semisimple simply connected group. Any spherical variety Y under G is equipped with a canonical equivariant rational map $\varphi\colon Y \dashrightarrow X$, where X is a wonderful variety under the same group G; see [76, Sec. 4.3]. This yields an equivariant ring homomorphism $\varphi^*\colon \mathcal{R}(X) \to \mathcal{R}(Y)$. By [76, Thm. 4.3.2], one obtains an induced isomorphism

$$\mathcal{R}(X) \otimes_{\mathcal{R}(X)^G} \mathcal{R}(Y)^G \to \mathcal{R}(Y).$$

We illustrate Theorem 4.5.5.4 with a glance at the wonderful compactification of an adjoint semisimple group; the construction is due to De Concini and Procesi [106], we refer also to [77, Sec. 6.1] for details. We first recall the necessary notation around roots of reductive groups.

Consider a reductive group G with Borel subgroup $B \subseteq G$ and maximal torus $T \subseteq B$. The *roots* of G with respect to T are the nonzero weights of the representation of T on the Lie algebra $\mathfrak{g}$ of G obtained by differentiating the action $t \cdot g = tgt^{-1}$ of T on G at the fixed point $e_G \in G$. A *positive root* with respect to B is a root of G with respect to T having an eigenvector in the Lie subalgebra $\mathfrak{b} \subseteq \mathfrak{g}$ of $B \subseteq G$. For $\lambda, \mu \in \mathbb{X}(T)$ one writes $\lambda \le \mu$ if $\mu - \lambda$ is a nonnegative linear combination of positive roots. The *simple roots* are the indecomposable members of the monoid generated by the positive roots; they freely generate this monoid.

Example 4.5.5.7 Let G be a semisimple simply connected group with maximal torus $T \subseteq G$. Fix a regular dominant weight $\eta \in \Lambda = \mathbb{X}(T)$ and let $V(\eta)$ be the associated simple G-module. Then $G \times G$-acts on $\operatorname{End}(V(\eta))$ via

$$(g_1, g_2) \cdot \varphi = g_1 \circ \varphi \circ g_2^{-1},$$

where we regard the g_i as endomorphisms via the G-action on $V(\lambda)$. This induces an action of $G \times G$ on the projective space $\mathbb{P} := \mathbb{P}(\operatorname{End}(V(\eta)))$. Let $X \subseteq \mathbb{P}$ be the closure of the $(G \times G)$-orbit through the class $x_0 \in \mathbb{P}$ of the identity of $V(\eta)$. The orbit through x_0 is open in X and, denoting by $C \subseteq G$ the center; one has

$$(G \times G) \cdot x_0 \cong (G \times G)/((C \times C)\operatorname{diag}(G)) \cong G/C = G_{\mathrm{ad}},$$

where G_{ad} is called the adjoint group of G. The variety X is the *wonderful compactification of G_{ad}* and is in fact wonderful with respect to the action of $G \times G$ and the induced action of $G_{\mathrm{ad}} \times G_{\mathrm{ad}}$; see [77, Thm. 6.1.8]. Now, take a pair $B^+, B^- \subseteq G$ of opposite Borel subgroups for T; that means that $T = B^+ \cap B^-$ holds. Then we have the Borel subgroup

$$B^+ \times B^- \subseteq G \times G.$$

There are as many boundary divisors E_k as colors D_i and the latter ones are the closures of the one-codimensional Bruhat cells. The $(B^+ \times B^-)$-weight of

$1_{D_i} \in \mathcal{R}(X)$ is $(\varpi_i, -\varpi_i)$, where $\varpi_1, \ldots, \varpi_r$ are the fundamental weights of G, which in turn generate $\mathbb{X}(T)$. In particular, we have a canonical identification

$$\mathrm{Cl}(X) = \mathbb{X}(T).$$

The spherical roots with respect to $T \times T$ are $\gamma_i = (\alpha_i, -\alpha_i)$, where $\alpha_1, \ldots, \alpha_r$ are the simple roots of G with respect to T. As the isotropy group $(G \times G)_0$ equals $(C \times C)\mathrm{diag}(G)$, the group $(G \times G)_1$ is the diagonal $\mathrm{diag}(G) \cong G$. Theorem 4.5.5.4 gives us a description of the Cox ring of X as

$$\mathcal{R}(X) \cong \bigoplus_{\mu \in \Lambda} \bigoplus_{\substack{\lambda \in \Lambda^+ \\ \lambda \leq \mu}} \Gamma(G, \mathcal{O})_{(\lambda)} \mu \cong \bigoplus_{\mu \in \Lambda} \bigoplus_{\substack{\lambda \in \Lambda^+ \\ \lambda \leq \mu}} \mathrm{End}(V(\lambda)) \mu,$$

where the algebra $\Gamma(G, \mathcal{O})$ is understood as a $(G \times G)$-module with respect to the action from left and right as $(g_1, g_2) \cdot g = g_1 g g_2^{-1}$ and, as before, $V(\lambda)$ denotes the simple G-module with highest weight $\lambda \in \Lambda^+$.

We conclude with relations to the theory of affine algebraic monoids. Recall that an *affine algebraic semigroup* is an affine algebraic variety S together with an associative multiplication

$$\mu \colon S \times S \to S,$$

which is a morphism of algebraic varieties. Examples are given by the semigroup $End(V)$ of endomorphisms of finite-dimensional vector spaces V, or, more generally, the semigroup of endomorphisms of V of rank at most k for some nonnegative integer k.

An *affine algebraic monoid* is an affine algebraic semigroup S with unit element. The group $S^* \subseteq S$ of invertible elements of an affine algebraic monoid S is an open subset and thus an algebraic group. Moreover, S comes with a natural action

$$(S^* \times S^*) \times S \to S, \qquad (g_1, g_2) \cdot s = g_1 s g_2^{-1}.$$

Observe that the open orbit S^* of this action is isomorphic as a homogeneous space to $(S^* \times S^*)/\mathrm{diag}(S^*)$.

Remark 4.5.5.8 If G is a connected affine algebraic group and S an irreducible affine $(G \times G)$-variety with an open orbit isomorphic to $(G \times G)/\mathrm{diag}(G)$, then the group law on the open orbit extends to an associative multiplication on S turning it into an affine algebraic monoid with $S^* = G$; see [259, 299].

In 1995, Vinberg classified affine algebraic monoids with prescribed reductive group of invertible elements. With any connected semisimple group G he associated the *enveloping monoid* $\mathrm{Env}(G)$ whose group of invertible elements is $G \times^C T$, where $T \subseteq G$ is a maximal torus and the center $C \subseteq G$ acts diagonally on $G \times T$. The monoid $\mathrm{Env}(G)$ can be characterized by a universal property among all reductive monoids S such that the commutator group of S^* is isomorphic to G; see [299, Thm. 5]. The relation to Cox rings of wonderful varieties is given by [299, Thm. 8].

Theorem 4.5.5.9 *Let G be a connected semisimple group and $T \subseteq G$ a maximal torus. Then the T-variety $\mathrm{Env}(G)$ is the total coordinate space of the wonderful compactification X of the group G_{ad}. In particular, X is the geometric quotient of a big open subset of $\mathrm{Env}(G)$ by the torus T.*

See also [260, Thm. 3], [76, Example 3.2.4], and [258] for related results. An explicit description of the enveloping monoid of the special linear group following [296, Example 27.30] is now provided.

Example 4.5.5.10 Consider the special linear group SL_n and the maximal torus $T \subseteq \mathrm{SL}_n$ consisting of all diagonal matrices $\mathrm{diag}(t_1, \ldots, t_n)$ in SL_n. Then we have a representation of $\mathrm{SL}_n \times T$ on the vector space

$$V = \left(\bigoplus_{k=1}^{n-1} \bigwedge^k \mathbb{K}^n \right) \oplus \mathbb{K}^{n-1}$$

where SL_n acts on $\wedge^k \mathbb{K}^n$ via the standard action on $\mathbb{K}^n$ and trivially on $\mathbb{K}^{n-1}$, and $\mathrm{diag}(t_1, \ldots, t_n) \in T$ acts on $\wedge^k \mathbb{K}^n$ by scalar multiplication with $t_1 \cdots t_k$ and on $\mathbb{K}^n$ as $\mathrm{diag}(t_1/t_2, \ldots, t_{n-1}/t_n)$.

The enveloping monoid $\mathrm{Env}(\mathrm{SL}_n)$ is the closure of the image of $\mathrm{SL}_n \times T$ in $\mathrm{End}(V)$. More explicitly, $\mathrm{Env}(\mathrm{SL}_n)$ is the closure in $\mathrm{End}(V)$ of the set of all tuples of the form

$$(t_1 g, \ldots, t_1 \cdots t_n \wedge^n g, t_1/t_2, \ldots, t_{n-1}/t_n),$$

where $g \in \mathrm{SL}_n$ and $t = \mathrm{diag}(t_1, \ldots, t_n) \in T$; that means that $t_1 \cdots t_n = 1$ holds. For example, looking at $n = 2$, the enveloping monoid is

$$\mathrm{Env}(\mathrm{SL}_2) = \{(A, z); \ A \in \mathrm{Mat}(2, 2; \mathbb{K}), \ z \in \mathbb{K}, \ \det A = z\} \cong \mathrm{Mat}(2, 2; \mathbb{K}).$$

This is the total coordinate space $\overline{X}$ of $X = \mathbb{P}(\mathrm{Mat}(2, 2; \mathbb{K}))$, which in turn is the wonderful completion of the adjoint group $\mathrm{PSL}_2 = \mathrm{SL}_2/\{\pm E_2\}$ of SL_2. For $n = 3$, we identify

$$\bigwedge^2 \mathbb{K}^3 \cong (\mathbb{K}^3)^* \cong \mathbb{K}^3.$$

Then we can realize the enveloping monoid $\mathrm{Env}(\mathrm{SL}_3)$ as the variety of tuples (A_1, A_2, z_1, z_2), where $A_i \in \mathrm{Mat}(3, 3; \mathbb{K})$ and $z_i \in \mathbb{K}$ such that

$$\wedge^2 A_i = z_i A_j \text{ for } i \neq j, \qquad A_1^t A_2 = A_1 A_2^t = z_1 z_2 E_3.$$

Here $\mathrm{Env}(\mathrm{SL}_3)$ is the total coordinate space $\overline{X}$ of the wonderful compactification X of PSL_3. Observe that the characteristic space $\widehat{X}$ is given by

$$\widehat{X} = \{(A_1, A_2, z_1, z_2) \in \overline{X};\ A_1 \neq 0 \text{ or } A_2 \neq 0\} \subseteq \overline{X}.$$

Other realizations of the enveloping monoid $\mathrm{Env}(G)$ of an arbitrary semisimple group G can be found in [260, 299]. Guay [152] generalized Vinberg's construction of the enveloping monoid to the case of any symmetric homogeneous space and thereby provided a realization of the total coordinate space of a complete symmetric variety.

Exercises to Chapter 4

Exercise 4.1 Let $n \geq 3$ and consider an n-dimensional irreducible, smooth complete intersection $Y \subseteq \mathbb{P}^{n+k}$. Then the pullback $\mathrm{Cl}(\mathbb{P}^{n+k}) \to \mathrm{Cl}(Y)$ is an isomorphism and the affine cone $C(Y) \subseteq \mathbb{K}^{n+k+1}$ with its $\mathbb{K}^*$-action is the total coordinate space of Y; see Corollary 4.1.1.7. Now assume $2n + 1 < n + k$ and let $X \subseteq \mathbb{P}^{n+k-1}$ be the projection of Y from a generic point. Then X is smooth and isomorphic to Y. The pullback $\mathrm{Cl}(\mathbb{P}^{n+k-1}) \to \mathrm{Cl}(X)$ is an isomorphism and with the affine cone $C(X) \subseteq \mathbb{K}^{n+k}$ we obtain that $C(X) \setminus \{0\}$ is the characteristic space over X; see Theorem 4.1.1.1. However, $C(X)$ is not normal and thus it is not the total coordinate space of X. Hint: Compare the tangent spaces of the cones $C(X)$ and $C(Y)$ at the respective apexes.

Exercise 4.2 Every irreducible, smooth rational projective surface X can be obtained by a series of blow-ups either from the projective plane $\mathbb{P}^2$ or the ath Hirzebruch surface $\mathbb{F}_a$. Recall that $\mathbb{P}^2$ and $\mathbb{F}_a$ are the toric surfaces arising from the complete fans in $\mathbb{Z}^2$ with the generator sets

for $\mathbb{P}^2$: $\{(1,0), (0,1), (-1,-1)\}$,   for $\mathbb{F}_a$: $\{(1,0), (0,1), (-1,-a), (0,-1)\}$.

Compute the Cox rings of rational surfaces up to Picard number $\varrho(X) \leq 4$. For blowing up a nonfixed point, proceed similarly as in Example 4.1.4.1. Conclude that all smooth rational surfaces with $\varrho(X) \leq 3$ are toric and for $\varrho(X) \leq 4$ they admit at least a $\mathbb{K}^*$-action. Moreover, observe that for $\varrho(X) \leq 4$ there are no moduli.

Exercise 4.3 Show that the Picard groups of the affine algebraic groups GL_n, SL_n, SO_n, Sp_{2n} and PGL_n are given as

$$\mathrm{Pic}(\mathrm{GL}_n) = 0, \qquad \mathrm{Pic}(\mathrm{SL}_n) = 0, \qquad \mathrm{Pic}(\mathrm{SO}_n) \cong \mathbb{Z}/2\mathbb{Z}\ (n \geq 3),$$
$$\mathrm{Pic}(\mathrm{Sp}_{2n}) = 0, \qquad \mathrm{Pic}(\mathrm{PGL}_n) \cong \mathbb{Z}/n\mathbb{Z}.$$

Moreover, show that every connected solvable affine algebraic group has trivial Picard group.

Exercise 4.4 The group $G = \mathrm{PGL}_2$ acts on the projective line $\mathbb{P}^1$ transitively, and the line bundle $\mathcal{O}(1)$ on $\mathbb{P}^1$ is not G-linearizable: the space $\Gamma(\mathbb{P}^1, \mathcal{O}(1))$ is two-dimensional, but the smallest nontrivial PGL_2-module is of dimension 3. However, the bundle $\mathcal{O}(1)$ is SL_2-linearizable with respect to the SL_2-action given by the double covering $\mathrm{SL}_2 \twoheadrightarrow \mathrm{PGL}_2$.

Exercise 4.5 For a finite group G it can happen that for a line bundle L on a G-variety X the line bundles $\mathrm{pr}_X^*(L)$ and $\mu^*(L)$ are isomorphic, but L is not G-linearizable. Take $X = \mathbb{P}^1$ and a noncyclic group G of order 4. Consider the action of G on X obtained by projectivizing a two-dimensional irreducible representation of the dihedral group D_4. If $L = \mathcal{O}(1)$, then the condition $\mathrm{pr}_X^*(L) \cong \mu^*(L)$ is fulfilled, $\mathrm{Aut}(X)$ acts trivially on $\mathrm{Pic}(X)$. But L admits no G-linearization: there is no representation of the group G on the space dual to $\Gamma(X, L) \cong \mathbb{K}^2$ whose projectivization defines the given action of G on X.

Exercise 4.6 Consider the projective toric surface X arising from the fan Σ in $\mathbb{Z}^2$ with the rays generated by the vectors $(-3, 2)$, $(1, 0)$, and $(1, 2)$. The one-parameter subgroup $t \mapsto (1, t)$ of the acting torus $\mathbb{T}^2$ defines an action $\mu \colon \mathbb{K}^* \times X \to X$. This action admits a lifting (ε, μ') to the characteristic space $\widehat{X} \to X$ in the sense of Definition 4.2.3.1 with $\varepsilon(t) = t^2$ but there is no such lifting with $\varepsilon(t) = t$.

Exercise 4.7 (Chamber structure of the G-ample cone) Let X be a Mori dream space with an action of a connected reductive group G. Fix a lifting of the G-action to the total coordinate space $\overline{X}$. The cone of ample G-linearized divisor classes of X is

$$\alpha(X, G) = (\mathrm{Ample}(X) \times \mathbb{X}_\mathbb{Q}(G)) \cap \omega_{\overline{X}} \subseteq \mathrm{Cl}_\mathbb{Q}(X) \times \mathbb{X}_\mathbb{Q}(G),$$

where $\omega_{\overline{X}}$ refers to the weight cone of the action of $G \times H_X$ on $\overline{X}$. Let $\Lambda(\overline{X}, G \times H_X)$ be the GIT fan in the sense of Exercise 3.3. Then the sets of semistable points of G-linearized ample line bundles of X are in order-reversing one-to-one correspondence with the cones of the partial fan

$$\{\lambda \cap \alpha(X, G); \ \lambda \in \Lambda(\overline{X}, G \times H_X)\}.$$

See [20] for more background and a solution. The decomposition of the G-ample cone into partially open polyhedral GIT chambers exists even in the non-Mori dream case and goes back to Dolgachev/Hu [128] and Thaddeus [295].

Exercise 4.8 (Picard quotient presentation) Let X be a not necessarily normal irreducible prevariety that has only constant invertible global functions. Regard the Cartier divisors X as global sections of the sheaf $\mathcal{K}^*/\mathcal{O}^*$ of nonzero rational modulo invertible functions, that is, a Cartier divisor is given by a collection (U_i, f_i) with $f_i \in \mathbb{K}(X)$ such that $f_i/f_j \in \Gamma(U_i \cap U_i, \mathcal{O}^*)$; see [160, Sec. II.6] for basic background and [42] for the solution of (ii) to (vi) in terms of line bundles.

(1) The Cartier divisors form a multiplicative group $\mathrm{CDiv}(X)$ and the Picard group is $\mathrm{Pic}(X) = \mathrm{CDiv}(X)/\mathrm{PDiv}(X)$, where $\mathrm{PDiv}(X)$ is the subgroup of the Cartier divisors (U_i, f), that is, the functions f_i do not depend on i.

(2) Every Cartier divisor D on X defines a sheaf $\mathcal{O}(D)$ that is locally free of rank 1 and, if D is given as (U_i, f_i), then one has $\Gamma(U_i, \mathcal{O}(D)) = \Gamma(U_i, \mathcal{O})f_i^{-1}$. Reformulate the concept of a sheaf of divisorial algebras in this setting.

(3) Assume that $\mathrm{Pic}(X)$ is finitely generated. Follow the principles of Construction 1.4.2.1 and define a $\mathrm{Pic}(X)$-graded sheaf $\mathcal{L}$ on X with

$$\mathcal{L} := \bigoplus_{[D] \in \mathrm{Pic}(X)} \mathcal{L}_{[D]}, \qquad \mathcal{L}_{[D]} \cong \mathcal{O}(D)$$

such that $\mathcal{L}$ is a quotient of a sheaf of divisorial algebras corresponding to a subgroup $K \subseteq \mathrm{CDiv}(X)$ mapping onto $\mathrm{Pic}(X)$.

(4) Assume that $\mathrm{Pic}(X)$ is finitely generated. Show that $\mathcal{L}$ is locally of finite type and $\widetilde{X} := \mathrm{Spec}_X \mathcal{L}$ with the canonical map $q_X : \widetilde{X} \to X$ is an étale G_X-principal bundle, where the action of $G_X := \mathrm{Spec}\,\mathbb{K}[\mathrm{Pic}(X)]$ on X is given by the $\mathrm{Pic}(X)$-grading of $\mathcal{L}$.

(5) Assume that $\mathrm{Pic}(X)$ is finitely generated. Show that $\widetilde{X}$ is quasiaffine if and only if X is divisorial in the sense that it is covered by affine open sets $X \setminus \mathrm{Supp}(D)$, where D is an effective Cartier divisor.

(6) Show that every morphism $\varphi : X \to Y$ of irreducible prevarieties X and Y with finitely generated Picard groups lifts to a morphism $\widetilde{X} \to \widetilde{Y}$ that is equivariant with respect to the actions of the quasitori G_X and G_Y.

(7) Let $q_X : \widetilde{X} \to X$ be the Picard quotient presentation of an irreducible prevariety X. Then, for any $D \in \mathrm{CDiv}(X)$, the class $[D] \in \mathrm{Pic}(X)$ defines a character $\chi \in \mathbb{X}(G_X)$ and there is an action

$$G_X \times (\widetilde{X} \times \mathbb{K}) \to \widetilde{X} \times \mathbb{K}, \qquad h \cdot (\widetilde{x}, z) := (h^{-1} \cdot \widetilde{x}, \chi(h)z),$$

The quotient space $L := (\widetilde{X} \times \mathbb{K})/G_X$ is the total space of a geometric line bundle $\varphi : L \to X$ over X, where the projection φ is induced by the G_X-invariant map

$$\psi : \widetilde{X} \times \mathbb{K} \to X, \qquad (\widetilde{x}, z) \mapsto q_X(\widetilde{x}).$$

The isomorpism class of the line bundle $\varphi : L \to X$ in the Picard group equals the class of D we started with.

(8) Describe the Picard quotient presentation for a toric variety X in terms of the defining fan; see also [182].

Exercise 4.9 (Automorphisms of complete toric varieties)   The original references are [104, 108]. See also [85, 233] for more and [21] for a generalization to the case of complexity 1. Let X be a complete toric variety. Denote by $w_1, \ldots, w_s \in K := \mathrm{Cl}(X)$ the classes of the invariant prime divisors and let $D_{i1}, \ldots, D_{in_i}$ denote the ones with class w_i. Thus, the Cox ring of X is

$$\mathbb{K}[T_{ij}; \ 1 \le i \le s, \ 1 \le j \le n_i], \qquad \deg(T_{ij}) = w_i \in K.$$

Moreover, consider the coordinate subspaces $W_i := V(T_{kj}, k \ne i)$ of the total coordinate space $\mathbb{K}^n$, where $n = n_1 + \cdots + n_s$. Consider

$$\widehat{G}_i := \mathrm{GL}(W_i) \subseteq \mathrm{GL}_n, \qquad \widehat{G} := \widehat{G}_1 \times \cdots \times \widehat{G}_s \subseteq \mathrm{GL}_n.$$

Then the action of $\widehat{G}$ in $\mathbb{K}^n$ commutes with the action of the characteristic quasitorus H_X. Moreover, $\widehat{G}$ leaves the characteristic space $\widehat{X} \subseteq \mathbb{K}^n$ invariant and defines a subgroup

$$G := \widehat{G}/H_X \subseteq \mathrm{Aut}(X)^0.$$

This is the reductive part of $\mathrm{Aut}(X)$. Given i, j, let $e_{ij} \in \mathbb{K}^n$ be the ijth canonical basis vector. For any nonlinear monomial z^μ in degree w_i not depending on z_{ij} and each $t \in \mathbb{K}$, we have a map

$$\widehat{\eta}_{ij,\mu}(t) \colon \mathbb{K}^n \to \mathbb{K}^n, \qquad z \mapsto z + tz^\mu e_{ij},$$

which defines an additive one-parameter subgroup $\eta_{ij,\mu} \colon \mathbb{K} \to \mathrm{Aut}(X)^0$. The images $\eta_{ij,\mu}(\mathbb{K})$ generate the unipotent radical $U \subseteq \mathrm{Aut}(X)^0$ and G, U together generate $\mathrm{Aut}(X)^0$. Let (Σ, N) be the describing lattice fan of X. A *Demazure root* is an integral linear form u on N such that there exist i, j with

$$\langle u, v_{ij} \rangle = -1, \qquad \langle u, v_{kl} \rangle > 0 \quad \text{for } (k, l) \ne (i, j),$$

where $v_{ij} \in N$ is the primitive generator of the ray of Σ corresponding to the variable T_{ij} of the Cox ring. The Demazure roots are precisely the weights of the additive one-parameter subgroups $\eta \colon \mathbb{K} \to \mathrm{Aut}(X)$ normalized by the acting torus T of X. Finally, consider the finite group

$$\Gamma := \{\zeta \in \mathrm{GL}(N); \ \zeta(\sigma) \in \Sigma \text{ for all } \sigma \in \Sigma\} \subseteq \mathrm{GL}(N)$$

and let $W \subseteq \Gamma$ denote the subgroup stabilizing the sets $\{v_{i1}, \ldots, v_{in_i}\}$. Then Γ/W is isomorphic to the factor group $\mathrm{Aut}(X)/\mathrm{Aut}(X)^0$ and W is isomorphic to the Weyl group of G.

Exercise 4.10 (Homogeneous toric varieties)   See [17] for more background and a solution.

(1) A complete toric variety admits a transitive action of a semisimple algebraic group if and only if it is a product of projective spaces.

(2) Find all not necessarily complete toric varieties that admit a transitive action of a semisimple algebraic group.

Exercise 4.11 (Cox sheaves that are not locally of finite type) Let X be an irreducible, normal projective variety with finitely generated divisor class group $\mathrm{Cl}(X)$ but nonfinitely generated Cox ring $\mathcal{R}(X)$. Moreover, let Y be any affine variety with a torus action $T \times Y \to Y$ such that X is the geometric quotient space of a T-invariant open subset $V \subseteq Y$. Then Y has finitely generated divisor class group $\mathrm{Cl}(Y)$ but the Cox sheaf $\mathcal{R}_Y$ is not locally of finite type. Examples are provided by any affine cone Y over $\mathbb{P}^2$ blown up at nine points in very general position; see Proposition 4.3.4.5.

Exercise 4.12* Let X be an irreducible, normal variety with finitely generated divisor class group $\mathrm{Cl}(X)$. Characterize local finite generation of the Cox sheaf $\mathcal{R}_X$ in terms of the singularities of X.

Exercise 4.13* Let X be an irreducible normal variety, $K \subseteq \mathrm{WDiv}(X)$ a finitely generated subgroup mapping onto $\mathrm{Cl}(X)$, and $\mathcal{S}$ the sheaf of divisorial algebras associated with K. Suppose that $\Gamma(X, \mathcal{S})$ is finitely generated for all open subsets $U \subseteq X$. Does X admit a normal completion X' such that the Cox ring $\mathcal{R}(X')$ is finitely generated?

Exercise 4.14 (Isomorphy of varieties with torus action of complexity 1) Recall the notation of Construction 3.4.3.1 and consider a pair (A, P) of defining matrices as used there. The columns of P are organized in blocks $v_{i1}, \ldots, v_{in_i}$ and $v_1, \ldots, v_m$. By an *admissible operation* on P we mean

- Adding a multiple of one of the upper r row to one of the lower s rows
- Applying an elementary row operation among the last s rows
- Switching two columns v_{ij_1} and v_{ij_2} inside a block $v_{i1}, \ldots, v_{in_i}$
- Switching two blocks $v_{i1}, \ldots, v_{in_i}$ and $v_{j1}, \ldots, v_{jn_j}$ and rearranging the L-block by elementary row operations into its required shape
- Switching two columns v_{k_1} and v_{k_2} of the d'-block

We say that two pairs (A, P) and (A', P') are *isomorphic* if $A' = BAD$ and $P' = SPU$ hold with $A \in \mathrm{GL}_2(\mathbb{K})$, a diagonal matrix $D \in \mathrm{GL}_{r+1}(\mathbb{K})$, and unimodular matrices S, U describing a series of admissible operations on P. If (A, P) and (A', P') define minimal presentations in the sense that $r, r' \geq 2$ and always $n_i l_{ij}, n_i' l_{ij}' > 1$ holds, then the following statements are equivalent.

(1) The graded rings $R(A, P)$ and $R(A', P')$ are isomorphic to each other.
(2) The pairs (A, P) and (A', P') are isomorphic to each other.

Now assume that in addition to the minimally presenting pairs (A, P) and (A', P') we have two bunches Φ and Φ' as in Construction 3.4.3.6 defining complete varieties $X(A, P, \Phi)$ and $X(A', P', \Phi')$. Then the following statements are equivalent.

(3) There is an isomorphism $(\psi, \widetilde{\psi})$ of the graded rings $R(A, P)$ and $R(A', P')$ such that $\widetilde{\psi}$ sends Φ to Φ'.

(4) There is a torus-equivariant isomorphism $(\varphi, \widetilde{\varphi})$ between $X(A, P, \Phi)$ and $X(A', P', \Phi')$.

(5) There is an isomorphism φ of algebraic varieties between $X(A, P, \Phi)$ and $X(A', P', \Phi')$.

Conclude that two $\mathbb{K}^*$-surfaces $X(A, P)$ and $X(A', P')$ are isomorphic to each other as varieties if and only if the defining pairs (A, P) and (A', P') are isomorphic to each other. Hint: For "(ii)$\Rightarrow$(i)" consider the induced isomorphism on the one-dimensional quotient spaces and for "(v)$\Rightarrow$(iv)" use the fact that the (nontoric) varieties in question have affine algebraic automorphism groups due to Corollary 4.2.4.2. See [177] for a solution.

Exercise 4.15 Let X be a projective toric surface arising from a fan Σ, consider a torus invariant curve $C \subseteq X$ corresponding to a ray $\varrho \in \Sigma$, and let $x_1, \ldots, x_r \in C$ be a collection of pairwise different points. Then the blow-up X' of X at $x_1, \ldots, x_r$ is a $\mathbb{K}^*$-surface. Describe defining data (A, P) of X' in the sense of Construction 3.4.3.1 in terms of Σ, ϱ and $[a_i, b_i] = \alpha(x_i)$, where $\alpha \colon C \to \mathbb{P}^1$ is any isomorphism. Generalize your observations to toric varieties X of higher dimension.

Exercise 4.16* Expand the approach of Example 4.4.1.8 to an explicit algorithm for computing Cox rings of trinomial hypersurfaces in the projective space $\mathbb{P}^n$.

Exercise 4.17* Let X be an irreducible variety and $T \times X \to X$ a torus action of complexity 1. Consider the Picard quotient presentation $\widetilde{X} \to X$ of Exercise 4.8 and describe the Pic(X)-graded algebra $\Gamma(\widetilde{X}, \mathcal{O})$ explicitly in terms of homogeneous generators and relations.

Exercise 4.18 (Danilov-Gizatullin surfaces) Let $S \subseteq \mathbb{F}_a$ be an ample divisor in a Hirzebruch surface $\mathbb{F}_a$ and set $n := S^2$. The nth *Danilov–Gizatullin* surface is the affine surface $V_n := \mathbb{F}_a \setminus S$. Here, the isomorphism class of V_n depends indeed only on n; see [147, Thm. 5.8.1] and also [137]. Now, let $n \geq 2$ and consider the affine hypersurface

$$F_n := V(T_1 T_4 - T_2^{n-1} T_3 - 1) \subseteq \mathbb{K}^4.$$

Then F_n is smooth, factorial, and $\Gamma(F_n, \mathcal{O}^*) = \mathbb{K}^*$ holds. Moreover, the action of $\mathbb{K}^*$ on F_n given by

$$t \cdot z = (t^{-1}z_1, tz_2, t^{1-n}z_3, tz_4)$$

is free. Show that the quotient morphism $F_n \to F_n/\mathbb{K}^*$ is the characteristic space of the Danilov–Gizatullin surface V_n. Compare also [130].

Exercise 4.19 (Affine SL_2-embeddings) We describe here the characteristic space over normal affine SL_2-embeddings following [31]. Let $p \le q$ and m be positive integers with p, q coprime. Set

$$k := \gcd(q - p, m), \qquad a := \frac{m}{k}, \qquad b := \frac{q-p}{k}.$$

In the case $p = q = 1$, we set $k := 1$, $a := m$ and $b := 0$. Now consider the affine hypersurface

$$C_b := V(T_1 T_4 - T_2 T_3 - T_5^b) \subseteq \mathbb{K}^5$$

Then C_b is factorial and $\Gamma(C_b, \mathcal{O}^*) = \mathbb{K}^*$ holds. Let $\xi \in \mathbb{K}^*$ be a primitive ath root of unity and set $Z_a = \langle \xi \rangle \subseteq \mathbb{K}^*$. Then the quasitorus $H := \mathbb{K}^* \times Z_a$ acts on C_b via

$$(t, \xi) \cdot z := (t^{-p}\xi^{-1}z_1, t^{-p}\xi^{-1}z_2, t^q\xi z_3, t^q\xi z_4, t^k z_5)$$

and this action is strongly stable. Thus, C_b is the characteristic space of $X := C_b /\!\!/ H$. We endow X with an almost homogeneous SL_2-action. The group SL_2 acts on C_b via left multiplication on the matrix

$$\begin{bmatrix} z_1 & z_3 \\ z_2 & z_4 \end{bmatrix}$$

and leaving z_5 fixed. This action commutes with the H-action and descends to the desired action on X. The stabilizer in SL_2 of a generic point on X is a cyclic subgroup of order m. The complement of the open SL_2-orbit is a two-dimensional orbit for $p = q = 1$ and the union of two-dimensional orbit and a fixed point otherwise. Every nontransitive SL_2-action with an open orbit on an irreducible normal affine threefold can be realized this way [31, 197, 255]. An affine SL_2-embedding defined by a triple (p, q, m) is a toric variety if and only if $q - p$ divides m; see [145].

Exercise 4.20* Extend the description of varieties X with a torus action of complexity 1 given in Theorem 4.4.1.6 to the case $\Gamma(X, \mathcal{O}^*) = \mathbb{K}^*$.

Exercise 4.21 (Spherical varieties of semisimple rank 1) Recall that the semisimple rank of an affine algebraic group G is the dimension of a maximal torus of $G/R(G)$, where $R(G) \subseteq G$ is the radical, that is, the largest normal connected solvable subgroup. For a nontoric complete normal variety X, the following statements are equivalent.

(1) X is spherical with respect to an action of a reductive group of semisimple rank one.

(2) X allows a torus action of complexity 1 and an almost homogeneous reductive group action.

Moreover, if one of these statements holds, then the Cox ring of X is a ring $R(A, P)$ with precisely one relation g and this is one of the following:

- $g = T_{01}T_{02} + T_{11}T_{12} + T_2^{l_2}$ with $w_{01} = w_{11}$ and $w_{02} = w_{12}$,
- $g = T_{01}T_{02} + T_{11}^2 + T_2^{l_2}$ with $w_{01} = w_{02} = w_{11}$,

where $w_{ij} = \deg_K(T_{ij}) \in K \cong \mathrm{Cl}(X)$ denotes the degrees of the generators of the Cox ring. Describe the defining matrix P. We refer to [21] for more details.

Exercise 4.22 Every affine toric variety X is a commutative algebraic monoid. The idempotent elements of X are precisely the distinguished points.

5

Surfaces

In this chapter we study Cox rings of surfaces. In Section 5.1 we figure out the smooth Mori dream surfaces of positive anticanonical Iitaka dimension. In the cases of rational elliptic surfaces, K3 surfaces, and Enriques surfaces, we obtain a detailed picture. The subsequent two sections are more concrete; there we aim for explicit descriptions of the Cox ring. In Section 5.2, we show that the Cox ring of a del Pezzo surface is generated by sections of anticanonical degree 1 and prove that its ideal of relations is generated by quadrics. Section 5.3 treats Cox rings of K3 surfaces. It includes a detailed study in the case of Picard number 2, and complete results are obtained for double covers of del Pezzo surfaces and of blow-ups of Hirzebruch surfaces at $n \leq 3$ points. Finally, in Section 5.4 we develop the theory of rational $\mathbb{K}^*$-surfaces. Here we allow singularities and show how their minimal resolution is encoded in the Cox ring. As an example class, we present the Gorenstein log del Pezzo $\mathbb{K}^*$-surfaces in terms of Cox rings.

5.1 Mori dream surfaces

5.1.1 Basic surface geometry

We recall the necessary concepts and facts from the theory of smooth projective surfaces and provide some basic facts needed later. As in most of the subsequent sections, we work here over the field $\mathbb{C}$ of complex numbers. By a *surface* we always mean an irreducible variety of dimension two. Basic references are the textbooks [29, 39].

We begin with the intersection theory on a smooth projective surface X. As usual, we call an effective Weil divisor on X also a *curve* on X. So, the prime divisors are precisely the irreducible and reduced curves. Given two distinct prime divisors D_1 and D_2 on X, their *intersection number* is defined as the degree, that is, the sum over all coefficients, of the pullback divisor of D_2 to the normalization D_1' of D_1:

$$D_1 \cdot D_2 := \deg(\nu^* \imath^* D_2) \in \mathbb{Z},$$

where $\imath\colon D_1 \to X$ denotes the inclusion and $\nu\colon D_1' \to D_1$ the normalization map. The intersection number can be expressed as the sum of the *intersection multiplicities* m_x at the points $x \in D_1 \cap D_2$:

$$D_1 \cdot D_2 = \sum_{D_1 \cap D_2} m_x, \qquad m_x := \dim_{\mathbb{C}} \left(\mathcal{O}_{X,x} / \langle f_x^{(1)}, f_x^{(2)} \rangle \right),$$

where $f_x^{(i)} \in \mathcal{O}_{X,x}$ is a generator for the vanishing ideal of D_i near x. The intersection number extends to a symmetric bilinear form with integer values on $\mathrm{WDiv}(X)$ and, as it depends only on linear equivalence classes, it defines a symmetric bilinear form on the divisor class group, called the *intersection product*:

$$\mathrm{Cl}(X) \times \mathrm{Cl}(X) \to \mathbb{Z}, \qquad ([D_1], [D_2]) \mapsto D_1 \cdot D_2.$$

The *self-intersection number* $C^2 = C \cdot C$ can be used to characterize contractibility of curves. By a $(-k)$-*curve* we mean a reduced irreducible curve $C \subseteq X$ with $C \cong \mathbb{P}^1$ and $C^2 = -k$. The *Castelnuovo criterion*, see [39, Thm. II.17], states that a reduced irreducible curve C on X is a (-1)-curve if and only if there is a contracting morphism $\pi_C\colon X \to Y$ onto a smooth projective surface, where contracting means that $C \subseteq X$ maps to a point $y \in Y$ and $X \setminus C$ maps isomorphically onto $Y \setminus \{y\}$. Moreover, by *Artin's criterion*, see [12, Thm. 2.3], a reduced connected curve C with irreducible components $C_1, \ldots, C_k$ admits a contracting morphism $\pi_C\colon X \to Y$ onto a normal projective surface Y if the matrix $(C_i \cdot C_j)$ is negative definite and $D^2 + D \cdot K_X < 0$ holds for any divisor D supported inside C. In Exercise 5.1 we give another criterion for Mori dream surfaces.

Two divisors $D_1, D_2 \in \mathrm{WDiv}(X)$ are *numerically equivalent*, denoted by $D_1 \equiv D_2$, if $(D_1 - D_2) \cdot C = 0$ holds for any prime divisor C on X. The *Néron–Severi group* of X is the factor group $\mathrm{N}^1(X)$ of $\mathrm{WDiv}(X)$ modulo the subgroup of divisors being numerically equivalent to zero. Because linearly equivalent divisors are numerically equivalent, one has a canonical epimorphism

$$\mathrm{Cl}(X) \to \mathrm{N}^1(X)$$

and, by the definition of numerical equivalence, the intersection product descends to the Néron–Severi group. The Néron–Severi group is finitely generated, free, abelian [202, Prop. 1.1.14], and its rank $\varrho(X)$ is called the *Picard number* of X. A divisor D on X is called *numerically effective (nef)* if $D \cdot C \geq 0$

holds for any prime divisor C on X. The classes of numerically effective divisors generate the *nef cone* $\text{Nef}(X) \subseteq N^1_{\mathbb{Q}}(X)$ in the rational Néron–Severi group of X.

The *Hodge index theorem* [160, Thm. V.1.9] states that the intersection form on $N^1(X)$ is represented by a symmetric matrix of signature $(1, \varrho - 1)$, that is, with one positive eigenvalue and $\varrho - 1$ negative eigenvalues, where $\varrho := \varrho(X)$ stands for the Picard number. Another way to state this theorem is to say that if D and E are two not numerically trivial divisors with $D^2 > 0$ and $D \cdot E = 0$, then $E^2 < 0$ holds.

Let us fix our notation for the cohomology on a complete variety X. As usual, we denote the ith cohomology group of a sheaf $\mathcal{F}$ of $\mathcal{O}_X$-modules by $H^i(X, \mathcal{F})$ and write $h^i(X, \mathcal{F})$ for its dimension as a complex vector space. For the sheaf $\mathcal{O}_X(D)$ associated with a divisor D on X we abbreviate

$$H^i(X, D) := H^i(X, \mathcal{O}_X(D)), \qquad h^i(X, D) := \dim_{\mathbb{C}}(H^i(X, D)).$$

In particular, we will also write $H^0(X, D)$ for $\Gamma(X, \mathcal{O}_X(D))$. The *Euler characteristic* $\chi(\mathcal{O}_X)$ of the structure sheaf and the *arithmetic genus* $p_a(X)$ of a complete variety X are defined as

$$\chi(\mathcal{O}_X) := \sum_{i=0}^{\dim(X)} (-1)^i h^i(X, \mathcal{O}_X), \qquad p_a(X) := (-1)^{\dim(X)}(\chi(\mathcal{O}_X) - 1).$$

Let K_X denote a canonical divisor on X. We recall the following weak form of *Serre's duality* [160, Cor. III.7.7]: given a divisor D of X we have

$$h^i(X, D) := h^{\dim(X)-i}(X, K_X - D) \qquad \text{for } 0 \leq i \leq \dim(X).$$

The arithmetic genus $p_a(C)$ of a reduced irreducible curve $C \subseteq X$ can be computed by means of intersection numbers using the *genus formula* [39, I.15]:

$$p_a(C) = \frac{1}{2}(C^2 + C \cdot K_X) + 1.$$

Recall that $p_a(C)$ is greater than or equal to the topological genus $g(\widetilde{C}) := \frac{1}{2}h^1(\widetilde{C}, \mathbb{Q})$ of the resolution $\widetilde{C}$ of the curve C. In particular $p_a(C) = 0$ holds if and only if C is smooth rational [39, Rem. I.16 (i)] and in this case we have $C \cdot K_X = -2 - C^2$. If the curve C is smooth then a canonical divisor for C can be calculated by means of the following *adjunction formula*:

$$K_C = (K_X + C)|_C.$$

Taking degrees of both sides we get the genus formula again. The *Riemann–Roch formula* [39, Thm. I.12] relates the Euler characteristic of a sheaf $\mathcal{O}_X(D)$

of a divisor D on X to that of the structure sheaf $\mathcal{O}_X$:

$$\chi(\mathcal{O}_X(D)) = \frac{1}{2}(D^2 - D \cdot K_X) + \chi(\mathcal{O}_X).$$

This formula is often used to provide a bound or to explicitly calculate the dimension $h^0(X, D)$. For these purposes also the *Kawamata–Viehweg vanishing theorem* [202, Thm. 4.3.1] is useful: it states that for any nef and big divisor D on X we have

$$h^i(X, K_X + D) = 0 \qquad \text{for all } i > 0.$$

The *irregularity* of a normal complex projective variety is $q(X) := h^1(X, \mathcal{O}_X)$. We have $q(X) = 0$ if and only if $\text{Pic}(X)$ is finitely generated. In particular, $q(X) = 0$ holds for every complex Mori dream space X. Moreover, if $q(X) = 0$ holds for a smooth X, then the kernel of the epimorphism $\text{Cl}(X) \to \text{N}^1(X)$ is finite and coincides with the torsion subgroup of $\text{Cl}(X)$; see [202, Rem. 1.1.20]. In particular, we then have a canonical identification $\text{Cl}_\mathbb{Q}(X) = \text{N}^1_\mathbb{Q}(X)$ of the associated rational vector spaces. Examples for $\text{Cl}(X) \to \text{N}^1(X)$ with a nontrivial finite kernel are provided by Enriques surfaces; see Section 5.1.6.

Proposition 5.1.1.1 *Let X be a smooth projective surface with $q(X) = 0$ and let D and E be two nef, not numerically trivial, divisors of X such that $D \cdot E = 0$. Then $D^2 = E^2 = 0$ and $E \sim \alpha D$ holds for some positive $\alpha \in \mathbb{Q}_{>0}$.*

Proof As they are nef, D and E are limits of ample divisors; see [202, Cor. 1.4.10]. In particular, $D^2, E^2 \geq 0$ holds. By the Hodge index theorem we have $\text{N}^1(X)_\mathbb{R} = U \oplus V$, where the intersection form is positive definite on $U = \mathbb{R}u$ and negative definite on V. Scaling with positive real numbers shifts the (with nonzero) classes of D and E to $w_D = u + v_D$ and $w_E = u + v_E$ with $v_D, v_E \in V$. Observe

$$0 \leq w_D^2 + w_E^2 = w_D^2 + w_E^2 - 2w_D \cdot w_E = (v_D - v_E)^2 \leq 0.$$

This implies $v_D = v_E = 0$. Due to $q(X) = 0$, the epimorphism $\text{Cl}(X) \to \text{N}^1(X)_\mathbb{R}$ has a finite kernel. Consequently, the classes of D and E in $\text{Cl}(X)$ differ by a positive rational number. $\qquad\square$

Recall that the complete linear system $|D|$ defined by a divisor D of X is the (possibly empty) set of all effective divisors D' on X being linearly equivalent to D; we can naturally identify $|D|$ with the projective space $\mathbb{P}(H^0(X, D))$. The base locus and stable base locus of $|D|$ have been defined in Section 3.3.2.

Proposition 5.1.1.2 *Let X be a smooth projective surface, D an effective divisor, and E a prime divisor of X. If $D \cdot E < 0$, then E is contained in the stable base locus of $|D|$.*

Proof For any positive integer n, the restriction of nD to E has degree $nD \cdot E < 0$. The Riemann–Roch formula for curves yields $h^0(E, nD|_E) = 0$. Thus, from the long exact cohomology sequence of the exact sequence of sheaves

$$0 \longrightarrow \mathcal{O}_X(nD - E) \longrightarrow \mathcal{O}_X(nD) \longrightarrow \mathcal{O}_X(nD)|_E \longrightarrow 0$$

we deduce that the inclusion map $H^0(X, nD - E) \to H^0(X, nD)$ is an isomorphism. It follows that E is contained in the base locus of $|nD|$ for any positive integer n and thus lies in the stable base locus of $|D|$.   □

Corollary 5.1.1.3 *Let X be a smooth projective surface. Then we have the following inclusions of cones in the rational Néron–Severi group:*

$$\mathrm{SAmple}(X) \subseteq \mathrm{Mov}(X) \subseteq \mathrm{Nef}(X) \subseteq \overline{\mathrm{Eff}(X)} \subseteq \mathrm{N}^1_{\mathbb{Q}}(X).$$

Proof By definition, $\mathrm{SAmple}(X)$ is contained in $\mathrm{Mov}(X)$ and both lie in $\mathrm{Eff}(X)$. Proposition 5.1.1.2 gives the second inclusion. The third one is due to the facts that the nef cone is the closure of the ample cone, see [202, Cor. 1.4.10], and that each ample class admits an effective multiple.   □

The *base point free theorem* tells us that a nef divisor D on a smooth projective variety X is semiample provided that $D - K_X$ is nef and big; see [275].

Definition 5.1.1.4 Let X be a smooth projective surface and let D be a divisor whose class in $\mathrm{N}^1_{\mathbb{R}}(X)$ is in the closure of the effective cone of X. A presentation $D = P + N$ with rational divisors P and N is a *Zariski decomposition* if P is nef, N is effective, $P \cdot N = 0$ holds, and the intersection matrix of the prime components of N is negative definite. We call P the *positive part* of the Zariski decomposition of D.

According to [202, Thm. 2.3.19], Zariski decompositions in the above sense exist and are unique. Let us compare this with the definition given earlier for Mori dream spaces.

Remark 5.1.1.5 If X is a Mori dream surface, then the above definition of Zariski decomposition is equivalent to Definition 3.3.4.6. Indeed both decompositions exist and are unique. So it is enough to show that one definition implies the other. If $D = P + N$ is a Zariski decomposition with positive part P, according to Definition 5.1.1.4, then we only have to show that the natural map $S^+(X, P) \to S^+(X, D)$ is an isomorphism. Let n be a positive integer such that nN is a divisor with integer coefficients. Because $N^2 < 0$ and N is effective, there exists a prime divisor C in the support of N such that $N \cdot C < 0$. Then

$$nD \cdot C = (nP + nN) \cdot C = nN \cdot C < 0.$$

Hence C is contained in the base locus of the linear system $|nD|$. Repeating the same argument with $nN - C$ instead of nN, we see that the whole divisor nN is contained in the base locus of $|nD|$ and thus the map $H^0(X, nP) \to H^0(X, nD)$ is an isomorphism.

Consider a divisor D on a projective variety X. Let $N(D)$ be the set of nonnegative integers n with $h^0(X, nD) \neq 0$ and denote by $\varphi_{|nD|}$ the rational map $X \dashrightarrow \mathbb{P}(H^0(X, nD))$ given by $|nD|$. The *Iitaka dimension* of D is

$$\kappa(D) := \begin{cases} \max_{n \in N(D)} \dim \varphi_{|nD|}(X), & \text{if } N(D) \neq \{0\} \\ -\infty & \text{otherwise.} \end{cases}$$

According to this definition, the divisor D is big if and only if $\kappa(D) = \dim X$. Moreover, if X is a smooth projective surface and $D = P + N$ is the Zariski decomposition of D with positive part P, then one has $\kappa(D) = \kappa(P)$.

Proposition 5.1.1.6 *Let X be a smooth projective surface with finitely generated divisor class group of rank at least 3 and polyhedral effective cone* $\mathrm{Eff}(X)$. *Then the extremal rays of* $\mathrm{Eff}(X)$ *are spanned by classes of prime divisors E with $E^2 < 0$.*

Proof The subset $\{w \in \mathrm{Cl}_{\mathbb{Q}}(X); \; w^2 \geq 0\}$ of $\mathrm{Cl}_{\mathbb{Q}}(X)$ is union of two cones Q and Q'. Let Q be the cone that contains the ample cone. The interior of Q is contained in the effective cone $\mathrm{Eff}(X)$ by the Riemann–Roch formula. The effective cone is closed because it is polyhedral, so that $Q \subseteq \mathrm{Eff}(X)$. Because the Picard number is at least 3, the boundary ∂Q of Q is a circular cone because the intersection form on $\mathrm{Cl}(X)_{\mathbb{Q}}$ is hyperbolic with signature $(1, \varrho - 1)$, by the Hodge index theorem. Thus an element $w \in \partial Q$ cannot span an extremal ray $\mathbb{Q}_{\geq 0} \cdot w$ of $\mathrm{Eff}(X)$ because otherwise $\mathrm{Eff}(X)$ would not be locally polyhedral in a neighborhood of that ray. Hence all the extremal rays of the effective cone are spanned by classes of effective divisors E with $E^2 < 0$. Moreover, any such E must be irreducible because otherwise its class would not span an extremal ray of the effective cone. Thus, after possibly rescaling E, we can assume that it is a prime divisor.

Assume now that E is a prime divisor with $E^2 < 0$. Then E is contained in the base locus of $|nE|$ for any positive integer n, by Proposition 5.1.1.2. Thus $H^0(X, nE)$ is one-dimensional for any positive integer n, so that the class of E spans an extremal ray of $\mathrm{Eff}(X)$. $\qquad \square$

5.1.2 Nef and semiample cones

We consider smooth projective complex surfaces with finitely generated divisor class group and ask when the semiample cone coincides with the nef cone; recall from Corollary 4.3.3.6 that given this equality, Mori dreamness is equivalent to the effective cone being polyhedral. The main results concern the case of a nef anticanonical divisor (Theorem 5.1.2.1) and the case of positive anticanonical Iitaka dimension (Theorem 5.1.2.4). The original references for this section are [11, 198, 292].

We begin with smooth projective surfaces X with $q(X) = 0$ and nef anti-canonical divisor $-K_X$. Recall that $q(X) = 0$ merely means that $Cl(X)$ is finitely generated. Moreover, by the base point free theorem, assuming $-K_X$ to be nef gives that any nef and big divisor on X is semiample.

Theorem 5.1.2.1 *Let X be a smooth projective surface with $q(X) = 0$ and nef anticanonical divisor $-K_X$. Then the following statements are equivalent.*

(i) $\operatorname{Nef}(X) = \operatorname{SAmple}(X)$.
(ii) $-K_X$ *is semiample.*

Lemma 5.1.2.2 *Let X be a smooth projective surface with $q(X) = 0$ and let D be a nef not numerically trivial divisor on X with $D^2 = 0$. Then either D is semiample or $\kappa(D) = 0$.*

Proof First observe that $D^2 = 0$ implies $\kappa(D) \leq 1$. If $\kappa(D) = 0$, then D is not semiample. If $\kappa(D) = 1$ there exists a positive integer n such that $nD \sim M + F$ with $|M|$ positive dimensional, fixed component free and F effective. In particular M is nef by Proposition 5.1.1.2. Because D and M are nef and F is effective, $0 \leq D \cdot M \leq D \cdot (M + F) = 0$ gives $D \cdot M = 0$. Hence $D^2 = M^2 = 0$ and $D \sim \alpha M$ for some nonnegative rational number α, by Proposition 5.1.1.1. Thus D is semiample if and only if M is semiample. Because $|M|$ is fixed component free and $M^2 = 0$, it is base point free, so that M is semiample. $\square$

Proof of Theorem 5.1.2.1 Only for "(ii)$\Rightarrow$ (i)" is there something to show. Let D be a not numerically trivial nef divisor of X. If $D - K_X$ is nef and big then D is semiample by the base point free theorem. So we assume now that $D - K_X$ is not big. This implies $(D - K_X)^2 = 0$, so that $D^2 = D \cdot (-K_X) = 0$, being both D and $-K_X$ nef divisors. If $-K_X$ is not numerically trivial then $D \sim \alpha(-K_X)$ for some positive rational number α, by Proposition 5.1.1.1. Hence D is semiample if and only if $-K_X$ is semiample. If K_X is numerically trivial, then X is either a K3 surface or an Enriques surface. In the first case, by the Riemann–Roch formula and Serre's duality $h^0(X, D) > 1$. Thus we have $\kappa(D) > 0$ and D is semiample by Proposition 5.1.2.2. In the second case the universal cover $\pi : Y \to X$ is a double covering where Y is a K3 surface. The pullback $\pi^* D$ of a nef divisor D of X is nef. Thus it is semiample by our previous discussion; hence $|\pi^* nD|$ is base point free for some positive integer n. From [29, Lemma 17.2] we infer

$$\mathcal{O}_Y(\pi^* nD) = \pi^* \mathcal{O}_X(nD) \oplus \pi^* \mathcal{O}_X(nD + K_X).$$

If $p \in X$ is a base point of $|nD|$, then there is a $g \in H^0(X, nD + K_X)$ such that $g(p) \neq 0$ holds; otherwise $\pi^{-1}(p)$ would be in the base locus of $|\pi^*nD|$, which is a contradiction. Because $2K_X$ is trivial, we deduce that $|2nD|$ has no base points so that D is semiample. $\qquad\square$

Using the general classification theory of surfaces, we can figure out more precisely the ones satisfying the hypothesis of Theorem 5.1.2.1. Note that basic examples of case (ii) below are del Pezzo surfaces; these are studied in detail in the next section.

Theorem 5.1.2.3 *Let X be a smooth projective surface with $q(X) = 0$ and nef anticanonical divisor $-K_X$. Then X is one of the following:*

(i) *A K3 surface or an Enriques surface*
(ii) *A rational surface with $\varrho(X) \leq 10$ whose unique negative curves are (-1)- and (-2)-curves*

Moreover, $\mathrm{Nef}(X) = \mathrm{SAmple}(X)$ unless we are in case (ii) with $\varrho(X) = 10$ and the intersection graph of the (-2)-curves of X is a Coxeter graph of type $\tilde{A}_8$, $\tilde{D}_8$, $\tilde{E}_8$.

Proof If $-K_X$ is trivial, then by the classification theory of smooth projective surface either X is a K3 surface, or it is an Enriques surface or an abelian surface [39]. The last case cannot occur because $q(X) = 0$.

If $-K_X$ is nontrivial, then X has negative Kodaira dimension so that it is rational due to $q(X) = 0$. By the genus formula and the fact that $-K_X$ is nef, any curve C of X has self-intersection $C^2 \geq -2$. Moreover, if C^2 is negative, then still by the genus formula the curve C must be smooth rational, so that it is either a (-1)-curve or a (-2)-curve. By Theorem 5.1.2.1 the nef and the semiample cones of X coincide if and only if $-K_X$ is semiample. The last condition holds in cases (i), (ii) and in case (iii) if $\varrho(X) < 10$. The case $\varrho(X) = 10$ is a consequence of [232, Ex. 1.4.1]. $\qquad\square$

We now consider the case of rational surfaces of positive anticanonical Iitaka dimension $\kappa(-K_X)$. An example of such surface is the blow-up of $\mathbb{P}^2$ at the intersection points of five general lines; note that this is not a del Pezzo surface.

Theorem 5.1.2.4 *Let X be a rational smooth projective surface with $\kappa(-K_X) \geq 1$. Then $\mathrm{Nef}(X) = \mathrm{SAmple}(X)$ holds.*

For the proof, we discuss the case $-K_X$ not nef with $\kappa(-K_X) \geq 1$. If D is a nef and big divisor of X such that $D \cdot C > 0$ for any prime component C of an effective multiple of $-K_X$, then $nD - K_X$ is nef and big for $n > 0$ sufficiently big. Hence D is semiample by the base point free theorem We will establish a more efficient statement in our setting. For a divisor D on a surface X we set in the sequel

$$D^\perp := \{w \in \mathrm{Cl}(X); \; w \cdot D = 0\} \subseteq \mathrm{Cl}(X).$$

Let D be a nef and big divisor on a smooth projective surface X. A *block* for D is an effective divisor B such that $B \cdot C < 0$ for any prime divisor C with $C \cdot D = 0$. Observe that in general a block B is nonreduced. Moreover, by Proposition 5.1.1.2 any prime divisor C with $C \cdot D = 0$ is a component of block B. Thus for our definition to make sense we have to show that there are only finitely many such C. By the Hodge index theorem the intersection form on $D^\perp$ is negative definite. Hence if $\{C_1, \ldots, C_n\}$ is a set of distinct prime divisors whose classes are in $D^\perp$, then their classes are linearly independent in $\mathrm{Cl}(X)$. Indeed if $M := \sum_i n_i C_i \sim 0$, with $n_i \in \mathbb{Z}$ for any i, then by writing $M = M_1 - M_2$, with both M_1 and M_2 effective and without common components, we get $0 \le M_1 \cdot M_2 = M_1^2 < 0$, a contradiction.

In what follows, given a Cartier divisor D and a (possibly nonreduced) curve B on a surface X, we denote by $D|_B$ the restriction of D to B, that means $D|_B = \iota^*(D)$, where $\iota : B \to X$ is the inclusion and we regard D as given by local equations.

Proposition 5.1.2.5 *Let D be a nef and big divisor on a smooth projective surface X and let B be a block for D. Then the stable base locus of $|D|$ equals the stable base locus of $|D|_B|$.*

Proof Let E be a prime divisor with $E \cdot D > 0$ and let B be a block for D. After possibly replacing B with a positive multiple nB we can assume that

$$(B + E + K_X) \cdot C < 0 \qquad (5.1.1)$$

for any component C of B. Now put $Z = B + E$ and consider the exact sequence

$$0 \longrightarrow H^0(X, nD - Z) \longrightarrow H^0(X, nD) \overset{r}{\longrightarrow} H^0(X, nD|_Z) \longrightarrow H^1(X, nD - Z).$$

We will show that $H^1(X, nD - Z)$ vanishes for n big enough, proving in this way that the map r is surjective. We deduce that the restriction of nD to Z admits global sections not vanishing identically on E, and hence that E is not contained in the stable base locus of N. Therefore the support of the stable base locus of N is contained in the support of B. Moreover, when $E = 0$ we can choose $Z = B$, establishing the main statement of the lemma.

To prove the claimed vanishing we apply the Kawamata–Viehweg vanishing theorem. Hence it is enough to show that the divisor $A_n := nD - Z - K_X$ is ample for n sufficiently big. Because D is big, there exists an n_0 such that $A_n^2 > 0$ for any $n \ge n_0$. In particular the set $\mathcal{E}$ of integral curves having negative intersection with A_{n_0} is finite. If $C \in \mathcal{E}$ is such that $C \cdot D = 0$, then C is a component of B so that $A_n \cdot C > 0$ by (5.1.1), for any $n \geq 0$. If $C \in \mathcal{E}$ is such that $C \cdot D > 0$, then again $A_n \cdot C > 0$ for n sufficiently big. Hence the statement is proved. $\qquad\square$

Remark 5.1.2.6 As a consequence of Proposition 5.1.2.5 we observe that if R is a connected component of the support of a block B of a nef and big divisor D such that $H^1(R, \mathcal{O}_R) = \langle 0 \rangle$, then the stable base locus of $|D|$ is disjoint from R, because in this case R behaves like a rational curve by [207, Prop. 11.1] and $D|_R$ is a divisor of nonnegative degree.

Proposition 5.1.2.7 *Let X be a smooth projective rational surface and let D be a nef divisor of X such that $D \cdot (-K_X) > 0$. Then D is semiample.*

Proof We first consider the case D big. Let B be a block for D and let R be a connected component of the support of B. Because $(R + K_X) \cdot D < 0$ and D is a nef divisor, $R + K_X$ cannot be linearly equivalent to an effective divisor or equivalently $H^0(X, R + K_X) = \langle 0 \rangle$. By Serre's duality [39, Thm. I.11] we get $H^2(X, -R) = \langle 0 \rangle$. Hence taking the long exact sequence in cohomology of the exact sequence of sheaves

$$0 \longrightarrow \mathcal{O}_X(-R) \longrightarrow \mathcal{O}_X \longrightarrow \mathcal{O}_R \longrightarrow 0$$

and recalling that $H^1(X, \mathcal{O}_X) = H^2(X, \mathcal{O}_X) = \langle 0 \rangle$, because X is rational, we obtain that $H^1(R, \mathcal{O}_R) = \langle 0 \rangle$, so that R is disjoint from the stable base locus of D by Proposition 5.1.2.5 and Remark 5.1.2.6. Hence D is semiample.

Suppose now that D is not big, and hence that $D^2 = 0$. From $D \cdot (K_X - D) < 0$, the Riemann–Roch formula and Serre's duality [39, Thm. I.11] we deduce that D is linearly equivalent to an effective divisor. To avoid introducing more notation, we assume that D itself is effective. Because $D \cdot (K_X + D) < 0$, then $K_X + D$ is not linearly equivalent to an effective divisor. Hence from the exact sequence

$$0 \longrightarrow \mathcal{O}_X(-D) \longrightarrow \mathcal{O}_X \longrightarrow \mathcal{O}_D \longrightarrow 0$$

Serre's duality [39, Thm. I.11] and the fact that X is rational, we deduce that $H^1(D, \mathcal{O}_D) = \langle 0 \rangle$. Tensoring the previous sequence by $\mathcal{O}_X(D)$ and taking cohomomology we get a surjective map $H^0(X, D) \to H^0(D, D|_D)$. We deduce that the base locus of $|D|$ is contained in D. Because the divisor $D|_D$ is trivial and $H^1(D, \mathcal{O}_D) = \langle 0 \rangle$ we conclude that the base locus of $|D|$ is empty. $\quad\square$

Proof of Theorem 5.1.2.4 Let D be a nef divisor of X. If $D \cdot (-K_X) > 0$, then D is semiample by Proposition 5.1.2.7. So, assume $D \cdot (-K_X) = 0$ and let $-K_X = P + N$ be the Zariski decomposition of $-K_X$ with positive part P.

By $D \cdot (P + N) = D \cdot (-K_X) = 0$, the fact that P is nef and N is effective, we deduce $D \cdot P = 0$. Because both D and P are nef nontrivial divisors, $D^2 = P^2 = 0$ and $D \sim \alpha P$ for some nonnegative rational number α, by Proposition 5.1.1.1. Hence $1 \le \kappa(-K_X) = \kappa(P) = \kappa(D) \le 1$, where the first equality is a property of the Zariski decomposition and the second inequality is due to $D^2 = 0$ and D nef. Hence $\kappa(D) = 1$ so that D is semiample by Proposition 5.1.2.2. □

We conclude the section with a glimpse beyond the cases treated so far. We begin with a result on the case $\kappa(-K_X) = 0$; there, if the linear system $| - K_X|$ is nonempty and E is the support of the unique effective divisor of $| - K_X|$, we denote by $j \colon \operatorname{Pic}(X) \to \operatorname{Pic}(E)$ the pullback homomorphism defined by the inclusion map $E \to X$. The following results are [198, Thm. 4.1, Thm. 5.4].

Theorem 5.1.2.8 *Let X be a rational smooth projective surface with anticanonical divisor of Iitaka dimension zero. If $| - K_X|$ is nonempty let E be the unique element of the linear system. Then the following are equivalent.*

(i) $\operatorname{Nef}(X) = \operatorname{SAmple}(X)$.
(ii) *Either $| - K_X|$ is empty or the image of the pullback homomorphism $K_X^\perp \to \operatorname{Pic}(E)$ is a finite group.*

Example 5.1.2.9 We show how to construct a non-Mori dream smooth complex projective surface X with polyhedral effective cone $\operatorname{Eff}(X)$ and $\operatorname{SAmple}(X) \ne \operatorname{Nef}(X)$.

Let $\pi \colon X \to S$ be the blow-up of a smooth very general quartic-surface $S \subseteq \mathbb{P}^3$ at a very general point $p \in S$. Denote by C the strict transform of the genus two singular curve $\Gamma := T_p S \cap S$ cut out on S by the tangent plane at p. By the adjunction formula [39, Rem. I.16 (ii)], we have

$$C|_C \sim K_C - E|_C = K_C - p_1 - p_2,$$

where E is the exceptional divisor of π and the set $\{p_1, p_2\} = \pi^{-1}(p) \cap C$ is the preimage in C of the node of Γ. It is possible to show [10, Sec. 6] that, due to the very generality assumptions made on S and p, the degree 0 divisor $C|_C$ of C is nontorsion in $\operatorname{Pic}^0(C)$ so that the fundamental sequence of C shows that $H^0(X, nC)$ is one-dimensional for any positive integer n and thus C is not semiample.

On the other hand, the divisor C is nef because it is prime and $C^2 = 0$. Hence X is not a Mori dream surface, by Corollary 4.3.3.6.

Remark 5.1.2.10 Examples of Mori dream surfaces with $\kappa(-K_X) = -\infty$ are known, even for surfaces of general type. Indeed, any normal projective variety with divisor class group of rank 1 is Mori dream. Further examples with higher rank divisor class group can be obtained by means of Lefschetz type theorems for Mori dream spaces; see Remark 4.1.1.4.

5.1.3 Rational surfaces

We consider rational smooth complex projective surfaces X; recall that rational means that the function field $\mathbb{C}(X)$ is isomorphic to $\mathbb{C}(T_1, T_2)$. The first result, Theorem 5.1.3.1, concerns the case of nonnegative anticanonical Iitaka dimension and characterizes polyhedrality of the effective cone of X. The second result, Theorem 5.1.3.7, characterizes the Mori dream surfaces X for the case of positive anticanonical Iitaka dimension. The original references for this section are [10, 198, 293].

Theorem 5.1.3.1 *Let X be a rational smooth complex projective surface with Iitaka dimension $\kappa(-K_X) \geq 0$. Then the following statements are equivalent.*

(i) *The effective cone $\mathrm{Eff}(X) \subseteq \mathrm{Cl}_{\mathbb{Q}}(X)$ is polyhedral.*
(ii) *The surface X contains only finitely many (-1)- and (-2)-curves.*

In a first step, we consider the cone $E(X) \subseteq \mathrm{Cl}_{\mathbb{Q}}(X)$ generated by the irreducible curves $E \subseteq X$ with self-intersection $E^2 < 0$. Note that we have $E(X) \subseteq \mathrm{Eff}(X)$ by definition.

Proposition 5.1.3.2 *Let X be a rational smooth complex projective surface with anticanonical Iitaka dimension $\kappa(-K_X) \geq 0$ and Picard number $\varrho(X) \geq 3$. Then $\overline{\mathrm{Eff}(X)} = \overline{E(X)}$ holds in $\mathrm{Cl}_{\mathbb{Q}}(X)$.*

Lemma 5.1.3.3 *Let X be a smooth projective rational surface and let C be an irreducible smooth rational curve of X with $C^2 \in \{0, 1\}$. If the class $w = [C]$ of C spans an extremal ray of the effective cone $\mathrm{Eff}(X)$, then X is either a Hirzebruch surface or the projective plane.*

Proof From the fundamental sequence of C we get the following long exact cohomology sequence

$$0 \longrightarrow H^0(X, \mathcal{O}_X) \longrightarrow H^0(X, C) \overset{r}{\longrightarrow} H^0(C, C|_C) \longrightarrow 0,$$

where the last zero is due to the rationality of X. Hence $|C|$ is a base point free linear system of dimension $C^2 + 1$ so that it defines a morphism $\varphi\colon X \to \mathbb{P}^n$. Because w spans an extremal ray of the effective cone of X and C is irreducible, all the elements of the linear system $|C|$ are irreducible. If $C^2 = 0$, then $n = 1$

and the fibers of φ are all isomorphic to $\mathbb{P}^1$ so that X is a Hirzebruch surface $\mathbb{F}_n = \mathbb{P}(\mathcal{O}_{\mathbb{P}^1} \oplus \mathcal{O}_{\mathbb{P}^1}(n))$, by [39, Lemma III.8]. If $C^2 = 1$, then $n = 2$ and φ is an embedding, being a degree 1 map between normal varieties, so that X is isomorphic to the projective plane. $\qquad\qquad\qquad\qquad\qquad\qquad\qquad\square$

Proof of Proposition 5.1.3.2   Let $w \in \mathrm{Cl}(X)$ be a class that lies on an extremal ray of $\overline{\mathrm{Eff}}(X)$. It is enough to show that $w \in \overline{E(X)}$.

Assume $w \cdot K_X < 0$. Then by the cone theorem [202, Thm. 1.5.33] w is the class of a rational curve C with $C \cdot K_X \geq -3$, so that $C^2 \in \{-1, 0, 1\}$ by the genus formula. By Lemma 5.1.3.3 and the hypothesis on the Picard number of X we conclude $C^2 = -1$ so that $w \in E(X)$.

Assume $w \cdot K_X > 0$. Let D be a divisor of X whose class is w. Because D is pseudoeffective, that means that its class lies in the closure of the effective cone; there is a Zariski decomposition $D = P + N$ with positive part P and N effective; see [202, Thm. 2.3.19]. Because the classes of P and N lie in $\overline{\mathrm{Eff}}(X)$ and w lies on an extremal ray, w is either the class of P or that of N. The first case cannot occur because $w \cdot K_X > 0$ and $P \cdot K_X < 0$, being P nef and $-K_X$ pseudoeffective. Thus the second case holds and so $w \in E(X)$.

Finally, assume $w \cdot K_X = 0$. We prove by contradiction that w lies on an extremal ray of $\overline{E(X)}$. Indeed if this were not the case there would be an open neighborhood U of $\mathbb{Q}_{>0} \cdot w$ that does not intersect $\overline{E(X)}$. Because U is open there exists a nontrivial class $w' \in U$ that lies on an extremal ray of $\overline{\mathrm{Eff}}(X)$ and such that $w' \cdot K_X \neq 0$, a contradiction. $\qquad\qquad\qquad\qquad\square$

Proof of Theorem 5.1.3.1   If $\varrho(X) \leq 2$, then both conditions are obviously satisfied because X is either $\mathbb{P}^2$ or an Hirzebruch surface $\mathbb{F}_n$. Assume now that $\varrho(X) \geq 3$.

We show "(i)$\Rightarrow$(ii)." By Proposition 5.1.1.2, if E is a negative curve, then $h^0(X, nE) = 1$ for any positive n. Hence the classes of negative curves span extremal rays of the effective cone.

We prove "(ii)$\Rightarrow$(i)." By Proposition 5.1.3.2 it is enough to prove that $E(X)$ is polyhedral, or equivalently that X contains finitely many classes of prime divisors D with $D^2 < 0$. If $D \cdot (-K_X) < 0$, then D is contained in the stable base locus of $|-K_X|$, by Proposition 5.1.1.2. Hence there are only finitely many divisors with this property. If on the other hand $D \cdot (-K_X) \geq 0$, then by genus formula D is a smooth rational curve with $D^2 \in \{-1, -2\}$. By (ii) there are finitely many such curves. Hence the statement follows. $\qquad\qquad\square$

We come to the characterization of rational smooth complex projective Mori dream surfaces X of positive anticanonical Iitaka dimension $\kappa(-K_X)$. For the case $\kappa(-K_X) = 1$, the following class of fibrations is central.

Definition 5.1.3.4 Let X be a smooth projective complex surface. An *elliptic fibration* on X is a morphism $\pi : X \to B$ onto a smooth curve B such that the general fiber of π is a smooth irreducible curve of genus 1. We say that an elliptic fibration $\pi : X \to B$ is *relatively minimal* if it does not contract (-1)-curves.

Note that an elliptic fibration $\pi : X \to B$ on a rational surface X necessarily has base curve $B \cong \mathbb{P}^1$, because otherwise the pullback of a nontrivial holomorphic 1-form of B would be a nontrivial holomorphic 1-form on X, which contradicts rationality. Moreover, if $\pi : X \to \mathbb{P}^1$ is a relatively minimal elliptic fibration on a smooth complex projective surface X, then $-K_X$ is linearly equivalent to a positive rational multiple of a π-fiber by [29, Cor. V.12.3]. In particular, X is then of anticanonical Iitaka dimension 1. A partial converse now follows.

Construction 5.1.3.5 Let X be a rational smooth complex projective surface with anticanonical Iitaka dimension $\kappa(-K_X) = 1$. Then there are a unique elliptic fibration $\pi : X \to \mathbb{P}^1$ and a unique commutative diagram

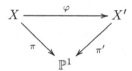

where $\varphi : X \to X'$ is a birational morphism onto a smooth (rational) surface X' with nef anticanonical divisor $-K_{X'}$ and relatively minimal elliptic fibration $\pi' : X' \to \mathbb{P}^1$. We refer to π' as the *relatively minimal model* of π.

Proof Let $-K_X = P + N$ be the Zariski decomposition of the anticanonical divisor with positive part P. Then P is semiample due to $\kappa(P) = \kappa(-K_X) = 1$ and Proposition 5.1.2.2. Let $\pi : X \to B$ be the morphism defined by P according to Remark 3.3.4.3. Observe that B is a curve because $\kappa(P) = \kappa(-K_X) = 1$. As X is rational, we have $B \cong \mathbb{P}^1$. Using Stein factorization, we can assume that π has connected fibers. By the genus formula the fibers of π have arithmetic genus 1 and by Bertini's theorem [160, Cor. III.10.9] the general fiber is smooth. Thus, π is an elliptic fibration.

To show the uniqueness of π observe that if F is a smooth fiber of an elliptic fibration on X, then $F^2 = 0$ and $p_a(F) = 1$, so that $F \cdot (-K_X) = 0$ by the genus formula. Moreover F is nef, being irreducible of nonnegative self-intersection. Thus we deduce $0 = F \cdot (P + N) \geq F \cdot P \geq 0$ so that $F \cdot P = 0$. By Proposition 5.1.1.1 and the fact that both F and a general fiber of π are irreducible we conclude that F is a fiber of π.

The birational morphism φ is now obtained by contracting all the (-1)-curves contained into the fibers of π, so that the resulting elliptic fibration π' is relatively minimal. By [29, Thm. 12.1] we conclude that $-K_{X'}$ is linearly equivalent to a positive rational multiple of a fiber of π' and thus is nef. □

Definition 5.1.3.6 An elliptic fibration $\pi : X \to \mathbb{P}^1$ on a rational surface X is called *extremal* if the components of its reducible fibers generate in $\mathrm{Cl}(X)$ a subgroup of corank 1.

Theorem 5.1.3.7 *Let X be a rational smooth complex projective surface of anticanonical Iitaka dimension $\kappa(-K_X) \geq 1$. Then X is a Mori dream surface exactly in the following cases:*

(i) $\kappa(-K_X) = 2$.
(ii) $\kappa(-K_X) = 1$ *and a relatively minimal model of π is extremal.*

An example class are the weak del Pezzo surfaces treated in Proposition 5.2.1.10. Let us prepare the proof of Theorem 5.1.3.7. Below, for a divisor D on X, we write $D^{\perp} \subseteq \mathrm{Cl}(X)$ for the subgroup of elements orthogonal to the class of D.

Proposition 5.1.3.8 *Let X be a rational smooth complex surface with $-K_X$ nef and $K_X^2 = 0$. Then the following statements are equivalent.*

(i) *X contains finitely many (-1)-curves.*
(ii) *$K_X{}^{\perp} \subseteq \mathrm{Cl}(X)$ is spanned by classes of (-2)-curves.*

Proof We show "(i)⇒(ii)." First of all we observe that X contains finitely many (-2)-curves. Indeed if C is a (-2)-curve of X, then $-K_X \cdot C = 0$. By the Hodge index theorem the intersection matrix on $K_X{}^{\perp}$ is negative semidefinite. By hypothesis the class of K_X is in $K_X{}^{\perp}$ so that the intersection form descends to a negative definite intersection form on the quotient group $V := K_X^{\perp}/\langle K_X \rangle$. Hence there is a finite number of classes $w \in V$ with $w^2 = -2$. For each such class there are infinitely many classes of the form $w + n[K_X] \in \mathrm{Cl}(X)$ where $n \in \mathbb{Z}$. We claim that only one of these classes is the class of a (-2)-curve C. Indeed if $C + nK_X$ would be linearly equivalent to a (-2)-curve E, then from

$$0 = (nK_X)^2 = (E - C)^2 = -4 - 2E \cdot C$$

we would deduce $E \cdot C = -2$, so that $E = C$, a contradiction. Hence X contains finitely many (-2)-curves. Because by (i) it contains also finitely many (-1)-curves, $\mathrm{Eff}(X)$ is polyhedral by Corollary 5.1.3.1. Observe that $-K_X$ is an extremal ray of the nef cone $\mathrm{Nef}(X)$. Because the effective cone is dual to the nef cone, the face $-K_X^{\perp} \cap \mathrm{Eff}(X)$ of the effective cone is polyhedral of

dimension 9. By the genus formula its extremal rays are spanned by classes of
(-2)-curves.

We sketch a proof of "(ii)$\Rightarrow$(i)." By hypothesis $-K_X^\perp \cap \mathrm{Eff}(X)$ is polyhedral
of codimension 1, then $\mathrm{Nef}(X)$ is locally polyhedral in a neighborhood of
$[-K_X] \in \mathrm{Cl}(X)$. As a consequence of the cone theorem [202, Thm. 1.5.33] it
is possible to show [232, p. 82] that the nef cone is polyhedral. This implies
that X contains finitely many (-1)-curves. $\qquad\square$

Proposition 5.1.3.9 *Let X be a smooth projective rational surface with $-K_X$
of Iitaka dimension 1 and let X' be as in Construction 5.1.3.5. Then the following
statements are equivalent.*

 (i) *X contains finitely many (-1)-curves.*
(ii) *X' contains finitely many (-1)-curves.*

Proof For a proof of the implication "(ii)$\Rightarrow$(i)," we refer to [10, Lemma 4.7,
Thm. 4.8]. We only present a proof of "(i)$\Rightarrow$(ii)" here.

If E is a (-1)-curve of X', then its strict transform via the birational morphism
$\phi\colon X \to X'$ is a smooth rational curve C of X of negative self-intersection.
Hence either C is a (-1)-curve or $C \cdot (-K_X) \le 0$ by the genus formula. In the
first case there are finitely many such curves by hypothesis. In the second case,
given the Zariski decomposition $-K_X = P + N$ with positive part P, either
$C \cdot P = 0$ or $C \cdot N < 0$. Both possibilities lead to a finite number of cases. In
the first case C is contracted by the morphism $\varphi(P)\colon X \to \mathbb{P}^1$, while in the
second case C is a prime component of the support of N by Proposition 5.1.1.2.
This proves the statement. $\qquad\square$

Proof of Theorem 5.1.3.7 Let X be a smooth projective rational surface with
$\kappa(-K_X) \ge 1$. By Theorem 5.1.2.4 we have $\mathrm{Nef}(X) = \mathrm{SAmple}(X)$. Hence X
is Mori dream if and only if $\mathrm{Nef}(X)$ is polyhedral, or equivalently if and only if
$\mathrm{Eff}(X)$ is polyhedral, as the two cones are dual with respect to the intersection
form. Hence X is Mori dream if and only if it contains finitely many (-1)-
and (-2)-curves, by Corollary 5.1.3.1. Observe that for any such curve C we
have $-K_X \cdot C \in \{1, 0\}$, by the genus formula. Moreover, because $H^0(X, C)$
is one-dimensional for any negative curve C, there are finitely many negative
curves if and only if there are finitely many negative classes.

If $\kappa(-K_X) = 2$, then $-K_X \sim A + E$, where A and E are $\mathbb{Q}$-divisors with
A ample and E effective. By the cone theorem [202, Thm. 1.5.33] there are
finitely many classes of curves C such that $C \cdot A \le 1$. If C is a (-1)- or (-2)-
curve such that $C \cdot A > 1$ then $C \cdot E < 0$, by the adjuncion formula. Hence C
is contained in the support of E, by Proposition 5.1.1.2, so that there are finitely
many such curves.

If $\kappa(-K_X) = 1$, consider the Zariski decomposition $-K_X = P + N$ with positive part P. Then $\kappa(P) = \kappa(-K_X) = 1$, so that P is semiample, by Proposition 5.1.2.2. Let $\pi \colon X \to \mathbb{P}^1$ be the morphism defined by a multiple of P. If C is a (-2)-curve of X, then $-K_X \cdot C = 0$, by the genus formula. Hence either $C \cdot N < 0$ or $C \cdot P = 0$. In the first case C is contained in the support of N, by Proposition 5.1.1.2. In the second case C is a component of a fiber of π. In both cases there are finitely many such curves. Hence X is Mori dream if and only if it contains finitely many (-1)-curves. By Propositions 5.1.3.8 and 5.1.3.9 the last condition is equivalent to ask that $\mathrm{Cl}(X')$ contains a rank 9 subgroup spanned by classes of (-2)-curves. $\square$

Remark 5.1.3.10 If $-K_X$ is nef but not semiample, then $K_X^2 = 0$. In this case the Cox ring $\mathcal{R}(X)$ is not finitely generated by Corollary 4.3.3.6. There are exactly three families of such rational surfaces whose effective cone is polyhedral. These are classified in [232, Ex. 1.4.1].

Remark 5.1.3.11 Examples of smooth Mori dream rational surfaces X with $\kappa(-K_X) < 0$ can be obtained by constructing smooth surfaces of complexity 1 with large Picard number. Consider, for example, the minimal resolution X of the singular rational $\mathbb{K}^*$-surface $Y \subseteq \mathbb{P}(33, 22, 6, 1)$ of equation $x_0^2 + x_1^3 + x_2^{11} = 0$. The canonical class of Y is ample because $\mathrm{Cl}(Y) \cong \mathbb{Z}$ and $-K_Y$ has degree 4 by adjunction formula. Thus $-K_X$ has negative Iitaka dimension. Moreover, X is Mori dream because any rational $\mathbb{K}^*$-surface is Mori dream. A presentation of the Cox ring of X is given in Example 5.4.4.6. Another way to obtain smooth Mori dream rational surfaces with anticanonical class of negative Iitaka dimension is via the theory of redundant blow-ups [179].

We conclude this section with Totaro's characterization [297, Cor. 5.1] of Mori dreamness for klt Calabi–Yau pairs in the surface case; see Section 5.1.5 for a detailed discussion of the special case of K3 surfaces.

Theorem 5.1.3.12 *Let (X, Δ) be a klt Calabi–Yau pair, where X is a complex projective surface. Then the following statements are equivalent.*

(i) *X is a Mori dream surface.*
(ii) *The image of $\mathrm{Aut}(X) \to \mathrm{GL}(\mathrm{Cl}(X))$ is a finite group.*

If the pair (X, Δ) is not a klt Calabi–Yau pair then the implication (ii)$\Rightarrow$(i) of Theorem 5.1.3.12 no longer holds as shown in Exercise 5.5.

5.1.4 Extremal rational elliptic surfaces

As shown in Theorem 5.1.3.7, a rational smooth complex projective surface of anticanonical Iitaka dimension 1 is Mori dream if and only if its relatively minimal model is an extremal elliptic fibration. Here, we classify the latter ones. The following type of fibrations plays a central role.

Definition 5.1.4.1 An elliptic fibration $\pi: X \to B$ is called *jacobian* if it admits a section $\sigma: B \to X$.

Jacobian elliptic fibrations on rational surfaces have been classified by Cossec–Dolgachev [103] and Miranda-Persson [220]. Basing on these results, we obtain a first statement, characterizing the Mori dream property in terms of the possible singular fibers; for the latter, we use Kodaira's notation [29, Sec. V.7].

Theorem 5.1.4.2 *Let X be a rational smooth complex projective surface admitting a relatively minimal elliptic fibration $\pi: X \to \mathbb{P}^1$. Then the following statements are equivalent.*

(i) *X is a Mori dream surface.*
(ii) *The configuration of singular nonmultiple fibers of π is one the following.*

Type	Fibers	Type	Fibers
X_{22}	$\mathrm{II}^* \, \mathrm{II}$	X_{431}	$\mathrm{IV}^* \, \mathrm{I}_3 \, \mathrm{I}_1$
X_{211}	$\mathrm{II}^* \, 2\mathrm{I}_1$	X_{222}	$\mathrm{I}_2^* \, 2\mathrm{I}_2$
X_{411}	$\mathrm{I}_4^* \, 2\mathrm{I}_1$	X_{141}	$\mathrm{I}_1^* \, \mathrm{I}_4 \, \mathrm{I}_1$
X_{9111}	$\mathrm{I}_9 \, 3\mathrm{I}_1$	X_{6321}	$\mathrm{I}_6 \, \mathrm{I}_3 \, \mathrm{I}_2 \, \mathrm{I}_1$
X_{33}	$\mathrm{III}^* \, \mathrm{III}$	$X_{11}(a)$	$2\mathrm{I}_0^*$
X_{321}	$\mathrm{III}^* \, \mathrm{I}_2 \, \mathrm{I}_1$	X_{5511}	$2\mathrm{I}_5 \, 2\mathrm{I}_1$
X_{8211}	$\mathrm{I}_8 \, \mathrm{I}_2 \, 2\mathrm{I}_1$	X_{4422}	$2\mathrm{I}_4 \, 2\mathrm{I}_2$
X_{44}	$\mathrm{IV}^* \, \mathrm{IV}$	X_{3333}	$4\mathrm{I}_3$

If one of these statements holds, then the elliptic fibration $\pi: X \to \mathbb{P}^1$ is extremal and the divisor class group $\mathrm{Cl}(X)$ is free of rank 10.

Proof We show "(i)$\Rightarrow$(ii)." Because π is relatively minimal and X is rational, $-K_X$ is linearly equivalent to a positive rational multiple of a fiber of π by [29, Thm. 12.1]. In particular $-K_X$ is semiample with $K_X^2 = 0$. Let $\mathrm{Jac}(\pi): \mathrm{Jac}(X) \to \mathbb{P}^1$ be the jacobian elliptic fibration associated with π as defined in [103, Prop. 5.2.5]. The fibrations π and $\mathrm{Jac}(\pi)$ have the same configuration of nonmultiple singular fibers by [103, Thm. 5.3.1]. Because X is Mori dream, we infer from Proposition 5.1.3.8 that the (-2)-curves of X span the subgroup $K_X^\perp$. Thus, π and hence also $\mathrm{Jac}(\pi)$ are extremal. By the classification of extremal jacobian elliptic fibrations [103, 220] and [131, (ii) p. 431] we conclude that the singular nonmultiple fibers of π are those given in the statement.

For the implication "(ii)⇒(i)," observe that surface X in the table contains a set of (-2)-curves (the components of the reducible fibers of π) whose classes span $K_X^\perp$. Hence X is Mori dream by Theorem 5.1.3.7.

Because X is a rational surface with $K_X^2 = 0$, the class group $\mathrm{Cl}(X)$ has rank 10. Moreover, the intersection form on the prime components of the π-fibers is negative semidefinite by [29, Lemma III.8.2]; thus such components span a subgroup of $\mathrm{Cl}(X)$ whose rank equals the number of prime components minus the number of reducible fibers plus 1. A direct verification shows that this rank is 9 for all the types listed in (ii). Thus each such fibration is extremal. □

Recall that for a jacobian elliptic fibration $\pi : X \to B$, the set of sections becomes an abelian group with respect to pointwise addition, isomorphic to the *Mordell–Weil group* $\mathrm{MW}(\pi)$, which consists of the $\mathbb{C}(B)$-rational points of the generic fiber of π.

Remark 5.1.4.3 Among the Mori dream surfaces listed in Theorem 5.1.4.2 (ii), there are jacobian extremal elliptic fibrations $\pi : X \to \mathbb{P}^1$ obtained by blowing up $\mathbb{P}^2$ at the nine base points of one of the following pencils of cubic curves.

Type	Pencil of cubics	$\mathrm{MW}(\pi)$
X_{22}	$y^3 + x^2 z + \alpha z^3$	$\langle 0 \rangle$
X_{211}	$27(y^3 + x^2 z + y^2 z) - 4\alpha z^3$	$\langle 0 \rangle$
X_{411}	$x^2 y + z^3 + y^2 z - 2yz^2 + 4\alpha\, yz^2$	$\mathbb{Z}/2\mathbb{Z}$
X_{9111}	$x^2 y + y^2 z + z^2 x - 3\alpha\, xyz$	$\mathbb{Z}/3\mathbb{Z}$
X_{33}	$x(xz - y^2) + \alpha z^3$	$\mathbb{Z}/2\mathbb{Z}$
X_{321}	$64x(xz - y^2 + yz) + \alpha z^3$	$\mathbb{Z}/2\mathbb{Z}$
X_{8211}	$(x - y)(xy - z^2) + 4\alpha\, xyz$	$\mathbb{Z}/4\mathbb{Z}$
X_{44}	$yz(y - z) + \alpha x^3$	$\mathbb{Z}/3\mathbb{Z}$
X_{431}	$27yz(x + y + z) - \alpha x^3$	$\mathbb{Z}/3\mathbb{Z}$
X_{222}	$y^3 + z(2y - z)(x - y) + \alpha\, xy(x - y)$	$\mathbb{Z}/2\mathbb{Z} \oplus \mathbb{Z}/2\mathbb{Z}$
X_{141}	$16z(xy - y^2 + xz) + \alpha\, xy(x - y)$	$\mathbb{Z}/4\mathbb{Z}$
X_{6321}	$(x + y)(y + z)(x + z) + \alpha\, xyz$	$\mathbb{Z}/6\mathbb{Z}$
$X_{11}(a)$	$yz(y - z) + \alpha\, x^2(y - az)$	$\mathbb{Z}/2\mathbb{Z} \oplus \mathbb{Z}/2\mathbb{Z}$
X_{5511}	$(y + z)(x + y)(x + y + z) + \alpha\, xyz$	$\mathbb{Z}/5\mathbb{Z}$
X_{4422}	$(x - z)(x - 2y + z)(x + 2y + z) + 8\alpha\, xyz$	$\mathbb{Z}/4\mathbb{Z} \oplus \mathbb{Z}/2\mathbb{Z}$
X_{3333}	$x^3 + y^3 + z^3 - 3\alpha\, xyz$	$\mathbb{Z}/3\mathbb{Z} \oplus \mathbb{Z}/3\mathbb{Z}$

Here, the computation of the Mordell–Weil group is done via the Shioda–Tate–Wazir exact sequence [305, Lemma 3.1] of a jacobian elliptic fibration π involving $\mathrm{MW}(\pi)$ and the subgroup $T \subseteq \mathrm{Cl}(X)$ generated by the classes of the irreducible and reduced components of π-fibers and the class of a section $\sigma(\mathbb{P}^1)$:

$$0 \longrightarrow T \longrightarrow \mathrm{Cl}(X) \longrightarrow \mathrm{MW}(\pi) \longrightarrow 0.$$

Theorem 5.1.4.4 *Let X be a rational smooth projective surface and $\pi : X \to \mathbb{P}^1$ a jacobian, relatively minimal elliptic fibration. Then the following statements are equivalent.*

(i) *X is a Mori dream surface.*
(ii) *The Mordell–Weil group $\mathrm{MW}(\pi)$ is finite.*
(iii) *X is isomorphic to one of the surfaces of Remark 5.1.4.3.*

Moreover, if one of the above conditions holds, then the Cox ring $\mathcal{R}(X)$ is generated by sections in the degrees $[-K_X]$ and $[C]$, where C is a smooth irreducible rational curve with $C^2 \in \{-2, -1, 0, 1\}$.

Proof The Shioda–Tate–Wazir exact sequence tells us that $\mathrm{MW}(\pi)$ is finite if and only if π is extremal. By Theorem 5.1.3.7, the latter property is equivalent to X being Mori dream. This proves "(i)⇔(ii)."

We sketch the proof of "(ii)⇒(iii)." Because π is relatively minimal and X is rational, $-K_X$ is linearly equivalent to a positive rational multiple of a fiber F of π by [29, Thm. 12.1]. If $E := \sigma(\mathbb{P}^1)$ is a section of π, then $E \cdot F = 1$, so that $H^0(X, E) = 1$ because otherwise the restriction of the linear system $|E|$ to F would be a linear system of degree 1 and positive dimension on a curve of positive genus, a contradiction. Thus, by the Riemann–Roch formula and Serre's duality we deduce $E^2 - E \cdot K_X \leq 0$. On the other hand, by the genus formula we have $E^2 + E \cdot K_X = -2$. Because $-K_X$ is nef and proportional to a positive multiple of F, we deduce $-K_X \cdot E = 1$ so that $-K_X$ is linearly equivalent to F. Moreover E is a (-1)-curve by the genus formula.

By the Shioda–Tate–Wazir exact sequence we know the Mordell–Weil group of π for each of the 16 possibilities listed in Theorem 5.1.4.2. By the Shioda height pairing we know how each section $E := \sigma(\mathbb{P}^1)$ of π intersects each prime component of a fiber of π. This information allows one to perform nine contractions of (-1)-curves whose composition is a birational morphism $\varphi \colon X \to \mathbb{P}^2$. One directly verifies that the image of the linear system $|-K_X|$ in each case is one of the pencils of cubic curves listed in (iii). Moreover it is possible to show that, a part $X_{11}(a)$, each such surface is unique up to isomorphism [103].

The implication "(iii)⇒(ii)" is clear. For a proof of the last statement and an explicit presentation of the Cox rings of the 16 families of surfaces see [8]. $\quad\square$

Remark 5.1.4.5   Whenever the fibration has two or three singular fibers, the corresponding values of α are 0, ∞ and 0, 1, ∞ respectively. The surfaces X_{9111} and X_{3333} have singular fibers at 1, ε, ε^2, ∞; the surface X_{8211} at $-1, 0, 1, \infty$; the surface X_{6321} at $-8, 0, 1, \infty$; and the surface X_{5511} at 0, $\frac{1}{2}(11 \pm 5\sqrt{5})$, ∞. The three surfaces X_{22}, X_{33}, X_{44} and the family $X_{11}(a)$, where $a \in \mathbb{C} - \{0, 1\}$, come with a $\mathbb{C}^*$-action, see [131, Prop. 9.2.17], given on $\mathbb{P}^2$ as follows:

$$X_{22}: \quad t \cdot [x, y, z] = [tx, y, t^{-2}z]$$
$$X_{33}: \quad t \cdot [x, y, z] = [tx, y, t^{-1}z]$$
$$X_{44} \text{ and } X_{11}(a): \quad t \cdot [x, y, z] = [tx, y, z].$$

Relatively minimal rational elliptic surfaces are always blow-ups of the projective plane at nine, possibly infinitely near, points [103].

Proposition 5.1.4.6   *The configurations of nine distinct points whose blow-up is Mori dream are dense with respect to the Zariski topology of the configuration space of nine points in $\mathbb{P}^2$.*

Proof   To prove the statement we start with the *Hesse surface* $X := X_{3333}$ of Theorem 5.1.4.4. The elliptic fibration $\pi \colon X \to \mathbb{P}^1$ is obtained by resolving the indeterminacy of the map $[x, y, z] \mapsto [x^3 + y^3 + z^3, xyz]$. The fibration π is relatively minimal because it is defined by $|-K_X|$ and $-K_X$ is nef. Moreover the only singular fibers of π are four fibers of type I_3. These are attained at the values $\alpha \in \{\infty, 1, \varepsilon, \varepsilon^2\}$, where ε is a primitive third root of unity. The four singular fibers are the strict transforms of the following reducible plane cubics:

$$xyz = 0$$
$$(x + y + z)(x + \varepsilon^2 y + \varepsilon z)(x + \varepsilon y + \varepsilon^2 z) = 0$$
$$(x + y + \varepsilon z)(x + \varepsilon^2 y + \varepsilon^2 z)(x + \varepsilon y + z) = 0$$
$$(x + y + \varepsilon^2 z)(x + \varepsilon^2 y + z)(x + \varepsilon y + \varepsilon z) = 0$$

Now let C be a smooth fiber of π, E the exceptional divisor over one of the base points, $p = E \cap C$, and q a point of C such that $q - p$ is a nontrivial m-torsion point of $\mathrm{Pic}^0(C)$. Denote by X_q the surface obtained by contracting E and blowing up the image of q, as in the diagram

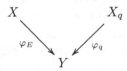

This operation is called *Halphen transform* [103]. If we denote by $C_q \cong C$ the strict transform of $\varphi_E(C)$ via φ_q, then a standard calculation shows that $-K_{X_q} \sim C_q$ and moreover its restriction to C_q is linearly equivalent to the

image of $p - q$, so that $-mK_{X_q}|_{C_q} \sim 0$. For any positive integer n there is an exact sequence of sheaves

$$0 \longrightarrow \mathcal{O}_{X_q}(-(n-1)K_{X_q}) \longrightarrow \mathcal{O}_{X_q}(-nK_{X_q}) \longrightarrow \mathcal{O}_{X_q}(-nK_{X_q})|_{C_q} \longrightarrow 0.$$

Because X_q is rational, C_q is an elliptic curve and $-K_{X_q}|_{C_q}$ is a nontrivial torsion divisor, then by putting $n = 1$ and passing to the long exact sequence in cohomology one obtains that the first and third h^1 vanish. Hence also $h^1(X_q, -K_{X_q}) = 0$. By induction on n one gets $h^1(X_q, -nK_{X_q}) = 0$ for $n \in \{1, \ldots, m - 1\}$. Hence for $n = m$ we get a surjective map

$$H^0(X_q, -mK_{X_q}) \rightarrow H^0(C_q, -mK_{X_q}|_{C_q}) \cong \mathbb{C}.$$

This shows that the linear system $|-mK_{X_q}|$ is base point free. Moreover, because $h^0(X_q, -(m-1)K_{X_q}) = 1$, we deduce that $|-mK_{X_q}|$ is one-dimensional and so it defines a fibration $\pi_q : X_q \rightarrow \mathbb{P}^1$. It is straightforward to see that π is a relatively minimal elliptic fibration and mC_q is one of its fibers. Observe that the prime components of the reducible fibers of π that do not intersect E are still components of the fibers of π_q. Hence π_q is extremal because the prime components of its reducible fibers span $K_{X_q}^\perp$. Now varying the curve C in the pencil and the point q on C we get our statement. $\qquad\square$

Example 5.1.4.7 Let C be the Fermat cubic, let $q = [-\sqrt[3]{2}, 1, 1]$ and $p = [0, 1, -1]$ be two points of C. The divisor $q - p$ is 2-torsion in $\mathrm{Pic}^0(C)$ because the tangent line $\sqrt[3]{4}x + y + z = 0$ of C at q passes through the flex p of C. Let S be the set of the nine base points of the Hesse pencil (so that $p \in S$) and let X be the blow-up of $\mathbb{P}^2$ at the nine points $S \cup \{q\} - \{p\}$. The surface X is the Halphen transform of the Hesse surface obtained by blowing up S. The elliptic fibration $\pi : X \rightarrow \mathbb{P}^1$ is defined by $|-2K_X|$, which comes from the pencil of sextics singular at the base points. Two such sextics are defined by $(x^3 + y^3 + z^3)^2$ and by

$$yz\left(x^4 + xy^3 + xz^3 + \sqrt[3]{2}\left(x^3y + y^4 + x^3z - 2y^3z + 3y^2z^2 - 2yz^3 + z^4\right) + \frac{3}{\sqrt[3]{2}}x^2yz\right).$$

The second sextic gives one of the fibers of type I_3 of π: two components are the strict transforms of the lines $y = 0$ and $z = 0$, which exist also in the Hesse fibration, while the new component is the rational plane quartic which has three singularities, $[0, \varepsilon, -1]$, $[0, \varepsilon^2, -1]$ and q. In a similar way one constructs the remaining three fibers of type I_3 and concludes that X is Mori dream by Theorem 5.1.4.2.

5.1.5 K3 surfaces

By a *K3 surface* we mean a smooth complex projective surface X of irregularity $q(X) = 0$ and canonical divisor K_X linearly equivalent to zero. We first characterize the Mori dreamness for K3 surfaces following [9, 297] and then give a detailed picture of the Mori dream K3 surfaces in terms of Picard lattices.

Theorem 5.1.5.1 *Let X be a K3 surface. Then the following statements are equivalent.*

(i) *X is a Mori dream surface.*
(ii) *The effective cone $\mathrm{Eff}(X) \subseteq \mathrm{Cl}_{\mathbb{Q}}(X)$ is polyhedral.*
(iii) *The automorphism group of X is finite.*

Proof The equivalence of the first two statements is an immediate consequence of Theorem 5.1.2.1. The implication "(i)$\Rightarrow$(iii)" is clear by Corollary 4.2.4.2. The implication "(iii)$\Rightarrow$(ii)" is sketched at the end of the section; for the full proof refer to [251, Sec. 7, Cor.]. $\qquad\square$

These equivalences allow us to specify the Mori dream K3 surfaces using the classification of K3 surfaces with finite automorphism group via their Picard lattices given in a series of articles by Piatetski-Shapiro, Nikulin, and Vinberg [230, 231, 232, 251, 300]. For the statement, we first recall first the basic concepts on lattices and indicate connections between lattices and K3 surfaces; a general reference is [29, Sec. VIII].

In the context of K3 surfaces, a *lattice* is a finitely generated free abelian group Λ equipped with an integer symmetric bilinear form $\beta \colon \Lambda \times \Lambda \to \mathbb{Z}$; we will mostly write $w \cdot w' := \beta(w, w')$ and $w^2 := \beta(w, w)$. Such a lattice Λ is called

(i) *Nondegenerate* if β is of rank $\mathrm{rk}\,(\beta) = \mathrm{rk}\,(\Lambda)$
(ii) *Even* if $w^2 \in 2\mathbb{Z}$ holds for all $w \in \Lambda$
(iii) *Hyperbolic* if β has signature $(1, \varrho - 1)$ where $\varrho = \mathrm{rk}\,(\beta)$

We briefly recall from [29, Sec. I.2] the basic examples of even hyperbolic lattices. There is the lattice $U = \mathbb{Z}^2$ of rank 2 with the intersection form given by the Gram matrix

$$\begin{bmatrix} 0 & 1 \\ 1 & 0 \end{bmatrix}.$$

Moreover, there are the lattices A_n, D_n, E_6, E_7, and E_8. They are characterized by means of their (equally named) Coxeter–Dynkin graphs: the vertices are the *simple roots* of the lattice, that is, the finitely many elements w with $w^2 = -2$ and two roots w, w' are joined by an edge if $w \cdot w' = 1$ (if the are not joined then $w \cdot w' = 0$ holds).

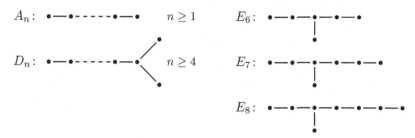

Given two lattices Λ_1, Λ_2 we denote by $\Lambda_1 \oplus \Lambda_2$ the direct sum lattice and by $n\Lambda$ the n-fold direct sum of Λ. Moreover, for a lattice Λ and an integer n, we write $\Lambda(n)$ for Λ with the stretched product $n\langle *, * \rangle$. Specially, we denote by (n) the lattice $\mathbb{Z}$ with $a \cdot b = nab$.

Remark 5.1.5.2 The above lattices admit the following geometric interpretation. Let $S := \{C_1, \ldots, C_n\}$ be a set of (-2)-curves on a smooth projective surface X. Assume that the intersection matrix $(C_i \cdot C_j)$ is negative definite. Then the sublattice of $\mathrm{Cl}(X)$ spanned by the classes of the curves in S is isometric to a direct sum of lattices of types A, D, and E. Given a divisor D whose support is contained in S we have $D^2 + D \cdot K_X = D^2 < 0$ by the genus formula and the hypothesis on S. Thus by Artin's criterion there exists a birational morphism $\pi\colon X \to Y$, with Y normal projective surface, which contracts exactly the curves in S. The singular points of Y are images of the connected curves supported on S and are called *rational double points*.

Let us briefly indicate the central role of lattices in the theory of K3 surfaces. The basic lattice associated with a K3 surface X is its second singular cohomology group $H^2(X, \mathbb{Z})$, which is torsion free of rank 22 and comes with the nondegenerate cup product. There is an isomorphism of lattices

$$H^2(X, \mathbb{Z}) \cong E_8 \oplus E_8 \oplus U \oplus U \oplus U =: \Lambda^{\mathrm{K3}}.$$

Because of $q(X) = h^1(X, \mathcal{O}) = 0$, the divisor class group $\mathrm{Cl}(X) \cong H^1(X, \mathcal{O}^*)$ embeds via the exponential sequence into $H^2(X, \mathbb{Z})$. This embedding is an isometry of lattices, where $\mathrm{Cl}(X)$ is endowed with the intersection product. Thus, $\mathrm{Cl}(X)$ is an even lattice and, by the Hodge index theorem it is hyperbolic.

The inclusion $\mathbb{Z} \subseteq \mathbb{C}$ gives an inclusion $H^2(X, \mathbb{Z}) \subseteq H^2(X, \mathbb{C})$. the latter vector space is naturally identified with the closed differentiable complex 2-forms on X modulo exact ones. From this point of view, the bilinear form on $H^2(X, \mathbb{C})$ induced by the cup product on $H^2(X, \mathbb{C})$ is given by

$$\omega_1 \cdot \omega_2 := \int_X \omega_1 \wedge \omega_2.$$

The Hodge decomposition of $H^2(X, \mathbb{C})$ into Dolbeault cohomology groups of (p, q)-forms provides further structure:

$$H^2(X, \mathbb{Z}) \subseteq H^2(X, \mathbb{C}) = H^{2,0}(X) \oplus H^{1,1}(X) \oplus H^{0,2}(X).$$

Note that $H^{2,0}(X)$ is spanned by ω_X and $H^{0,2}(X)$ by the complex conjugate $\overline{\omega}_X$ for any nowhere vanishing closed holomorphic 2-form ω_X on X. Moreover, via the Hodge structure, we can recover the divisor class group $\mathrm{Cl}(X)$ in $H^2(X, \mathbb{C})$ as

$$\mathrm{Cl}(X) = H^{1,1}(X) \cap H^2(X, \mathbb{Z}) = \omega_X^{\perp} \cap H^2(X, \mathbb{Z}),$$

where the first equality is due to the Lefschetz theorem on $(1, 1)$-classes [29, Thm. IV.2.13]. In particular, we see that $\mathrm{Cl}(X)$ is an even hyperbolic sublattice of rank at most 20 in $H^2(X, \mathbb{Z})$. The *transcendental lattice* of the K3 surface X is the lattice $H^2(X, \mathbb{Z}) \cap \mathrm{Cl}(X)^{\perp}$ in $H^2(X, \mathbb{C})$ with the induced bilinear form. Now, via a marking, that is, an isometry $\Phi\colon H^2(X, \mathbb{Z}) \to \Lambda^{K3}$, one associates with the K3 surface X a period point $[\Phi(\omega_X)]$ in the period domain

$$\mathcal{Q} := \{[\omega] \in \mathbb{P}(\Lambda_{\mathbb{C}}^{K3});\ \omega \cdot \omega = 0,\ \omega \cdot \overline{\omega} > 0\} \subseteq \mathbb{P}(\Lambda_{\mathbb{C}}^{K3}),$$

where $\Lambda_{\mathbb{C}}^{K3} := \Lambda^{K3} \otimes_{\mathbb{Z}} \mathbb{C}$ is the complexified K3 lattice. The *global Torelli theorem* basically says that two K3 surfaces X and X' are isomorphic if and only if they have the same period points for suitable markings. Moreover, Todorov showed that every element of $\mathcal{Q}$ is the period point of a K3 surface.

Building on this, one constructs moduli spaces of K3 surfaces X with $\mathrm{Cl}(X)$ containing a given hyperbolic sublattice $S \subseteq \Lambda^{K3}$. Set $T := S^{\perp}$. Then for any $[\omega] \in \mathcal{Q} \cap \mathbb{P}(T_{\mathbb{C}})$, there exists a K3 surface X with period point $[\omega]$ and, because $S \subseteq \omega^{\perp}$ holds, $\mathrm{Cl}(X)$ contains an isometric copy of S. In fact, every K3 surface with $\mathrm{Cl}(X)$ containing an isometric copy of S is obtained in this way. For very general points $[\omega] \in \mathcal{Q} \cap \mathbb{P}(T_{\mathbb{C}})$, the corresponding K3 surfaces satisfy $\mathrm{Cl}(X) \cong S$ and have T as transcendental lattice.

Now the classification of Mori dream K3 surfaces is based on the fact that for a K3 surface with a finite automorphism group, the lattice $\mathrm{Cl}(X)$ is *2-reflective*. The latter means that the Weyl group $W(\mathrm{Cl}(X)) \subseteq O(\mathrm{Cl}(X))$ generated by the reflections $u \mapsto u + (u \cdot w)u$, where w runs through the roots, has finite index in $O(\mathrm{Cl}(X))$. The classification results on even hyperbolic 2-reflective lattices obtained in [230, 231, 232, 251, 300] then lead to the following.

Theorem 5.1.5.3 *Let X be an algebraic K3 surface with Picard number $\varrho(X)$. Then X is Mori dream if and only if one of the following occurs.*

(i) $\varrho(X) = 2$ and $\mathrm{Cl}(X)$ contains a class w with $w^2 \in \{-2, 0\}$.

(ii) $\varrho(X) = 3$ and $\mathrm{Cl}(X)$ is isometric to one of the 26 lattices of [231]:

$$\langle 2e_1 + e_3, e_2, 2e_3 \rangle, \qquad \langle ke_1, e_2, e_3 \rangle,\ k \in \{4, 5, 6, 7, 8, 10, 12\},$$

$$\langle e_1, ke_2, e_3 \rangle,\ k \in \{2, 3, 4, 5, 6, 9\}, \qquad \langle e_1, e_2, ke_3 \rangle,\ k \in \{1, 2, 3, 4, 6, 8\},$$

where the intersection matrix of e_1, e_2, e_3 is $\begin{bmatrix} -2 & 0 & 1 \\ 0 & -2 & 2 \\ 1 & 2 & -2 \end{bmatrix}$, and

$$(6) \oplus 2A_1, \qquad (k) \oplus A_2, \ k \in \{4, 12, 36\},$$

$$\begin{bmatrix} 6 & 0 & -1 \\ 0 & -2 & 1 \\ -1 & 1 & -2 \end{bmatrix}, \qquad \begin{bmatrix} 14 & 0 & -1 \\ 0 & -2 & 1 \\ -1 & 1 & -2 \end{bmatrix}.$$

(iii) $\varrho(X) = 4$ and $\mathrm{Cl}(X)$ is isometric to one of the 14 lattices of [300, Thm. 1]:

$$(8) \oplus 3A_1, \qquad (-4) \oplus (4) \oplus A_2, \qquad (4) \oplus A_3,$$

$$U(k) \oplus 2A_1, \ k \in \{1, 2, 3, 4\}, \qquad U(k) \oplus A_2, \ k \in \{1, 2, 3, 6\},$$

$$\begin{bmatrix} 0 & -3 \\ -3 & 2 \end{bmatrix} \oplus A_2, \quad \begin{bmatrix} 2 & -1 & -1 & -1 \\ -1 & -2 & 0 & 0 \\ -1 & 0 & -2 & 0 \\ -1 & 0 & 0 & -2 \end{bmatrix}, \quad \begin{bmatrix} 12 & -2 & 0 & 0 \\ -2 & -2 & -1 & 0 \\ 0 & -1 & -2 & -1 \\ 0 & 0 & -1 & -2 \end{bmatrix}.$$

(iv) $5 \le \varrho(X) \le 19$ and $\mathrm{Cl}(X)$ is isometric to one of the following lattices:

ϱ	Λ
5	$U \oplus 3A_1$, $U(2) \oplus 3A_1$, $U \oplus A_1 \oplus A_2$, $U \oplus A_3$, $U(4) \oplus 3A_1$, $(6) \oplus 2A_2$, $(4) \oplus D_4$, $(8) \oplus D_4$, $(16) \oplus D_4$
6	$U \oplus D_4$, $U(2) \oplus D_4$, $U \oplus 4A_1$, $U(2) \oplus 4A_1$, $U \oplus 2A_1 \oplus A_2$, $U \oplus 2A_2$, $U \oplus A_1 \oplus A_3$, $U \oplus A_4$, $U(4) \oplus D_4$, $U(3) \oplus 2A_2$
7	$U \oplus D_4 \oplus A_1$, $U \oplus 5A_1$, $U(2) \oplus 5A_1$, $U \oplus A_1 \oplus 2A_2$, $U \oplus 2A_1 \oplus A_3$, $U \oplus A_2 \oplus A_3$, $U \oplus A_1 \oplus A_4$, $U \oplus A_5$, $U \oplus D_5$
8	$U \oplus D_6$, $U \oplus D_4 \oplus 2A_1$, $U \oplus 6A_1$, $U(2) \oplus 6A_1$, $U \oplus 3A_2$, $U \oplus 2A_3$, $U \oplus A_2 \oplus A_4$, $U \oplus A_1 \oplus A_5$, $U \oplus A_6$, $U \oplus A_2 \oplus D_4$, $U \oplus A_1 \oplus D_5$, $U \oplus E_6$
9	$U \oplus E_7$, $U \oplus D_6 \oplus A_1$, $U \oplus D_4 \oplus 3A_1$, $U \oplus 7A_1$, $U(2) \oplus 7A_1$, $U \oplus A_7$, $U \oplus A_3 \oplus D_4$, $U \oplus A_2 \oplus D_5$, $U \oplus D_7$, $U \oplus A_1 \oplus E_6$
10	$U \oplus E_8$, $U \oplus D_8$, $U \oplus E_7 \oplus A_1$, $U \oplus D_4 \oplus D_4$, $U \oplus D_6 \oplus 2A_1$, $U(2) \oplus D_4 \oplus D_4$, $U \oplus D_4 \oplus 4A_1$, $U \oplus 8A_1$, $U \oplus A_2 \oplus E_6$
11	$U \oplus E_8 \oplus A_1$, $U \oplus D_8 \oplus A_1$, $U \oplus D_4 \oplus D_4 \oplus A_1$, $U \oplus D_4 \oplus 5A_1$
12	$U \oplus E_8 \oplus 2A_1$, $U \oplus D_8 \oplus 2A_1$, $U \oplus D_4 \oplus D_4 \oplus 2A_1$, $U \oplus A_2 \oplus E_8$
13	$U \oplus E_8 \oplus 3A_1$, $U \oplus D_8 \oplus 3A_1$, $U \oplus A_3 \oplus E_8$
14	$U \oplus E_8 \oplus D_4$, $U \oplus D_8 \oplus D_4$, $U \oplus E_8 \oplus 4A_1$
15	$U \oplus E_8 \oplus D_4 \oplus A_1$
16	$U \oplus E_8 \oplus D_6$
17	$U \oplus E_8 \oplus E_7$
18	$U \oplus 2E_8$
19	$U \oplus 2E_8 \oplus A_1$

Given a nondegenerate lattice Λ, one can define its dual as $\Lambda^* = \mathrm{Hom}(\Lambda, \mathbb{Z})$ equipped with the quadratic form, with values in $\mathbb{Q}$, coming from Λ. In this way there is an embedding of lattices

$$\Lambda \to \Lambda^*, \qquad v \mapsto [u \mapsto u(v)].$$

The *discriminant group* of a nondegenerate even lattice Λ is the finite abelian group $d(\Lambda) := \Lambda^*/\Lambda$. The lattice Λ is called 2-*elementary* if its discriminant group is isomorphic to $(\mathbb{Z}/2\mathbb{Z})^a$ for some positive integer a.

Remark 5.1.5.4   The discriminant groups of the lattices $\Lambda = U, A_n, D_{2n+1}$, D_{2n}, E_k, where $n \geq 1$ and $k = 6, 7, 8$ can be directly computed:

Λ	U	A_n	D_{2n+1}	D_{2n}	E_k
$d(\Lambda)$	0	$\mathbb{Z}/(n+1)\mathbb{Z}$	$\mathbb{Z}/4\mathbb{Z}$	$\mathbb{Z}/2\mathbb{Z} \oplus \mathbb{Z}/2\mathbb{Z}$	$\mathbb{Z}/(9-k)\mathbb{Z}$

In particular, the only 2-elementary lattices occuring in Theorem 5.1.5.3 are the direct sums of copies of A_1, E_7, E_8, and D_{2n}, with n positive integer.

It is possible to show, as a consequence of the global Torelli theorem for K3 surfaces, that if X is a K3 surface and $\mathrm{Cl}(X)$ is a 2-elementary lattice, then X admits an automorphism $\sigma \in \mathrm{Aut}(X)$, with $\sigma^2 = \mathrm{id}$, that acts as the identity on $\mathrm{Cl}(X)$. The quotient surface $Y = X/\langle\sigma\rangle$ is either smooth rational or it is an Enriques surface. In Section 5.3.4 we use such double coverings $X \to Y$ for describing the Cox ring of X in terms of that of Y.

We conclude with indicating why a K3 surface with a finite automorphism group has a polyhedral effective cone. If X is a K3 surface, a class $w \in \mathrm{Cl}(X)$ is a *root* if $w^2 = -2$. Given a root w define the *Picard–Lefschetz reflection* associated with w as the isometry $s_w \colon \mathrm{Cl}(X) \to \mathrm{Cl}(X)$ given by $x \mapsto x + (x \cdot w)\, w$. The *Weyl group* of $\mathrm{Cl}(X)$ is:

$$\mathrm{W}(X) := \langle s_w;\ w \text{ is a root of } \mathrm{Cl}(X)\rangle \subseteq \mathrm{Aut}(\mathrm{Cl}(X)).$$

Observe that no element of $\mathrm{W}(X)$ is induced by an automorphism of X because $s_w(w) = -w$ and either w or $-w$ is an effective class by the Riemann–Roch formula. The effect of applying a Picard–Lefschetz reflection with respect to a root w is to make a reflection with respect to the hyperplane $w^\perp$ of $\mathrm{Cl}_{\mathbb{Q}}(X)$. This reflection moves the whole ample cone into another chamber. To see this observe that if h is an ample class of X and w is an effective root, then $h \cdot w > 0$, so that $s_w(h) \cdot w < 0$ and $s_w(h)$ is not ample.

It can be proved that the nef cone Nef(X) is a fundamental chamber for the action of W(X) on Cl$_\mathbb{Q}$(X), meaning with this that W(X) · Nef(X) gives a decomposition of the positive light cone

$$Q = \{w \in \mathrm{Cl}_\mathbb{Q}(X);\ w^2 > 0 \text{ and } w \cdot h > 0 \text{ with } h \text{ ample}\} \subseteq \mathrm{Cl}_\mathbb{Q}(X)$$

into chambers that are congruent to Nef(X), and W(X) acts freely and transitively on this set of chambers. On the other hand an isometry of Cl(X) coming from an automorphism clearly preserves the nef cone. Hence if we denote by $O(\mathrm{Nef}(X))$ the group of isometries of Cl(X) that preserve the nef cone, we have a homomorphism Aut(X) → $O(\mathrm{Nef}(X))$. As a consequence of the global Torelli theorem this homomorphism has finite kernel and cokernel.

Observe that W(X) is a normal subgroup of $O(\mathrm{Cl}(X))$ because the conjugate of a Picard–Lefschetz reflection is a Picard–Lefschetz reflection. Hence for any isometry σ of Cl(X) there is an $s \in W(X)$ such that $\sigma(\mathrm{Nef}(X)) = s(\mathrm{Nef}(X))$. In other words $O(\mathrm{Cl}(X))$ is a semidirect product of W with $O(\mathrm{Nef}(X))$, so that

$$O(\mathrm{Cl}(X)) \doteq W(X) \rtimes \mathrm{Aut}(X),$$

where the symbol $\doteq$ means isomorphism "up to a finite group." In particular Aut(X) is finite if and only if $W(X)$ has finite index in $O(\mathrm{Cl}(X))$. If this is the case, the lattice Cl(X) is 2-reflective. Hyperbolic, even, 2-reflective lattices have been classified in [230, 231, 232, 251, 300]. The results of these papers, together with Theorem 5.1.5.1, give Theorem 5.1.5.3.

5.1.6 Enriques surfaces

An *Enriques surface* is a smooth complex projective surface X with $q(X) = 0$ and $2K_X \sim 0$ but $K_X \not\sim 0$. We characterize here the (families) of Mori dream Enriques surfaces and study limits of one of these families, so-called Coble surfaces, in detail.

First recall from [103], that the divisor class group Cl(X) of an Enriques surface X is isomorphic to $\cong \mathbb{Z}^{10} \oplus \mathbb{Z}/2\mathbb{Z}$. Moreover, as a lattice, the free part of Cl(X) is isomorphic to $U \oplus E_8$. We denote by $\tilde{X} \to X$ the universal covering of an Enriques surface X; this is always a map of degree 2 and $\tilde{X}$ is a K3 surface with $U(2) \oplus E_8(2) \subseteq \mathrm{Cl}(\tilde{X})$. We write $T_{\tilde{X}}$ for the transcendental lattice of $\tilde{X}$.

The following result is a direct consequence of Theorem 5.1.3.12 and the classification of Enriques surfaces with finite automorphism groups [126] and [194, p. 193].

Theorem 5.1.6.1 *An Enriques surface X is Mori dream if and only if it belongs to one of the following families.*

Family	Number of (-2)-curves of X	$T_{\tilde{X}}$
I	12	$U \oplus \langle 4 \rangle$
II	12	$U \oplus \langle 8 \rangle$
III	20	$\langle 4 \rangle \oplus \langle 4 \rangle$
IV	20	$\langle 4 \rangle \oplus \langle 4 \rangle$
V	20	$\begin{bmatrix} 4 & 2 \\ 2 & 4 \end{bmatrix}$
VI	20	$\begin{bmatrix} 4 & 1 \\ 1 & 4 \end{bmatrix}$
VII	20	$\begin{bmatrix} 4 & 2 \\ 2 & 6 \end{bmatrix}$

Families I and II are one-dimensional inside the moduli space of Enriques surfaces, while families III–VII have no moduli. We show in Theorem 5.1.6.4 that the limit of Enriques surfaces of type I are Mori dream rational surfaces whose minimal resolution is a Mori dream Coble surface.

We now recall, following [194, Sec. 3], how to describe geometrically the general element X_t of the family I. Let φ be the involution of $\mathbb{P}^1 \times \mathbb{P}^1$ defined by

$$\varphi([x_0, x_1], [y_0, y_1]) = ([x_0, -x_1], [y_0, -y_1]).$$

Let $\{C_t\}_{t \in \mathbb{P}^1}$ be the pencil of curves of degree $(2, 2)$ in $\mathbb{P}^1 \times \mathbb{P}^1$ given by

$$(2x_0^2 - x_1^2)(y_0^2 - y_1^2) + (2ty_0^2 + (1 - 2t)y_1^2)(x_1^2 - x_0^2) = 0$$

and let L_1, L_2, L_3, L_4 be the four lines of equations

$$x_0 - x_1 = 0, \qquad x_0 + x_1 = 0, \qquad y_0 - y_1 = 0, \qquad y_0 + y_1 = 0$$

An elementary calculation shows that C_t is smooth irreducible for $t \neq 1, \frac{1}{2}, \frac{3}{2}, \infty$. Moreover, the curve C_t has an ordinary double point, for $t \in \{\frac{1}{2}, \frac{3}{2}\}$; it is the union of two irreducible curves of degree $(1, 1)$ for $t = 1$; and is the union of $L_1, \ldots, L_4$ for $t = \infty$. Note that the base locus of the pencil C_t consists of the four points

$$([1, 1], [1, 1]) \quad ([1, 1], [1, -1]), \quad ([1, -1], [1, 1]), \quad ([1, -1], [1, -1]),$$

and also note that each of these points is contained in two of the curves $L_1, \ldots, L_4$. For each $t \in \mathbb{P}^1$ the birational map $\pi_t \colon S_t \to \mathbb{P}^1 \times \mathbb{P}^1$ is obtained by first blowing up the four base points of the pencil C_t and then blowing up the 12 points of intersections of any exceptional divisor with the strict transform of C_t and of $L_1, \ldots, L_4$. Thus the surface S_t is rational with Picard group of rank 18. The reducible curve B_t of $\mathbb{P}^1 \times \mathbb{P}^1$ defined by

$$B_t := C_t + L_1 + L_2 + L_3 + L_4$$

is a divisor of degree $(4, 4)$, invariant with respect to the involution φ. Dropping the subscript t from the morphisms to simplify the notation, this leads us to the diagram

$$\tilde{X}_t \xrightarrow[2:1]{\varphi} S_t \xrightarrow{\pi} \mathbb{P}^1 \times \mathbb{P}^1$$
$$\psi \downarrow 2:1$$
$$X_t$$

where the morphism π is birational; the surface S_t is rational with Picard group of rank 18; the morphism φ is a double cover; the branch locus of φ consists of the strict transform B_t' of B_t together with a union Γ of disjoint curves in the exceptional locus of π; and finally ψ is the double cover $\tilde{X}_t \to X_t = \tilde{X}_t / \langle \sigma \rangle$, where $\sigma \in \text{Aut}(\tilde{X}_t)$ is the involution induced by φ.

Kondo proves that X_t is an Enriques surface for all t different from $1, \frac{1}{2}, \frac{3}{2}, \infty$ and that $\text{Aut}(X_t) \cong D_4$, the dihedral group of order 8.

Let $t_0 \in \mathbb{P}^1$ be such that C_{t_0}' is irreducible and has an ordinary double point and let $\tilde{X}_0 := \tilde{X}_{t_0}$ be the corresponding surface. Then $\tilde{X}_0$ has a singularity of type A_1 at a point p. Because p is the only singular point of $\tilde{X}_0$, it must be stable with respect to the involution σ, and thus X_0 is singular at one point q as well. This leads to the commutative diagram

$$\tilde{X} \xrightarrow{\pi_p} \tilde{X}_0$$
$$\psi \downarrow \qquad \downarrow \psi_0$$
$$X \xrightarrow{\pi_q} X_0$$

where π_p and π_q are minimal resolution of singularities, $\tilde{X}$ is a K3 surface, and ψ is a double cover branched along the exceptional divisor E of π_q. A *Coble surface* is a smooth rational surface X with $|-K_X|$ empty and $|-2K_X|$ nonempty.

Proposition 5.1.6.2 *The surface X is a Coble surface.*

Proof Let $\sigma_0 \in \text{Aut}(\tilde{X}_0)$ be the involution induced by φ. As we noted earlier, $\sigma_0(p) = p$ so that σ_0 lifts to an involution $\sigma \in \text{Aut}(\tilde{X})$. The involution σ is nonsymplectic because σ_0 is nonsymplectic and the exceptional divisor $R := \pi_p^{-1}(p)$ is fixed by σ_0. Thus the quotient surface $X = \tilde{X}/\langle \sigma \rangle$ is a smooth projective rational surface. Observe that the branch divisor of ψ is contained in $|-2K_X|$, and in particular $|-2K_X|$ is not empty. Moreover, because ψ_0 does not ramify along a divisor, the branch divisor of ψ is exactly E. Because X_0 has an A_1-singularity at p, it follows that $F = \pi_p^{-1}(p)$ is a (-2)-curve on $\tilde{X}$. In particular, the curve $E = \psi(F)$ is irreducible and reduced with $E^2 < 0$. Thus $|-K_X|$ is empty, because otherwise $E \in |-2K_X|$ would be nonreduced. $\qquad \square$

Remark 5.1.6.3 Coble surfaces have been introduced in [129], where the authors show that Coble surfaces are projective degenerations of Enriques surfaces.

Theorem 5.1.6.4 *The Coble surface X is Mori dream.*

Proof We want to use [194, Cor. 5.7] to prove that the group $\mathrm{Aut}(X_0)$ is finite, even though X_0 is a rational surface and not an Enriques surface. The reason we can still apply Kondo's result is that the Néron–Severi lattice of Y_0 is isomorphic to the Néron–Severi lattice of an Enriques surface, and moreover the minimal resolution of the limit $\tilde{X}_0$ of the K3 surfaces $\tilde{X}_t$ is still a K3 surface and therefore it still corresponds to a point in the period space. Moreover, X_0 is Kawamata log terminal, klt for short, because

$$K_X = \pi_q^* K_{X_0} - \frac{1}{2}E$$

and π_q is a resolution of singularities of X_0. Observe that indeed X_0 is a klt Calabi–Yau surface because its canonical divisor is numerically trivial. Let $\Delta := \frac{1}{2}E$. Because the pair (X, Δ) is the terminal model of X_0, it follows that $\mathrm{Aut}(X_0) = \mathrm{Aut}(X, \Delta)$. In particular the last group is finite so that the image of the map $\mathrm{Aut}(X, \Delta) \to \mathrm{GL}\big(\mathrm{Cl}(X)\big)$ is finite as well. Because 2Δ is linearly equivalent to $-K_X$ and $h^0(X, -K_X) = 1$, being $-K_X$ of Iitaka dimension zero, any automorphism of X must preserve Δ. Thus $\mathrm{Aut}(X) = \mathrm{Aut}(X, \Delta)$ and we conclude that X is Mori dream by Theorem 5.1.3.12. □

Remark 5.1.6.5 In general a Coble surface is not a Mori dream space. To see this let $p_1, \ldots, p_9$ denote the nine intersection points of two general plane cubic curves C_1 and C_2. Let $q \in C_1$ be such that the divisor class of $p_9 - q$ is a 2-torsion point of $\mathrm{Cl}(C_1)$. Then the blow-up Z of the plane at $p_1, \ldots, p_8, q$ admits an elliptic fibration defined by $|-2K_Z|$. Due to the generality assumption on C_1 and C_2, it is easy to see that the linear system $|-2K_Z|$ does not contain reducible elements. Equivalently Z does not contain (-2)-curves, so that it is not a Mori dream space by [10]. By an Euler characteristic calculation the fibration induced by $|-2K_Z|$ contains 12 nodal curves. The blow-up X of Z at one of these nodes p is a Coble surface because $|-K_X|$ is empty, but $|-2K_X|$ contains the strict transform of the fiber through p. Moreover, X is not a Mori dream surface because Z is not a Mori dream surface.

In the following remark, we obtain a description of the Coble surface X as an iterated blow-up of $\mathbb{P}^2$.

Remark 5.1.6.6 The Coble surface X constructed above as the minimal resolution of a limit of a family of Mori dream Enriques surfaces contains (at least) the configuration of (-2)-curves of the general element of such family.

The intersection graph of this surface is the following (see [194, Ex. 1]), where double edges mean that the corresponding curves have intersection equal to 2.

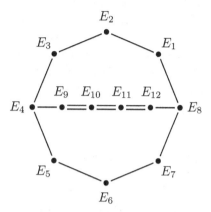

It is not difficult to see that $F := E_1 + E_2 + E_3 + E_4 + E_5 + E_6 + E_7 + E_8$ has $F^2 = 0$ and $|2F|$ is an elliptic pencil on X with two reducible fibers one of type I_8 and one of type I_2. Thus X is the blow-up of a Mori dream surface Z with Picard group of rank 10, admitting an elliptic fibration whose dual graph of singular fibers contains I_8 and I_2. The surface Z' defined by the jacobian fibration of that induced by $|-2K_Z|$ has finite Mordell–Weil group by [10]. Moreover, Z' is unique up to isomorphism and its unique elliptic fibration, given by $|-K_{Z'}|$, admits exactly four singular fibers of type I_8, I_2, I_1, I_1, by [131, Prop. 9.2.19]. This is the same configuration of singular fibers of the elliptic pencil defined by the linear system $|-2K_Z|$. Explicitly, let $[x, y, z]$ be homogeneous coordinates on $\mathbb{P}^2$ and set

$$c_0 := x^2(x^2 - yz)^2,$$

$$c_\infty := (2x^2y - x^2z + y^3 - 2y^2z + yz^2)(2x^2y + x^2z - y^3 - 2y^2z - yz^2).$$

The surface Z is the rational elliptic surface associated with the pencil generated by the two plane sextics $C_0 = V(c_0)$ and $C_\infty = V(c_\infty)$. The fiber corresponding to the curve C_0 has multiplicity 2 and is of type I_8. The fiber corresponding to the curve C_∞ is reduced and is of type I_2. The only two remaining singular fibers are both of type I_1 and correspond to the plane sextics $V(c_0 + c_\infty)$ and $V(c_0 - c_\infty)$. To construct the jacobian surface Z' associated with Z, set

$$d_0 := 4x(yz - x^2), \qquad d_\infty := z(y + z)^2 - x^2(y + 2z).$$

The surface Z' is the rational elliptic surface determined by the pencil generated by the plane cubics $D_0 = V(d_0)$ and $D_\infty = V(d_\infty)$. As before, the fiber corresponding to D_0 is of type I_8; the fiber corresponding to D_∞ is of type I_2; the only remaining singular fibers are of type I_1 and correspond to the cubics $V(d_0 + d_\infty)$ and $V(d_0 - d_\infty)$. Observe that the singularity of the pencil defining Z' at the point $[0, 1, -1]$ is resolved only after two successive blow-ups. To obtain Z from Z' it suffices to contract the exceptional curve lying above the point $[0, 1, -1]$ (the one introduced by the second blow-up) and to blow-up the point $[0, 1, 1]$. Finally, the Coble surface $\tilde{Y}$ is obtained from Z by blowing up the singular point of one of the two I_1 fibers. The two surfaces obtained by the choice of the last blown up point are isomorphic: the substitution $([x, y, z], t) \mapsto ([ix, y, -z], -t)$ in the pencil $c_0 + tc_\infty$ defining Z determines an automorphism of order 4 on Z exchanging the two I_1 fibers.

5.2 Smooth del Pezzo surfaces

5.2.1 Preliminaries

We recall the basic geometric properties of del Pezzo and weak del Pezzo surfaces. General references are [127, 214]. The ground field $\mathbb{K}$ is algebraically closed and of characteristic zero.

Definition 5.2.1.1 A *del Pezzo surface* is a normal projective surface with ample anticanonical divisor.

So, by definition, a del Pezzo surface is a two-dimensional Fano variety. Let us recall from [214, Thm. 24.4] the characterization of *smooth del Pezzo surfaces* as blow-ups of minimal smooth rational surfaces.

Definition 5.2.1.2 We say that $1 \le r \le 8$ distinct points $p_1, \ldots, p_r$ in the projective plane $\mathbb{P}^2$ are in *general position* if

- No three of them lie on a line.
- No six of them lie on a conic.
- No eight of them lie on a cubic with a singularity at some of the p_i.

Theorem 5.2.1.3 *Up to isomorphy, the smooth del Pezzo surfaces are* $\mathbb{P}^1 \times \mathbb{P}^1$ *and the blow-ups of* $\mathbb{P}^2$ *at* $0 \le r \le 8$ *points in general position.*

As a consequence, we obtain a precise picture about the negative curves on X; recall that for $k \in \mathbb{Z}_{\ge 0}$ a $(-k)$-*curve* on a smooth projective surface X is a curve C isomorphic to $\mathbb{P}^1$ with $C^2 = -k$.

Remark 5.2.1.4 If X is a smooth del Pezzo surface, then every irreducible curve with negative self intersection number is a (-1)-curve. The total number M_r of (-1)-curves on a blow-up X of $\mathbb{P}^2$ at r points in general position is given by

r	1	2	3	4	5	6	7	8
M_r	1	3	6	10	16	27	56	240

see [214, Thm. 26.2]. More explicitly, let $E_i \subseteq X$ denote the exceptional divisor over $p_i \in \mathbb{P}^2$ and let H be the total transform of any line in $\mathbb{P}^2$. Then, up to permutation of the indices, the (-1)-curves of X are linearly equivalent to divisors of the form

$$
\begin{aligned}
&E_1 \\
&H - E_1 - E_2 \\
&2H - E_1 - E_2 - E_3 - E_4 - E_5 \\
&3H - 2E_1 - E_2 - E_3 - E_4 - E_5 - E_6 - E_7 \\
&4H - 2E_1 - 2E_2 - 2E_3 - E_4 - E_5 - E_6 - E_7 - E_8 \\
&5H - 2E_1 - 2E_2 - 2E_3 - 2E_4 - 2E_5 - 2E_6 - E_7 - E_8 \\
&6H - 3E_1 - 2E_2 - 2E_3 - 2E_4 - 2E_5 - 2E_6 - 2E_7 - 2E_8
\end{aligned}
$$

In other words the (-1)-curves on X are the exceptional divisors, the strict transforms of lines in $\mathbb{P}^2$ passing through two of $p_1, \ldots, p_r$, and so on, up to the strict transforms of sextics in $\mathbb{P}^2$ passing through eight of $p_1, \ldots, p_r$ having one as a triple point and the remaining seven as double points.

Definition 5.2.1.5 A *weak del Pezzo surface* is a smooth projective surface whose anticanonical divisor is nef and big.

Of course, every smooth del Pezzo surface is a weak del Pezzo surface. A simple example of an honestly weak del Pezzo surface is the second Hirzebruch surface $\mathbb{F}_2$: its anticanonical divisor is nef and big but not ample.

The weak del Pezzo surfaces can be characterized as well as blow-ups of minimal smooth rational surfaces. This time iterated blow-ups are involved: starting with $X_0 = \mathbb{P}^2$, we blow up a point $p_1 \in X_0$, then a point $p_2 \in X_1$ that might lie on the exceptional divisor of p_1, and so on, until we reach $X = X_r$:

$$
X = X_r \longrightarrow X_{r-1} \longrightarrow \cdots\cdots \longrightarrow X_1 \longrightarrow X_0 = \mathbb{P}^2.
$$

We denote by $E_i \subseteq X$ the total transform of the exceptional divisor over $p_i \in X_{i-1}$. Observe that $E_i \cdot E_j = 0$ for any $i \neq j$ and $E_i^2 = -1$ for any i. Moreover each E_i is an *extended* (-1)-*curve*: either a (-1)-curve or a chain of rational curves whose last component is a (-1)-curve and all the remaining components are (-2)-curves.

Definition 5.2.1.6 Let $X_0 = \mathbb{P}^2, \ldots, X_r = X$ be a sequence of blow-ups of points $p_i \in X_{i-1}$, where $1 \leq r \leq 8$. We say that $p_1, \ldots, p_r$ are in *almost general position* if

- No four of them are mapped to the same line in $\mathbb{P}^2$.
- No seven of them are mapped to the same conic in $\mathbb{P}^2$.
- The total transform E_i of the exceptional divisor over $p_i \in X_{i-1}$ is an extended (-1)-curve for any i.

Equivalently the points, $p_1, \ldots, p_r$, with $1 \le r \le 8$, are in almost general position if and only if no $p_i \in X_{i-1}$ lies on a (-2)-curve of X_{i-1}. Because $-K_X$ is semiample and big there is a birational morphism

$$\varphi \colon X \ \to \ Y := \mathrm{Proj}(S^+(X, -K_X)).$$

This is a minimal resolution of singularities for Y, because, by the genus formula φ contracts exactly the (-2)-curves. The surface Y is called the *anticanonical model* of X. The following statements characterize weak del Pezzo surfaces and their anticanonical models; see [127, Thm. 8.1.15, Cor. 8.1.24].

Theorem 5.2.1.7 *Let X be a weak del Pezzo surface. Then the following hold.*

(i) *Up to isomorphy X is either $\mathbb{P}^1 \times \mathbb{P}^1$ or the Hirzebruch surface $\mathbb{F}_2$ or a blow-up of $\mathbb{P}^2$ at $0 \le r \le 8$ points in almost general position.*
(ii) *The anticanonical model of X is a del Pezzo surface whose singularities, if any, are rational double points.*

Conversely, for any del Pezzo surface with at most rational double points as singularities, the minimal resolution is a weak del Pezzo surface.

Remark 5.2.1.8 Let X be a weak del Pezzo surface arising from $\mathbb{P}^2$ as a blow-up of $1 \le r \le 8$ points almost general position. Then we have

$$\mathrm{Pic}(X) \ = \ \mathrm{Cl}(X) \ = \ \mathbb{Z}[H] \oplus \mathbb{Z}[E_1] \oplus \ldots \oplus \mathbb{Z}[E_r],$$

where $H \subseteq X$ is the total transform of any line in $\mathbb{P}^2$ and the E_i are the total transforms of the exceptional divisors. An anticanonical divisor is given by

$$-K_X = 3H - E_1 - \cdots - E_r.$$

Observe that the linear system $|-K_X|$ is base point free for $r < 8$ and it has just one base point for $r = 8$.

Remark 5.2.1.9 The *degree* of a weak del Pezzo surface X is the self-intersection number of the anticanonical divisor $-K_X$. The surfaces $\mathbb{P}^1 \times \mathbb{P}^1$ and $\mathbb{F}_2$ are of degree 8 and for a surface X arising from $\mathbb{P}^2$ as a blow-up of $1 \le r \le 8$ points in almost general position we have

$$\deg(X) \ = \ (-K_X)^2 \ = \ 9 - r.$$

Proposition 5.2.1.10 *Let X be a weak del Pezzo surface. Then the following statements hold.*

(i) *The surface X is Mori dream. In particular the effective cone $\mathrm{Eff}(X)$ is polyhedral.*
(ii) *If $\varrho(X) \ge 3$ then the effective cone $\mathrm{Eff}(X)$ is generated by the classes of (-1)-curves and (-2)-curves.*

Proof Weak del Pezzo surfaces are Mori dream surfaces by Theorem 5.1.3.7. Thus $\mathrm{Eff}(X)$ is polyhedral by Corollary 4.3.3.6. If $\varrho(X) \geq 3$ holds, then the effective cone is generated by all the classes of prime divisors C with $C^2 < 0$ by Proposition 5.1.1.6. If C is such a divisor then $2p_a(C) - 2 = C^2 + C \cdot K_X < 0$, because $-K_X$ is nef. Thus $p_a(C) = 0$, so that C is smooth rational. Moreover, we have $C^2 \geq -2$, which proves the statement. $\qquad\square$

Remark 5.2.1.11 If X is a blow-up of $\mathbb{P}^2$ at $2 \leq r \leq 8$ points in almost general position then each (-2)-curve of X is either of the form $E_i - E_{i+1}$, if E_i is reducible, or it is linearly equivalent to one of the following (modulo permutation of the indices):

$$H - E_1 - E_2 - E_3$$
$$2H - E_1 - E_2 - E_3 - E_4 - E_5 - E_6$$
$$3H - 2E_1 - E_2 - E_3 - E_4 - E_5 - E_6 - E_7 - E_8.$$

5.2.2 Generators of the Cox ring

The aim is to figure out generators for the Cox ring $\mathcal{R}(X)$ of a smooth del Pezzo surface X. The ground field $\mathbb{K}$ is algebraically closed of characteristic zero. The original reference for this section is [33]. The main result now follows.

Theorem 5.2.2.1 *Let X be a smooth del Pezzo surface with $\varrho(X) \geq 3$; let $f_1, \ldots, f_k \in \mathcal{R}(X)$ denote the canonical sections of the (-1)-curves; and, for $\varrho(X) = 9$, let $h_1, h_2 \in \mathcal{R}(X)_{[-K_X]}$ be a vector space basis for the anticanonical degree.*

(i) *If X is of Picard number $\varrho(X) \leq 8$, then the Cox ring $\mathcal{R}(X)$ is generated by $f_1, \ldots, f_k$.*
(ii) *If X is of Picard number $\varrho(X) = 9$, then the Cox ring $\mathcal{R}(X)$ is generated by $f_1, \ldots, f_k, h_1, h_2$.*

In both cases the system of generators is irredundant, and conversely, every irredundant system of homogeneous generators is of this form.

The idea for the proof of Theorem 5.2.2.1 is to construct inductively generators for the Cox ring $\mathcal{R}(X)$ from generators of the Cox rings $\mathcal{R}(Y)$ of the surfaces Y obtained by contracting (-1)-curves. This works more generally for weak del Pezzo surfaces, and the precise formulation is the following.

Definition 5.2.2.2 Let X be a weak del Pezzo surface. A *distinguished section* is a homogeneous element $f \in \mathcal{R}(X)$ of one of the following types:

(i) A canonical section $f = 1_{[E]} \in \mathcal{R}(X)_{[E]}$, where $E \subseteq X$ is a (-1)-curve or a (-2)-curve
(ii) A section $f \in \mathcal{R}(X)_{[-K_X]}$, where $[-K_X]$ is the anticanonical class

(iii) A pullback section $f = \pi_E^*(g)$, where $\pi_E \colon X \to Y$ contracts a (-1)-curve $E \subseteq X$ and $g \in \mathcal{R}(Y)_{[D]}$ is homogeneous with $\pi_E(E) \notin \mathrm{div}_{[D]}(g)$

We say that a homogeneous element $f \in \mathcal{R}(X)$ is a *distinguished polynomial* if it is a polynomial in distinguished sections.

Theorem 5.2.2.3 *Let X be a weak del Pezzo surface with $\varrho(X) \geq 3$ such that $\mathrm{Bs}\,|-K_X|$ is not contained in any (-1)-curve of X. Then the Cox ring $\mathcal{R}(X)$ is generated by finitely many distinguished sections.*

For the proof of Theorem 5.2.2.3, we follow [33]. With any effective divisor D on a weak del Pezzo surface X, we associate the integers

$$m(D) := \min(D \cdot E; \ E \subseteq X \text{ a } (-1)\text{-curve}), \qquad \deg(D) := -K_X \cdot D.$$

Proposition 5.2.2.4 *Let D be a nef divisor on a weak del Pezzo surface X. Then either $|D|$ is base point free or $\varrho(X) = 9$ holds, D is an anticanonical divisor, and the base locus $\mathrm{Bs}\,|D|$ consists of exactly one point.*

Proof We proceed by induction on the Picard number. For $\varrho(X) = 1$, we have $X \cong \mathbb{P}^2$ and the assertion is obvious. Assume now that the assertion holds for all weak del Pezzo surfaces Y with $\varrho(Y) \leq n - 1 \leq 8$. Consider a weak del Pezzo surface X with $\varrho(X) = n$ and a nef divisor D on X. Let $E \subseteq X$ be a (-1)-curve with $D \cdot E = m := m(D)$ and $\pi_E \colon X \to Y$ the contraction of E. The divisor $D + mK_X$ is linearly equivalent to the pullback of a nef divisor N of Y so that

$$\mathrm{Bs}\,|D| \subseteq \mathrm{Bs}\,|-mK_X| \ \cup \ \mathrm{Bs}\,|\pi^*N|.$$

By the induction hypothesis, $|N|$ is base point free. Thus $\mathrm{Bs}\,|\pi^*N|$ is empty. Moreover, $\mathrm{Bs}\,|-mK_X|$ is nonempty if and only if $m = 1$ and $\varrho(X) = 9$ hold; in this case, it consists of a single point. It remains to show that in the latter case, either $|D|$ is base point free or N is principal. Let C be an element of $|-K_X|$ and consider the corresponding exact sequence

$$0 \longrightarrow H^0(X, \pi^*N) \longrightarrow H^0(X, D) \overset{\varrho}{\longrightarrow} H^0(C, D|_C) \longrightarrow H^1(X, \pi^*N).$$

Because $\pi^*N - K_X$ is nef and big, the last cohomology group is trivial by the Kawamata–Viehweg vanishing theorem. Thus, the map ϱ is surjective. Because $(\pi^*N)^2 \geq 0$ and $C^2 > 0$ hold, either π^*N is principal or $\pi^*N \cdot C > 0$, by the Hodge index theorem. In the second case $D|_C$ is a divisor of degree $D \cdot C = (\pi^*N + C) \cdot C \geq 2$ on the elliptic curve C. Hence $|D|_C|$ is base point free and we conclude that $|D|$ is base point free because ϱ is surjective. $\square$

Proof of Theorem 5.2.2.3 First observe, for example, by induction on $\varrho(X)$, that the subalgebra of distinguished polynomials of $\mathcal{R}(X)$ is finitely generated for all weak del Pezzo surfaces X. Thus, given an effective divisor D on X, we only have to show that any element of $H^0(X, D)$ is a distinguished polynomial.

We reduce to the case that D is nef. Let E be a negative curve such that $D \cdot E < 0$. Then $E \subseteq \mathrm{Bs}\,|D|$ by Proposition 5.1.1.2. Thus, any section $f \in H^0(X, D)$ is of the form $f = f_E \cdot f'$ with $f_E \in H^0(X, E)$ and $f' \in H^0(X, D - E)$. Thus it is enough to prove that any element of $H^0(X, D - E)$ is a distinguished polynomial. If $D - E$ in not nef, then we iterate this process until we end up with a nef divisor.

Now, let D be nef. We proceed by induction on $\deg(D)$. If $\deg(D) = 0$ holds, then the Hodge index theorem gives $D = 0$ and we have nothing to prove. So, suppose that for any nef divisor D' with $\deg(D') \leq d - 1$ the elements of $f' \in H^0(X, D')$ are distinguished polynomials and let D be a nef divisor with $\deg(D) = d$.

Consider the case $m(D) = 0$. Choose a (-1)-curve $E \subseteq X$ with $D \cdot E = 0$ and let $\pi_E \colon X \to Y$ be the corresponding contraction. Then $D = \pi_E^* D_Y$ holds for an effective divisor D_Y of Y and we have an isomorphism of vector spaces

$$H^0(Y, D_Y) \ \to \ H^0(X, D), \qquad g \ \mapsto \ \pi_E^*(g).$$

Thus, it is enough to show that for any $g \in H^0(Y, D_Y)$, the pullback $\pi_E^*(g)$ is a distinguished polynomial. Let $D_g := \mathrm{div}_{D_Y}(g) \sim D_Y$ be the divisor defined by g and let $\tilde{D}_g$ be its strict transform. Then there exist $l \geq 0$ and an element $\tilde{g} \in H^0(Y, \tilde{D}_g)$ such that

$$\pi_E^* D_g \ = \ lE + \tilde{D}_g, \qquad \pi_E^*(g) \ = \ f_E^l \cdot \tilde{g}.$$

For $l > 0$, we obtain $\deg(\tilde{D}_g) < \deg(D)$ so that $\tilde{g}$, and consequently $\pi_E^*(g)$, are distinguished polynomials by the induction hypothesis. If $l = 0$ holds, then D_g does not contain $p = \pi_E(E)$. If D_g is a negative curve, then also $\pi_E^* D_g$ is a negative curve so that $\pi_E^*(g)$ is a distinguished section. If D_g is not a negative curve, then $\pi_E^*(g)$ is a distinguished polynomial by definition.

Now consider the case $m := m(D) \geq 1$. Again, let E be a (-1)-curve with $D \cdot E = m$. Consider the exact sequence

$$0 \ \longrightarrow H^0(X, D - E) \ \overset{j}{\longrightarrow} H^0(X, D) \ \overset{r}{\longrightarrow} H^0(E, D|_E).$$

Because $\deg(D - E) < \deg(D)$ holds, any element of $H^0(X, D - E)$ is a distinguished polynomial by hypothesis. Thus the image $im(j)$ is generated by distinguished polynomials. Let $D' := D + mK_X$. Observe that D' has nonnegative intersection with any (-1)- and (-2)-curve, so that it is nef by Proposition 5.2.1.10. Thus the linear system $|D'|$ does not have fixed components by Proposition 5.2.2.4. Moreover, $\deg(D') < \deg(D)$ because $m(D) > 0$. Let

$s \in H^0(X, D')$ be a distinguished polynomial whose restriction $s|_E$ is constant with nonzero. Such a section exists because $D' \cdot E = 0$ and $|D'|$ does not have fixed components. We have a commutative diagram

$$
\begin{array}{ccccc}
\mathrm{Sym}^m H^0(X, -K_X) & \xrightarrow{\ \iota\ } & H^0(X, -mK_X) & \xrightarrow{\ \varphi\ } & H^0(X, D) \\
{\scriptstyle r_1}\downarrow & & {\scriptstyle r_m}\downarrow & & \downarrow{\scriptstyle r} \\
\mathrm{Sym}^m H^0(E, -K_X|_E) & \xrightarrow{\ \iota_E\ } & H^0(E, -mK_X|_E) & \xrightarrow{\ \varphi_E\ } & H^0(E, D|_E)
\end{array}
$$

where $\varphi(f) = s \cdot f$, the maps r_1, r_m and r are restrictions, ι is an inclusion, and ι_E as well as φ_E are isomorphisms. The map r_1 is surjective because the base locus of the linear system $|-K_X|$ is either empty or it is a point p that does not lie on any (-1)-curve E by hypothesis. Thus $\varphi_E \circ r_m \circ \iota = \varphi_E \circ \iota_E \circ r_1$ is surjective. This proves that $r \circ \varphi$ is surjective. Let $h_1, h_2 \in H^0(X, -K_X)$ be two linearly independent distinguished sections. For each $0 \leq k \leq m$, the following sections

$$
g_k := (\varphi \circ \iota)(h_1^k \cdot h_2^{m-k}) \in H^0(X, D)
$$

are distinguished polynomials. Because $H^0(X, D)$ is generated as a vector space by the g_k and $\mathrm{im}(j)$, we obtain the assertion. $\qquad\square$

Lemma 5.2.2.5 *Let X be a smooth del Pezzo surface with $\varrho(X) = 9$ and* $\mathrm{Bs}\,|-K_X| = \{p\}$. *Then p does not lie on any (-1)-curve of X.*

Proof Assume that there is a (-1)-curve $E \subseteq X$ through p and let $q \in E$ be a point distinct from p. By hypothesis, the linear system $|-K_X|$ is at least one-dimensional, so that there exists an element $D \in |-K_X|$ that contains q. Because D and E intersect at least in two points and $D \cdot E = 1$ holds, E must be a component of D. But then $D - E$ is an effective divisor that has zero intersection with $-K_X$. This contradicts ampleness of $-K_X$. $\qquad\square$

Proof of Theorem 5.2.2.1 We proceed by induction on the Picard number $\varrho(X)$. In the cases $\varrho(X) = 3, 4$, the surface X is the toric blow-up of $\mathbb{P}^2$ in two or three points and we directly see that $\mathcal{R}(X)$ is generated by the canonical sections of the (-1)-curves.

Now let $\varrho(X) \leq 8$. By Lemma 5.2.2.5, we are in the situation of Theorem 5.2.2.3. Thus, our task is to show that any distinguished section $f \in \mathcal{R}(X)$ is a polynomial in the canonical sections of the (-1)-curves. This is obvious for the distinguished sections f of type (i) and (iii). So, we are left with type (ii), which means that we have $f \in H^0(X, -K_X)$. Let $E \subseteq X$ be a (-1)-curve and consider the exact sequence

$$0 \longrightarrow H^0(X, -K_X - E) \xrightarrow{\cdot f_E} H^0(X, -K_X) \xrightarrow{r} H^0(E, -K_X|_E).$$

The left-hand side vector space is generated by distinguished sections by Theorem 5.2.2.3. Because $\deg(-K_X - E) < \deg(-K_X)$ holds, $H^0(X, -K_X - E)$ contains no sections of $H^0(X, -K_X)$. We conclude that $H^0(X, -K_X - E)$ is generated by (monomials in) the canonical sections of (-1)-curves. Because $-K_X|_E \cong \mathcal{O}_{\mathbb{P}^1}(1)$, it is enough to show that there are elements $g_1, g_2 \in H^0(X, -K_X)$ that are polynomials in the canonical sections of (-1)-curves and restrict to generators of the two-dimensional vector space $H^0(E, -K_X|_E)$. In this way we prove that r is surjective and construct a section for r.

We can always assume that $E = E_1$, the first exceptional divisor of a blow-up $\pi: X \to \mathbb{P}^2$. Then the following (-1)-curves correspond to pairs of linearly independent sections of $H^0(X, -K_X)$ whose images via r span $H^0(E, -K_X|_E)$.
If $\varrho(X) = 5$:

$$-K_X \sim 3H - E_1 - \cdots - E_4$$
$$\sim (H - E_1 - E_2) + (H - E_3 - E_4) + (H - E_2 - E_3) + E_2 + E_3$$
$$\sim (H - E_1 - E_3) + (H - E_2 - E_4) + (H - E_2 - E_3) + E_2 + E_3.$$

If $\varrho(X) = 6$:

$$-K_X \sim 3H - E_1 - \cdots - E_5$$
$$\sim (H - E_1 - E_2) + (H - E_3 - E_4) + (H - E_4 - E_5) + E_4$$
$$\sim (H - E_1 - E_5) + (H - E_2 - E_3) + (H - E_3 - E_4) + E_3.$$

If $\varrho(X) = 7$:

$$-K_X \sim 3H - E_1 - \cdots - E_6$$
$$\sim (H - E_1 - E_2) + (H - E_3 - E_4) + (H - E_5 - E_6)$$
$$\sim (H - E_1 - E_5) + (H - E_2 - E_4) + (H - E_3 - E_6).$$

If $\varrho(X) = 8$:

$$-K_X \sim 3H - E_1 - \cdots - E_7$$
$$\sim (2H - E_1 - E_2 - E_3 - E_4 - E_5) + (H - E_6 - E_7)$$
$$\sim (2H - E_3 - E_4 - E_5 - E_6 - E_7) + (H - E_1 - E_2).$$

If $\varrho(X) = 9$, then $-K_X$ is not linearly equivalent to a sum of (-1)-curves with positive integer coefficients because $-K_X$ is not a (-1)-curve, $\deg(-K_X) = 1$

and any (-1)-curve has degree 1. Thus any set of generators of the Cox ring must contain a basis h_1, h_2 of $H^0(X, -K_X)$.

To conclude the proof, observe that, modulo multiplication by nonzero scalars, any generating set of $\mathcal{R}(X)$ must contain the canonical sections of the (-1)-curves, because the corresponding classes span an extremal rays of the effective cone $\mathrm{Eff}(X)$. Moreover, if $\varrho(X) = 9$, a basis of $H^0(X, -K_X)$ must appear in any generating set of $\mathcal{R}(X)$ as seen before.   □

Remark 5.2.2.6   As a consequence of the proof of Theorem 5.2.2.1 the anti-canonical divisor $-K_X$ of a del Pezzo surface X is linearly equivalent to a sum of (-1)-curves whose intersection graph is given below for each value of $\varrho(X)$ between 5 and 8.

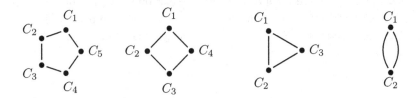

5.2.3 The ideal of relations

We consider a smooth del Pezzo surface of Picard number at least 5 and discuss the ideal of relations between the generators of the Cox ring. As before, we work over an algebraically closed field $\mathbb{K}$ of characteristic zero. The original references for this section are [33, 199, 284, 292].

Let X be a smooth del Pezzo surface and let $(f_1, \ldots, f_s)$ be a minimal system of homogeneous generators of the Cox ring $\mathcal{R}(X)$ as provided by Theorem 5.2.2.1. Then we have a $\mathrm{Cl}(X)$-graded presentation

$$0 \longrightarrow \mathfrak{J} \longrightarrow \mathbb{K}[T] \xrightarrow{\;T_i \mapsto f_i\;} \mathcal{R}(X) \longrightarrow 0,$$

where $\mathbb{K}[T]$ is a $\mathrm{Cl}(X)$-graded polynomial ring in s variables with $\deg(T_i) = \deg(f_i)$ for all i and $\mathfrak{J}$ is the homogeneous ideal of relations. As usual, a *quadric* in $\mathbb{K}[T]$ is a polynomial whose monomials are of the form $T_i T_j$. By a *conic* of X, we mean a divisor Q of X with $Q^2 = 0$ and $-K_X \cdot Q = 2$.

Theorem 5.2.3.1   *Let X be a smooth del Pezzo surface of Picard number $\varrho(X) \geq 4$ with Cox ring $\mathcal{R}(X) \cong \mathbb{K}[T]/\mathfrak{J}$ presented as above. Then the ideal of relations $\mathfrak{J}$ is generated by $\mathrm{Cl}(X)$-homogeneous quadrics in the following degrees $w \in \mathrm{Cl}(X)$, where Q denotes a conic of X, E a (-1)-curve of X and $\#w$ the number of classes of type w.*

$\varrho(X)$	w	$\dim \mathbb{K}[T]_w$	$\dim \mathcal{R}(X)_w$	$\dim(\mathfrak{I}_w)$	#w
5	$[Q]$	3	2	1	5
6	$[Q]$	4	2	2	10
7	$[Q]$	5	2	3	27
8	$[Q]$	6	2	4	126
	$[-K_X]$	28	3	25	1
9	$[Q]$	7	2	5	2160
	$[-K_X + E]$	30	3	27	240
	$[-2K_X]$	123	4	119	1

Moreover any set of generators of $\mathfrak{I}$ must contain at least $\dim(\mathfrak{I}_w)$ homogeneous elements for any degree w in the above table.

We will prove Theorem 5.2.3.1 in the cases $\varrho(X) \le 7$ and refer to [199, 284, 292] for the remaining cases. We first provide the necessary techniques introduced in [199, 200].

Construction 5.2.3.2 Let X be a Mori dream space, $(f_1, \ldots, f_n)$ a minimal system of homogeneous generators for the Cox ring $\mathcal{R}(X)$, and set $w_i := \deg(f_i)$ and $E_i = \operatorname{div}_{w_i}(f_i)$. Moreover, fix a subgroup $K \subseteq \operatorname{WDiv}(X)$ as in Construction 1.4.2.1. Consider the $\operatorname{Cl}(X)$-graded presentation:

$$0 \longrightarrow \mathfrak{I} \longrightarrow \mathbb{K}[T] \xrightarrow{T_i \mapsto f_i} \mathcal{R}(X) \longrightarrow 0 \qquad (5.2.1)$$

where $\mathbb{K}[T] := \mathbb{K}[T_1, \ldots, T_n]$ is a $\operatorname{Cl}(X)$-graded polynomial ring and $\mathfrak{I}$ is the homogeneous ideal of relations. We say that $\mathcal{R}(X)$ has *a relation in degree w* if for any such presentation of the Cox ring any set of homogeneous generators of $\mathfrak{I}$ contains an element of degree w. Now, consider a minimal $\operatorname{Cl}(X)$-graded resolution

$$\cdots \longrightarrow \bigoplus_{w \in \operatorname{Cl}(X)} \mathbb{K}[T](-w)^{b_{2,w}} \longrightarrow \bigoplus_{w \in \operatorname{Cl}(X)} \mathbb{K}[T](-w)^{b_{1,w}} \xrightarrow{g} \mathbb{K}[T] \longrightarrow 0,$$

where g is given by a row matrix whose entries are a set of minimal generators of the ideal $\mathfrak{I}$. The $b_{i,w}$ are the *Picard-graded Betti numbers* of $\mathcal{R}(X)$. The notation has to be understood in the following way: If $b_{i,w} = 0$, then the corresponding shifted polynomial ring is not considered into the sum. Observe that $\mathfrak{I}_w$ has exactly $b_{1,w}$ generators.

Denote by $\mathbb{K}^\bullet$ the Koszul complex of $\mathbb{K}$ with respect to the variables $T_1, \ldots T_n$, or equivalently $\mathbb{K}^\bullet$ is the minimal free resolution of the field $\mathbb{K}$

over $\mathbb{K}[T]$. We recall the definition of the first two maps of the Koszul complex $\mathbb{K}^\bullet$

$$\bigoplus_{1 \le i < j \le n} \mathbb{K}[T](-w_i - w_j) \xrightarrow{d_2} \bigoplus_{i=1}^{n} \mathbb{K}[T](-w_i) \xrightarrow{d_1} \mathbb{K}[T] \longrightarrow 0.$$

Denoting by e_i the ith coordinate vector of the middle direct sum and by e_{ij} the ijth coordinate vector of the left-hand side direct sum we have $d_1(e_i) = x_i$ and $d_2(e_{ik}) = x_i e_k - x_i e_i$. Then d_1 and d_2 are completely defined by $\mathbb{K}[T]$ linearity. Consider now the tensor product complex $B^\bullet := \mathcal{R}(X) \otimes_{\mathbb{K}[T]} \mathbb{K}^\bullet$. Then its degree w part $(B^\bullet)_w$ is

$$\bigoplus_{1 \le i < j \le n} H^0(X, D - E_i - E_j) \xrightarrow{\partial_2} \bigoplus_{i=1}^{n} H^0(X, D - E_i) \xrightarrow{\partial_1} H^0(X, D),$$

where D is such that $[D] = w$. The maps are explicitly given by: $\partial_1(e_i) = f_i$ and $\partial_2(e_{ik}) = f_i e_k - f_k e_i$.

Theorem 5.2.3.3 *Consider the situation of Construction 5.2.3.2.*

(i) $\mathcal{R}(X)$ *has a relation in degree w if and only if $b_{1,w} \ne 0$.*
(ii) *One has $b_{1,w} = \dim_{\mathbb{K}} H_1(B_w^\bullet)$.*

Proof The first statement is due to the fact that we are taking a minimal resolution of $\mathcal{R}(X)$. To prove the second statement observe that the ith homology group of the complex $B^\bullet$ is, by the definition of Tor, isomorphic to $\mathrm{Tor}_i^{\mathbb{K}[T]}(\mathcal{R}(X), \mathbb{K})$. Taking the resolution of $\mathcal{R}(X)$ and tensoring with the $\mathbb{K}[T]$-module $\mathbb{K}$, by [218, Lemma 1.32] we obtain the following which proves the statement:

$$(\mathrm{Tor}_i^{\mathbb{K}[T]}(\mathcal{R}(X), \mathbb{K}))_w \cong \mathbb{K}^{b_{i,w}}.$$

$\square$

By Theorem 5.2.3.3, to decide if $\mathcal{R}(X)$ has relations in degree w one has to compute the first homology group of $(B^\bullet)_w$. As usual a *cycle* is an element of $\ker(\partial_1)$ and a *boundary* is an element of $\mathrm{im}(\partial_2)$. The *size* of a cycle $\sigma = \sum_i \sigma_i e_i$ is the cardinality of the set $\{i : \sigma_i \ne 0\}$.

Lemma 5.2.3.4 *Any cycle of size at most 2 is a boundary.*

Proof The only cycle σ of size at most 1 is $\sigma = 0$ because $\mathcal{R}(X)$ is an integral domain. If σ is a cycle of size 2 then $\sigma = \sigma_1 e_1 + \sigma_2 e_2$ with $\sigma_1 f_1 + \sigma_2 f_2 = 0$. Because $\mathcal{R}(X)$ is factorially $\mathrm{Cl}(X)$-graded, by Theorem 1.5.3.7, and because f_1, f_2 are prime elements, we conclude that $\sigma_1 = -\sigma_{12} f_2$ and $\sigma_2 = \sigma_{12} f_1$

hold with an element $\sigma_{12} \in H^0(X, D - E_1 - E_2)$. Thus $\sigma = \partial_2(\sigma_{12} e_{12})$ is a boundary. $\qquad\square$

Definition 5.2.3.5 Let $\mathcal{S} \subseteq \{E_1, \ldots, E_n\}$, and $E_k \in \{E_1, \ldots, E_n\} - \mathcal{S}$. We say that E_k is *capturable by* $\mathcal{S}$ if the following natural map is surjective:

$$\bigoplus_{E_i \in \mathcal{S}} H^0(X, D - E_i - E_k) \otimes \langle f_i \rangle \ \to \ H^0(X, D - E_k).$$

Proposition 5.2.3.6 *Let* $\mathcal{S}_1 \subseteq \cdots \subseteq \mathcal{S}_k = \{E_1, \ldots, E_n\}$ *be a sequence of proper inclusions such that any element of* $\mathcal{S}_i - \mathcal{S}_{i-1}$ *is capturable by* $\mathcal{S}_{i-1}$, *for any* i. *If* $\mathcal{S}_1$ *has cardinality 2, then* $\mathcal{R}(X)$ *does not have relations in degree* $[D]$.

Proof Let σ be a cycle and let $E_k \in \mathcal{S}_i - \mathcal{S}_{i-1}$. Because E_k is capturable by $\mathcal{S}_{i-1}$ and $\sigma_k \in H^0(X, D - E_k)$, then $\sigma_k = \sum_{E_j \in \mathcal{S}_{i-1}} x_j f_j$, with $x_j \in H^0(X, D - E_j - E_k)$. Thus we obtain

$$\sigma_k e_k = \sum_{E_j \in \mathcal{S}_{i-1}} x_j f_k e_j + \partial_2 \left(\sum_{E_j \in \mathcal{S}_{i-1}} x_j e_{hk} \right).$$

Hence σ is homologous to a cycle σ' whose support does not contain E_k. Proceeding in the same way, after a finite number of steps, one shows that σ is homologous to a cycle σ'' whose support is contained in $\mathcal{S}_1$ and we conclude by Lemma 5.2.3.4 and Theorem 5.2.3.3. $\qquad\square$

We now re-enter the setting of Theorem 5.2.3.1 and prepare the proof. We write X_r for the blow-up of $\mathbb{P}^2$ at r points in general position; note that we have $\varrho(X_r) = r + 1$.

Proposition 5.2.3.7 *Let* X *be a del Pezzo surface and let* Q *be a conic of* X. *Then* Q *is a nef divisor, and* $H^0(X, Q)$ *is a two-dimensional vector space that contains* $\varrho(X) - 2$ *sections of the form* $f_E \cdot f_{E'}$, *where* E *and* E' *are* (-1)-*curves with* $E \cdot E' = 1$.

Proof We first show that Q is nef. Because we have $-K_X \cdot (K_X - Q) < 0$ and $-K_X$ is ample, $K_X - Q$ is not effective. Thus, the Riemann–Roch formula and Serre's duality [39, Thm. I.11] yield $h^0(X, Q) - h^1(X, Q) = 2$. So, without loss of generality, we can assume Q to be effective. By Proposition 5.2.1.10 and the fact that Q is a conic, we deduce that either Q is irreducible or $Q = E + E'$, where E and E' are (-1)-curves with $E \cdot E' = 1$. In particular Q is nef and $|Q|$ is base point free by Proposition 5.2.2.4.

Because $Q - K_X$ is nef and big, the Kawamata–Viehweg vanishing theorem gives us $h^1(X, Q) = 0$. Thus $|Q|$ has dimension 1 and it defines a morphism

$\varphi \colon X \to \mathbb{P}^1$. We have a commutative diagram of morphisms

$$\begin{array}{ccc} X & \xrightarrow{\ \pi\ } & Y \\ {\scriptstyle\varphi}\downarrow & & \downarrow{\scriptstyle\psi} \\ \mathbb{P}^1 & \xrightarrow[\text{id}]{} & \mathbb{P}^1 \end{array}$$

where π contracts the (-1)-curve E' for each pair E, E' such that $E + E' \sim Q$. The (-1)-curves contracted by π are contained in the reducible fibers of φ. Hence all the fibers of ψ are smooth rational. Thus, Y is a Hirzebruch surface [39]. This implies $\varrho(Y) = 2$. Hence the number of (-1)-curves contracted by π is $\varrho(X) - 2$. $\qquad\square$

Proposition 5.2.3.8 *Let E, E' be two distinct (-1)-curves on a del Pezzo surface X with $E \cdot E' > 0$. Then one of the following holds.*

(i) $E \cdot E' = 1$.
(ii) $E \cdot E' = 2$, $\varrho(X) = 8$ and $E + E' \sim -K_X$.
(iii) $E \cdot E' = 2$, $\varrho(X) = 9$ and $E + E' \sim -K_X + E''$ for some (-1)-curve E''.
(iv) $E \cdot E' = 3$, $\varrho(X) = 9$ and $E + E' \sim -2K_X$.

Proof We can always assume that E is an exceptional divisor and E' is one of the curves listed in Remark 5.2.1.4 (distinct from E_1). Thus the statement follows by direct calculation. $\qquad\square$

Proposition 5.2.3.9 *Let X be a smooth nontoric del Pezzo surface and let $w := [E + E']$ where E, E' are two (-1)-curves with $E \cdot E' > 0$. Then $\mathcal{R}(X)$ has quadratic relations in degree w.*

Proof First observe that w is a nef class because $E + E'$ has nonnegative intersection product with both E and E'. Hence by the Riemann–Roch formula and the Kawamata–Viehweg vanishing theorem the space $\mathcal{R}(X)_w \cong H^0(X, E + E')$ has dimension $E \cdot E' + 1$. There are four possibilities for the value of $E \cdot E'$ according to Proposition 5.2.3.8. In the first case $\mathcal{R}(X)_w$ contains $\varrho(X) - 2 \geq 3$ quadratic monomials by Proposition 5.2.3.7, where the inequality is due to the fact that X is nontoric. Thus the Cox ring has quadratic relations in degree w.

In cases (ii) and (iv) of Proposition 5.2.3.8 the class w is invariant under the action of the Weyl group of $\mathrm{Cl}(X)$; see Section 5.2.4 for the definition. The action of the Weyl group is transitive on the set of (-1)-classes and moreover given a pair of (-1)-curves E and E' with $[E + E'] = w$ there is an element of the Weyl group that permutes E and E'. Hence the number of such pairs equals the order of the Weyl group divided by two times the order of the stabilizers of a (-1)-curve. This number is given in the statement of Theorem 5.2.3.1

and again one gets the statement. Finally in case (iii) of Proposition 5.2.3.8 a similar argument applies but this time one has to look at the subgroup of the Weyl group which stabilizes the (-1)-curve C given in (iii). $\qquad\square$

Lemma 5.2.3.10 *Let D be a nonnef divisor of X. Then the Cox ring $\mathcal{R}(X)$ does not have relations in degree $[D]$.*

Proof Without loss of generality one can assume D to be effective. Because D is not nef, it has negative intersection with some (-1)-curve E. Hence by Proposition 5.1.1.2 the curve E is contained in the base locus of $|D|$, or equivalently all the sections of $H^0(X, D)$ are divisible by the generator $f_E \in \mathcal{R}(X)_{[E]}$. Thus the Cox ring has no relations in degree $[D]$, because any such relation would be made by a reducible polynomial. $\qquad\square$

Given a smooth non toric del Pezzo surface X_r with $r \le 6$, let $\mathbb{K}[X_r]$ be the $\mathrm{Cl}(X_r)$-graded polynomial ring whose variables correspond to the (-1)-curves of X_r. Given a conic Q of X_r, according to Proposition 5.2.3.9, the Cox ring $\mathcal{R}(X_r)$ admits quadratic relations in degree $[Q]$. Denote by Q_r the homogeneous ideal of $\mathbb{K}[X_r]$ generated by all such quadrics as Q varies over all the conics of X_r.

Lemma 5.2.3.11 *Let D be a nef, nonample divisor of X_r such that $-K_{X_r} \cdot D \ge 3$ and $r \le 6$. Assume that the Cox ring of X_{r-1} is isomorphic to $\mathbb{K}[X_{r-1}]/Q_{r-1}$; then the Cox ring $\mathcal{R}(X_r)$ does not have relations in degree $[D]$.*

Proof If $r \le 3$, then X_r is isomorphic to a toric variety and the statement holds. Assume now $4 \le r \le 6$. Because D is nef not-ample, there exists a (-1)-curve E with $D \cdot E = 0$. Let $\pi : X_r \to X_{r-1}$ be the blow-down of E. Because the set of (-1)-curves of X_r that do not intersect E is in bijection with the set of (-1)-curves of X_{r-1}, we obtain $Q_{r-1} \subseteq Q_r \cap \mathbb{K}[X_{r-1}]$. Thus, we have an homomorphism of $\mathrm{Cl}(X_r)$-graded $\mathbb{K}$-algebras:

$$\mathbb{K}[X_{r-1}]/Q_{r-1} \to \mathbb{K}[X_r]/Q_r.$$

We want to prove that this map is surjective in degree $[D]$. Let $\{E_1, \dots, E_k\}$ be the set of (-1)-curves of X_r that intersect E and let $\{E_{k+1}, \dots, E_s\}$ be the set of those that do not intersect E. Because $r \le 6$, then $E \cdot E_i = 1$ for any i, as can be easily deduced by using the list of Remark 5.2.1.4. Denote by T_i the variable of $\mathbb{K}[X_r]$ that corresponds to E_i and by T_E be the variable that corresponds to E. Given a degree w monomial $g \in \mathbb{K}[X_r]_w$ we have

$$g = T_E^m \cdot T_1^{m_1} \cdots T_s^{m_s}$$

$$= (T_E T_1)^{m_1} \cdots (T_E T_k)^{m_k} \cdot T_{k+1}^{m_{m}+1} \cdots T_s^{m_s}$$

$$= p(T_{k+1}, \dots, T_s),$$

where the second equality is due to $m = m_1 + \cdots + m_k$ because $D \cdot E = 0$ and $E \cdot E_i = 1$ for any $i \in \{1, \ldots, k\}$, while the third equality is due to the relations contained in Q_r of the form $T_E T_i = \sum \alpha_{jh} T_j T_h$ where $i \in \{1, \ldots, k\}$ and $i, h \in \{k + 1, \ldots, s\}$, coming from the conic $E_i + E$. This proves the claimed surjectivity.

By the hypothesis and the fact that $\mathbb{K}[X_r]/Q_r$ surjects onto the Cox ring $\mathcal{R}(X_r)$, we obtain

$$\dim_{\mathbb{K}} \mathcal{R}(X_{r-1})_{[\pi(D)]} \geq \dim_{\mathbb{K}} (\mathbb{K}[X_r]/Q_r)_{[D]} \geq \dim_{\mathbb{K}} \mathcal{R}(X_r)_{[D]}.$$

Because the first and the last dimensions are equal, then we conclude that the degree $[D]$ part of $\mathbb{K}[X_r]/Q_r$ equals that of $\mathcal{R}(X_r)$, which proves the statement because $\mathbb{K}[X_r]/Q_r$ does not have relations in degree $[D]$. $\qquad\square$

Lemma 5.2.3.12 *Let Q be a nef divisor of X_r with $-K_{X_r} \cdot Q \leq 2$. Then Q is a conic.*

Proof Because $\mathcal{R}(X_r)$ is generated by sections whose degrees are (-1)-classes, by Theorem 5.2.2.1, Q is linearly equivalent to a sum, with nonnegative integer coefficients, of (-1)-curves. From $-K_{X_r} \cdot Q \leq 2$ this sum contains at most two summands. Because Q is nef, the only possibility is $Q \sim E + E'$ with $E \cdot E' > 0$. Hence $E \cdot E' = 1$, by Proposition 5.2.3.8, so that Q is a conic. $\qquad\square$

Lemma 5.2.3.13 *Let D be an ample divisor of X_r and let A, B, and C be three (-1)-curves with $A \cdot B = 0$, $B \cdot C = 1$, and $A \cdot C = 0$ if $r = 4, 5$ and $A \cdot C = 1$ if $r = 6$. Then $h^1(X, D - A - B - C) = 0$.*

Proof To prove the statement it is enough to show that the divisor $L_n := D - A - B - C - K_{X_r}$, where $-K_{X_r} = 3H - E_1 - \cdots - E_n$, is nef and big and conclude by the Kawamata–Viehweg vanishing theorem. If $r \in \{4, 5\}$ we can assume, without loss of generality, $A = H - E_1 - E_2$, $B = H - E_1 - E_3$ and $C = H - E_2 - E_4$. Hence both

$$L_4 = D + E_1 + E_2 \quad \text{and} \quad L_5 = (D - E_5) + E_1 + E_2$$

are nef because D is ample and $E_i \cdot E \leq 1$ for any (-1)-curve E of X_r. Moreover, both the divisors have positive self-intersection so that they are big. If $r = 6$, by taking $A = H - E_1 - E_2$, $B = H - E_1 - E_3$ and $C = H - E_4 - E_5$ we get $L_6 = (D - E_6) + E_1$ which is again nef and big. $\qquad\square$

Proof of Theorem 5.2.3.1 For $r \leq 6$ If w is the class of a conic Q of X_r, then, by Lemma 5.2.3.7, $Q \sim E + E'$ for some pair of (-1)-curves E, E' that intersect in one point. Then $\mathcal{R}(X_r)$ has relations in degree w, by Proposition 5.2.3.9.

We will show now the second part of the statement by induction on n. If $n \leq 3$, then X_r is toric and the statement holds. So we assume now that the statement holds for X_{r-1} and prove it for X_r.

We already know that $\mathcal{R}(X_r)$ is not generated in degree w if w is not nef, by Lemma 5.2.3.10. Hence assume now that D is a nef divisor. Let $E_1, \ldots, E_r$ be the exceptional divisors of $\pi : X \to \mathbb{P}^2$ and let $p_i := \pi(E_i)$ be the corresponding points for $1 \leq i \leq n$. Denote by E_{ij} the strict transform of the line through p_i, p_j, by C the strict transform of the conic through $p_1, \ldots, p_5$ if $r = 5$, and by C_i the strict transform of the conic through five points avoiding p_i, if $r = 6$.

If D is ample we construct subsets $\mathcal{S}_1 \subseteq \cdots \subseteq \mathcal{S}_m$ of (-1)-curves of X_r such that any element of $\mathcal{S}_{i+1} - \mathcal{S}_i$ is capturable by $\mathcal{S}_i$. according to Definition 5.2.3.5. On X_4 we have:

$$\mathcal{S}_1 := \{E_1, E_2\}$$
$$\mathcal{S}_2 := \mathcal{S}_1 \cup \{E_{13}, E_{14}, E_{23}, E_{24}\}$$
$$\mathcal{S}_3 := \mathcal{S}_2 \cup \{E_3, E_4, E_{12}\}$$
$$\mathcal{S}_4 := \mathcal{S}_3 \cup \{E_{34}\}.$$

On X_5 we have:

$$\mathcal{S}_1 := \{E_1, E_2\}$$
$$\mathcal{S}_2 := \mathcal{S}_1 \cup \{E_{13}, E_{14}, E_{15}, E_{23}, E_{24}, E_{25}\}$$
$$\mathcal{S}_2 := \mathcal{S}_2 \cup \{E_{23}, E_{24}, E_{25}, E_{34}, E_{35}, E_{45}, C\}.$$

On X_6 we have:

$$\mathcal{S}_1 := \{E_1, E_2\}$$
$$\mathcal{S}_2 := \mathcal{S}_1 \cup \{E_{12}, C_3, C_4, C_5, C_6\}$$
$$\mathcal{S}_3 := \mathcal{S}_2 \cup \{E_{34}, E_{35}, E_{36}, E_{45}, E_{46}, E_{56}, E_3, E_4, E_5, E_6\}$$
$$\mathcal{S}_4 := \mathcal{S}_3 \cup \{E_{13}, \ldots E_{16}, E_{23}, \ldots, E_{26}, C_1, C_2\}.$$

Hence $\mathcal{R}(X_r)$ does not have relations in degree $[D]$ by Proposition 5.2.3.6. Suppose now that D is not ample. Then $\mathcal{R}(X_r)$ does not have relations in degree $[D]$ if D is nef and such that $-K_{X_r} \cdot D \geq 3$, by Lemma 5.2.3.11. Hence by Lemma 5.2.3.12 we conclude that D is a conic. $\qquad\square$

5.2.4 Del Pezzo surfaces and flag varieties

In this section we discuss relations between the geometry of del Pezzo surfaces and representations of certain simple affine algebraic groups following [33] and [268]; see also [267].

For $3 \leq r \leq 8$, denote by X_r the del Pezzo surface of degree $9 - r$ obtained by blowing up $\mathbb{P}^2$ at points $p_1, \ldots, p_r$ in general position and let $\pi : X_r \to \mathbb{P}^2$ be the contraction morphism. Then $\mathrm{Cl}(X_r)$ comes with a basis $(l_0, \ldots, l_r)$, where l_0 is the class of the total transform of a line in $\mathbb{P}^2$ and l_i is the class of the exceptional divisor over p_i. The associated intersection matrix is diagonal with $l_0^2 = 1$ and $l_i^2 = -1$ for any $i > 0$. It turns out that the set

$$R_r := \{\alpha \in \text{Cl}(X_r); \ \alpha^2 = -2, \ \alpha \cdot K_X = 0\}$$

is a root system. Note that R_r comprises exactly the classes $\alpha = [E] - [E']$, where E and E' are two disjoint (-1)-curves on X_r. The corresponding Weyl group W_r is the subgroup of $\text{Aut}(\text{Cl}(X_r))$ generated by the reflections

$$\sigma: \text{Cl}(X_r) \ \to \ \text{Cl}(X_r), \qquad u \mapsto u + (u \cdot \alpha)\alpha,$$

where $\alpha \in R_r$. For any set of simple roots in R_r, the corresponding reflections form a minimal generating subset of W_r. As a simple roots we can take

$$\alpha_1 = l_1 - l_2, \quad \alpha_2 = l_2 - l_3, \quad \alpha_3 = l_0 - l_1 - l_2 - l_3, \quad \alpha_i = l_{i-1} - l_i, \ i \geq 4.$$

Via pullback, the blow-up morphism $X_r \to X_{r-1}$ determines an isometric embedding of lattices $\text{Cl}(X_{r-1}) \to \text{Cl}(X_r)$ which in turn induces embeddings of root systems $R_{r-1} \to R_r$. From this point of view, the Dynkin diagram of R_r, where $3 \leq r \leq 8$, can be considered as the subgraph on the vertices α_i with $i \leq r$ of the graph

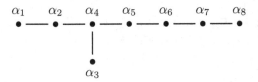

In particular, the root systems R_r are given as

$$R_3 = A_2 + A_1, \quad R_4 = A_4, \quad R_5 = D_5, \quad R_6 = E_6, \quad R_7 = E_7, \quad R_8 = E_8.$$

Now let G_r be a connected, simply connected, semisimple algebraic group having R_r as its root system; recall that G_r is unique up to isomorphism. Denote by $\varpi_1, \ldots, \varpi_r$ the dual basis to the basis $\alpha_1, \ldots, \alpha_r$. Each ϖ_i is the highest weight of an irreducible representation of G_r, called the *fundamental representation*; see Section 3.2.3 for a little more background. We denote by $V(\varpi)$ the representation space of G_r with the highest weight ϖ.

Remark 5.2.4.1 The dimension d_r of the representation space $V(\varpi_r)$ of G_r is given by the following table:

r	4	5	6	7	8
d_r	10	16	27	56	248

Comparing with Remark 5.2.1.4, we see that for $r \leq 7$ the dimension d_r equals the number M_r of negative curves on the del Pezzo surface X_r. The following result, conjectured by Batyrev and proven in [33, 110, 254, 268, 277], explains this observation (we discuss later the reason for $d_8 \neq M_8$).

Theorem 5.2.4.2 *Fix $3 \le r \le 7$. Let $H_r \subseteq G_r$ be a maximal torus and $P_r \subseteq G_r$ the stabilizer of the line $\mathbb{K}\,v(r)$ through the highest weight vector $v(r) \in V(\varpi_r)$.*

(i) *We have $\mathrm{Cl}(X_r) = \mathbb{X}(H_r) \oplus \mathbb{Z}$, where $\mathbb{X}(H_r) \subseteq \mathrm{Cl}(X_r)$ is the kernel of the degree; in particular, H_{X_r} is isomorphic to $H_r \times \mathbb{K}^*$.*

(ii) *The vector space $V(\varpi_r)$ admits a basis $v_1, \ldots, v_{d_r}$ of H_r-eigenvectors with pairwise different weights.*

(iii) *For a suitable numbering of the (-1)-curves $E_1, \ldots, E_{d_r}$ on X_r one obtains a surjection of graded algebras*

$$\mathbb{K}[V(\varpi_r)] \;\to\; \mathcal{R}(X_r), \qquad v_i^* \mapsto 1_{E_i},$$

the associated embedding $\overline{X}_r \to V(\varpi_r)$ maps $\overline{X}_r$ into $G_r \cdot \mathbb{K}\,v(r)$ and, with $\mathbb{P}(X_r) := \mathrm{Proj}(\mathcal{R}(X_r))$, this gives rise to a commutative diagram

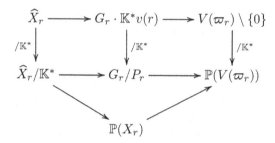

where $\widehat{X}_r/\mathbb{K}^* \subseteq \mathbb{P}(X_r)$ is an open embedding and the flag variety G_r/P_r is identified with the G_r-orbit through $v(r)$ in $\mathbb{P}(V(\varpi_r))$.

(iv) *The (-1)-curves of X_r are precisely the images under $p_X \colon \widehat{X}_r \to X_r$ of the intersections of $\widehat{X}_r$ with the coordinate hyperplanes $L_i = \mathrm{lin}(v_j; \; j \ne 0)$ of $V(\varpi_r)$.*

(v) *For every (-1)-curve $E_i \subseteq X_r$, the localization $\mathbb{P}(X_r)_{v_i^*}$ with respect to the corresponding linear form v_i^* is isomorphic to the affine cone $C(X_{r-1}) \subseteq V(\varpi_{r-1})$ over $\mathbb{P}(X_{r-1}) \subseteq \mathbb{P}(V(\varpi_{r-1}))$, where X_{r-1} is obtained from X_r by contracting E_i.*

The case $r = 4$ was proven by Skorobogatov [277] and is discussed in Example 5.2.4.3. The proof for $r = 5$ is due to Oleg Popov; see also Remark 5.2.4.5. For $r = 6, 7$, the theorem is proven in [110]; the idea is to compare by suitably scaling the generators of the Cox ring its relations with equations of the cone over G_r/P_r. Serganova and Skorobogatov presented another proof for $r = 5, 6, 7$, see [268, 269] and Remark 5.2.4.5.

Example 5.2.4.3 See also [277]. Consider X_4, the blow-up of $\mathbb{P}^2$ at four points in general position. Applying a suitable linear transformation, we achieve that the points to be blown up are

$$p_1 = [1, 0, 0], \qquad p_2 = [0, 1, 0], \qquad p_3 = [0, 0, 1], \qquad p_4 = [1, 1, 1].$$

The root system associated with X_4 is $R_4 = A_4$ and thus we have $G_4 = SL_5$. The fundamental representation is the action of SL_5 on $V(\varpi_4) = \wedge^2 \mathbb{K}^5$ which is explicitly given by

$$A \cdot (e_i \wedge e_j) = (Ae_i \wedge Ae_j), \qquad 1 \le i < j \le 5.$$

The 10 basis vectors $e_i \wedge e_j$, where $1 \le i < j \le 5$, of $\wedge^2 \mathbb{K}^5$ are eigenvectors of the diagonal maximal torus $H_4 \cong \mathbb{T}^4$ of $G_4 = SL_5$. Moreover, P_4 is the second maximal parabolic subgroup and the highest weight vector is $v(4) = e_1 \wedge e_2$. We have

$$G_4 \cdot \mathbb{K} v(4) = \{v_1 \wedge v_2; \ v_1, v_2 \in \mathbb{K}^5\} = C(G(2, 5)),$$

where $C(G(2, 5))$ denotes the cone over the Grassmannian $G(2, 5)$. Now, let us see how the total coordinate space $\overline{X}_4$ embeds into $G_4 \cdot \mathbb{K}\varpi_4 = C(G(2, 5))$. The Cox ring of X_4 is generated by the canonical sections of the (-1)-curves. The latter ones are

$$E_i := \pi^{-1}(p_i), \quad 1 \le i \le 4, \qquad\qquad \pi^\sharp(\overline{p_i p_j}), \quad 1 \le i < j \le 4,$$

where $\pi \colon X_4 \to \mathbb{P}^2$ is the contraction, $\overline{p_i p_j} \subseteq \mathbb{P}^2$ the line through $p_i, p_j \in \mathbb{P}^2$ and $\pi^\sharp(\overline{p_i p_j})$ the strict transform. Then, in the Cox ring $\mathbb{K}[T_1, T_2, T_3]$ of $\mathbb{P}^2$, every point p_i defines a linear relation involving the canonical sections f_{ij} of the three lines $\overline{p_i p_j}$ through p_i, and we find a quadratic relation:

$$f_{12} - f_{13} + f_{14} = 0, \quad f_{12} - f_{23} + f_{24} = 0, \quad f_{13} - f_{23} + f_{34} = 0,$$

$$f_{14} - f_{24} + f_{34} = 0, \qquad f_{12}f_{34} - f_{13}f_{24} + f_{14}f_{23} = 0.$$

Write g_i for the canonical section of E_i and h_{ij} for the canonical section of $\pi^\sharp(\overline{p_i p_j})$. Then f_{ij} pulls back to $g_i g_j h_{ij}$. Pulling back the relations to the Cox ring $\mathcal{R}(X_4)$ and saturating with respect to $g_1 g_2 g_3 g_4$ gives the relations

$$g_2 h_{12} - g_3 h_{13} + g_4 h_{14} = 0, \quad g_1 h_{12} - g_3 h_{23} + g_4 h_{24} = 0,$$

$$g_1 h_{13} - g_2 h_{23} + g_4 h_{34} = 0, \quad g_1 h_{14} - g_2 h_{24} + g_3 h_{34} = 0,$$

$$h_{12} h_{34} - h_{13} h_{24} + h_{14} h_{23} = 0.$$

These are in fact the Plücker relations which generate the ideal of relations between the generators g_i and h_{ij} of the Cox ring of X_4; compare also Example 4.1.4.1. Thus, we see that in the case of X_4, the embedding $\mathbb{P}(X_4) \to G(2, 5)$ and the open embedding $\widehat{X}_4/\mathbb{K}^* \to \mathbb{P}(X_4)$ realizes X_4 as a quotient of the maximal torus action of the Grassmannian $G(2, 5)$. In fact, $\widehat{X}_4/\mathbb{K}^*$ is the unique big open subset of $G(2, 5)$ having a geometric quotient with projective quotient space.

Remark 5.2.4.4 Theorem 5.2.4.2 (iv) allows inductive approches. For example, Batyrev and Popov [33] showed this way that the singular loci of $\mathbb{P}(X_r)$ and $C(X_r)$ have codimension 7. Moreover, they gave a simple proof of the fact that the radical of the ideal of relations of the Cox ring $\mathcal{R}(X_r)$, for $4 \le r \le 6$, is generated by quadrics.

Remark 5.2.4.5 For their inductive proof of Theorem 5.2.4.2 Serganova and Skorobogatov [268] show that there is a projection $\varphi \colon V(\varpi_r) \to V(\varpi_{r-1})$ and an open set $U \subseteq C(G_r/P_r)$ such that the restriction of φ to U is a composition of a $\mathbb{K}^*$-torsor and the blow-up morphism of $V(\varpi_{r-1}) \setminus \{0\}$ at $C(G_{r-1}/P_{r-1}) \setminus \{0\}$. Then $\widehat{X}_r$ is recovered as the strict transform with respect to φ of a suitably moved $\widehat{X}_{r-1}$. For example, for $r = 5$, then we have $G_5 = \mathrm{Spin}_{10}$ and $V(\varpi_5)$ is the 16-dimensional half-spin module. It admits a decomposition into SL_5-modules

$$V(\varpi_5) = V(\varpi_4) \oplus \mathbb{K}^5 \oplus \mathbb{K},$$

where $\mathbb{K}^5$ and $\mathbb{K}$ are the dual to the tautological and the trivial modules, respectively. The 11-dimensional cone $C(G_5/P_5)$ over the isotropic Grassmannian contains an open subset U mapping onto 10-dimensional space $(\wedge^2 \mathbb{K}^5) \setminus \{0\}$ blown up at the 7-dimensional cone $C(G_4/P_4) \setminus \{0\}$ of decomposable 2-vectors. The characteristic space $\widehat{X}_5$ is contained in $C(G_5/P_5)$ as a locally closed 8-dimensional subvariety.

Remark 5.2.4.6 For $r = 8$, the module $V(\varpi_8)$ is the adjoint module of the exceptional group E_8. It contains an 8-dimensional zero-weight subspace, which is the Cartan subalgebra. Another 240 weights of T_8 in $V(\varpi_8)$ are roots of the Lie algebra E_8 and they are pairwise distinct. This explains why $M_8 = 240$, but $d_8 = 248$. In fact, the case $r = 8$ can be settled following the same ideas, but technical difficulties here are much harder; see [270] and also [311].

5.3 K3 surfaces

5.3.1 Abelian coverings

In this preparatory section, we consider certain finite coverings $\pi \colon X \to Y$, for example, cyclic ones, and provide basic statements relating the Cox rings of X and Y to each other. For example, Proposition 5.3.1.4 states that finite generation is preserved, provided that $\pi^*(\mathrm{Cl}(Y))$ is of finite index in $\mathrm{Cl}(X)$. We work over an algebraically closed field $\mathbb{K}$ of characteristic zero. First, we make precise which type of coverings we will treat.

Construction 5.3.1.1 Let Y be a normal variety and $D_1, \ldots, D_r$ be a linearly independent collection of Cartier divisors. Denote by $M^+ \subseteq \mathrm{CDiv}(Y)$ the semigroup generated by the divisors $D_1, \ldots, D_r$ and set

$$Y(D_1, \ldots, D_r) := \mathrm{Spec}_Y(\mathcal{A}), \qquad \mathcal{A} := \bigoplus_{D \in M^+} \mathcal{O}_Y(-D).$$

Then the inclusion $\mathcal{O}_Y \to \mathcal{A}$ defines a morphism $\alpha \colon Y(D_1, \ldots, D_r) \to Y$, which is a (split) vector bundle of rank r over Y. Similarly, with $n_1, \ldots, n_r \in \mathbb{Z}_{>0}$ and $E_i := n_i D_i$, denote by $N^+ \subseteq \mathrm{CDiv}(Y)$ the semigroup generated by $E_1, \ldots, E_r$. Setting

$$Y(E_1, \ldots, E_r) := \mathrm{Spec}_Y(\mathcal{B}), \qquad \mathcal{B} := \bigoplus_{E \in N^+} \mathcal{O}_Y(-E)$$

gives a further (split) vector bundle $\beta \colon Y(E_1, \ldots, E_r) \to Y$ of rank r over Y. The inclusion $\mathcal{B} \subseteq \mathcal{A}$ defines a morphism $\kappa \colon Y(D_1, \ldots, D_r) \to Y(E_1, \ldots, E_r)$. Now, let $\sigma \colon Y \to Y(E_1, \ldots, E_r)$ be a section such that all projections of σ to the factors $Y(E_i)$ are nontrivial. Then we define

$$X := \kappa^{-1}(\sigma(Y)) \subseteq Y(D_1, \ldots, D_r).$$

Restricting α gives a morphism $\pi \colon X \to Y$, which we call an *abelian covering* of Y. Note that $\pi \colon X \to Y$ is the quotient for the action of the abelian group $\Gamma := \mathbb{Z}/n_1\mathbb{Z} \oplus \ldots \oplus \mathbb{Z}/n_r\mathbb{Z}$ on X defined by the inclusion $\mathcal{B} \subseteq \mathcal{A}$ of sheaves of graded algebras.

Remark 5.3.1.2 Consider a smooth variety Y, Cartier divisors D and $E := nD$ on Y, and a section $\sigma \colon Y \to Y(E)$ with a nontrivial reduced divisor

$$B := \mathrm{div}(\sigma) \in \mathrm{WDiv}(Y).$$

Then Construction 5.3.1.1 gives an *n-cyclic covering* $\pi \colon X \to Y$ *with branch divisor* B, in the sense of [29, Sec. I.17]. Note that $\pi^{-1}(B)$ is precisely the fixed point set of $\Gamma = \mathbb{Z}/n\mathbb{Z}$ and Γ acts freely outside $\pi^{-1}(B)$.

We need a pullback for Weil divisors under an abelian covering $\pi \colon X \to Y$ of normal varieties. Given $D \in \mathrm{WDiv}(Y)$, consider the restriction D' of D to the set $Y' \subseteq Y$ of smooth points and take the usual pullback $\pi^*(D')$ on $\pi^{-1}(Y')$. Because π is finite, the complement $X \setminus \pi^{-1}(Y')$ is of codimension at least 2 in X and hence $\pi^*(D')$ uniquely extends to a Weil divisor $\pi^*(D)$ on X.

Proposition 5.3.1.3 *Let* $\pi : X \to Y$ *be an abelian covering as in Construction 5.3.1.1, assume that* X *is normal, and let* $K \subseteq \mathrm{WDiv}(Y)$ *be a subgroup containing* $D_1, \ldots, D_r$ *of 5.3.1.1. Set*

$$\mathcal{S}_Y := \bigoplus_{D \in K} \mathcal{O}_Y(D), \qquad \mathcal{S}_X := \bigoplus_{D \in K} \mathcal{O}_X(\pi^*D).$$

Then setting $\deg(T_i) := D_i$ *turns* $\mathcal{S}_Y[T_1, \ldots, T_r]$ *into a* K-*graded sheaf of* $\mathcal{O}_Y$-*algebras and there is a* K-*graded isomorphism of sheaves*

$$\pi_*\mathcal{S}_X \cong \mathcal{S}_Y[T_1, \ldots, T_r]/\langle T_1^{n_1} - g_1, \ldots, T_r^{n_r} - g_r \rangle,$$

where $g_i \in \Gamma(Y, \mathcal{O}(E_i))$ *are sections such that the branch divisor* B *of the covering* $\pi : X \to Y$ *is given as*

$$B = \mathrm{div}(g_1) + \cdots + \mathrm{div}(g_r).$$

Proof Note that for any open set $V \subseteq Y$ and its intersection $V' := V \cap Y'$ with the set $Y' \subseteq Y$ of smooth points, the sections of $\mathcal{S}_Y$ over V and V' coincide and also the sections of $\pi_*\mathcal{S}_X$ over V and V' coincide. Hence, we may assume that $K \subseteq \mathrm{WDiv}(Y)$ consists of Cartier divisors.

A first step is to express the direct image $\pi_*\mathcal{S}_X$ in terms of $\pi_*\mathcal{O}_X$ and data living on Y. Using the projection formula, we obtain

$$\pi_*\mathcal{S}_X = \pi_* \bigoplus_{D \in K} \mathcal{O}_X(\pi^*D) \cong \bigoplus_{D \in K} \mathcal{O}_Y(D) \otimes_{\mathcal{O}_Y} \pi_*\mathcal{O}_X \cong \mathcal{S}_Y \otimes_{\mathcal{O}_Y} \pi_*\mathcal{O}_X.$$

(5.3.1)

Now we have to investigate $\pi_*\mathcal{O}_X$. Denote by $q : \widetilde{Y} \to Y$ the torsor associated with $\mathcal{S}_Y$, that is, we have $\widetilde{Y} = \mathrm{Spec}_Y(\mathcal{S}_Y)$. Moreover, in Construction 5.3.1.1 we constructed the rank r vector bundles

$$\alpha : Y(D_1, \ldots, D_r) \to Y, \qquad \beta : Y(E_1, \ldots, E_r) \to Y.$$

Using the pullback divisors $q^*(D_i)$ and $q^*(E_i)$, we obtain the respective pullback bundles

$$\widetilde{\alpha} : \widetilde{Y}(q^*D_1, \ldots, q^*D_r) \to \widetilde{Y}, \qquad \widetilde{\beta} : \widetilde{Y}(q^*E_1, \ldots, q^*E_r) \to \widetilde{Y}.$$

Set for short $Y(\mathbf{D}) := Y(D_1, \ldots, D_r)$ and $\widetilde{Y}(q^*\mathbf{D}) := \widetilde{Y}(q^*D_1, \ldots, q^*D_r)$. Similarly, define $Y(\mathbf{E})$ and $\widetilde{Y}(q^*\mathbf{E})$. Then we have a commutative diagram

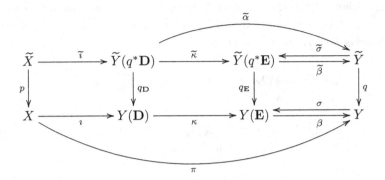

where p, $q_{\mathbf{D}}$, and $q_{\mathbf{E}}$ are the canonical morphisms; we set $\widetilde{X} := q_{\mathbf{D}}^{-1}(X)$ and $\widetilde{\sigma} := q^*\sigma$ is the pullback section.

Recall that $q\colon \widetilde{Y} \to Y$ is the quotient for the free action of the torus $H :=$ $\mathrm{Spec}(\mathbb{K}[K])$ defined by the grading of $q_*\mathcal{O}_{\widetilde{Y}} = \mathcal{S}_Y$. Thus, $\widetilde{Y}(\mathbf{D})$ and $\widetilde{X}$ inherit free H-actions having $q_{\mathbf{D}}$ and p as their respective quotients. Moreover, let $\widetilde{\mathcal{I}}$ denote the ideal sheaf of $\widetilde{X}$ in $\widetilde{Y}(q^*\mathbf{D})$. Then $\widetilde{\mathcal{I}}$ is homogeneous, and we have

$$\pi_*\mathcal{O}_X \cong \left(\pi_* p_* \mathcal{O}_{\widetilde{X}}\right)_0 \cong \left(\pi_* p_* \widetilde{\imath}^*\left(\mathcal{O}_{\widetilde{Y}(q^*\mathbf{D})}/\widetilde{\mathcal{I}}\right)\right)_0 = \left(q_* \widetilde{\alpha}_*\left(\mathcal{O}_{\widetilde{Y}(q^*\mathbf{D})}/\widetilde{\mathcal{I}}\right)\right)_0. \tag{5.3.2}$$

To proceed, we need a suitable trivialization of the bundle $\widetilde{\alpha}\colon \widetilde{Y}(q^*\mathbf{D}) \to \widetilde{Y}$. For this, consider an open affine subset $V \subseteq Y$ such that on V we have $D_i = \mathrm{div}(h_{i,V})$ for $1 \le i \le r$. This gives us sections

$$\eta_{i,V} := h_{i,V}^{-1} \in \Gamma(V, \mathcal{O}_Y(D_i)) \subseteq \Gamma(q^{-1}(V), \mathcal{O}_{\widetilde{Y}})_{D_i},$$

$$q_{\mathbf{D}}^*(h_{i,V}) \in q_{\mathbf{D}}^*\left(\Gamma(\alpha^{-1}(V), \mathcal{O}_{Y(\mathbf{D})})\right) \subseteq \Gamma(\widetilde{\alpha}^{-1}(q^{-1}(V)), \mathcal{O}_{\widetilde{Y}(q^*\mathbf{D})})_0$$

Given another open affine subset $W \subseteq Y$ such that on W we have $D_i = \mathrm{div}(h_{i,W})$ for $1 \le i \le r$, we obtain over $V \cap W$ for the corresponding sections:

$$\frac{\widetilde{\alpha}^*(\eta_{i,W})}{\widetilde{\alpha}^*(\eta_{i,V})} = \widetilde{\alpha}^* q^*\left(\frac{\eta_{i,W}}{\eta_{i,V}}\right) = q_{\mathbf{D}}^* \alpha^*\left(\frac{h_{i,V}}{h_{i,W}}\right) = \frac{q_{\mathbf{D}}^*(h_{i,V})}{q_{\mathbf{D}}^*(h_{i,W})}.$$

Covering Y with V's as above, we obtain that the functions $\widetilde{\alpha}^*(\eta_{i,V}) \cdot q_{\mathbf{D}}^*(h_{i,V})$ living on $\widetilde{\alpha}^{-1}(q^{-1}(V))$ glue together to a global regular function f_i of degree D_i on $\widetilde{Y}(q^*D_i)$ generating $\mathcal{O}_{\widetilde{Y}}(q^*D_i)$ over $\mathcal{O}_{\widetilde{Y}}$. Thus, the f_i define a trivialization

$$\begin{array}{ccc}
\widetilde{Y} \times \mathbb{K}^r & \xrightarrow{(\widetilde{y},z) \mapsto (\widetilde{y},z^n)} & \widetilde{Y} \times \mathbb{K}^r
\end{array}$$

where we write z for $(z_1, \ldots, z_r)$ and z^n for $(z_1^{n_1}, \ldots, z_r^{n_r})$ etc. Because $\widetilde{\sigma} = q^*\sigma$ is H-equivariant, each component g_i of g is homogeneous with $\deg(g_i) = E_i$. Note that the divisors $\mathrm{div}(g_i)$ describe the branch divisor as claimed.

Denote by $\widetilde{\mathcal{J}} \subseteq \mathcal{O}_{\widetilde{Y}}[T_1, \ldots, T_r]$ the ideal sheaf of the image of $\widetilde{X}$ in $\widetilde{Y} \times \mathbb{K}^r$. Then, using $\widetilde{X} = \widetilde{\kappa}^{-1}(\widetilde{\sigma}(\widetilde{Y}))$, we obtain

$$\widetilde{\mathcal{I}} = \langle f_1^{n_1} - \widetilde{\alpha}^* g_1, \ldots, f_r^{n_r} - \widetilde{\alpha}^* g_r \rangle, \qquad \widetilde{\mathcal{J}} = \langle T_1^{n_1} - g_1, \ldots, T_r^{n_r} - g_r \rangle.$$

Thus, using the isomorphism $q_* \widetilde{\alpha}_* \mathcal{O}_{\widetilde{Y}(q^*\mathbf{D})} \cong \mathcal{S}_Y[T_1, \ldots, T_r]$ established by the above commutative diagram, we may continue (5.3.2) as

$$\pi_* \mathcal{O}_X \cong \big(\mathcal{S}_Y[T_1, \ldots, T_r]/\widetilde{\mathcal{J}}\big)_0 \cong \mathcal{S}_Y[T_1, \ldots, T_r]_0/\widetilde{\mathcal{J}}_0 \qquad (5.3.3)$$

The homogeneous ideal sheaf $\widetilde{\mathcal{I}}$ is locally, over Y, generated in degree zero in the sense that we have $\widetilde{\mathcal{I}} = \widetilde{\alpha}^* \mathcal{O}_{\widetilde{Y}} \cdot \widetilde{\mathcal{I}}_0$. The same holds for $\widetilde{\mathcal{J}}$, and we obtain

$$\pi_* \mathcal{S}_X \cong \mathcal{S}_Y \otimes_{\mathcal{O}_Y} \pi_* \mathcal{O}_X \cong \mathcal{S}_Y \otimes_{\mathcal{O}_Y} \mathcal{S}_Y[T_1, \ldots, T_r]_0/\widetilde{\mathcal{J}}_0 \cong \mathcal{S}_Y[T_1, \ldots, T_r]/\widetilde{\mathcal{J}}.$$

$\square$

Proposition 5.3.1.4 *Let $\pi: X \to Y$ be an abelian covering of normal varieties with finitely generated free divisor class groups such that $\pi^*(\mathrm{Cl}(Y))$ is of finite index in $\mathrm{Cl}(X)$. Then the following statements are equivalent.*

(i) *The Cox ring $\mathcal{R}(X)$ is a finitely generated $\mathbb{K}$-algebra.*
(ii) *The Cox ring $\mathcal{R}(Y)$ is a finitely generated $\mathbb{K}$-algebra.*

Proof Let $M \subseteq \mathrm{WDiv}(Y)$ and $K \subseteq \mathrm{WDiv}(X)$ be subgroups mapping isomorphically to the respective divisor class groups $\mathrm{Cl}(Y)$ and $\mathrm{Cl}(X)$. Then the Cox rings are given as

$$\mathcal{R}(Y) = \bigoplus_{E \in M} \Gamma(Y, \mathcal{O}_Y(E)), \qquad \mathcal{R}(X) = \bigoplus_{D \in K} \Gamma(X, \mathcal{O}_X(D)).$$

Because $\mathrm{Cl}(Y)$ is free and $\pi: X \to Y$ is the quotient for a finite group action, the pullback $\pi^*: \mathrm{Cl}(Y) \to \mathrm{Cl}(X)$ is injective, see [142, Ex. 1.7.6]. Consequently, there are a unique subgroup $L \subseteq K$ and an isomorphism $\pi^*(M) \to L$ inducing the identity on $\pi^*(\mathrm{Cl}(Y))$. By our assumption, L is of finite index in K. Moreover, we have canonical identifications

$$\mathcal{R}(Y) \subseteq \bigoplus_{E \in M} \Gamma(X, \mathcal{O}_X(\pi^*(E))) = S := \bigoplus_{D \in L} \Gamma(X, \mathcal{O}_X(D)) \subseteq \mathcal{R}(X).$$

Suppose that $\mathcal{R}(X)$ is finitely generated over $\mathbb{K}$. Then also the Veronese subalgebra $S \subseteq \mathcal{R}(X)$ is finitely generated over $\mathbb{K}$. Moreover, by Proposition 5.3.1.3, the algebra S is a finite module over $\mathcal{R}(Y)$. Thus, the tower $\mathbb{K} \subseteq \mathcal{R}(Y) \subseteq S$ fullfills the assumptions of the Artin–Tate lemma [22, Prop. 7.8], and we obtain that $\mathcal{R}(Y)$ is a finitely generated $\mathbb{K}$-algebra.

Now let $\mathcal{R}(Y)$ be finitely generated over $\mathbb{K}$. Then Proposition 5.3.1.3 tells us that S is finitely generated over $\mathbb{K}$. Thus Proposition 1.1.2.4 shows that $\mathcal{R}(X)$ is finitely generated over $\mathbb{K}$. $\qquad\square$

5.3.2 Picard numbers 1 and 2

Recall that a K3 surface is a smooth compact complex surface X with $K_X \sim 0$ and $q(X) = 0$. Here, we study the Cox ring of a K3 surface X in the cases $\varrho(X) = 1, 2$. The original references for this section are [9, 240].

Theorem 5.3.2.1 *Let X be a K3 surface with $\varrho(X) = 1$ and let C be a smooth curve whose class generates the ample cone of X. Let $\varphi \colon X \to \mathbb{P}^n$ be the morphism defined by the complete linear system $|C|$. Then precisely one of the following statements holds.*

(i) *The Cox ring $\mathcal{R}(X)$ is isomorphic to the classical homogeneous coordinate ring of $X \subseteq \mathbb{P}^n$.*

(ii) *We have $n = 2$ and $\varphi \colon X \to \mathbb{P}^2$ is a double covering branched over a sextic curve $V(f) \subseteq \mathbb{P}^2$ and the Cox ring of X is the $\mathbb{Z}$-graded ring*

$$\mathcal{R}(X) \cong \mathbb{K}[T_1, \ldots, T_4]/\langle T_4^2 - f(T_1, T_2, T_3)\rangle,$$

$$\deg(T_i) = 1, \ i = 1, 2, 3, \quad \deg(T_4) = 3.$$

In the proof, we work with the section ring of a single divisor. Recall that if X is a smooth projective variety and D is a divisor of X then we can define the graded algebra

$$S^+(X, D) := \bigoplus_{n \in \mathbb{N}} H^0(X, nD).$$

An interesting fact about K3 surfaces is that $S^+(X, D)$ is finitely generated for any divisor D of X, even if X is not a Mori dream surface [262, Thm. 6.1]. Moreover, one has the following.

Proposition 5.3.2.2 *Let X be a K3 surface and let C be a smooth irreducible curve of genus g on X. Then the algebra $S^+(X, C)$ is generated in degrees*

(i) *1 if $g \neq 2$ and C is not hyperelliptic*

(ii) *1, 2 if $g > 2$ and C is hyperelliptic*

(iii) *1, 3 if $g = 2$*

Proof If $g = 0$ holds, then $S^+(X, C) = \mathbb{C}[s]$ with $s \in H^0(X, C)$, because D is irreducible with negative self-intersection. Thus $S^+(X, C)$ is generated in degree one.

For a nonrational curve C, the canonical algebra $S^+(C, K_C)$is generated in degree 1 if C is not hyperelliptic or $g = 1$, in degrees 1 and 3 if $g = 2$ and in degrees 1 and 2 if C is hyperelliptic and $g \geq 3$, see [7, p. 117]. By the genus formula we have $\mathcal{O}_X(C)|_C \cong K_C$. Thus, we obtain the exact sequence

$$0 \longrightarrow H^0(X, (n-1)C) \longrightarrow H^0(X, nC) \longrightarrow H^0(C, nK_C) \longrightarrow 0,$$

where the last zero is due to the Kawamata–Viehweg vanishing theorem. This gives the assertion. □

Proof of Theorem 5.3.2.1 As a consequence of Proposition 5.3.2.2 the Cox ring of X is generated in degree 1 if and only if the curve C is not hyperelliptic. By [262, Thm. 5.2] and the hypothesis this holds exactly when $C^2 > 2$. In this case $\varphi \colon X \to \mathbb{P}^n$ is an embedding. By Bertini's theorem [160, Thm. II.8.18] a general hyperplane section of X is a prime divisor; hence, after possibly applying a linear change of coordinates, we can assume that the $n + 1$ generators of $\mathcal{R}(X)$ are pairwise nonassociated primes; thus we obtain statement (i).

If $C^2 = 2$, the map defined by $|C|$ is a double covering $\pi \colon X \to \mathbb{P}^2$ branched along a smooth degree 6 curve B such that C is the preimage of a line via π. Because $\pi^* \colon \mathrm{Cl}(\mathbb{P}^2) \to \mathrm{Cl}(X)$ is an isomorphism, Proposition 5.3.1.3, yields statement (ii). Observe that the degrees of the T_i are $[1, 1, 1, 3]$. The generator of the ideal of relations of the Cox ring of X can be thought as the equation of X in the weighted projective space $\mathbb{P}(1, 1, 1, 3)$. □

We turn to K3 surfaces X of Picard number $\varrho(X) = 2$. So, we have $\mathrm{Cl}(X) \cong \mathbb{Z}^2$. According to Theorem 5.1.5.3 such an X is a Mori dream surface if and only if there exists a class $w \in \mathrm{Cl}(X)$ with $w^2 \in \{0, -2\}$. Note that this is a purely arithmetic condition on the intersection form.

Remark 5.3.2.3 By [222, Cor. 2.9 (i)] any even lattice of rank 2 with signature $(1, 1)$ is the Picard lattice of an algebraic K3 surface. For example,

$$\begin{bmatrix} -4 & 0 \\ 0 & 12 \end{bmatrix}$$

is the intersection matrix of a K3 surface X with $\mathrm{Cl}(X) \cong \mathbb{Z}^2$. Because $w^2 \neq 0, -2$ for any class $w \in \mathrm{Cl}(X)$ holds, then X is not a Mori dream surface.

Proposition 5.3.2.4 *Let X be a K3 surface with $\mathrm{Cl}(X) \cong \mathbb{Z}w_1 \oplus \mathbb{Z}w_2$. Then the following are equivalent.*

(i) *The intersection matrix of w_1, w_2 is one of the following, where $k \geq 2$ in case I and $k \geq 1$ in the remaining cases:*

$$I: \begin{bmatrix} 0 & k \\ k & 0 \end{bmatrix} \qquad II: \begin{bmatrix} -2 & k \\ k & 0 \end{bmatrix} \qquad III: \begin{bmatrix} -2 & k \\ k & -2 \end{bmatrix}$$

(ii) *The effective cone $\mathrm{Eff}(X)$ is polyhedral and, after possibly substituting w_i with $-w_i$, it is generated as a cone by w_1, w_2.*

Proof We show (i)$\Rightarrow$(ii). By the Riemann–Roch formula and $w_i^2 \in \{-2, 0\}$, $\mathrm{Eff}(X)$ is polyhedral and either w_i or $-w_i$ is an effective class. So from now on we can assume w_1, w_2 to be effective. Suppose that $\mathrm{cone}(w_1, w_2) \subsetneq \mathrm{Eff}(X)$ holds. Then we may assume that w_1 does not lie on the boundary of $\mathrm{Eff}(X)$. Thus, one of the generatorsof $\mathrm{Eff}(X)$ is of the form $w = aw_1 - bw_2$ for some $a, b \in \mathbb{N}$, where $a > 0$. Hence

$$w^2 = a^2 w_1^2 + b^2 w_2^2 - 2ab \, w_1 \cdot w_2 \in \{0, -2\}.$$

This can be realized only for $b = 0$, because we assumed $w_i^2 \in \{0, -2\}$ and $w_1 \cdot w_2 \geq 2$. Thus, $w = aw_1$ holds and, consequently, w_1 spans an extremal ray of $\mathrm{Eff}(X)$, a contradiction.

We prove (ii)$\Rightarrow$(i). Because the signature of the quadratic form on $\mathrm{Cl}(X)$ is $(1, 1)$, the determinant of the intersection matrix is negative: $w_1^2 w_2^2 - (w_1 \cdot w_2)^2 < 0$ and in particular $k := w_1 \cdot w_2 \geq 1$ holds. Moreover if in case I we take $k = 1$, then the intersection matrix of the basis $\{w_1 - w_2, w_2\}$ is already considered in case II with $k = 1$. $\qquad\square$

We investigate the possible degrees of generators and relations for $\mathcal{R}(X)$. An explicit computation of $\mathcal{R}(X)$ for the cases $k := w_1 \cdot w_2 = 1, 2$ is given in Subsection 5.3.3.

Theorem 5.3.2.5 *Let X be a K3 surface such that $\mathrm{Eff}(X) = \mathrm{cone}(w_1, w_2)$ with a $\mathbb{Z}$-basis w_1, w_2 of $\mathrm{Cl}(X)$. Then we have three possibilities for the intersection matrix with respect to w_1, w_2:*

$$I: \begin{bmatrix} 0 & k \\ k & 0 \end{bmatrix} \qquad II: \begin{bmatrix} -2 & k \\ k & 0 \end{bmatrix} \qquad III: \begin{bmatrix} -2 & k \\ k & -2 \end{bmatrix}$$

where $k \geq 2$ in case I and $k \geq 1$ in the remaining cases. According to the cases, the following statements hold:

(i) *In case I the Cox ring is generated in degrees w_1, w_2, $w_1 + w_2$.*

(ii) *In case II the Cox ring has generators in degrees w_1, $aw_1 + w_2$, where $0 \leq a \leq \lfloor k/2 \rfloor$. Moreover, if $k > 1$ holds and k is odd, then the Cox ring has also generators in degree $kw_1 + 2w_2$.*

(iii) *In case* III *the Cox ring has generators* $aw_1 + w_2$ *and* $w_1 + aw_2$, *where* $0 \leq a \leq \lfloor k/2 \rfloor$. *Moreover, if* $k > 1$ *holds and* k *is odd, then the Cox ring* $\mathcal{R}(X)$ *has also generators in degree* $kw_1 + 2w_2$ *and* $2w_1 + kw_2$.

The following figure indicates degrees of generators of the Cox ring, effective and semiample cones according to the cases I, II, III when $k = 3$.

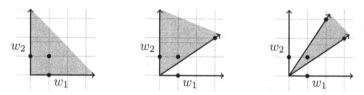

Lemma 5.3.2.6 *Let* X *be a smooth surface and* D, E_1, E_2 *divisors on* X *and* $f_i \in H^0(X, E_i)$ *with* $\mathrm{div}_{E_1}(f_1) \cap \mathrm{div}_{E_2}(f_2) = \emptyset$. *If* $h^1(X, D - E_1 - E_2) = 0$ *holds, then we have a surjective map*

$$H^0(X, D - E_1) \oplus H^0(X, D - E_2) \to H^0(X, D) \qquad (g_1, g_2) \mapsto g_1 f_1 + g_2 f_2.$$

Proof First note that due to the assumptions, the maps $\iota\colon h \mapsto (hf_2, -hf_1)$ and $\varphi\colon (h_1, h_2) \mapsto (h_1 f_1 + h_2 f_2)$ give rise to an exact sequence of sheaves:

$$0 \longrightarrow \mathcal{O}_X(-E_1 - E_2) \overset{\iota}{\longrightarrow} \mathcal{O}_X(-E_1) \oplus \mathcal{O}_X(-E_2) \overset{\varphi}{\longrightarrow} \mathcal{O}_X \longrightarrow 0.$$

Tensoring this sequence with $\mathcal{O}_X(D)$ and passing to the associated long exact sequence in cohomology gives the assertion. $\qquad\square$

Proof of Theorem 5.3.2.5 The first part of the statement is a direct consequence of Proposition 5.3.2.4. We provide here a proof of (i); the proofs of (ii) and (iii) are similar and can be found in [9].

By the Riemann–Roch formula either w_i or $-w_i$ is effective. Hence we can assume that both w_1, w_2 are classes of effective curves C_1, C_2. Observe that both the C_i are irreducible, because otherwise $C_i = C' + C''$ and $C_i^2 = 0$ would imply $C'^2 < 0$, a contradiction because by hypothesis X does not contain negative curves.

By the genus formula the linear system $|C_i|$ are elliptic pencils as in Definition 5.1.3.4. Let $\{f_{i1}, f_{i2}\}$ be a basis of $H^0(X, C_i)$, for $i = 1, 2$.

Now we prove the statement. Consider a divisor $D = a C_1 + b C_2$. If $a \geq 3$ and $b \geq 1$ or $(a, b) = (2, 1)$ holds, then we have $h^1(X, D - 2C_1) = 0$ by the Kawamata–Viehweg vanishing theorem. Thus, Lemma 5.3.2.6 provides a surjective map

$$\varphi\colon H^0(X, D - C_1) \oplus H^0(X, D - C_2) \to H^0(X, D),$$

$$(g_1, g_2) \mapsto g_1 f_{11} + g_2 f_{12}.$$

If $b = 0$ holds, then $|D| = |aC_1|$ is composed with the pencil $|C_1|$ and again φ is surjective. Iterating this procedure and possibly exchanging the roles of a and b we see that the elements of $H^0(X, D)$ are polynomials in $f_{11}, f_{12}, f_{21}, f_{22}$ and elements of $H^0(X, n(C_1 + C_2))$, where $n \le 2$. Because $(C_1 + C_2) \cdot C_i = k \ge 3$, then the general element of $|C_1 + C_2|$ is a nonhyperelliptic genus $g > 3$ curve, by [262, Thm. 5.2]. Hence the algebra $\mathcal{R}(X, w_1 + w_2)$ is generated in degree one by Proposition 5.3.2.2. Thus, we obtained that $\mathcal{R}(X)$ is generated in the degrees w_1, w_2 and $w_1 + w_2$. $\qquad\square$

Example 5.3.2.7 Two classical examples for cases II and III of Theorem 5.3.2.5 are given by the following intersection matrices:

$$\begin{bmatrix} -2 & 3 \\ 3 & 0 \end{bmatrix} \quad \begin{bmatrix} -2 & 4 \\ 4 & -2 \end{bmatrix}.$$

In both cases the class $w := w_1 + w_2$ is ample with $w^2 = 4$. The projective models for such surfaces are respectively: a smooth quartic surface of $\mathbb{P}^3$ that contains a line (corresponding to w_1) and a smooth quartic surface of $\mathbb{P}^3$ that admits a hyperplane section that is union of two irreducible conics (corresponding to w_1 and w_2). A presentation for the Cox ring of the first type of surface is given in [240, Thm. 3.1] and [11, Thm. 4.1], while for the second type of surface see [240, Prop. 3.2].

Remark 5.3.2.8 The assumptions $w_i^2 \in \{0, -2\}$ made in Theorem 5.3.2.5 imply that the primitive generators w_1, w_2 of the class group span $\mathrm{Cl}(X)$. This is not always the case as, for example, when the intersection form of X with respect to the basis $\{w_1, w_2\}$ is:

$$\begin{bmatrix} -4 & 6 \\ 6 & 0 \end{bmatrix}.$$

Indeed in this case the classes $3w_1 + w_2$, w_2 have zero self-intersection and, after possibly a change of sign, they span the effective cone. Their intersection matrix is of type I, but $\mathbb{Z}(3w_1 + w_2) \oplus \mathbb{Z}w_2$ has index two in $\mathrm{Cl}(X)$. The effective cone, which coincides with the semiample one, is the shaded region in the following figure.

Observe that the Cox ring has generators in degrees w_1, $w_1 + w_2$, $2w_1 + w_2$ and $3w_1 + w_2$ because these degrees form a Hilbert basis of the effective cone.

The techniques of the proof of Theorem 5.3.2.5 can be used also to treat such cases; see, for example, [9].

We now consider the ideal of relations of the Cox ring for K3 surfaces as in Case I of Theorem 5.3.2.5. Cases II and III are still open.

Theorem 5.3.2.9 *Let X be a K3 surface as in Case* I *of Theorem 5.3.2.5, that is, $\mathrm{Cl}(X) \cong \mathbb{Z}w_1 \oplus \mathbb{Z}w_2$ and intersection form given by $w_1^2 = w_2^2 = 0$ and $w_1 \cdot w_2 = k$.*

(i) *For $k = 3$ the Cox ring is isomorphic to $\mathbb{C}[T_1, \ldots, T_5]/\langle f \rangle$, where f has degree $3w_1 + 3w_2$ and the degree matrix of the generators is*

$$\begin{bmatrix} 1 & 1 & 1 & 0 & 0 \\ 0 & 0 & 1 & 1 & 1 \end{bmatrix}.$$

(ii) *For $k \geq 4$ the ideal of relations of the Cox ring is generated in degree $2w_1 + 2w_2$ and we have*

$$\dim(\mathcal{I}(X)_{2w_1+2w_3}) = \frac{k(k-3)}{2}.$$

Moreover with the ample class $w := w_1 + w_2$, the Cox ring $\mathcal{R}(X)$ admits a minimal resolution

$$0 \longrightarrow R(-kw) \longrightarrow R((-k+2)w)^{\beta_{k-3}} \longrightarrow \cdots$$

$$\cdots \longrightarrow R(-2w)^{\beta_1} \longrightarrow R \longrightarrow \mathcal{R}(X) \longrightarrow 0,$$

where $\beta_i = i\binom{k-1}{i+1} - \binom{k-2}{i-1}$ and R is the polynomial ring $\mathbb{K}[T_1, \ldots, T_{k+2}]$ with generator degrees $1, 1, 1, 1, 2, \ldots, 2$.

Proof By the proof of Theorem 5.3.2.5 there are elliptic curves C_1 and C_2 of X whose classes are w_1 and w_2, respectively. Let $\{f_{i1}, f_{i2}\}$ be a basis of $H^0(X, C_i)$, for $i = 1, 2$ and let $D = a\,C_1 + b\,C_2$. If $a \geq 4$ and $b \geq 2$ or $(a, b) = (3, 2)$ holds, then we have

$$h^1(X, D - 2C_1 - C_2) = h^1(X, D - 3C_1 - C_2) = 0,$$

by the Kawamata–Viehweg vanishing theorem. Thus, taking $f_1 = f_{11}$ and $f_2 = f_{12}$ in Lemma 5.3.2.6 and using Proposition 5.2.3.6, we see that $\mathcal{R}(X)$ has no relations in degree $w = [D]$. If $b = 1$ holds, then

$$H^0(X, (n-1)C_1) \oplus H^0(X, (n-1)C_1) \longrightarrow H^0(X, n\,C_1)$$

$$(g_1, g_2) \mapsto g_1 f_{11} + g_2 f_{12}$$

is surjective for $n = a$, $a - 1$, because $\mathcal{R}(X)$ is generated in degrees w_1, w_2 and $w_1 + w_2$. Thus $\mathcal{R}(X)$ has no relations in degree w for $b = 1$, by Proposition 5.2.3.6 and Lemma 5.3.2.6. Eventually, there are no relations of degree w for $b = 0$. In fact, then $\mathcal{R}(X)_w$ is generated by f_{11} and f_{12} and hence any such relation defines a relation among f_{11} and f_{12}, which contradicts the fact that f_{11}, f_{12} define a surjection $X \to \mathbb{P}^1$.

Exchanging the roles of w_1 and w_2 in this consideration, we obtain that essential relations can occur only in degrees $2u$ and $3u$, where $u = w_1 + w_2$.

In the case $k = 3$, the statements proven so far give that any minimal system of generators has five members and their degrees are w_1, w_1, w_2, w_2, and u. Hence there must be exactly one relation in $\mathcal{R}(X)$. The degree of this relation minus the sum of the degrees of the generators gives the canonical class, by Proposition 3.3.3.2, and hence vanishes. Thus our relation must have degree $3u$.

Finally, let $k \geq 4$. As observed before, $\mathcal{R}(X)_{nu}$ is generated by $\mathcal{R}(X)_u$. Hence, any relation in degree nu is also a relation of

$$\bigoplus_{n \in \mathbb{N}} \mathcal{R}(X)_{nu}.$$

Because $u \cdot w_i > 3$ holds and X does not contain smooth rational curves, [262, Thm. 7.2] tells us that the ideal of relations of this algebra is generated in degree 2. Thus, there are only essential relations of degree $2u$ in $\mathcal{R}(X)$.

To determine the dimension of $\mathcal{I}(X)_{2u}$ for a minimal ideal of relations $\mathcal{I}(X)$, note that we have the four generators f_{ij}, where $1 \leq i, j \leq 2$, of degree w_i, where $1 \leq i \leq 2$, and $k - 2$ generators of degree $u = w_1 + w_2$. Using the Riemann–Roch formula we obtain that $\mathcal{R}(X)_{2u}$ is of dimension $4k + 2$. Thus, denoting by $\mathbb{C}[T]$ the polynomial ring in the above generators and by $V \subseteq \mathcal{R}(X)_u$ the vector space spanned by the $k - 2$ generators of degree u, we obtain

$$\dim(\mathcal{I}(X)_{2u}) = \dim(\mathbb{C}[T]_{2u}) - \dim(\mathcal{R}(X)_{2u})$$
$$= \dim(\mathrm{Sym}^2 V) + 4(k - 2) + 9$$
$$= \frac{k(k - 3)}{2}.$$

For the last part of the proof see [240, Thm. 4.3]. $\qquad\square$

Note that for $k = 3, 4$ the Cox ring of X is a complete intersection, while for $k \geq 5$ this no longer holds.

Example 5.3.2.10 Let X be a K3 surface with $\mathrm{Cl}(X) \cong \mathbb{Z}w_1 \oplus \mathbb{Z}w_2$ and intersection form given by the matrix

$$\begin{bmatrix} -2 & 3 \\ 3 & -2 \end{bmatrix},$$

that is $w_1^2 = w_2^2 = -2$ and $w_1 \cdot w_2 = 3$. Then the Cox ring $\mathcal{R}(X)$ has generators

$$f_{1,0}, \ f_{0,1}, \ f_{1,1}, \ g_{1,1}, \ f_{2,3}, \ f_{3,2}$$

in the corresponding degrees by Theorem 5.3.2.5. Let $u = w_1 + w_2$. A monomial basis of $\mathrm{Sym}^3 \, \mathcal{R}(X)_u$, plus $f_{2,3} f_{0,1}$ and $f_{3,2} f_{1,0}$, give 12 linearly dependent elements of $\mathcal{R}(X)_{3u}$ because this space has dimension 11 by the Riemann–Roch formula. This means that $\mathcal{R}(X)$ has a relation in degree $3u$.

Similarly, a monomial basis of $\mathrm{Sym}^5 \, \mathcal{R}(X)_u$, plus $f_{2,3} f_{3,2}$ and the product of $f_{2,3} f_{1,0}$ for a monomial basis of $\mathrm{Sym}^2 \, \mathcal{R}(X)_u$ give 28 monomials. These are linearly dependent because the dimension of $\mathcal{R}(X)_{5u}$ is 27 by the Riemann–Roch formula. This means that $\mathcal{R}(X)$ has a relation in degree $5u$.

We now give a geometric interpretation for generators and relations. Because $w_1 + w_2$ is ample with $(w_1 + w_2)^2 = 2$ then the map

$$\pi : X \to \mathbb{P}^2$$

associated with $w_1 + w_2$ is a double cover branched along a smooth plane sextic, see [262, Sec. 5] Observe that $f_{1,0} f_{0,1} = \pi^*(s)$ and $f_{2,3} f_{3,2} = \pi^*(t)$, where $s = 0$ is a line and $t = 0$ a quintic in $\mathbb{P}^2$. The second equality gives a relation in degree $5w_1 + 5w_2$.

5.3.3 Nonsymplectic involutions

We consider K3 surfaces coming with a nonsymplectic involution and provide the necessary statements needed for the Cox ring computations in the subsequent section. We work over the field $\mathbb{C}$ of complex numbers. The original reference for the present section is [9].

A *nonsymplectic involution* on X is an automorphism $\sigma : X \to X$ of order 2 such that $\sigma^* \omega_X = -\omega_X$ holds, where ω_X is a nonzero holomorphic 2-form of X. Note that the fixed lattice sits inside the divisor class group:

$$L^\sigma := \{ u \in H^2(X, \mathbb{Z}); \ \sigma^*(u) = u \} \subseteq \mathrm{Cl}(X) = H^2(X, \mathbb{Z}) \cap \omega_X^\perp.$$

It is known, see [230, p. 1424] that the moduli space of K3 surfaces X with $L^\sigma \subseteq \mathrm{Cl}(X)$, has dimension $20 - \mathrm{rk}\,(L^\sigma)$. A K3 surface X with $\mathrm{Cl}(X) = L^\sigma$ is called *generic*.

For any K3 surface with a nonsymplectic involution $\sigma : X \to X$, one has a quotient surface $Y := X/\langle \sigma \rangle$ and the quotient map $\pi : X \to Y$. We recall that Y is an Enriques surface if and only if π is unramified; otherwise Y is rational. Moreover, in this last case the branch divisor $B \in \mathrm{WDiv}(Y)$ satisfies

$$\pi^*(B) = 2\pi^{-1}(B), \qquad B \sim -2K_Y,$$

where $\pi^{-1}(B)$ refers to the set-theoretical inverse image and K_Y is the canonical divisor of Y; see [310, Lemma 1.2] for a proof.

Proposition 5.3.3.1 *Let X be a generic K3 surface with a nonsymplectic involution $\sigma \in \mathrm{Aut}(X)$ and quotient map $\pi \colon X \to Y = X/\langle\sigma\rangle$ with branch divisor B. Then the pullback $\pi^* \colon \mathrm{Cl}(Y) \to \mathrm{Cl}(X)$ is injective and $\pi^*(\mathrm{Cl}(Y))$ is of index 2^{n-1} in $\mathrm{Cl}(X)$, where n is the number of components of B.*

Proof Note first that $\mathrm{Cl}(Y)$ is free, because Y arises by blowing up points from $\mathbb{P}^2$ or a Hirzebruch surface. According to [142, Ex. 1.7.6], we have

$$\pi^*(\mathrm{Cl}_{\mathbb{Q}}(Y)) = \mathrm{Cl}_{\mathbb{Q}}(X)^{\sigma}, \qquad \pi_*\pi^*(\mathrm{Cl}(Y)) = 2\,\mathrm{Cl}(Y).$$

Because X is generic, the first equation tells us that $\pi^*(\mathrm{Cl}(Y))$ is of finite index in $\mathrm{Cl}(X)$. The second one shows that π^* is injective. Moreover, by [29, Lemma 2.1], we have

$$[\pi^*(\mathrm{Cl}(Y)) : \mathrm{Cl}(X)]^2 = \frac{\det \pi^*(\mathrm{Cl}(Y))}{\det \mathrm{Cl}(X)}.$$

Because $\mathrm{Cl}(Y)$ is unimodular, see [29], the numerator equals $2^{\varrho(Y)}$. By [230, Thm. 4.2.2], the lattice $L^{\sigma} = \mathrm{Cl}(X)$ has determinant 2^l and the difference $\varrho(Y) - l$ equals $2(n - 1)$, where n is the number of connected components of the branch divisor. $\qquad\square$

We now determine the quotient surfaces $Y = X/\langle\sigma\rangle$ of generic K3 surfaces X with small Picard number $\varrho(X)$. In the sequel, we denote by $\mathrm{Bl}_k(Z)$ the blow-up of a variety Z in k general points. Moreover, we adopt the standard notation for integral lattices, see [29, Sec. I.2], and $L(k)$ denotes the lattice obtained from L by multiplying the intersection matrix by k.

Proposition 5.3.3.2 *Let X be a generic K3 surface X with a nonsymplectic involution and associated double cover $\pi \colon X \to Y$. For $2 \le \varrho(X) \le 5$, the table*

$\varrho(X)$	$\mathrm{Cl}(X)$	Y	B
2	$U, U(2), (2) \oplus A_1$	$\mathbb{F}_4, \mathbb{F}_0, \mathrm{Bl}_1(\mathbb{P}^2)$	$\mathbb{P}^1 + C_{10}, C_9, C_9$
$3 \le k \le 5$	$U \oplus A_1^{k-2}, U(2) \oplus A_1^{k-2}$	$\mathrm{Bl}_{k-2}(\mathbb{F}_4), \mathrm{Bl}_{k-2}(\mathbb{F}_0)$	$\mathbb{P}^1 + C_{12-k}, C_{11-k}$

describes the intersection form of X, the quotient surface Y, and the branch divisor B of π, where C_g denotes a smooth irreducible curve of genus g.

Observe that Y is not a del Pezzo surface only if $Y \cong \mathbb{F}_4$ or $\mathrm{Bl}_{k-2}(\mathbb{F}_4)$, which are exactly the cases when the branch divisor B has two components.

Lemma 5.3.3.3 *Let X be a generic K3 surface with a nonsymplectic involution such that the associated double cover $\pi: X \to Y$ has branch divisor $B = C_1 + C_B$, where $C_1 \subseteq Y$ is a smooth rational curve and $C_B \subseteq Y$ is any irreducible curve.*

(i) *Let $w_1 \in \text{Cl}(Y)$ be the class of $C_1 \subseteq Y$. Then $(w_1, w_2, \ldots, w_r)$ is a basis of $\text{Cl}(Y)$ if and only if $(\pi^*(w_1)/2, \pi^*(w_2), \ldots, \pi^*(w_r))$ is a basis of $\text{Cl}(X)$.*

(ii) *With respect to bases as in (i), the homomorphism $\pi^*: \text{Cl}(Y) \to \text{Cl}(X)$ is given by the diagonal matrix*

$$
A := \begin{bmatrix}
2 & 0 & \cdots & 0 \\
0 & 1 & \cdots & 0 \\
\vdots & \vdots & \ddots & \vdots \\
0 & 0 & \cdots & 1
\end{bmatrix}.
$$

(iii) *Let $C \subseteq X$ be any smooth rational curve and let $w \in \text{Cl}(X)$ be its class. Then precisely one of the following statements holds.*

    a. *$\pi(C)$ is a component of B and $\pi(C)^2 = -4$.*

    b. *$w \in \pi^*(\text{Cl}(Y))$ and $\pi(C)^2 = -1$.*

Proof Because $\pi^*: \text{Cl}(Y) \to \text{Cl}(X)$ is injective, $(w_1, \ldots, w_r)$ is a basis of $\text{Cl}(Y)$ if and only if $(\pi^*(w_1), \ldots, \pi^*(w_r))$ is a basis of $\pi^*(\text{Cl}(Y))$. By Proposition 5.3.3.1 (ii), we have $\pi^*(w_1) = 2u_1$ with some $u_1 \in \text{Cl}(X)$. Moreover, also by Proposition 5.3.3.1 (ii), the pullback $\pi^*(\text{Cl}(Y))$ is of index 2 in $\text{Cl}(X)$. This gives (i) and (ii).

To prove (iii), let $w \in \text{Cl}(X)$ denote the class of $C \subseteq X$. The genus formula and the Riemann–Roch formula give $w^2 = -2$ and $h^0(X, w) = 1$. Because the elements of $\text{Cl}(X)$ are fixed under the involution $\sigma: X \to X$, we can conclude $\sigma(C) = C$. If $\sigma = \text{id}$ holds on C, then C is contained in the ramification divisor. By Proposition 5.3.3.1 (ii), we have $2C = \pi^{-1}(\pi(C))$, which implies $\pi(C)^2 = -4$. If $\sigma \neq \text{id}$ on C, then the restriction $\pi: C \to \pi(C)$ is a double cover. This implies $C = \pi^{-1}(\pi(C))$ and $\pi(C)^2 = 1/2 \cdot C^2 = -1$. $\qquad\square$

Proof of Proposition 5.3.3.2 According to [230, Sec. 4], a lattice L of rank at most 5 is the fixed lattice L^σ of an involution σ on a K3 surface if and only if it is an even lattice of signature $(1, k - 1)$ which is 2-elementary, that is, satisfies $\text{Hom}(L, \mathbb{Z})/L = \mathbb{Z}_2^a$, where $2^a = |\det(L)|$. Such lattices are classified up to isometries by three invariants: the rank k, the integer a, and an invariant δ defined as

$$
\delta(L) = \begin{cases}
0 & \text{if } u^2 \in \mathbb{Z} \text{ for all } u \in \text{Hom}(L, \mathbb{Z}), \\
1 & \text{otherwise.}
\end{cases}
$$

It is easy to check that the lattices in the table are the only 2-elementary even lattices of signature $(1, k - 1)$ with $2 \le k \le 5$ because they cover all possible triples (k, a, δ), see [230, Thm. 4.3.1] and also [5, Sec. 2.3].

Now, suppose that the intersection form on $\mathrm{Cl}(X)$ is $U \oplus A_1^{k-2}$. Then it is known that there is an elliptic fibration $p \colon X \to \mathbb{P}^1$ with a section E and $k - 2$ reducible fibers, see [195, Lemma 3.1]. In fact, if e, f is the natural basis of U and $v_1, \ldots, v_{k-2}$ is an orthogonal basis of A_1^{k-2}, we can assume that the class of E is $f - e$ and v_i are represented by components of the reducible fibers not intersecting E.

By [230, Thm. 4.2.2] the ramification divisor of σ is the disjoint union of a smooth irreducible curve of genus $12 - k$ and a smooth irreducible rational curve. This implies that C is transverse to the fibers of p; hence any fiber is preserved by σ and the section E is the rational curve in the ramification divisor.

A basis of $\mathrm{Cl}(X)$ is given by $e, f - e, v_1, \ldots, v_{k-2}$. It follows from Lemma 5.3.3.3 (ii) and (iii) that the Picard lattice of Y has intersection form

$$\begin{bmatrix} 0 & 1 \\ 1 & -4 \end{bmatrix} \oplus (-1)^{k-2}.$$

Consequently, the classification of minimal rational surfaces yields that Y is the blow-up of the Hirzebruch surface $\mathbb{F}_4$ at $k - 2$ points.

Now assume that the intersection form on $\mathrm{Cl}(X)$ is $U(2) \oplus A_1^{k-2}$. Then, by [230, Thm. 4.2.2], the ramification divisor has only one connected component and this is a smooth irreducible curve of genus $11 - k$. Thus, Proposition 5.3.3.1 (ii) gives $\mathrm{Cl}(X) = \pi^*(\mathrm{Cl}(Y))$. It follows that the intersection form on $\mathrm{Cl}(Y)$ is $U \oplus (-1)^{k-2}$. Hence, as before, we can conclude that Y is the blow-up of $\mathbb{F}_0$ at $k - 2$ points.

Similarly, if the intersection form on $\mathrm{Cl}(X)$ is $(2) \oplus A_1$, then we obtain that the intersection form on $\mathrm{Cl}(Y)$ is $(1) \oplus (-1)$, the ramification divisor is a smooth irreducible curve of genus 9 and we conclude that Y is the blow-up of $\mathbb{P}^2$ at one point. □

5.3.4 Cox rings of K3 surfaces

We consider generic K3 surfaces X with a nonsymplectic involution. Here we determine the Cox ring of such surfaces when the Picard number is $2 \le \varrho(X) \le 5$. Our main results for this section are the following two propositions.

Proposition 5.3.4.1 *Let X be a generic K3 surface with a nonsymplectic involution σ and associated double cover $\pi \colon X \to Y$ and intersection form of type (2) or $U(2) \oplus A_1^{k-2}$, where $2 \le k \le 9$. Then Y is a del Pezzo surface of*

Picard number k and there is an isomorphism of $\text{Cl}(X)$*-graded rings*

$$\mathcal{R}(X) \cong \mathcal{R}(Y)[T]/\langle T^2 - f \rangle,$$

where f *is the pullback of the canonical section of the branch divisor and* $\deg(T) = \deg(f)/2$.

Proposition 5.3.4.2 *Let* X *be a generic K3 surface with a nonsymplectic involution* σ *and associated double cover* $\pi : X \to Y$ *and intersection form of type* $U \oplus A_1^k$, *where* $0 \le k \le 3$. *Then* Y *is the blow-up of* $\mathbb{F}_4$ *along* k *general points. The folllowing cases can occur.*

(i) *The surface* Y *is* $\mathbb{F}_4$ *and the Cox ring of* X *is*

$$\mathcal{R}(X) = \mathbb{C}[T_1, \ldots, T_5]/\langle T_5^2 - f \rangle$$

where $f \in \mathbb{C}[T_1, \ldots, T_5]$ *is a prime polynomial and the degree of* $T_i \in \mathcal{R}(X)$ *is the* i*th column of*

$$Q = \begin{bmatrix} 0 & 1 & 1 & 4 & 6 \\ 1 & 0 & 0 & 2 & 3 \end{bmatrix}.$$

(ii) *The surface* Y *is the blow-up of* $\mathbb{F}_4$ *at one general point and the Cox ring of* X *is*

$$\mathcal{R}(X) = \mathbb{C}[T_1, \ldots, T_6]/\langle T_6^2 - f \rangle$$

where $f \in \mathbb{C}[T_1, \ldots, T_6]$ *is a prime polynomial and the degree of* $T_i \in \mathcal{R}(X)$ *is the* i*th column of*

$$Q = \begin{bmatrix} 0 & 1 & 1 & 0 & 4 & 6 \\ 1 & 0 & 0 & 0 & 2 & 3 \\ 0 & -1 & 0 & 1 & -1 & -1 \end{bmatrix}.$$

(iii) *The surface* Y *is the blow-up of* $\mathbb{F}_4$ *at two general points and the Cox ring of* X *is*

$$\mathcal{R}(X) = \mathbb{C}[T_1, \ldots, T_7]/\langle T_7^2 - f \rangle$$

where $f \in \mathbb{C}[T_1, \ldots, T_7]$ *is a prime polynomial and the degree of* $T_i \in \mathcal{R}(X)$ *is the* i*th column of*

$$Q = \begin{bmatrix} 0 & 1 & 1 & 0 & 0 & 4 & 6 \\ 1 & 0 & 0 & 0 & 0 & 2 & 3 \\ 0 & -1 & 0 & 1 & 0 & -1 & -1 \\ 0 & 0 & -1 & 0 & 1 & -1 & -1 \end{bmatrix}.$$

(iv) *The surface* Y *is the blow-up of* $\mathbb{F}_4$ *at three general points and the Cox ring of* X *is*

$$\mathcal{R}(X) = \mathbb{C}[T_1, \ldots, T_9]/\langle T_2 T_5 + T_3 T_6 + T_4 T_7, \, T_9^2 - f \rangle$$

where $f \in \mathbb{C}[T_1, \ldots, T_8]$ is a prime polynomial and the degree of $T_i \in \mathcal{R}(X)$ is the ith column of

$$Q = \begin{bmatrix} 0 & 1 & 1 & 1 & 0 & 0 & 0 & 4 & 6 \\ 1 & 0 & 0 & 0 & 0 & 0 & 0 & 2 & 3 \\ 0 & -1 & 0 & 0 & 1 & 0 & 0 & -1 & -1 \\ 0 & 0 & -1 & 0 & 0 & 1 & 0 & -1 & -1 \\ 0 & 0 & 0 & -1 & 0 & 0 & 1 & -1 & -1 \end{bmatrix}.$$

We enter the proof of the above results. A key step is to relate the Cox ring of X to that of Y for a double cover $X \to Y$ as it occurs in our setting.

Proposition 5.3.4.3 *Let X be a generic K3 surface with a nonsymplectic involution. Suppose that*

- *The branch divisor of the associated double cover $\pi : X \to Y$ is of the form $B = C_1 + C_B$ with $C_1, C_B \subseteq Y$ irreducible and C_1 rational.*
- *The Cox ring of Y is a polynomial ring $S = S'[t_1]$ with the canonical section t_1 of C_1 and a finitely generated $\mathbb{C}$-algebra S'.*

Moreover, denote by $f \in S'$ the canonical section of C_B. Then the Cox ring of X is given as

$$R = \pi^*(S')[T_1, T_2]/\langle T_2^2 - \pi^*(f)\rangle,$$

with the $\mathrm{Cl}(X)$-grading defined by $\deg(\pi^*(g)) := \pi^*(\deg(g))$ for any homogeneous $g \in S'$ and

$$\deg(T_1) := \frac{\pi^*(w_1)}{2}, \qquad \deg(T_2) := -\frac{\pi^*(2K_Y + w_1)}{2}.$$

Moreover, the pullback homomorphism $\pi^*: S \to R$ of graded rings is given on the grading groups by $\mathbb{Z}^r \to \mathbb{Z}^r$, $w \mapsto Aw$ and as a ring homomorphism by

$$t_1 \mapsto T_1^2, \qquad g \mapsto \pi^*(g) \text{ for any homogeneous } g \in S'.$$

Proof First note that, by Proposition 5.3.1.4, the Cox ring R of X inherits finite generation from the Cox ring S of Y. Consider the pullback group of $\mathrm{Cl}(Y)$ and the corresponding Veronese subalgebra

$$L := \pi^*(\mathrm{Cl}(Y)) \subseteq \mathrm{Cl}(X), \qquad R_L := \bigoplus_{w \in L} R_w.$$

Write, for the moment $B = B_1 + B_2$, and let r and b_i denote the canonical sections of $\pi^{-1}(B)$ and B_i respectively. We claim that there is a commutative diagram of finite ring homomorphisms

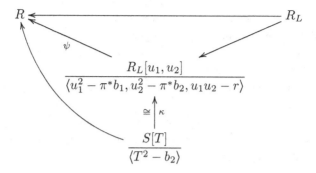

where, denoting by r_1 and r_2 the canonical sections of the reduced divisors $\pi^{-1}(B_1)$ and $\pi^{-1}(B_2)$ respectively, the homomorphism ψ is induced by $u_i \mapsto r_i$.

In this claim, everything is straightforward except the definition of the isomorphism κ. By Proposition 5.3.1.3 we know that R_L is generated as a $\pi^*(S)$-module by 1 and a section $s \in R_L$ satisfying $s^2 = \pi^*(b)$, where b denotes the canonical section of B. According to Lemma 5.3.3.3, we may choose s to be the canonical section r of the ramification divisor $\pi^{-1}(B)$. Thus, we obtain isomorphisms

$$R_L[u_1, u_2]/\langle u_1^2 - b_1, u_2^2 - b_2, u_1 u_2 - r\rangle$$
$$\cong \pi^*(S)[y, u_1, u_2]/\langle y^2 - b, u_1^2 - b_1, u_2^2 - b_2, u_1 u_2 - y\rangle$$
$$\cong \pi^*(S)[u_1, u_2]/\langle u_1^2 - b_1, u_2^2 - b_2\rangle.$$

Now we use our assumption $S = S'[t_1]$. This enables us to define a ring homomorphism

$$\widetilde{\kappa} : S[T] \rightarrow \pi^*(S)[u_1, u_2], \qquad S' \ni g \mapsto \pi^*(g) \in \pi^*(S'),$$

$$t_1 \mapsto u_1, \quad T \mapsto u_2.$$

It sends t_1^2 to u_1^2, which defines the same element in $\pi^*(S)[u_1, u_2]/\langle u_1^2 - b_1, u_2^2 - b_2\rangle$ as $\pi^*(t_1)$. Consequently, $\widetilde{\kappa}$ induces the desired isomorphism

$$\kappa : S[T]/\langle T^2 - b_2\rangle \rightarrow \pi^*(S)[u_1, u_2]/\langle u_1^2 - b_1, u_2^2 - b_2\rangle.$$

The next step is to show that the homomorphism ψ of the above diagram is an isomorphism. For this, it is enough to show that $S[T]/\langle T^2 - b_2\rangle$ is a normal ring. Indeed, $R_L \rightarrow R$ is of degree 2,

$$R_L \rightarrow R_L[u_1, u_2]/\langle u_1^2 - b_1, u_2^2 - b_2, u_1 u_2 - r\rangle$$

is of degree at least 2 and thus ψ is a finite morphism of degree one. If we know that $S[T]/\langle T^2 - b_2\rangle$ is normal, we can conclude that ψ is an isomorphism.

To show that $S[T]/\langle T^2 - b_2 \rangle$ is normal, note that S can be made into a $\mathbb{Z}$-graded ring by assigning to each $\mathbb{Z}^r$-homogeneous element the w_1-component of its $\mathbb{Z}^r$-degree. In particular, then $\deg(b_2)$ is odd. Morever, $b_2 \in S$ is a prime element. Thus, we can apply the result [266, p. 45] and obtain that $S[T]/\langle T^2 - b_2 \rangle$ is even factorial. In particular it is normal.

Having verified that ψ is an isomorphism, the commutative diagram tells us that the Cox ring R of X is isomorphic to $S[T]/\langle T^2 - b_2 \rangle$. Consequently, R is the polynomial ring $\pi^*(S')[T_1, T_2]$ divided by the relation $T_2^2 - \pi^*(b_2)$, where $\pi^*(b_2)$ only depends on the first variable. The degrees of the generators T_i are easily computed using Lemma 5.3.3.3 (ii). $\square$

If X is a generic K3 surface with a nonsymplectic involution σ such that the associated double cover has an irreducible branch divisor B, then the quotient surface $Y = X/\langle \sigma \rangle$ is smooth rational. Moreover, by [230, Thm. 4.2.2] one has $-1 \leq K_Y^2 \leq 9$ and, due to the generality assumption on X, the surface Y is a del Pezzo surface exactly when $K_Y^2 \geq 1$. Observe that $\mathrm{Cl}(X)$ is isometric to one of the following lattices:

$$(2) \qquad U(2) \oplus A_1^{k-2} \text{ with } 2 \leq k \leq 9,$$

where k is the Picard rank of X. We now proceed to compute the Cox rings of such K3 surfaces.

Proof of Proposition 5.3.4.1   As in the proof of Proposition 5.3.3.2, we use [230, Thm. 4.2.2] to see that the ramification divisor of $\pi: X \to Y$ is irreducible. Then Proposition 5.3.3.1 (ii) yields $\mathrm{Cl}(X) = \pi^*(\mathrm{Cl}(Y))$. It follows that the intersection form on $\mathrm{Cl}(Y)$ is either (1) or $U \oplus (-1)^{k-2}$. Consequently, Y is the blow-up of $\mathbb{F}_0$ at $k - 2$ general points and hence is a del Pezzo surface of Picard number k. Thus the statement follows from Proposition 5.3.1.3. $\square$

We are ready to compute the Cox rings of generic K3 surfaces X admitting a nonsymplectic involution and satisfying $2 \leq \varrho(X) \leq 5$. According to Propositions 5.3.3.2 and 5.3.4.1 it is enough to work out the case where the quotient surface Y is the blow-up of $\mathbb{F}_4$ at k general points with $k \in \{0, 1, 2, 3\}$. If $k < 3$, then the surface Y is isomorphic to a toric surface whose fan is one of the following:

The degree matrix of the Cox ring of such a Y is immediately obtained by Gale duality.

Proof of Proposition 5.3.4.2 The assertion is a direct consequence of Proposition 5.3.4.3, Lemma 5.3.3.3, and, in case (iv), of Example 5.4.2.3. Note that the canonical class of Y can be determined according to Proposition 3.3.3.2 as the degree of the relation minus the sum of the degrees of the generators of the Cox ring of Y. □

5.4 Rational $\mathbb{K}^*$-surfaces

5.4.1 Defining data and their surfaces

Normal complete rational $\mathbb{K}^*$-surfaces X arise from the general Construction 3.4.3.1 of rational varieties with a torus action of complexity 1. Here we re-enter the construction in the simpler surface case for a more detailed discussion and show how the defining data are reflected in basic geometric properties of the action. The ground field $\mathbb{K}$ is algebraically closed and of characteristic zero.

Definition 5.4.1.1 Fix $r \in \mathbb{Z}_{\geq 1}$ and $n_0, \ldots, n_r \in \mathbb{Z}_{\geq 1}$. Consider integral tuples $l_i := (l_{i1}, \ldots, l_{in_i}) \in \mathbb{Z}_{\geq 1}^{n_i}$ and $d_i := (d_{i1}, \ldots, d_{in_i}) \in \mathbb{Z}^{n_i}$ such that

$$\frac{d_{i1}}{l_{i1}} > \cdots > \frac{d_{in_i}}{l_{in_i}} \text{ for all } i, \qquad \gcd(l_{ij}, d_{ij}) = 1 \text{ for all } i, j.$$

Set $n := n_0 + \cdots + n_r$. Define an integral $r \times n$ matrix L and an integral $1 \times n$ matrix d by

$$L := \begin{bmatrix} -l_0 & l_1 & \cdots & 0 \\ \vdots & & \ddots & \vdots \\ -l_0 & 0 & \cdots & l_r \end{bmatrix}, \qquad d := \begin{bmatrix} d_0 & d_1 & \cdots & d_r \end{bmatrix}.$$

Then we define four types of integral $(r + 1) \times (n + m)$ matrices P, where $m = 0, 2, 1, 1$ according to the cases:

$$\text{(e-e)} \quad P = \begin{bmatrix} L \\ d \end{bmatrix}, \qquad \text{(p-p)} \quad P = \begin{bmatrix} L & 0 & 0 \\ d & 1 & -1 \end{bmatrix},$$

$$\text{(p-e)} \quad P = \begin{bmatrix} L & 0 \\ d & 1 \end{bmatrix}, \qquad \text{(e-p)} \quad P = \begin{bmatrix} L & 0 \\ d & -1 \end{bmatrix},$$

where we require in all cases that the (pairwise different and primitive) columns of P generate $\mathbb{Q}^{r+1}$ as a cone.

Remark 5.4.1.2 Consider the canonical basis vectors $e_i \in \mathbb{Z}^r$ and the vector $e_0 := -e_1 - \cdots - e_r \in \mathbb{Z}^r$. Then the possible columns of the matrices P of Definition 5.4.1.1 are of the form

$$v_{ij} := \begin{bmatrix} l_{ij} e_i \\ d_{ij} \end{bmatrix}, \qquad v^+ := \begin{bmatrix} 0 \\ 1 \end{bmatrix}, \qquad v^- := \begin{bmatrix} 0 \\ -1 \end{bmatrix}.$$

For each i, we have an ith vertical block $[v_{i1}, \ldots, v_{in_i}]$ in P. The projection $P_1 \colon \mathbb{Z}^{r+1} \to \mathbb{Z}^r$ onto the first r coordinates maps v^+, v^- to zero and the v_{ij} are blockwise distributed over the rays through the vectors e_i:

$$P_1(v_{0j}) \in \mathrm{cone}(e_0), \qquad P_1(v_{1j}) \in \mathrm{cone}(e_1), \qquad \ldots, \qquad P_1(v_{rj}) \in \mathrm{cone}(e_r).$$

Inside each block $[v_{i1}, \ldots, v_{in_i}]$, the conditions $d_{i1}/l_{i1} > \ldots > d_{in_i}/l_{in_i}$ on the slopes ensure that the rays through $v_{i1}, \ldots, v_{in_i}$ are enumerated from above to below with respect to the last coordinate in $\mathbb{Q}^{r+1}$.

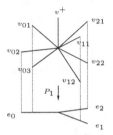

Construction 5.4.1.3 Let $r \in \mathbb{Z}_{\geq 1}$, let $A = [a_0, \ldots, a_r]$ be a matrix with pairwise linearly independent columns $a_0, \ldots, a_r \in \mathbb{K}^2$ and P a matrix as in 5.4.1.1. Define a $\mathbb{K}$-algebra

$$R(A, P) := \mathbb{K}[T_{ij}, S_k; \ 0 \leq i \leq r, \ 1 \leq j \leq n_i, \ 1 \leq k \leq m] / \langle g_I; \ I \in \mathfrak{J} \rangle,$$

where $\mathfrak{J}$ is the set of all triples $I = (i_1, i_2, i_3)$ with $0 \leq i_1 < i_2 < i_3 \leq r$ and for each such I the associated trinomial g_I is given by

$$g_I := g_{i_1, i_2, i_3} := \det \begin{bmatrix} T_{i_1}^{l_{i_1}} & T_{i_2}^{l_{i_2}} & T_{i_3}^{l_{i_3}} \\ a_{i_1} & a_{i_2} & a_{i_3} \end{bmatrix}, \qquad T_i^{l_i} := T_{i1}^{l_{i1}} \cdots T_{in_i}^{l_{in_i}}.$$

Consider the transpose P^* of P and the projection $Q \colon \mathbb{Z}^{n+m} \to K := \mathbb{Z}^{n+m}/\mathrm{im}(P^*)$. We endow $R(A, P)$ with the K-grading inherited from the one on $\mathbb{K}[T_{ij}, S_k]$ given by

$$\deg(T_{ij}) := Q(e_{ij}), \qquad \deg(S_k) := Q(e_k).$$

Remark 5.4.1.4 As observed in Lemma 3.4.2.4, the ring $R(A, P)$ is a complete intersection and as generators of the ideal of relations, we can, for example, take

$$g_{0,1,2}, \quad g_{1,2,3}, \quad \ldots, \quad g_{r-3,r-2,r-1}, \quad g_{r-2,r-1,r}.$$

Moreover, applying a suitable homothety to the variables, we see that the ring $R(A, P)$ is isomorphic to the factor ring of $\mathbb{K}[T_{ij}, S_k]$ by the ideal

$$\langle T_0^{l_0} + T_1^{l_1} + T_2^{l_2}, \; \alpha_1 T_1^{l_1} + T_2^{l_2} + T_3^{l_3}, \; \ldots, \; \alpha_{r-2} T_{r-2}^{l_{r-2}} + T_{r-1}^{l_{r-1}} + T_r^{l_r} \rangle$$

with factors $1 \neq \alpha_i \in \mathbb{K}^*$ indicating parameters for $R(A, P)$; however, note that for $r \geq 4$, this presentation need not stem from data A and P.

Theorem 5.4.1.5 *The K-graded $\mathbb{K}$-algebra $R(A, P)$ provided by Construction 5.4.1.3 is the Cox ring of a $\mathbb{Q}$-factorial projective $\mathbb{K}^*$-surface $X(A, P)$ and the $\mathbb{K}^*$-surface $X(A, P)$ is determined up to isomorphy by A and P. Moreover, every rational normal complete $\mathbb{K}^*$-surface is isomorphic to some $X(A, P)$.*

Proof According to Theorems 3.4.3.4 and 3.4.3.7 the K-graded $\mathbb{K}$-algebra $R(A, P)$ is the Cox ring of some $\mathbb{K}^*$-surface $X(A, P)$. As any other Mori dream surface, also $X(A, P)$ is determined up to isomorphy by its Cox ring and it is $\mathbb{Q}$-factorial projective; see Theorem 4.3.3.5. In fact, taking the unique $\mathfrak{F}$-bunch Φ arising from any $w \in \mathrm{Mov}(X(A, P))^\circ$, where $\mathfrak{F} = (T_{ij}, S_k)$, we have $X(A, P) = X(A, P, \Phi)$. This proves the first statement. The second statement is a special case of Theorem 4.4.1.6. $\qquad\square$

To make the construction of $X(A, P)$ more explicit, we realize the surface via the canonical toric embedding 3.2.5.3 as a $\mathbb{K}^*$-invariant closed subvariety of a toric variety. The ambient toric variety will be directly constructed from the matrix P.

Construction 5.4.1.6 Given a matrix P as in Definition 5.4.1.1, denote by v_{ij} the first n columns of P. Moreover, write $v^+ := (0, \ldots, 0, 1)$ and $v^- := (0, \ldots, 0, -1)$ for the possible remaining columns of P. According to the four types, we associate with P a fan $\Sigma(P)$ in the lattice $\mathbb{Z}^{r+1}$.

(e-e) The fan $\Sigma(P)$ has the rays $\varrho_{ij} := \mathrm{cone}(v_{ij})$. The maximal cones of $\Sigma(P)$ are

$$
\begin{aligned}
\sigma^+ &:= \mathrm{cone}(v_{01}, \ldots, v_{r1}), \\
\tau_{ij} &:= \mathrm{cone}(v_{ij}, v_{ij+1}), \quad \text{for } 0 \leq i \leq r, \; 1 \leq j \leq n_i - 1, \\
\sigma^- &:= \mathrm{cone}(v_{0n_0}, \ldots, v_{rn_r}).
\end{aligned}
$$

(p-p) The fan $\Sigma(P)$ has the rays $\varrho_{ij} := \mathrm{cone}(v_{ij})$ and $\varrho^+ := \mathrm{cone}(v^+)$ and $\varrho^- := \mathrm{cone}(v^-)$. The maximal cones of $\Sigma(P)$ are

$$
\begin{aligned}
\tau_i^+ &:= \mathrm{cone}(v^+, v_{i1}), \quad \text{for } 0 \leq i \leq r, \\
\tau_{ij} &:= \mathrm{cone}(v_{ij}, v_{ij+1}), \quad \text{for } 0 \leq i \leq r, \; 1 \leq j \leq n_i - 1, \\
\tau_i^- &:= \mathrm{cone}(v_{in_i}, v^-), \quad \text{for } 0 \leq i \leq r.
\end{aligned}
$$

(p-e) The fan $\Sigma(P)$ has the rays $\varrho_{ij} := \mathrm{cone}(v_{ij})$ and $\varrho^+ := \mathrm{cone}(v^+)$. The maximal cones of $\Sigma(P)$ are

$$
\begin{aligned}
\tau_i^+ &:= \mathrm{cone}(v_{i1}, v^+), \quad \text{for } 0 \le i \le r, \\
\tau_{ij} &:= \mathrm{cone}(v_{ij}, v_{ij+1}), \quad \text{for } 0 \le i \le r,\ 1 \le j \le n_i - 1, \\
\sigma^- &:= \mathrm{cone}(v_{0n_0}, \dots, v_{rn_r}).
\end{aligned}
$$

(e-p) The fan $\Sigma(P)$ has the rays $\varrho_{ij} := \mathrm{cone}(v_{ij})$ and $\varrho^- := \mathrm{cone}(v^-)$. The maximal cones of $\Sigma(P)$ are

$$
\begin{aligned}
\sigma^+ &:= \mathrm{cone}(v_{01}, \dots, v_{r1}), \\
\tau_{ij} &:= \mathrm{cone}(v_{ij}, v_{ij+1}), \quad \text{for } 0 \le i \le r,\ 1 \le j \le n_i - 1, \\
\tau_i^- &:= \mathrm{cone}(v^-, v_{in_i}), \quad \text{for } 0 \le i \le r.
\end{aligned}
$$

Remark 5.4.1.7 Let P be a matrix as in Definition 5.4.1.1. Then, according to the four posssible cases, we obtain for the fans $\Sigma(P)$ associated with P pictures of the following shape:

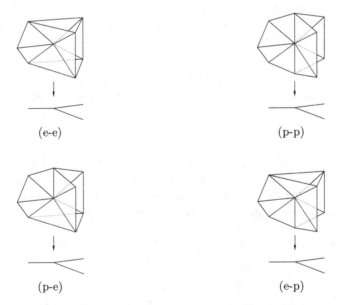

(e-e)   (p-p)

(p-e)   (e-p)

where the downwards arrows indicate the projection $P_1 : \mathbb{Z}^{r+1} \to \mathbb{Z}^r$ onto the first r coordinates and on the bottom we draw the rays through the canonical basis vectors $e_1 \dots, e_r \in \mathbb{Z}^r$ and $e_0 = -e_1 - \dots - e_r$; here for $r = 2$.

Proposition 5.4.1.8 *Let A and P be data as in Construction 5.4.1.3, $\Sigma(P)$ the fan introduced in Construction 5.4.1.6, and $Z(P)$ the toric variety associated with $\Sigma(P)$. Then we have a commutative diagram*

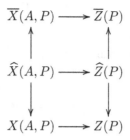

with the characteristic spaces $\widehat{X}(A, P) \to X(A, P)$ and $\widehat{Z}(P) \to Z(P)$, the total coordinate spaces

$$\overline{X}(A, P) = V(g_I; \; I \in \mathfrak{J}) \subseteq \mathbb{K}^{n+m} = \overline{Z}(P)$$

and the canonical toric embedding $X(A, P) \to Z(P)$. The $\mathbb{K}^*$-action on $X(A, P)$ arises from the one-parameter subgroup $t \mapsto (1, \ldots, 1, t)$ of the acting torus $\mathbb{T}^{r+1}$ of $Z(P)$.

Proof We have $X(A, P) = X(A, P, \Phi)$ with the $\mathfrak{F}$-bunch Φ arising from any $w \in \mathrm{Mov}(X)^\circ$, where $\mathfrak{F} = (T_{ij}, S_k)$. One directly verifies that $\Sigma(P)$ is the fan Gale dual to the envelope of Φ in the sense of Construction 3.2.5.3. Thus, the induced morphism $X(A, P) \to Z(P)$ is the canonical toric embedding. $\qquad\square$

We indicate how to read off basic properties of the $\mathbb{K}^*$-action on $X(A, P)$ from the matrix P. Let us recall the necessary notions on complete $\mathbb{K}^*$-varieties X. For any $x \in X$ the orbit map $\mu_x \colon \mathbb{K}^* \to X, t \mapsto t \cdot x$ extends to a morphism $\overline{\mu}_x \colon \mathbb{P}^1 \to X$ and one sets

$$\lim_{t \to 0} t \cdot x := \overline{\mu}_x(0), \qquad \lim_{t \to \infty} t \cdot x := \overline{\mu}_x(\infty).$$

The image of $\overline{\mu}_x$ is the closure of the orbit $\mathbb{K}^* \cdot x$ and the two limit points are always fixed points. If X is normal projective, then the two limits of a point with nontrivial orbit are necessarily distinct. A fixed point of a $\mathbb{K}^*$-surface is called

- *Elliptic* if it is isolated and lies in the closure of infinitely many $\mathbb{K}^*$-orbits
- *Parabolic* if it belongs to a fixed point curve
- *Hyperbolic* if it is isolated and lies in the closure of two $\mathbb{K}^*$-orbits

These are in fact the only possible types of fixed points for a normal $\mathbb{K}^*$-surface. A normal projective $\mathbb{K}^*$-surface X has a *source* $F^+ \subseteq X$ and a *sink* $F^- \subseteq X$. They are characterized by the behavior of general points: there is a nonempty open set $U \subseteq X$ with

$$\lim_{t \to 0} t \cdot x \in F^+, \qquad \lim_{t \to \infty} t \cdot x \in F^- \qquad \text{for all } x \in U.$$

The source can either consist of an elliptic fixed point or it is a curve of parabolic fixed points; the same holds for the sink. Any fixed point outside the source or the sink is hyperbolic.

We turn to $X = X(A, P)$. As in Construction 3.2.1.3, we denote by $D_X^{ij} \subseteq X$ the prime divisor corresponding to the variable T_{ij} of $R(A, P)$. Thus, D_X^{ij} is the intersection of X with the toric prime divisor of $Z(P)$ corresponding to the ray ϱ_{ij}. Similarly, we write D_X^+, D_X^- for the prime divisors on X obtained as intersecions of X with the toric prime divisor of $Z(P)$ corresponding to the rays ϱ^+, ϱ^- if present.

Proposition 5.4.1.9 *Let A and P be data as in Construction 5.4.1.3 and $X = X(A, P)$ the associated $\mathbb{K}^*$-surface. According to the four types of P we have the following.*

(e-e) *The source $F^+ = \{x_{\sigma^+}\}$ and the sink $F^- = \{x_{\sigma^-}\}$ both consist of an elliptic fixed point.*

(p-p) *The source $F^+ = D_X^+$ and the sink $F^- = D_X^-$ both are smooth rational curves.*

(p-e) *The source $F^+ = D_X^+$ is a smooth rational curve and the sink $F^- = \{x_{\sigma^-}\}$ consists of an elliptic fixed point.*

(e-p) *The source $F^+ = \{x_{\sigma^+}\}$ consists of an elliptic fixed point and the sink $F^- = D_X^-$ is a smooth rational curve.*

Moreover, in all four cases, the divisors D_X^{ij} are rational curves and we have the following statements on the hyperbolic fixed points and isotropy groups of the nontrivial $\mathbb{K}^*$-orbits.

(i) *The toric orbit corresponding to a ray $\varrho_{ij} \in \Sigma(P)$ cuts out a nontrivial $\mathbb{K}^*$-orbit $B^{ij} \subseteq D_X^{ij}$.*

(ii) *For any $x \in B^{ij}$, the isotropy group $\mathbb{K}_x^*$ is of order l_{ij} and $\overline{d}_{ij} \in \mathbb{Z}/l_{ij}\mathbb{Z}$ is the weight of the cotangent representation of $\mathbb{K}_x^*$.*

(iii) *The toric orbit corresponding to a cone $\tau_{ij} \in \Sigma(P)$ cuts out a hyperbolic fixed point $x_{ij} \in X$.*

(iv) *For every $0 \leq i \leq r$, the divisors $D_X^{ij} \subseteq X$ form a chain of rational curves connecting the source with the sink in the sense that picking points $b_{ij} \in B_{ij}$ we have*

$$\lim_{t \to 0} t \cdot b_{i1} \in F^+, \quad \lim_{t \to \infty} t \cdot b_{ij} = x_{ij+1} = \lim_{t \to 0} t \cdot b_{ij+1}, \quad \lim_{t \to \infty} t \cdot b_{in_i} \in F^-.$$

(v) *For any $x \in X$ not lying in F^+, F^- or one of the D_X^{ij}, the isotropy group $\mathbb{K}_x^*$ is trivial and we have*

$$\lim_{t \to 0} t \cdot x \in F^+, \quad \lim_{t \to \infty} t \cdot x \in F^-.$$

Finally, with the subfan $\Sigma_0(P) \preceq \Sigma(P)$ having the rays ϱ_{ij} as its maximal cones and the associated open toric subvariety $Z_0(P) \subseteq Z(P)$, we have

$$X_0 := \{x \in X; \, \mathbb{K}_x^* \text{ is finite}\} = X \cap Z_0(P)$$

and the projection $P_1 \colon \mathbb{Z}^{r+1} \to \mathbb{Z}^r$ defines a toric morphism $Z_0(P) \to \mathbb{P}^r$. The restriction to X_0 is the geometric quotient followed by the separation and its image in $\mathbb{P}^r$ is the line given by

$$V(h_I; \, I \in \mathfrak{J}) \subseteq \mathbb{P}^r, \qquad h_I := h_{i_1, i_2, i_3} := \det \begin{bmatrix} z_{i_1} & z_{i_2} & z_{i_3} \\ a_{i_1} & a_{i_2} & a_{i_3} \end{bmatrix}.$$

Proof Because X inherits the $\mathbb{K}^*$-action from $Z(P)$, where it is given by the one-parameter subgroup corresponding to $(0, \ldots, 0, 1) \in \mathbb{Z}^{r+1}$, we can derive the statements directly from toric geometry. In particular, we see that assertion (iv) reflects the fact that for fixed i, the rays ϱ_{ij} are numbered according to their slopes. Moreover, we can conclude that parabolic fixed point curves are smooth rational, because they are mapped isomorphically onto the image by a suitable extension of $\pi \colon X_0 \to \mathbb{P}^1$. $\qquad\square$

5.4.2 Intersection numbers

We describe the intersection theory of a $\mathbb{K}^*$-surface $X = X(A, P)$. Recall that the invariant divisors D_X^{ij} and D_X^+, D_X^- if present, generate the divisor class group of X. Thus, the intersection theory of X is determined by the intersection numbers of its invariant divisors. In Proposition 5.4.2.1 and Corollary 5.4.2.2 we compute these numbers in terms of the defining matrix P. We work over an algebraically closed field $\mathbb{K}$ of characteristic zero. The original reference for this section is [177].

Proposition 5.4.2.1 *Consider a $\mathbb{K}^*$-surface $X(A, P)$. For the slopes of the vectors v_{ij} and the upper and lower slope sums write*

$$m_{ij} := \frac{d_{ij}}{l_{ij}}, \qquad m^+ := m_{01} + \cdots + m_{r1}, \qquad m^- := m_{0n_0} + \cdots + m_{rn_r}.$$

Then we have $m_{ij} > m_{ij+1}$ and $m^+ > m^-$. The intersection numbers of the invariant curves D_X^{ij} and D_X^+, D_X^-, if present, are given as follows.

(i) *If D_X^{ij} and D_X^{ik} do not intersect, that is, $|k - j| > 1$, then we have $D_X^{ij} \cdot D_X^{ik} = 0$. For $k = j + 1$, we have*

$$D_X^{ij} \cdot D_X^{ij+1} = \frac{1}{l_{ij} l_{ij+1}} \cdot \frac{1}{m_{ij} - m_{ij+1}}.$$

(ii) *For curves D_X^{i1}, D_X^{k1} or $D_X^{in_i}$, $D_X^{kn_k}$ with $i \neq k$ intersecting in an elliptic fixed point, we obtain*

$$D_X^{i1} \cdot D_X^{k1} = \begin{cases} \frac{1}{l_{i1}l_{k1}}\left(\frac{1}{m^+} - \frac{1}{m^-}\right), & \text{(e-e) with } n_i n_k = 1, \\ \frac{1}{l_{i1}l_{k1}m^+}, & \text{(e-e) with } n_i n_k \neq 1 \text{ or (e-p)}, \end{cases}$$

$$D_X^{in_i} \cdot D_X^{kn_k} = \begin{cases} \frac{1}{l_{in_i}l_{kn_i}}\left(\frac{1}{m^+} - \frac{1}{m^-}\right), & \text{(e-e) with } n_i n_k = 1, \\ \frac{-1}{l_{in_i}l_{kn_k}m^-}, & \text{(e-e) with } n_i n_k \neq 1 \text{ or (p-e)}. \end{cases}$$

(iii) *For curves D_X^{i1}, $D_X^{in_i}$ and the parabolic fixed point curves D_X^+, D_X^-, if present, we obtain*

$$D_X^{i1} \cdot D_X^+ = \frac{1}{l_{i1}}, \qquad D_X^{in_i} \cdot D_X^- = \frac{1}{l_{in_i}}, \qquad D_X^+ \cdot D_X^- = 0.$$

Proof Recall that $X = X(A, P)$ is a complete intersection in its $(r + 1)$-dimensional ambient toric variety $Z = Z(P)$, which we may assume to be projective and simplicial; use Construction 3.2.5.7 for the latter. Thus, we can compute the intersection number of curves $C_X = \iota^*(D_Z)$ and $C_X' = \iota^*(D_Z')$ obtained by restricting toric divisors as

$$C_X \cdot C_X' = D_Z \cdot D_Z' \cdot E_1 \cdots E_{r-1},$$

where $E_1, \ldots, E_{r-1}$ are any toric representatives of the degree of the $r - 1$ relations of $R(A, P)$. Note that the degree of the relations is represented by $l_{i1}D_Z^{i1} + \cdots + l_{in_i}D_Z^{in_i}$ for any i.

We prove (i). To intersect curves D_X^{ij}, D_X^{ij+1}, take a cone σ of the fan of Z having primitive generators $v_{ij}, v_{ij+1}, v_{s_1 t_1}, \ldots, v_{s_{r-1}t_{r-1}}$, oriented positively. With Proposition 2.1.2.8 we obtain

$$D_X^{ij} \cdot D_X^{ij+1} = D_Z^{ij} \cdot D_Z^{ij+1} \cdot \left(\sum_{j=1}^{n_{s_1}} l_{s_1 j}D_Z^{s_1 j}\right) \cdots \left(\sum_{j=1}^{n_{s_{r-1}}} l_{s_{r-1} j}D_Z^{s_{r-1} j}\right)$$

$$= \frac{l_{s_1 t_1} \cdots l_{s_{r-1}t_{r-1}}}{\det(v_{ij}, v_{ij+1}, v_{s_1 t_1}, \ldots, v_{s_{r-1}t_{r-1}})}$$

$$= \frac{1}{d_{ij}l_{ij+1} - d_{ij+1}l_{ij}},$$

where for the computation of the determinant one uses the special shape of the columns $v_{ij}, v_{ij+1}, v_{s_1 t_1}, \ldots, v_{s_{r-1}t_{r-1}}$ of P and the fact that, by simpliciality of σ, the indices $i, s_1, \ldots, s_{r-1}$ are pairwise different.

For, for example, the first formula of (ii), consider $i \neq k$ and let $s_1, \ldots, s_{r-1}$ be the indices different from i, k such that $v_{i1}, v_{k1}, v_{s_1 1}, \ldots, v_{s_{r-1} 1}$ are oriented positively. Then the assertion follows up to a direct computation of determinants from

$$
D_X^{i1} \cdot D_X^{k1} = D_Z^{i1} \cdot D_Z^{k1} \cdot \left(\sum_{j=1}^{n_{s_1}} l_{s_1 j} D_Z^{s_1 j} \right) \cdots \left(\sum_{j=1}^{n_{s_{r-1}}} l_{s_{r-1} j} D_Z^{s_{r-1} j} \right)
$$

$$
= \begin{cases} \dfrac{l_{s_1 1} \cdots l_{s_{r-1} 1}}{\det(v_{01}, \ldots, v_{r1})} - \dfrac{l_{s_1 n_{s_1}} \cdots l_{s_{r-1} n_{s_{r-1}}}}{\det(v_{0 n_0}, \ldots, v_{r n_r})}, & \text{(e-e) with } n_i n_k = 1, \\[4mm] \dfrac{l_{s_1 1} \cdots l_{s_{r-1} 1}}{\det(v_{01}, \ldots, v_{r1})}, & \text{(e-e) with } n_i n_k \neq 1 \text{ or (e-p).} \end{cases}
$$

To see (iii), consider exemplarily D_X^+, the restriction of the toric divisor D_Z^+ associated with $v^+ = e_{r+1}$. To intersect D_X^{i1} and D_X^+, take a cone σ of the fan of Z having primitive generators $v_{i1}, v^+, v_{s_1 1}, \ldots, v_{s_{r-1} 1}$, oriented positively, where all s_k differ from i. This gives

$$
D_X^{i1} \cdot D_X^+ = D_Z^{i1} \cdot D_Z^+ \cdot \left(\sum_{j=1}^{n_{s_1}} l_{s_1 j} D_Z^{s_1 j} \right) \cdots \left(\sum_{j=1}^{n_{s_{r-1}}} l_{s_{1-j} j} D_Z^{s_{r-1} j} \right)
$$

$$
= \frac{l_{s_1 t_1} \cdots l_{s_{r-1} t_{r-1}}}{\det(v_{i1}, v^+, v_{s_1 1}, \ldots, v_{s_{r-1} 1})}
$$

$$
= \frac{1}{l_{i1}}. \qquad \square
$$

Corollary 5.4.2.2 *Consider a $\mathbb{K}^*$-surface $X(A, P)$, let m_{ij}, m^+ and m^- be defined as in Proposition 5.4.2.1 and set*

$$
l^+ := l_{01} \cdots l_{r1}, \qquad\qquad l^- := l_{0 n_0} \cdots l_{r n_r}.
$$

Then the self-intersection numbers of the invariant prime divisors on $X = X(A, P)$ are given as follows.

(i) *For a curve D_X^{i1} with $n_i = 1$, we have*

$$
\left(D_X^{i1} \right)^2 = \begin{cases} \dfrac{1}{l_{i1}^2 (m^+ - m^-)}, & \text{(e-e)}, \\[3mm] 0 & \text{(p-p)}, \\[3mm] \dfrac{1}{l_{i1}^2 m^+}, & \text{(e-p)}, \\[3mm] \dfrac{-1}{l_{i1}^2 m^-}, & \text{(p-e)}. \end{cases}
$$

(ii) *For a curve D_X^{ij} with $n_i > 1$, we have*

$$
\left(D_X^{ij}\right)^2 = \begin{cases}
\frac{1}{l_{ij}^2}\left(\frac{1}{m^+} - \frac{1}{m_{ij}-m_{ij+1}}\right), & j = 1 \text{ and } (e\text{-}e) \text{ or } (e\text{-}p), \\[2ex]
\frac{-1}{l_{ij}^2(m_{ij}-m_{ij+1})}, & j = 1 \text{ and } (p\text{-}p) \text{ or } (p\text{-}e), \\[2ex]
\frac{-(m_{ij-1}-m_{ij+1})}{l_{ij}^2(m_{ij-1}-m_{ij})(m_{ij}-m_{ij+1})}, & 1 < j < n_i, \\[2ex]
\frac{-1}{l_{ij}^2(m_{ij-1}-m_{ij})}, & j = n_i \text{ and } (p\text{-}p) \text{ or } (e\text{-}p), \\[2ex]
\frac{1}{l_{ij}^2}\left(-\frac{1}{m^-} - \frac{1}{m_{ij}-m_{ij-1}}\right), & j = n_i \text{ and } (e\text{-}e) \text{ or } (p\text{-}e).
\end{cases}
$$

(iii) *For fixed point curves D_X^+ and D_X^-, we have*

$$
\left(D_X^+\right)^2 = -l^+ m^+, \qquad \left(D_X^-\right)^2 = l^- m^-.
$$

Proof The self-intersection number $(D_X^{ij})^2$ equals the intersection number $D_X^{ij} \cdot E_X^{ij}$ for some linearly equivalent divisor E_X^{ij}. We take

$$
E_X^{ij} := \frac{1}{l_{ij}}\left(\sum_j l_{tj} D_X^{tj} - \sum_{j \neq i} l_{ij} D_X^{ij}\right).
$$

The computation then is done by using Proposition 5.4.2.1. Similarly, to obtain $(D_X^+)^2$ and $(D_X^-)^2$, we just compute $D_X^+ \cdot E^+$ and $D_X^- \cdot E^-$ with

$$
E^+ := -\sum_{i=0}^r \sum_{j=1}^{n_i} d_{ij} D_X^{ij}, \qquad E^- := \sum_{i=0}^r \sum_{j=1}^{n_i} d_{ij} D_X^{ij}. \qquad \square
$$

Example 5.4.2.3 Fix an integer $a \in \mathbb{Z}_{\geq 0}$. We investigate the geometry of the $\mathbb{K}^*$-surface $X_a = X(A, P_a)$ associated with the matrices

$$
A := \begin{bmatrix} 0 & -1 & 1 \\ 1 & -1 & 0 \end{bmatrix}, \qquad
P_a := \begin{bmatrix} -1 & -1 & 1 & 1 & 0 & 0 & 0 & 0 \\ -1 & -1 & 0 & 0 & 1 & 1 & 0 & 0 \\ 1-a & -a & 1 & 0 & 1 & 0 & 1 & -1 \end{bmatrix}.
$$

Proposition 5.4.1.9 tells us that we have a parabolic source $D_{X_a}^+$, a parabolic sink $D_{X_a}^-$, three hyperbolic fixed points and all nontrivial orbits are free.

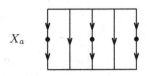

X_a

The surface X_a is smooth, its Cox ring $\mathcal{R}(X_a) = R(A, P_a)$ with the grading by $\mathrm{Cl}(X_a) = \mathbb{Z}^4$ is explicitly given as

$$R(A, P_a) = \mathbb{K}[T_{01}, T_{02}, T_{11}, T_{12}, T_{21}, T_{22}, S_1, S_2] / \langle T_{01}T_{02} + T_{11}T_{12} + T_{21}T_{22} \rangle,$$

$$Q_a = \begin{bmatrix} 1 & 0 & 0 & 1 & 0 & 1 & 0 & 1-a \\ 0 & 1 & 0 & 1 & 0 & 1 & 0 & -a \\ 0 & 0 & 1 & -1 & 0 & 0 & 0 & 1 \\ 0 & 0 & 0 & 0 & 1 & -1 & 0 & 1 \\ 0 & 0 & 0 & 0 & 0 & 0 & 1 & 1 \end{bmatrix}.$$

Let us compute the intersection numbers of the curves $D_{X_a}^{ij}$ and $D_{X_a}^+$, $D_{X_a}^-$. The slopes $m_{ij} = d_{ij}/l_{ij}$ are

$$m_{01} = 1-a, \quad m_{02} = -a, \quad m_{11} = 1, \quad m_{12} = 0, \quad m_{21} = 1, \quad m_{22} = 0.$$

Proposition 5.4.2.1 provides us with the intersection numbers of any two different curves intersecting nontrivially:

$$D_{X_a}^{i1} \cdot D_{X_a}^{i2} = 1, \quad D_{X_a}^{i1} \cdot D_{X_a}^+ = 1, \quad D_{X_a}^{i2} \cdot D_{X_a}^- = 1.$$

For the slope sums, we have $m^+ = 3 - a$ and $m^- = -a$. Moreover, $l^+ = l^- = 1$ holds. Corollary 5.4.2.2 gives us the self-intersection numbers:

$$\left(D_{X_a}^{ij}\right)^2 = -1 \quad \text{for all } i, j, \quad \left(D_{X_a}^+\right)^2 = a - 3, \quad \left(D_{X_a}^-\right)^2 = -a.$$

In particular, we can conclude that X_a is the blow-up of the ath Hirzebruch surface at three distinct points on the zero section. For a direct verification, consider

$$A' := \begin{bmatrix} 0 & -1 & 1 \\ 1 & -1 & 0 \end{bmatrix}, \qquad P'_a := \begin{bmatrix} -1 & 1 & 0 & 0 & 0 \\ -1 & 0 & 1 & 0 & 0 \\ -a & 0 & 0 & 1 & -1 \end{bmatrix}.$$

As P'_a arises from P_a by omitting the columns v_{01}, v_{11}, v_{21}, we obtain a map of the fans $\Sigma(P_a)$ and $\Sigma(P'_a)$ constructed in 5.4.1.6. This leads to a commutative diagram

$$\begin{array}{ccc} X_a & \longrightarrow & Z(P_a) \\ \downarrow & & \downarrow \\ X'_a & \longrightarrow & Z(P'_a) \end{array}$$

involving $X'_a = X(A, P'_a)$ and the toric ambient varieties. The map $X_a \to X'_a$ contracts the divisors $D^{i1}_{X_a}$, where $i = 0, 1, 2$. The Cox ring of X'_a is given by

$$R(A, P'_a) = \mathbb{K}[T_{01}, T_{11}, T_{21}, S^+, S^-]/\langle T_{01} + T_{11} + T_{21}\rangle,$$

$$Q'_a = \begin{bmatrix} 0 & 0 & 0 & 1 & 1 \\ 1 & 1 & 1 & a & 0 \end{bmatrix}.$$

We conclude that $R(A, P'_a)$ is a $\mathbb{Z}^2$-graded polynomial ring in four variables isomorphic to the Cox ring of the projective toric surface with the fan generated by

$$(1, 0), \qquad (0, 1), \qquad (-1, -a), \qquad (0, -1).$$

Remark 5.4.2.4 Consider a smooth $\mathbb{K}^*$-surface X with a parabolic source F^+ and a parabolic sink F^-. Orlik and Wagreich [238] associated with X a graph having besides F^+, F^- the exceptional curves as vertices, labeled by the respective self-intersection numbers, where two vertices are joined by an edge if and only if the corresponding curves intersect.

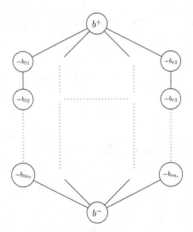

Note that for $X = X(A, P)$ the vertical chains of the graph reflect the vertical blocks of the matrix P and the position of an edge inside a chain reflects the position of a ray through a colunn of P. Moreover, according to [238, Sec. 3.5], the order l_{ij} of the generic isotropy group of D^{ij}_X equals the numerator of the canceled continued fraction

$$b_{i1} - \cfrac{1}{b_{i2} - \cfrac{1}{\cdots - \cfrac{1}{b_{ij-1}}}}.$$

5.4.3 Resolution of singularities

We consider the general procedure of resolving singularities of a variety with a torus action of complexity 1 in the surface setting. There, the combinatorics behind the procedure becomes very clear and allows to recover easily statements proven by Orlik and Wagreich [237, 238]. We work over an algebraically closed field $\mathbb{K}$ of characteristic zero.

Remark 5.4.3.1 Consider a $\mathbb{K}^*$-surface $X = X(A, P)$. Then every singular point of X is a fixed point. For the three different types of fixed points, we have the following statements.

(i) An elliptic fixed point $x^{\pm}$ is singular if and only if the corresponding cone $\sigma^{\pm} \in \Sigma(P)$ is singular or $x^{\pm} = p(z)$ with a singular point z of the coordinate space $\overline{X} = V(g_i; \; i \in \mathfrak{J})$.

(ii) A parabolic fixed point $x \in D_X^+$ ($x \in D_X^-$) is singular if and only if $x \in D_X^+ \cap D_X^{i1}$ holds with $l_{i1} > 1$ ($x \in D_X^- \cap D_X^{in_i}$ holds with $l_{in_i} > 1$).

(iii) A hyperbolic fixed point $x_{ij} \in D_X^{ij} \cap D_X^{ij+1}$ is singular if and only if the corresponding cone $\tau_{ij} = \mathrm{cone}(v_{ij}, v_{ij+1})$ is singular.

These are all direct applications of Corollary 3.3.1.12. Moreover, we obtain that $x \in X$ is a factorial singularity if and only if x is an elliptic fixed point, $x \in Z(P)$ is smooth, and $x = p(z)$ with a singular point $z \in \overline{X}$.

Now we apply the resolution of singularities provided by Theorem 3.4.4.9 to a $\mathbb{K}^*$-surface $X = X(A, P)$. The output is the so-called canonical resolution of singularities presented in [238].

Construction 5.4.3.2 (Canonical resolution of singularities)   Consider a $\mathbb{K}^*$-surface $X = X(A, P)$.

(i) *Tropical step.* Enlarge P to a matrix P' by adding e_{r+1} and $-e_{r+1}$, if not already present. Then the surface $X' := X(A, P')$ is quasismooth and there is a canonical morphism $X' \to X$.

(ii) *Toric step.* Let P'' be the matrix having the primitive generators of the regular subdivision of $\Sigma(P')$ as its columns, ordered as in Definition 5.4.1.1. Then $X'' := X(A, P'')$ is smooth and there is a canonical morphism $X'' \to X'$.

The composition $X'' \to X'$ of the tropical and the toric step is the *canonical resolution of singularities* of X.

Remark 5.4.3.3 Let $X = X(A, P)$ have elliptic fixed points. The tropical step in the resolution of singularities is a stellar subdivision of the fan $\Sigma(P)$

of the toric ambient variety $Z(P)$ at the cones $\sigma^{\pm}$ by inserting the ray through $\pm e_{r+1}$. It can be interpreted as the coarsest common refinement of $\Sigma(P)$ and the tropical variety of $X \cap \mathbb{T}^{r+1} \subseteq Z(P)$.

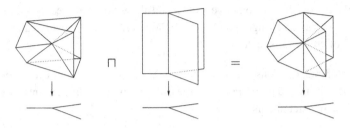

Geometrically this means replacing the elliptic fixed points with parabolic fixed point curves. Note that this process resolves the locus of indeterminacy of the rational quotient map $X \dashrightarrow \mathbb{P}^1$. By Proposition 3.4.4.6, the resulting surface is locally toric. The toric step is then done in the "sheets" of the fan $\Sigma(P')$, that means in the parts of $\Sigma(P')$ projecting via P_1 onto a ray.

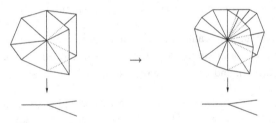

We see that each parabolic and each hyperbolic fixed point singularity has, as a toric surface singularity, a chain of rational curves as exceptional divisor. For the elliptic singularities, we look from above (below) what happened to σ^+ (σ^-) during the whole process and see a star-shaped exceptional divisor

as also observed by Orlik and Wagreich [237]. Note that the number of rays in the star shaped picture equals $r + 1$, that is, it is 2 plus the number of relations of the Cox ring of X.

For the discussion of examples, recall that a *rational double point*, also called *du Val singularity* or *ADE singularity*, of a normal surface X is a singular point $x_0 \in X$ admitting a resolution $\pi \colon X' \to X$ such that the exceptional divisor E over x_0 defines a graph of one of the following shapes

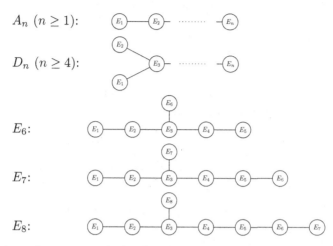

where the nodes represent the irreducible components $E_1, \ldots, E_n$ of E, the E_i are smooth rational curves of self-intersection $E_i^2 = -2$ with $E_i \cdot E_j = 0, 1$ and $E_i \cdot E_j = 1$ is indicated by joining the corresponding nodes with an edge.

Example 5.4.3.4 We consider once more the surface of Example 3.2.1.6, this time as a $\mathbb{K}^*$-surface. The defining matrices are

$$A := \begin{bmatrix} 0 & -1 & 1 \\ 1 & -1 & 0 \end{bmatrix}, \qquad P := \begin{bmatrix} -1 & -1 & 2 & 0 & 0 \\ -1 & -1 & 0 & 1 & 1 \\ 1 & 0 & -1 & 1 & 0 \end{bmatrix}.$$

To see that $X = X(A, P)$ is indeed the surface of 3.2.1.6, just compare the Cox rings; the one of $X(A, P)$ is by definition given as

$$R(A, P) = \mathbb{K}[T_{01}, T_{02}, T_{11}, T_{21}, T_{22}]/\langle T_{01}T_{02} + T_{11}^2 + T_{21}T_{22}\rangle,$$

$$Q = \begin{bmatrix} 1 & -1 & 0 & -1 & 1 \\ 1 & 1 & 1 & 0 & 2 \end{bmatrix},$$

where Q is a Gale dual of P, defining the grading of $R(A, P)$ by $K = \mathbb{Z}^4/\mathrm{im}\, P^* \cong \mathbb{Z}^2$. The fan $\Sigma(P)$ of the toric ambient variety Z has the maximal cones

$$\sigma^+ := \mathrm{cone}(v_{01}, v_{11}, v_{21}), \qquad \sigma^- := \mathrm{cone}(v_{02}, v_{11}, v_{22}),$$

$$\tau_{01} := \mathrm{cone}(v_{01}, v_{02}), \qquad \tau_{21} := \mathrm{cone}(v_{21}, v_{22}).$$

As we already observed in 3.3.1.14, the surface X comes with one singularity $x^+ = [0, 1, 0, 0, 1] \in X$; it is the toric fixed point corresponding to the cone σ^+. The tropical resolution step gives the matrix

$$P' = \begin{bmatrix} -1 & -1 & 2 & 0 & 0 & 0 & 0 \\ -1 & -1 & 0 & 1 & 1 & 0 & 0 \\ 1 & 0 & -1 & 1 & 0 & 1 & -1 \end{bmatrix}.$$

The resulting variety $X' = X(A, P')$ has two singularities, namely the hyperbolic fixed points corresponding to the cones

$$\tau_1^+ = \text{cone}(v_{11}, v^+), \qquad \tau_1^- = \text{cone}(v_{11}, v^-).$$

The toric step means passing to a regular subdivision of these cones. This is achieved by inserting the vectors $(1, 0, 0)$ and $(1, 0, -1)$. We arrive at the matrix

$$P'' = \begin{bmatrix} -1 & -1 & 1 & 2 & 1 & 0 & 0 & 0 & 0 \\ -1 & -1 & 0 & 0 & 0 & 1 & 1 & 0 & 0 \\ 1 & 0 & 0 & -1 & -1 & 1 & 0 & 1 & -1 \end{bmatrix}.$$

The associated $\mathbb{K}^*$-surface $X'' = X(A, P'')$ provides the canonical resolution. To obtain a minimal resolution, we figure out (-1)-curves among the new divisors. The slopes are

$$m_{01} = 1, \qquad m_{02} = 0, \qquad m_{11} = 0, \qquad m_{12} = -\frac{1}{2},$$

$$m_{13} = -1, \qquad m_{21} = 1, \qquad m_{22} = 0.$$

Thus, we have $m^+ = 2$ and $m^- = -1$. Moreover, $l^+ = l^- = 1$ holds. Computing the self intersection numbers according to Corollary 5.4.2.2 gives

$$\left(D_{X''}^+\right)^2 = -2, \qquad \left(D_{X''}^{11}\right)^2 = -2, \qquad \left(D_{X''}^{13}\right)^2 = -2, \qquad \left(D_{X''}^-\right)^2 = -1.$$

Thus, we may smoothly contract first $D_{X''}^-$ and then $D_{X''}^{13}$. In terms of matrices, this means passing to

$$\widetilde{P} = \begin{bmatrix} -1 & -1 & 1 & 2 & 0 & 0 & 0 \\ -1 & -1 & 0 & 0 & 1 & 1 & 0 \\ 1 & 0 & 0 & -1 & 1 & 0 & 1 \end{bmatrix}.$$

The associated $\mathbb{K}^*$-surface $\widetilde{X}$ comes with a map $\widetilde{X} \to X$ which is a minimal resolution. The Cox ring of $\widetilde{X}$ is given by

$$\mathcal{R}(\widetilde{X}) = \mathbb{K}[T_{ij}, S;\ i = 0, 1, 2\ j = 1, 2] / \langle T_{01}T_{02} + T_{11}T_{12}^2 + T_{21}T_{22} \rangle,$$

$$\widetilde{Q} := \begin{bmatrix} 0 & 1 & -1 & 1 & 1 & 0 & 0 \\ 1 & 0 & -1 & 1 & 0 & 1 & 0 \\ 0 & -1 & -1 & 0 & 0 & -1 & 0 \\ 0 & 0 & 0 & 0 & -1 & 1 & 1 \end{bmatrix}.$$

where the degrees of the generators in $\mathrm{Cl}(\widetilde{X}) = \mathbb{Z}^4$ are the columns of the Gale dual matrix $\widetilde{Q}$ of $\widetilde{P}$. The exceptional divisors $D_{\widetilde{X}}^{\pm}$ and $D_{\widetilde{X}}^{11}$ form a chain of smooth rational (-2)-curves and thus the singularity $x^+ \in X$ is of type A_2.

In the second example, we consider once more the E_6-singular cubic surface from Examples 3.4.3.3 and 3.4.4.2. The Cox ring of its minimal desingularization was first computed by Hassett and Tschinkel, without using the $\mathbb{K}^*$-action [162, Thm. 3.8].

Example 5.4.3.5 (The E_6-singular cubic V) Consider the $\mathbb{K}^*$-surface $X = X(A, P)$ defined by the two matrices

$$A := \begin{bmatrix} 0 & -1 & 1 \\ 1 & -1 & 0 \end{bmatrix}, \qquad P = \begin{bmatrix} -3 & -1 & 3 & 0 \\ -3 & -1 & 0 & 2 \\ -2 & -1 & 1 & 1 \end{bmatrix}.$$

The Cox ring $\mathcal{R}(X) = R(A, P)$ and its grading by $\mathrm{Cl}(X) = K = \mathbb{Z}$ are explicitly given as

$$R(A, P) = \mathbb{K}[T_{01}, T_{02}, T_{11}, T_{21}] / \langle T_{01}^3 T_{02} + T_{11}^3 + T_{21}^2 \rangle,$$

$$Q := \begin{bmatrix} 1 & 3 & 2 & 3 \end{bmatrix}.$$

Moreover, in terms of the columns $v_{01}, v_{02}, v_{11}, v_{21}$ of P, the fan $\Sigma(P)$ of the ambient toric variety $Z(P)$ has the maximal cones

$$\sigma^+ := \mathrm{cone}(v_{01}, v_{11}, v_{21}), \qquad \sigma^- := \mathrm{cone}(v_{02}, v_{11}, v_{21}),$$

$$\tau_{01} := \mathrm{cone}(v_{01}, v_{02}).$$

As observed in 3.4.4.2, the elliptic fixed point $x^+ = [0, 1, 0, 0]$ in the source is the only singular point in X. Consider the matrix

$$\widetilde{P} = \begin{bmatrix} -1 & -2 & -3 & -1 & 1 & 2 & 3 & 0 & 0 & 0 \\ -1 & -2 & -3 & -1 & 0 & 0 & 0 & 1 & 2 & 0 \\ 0 & -1 & -2 & -1 & 1 & 1 & 1 & 1 & 1 & 1 \end{bmatrix}.$$

obtained by adding to P the rays from the canonical resolution that lie in σ^+. Looking from above to σ^+, we see the shape of the resolution graph

where the black bullets indicate the new rays and hence the exceptional divisors of $\widetilde{X} := X(A, \widetilde{P})$ with respect to the minimal resolution $\widetilde{X} \to X$.

With the slopes

$$m_{01} = 0, \quad m_{02} = -\frac{1}{2}, \quad m_{03} = -\frac{2}{3}, \quad m_{04} = -1,$$

$$m_{11} = 1, \quad m_{12} = \frac{1}{2}, \quad m_{13} = \frac{1}{3},$$

$$m_{21} = 1, \quad m_{22} = \frac{1}{2},$$

the slope sum $m^+ = 2$ and $l^+ = 1$, we can compute the self-intersection numbers of the exceptional divisors according to Corollary 5.4.2.2. They are given as

$$\left(D_{\widetilde{X}}^{01}\right)^2 = \frac{-1}{l_{01}^2(m_{01} - m_{02})} = -2,$$

$$\left(D_{\widetilde{X}}^{02}\right)^2 = \frac{-(m_{01} - m_{03})}{l_{02}^2(m_{01} - m_{02})(m_{02} - m_{03})} = -2,$$

$$\left(D_{\widetilde{X}}^{11}\right)^2 = \frac{-1}{l_{11}^2(m_{11} - m_{12})} = -2,$$

$$\left(D_{\widetilde{X}}^{12}\right)^2 = \frac{-(m_{11} - m_{13})}{l_{12}^2(m_{11} - m_{12})(m_{12} - m_{13})} = -2,$$

$$\left(D_{\widetilde{X}}^{21}\right)^2 = \frac{-1}{l_{21}^2(m_{21} - m_{22})} = -2, \quad \left(D_{\widetilde{X}}^{+}\right)^2 = -l^+ m^+ = -2.$$

Thus, we see that the singularity $x^+ \in X$ is of type E_6. The Cox ring of the resolution $\widetilde{X}$ and its degree matrix (in row-echelon form) are given by

$$R(A, \widetilde{P}) = \frac{\mathbb{K}[T_{01}, T_{02}, T_{03}, T_{04}, T_{11}, T_{12}, T_{13}, T_{21}, T_{22}, S]}{\langle T_{01}T_{02}^2T_{03}^3T_{04} + T_{11}T_{12}^2T_{13}^3 + T_{21}T_{22}^2 \rangle},$$

$$\widetilde{Q} = \begin{bmatrix} 1 & 0 & 0 & 0 & 0 & 2 & -1 & 1 & 0 & 2 \\ 0 & 1 & 0 & 0 & 0 & 1 & 0 & 0 & 1 & 1 \\ 0 & 0 & 1 & 0 & 0 & 0 & 1 & 1 & 1 & 1 \\ 0 & 0 & 0 & 1 & 0 & 2 & -1 & 1 & 0 & 1 \\ 0 & 0 & 0 & 0 & 1 & 1 & -1 & 0 & 0 & 1 \\ 0 & 0 & 0 & 0 & 0 & 3 & -2 & 0 & 0 & 1 \\ 0 & 0 & 0 & 0 & 0 & 0 & 0 & 2 & -1 & 1 \end{bmatrix}.$$

5.4.4 Gorenstein log del Pezzo $\mathbb{K}^*$-surfaces

We present all Gorenstein log del Pezzo $\mathbb{K}^*$-surfaces X by listing their Cox rings and their defining matrices P. Recall that a *del Pezzo surface* is a normal projective surface with ample anticanonical divisor and a del Pezzo surface is *Gorenstein* if its anticanonical divisor is Cartier. Moreover, "log" means that X has at most *log terminal singularities* in the sense that

$$K_{\widetilde{X}} - \pi^* K_X = \sum_{i=1}^{r} \alpha_i E_i \qquad \text{with } \alpha_i \in \mathbb{Q}_{>-1}$$

holds for every resolution $\pi : \widetilde{X} \to X$ of singularities, where $E_1, \ldots, E_r$ are the irreducible components of the exceptional divisor and $K_{\widetilde{X}}$ as well as K_X are suitable representatives of the canonical divisor. Besides the general references [5, 229] on classification of log del Pezzo surfaces, the original references for this section are [110, 169, 177, 286]. The ground field $\mathbb{K}$ is algebraically closed and of characteristic zero.

Remark 5.4.4.1 Let $X = X(A, P)$ be a $\mathbb{K}^*$-surface arising from matrices A and P. Then an anticanonical divisor on X is given by

$$-K_X = \sum D_X^{ij} + \left(D_X^+ + D_X^- \right) - \sum_{j=1}^{n_{i_1}} l_{i_1 j} D_X^{i_1 j} - \cdots - \sum_{j=1}^{n_{i_{r-1}}} l_{i_{r-1} j} D_X^{i_{r-1} j},$$

where $0 \le i_1, \ldots, i_{r-1} \le r$ is any family of not necessarily distinct indices and we add D_X^+, D_X^- if they are present.

As it turns out, the Gorenstein del Pezzo $\mathbb{K}^*$-surfaces X have at most ADE-singularities and their Picard number is at most 4. The following statements provide a complete classification; for each X we list the Cox ring $\mathcal{R}(X)$, the divisor class group $\mathrm{Cl}(X)$, the degrees $(w_1, \ldots, w_t)$ of the generators, and the type $S(X)$ of the singularities; for example "$2A_1 A_3$" indicates that there are three singularities, two of type A_1 and one of type A_3.

Theorem 5.4.4.2 *The following table lists the Cox ring $\mathcal{R}(X)$ and the singularity type $S(X)$ of the Gorenstein log del Pezzo $\mathbb{K}^*$-surfaces X of Picard number 1.*

No.	$\mathcal{R}(X)$	$\mathrm{Cl}(X)$	$(w_1, \ldots, w_r)$	$S(X)$
1	$\mathbb{K}[T_1, T_2, T_3]$	$\mathbb{Z}$	$\begin{bmatrix} 1 & 1 & 1 \end{bmatrix}$	—
2	$\mathbb{K}[T_1, T_2, T_3]$	$\mathbb{Z}$	$\begin{bmatrix} 2 & 1 & 1 \end{bmatrix}$	A_1
3	$\mathbb{K}[T_1, T_2, T_3]$	$\mathbb{Z}$	$\begin{bmatrix} 3 & 2 & 1 \end{bmatrix}$	$A_1 A_2$
4	$\mathbb{K}[T_1, T_2, T_3]$	$\mathbb{Z} \oplus \mathbb{Z}/2\mathbb{Z}$	$\begin{bmatrix} 2 & 1 & 1 \\ \bar{1} & \bar{1} & \bar{0} \end{bmatrix}$	$2A_1 A_3$
5	$\mathbb{K}[T_1, T_2, T_3]$	$\mathbb{Z} \oplus \mathbb{Z}/3\mathbb{Z}$	$\begin{bmatrix} 1 & 1 & 1 \\ \bar{0} & \bar{1} & \bar{2} \end{bmatrix}$	$3A_2$

6	$\dfrac{\mathbb{K}[T_1,...,T_4]}{\langle T_1T_2+T_3^3+T_4^2\rangle}$	$\mathbb{Z}$	$\begin{bmatrix}1&5&2&3\end{bmatrix}$	A_4
7	$\dfrac{\mathbb{K}[T_1,...,T_4]}{\langle T_1^2T_2+T_3^3+T_4^2\rangle}$	$\mathbb{Z}$	$\begin{bmatrix}1&4&2&3\end{bmatrix}$	D_5
8	$\dfrac{\mathbb{K}[T_1,...,T_4]}{\langle T_1T_2+T_3^4+T_4^2\rangle}$	$\mathbb{Z}\oplus\mathbb{Z}/2\mathbb{Z}$	$\begin{bmatrix}1&3&1&2\\\bar1&\bar1&\bar0&\bar1\end{bmatrix}$	A_1A_5
9	$\dfrac{\mathbb{K}[T_1,...,T_4]}{\langle T_1^3T_2+T_3^3+T_4^2\rangle}$	$\mathbb{Z}$	$\begin{bmatrix}1&3&2&3\end{bmatrix}$	E_6
10	$\dfrac{\mathbb{K}[T_1,...,T_4]}{\langle T_1T_2+T_3^2+T_4^2\rangle}$	$\mathbb{Z}\oplus\mathbb{Z}/4\mathbb{Z}$	$\begin{bmatrix}1&1&1&1\\\bar1&\bar3&\bar2&\bar0\end{bmatrix}$	A_12A_3
11	$\dfrac{\mathbb{K}[T_1,T_2,T_3,S_1]}{\langle T_1^2+T_2^2+T_3^2\rangle}$	$\mathbb{Z}\oplus\mathbb{Z}/2\mathbb{Z}\oplus\mathbb{Z}/2\mathbb{Z}$	$\begin{bmatrix}1&1&1&1\\\bar1&\bar0&\bar1&\bar0\\\bar1&\bar1&\bar0&\bar0\end{bmatrix}$	$3A_1D_4$
12	$\dfrac{\mathbb{K}[T_1,...,T_4]}{\langle T_1^2T_2+T_3^4+T_4^2\rangle}$	$\mathbb{Z}\oplus\mathbb{Z}/2\mathbb{Z}$	$\begin{bmatrix}1&2&1&2\\\bar1&\bar0&\bar0&\bar1\end{bmatrix}$	A_1D_6
13	$\dfrac{\mathbb{K}[T_1,...,T_4]}{\langle T_1T_2+T_3^3+T_4^3\rangle}$	$\mathbb{Z}\oplus\mathbb{Z}/3\mathbb{Z}$	$\begin{bmatrix}1&2&1&1\\\bar1&\bar2&\bar2&\bar0\end{bmatrix}$	A_2A_5
14	$\dfrac{\mathbb{K}[T_1,...,T_4]}{\langle T_1^4T_2+T_3^3+T_4^2\rangle}$	$\mathbb{Z}$	$\begin{bmatrix}1&2&2&3\end{bmatrix}$	E_7
15	$\dfrac{\mathbb{K}[T_1,...,T_4]}{\langle T_1^3T_2+T_3^4+T_4^2\rangle}$	$\mathbb{Z}\oplus\mathbb{Z}/2\mathbb{Z}$	$\begin{bmatrix}1&1&1&2\\\bar0&\bar0&\bar1&\bar1\end{bmatrix}$	A_1E_7
16	$\dfrac{\mathbb{K}[T_1,...,T_4]}{\langle T_1^2T_2+T_3^3+T_4^3\rangle}$	$\mathbb{Z}\oplus\mathbb{Z}/3\mathbb{Z}$	$\begin{bmatrix}1&1&1&1\\\bar1&\bar1&\bar2&\bar0\end{bmatrix}$	A_2E_6
17	$\dfrac{\mathbb{K}[T_1,...,T_4]}{\langle T_1^5T_2+T_3^3+T_4^2\rangle}$	$\mathbb{Z}$	$\begin{bmatrix}1&1&2&3\end{bmatrix}$	E_8
18	$\dfrac{\mathbb{K}[T_1,...,T_5]}{\left(\begin{smallmatrix}T_1T_2+T_3^2+T_4^2,\\ \lambda T_3^2+T_4^2+T_5^2\end{smallmatrix}\right)}$	$\mathbb{Z}\oplus\mathbb{Z}/2\mathbb{Z}\oplus\mathbb{Z}/2\mathbb{Z}$	$\begin{bmatrix}1&1&1&1&1\\\bar1&\bar1&\bar0&\bar1&\bar0\\\bar0&\bar0&\bar1&\bar1&\bar0\end{bmatrix}$	$2D_4$

where $\lambda \in \mathbb{K}^* \setminus \{1\}$ in no. 18. The surfaces no. 1 to no. 5 are toric and the surfaces no. 6 to no. 18 are nontoric. Moreover, the defining matrices P of the latter ones are given as

$$P_{A_4} = \begin{bmatrix} -1 & -1 & 3 & 0 \\ -1 & -1 & 0 & 2 \\ 0 & -1 & 1 & 1 \end{bmatrix} \qquad P_{D_5} = \begin{bmatrix} -2 & -1 & 3 & 0 \\ -2 & -1 & 0 & 2 \\ -1 & -1 & 1 & 1 \end{bmatrix}$$

$$P_{A_1A_5} = \begin{bmatrix} -1 & -1 & 4 & 0 \\ -1 & -1 & 0 & 2 \\ 0 & -1 & 1 & 1 \end{bmatrix} \qquad P_{E_6} = \begin{bmatrix} -3 & -1 & 3 & 0 \\ -3 & -1 & 0 & 2 \\ -2 & -1 & 1 & 1 \end{bmatrix}$$

$$P_{A_12A_3} = \begin{bmatrix} -1 & -1 & 2 & 0 \\ -1 & -1 & 0 & 2 \\ 0 & -2 & 1 & 1 \end{bmatrix} \qquad P_{3A_1D_4} = \begin{bmatrix} -2 & 2 & 0 & 0 \\ -2 & 0 & 2 & 0 \\ -3 & 1 & 1 & 1 \end{bmatrix}$$

$$P_{A_1D_6} = \begin{bmatrix} -2 & -1 & 4 & 0 \\ -2 & -1 & 0 & 2 \\ -1 & -1 & 1 & 1 \end{bmatrix} \qquad P_{A_2A_5} = \begin{bmatrix} -1 & -1 & 3 & 0 \\ -1 & -1 & 0 & 3 \\ 0 & -1 & 1 & 1 \end{bmatrix}$$

$$P_{E_7} = \begin{bmatrix} -4 & -1 & 3 & 0 \\ -4 & -1 & 0 & 2 \\ -3 & -1 & 1 & 1 \end{bmatrix} \qquad P_{A_1E_7} = \begin{bmatrix} -3 & -1 & 4 & 0 \\ -3 & -1 & 0 & 2 \\ -2 & -1 & 1 & 1 \end{bmatrix}$$

$$P_{A_2E_6} = \begin{bmatrix} -2 & -1 & 3 & 0 \\ -2 & -1 & 0 & 3 \\ -1 & -1 & 1 & 1 \end{bmatrix} \qquad P_{E_8} = \begin{bmatrix} -5 & -1 & 3 & 0 \\ -5 & -1 & 0 & 2 \\ -4 & -1 & 1 & 1 \end{bmatrix}$$

$$P_{2D_4} = \begin{bmatrix} -1 & -1 & 2 & 0 & 0 \\ -1 & -1 & 0 & 2 & 0 \\ -1 & -1 & 0 & 0 & 2 \\ -1 & -2 & 1 & 1 & 1 \end{bmatrix}$$

Theorem 5.4.4.3 *The following table lists the Cox ring $\mathcal{R}(X)$ and the singularity type $S(X)$ of the Gorenstein log del Pezzo $\mathbb{K}^*$-surfaces X of Picard number 2.*

No.	$\mathcal{R}(X)$	$\mathrm{Cl}(X)$	$(w_1, \ldots, w_r)$	$S(X)$
1	$\mathbb{K}[T_1, \ldots, T_4]$	$\mathbb{Z}^2$	$\begin{bmatrix} 1 & 0 & 1 & 0 \\ 0 & 1 & 0 & 1 \end{bmatrix}$	—
2	$\mathbb{K}[T_1, \ldots, T_4]$	$\mathbb{Z}^2$	$\begin{bmatrix} 1 & -1 & 1 & 0 \\ 0 & 1 & 0 & 1 \end{bmatrix}$	—
3	$\mathbb{K}[T_1, \ldots, T_4]$	$\mathbb{Z}^2$	$\begin{bmatrix} 1 & 1 & 1 & 0 \\ 0 & 1 & -1 & 1 \end{bmatrix}$	A_1
4	$\mathbb{K}[T_1, \ldots, T_4]$	$\mathbb{Z}^2$	$\begin{bmatrix} 2 & 0 & 1 & 1 \\ 0 & 1 & 0 & 1 \end{bmatrix}$	$2A_1$
5	$\mathbb{K}[T_1, \ldots, T_4]$	$\mathbb{Z}^2 \oplus \mathbb{Z}/2\mathbb{Z}$	$\begin{bmatrix} 1 & 0 & 1 & 0 \\ 0 & 1 & 0 & 1 \\ \bar{1} & \bar{1} & \bar{0} & \bar{0} \end{bmatrix}$	$4A_1$

6	$\mathbb{K}[T_1,\ldots,T_4]$	$\mathbb{Z}^2$	$\begin{bmatrix} 1 & -1 & 1 & 0 \\ 0 & 2 & 1 & 1 \end{bmatrix}$	A_1A_2
7	$\mathbb{K}[T_1,\ldots,T_4]$	$\mathbb{Z}^2$	$\begin{bmatrix} 1 & 1 & 1 & 1 \\ 2 & 0 & 1 & -1 \end{bmatrix}$	$2A_1A_2$
8	$\dfrac{\mathbb{K}[T_1,\ldots,T_4,S_1]}{\langle T_1T_2+T_3^2+T_4^2\rangle}$	$\mathbb{Z}^2 \oplus \mathbb{Z}/2\mathbb{Z}$	$\begin{bmatrix} 1 & 1 & 1 & 1 & 1 \\ -1 & 1 & 0 & 0 & 1 \\ \bar{1} & \bar{1} & \bar{1} & \bar{0} & \bar{0} \end{bmatrix}$	$2A_1A_3$
9	$\dfrac{\mathbb{K}[T_1,\ldots,T_5]}{\langle T_1T_2+T_3T_4+T_5^2\rangle}$	$\mathbb{Z}^2$	$\begin{bmatrix} 1 & 3 & 1 & 3 & 2 \\ 1 & 1 & 0 & 2 & 1 \end{bmatrix}$	A_2
10	$\dfrac{\mathbb{K}[T_1,\ldots,T_5]}{\langle T_1T_2+T_3^2T_4+T_5^2\rangle}$	$\mathbb{Z}^2$	$\begin{bmatrix} 1 & 3 & 1 & 2 & 2 \\ 0 & 2 & 1 & 0 & 1 \end{bmatrix}$	A_3
11	$\dfrac{\mathbb{K}[T_1,\ldots,T_5]}{\langle T_1T_2+T_3T_4+T_5^3\rangle}$	$\mathbb{Z}^2$	$\begin{bmatrix} 1 & 2 & 1 & 2 & 1 \\ 1 & -1 & -1 & 1 & 0 \end{bmatrix}$	A_1A_3
12	$\dfrac{\mathbb{K}[T_1,\ldots,T_5]}{\langle T_1T_2+T_3^3T_4+T_5^2\rangle}$	$\mathbb{Z}^2$	$\begin{bmatrix} 1 & 3 & 1 & 1 & 2 \\ 1 & 1 & 0 & 2 & 1 \end{bmatrix}$	A_4
13	$\dfrac{\mathbb{K}[T_1,\ldots,T_5]}{\langle T_1^2T_2+T_3^2T_4+T_5^2\rangle}$	$\mathbb{Z}^2$	$\begin{bmatrix} 1 & 2 & 1 & 2 & 2 \\ 1 & 0 & 0 & 2 & 1 \end{bmatrix}$	D_4
14	$\dfrac{\mathbb{K}[T_1,\ldots,T_5]}{\langle T_1T_2+T_3T_4+T_5^2\rangle}$	$\mathbb{Z}^2$	$\begin{bmatrix} 1 & 1 & 1 & 1 & 1 \\ -1 & 1 & 2 & -2 & 0 \end{bmatrix}$	A_12A_2
15	$\dfrac{\mathbb{K}[T_1,\ldots,T_5]}{\langle T_1T_2+T_3^2T_4+T_5^3\rangle}$	$\mathbb{Z}^2$	$\begin{bmatrix} 1 & 2 & 1 & 1 & 1 \\ -1 & 1 & 1 & -2 & 0 \end{bmatrix}$	A_1A_4
16	$\dfrac{\mathbb{K}[T_1,\ldots,T_5]}{\langle T_1^3T_2+T_3^2T_4+T_5^2\rangle}$	$\mathbb{Z}^2$	$\begin{bmatrix} 1 & 1 & 1 & 2 & 2 \\ 1 & -1 & 0 & 2 & 1 \end{bmatrix}$	D_5
17	$\dfrac{\mathbb{K}[T_1,\ldots,T_5]}{\langle T_1^2T_2+T_3^2T_4+T_5^3\rangle}$	$\mathbb{Z}^2$	$\begin{bmatrix} 1 & 1 & 1 & 1 & 1 \\ -1 & 2 & 1 & -2 & 0 \end{bmatrix}$	A_1D_5
18	$\dfrac{\mathbb{K}[T_1,\ldots,T_5]}{\langle T_1^3T_2+T_3^3T_4+T_5^2\rangle}$	$\mathbb{Z}^2$	$\begin{bmatrix} 1 & 1 & 1 & 1 & 2 \\ 1 & -1 & 0 & 2 & 1 \end{bmatrix}$	E_6
19	$\dfrac{\mathbb{K}[T_1,\ldots,T_6]}{\left\langle \begin{array}{c} T_1T_2+T_3T_4+T_5^2, \\ \lambda T_3T_4+T_5^2+T_6^2 \end{array}\right\rangle}$	$\mathbb{Z}^2 \oplus \mathbb{Z}/2\mathbb{Z}$	$\begin{bmatrix} 1 & 1 & 1 & 1 & 1 & 1 \\ -1 & 1 & 1 & -1 & 0 & 0 \\ \bar{0} & \bar{0} & \bar{1} & \bar{1} & \bar{1} & \bar{0} \end{bmatrix}$	$2A_3$

where $\lambda \in \mathbb{K}^* \setminus \{1\}$ in no. 19. The surfaces no. 1 to no. 7 are toric and the surfaces no. 8 to no. 19 are nontoric. Moreover, the defining matrices P of the latter ones are given as

$$P_{2A_1A_3} = \begin{bmatrix} -1 & -1 & 2 & 0 & 0 \\ -1 & -1 & 0 & 2 & 0 \\ -1 & -2 & 1 & 1 & 1 \end{bmatrix} \qquad P_{A_2} = \begin{bmatrix} -1 & -1 & 1 & 1 & 0 \\ -1 & -1 & 0 & 0 & 2 \\ 0 & -1 & 1 & 0 & 1 \end{bmatrix}$$

$$P_{A_3} = \begin{bmatrix} -1 & -1 & 2 & 1 & 0 \\ -1 & -1 & 0 & 0 & 2 \\ 0 & -1 & 1 & 0 & 1 \end{bmatrix} \qquad P_{A_1A_3} = \begin{bmatrix} -1 & -1 & 1 & 1 & 0 \\ -1 & -1 & 0 & 0 & 3 \\ 0 & -1 & 1 & 0 & 1 \end{bmatrix}$$

$$P_{A_4} = \begin{bmatrix} -1 & -1 & 3 & 1 & 0 \\ -1 & -1 & 0 & 0 & 2 \\ 0 & -1 & 1 & 0 & 1 \end{bmatrix} \qquad P_{D_4} = \begin{bmatrix} -2 & -1 & 2 & 1 & 0 \\ -2 & -1 & 0 & 0 & 2 \\ -1 & -1 & 1 & 0 & 1 \end{bmatrix}$$

$$P_{A_12A_2} = \begin{bmatrix} -1 & -1 & 1 & 1 & 0 \\ -1 & -1 & 0 & 0 & 2 \\ 0 & -2 & 1 & 0 & 1 \end{bmatrix} \qquad P_{A_1A_4} = \begin{bmatrix} -1 & -1 & 2 & 1 & 0 \\ -1 & -1 & 0 & 0 & 3 \\ 0 & -1 & 1 & 0 & 1 \end{bmatrix}$$

$$P_{D_5} = \begin{bmatrix} -3 & -1 & 2 & 1 & 0 \\ -3 & -1 & 0 & 0 & 2 \\ -2 & -1 & 1 & 0 & 1 \end{bmatrix} \qquad P_{A_1D_5} = \begin{bmatrix} -2 & -1 & 2 & 1 & 0 \\ -2 & -1 & 0 & 0 & 3 \\ -1 & -1 & 1 & 0 & 1 \end{bmatrix}$$

$$P_{E_6} = \begin{bmatrix} -3 & -1 & 3 & 1 & 0 \\ -3 & -1 & 0 & 0 & 2 \\ -2 & -1 & 1 & 0 & 1 \end{bmatrix} \qquad P_{2A_3} = \begin{bmatrix} -1 & -1 & 1 & 1 & 0 & 0 \\ -1 & -1 & 0 & 0 & 2 & 0 \\ -1 & -1 & 0 & 0 & 0 & 2 \\ -1 & -2 & 1 & 0 & 1 & 1 \end{bmatrix}$$

Theorem 5.4.4.4 *The following table lists the Cox ring $\mathcal{R}(X)$ and the singularity type $S(X)$ of the Gorenstein log del Pezzo $\mathbb{K}^*$-surfaces X of Picard number 3.*

No.	$\mathcal{R}(X)$	$Cl(X)$	$(w_1, \ldots, w_r)$	$S(X)$
1	$\mathbb{K}[T_1, \ldots, T_5]$	$\mathbb{Z}^3$	$\begin{bmatrix} 1 & 0 & 1 & 0 & 0 \\ 1 & 1 & 0 & 1 & 0 \\ 0 & 1 & 0 & 0 & 1 \end{bmatrix}$	—
2	$\mathbb{K}[T_1, \ldots, T_5]$	$\mathbb{Z}^3$	$\begin{bmatrix} 1 & 0 & 1 & 0 & 0 \\ 1 & 1 & 0 & 1 & 0 \\ 0 & 1 & 1 & 0 & 1 \end{bmatrix}$	A_1
3	$\mathbb{K}[T_1, \ldots, T_5]$	$\mathbb{Z}^3$	$\begin{bmatrix} 1 & -1 & 1 & 0 & 0 \\ 1 & 0 & 0 & 1 & 0 \\ 0 & 1 & 1 & 0 & 1 \end{bmatrix}$	$2A_1$

4	$\dfrac{\mathbb{K}[T_1,...,T_6]}{\langle T_1T_2+T_3T_4+T_5T_6\rangle}$	$\mathbb{Z}^3$	$\begin{bmatrix} 1 & -1 & -1 & 1 & 0 & 0 \\ 0 & 1 & 0 & 1 & 1 & 0 \\ 1 & 0 & 0 & 1 & 0 & 1 \end{bmatrix}$	A_1
5	$\dfrac{\mathbb{K}[T_1,...,T_6]}{\langle T_1T_2+T_3T_4+T_5^2T_6\rangle}$	$\mathbb{Z}^3$	$\begin{bmatrix} 1 & 2 & 1 & 2 & 1 & 1 \\ 1 & 0 & 0 & 1 & 0 & 1 \\ 1 & -1 & -1 & 1 & 0 & 0 \end{bmatrix}$	A_2
6	$\dfrac{\mathbb{K}[T_1,...,T_6]}{\langle T_1T_2+T_3T_4+T_5T_6\rangle}$	$\mathbb{Z}^3$	$\begin{bmatrix} 1 & 1 & 1 & 1 & 1 & 1 \\ 1 & 0 & 0 & 1 & 0 & 1 \\ 0 & 0 & 1 & -1 & -1 & 1 \end{bmatrix}$	$3A_1$
7	$\dfrac{\mathbb{K}[T_1,...,T_5,S_1]}{\langle T_1T_2+T_3T_4+T_5^2\rangle}$	$\mathbb{Z}^3$	$\begin{bmatrix} 1 & 1 & 1 & 1 & 1 & 1 \\ -1 & 1 & 0 & 0 & 0 & 1 \\ -1 & 1 & 1 & -1 & 0 & 0 \end{bmatrix}$	A_1A_2
8	$\dfrac{\mathbb{K}[T_1,...,T_6]}{\langle T_1T_2+T_3^2T_4+T_5^2T_6\rangle}$	$\mathbb{Z}^3$	$\begin{bmatrix} 1 & 2 & 1 & 1 & 1 & 1 \\ 1 & 0 & 0 & 1 & 0 & 1 \\ 0 & 1 & 1 & -1 & 0 & 1 \end{bmatrix}$	A_3
9	$\dfrac{\mathbb{K}[T_1,...,T_6]}{\langle T_1^2T_2+T_3^2T_4+T_5^2T_6\rangle}$	$\mathbb{Z}^3$	$\begin{bmatrix} 1 & 1 & 1 & 1 & 1 & 1 \\ -1 & 1 & 0 & -1 & 0 & -1 \\ 0 & 1 & 1 & -1 & 0 & 1 \end{bmatrix}$	D_4
10	$\dfrac{\mathbb{K}[T_1,...,T_7]}{\left\langle \begin{array}{c} T_1T_2+T_3T_4+T_5T_6, \\ \lambda T_3T_4+T_5T_6+T_7^2 \end{array}\right\rangle}$	$\mathbb{Z}^3$	$\begin{bmatrix} 1 & 1 & 1 & 1 & 1 & 1 & 1 \\ 0 & 0 & -1 & 1 & 1 & -1 & 0 \\ 1 & -1 & -1 & 1 & 0 & 0 & 0 \end{bmatrix}$	$2A_2$

where $\lambda \in \mathbb{K}^* \setminus \{1\}$ in no. 10. The surfaces nos. 1, 2, and 3 are toric and the surfaces no. 4 to no. 10 are nontoric. Moreover, the defining matrices P of the latter ones are given as

$$P_{A_1} = \begin{bmatrix} -1 & -1 & 1 & 1 & 0 & 0 \\ -1 & -1 & 0 & 0 & 1 & 1 \\ 0 & -1 & 1 & 0 & 1 & 0 \end{bmatrix} \quad P_{A_2} = \begin{bmatrix} -1 & -1 & 1 & 1 & 0 & 0 \\ -1 & -1 & 0 & 0 & 2 & 1 \\ 0 & -1 & 1 & 0 & 1 & 0 \end{bmatrix}$$

$$P_{3A_1} = \begin{bmatrix} -1 & -1 & 1 & 1 & 0 & 0 \\ -1 & -1 & 0 & 0 & 1 & 1 \\ 0 & -2 & 1 & 0 & 1 & 0 \end{bmatrix} \quad P_{A_1A_2} = \begin{bmatrix} -1 & -1 & 1 & 1 & 0 & 0 \\ -1 & -1 & 0 & 0 & 2 & 0 \\ -1 & -2 & 1 & 0 & 1 & 1 \end{bmatrix}$$

$$P_{A_3} = \begin{bmatrix} -1 & -1 & 2 & 1 & 0 & 0 \\ -1 & -1 & 0 & 0 & 2 & 1 \\ 0 & -1 & 1 & 0 & 1 & 0 \end{bmatrix} \quad P_{D_4} = \begin{bmatrix} -2 & -1 & 2 & 1 & 0 & 0 \\ -2 & -1 & 0 & 0 & 2 & 1 \\ -1 & -1 & 1 & 0 & 1 & 0 \end{bmatrix}$$

$$P_{2A_2} = \begin{bmatrix} -1 & -1 & 1 & 1 & 0 & 0 & 0 \\ -1 & -1 & 0 & 0 & 1 & 1 & 0 \\ -1 & -1 & 0 & 0 & 0 & 0 & 2 \\ -1 & -2 & 1 & 0 & 1 & 0 & 1 \end{bmatrix}$$

Theorem 5.4.4.5 *The following table lists the Cox ring $\mathcal{R}(X)$ and the singularity type $S(X)$ of the Gorenstein log del Pezzo $\mathbb{K}^*$-surfaces X of Picard number 4.*

No.	$\mathcal{R}(X)$	$\mathrm{Cl}(X)$	$(w_1, \ldots, w_r)$	$S(X)$
1	$\mathbb{K}[T_1, \ldots, T_6]$	$\mathbb{Z}^4$	$\begin{bmatrix} 1 & -1 & 1 & 0 & 0 & 0 \\ 1 & 0 & 0 & 1 & 0 & 0 \\ 0 & 1 & 0 & 0 & 1 & 0 \\ 0 & 0 & 1 & 0 & 0 & 1 \end{bmatrix}$	—
2	$\dfrac{\mathbb{K}[T_1, \ldots, T_6, S_1]}{(T_1 T_2 + T_3 T_4 + T_5 T_6)}$	$\mathbb{Z}^4$	$\begin{bmatrix} 1 & 1 & 1 & 1 & 1 & 1 & 1 \\ 0 & 1 & 1 & 0 & 1 & 0 & 0 \\ 1 & 0 & 1 & 0 & 0 & 1 & 0 \\ -1 & 1 & 0 & 0 & 0 & 0 & 1 \end{bmatrix}$	A_1
3	$\dfrac{\mathbb{K}[T_1, \ldots, T_8]}{\left(\substack{T_1 T_2 + T_3 T_4 + T_5 T_6, \\ \lambda T_3 T_4 + T_5 T_6 + T_7 T_8}\right)}$	$\mathbb{Z}^4$	$\begin{bmatrix} 1 & 1 & 1 & 1 & 1 & 1 & 1 & 1 \\ 0 & 1 & 1 & 0 & 0 & 1 & 1 & 0 \\ 0 & 0 & 1 & -1 & -1 & 1 & 0 & 0 \\ -1 & 1 & 1 & -1 & 0 & 0 & 0 & 0 \end{bmatrix}$	$2A_1$

where $\lambda \in \mathbb{K}^* \setminus \{1\}$ in no. 3. The surface no. 1 is toric and the surfaces no. 2 and no. 3 are nontoric. Moreover, the defining matrices P of the latter ones are given as

$$P_{A_1} = \begin{bmatrix} -1 & -1 & 1 & 1 & 0 & 0 & 0 \\ -1 & -1 & 0 & 0 & 1 & 1 & 0 \\ -1 & -2 & 1 & 0 & 1 & 0 & 1 \end{bmatrix}$$

$$P_{2A_1} = \begin{bmatrix} -1 & -1 & 1 & 1 & 0 & 0 & 0 & 0 \\ -1 & -1 & 0 & 0 & 1 & 1 & 0 & 0 \\ -1 & -1 & 0 & 0 & 0 & 0 & 1 & 1 \\ -1 & -2 & 1 & 0 & 1 & 0 & 1 & 0 \end{bmatrix}$$

We conclude the section with an example of a smooth rational $\mathbb{K}^*$-surface having a noneffective anticanonical divisor.

Example 5.4.4.6 The surface X from Remark 5.1.3.11 is in fact a $\mathbb{K}^*$-surface of the form $X = X(A, P)$ with the matrices

$$A := \begin{bmatrix} 0 & -1 & 1 \\ 1 & -1 & 0 \end{bmatrix}, \qquad P = \begin{bmatrix} -2 & 3 & 0 & 0 \\ -2 & 0 & 11 & 0 \\ -3 & 2 & 9 & 1 \end{bmatrix}.$$

The Cox ring $\mathcal{R}(X) = R(A, P)$ and its grading by $\mathrm{Cl}(X) = K = \mathbb{Z}$ are explicitly given as

$$R(A, P) = \mathbb{K}[T_{01}, T_{11}, T_{21}, S_1] / \langle T_{01}^2 + T_{11}^3 + T_{21}^{11} \rangle,$$

$$Q = \begin{bmatrix} 33 & 22 & 6 & 1 \end{bmatrix}.$$

The $\mathbb{K}^*$-surface X comes with a parabolic source D_X^+ and the sink consists of an elliptic fixed point given in Cox coordinates as

$$x^- = [0, 0, 0, 1].$$

Note that x^- defines a smooth point in the ambient toric variety $Z(P) = \mathbb{P}_{33,22,6,1}$ but it comes from a singular point of the total coordinate space

$$\overline{X} = V(T_{01}^2 + T_{11}^3 + T_{21}^{11}) \subseteq \mathbb{K}^4.$$

Consequently x^- is a factorial singularity of X. There are three further singularities on X, given in Cox coordinates as

$$x_1 := [1, 0, 0, 0], \qquad x_2 := [0, 1, 0, 0], \qquad x_3 := [0, 0, 1, 0].$$

All three lie on D_X^+ and are inherited from the ambient toric variety $\mathbb{P}_{33,22,6,1}$. The anticanonical class of X is given by

$$-w_X^{\text{can}} = 33 + 22 + 6 + 1 - 66 = -4 \in \mathbb{Z} \cong \text{Cl}(X)$$

and is not effective. Now consider the canonical resolution $X'' \to X$. We have $X'' = X(A, P'')$ with the matrix

$$P'' = \begin{bmatrix} -1 & -2 & -1 & 1 & 3 & 2 & 1 & 0 & 0 & 0 & 0 & 0 & 0 & 0 & 0 & 0 \\ -1 & -2 & -1 & 0 & 0 & 0 & 1 & 6 & 11 & 5 & 4 & 3 & 2 & 1 & 0 & 0 \\ -2 & -3 & -1 & 1 & 2 & 1 & 0 & 1 & 5 & 9 & 4 & 3 & 2 & 1 & 0 & 1 & -1 \end{bmatrix}.$$

The Cox ring of the smooth $\mathbb{K}^*$-surface X'' and its grading by $\text{Cl}(X'') \cong \mathbb{Z}^{14}$ are given by

$$R(A, P'') = \frac{\mathbb{K}[T_{01}, T_{02}, T_{03}, T_{11}, T_{12} T_{13}, T_{14}, T_{21}, T_{22}, T_{23}, T_{24}, T_{25}, T_{26} T_{27}, T_{28}, S_1, S_2]}{\langle T_{01} T_{02}^2 T_{03} + T_{11} T_{12}^3 T_{13}^2 T_{14} + T_{21} T_{22}^6 T_{23}^{11} T_{24}^5 T_{25}^4 T_{26}^3 T_{27}^2 T_{28} \rangle},$$

$$Q'' = \begin{bmatrix}
1 & 0 & 0 & 0 & 0 & 0 & 1 & 0 & 0 & 0 & 0 & 0 & 0 & 0 & 1 & 0 & -2 \\
0 & 1 & 0 & 0 & 0 & 0 & 2 & 0 & 0 & 0 & 0 & 0 & 0 & 0 & 2 & 0 & -3 \\
0 & 0 & 1 & 0 & 0 & 0 & 1 & 0 & 0 & 0 & 0 & 0 & 0 & 0 & 1 & 0 & -1 \\
0 & 0 & 0 & 1 & 0 & 0 & -1 & 0 & 0 & 0 & 0 & 0 & 0 & 0 & 0 & 0 & 1 \\
0 & 0 & 0 & 0 & 1 & 0 & -3 & 0 & 0 & 0 & 0 & 0 & 0 & 0 & 0 & 0 & 2 \\
0 & 0 & 0 & 0 & 0 & 1 & -2 & 0 & 0 & 0 & 0 & 0 & 0 & 0 & 0 & 0 & 1 \\
0 & 0 & 0 & 0 & 0 & 0 & 0 & 1 & 0 & 0 & 0 & 0 & 0 & 0 & -1 & 0 & 1 \\
0 & 0 & 0 & 0 & 0 & 0 & 0 & 0 & 1 & 0 & 0 & 0 & 0 & 0 & -6 & 0 & 5 \\
0 & 0 & 0 & 0 & 0 & 0 & 0 & 0 & 0 & 1 & 0 & 0 & 0 & 0 & -11 & 0 & 9 \\
0 & 0 & 0 & 0 & 0 & 0 & 0 & 0 & 0 & 0 & 1 & 0 & 0 & 0 & -5 & 0 & 4 \\
0 & 0 & 0 & 0 & 0 & 0 & 0 & 0 & 0 & 0 & 0 & 1 & 0 & 0 & -4 & 0 & 3 \\
0 & 0 & 0 & 0 & 0 & 0 & 0 & 0 & 0 & 0 & 0 & 0 & 1 & 0 & -3 & 0 & 2 \\
0 & 0 & 0 & 0 & 0 & 0 & 0 & 0 & 0 & 0 & 0 & 0 & 0 & 1 & -2 & 0 & 1 \\
0 & 0 & 0 & 0 & 0 & 0 & 0 & 0 & 0 & 0 & 0 & 0 & 0 & 0 & 0 & 1 & 1
\end{bmatrix}.$$

The anticanonical class $-w_{X''}^{\text{can}} = (0, 0, 1, 1, 0, 0, 1, 0, -1, 0, 0, 0, 0, 2)$ does not belong to the effective cone $\text{Eff}(X'')$ which is spanned by the columns of Q''.

Exercises to Chapter 5

Exercise 5.1 (Contractibility in Mori dream surfaces)   Let X be a Mori dream surface and let E be a connected curve of X. Prove that the following statements are equivalent.

(1) E is contractible.
(2) The class of E is contained in a proper face of the effective cone $\mathrm{Eff}(X)$ and for all integers $n \geq 0$ one has $\dim \mathcal{R}_{[nE]} = 1$.
(3) The intersection matrix on the irreducible components of the support of E is negative definite.

Here, we say that a connected curve $E \subseteq X$ is contractible if there is a birational morphism $\pi : X \to X'$ of $\mathbb{Q}$-factorial complete surfaces mapping E to a point and $X \setminus E$ isomorphically to $X' \setminus \pi(E)$.

Exercise 5.2 (Nef and semiample cones)   Let X be a smooth projective rational surface with $K_X^2 = 0$ and $-K_X$ nef. Show that $-K_X$ is linearly equivalent to an effective divisor C. Assuming that C is a smooth irreducible curve prove that the following are equivalent.

(1) The class of $-K_X|_C$ in $\mathrm{Pic}(C)$ is torsion, that is, $n(-K_X|_C) \sim 0$ for some positive integer n.
(2) The class of C in $\mathrm{Pic}(X)$ is semiample.
(3) $\mathrm{Nef}(X) = \mathrm{SAmple}(X)$.

Exercise 5.3 ((-2)-curves)   With the same notation of Exercise 5.2 consider the homomorphism $\iota^* : \mathrm{Pic}(X) \to \mathrm{Pic}(C)$ induced by the inclusion $\iota : C \to X$. Prove the following.

(1) $\iota^*(K_X^{\perp}) \subseteq \mathrm{Pic}^0(C)$.
(2) Given a class $w \in \ker(\iota^*)$ with $w^2 = -2$ either w or $-w$ is effective.
(3) X contains a (-2)-curve if and only if there is an $w \in \ker(\iota^*)$ with $w^2 = -2$.
(4) If the class of $-K_X$ is semiample then a positive integer multiple of $-K_X$ defines an elliptic fibration $X \to \mathbb{P}^1$. This fibration is extremal if and only if $\ker(\iota^*)$ contains a root lattice of rank 9.

Exercise 5.4   Let Γ be a general smooth plane cubic curve, L be a line, and $p_1, \ldots, p_9 \in \Gamma$ be such that the classes of $3L|_\Gamma - p_1 - \cdots - p_9$ and $p_2 - p_1, \ldots, p_9 - p_1$ span a free subgroup of rank 9 of $\mathrm{Pic}^0(\Gamma)$. Denote by $\pi : X \to \mathbb{P}^2$ the blow-up of the projective plane at the nine points with exceptional divisor $E_i := \pi^{-1}(p_i)$ for any i. Prove that $\mathrm{Aut}(X)$ is trivial. Proceed as follows (see also [191]).

(1) Show that the linear system $| -K_X|$ is zero-dimensional and that it contains only the strict transform C of Γ.

(2) Let $\iota^*\colon \operatorname{Pic}(X) \to \operatorname{Pic}(C)$ be the pullback homomorphism induced by the inclusion $\iota\colon C \to X$, let G_1 be the subgroup of $\operatorname{Pic}(X)$ generated by the classes of $E_2 - E_1, \ldots, E_9 - E_1$ and let $G_2 := \iota^*(G_1)$. Show that

$$\iota^*|_{G_1}\colon G_1 \to G_2$$

is an isomorphism.

(3) Given $f \in \operatorname{Aut}(X)$ prove that $f(C) = C$ and that either $f|_C$ is a translation or it is a composition of a translation with a hyperelliptic involution. Use the above and $(f|_C)^* \circ \iota^*|_{G_1} = \iota^*|_{G_1} \circ f^*|_{G_1}$ to deduce that

$$(f|_C)^* = \pm\operatorname{id} \quad \text{and} \quad f^*|_{G_1} = \pm\operatorname{id}.$$

(4) Prove that $f(E_1) \cdot (E_i - E_1) = \pm 1$ for any $i > 1$, $f(E_1) \cdot K_X = -1$ and $f^*(K_X) \sim K_X$. Conclude that $f^* = \operatorname{id}$ on $\operatorname{Pic}(X)$.

(5) Show that the kernel of the map $\operatorname{Aut}(X) \to \operatorname{GL}(\operatorname{Pic}(X))$ is trivial and conclude that $\operatorname{Aut}(X)$ is trivial as well.

Exercise 5.5 Let X be a normal projective variety with finitely generated class group. Show that if $\operatorname{Eff}(X) \subseteq \operatorname{Cl}_{\mathbb{Q}}(X)$ is polyhedral then the image of the homomorphism $\varphi\colon \operatorname{Aut}(X) \to \operatorname{GL}(\operatorname{Cl}(X))$ is a finite group. To prove that the converse does not hold proceed as follows.

(1) Let X and C be as in Exercise 5.4. Use the exact sequence

$$0 \longrightarrow \mathcal{O}_X(-C) \longrightarrow \mathcal{O}_X \longrightarrow \mathcal{O}_C \longrightarrow 0$$

and the fact that the class of the divisor $C|_C$ is nontorsion, to show that $h^0(X, nC) = 1$ for any $n > 0$.

(2) Show that $[C]$ spans an extremal ray of the effective cone $\operatorname{Eff}(X) \subseteq \operatorname{Cl}_{\mathbb{Q}}(X)$. Use $C^2 = 0$ to conclude that the effective cone is not polyhedral.

Exercise 5.6 (Mori dream spaces have trivial irregularity)   Let X be a projective complex variety. Show that the irregularity $q(X) = h^1(X, \mathcal{O}_X)$ vanishes if and only if the Picard group $\operatorname{Pic}(X)$ is finitely generated. Proceed as follows.

(1) Consider the associated analytic variety X^{an} defined by X and show that the following *exponential sequence* is an exact sequence of sheaves:

$$0 \longrightarrow \mathbb{Z}_{X^{\mathrm{an}}} \longrightarrow \mathcal{O}_{X^{\mathrm{an}}} \xrightarrow{\;\exp\;} \mathcal{O}^*_{X^{\mathrm{an}}} \longrightarrow 0,$$

where $\mathbb{Z}_{X^{\mathrm{an}}}$ is the sheaf of locally constant integer values functions on X^{an} and exp is defined by $f \mapsto e^{2\pi i f}$.

(2) Consider the long exact sequence in cohomology associated with the exponential sequence:

$$H^1(X^{an}, \mathbb{Z}_{X^{an}}) \to H^1(X^{an}, \mathcal{O}_{X^{an}}) \to H^1(X^{an}, \mathcal{O}^*_{X^{an}}) \to H^2(X^{an}, \mathbb{Z}_{X^{an}}).$$

Show that the groups $H^i(X^{an}, \mathbb{Z}_{X^{an}})$ are isomorphic to the singular cohomology groups of X^{an} and that these are finitely generated abelian groups using that X^{an} is triangulable [173].

(3) Extract from Serre's GAGA [271], that $H^1(X^{an}, \mathcal{O}_{X^{an}})$ and $H^1(X^{an}, \mathcal{O}^*_{X^{an}})$ are isomorphic to $H^1(X, \mathcal{O}_X)$ and $H^1(X, \mathcal{O}^*_X)$, respectively.

(4) Show that $\mathrm{Pic}(X) \cong H^1(X, \mathcal{O}^*_X)$ holds [160, Ex. III.4.5] and thus $\mathrm{Pic}(X)$ is finitely generated if and only if $q(X) = 0$.

Exercise 5.7 (Smooth del Pezzo surfaces) Let X be a smooth del Pezzo surface and let $w \in \mathrm{Cl}(X)$ be a class such that $w \cdot K_X = 0$ and $w^2 = -2$. Show that the map $\sigma_w \colon \mathrm{Cl}(X) \to \mathrm{Cl}(X)$ defined by $x \mapsto x + (w \cdot x)w$ is $\mathbb{Z}$-linear of order 2. Define the *Weyl group* of $\mathrm{Cl}(X)$ as:

$$W(X) := \langle \sigma_w : w^2 = -2, \ w \cdot K_X = 0 \rangle.$$

Prove the following.

(1) $\sigma(\mathrm{Eff}(X)) = \mathrm{Eff}(X)$ for any $\sigma \in W(X)$.
(2) $W(X)$ acts transitively on the set of (-1)-classes of X if $\mathrm{Cl}(X)$ has rank > 3.
(3) If $\mathrm{Cl}(X)$ has rank ≤ 5 show that for any $\sigma \in W(X)$ there exists an automorphism $f \in \mathrm{Aut}(X)$ such that $f^* = \sigma$.

Exercise 5.8 (Root lattices on smooth del Pezzo surfaces) Let X be a smooth del Pezzo surface with Picard number $\varrho(X) \geq 3$. Show that the sublattice $K_X^\perp$ of $\mathrm{Cl}(X)$ is nondegenerate, even, and negative definite. Prove that it is isometric to the root lattice of one of the following root systems by providing an explicit Gram matrix in each case:

$\varrho(X)$	4	5	6	7	8	9
Lattice	$A_2 \times A_1$	A_4	D_5	E_6	E_7	E_8

Exercise 5.9 (Contraction of a (-1)-curve) Let X be a smooth del Pezzo surface of degree ≥ 3. Prove the following.

(1) Given any pair of (-1)-curves (E_1, E_2) of X we have $E_1 \cdot E_2 \leq 1$.
(2) If E is a (-1)-curve of X and $\pi \colon X \to Y$ is the contraction of E to a point $p \in Y$, then the image of any (-1)-curve of X passes through p with multiplicity at most 1.

(3) According to Proposition 4.1.3.8 the Cox ring of X is a saturated Rees algebra $R_1[I]^{\mathrm{sat}}$. Show that actually $R_1[I]^{\mathrm{sat}} = R_1[I]$, where $R_1 = \mathcal{R}(Y)$.

Exercise 5.10 (Curves on K3 surfaces) Let X be a K3 surface with class group of rank 2 and let $w \in \mathrm{Cl}(X)$ be a primitive class that spans an extremal ray of the effective cone of X. Prove that w is either the class of a smooth genus one curve of X or it is the class of a (-2)-curve of X.

Exercise 5.11 (K3 surfaces with class group of rank two) Let X be a K3 surface with class group of rank two. Prove the following.

(1) The group of isometries of $\mathrm{Cl}(X)$ is finite if there exists $w \in \mathrm{Cl}(X)$ with $w^2 = 0$.

(2) Assume that a Gram matrix for $\mathrm{Cl}(X)$ is $\begin{bmatrix} 4 & 0 \\ 0 & -4d \end{bmatrix}$ for some integer $d > 0$. Prove that the group of isometries $O(\mathrm{Cl}(X))$ of $\mathrm{Cl}(X)$ is finite if and only if there exists an $w \in \mathrm{Cl}(X)$ with $w^2 = 0$. Use the global Torelli theorem to conclude that $\mathrm{Aut}(X)$ is a finite group if and only if $O(\mathrm{Cl}(X))$ is finite.

Exercise 5.12 (2-elementary lattices) Let X be a Mori dream K3 surface with class group of rank $\varrho(X) \geq 4$. Assume that $\mathrm{Cl}(X)$ is a 2-elementary lattice.

(1) Determine all such lattices in Theorem 5.1.5.3.
(2) Use the global Torelli theorem to show that there exists an involution $\sigma \in \mathrm{Aut}(X)$ that acts like the identity on $\mathrm{Cl}(X)$.
(3) Construct the quotient surface $Y := X/\langle \sigma \rangle$, show that it is a rational surface with $\mathrm{rk}\,\mathrm{Cl}(Y) = \mathrm{rk}\,\mathrm{Cl}(X)$ and describe Y as a blow-up of $\mathbb{P}^2$.

Exercise 5.13 (Kummer surfaces) Let C be a genus 2 curve. Recall that the Kummer surface $\mathrm{Kum}(C) \subseteq \mathbb{P}^3$ is the quotient of $JC := \mathrm{Pic}^0(C)$ by the involution $x \mapsto -x$. The surface $\mathrm{Kum}(C)$ has degree 4 and 16 ordinary double points coming from the 16 fixed points of the involution. Prove the following.

(1) Let $\pi \colon X \to \mathrm{Kum}(C)$ be a minimal resolution of singularities. Show that the class group of X has rank at least 17 and that it is not isometric to any of the lattices of Theorem 5.1.5.3. Deduce that X is not Mori dream.
(2) Show that if C is very general, then $\mathrm{Kum}(C)$ has class group of rank 1 and thus it is Mori dream.

Exercise 5.14 (Picard group of a K3 surface) Let X be a K3 surface with holomorphic 2-form ω_X. Recall that the cup product in $H^2(X, \mathbb{C})$ is given by

$$\omega_1 \cdot \omega_2 := \int_X \omega_1 \wedge \omega_2.$$

Show that $\text{Pic}(X) = \omega_X^{\perp} \cap H^2(X, \mathbb{Z})$. Proceed as follows.

(1) Using the Hodge decomposition, show that $\omega_X^{\perp} = \mathbb{C}\,\omega_X \oplus H^{1,1}(X)$.
(2) Using the fact that $\overline{\omega} = \omega$ for any $\omega \in H^2(X, \mathbb{Z})$ deduce that $\omega_X^{\perp} \cap H^2(X, \mathbb{Z}) = H^{1,1}(X) \cap H^2(X, \mathbb{Z})$.
(3) Conclude by using the Lefschetz theorem on $(1, 1)$-classes [29, Thm. IV.2.13].

Exercise 5.15 (Rational elliptic $\mathbb{C}^*$-surfaces) Let X be a smooth complex projective rational $\mathbb{C}^*$-surface that admits an extremal jacobian elliptic fibration $\pi\colon X \to \mathbb{P}^1$. It is known that there are four families of such surfaces [131, Pro. 9.2.17] defined by the following pencils of plane cubic curves:

$$
\begin{array}{ll}
X_{22}: & x_1^3 + x_2 x_0^2 + t x_2^3 = 0 \\
X_{44}: & x_1 x_2(x_1 - x_2) + t x_0^3 = 0
\end{array}
\quad
\begin{array}{ll}
X_{33}: & (x_0 x_2 - x_1^2)x_0 + t x_2^3 = 0 \\
X_{11}(a): & x_1 x_2(x_1 - x_2) + t(x_1 - a x_2)x_0^2 = 0.
\end{array}
$$

We are going to find a presentation for the Cox ring of each such X (see also [8]). Proceed as follows.

(1) Determine the $\mathbb{C}^*$-action on each such X.
(2) Construct a $\mathbb{C}^*$-equivariant birational morphism $\varphi\colon \tilde{X} \to X$ such that $\tilde{X}$ has minimal Picard rank within all the complexity 1 surfaces that dominate X and has no elliptic fixed points for the $\mathbb{C}^*$-action.
(3) Show that the Orlik and Wagreich graph of $\tilde{X}$ is one of the following, where each white node represents a curve contracted by φ.

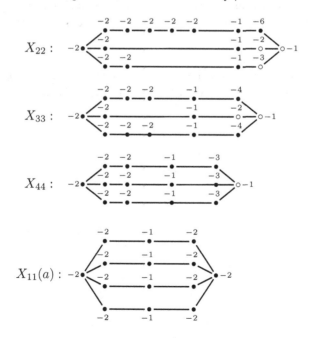

(4) Find the P-matrices of $\tilde{X}$ and X.

(5) Give a presentation for the Cox ring of X.

Exercise 5.16 Using the Orlik and Wagreich graphs of the surfaces X described in Exercise 5.15 determine which Gorenstein log del Pezzo $\mathbb{C}^*$-surfaces Y are dominated by one such X, that is, there exists a birational morphism $X \to Y$. Give a presentation for the Cox ring of Y.

Exercise 5.17* (Blow-ups of Mori dream surfaces) Let X be a Mori dream surface. Prove or disprove the following: There exists a point $p \in X$ such that the blow-up of X at p is again a Mori dream surface.

Exercise 5.18* (Families of del Pezzo surfaces) Let $\pi : \mathcal{X} \to U \subseteq \mathbb{A}^1$ be a flat family of smooth generalized del Pezzo surfaces. Assume that the general fiber of π is a smooth del Pezzo surface and that the fiber $\mathcal{X}_0 := \pi^{-1}(0)$ contains (-2)-curves. Study how the Cox ring of the general fiber deforms to the Cox ring of $\mathcal{X}_0$.

Exercise 5.19* (Enriques surfaces) Find a presentation for the Cox rings of Mori dream Enriques surfaces. Find a presentation for the Cox ring of the minimal resolution of the rational surface obtained as limit of Mori dream Enriques surfaces of type I or II.

Exercise 5.20* (Blow-ups of quartic surfaces) Give the equation of a smooth quartic surface S of $\mathbb{P}^3$ and a point $p \in S$ such that the blow-up of S at p is a Mori dream surface.

Exercise 5.21* (K3 surfaces with $\varrho(X) = 2$) Find the degrees of generators and relations of Cox rings of Mori dream K3 surfaces with divisor class group of rank 2.

6

Arithmetic applications

In Section 6.1, we give an overview of the theory of universal torsors over arbitrary fields of characteristic zero. Furthermore, we explore their connection to characteristic spaces and Cox rings. Section 6.2 deals with the existence of rational points over number fields. We discuss the Hasse principle and weak approximation. The failure of these principles is often explained by Brauer–Manin obstructions. We indicate how this can be approached via universal torsors and give an overview of the existing results. Section 6.3 is devoted to the distribution of rational points on varieties over number fields. Here, one introduces height functions on the set of rational points and is interested mainly in the asymptotic behavior of the number of rational points of bounded height. For Fano varieties, this behavior is predicted by Manin's conjecture. We indicate how universal torsors and Cox rings can be applied to prove Manin's conjecture. In Section 6.4, we specialize to del Pezzo surfaces. Here, Manin's conjecture is known in many cases, and a general strategy emerges. We discuss this strategy and some technical ingredients in detail. In Section 6.5, we show how it can be applied to prove Manin's conjecture for a singular cubic surface.

6.1 Universal torsors and Cox rings

6.1.1 Quasitori and principal homogeneous spaces

Here we summarize the basic concepts and facts on quasitori and their principal homogeneous spaces over nonclosed fields. The main references are [303, 304]. As an introduction to varieties over nonclosed fields and their rational points, we mention [253]. First we fix the setting.

Let $\mathbb{K}$ be a field of characteristic zero, not necessarily algebraically closed. We fix an algebraic closure $\overline{\mathbb{K}} \supseteq \mathbb{K}$ and denote by $G := \mathrm{Gal}(\overline{\mathbb{K}}/\mathbb{K})$ the Galois group. A *variety X over $\mathbb{K}$*, or $\mathbb{K}$-*variety*, is a reduced separated scheme of finite type over $\mathrm{Spec}(\mathbb{K})$, with structure morphism $X \to \mathrm{Spec}(\mathbb{K})$. For two $\mathbb{K}$-varieties X, X', we write

$$X \times_{\mathbb{K}} X' := X \times_{\mathrm{Spec}(\mathbb{K})} X'.$$

A *morphism* of $\mathbb{K}$-varieties, or $\mathbb{K}$-morphism, is a morphism of schemes over $\mathrm{Spec}(\mathbb{K})$. For any field extension $\mathbb{L} \supseteq \mathbb{K}$, we set

$$X_{\mathbb{L}} := X \times_{\mathbb{K}} \mathbb{L} := X \times_{\mathrm{Spec}(\mathbb{K})} \mathrm{Spec}(\mathbb{L}),$$

and denote by $X(\mathbb{L})$ the set of all $\mathbb{L}$-*rational points on* X, that is, the set of all $\mathbb{K}$-morphisms $\mathrm{Spec}(\mathbb{L}) \to X$. The *rational points* on a variety X over $\mathbb{K}$ are the $\mathbb{K}$-rational points on X; they can be identified with the closed points on X with residue field $\mathbb{K}$. The natural action of $g \in G$ on $x \in X(\overline{\mathbb{K}})$ is denoted by ^{g}x. We say that X is *geometrically irreducible* (resp. *geometrically connected, geometrically rational*, etc.) if $X_{\overline{\mathbb{K}}}$ is irreducible (resp. connected, rational, etc.). Here, we say that an irreducible variety X is rational over $\mathbb{K}$ if its function field $\mathbb{K}(X)$ is isomorphic to the function field $\mathbb{K}(T_1, \ldots, T_n)$ of $\mathbb{P}^n_{\mathbb{K}}$ for some n.

An *(affine) algebraic group over* $\mathbb{K}$ is an (affine) variety H over $\mathbb{K}$ together with a "group structure" given by $\mathbb{K}$-morphisms

$$e\colon \mathrm{Spec}(\mathbb{K}) \to H, \qquad \mathrm{m}\colon H \times_{\mathbb{K}} H \to H, \qquad \mathrm{i}\colon H \to H,$$

to be thought of as neutral element, multiplication, and inversion, that satisfy associativity, law of inverse, and law of identity in the following sense:

$$\mathrm{m} \circ (\mathrm{id}_H \times \mathrm{m}) = \mathrm{m} \circ (\mathrm{m} \times \mathrm{id}_H),$$
$$\mathrm{m} \circ (\mathrm{id}_H \times \mathrm{i}) \circ \Delta = \mathrm{m} \circ (\mathrm{i} \times \mathrm{id}_H) \circ \Delta = e \circ \pi,$$
$$\mathrm{m} \circ (e \times \mathrm{id}_H) = \mathrm{m} \circ (\mathrm{id}_H \times e) = \mathrm{id}_H,$$

where $\Delta\colon H \to H \times_{\mathbb{K}} H$ is the diagonal map and $\pi\colon H \to \mathrm{Spec}(\mathbb{K})$ the structure map [225, Defs. 0.1 and 0.2]. For $h, h' \in H$, we usually write $\mathrm{m}(h, h') = h \cdot h' = hh'$ and $\mathrm{i}(h) = h^{-1}$. A *homomorphism* of algebraic groups H and H' over $\mathbb{K}$ is a $\mathbb{K}$-morphism $\varphi\colon H \to H'$ that is compatible with the group structure in the sense that $\varphi \circ \mathrm{m} = \mathrm{m}' \circ (\varphi \times \varphi)$ holds with the multiplication m' on H'.

Remark 6.1.1.1 Let H be an algebraic group over $\mathbb{K}$ and $\mathbb{L} \supseteq \mathbb{K}$ a field extension. Then $H_{\mathbb{L}}$ is an algebraic group over $\mathbb{L}$. Moreover, the set $H(\mathbb{L})$ of $\mathbb{L}$-rational points comes naturally with a group structure.

The Definition 1.2.1.1 of a quasitorus over algebraically closed fields generalizes as follows to the nonclosed case; see also [303, Sec. 3.4].

Definition 6.1.1.2 A *quasitorus* over $\mathbb{K}$ is an affine algebraic group H over $\mathbb{K}$ such that $H_{\overline{\mathbb{K}}}$ is a quasitorus over the algebraic closure $\overline{\mathbb{K}}$. A *torus* over $\mathbb{K}$ is a geometrically connected quasitorus over $\mathbb{K}$.

Example 6.1.1.3 The *multiplicative group*, also called the *standard* 1-*torus*, over $\mathbb{K}$ is the $\mathbb{K}$-variety

$$\mathbb{G}_{m,\mathbb{K}} := \operatorname{Spec}(\mathbb{K}[\mathbb{Z}]) = \operatorname{Spec}(\mathbb{K}[T, T^{-1}])$$

where the neutral element is the rational point induced by $\mathbb{K}[\mathbb{Z}] \to \mathbb{K}$ with $T \mapsto 1$ and multiplication and inversion are induced by

$$\mathbb{K}[\mathbb{Z}] \to \mathbb{K}[\mathbb{Z}] \otimes_{\mathbb{K}} \mathbb{K}[\mathbb{Z}], \quad T \mapsto T \otimes T, \quad \mathbb{K}[\mathbb{Z}] \to \mathbb{K}[\mathbb{Z}], \quad T \mapsto T^{-1}.$$

For any field extension $\mathbb{L} \supseteq \mathbb{K}$, we have $\mathbb{G}_{m,\mathbb{K}}(\mathbb{L}) = \mathbb{L}^*$, and the induced group structure on the set $\mathbb{G}_{m,\mathbb{K}}(\mathbb{L})$ of $\mathbb{L}$-rational points coincides with the multiplication on $\mathbb{L}^*$.

Quasitori are also called *groups of multiplicative type* in the literature. The definition of the character group involves also the action of the Galois group G. By a finitely generated G-module M we mean a finitely generated abelian group M together with a *continuous* G-action, that is, the action of G factorizes via a finite quotient of G.

Definition 6.1.1.4 Let H be a quasitorus over $\mathbb{K}$. The *module of characters* of H is the finitely generated G-module $\mathbb{X}(H) := \operatorname{Hom}(H_{\overline{\mathbb{K}}}, \mathbb{G}_{m,\overline{\mathbb{K}}})$, where the action of $g \in G$ on $\chi \in \mathbb{X}(H)$ is given by $({}^{g}\chi)(h) := {}^{g}(\chi({}^{g^{-1}}h))$.

With this definition, one naturally generalizes Theorem 1.2.1.4, see [303, Sec. 3.4] and [304, Sec. 7.3] for details. For every G-module M, the associated group algebra $\overline{\mathbb{K}}[M]$ carries an induced action of G. We denote by $\overline{\mathbb{K}}[M]^{G}$ the $\mathbb{K}$-algebra of its G-invariant elements. Every G-equivariant homomorphism $\psi \colon M \to N$ of G-modules defines a $\overline{\mathbb{K}}$-algebra homomorphism $\overline{\mathbb{K}}[\psi] \colon \overline{\mathbb{K}}[M] \to \overline{\mathbb{K}}[N]$ and thus a homomorphism of $\mathbb{K}$-algebras

$$\overline{\mathbb{K}}[\psi]^{G} \colon \overline{\mathbb{K}}[M]^{G} \to \overline{\mathbb{K}}[N]^{G}, \qquad f \mapsto \psi(f).$$

Proposition 6.1.1.5 *We have contravariant exact functors being essentially inverse to each other:*

$$\{\text{finitely generated } G\text{-modules}\} \longleftrightarrow \{\text{quasitori over } \mathbb{K}\}$$

$$M \mapsto \operatorname{Spec}(\overline{\mathbb{K}}[M]^{G}),$$

$$\psi \mapsto \operatorname{Spec}(\overline{\mathbb{K}}[\psi]^{G}),$$

$$\mathbb{X}(H) \leftarrow\!\shortmid H,$$

$$\varphi^* \leftarrow\!\shortmid \varphi.$$

Under these equivalences, the free finitely generated G-modules correspond to the tori.

Example 6.1.1.6 Let $\mathbb{K} := \mathbb{R}$. Then we have the torus $\mathbb{G}_{m,\mathbb{R}} = \mathrm{Spec}(\mathbb{R}[T, T^{-1}])$ and another torus

$$H := \mathrm{Spec}(\mathbb{R}[Y, Z]/\langle Y^2 + Z^2 - 1 \rangle),$$

where for the latter, multiplication and inversion are given on the sets of $\mathbb{R}$-points by

$$((y_1, z_1), (y_2, z_2)) \mapsto (y_1 y_2 - z_1 z_2, y_1 z_2 + z_1 y_2), \quad (y, z) \mapsto (y, -z).$$

Note that H and $\mathbb{G}_{m,\mathbb{R}}$ are not isomorphic over $\mathbb{R}$, but become isomorphic after base change to $\mathbb{C}$, via the map $\chi : H_{\mathbb{C}} \to \mathbb{G}_{m,\mathbb{C}}$ induced by

$$\mathbb{C}[T, T^{-1}] \to \mathbb{C}[Y, Z]/\langle Y^2 + Z^2 - 1 \rangle, \quad T \mapsto Y + Z\sqrt{-1},$$

for $\sqrt{-1} \in \mathbb{C}$ with $(\sqrt{-1})^2 = -1$. Both H and $\mathbb{G}_{m,\mathbb{R}}$ are one-dimensional tori whose character groups are both isomorphic to $\mathbb{Z}$, but have different G-module structures. The Galois group

$$G = \mathrm{Gal}(\mathbb{C}/\mathbb{R}) \cong \mathbb{Z}/2\mathbb{Z}$$

acts trivially on the characters of $\mathbb{G}_{m,\mathbb{R}}$ and nontrivially on the characters of H. Indeed, χ as above is a character of H, in fact a generator of $\mathbb{X}(H)$, and for the complex conjugation $g \in G$, the character $^g\chi$ is induced by $T \mapsto Y - Z\sqrt{-1}$.

Example 6.1.1.7 Let $\mathbb{L} \supseteq \mathbb{K}$ be a field extension of finite degree n such that $\mathbb{L}$ is contained in $\overline{\mathbb{K}}$. Then one has the inclusion of Galois groups

$$G_{\mathbb{L}} := \mathrm{Gal}(\overline{\mathbb{K}}/\mathbb{L}) \subseteq \mathrm{Gal}(\overline{\mathbb{K}}/\mathbb{K}) = G,$$

and G acts naturally on the quotient set $G/G_{\mathbb{L}}$, which has cardinality n. Let $\mathbb{Z}[G/G_{\mathbb{L}}]$ be the free G-module of rank n with basis $G/G_{\mathbb{L}}$. The torus over $\mathbb{K}$ associated with $\mathbb{Z}[G/G_{\mathbb{L}}]$ is called the *Weil restriction* of $\mathbb{G}_{m,\mathbb{L}}$ and is denoted as $R_{\mathbb{L}/\mathbb{K}}(\mathbb{G}_{m,\mathbb{L}})$. For the rational points of the Weil restriction, we have $R_{\mathbb{L}/\mathbb{K}}(\mathbb{G}_{m,\mathbb{L}})(\mathbb{K}) = \mathbb{L}^*$.

We turn to principal homogeneous spaces; see also [303, Sec. 3.5] and [279, Sec. 2.1]. An *action* of an algebraic group H over $\mathbb{K}$ on a variety X over $\mathbb{K}$ is a $\mathbb{K}$-morphism $\mu \colon H \times_{\mathbb{K}} X \to X$ with the compatibility rules mimicking the axioms of a group action

$$\mu \circ (\mathrm{id}_H \times \mu) = \mu \circ (\mathrm{m} \times \mathrm{id}_X), \qquad \mu \circ (e \times \mathrm{id}_X) = \mathrm{id}_X,$$

where we refer to [225, Def. 0.3] for the details. For $x \in X$ and $h \in H$, we will usually write $\mu(h, x) = h \cdot x = hx$. By a *morphism* of varieties X and X' with actions μ and μ' of algebraic groups H and H', we mean a pair $(\varphi, \widetilde{\varphi})$, where $\varphi \colon X \to X'$ is a morphism of varieties over $\mathbb{K}$ and $\widetilde{\varphi} \colon H \to H'$ is a morphism of algebraic groups over $\mathbb{K}$ such that $\varphi \circ \mu$ equals $\mu' \circ (\widetilde{\varphi} \times \varphi)$. If $H = H'$ and $\widetilde{\varphi} = \mathrm{id}_H$, then we speak of an *equivariant morphism*.

Definition 6.1.1.8 Let H be a quasitorus over $\mathbb{K}$. A $\mathbb{K}$-*torsor*, also called *principal homogeneous space under H*, is a variety $\mathcal{H}$ over $\mathbb{K}$ together with an action of H on $\mathcal{H}$ such that $\mathcal{H}_{\overline{\mathbb{K}}}$ with the induced action of $H_{\overline{\mathbb{K}}}$ is equivariantly isomorphic to $H_{\overline{\mathbb{K}}}$ with its natural action on itself.

Example 6.1.1.9 The variety $\mathcal{H} := \mathrm{Spec}(\mathbb{R}[Y', Z']/\langle Y'^2 + Z'^2 + 1\rangle)$ is an $\mathbb{R}$-torsor under the torus H from Example 6.1.1.6 via the action $H \times_{\mathbb{R}} \mathcal{H} \to \mathcal{H}$ induced by

$$\mathbb{R}[Y', Z'] \rightarrow \mathbb{R}[Y, Z] \otimes_{\mathbb{R}} \mathbb{R}[Y', Z'],$$

$$Y' \mapsto YY' - ZZ', \quad Z' \mapsto YZ' + ZY'.$$

Note that $\mathcal{H}$ and H are not isomorphic over $\mathbb{R}$ because $\mathcal{H}(\mathbb{R}) = \emptyset$ and $H(\mathbb{R}) \neq \emptyset$. However, $\mathcal{H}$ and H become isomorphic after base change to $\mathbb{C}$.

Principal homogeneous spaces are classified in terms of Galois cohomology. We briefly recall the main facts on the latter, see [274, Sec. I.2] for details. Let H be a quasitorus over $\mathbb{K}$. A *continuous 1-cocycle* is a map

$$\sigma : G \rightarrow H(\overline{\mathbb{K}}), \qquad g \mapsto \sigma_g$$

that is continuous for the profinite topology on G and the discrete topology on $H(\overline{\mathbb{K}})$ and satisfies $\sigma_{gg'} = \sigma_g \cdot {}^g\sigma_{g'}$. Let $Z^1(\mathbb{K}, H)$ be the group of continuous 1-cocycles, where the group law is defined by $(\sigma \cdot \tau)_g := \sigma_g \cdot \tau_g$. Two 1-cocycles σ, τ are called *cohomologous* if and only if there is an $h \in H(\overline{\mathbb{K}})$ such that $\tau_g = h^{-1} \cdot \sigma_g \cdot {}^g h$ for all $g \in G$. This defines an equivalence relation on the set of continuous 1-cocycles. The continuous 1-cocycles cohomologous to the trivial cocycle $g \mapsto 1$ form the subgroup of 1-*coboundaries*. Let H be a quasitorus over $\mathbb{K}$. The *first Galois cohomology group* $H^1(\mathbb{K}, H) := H^1(G, H(\overline{\mathbb{K}}))$ is the quotient of $Z^1(\mathbb{K}, H)$ by the subgroup of 1-coboundaries.

Proposition 6.1.1.10 *Let H be a quasitorus over $\mathbb{K}$. There is a bijection between $H^1(\mathbb{K}, H)$ and the set of isomorphy classes of $\mathbb{K}$-torsors under H.*

Proof See [279, Sec. 2.1, Ex. 2]. For each $\mathbb{K}$-torsor $\mathcal{H}$ under H, a corresponding cocycle $\sigma \in H^1(\mathbb{K}, H)$ is defined as follows. Choose a point $x \in \mathcal{H}(\overline{\mathbb{K}})$. Then for each $g \in G$, we define σ_g as the unique element of $H(\overline{\mathbb{K}})$ such that ${}^g x = \sigma_g \cdot x$.

Conversely, given any $\mathbb{K}$-torsor $\mathcal{H}$ under H, for example, H itself with its natural action, we can define its *twist* $\mathcal{H}^\sigma$ by a 1-cocycle σ as follows. The usual action $(g, x) \mapsto {}^g x$ of $g \in G$ on $\mathcal{H}(\overline{\mathbb{K}})$ gives rise to a *twisted action* defined by $(g, x) \mapsto \sigma_g \cdot {}^g x$, inducing a twisted G-action on the coordinate ring $\Gamma(\mathcal{H}_{\overline{\mathbb{K}}}, \mathcal{O})$.

Then $\mathcal{H}^\sigma := \mathrm{Spec}(\Gamma(\mathcal{H}_{\overline{\mathbb{K}}}, \mathcal{O})^G)$, where the G-invariant functions $\Gamma(\mathcal{H}_{\overline{\mathbb{K}}}, \mathcal{O})^G$ are taken with respect to the twisted G-action. □

Lemma 6.1.1.11 *For a $\mathbb{K}$-torsor $\mathcal{H}$ under a quasitorus H, the set $\mathcal{H}(\mathbb{K})$ of rational points is nonempty if and only if the class of $\mathcal{H}$ in $H^1(\mathbb{K}, H)$ is trivial.*

Proof If $\mathcal{H}(\mathbb{K})$ is nonempty, then we can choose $x \in \mathcal{H}(\mathbb{K})$ when defining the cocycle σ in the proof of Proposition 6.1.1.10. Conversely, if the class is trivial, $\mathcal{H}$ is isomorphic to H over $\mathbb{K}$; hence $\mathcal{H}$ also has $\mathbb{K}$-rational points. □

Example 6.1.1.12 The $\mathbb{R}$-torsor $\mathcal{H}$ under H from Example 6.1.1.9 corresponds to the cocycle $\sigma \colon G \to H(\mathbb{C})$ that sends the nontrivial element $g \in \mathrm{Gal}(\mathbb{C}/\mathbb{R})$ to $(-1, 0) \in H(\mathbb{C})$. This cocycle is not a coboundary because for any $h = (y, z) \in H(\mathbb{C})$, we have

$$h^{-1}(-1, 0)^g h = (-y\overline{y} - z\overline{z}, -y\overline{z} + z\overline{y}) \neq (1, 0),$$

where $\overline{x}$ denotes the complex conjugate of $x \in \mathbb{C}$. This agrees with our earlier observation that $\mathcal{H}$ is a nontrivial $\mathbb{K}$-torsor under H. In fact, we have $H^1(\mathbb{R}, H) \cong \mathbb{Z}/2\mathbb{Z}$.

6.1.2 Universal torsors

Universal torsors were introduced and investigated by Colliot-Thélène and Sansuc to study rational points on geometrically rational varieties over number fields [96, 97, 98, 99, 100]. We present here the main concepts and outline the basic theory following [100, 279]; see also [157] for recent generalizations. The setting is the same as in Section 6.1.1; in particular, $\mathbb{K}$ is a field of characteristic zero and $G = \mathrm{Gal}(\overline{\mathbb{K}}/\mathbb{K})$.

Definition 6.1.2.1 Let H be a quasitorus over $\mathbb{K}$, and X a variety over $\mathbb{K}$. Then an *X-torsor under H* is a variety $\mathcal{X}$ over $\mathbb{K}$ with a faithfully flat $\mathbb{K}$-morphism $\pi \colon \mathcal{X} \to X$ and an action $\mu \colon H \times_{\mathbb{K}} \mathcal{X} \to \mathcal{X}$ of H on $\mathcal{X}$ satisfying $\pi \circ \mu = \pi \circ \mathrm{pr}_2$, such that the natural map $(\mu, \mathrm{pr}_2) \colon H \times_{\mathbb{K}} \mathcal{X} \to \mathcal{X} \times_X \mathcal{X}$ is an isomorphism.

The last property is equivalent to the requirement that $\mathcal{X}$ is locally, with respect to the étale topology, isomorphic to $H \times_{\mathbb{K}} X$, the *trivial X-torsor under H* with the natural action of H on the first factor and π the projection to the second factor. Therefore, an X-torsor under H is an étale H-principal bundle over X.

The X-torsors under the multiplicative group $\mathbb{G}_{\mathrm{m},\mathbb{K}}$ are line bundles over X with the zero section removed. These are classified by the Picard group

$\text{Pic}(X) = H^1_{\text{ét}}(X, \mathbb{G}_{m,\mathbb{K}})$, see [219, Prop. III.4.9]. This is a special case of the following result.

Proposition 6.1.2.2 *Let H be a quasitorus over $\mathbb{K}$, and X a variety over $\mathbb{K}$. Then X-torsors under H are classified up to isomorphism by the étale cohomology group $H^1_{\text{ét}}(X, H)$.*

Proof See [219, Sec. III.4] and [279, Sec. 2.2]. □

In the above statement, the trivial X-torsor under H corresponds to the trivial element of $H^1_{\text{ét}}(X, H)$. For the following two constructions, see [279, Lemma 2.2.3].

Construction 6.1.2.3 Let H be a quasitorus over $\mathbb{K}$, and X a variety over $\mathbb{K}$. Given a $\mathbb{K}$-torsor $\mathcal{H}$ under H and an X-torsor $\pi : \mathcal{X} \to X$ under H, consider the action

$$H \times_{\mathbb{K}} (\mathcal{H} \times_{\mathbb{K}} \mathcal{X}) \to \mathcal{H} \times_{\mathbb{K}} \mathcal{X} \qquad h \cdot (h', x) := (h^{-1}h', hx).$$

This action admits a geometric quotient and the quotient space $_{\mathcal{H}}\mathcal{X}$ is called the *twist* of $\mathcal{X}$ by $\mathcal{H}$. With its natural map $_{\mathcal{H}}\pi : {}_{\mathcal{H}}\mathcal{X} \to X$, this is an X-torsor under H. In $H^1_{\text{ét}}(X, H)$, we have

$$[_{\mathcal{H}}\mathcal{X}] = [\mathcal{X}] - [\mathcal{H} \times_{\mathbb{K}} X].$$

Let $\sigma \in Z^1(\mathbb{K}, H)$ be a cocycle corresponding to $\mathcal{H}$. Then $(g, x) \mapsto \sigma_g^{-1} \cdot {}^g x$ defines a twisted action of G on $\mathcal{X}_{\overline{\mathbb{K}}}$, and Galois descent gives a torsor $\pi^\sigma : \mathcal{X}^\sigma \to X$ that is isomorphic to $_{\mathcal{H}}\pi : {}_{\mathcal{H}}\mathcal{X} \to X$.

Construction 6.1.2.4 Let $\varphi : H \to H'$ be a morphism of quasitori over $\mathbb{K}$. Let X be a variety over $\mathbb{K}$, and $\mathcal{X}$ an X-torsor under H. Let $H' \times^H \mathcal{X}$ be the quotient of the action

$$H \times_{\mathbb{K}} (H' \times_{\mathbb{K}} \mathcal{X}) \to H' \times_{\mathbb{K}} \mathcal{X}, \qquad h \cdot (h', x) := (\varphi(h^{-1})h', hx).$$

Then $H' \times^H \mathcal{X}$ is a torsor over X under H', and we have $\varphi_*([\mathcal{X}]) = [H' \times^H \mathcal{X}]$ for the natural map $\varphi_* : H^1_{\text{ét}}(X, H) \to H^1_{\text{ét}}(X, H')$. We call $H' \times^H \mathcal{X}$ the *pushforward* of $\mathcal{X}$ via φ and use the notation $\varphi_*(\mathcal{X})$.

Note that the *geometric Picard group* $\text{Pic}(X_{\overline{\mathbb{K}}})$ of a $\mathbb{K}$-variety X is naturally a G-module. By [279, Lemma 2.3.1], the following definition makes sense.

Definition 6.1.2.5 Let H be a quasitorus over $\mathbb{K}$, and X a variety over $\mathbb{K}$. Let $\pi : \mathcal{X} \to X$ be an X-torsor under H. For each $\chi \in \mathbb{X}(H)$, we have the pushforward $\chi_*(\mathcal{X}_{\overline{\mathbb{K}}})$, an $X_{\overline{\mathbb{K}}}$-torsor under $\mathbb{G}_{m,\overline{\mathbb{K}}}$, whose class is an element of $H^1_{\text{ét}}(X_{\overline{\mathbb{K}}}, \mathbb{G}_{m,\overline{\mathbb{K}}}) = \text{Pic}(X_{\overline{\mathbb{K}}})$. This defines a G-module homomorphism

$$\text{type}(\mathcal{X}) : \mathbb{X}(H) \to \text{Pic}(X_{\overline{\mathbb{K}}})$$

that is called the *type* of the torsor $\pi \colon \mathcal{X} \to X$. Hence the type$(\mathcal{X}) \in$ $\mathrm{Hom}_G(\mathbb{X}(H), \mathrm{Pic}(X_{\overline{\mathbb{K}}}))$, the group of G-module homomorphisms $\mathbb{X}(H) \to$ $\mathrm{Pic}(X_{\overline{\mathbb{K}}})$.

The notion of universal torsors was introduced with the following definition by Colliot-Thélène and Sansuc [96], see also [100, (2.0.4)]. Note that Skorobogatov [279, Def. 2.3.3] defines universal torsors slightly more generally as those torsors whose type is an isomorphism.

Definition 6.1.2.6 Let X be a variety over $\mathbb{K}$ whose geometric Picard group $\mathrm{Pic}(X_{\overline{\mathbb{K}}})$ is finitely generated. Let $H_{\mathrm{Pic}(X_{\overline{\mathbb{K}}})}$ be the quasitorus over $\mathbb{K}$ whose module of characters $\mathbb{X}(H_{\mathrm{Pic}(X_{\overline{\mathbb{K}}})})$ is the G-module $\mathrm{Pic}(X_{\overline{\mathbb{K}}})$; we call $H_{\mathrm{Pic}(X_{\overline{\mathbb{K}}})}$ the *Picard quasitorus* of X. A *universal torsor* over X is an X-torsor $\pi \colon \mathcal{X} \to X$ under $H_{\mathrm{Pic}(X_{\overline{\mathbb{K}}})}$ whose type is the identity on $\mathrm{Pic}(X_{\overline{\mathbb{K}}})$.

If $\mathrm{Pic}(X_{\overline{\mathbb{K}}})$ coincides with the Néron–Severi group of $X_{\overline{\mathbb{K}}}$, then $H_{\mathrm{Pic}(X_{\overline{\mathbb{K}}})}$ is the *Néron–Severi torus* of X.

For a variety X over $\mathbb{K}$, we use in the following often the condition $\Gamma(X_{\overline{\mathbb{K}}}, \mathcal{O}^*) = \overline{\mathbb{K}}^*$, that is, the only invertible regular functions on $X_{\overline{\mathbb{K}}}$ are the constants. This holds for example when X is a proper geometrically connected variety over $\mathbb{K}$.

Theorem 6.1.2.7 *Let H be a quasitorus over $\mathbb{K}$, and X a variety over $\mathbb{K}$ with* $\Gamma(X_{\overline{\mathbb{K}}}, \mathcal{O}^*) = \overline{\mathbb{K}}^*$. *Then we have the* fundamental exact sequence

$$0 \to H^1(\mathbb{K}, H) \xrightarrow{p^*} H^1_{\mathrm{ét}}(X, H) \xrightarrow{\mathrm{type}} \mathrm{Hom}_G(\mathbb{X}(H), \mathrm{Pic}(X_{\overline{\mathbb{K}}}))$$

$$\xrightarrow{\partial} H^2(\mathbb{K}, H) \xrightarrow{p^*} H^2_{\mathrm{ét}}(X, H)$$

of abelian groups, where the second map sends an isomorphy class of X-torsors under H to the type of any representative, and the first and fourth map are induced by the structure morphism $p \colon X \to \mathrm{Spec}(\mathbb{K})$.

Proof See [100, (2.0.2)], [279, Cor. 2.3.9]. $\square$

The fundamental exact sequence in Theorem 6.1.2.7 shows: X-torsors under H of type $\lambda \in \mathrm{Hom}_G(\mathbb{X}(X), \mathrm{Pic}(X_{\overline{\mathbb{K}}}))$ exist if and only if $\partial(\lambda) \in H^2(\mathbb{K}, H)$ vanishes. In this case, the isomorphy classes of X-torsors under H of type λ are represented precisely by the twists $\mathcal{X}^\sigma$ from Construction 6.1.2.3 of one such torsor $\mathcal{X}$ for a set of representatives σ of $H^1(\mathbb{K}, H)$. In particular, over $\overline{\mathbb{K}}$, the $X_{\overline{\mathbb{K}}}$-torsors are determined uniquely up to isomorphism by their type.

See [157, Sec. 8.2] for a generalization of the following to torsors under quasitori.

Proposition 6.1.2.8 *Let H be a torus over $\mathbb{K}$, and X a smooth, geometrically irreducible variety over $\mathbb{K}$. Let $\mathcal{X}$ be an X-torsor under H of type $\lambda : \mathbb{X}(H) \to$*

$\text{Pic}(X_{\overline{\mathbb{K}}})$. *Then we have an exact sequence*

$$1 \to \Gamma(X_{\overline{\mathbb{K}}}, \mathcal{O}^*) \to \Gamma(\mathcal{X}_{\overline{\mathbb{K}}}, \mathcal{O}^*) \to \mathbb{X}(H) \to \text{Pic}(X_{\overline{\mathbb{K}}}) \to \text{Pic}(\mathcal{X}_{\overline{\mathbb{K}}}) \to 0,$$

where the map $\mathbb{X}(H) \to \text{Pic}(X_{\overline{\mathbb{K}}})$ coincides with λ up to sign.

Proof   See [100, Prop. 2.1.1] or [279, (2.9)]. $\qquad\square$

Corollary 6.1.2.9   *Let X be a smooth, geometrically irreducible variety over $\mathbb{K}$ with free and finitely generated $\text{Pic}(X_{\overline{\mathbb{K}}})$ and $\Gamma(X_{\overline{\mathbb{K}}}, \mathcal{O}^*) = \overline{\mathbb{K}}^*$, and let $\mathcal{X}$ be a universal torsor over X. Then $\Gamma(\mathcal{X}_{\overline{\mathbb{K}}}, \mathcal{O}^*) = \overline{\mathbb{K}}^*$ and $\text{Pic}(\mathcal{X}_{\overline{\mathbb{K}}}) = 0$.*

Proof   Because $\text{Pic}(X_{\overline{\mathbb{K}}})$ is free, $H_{\text{Pic}(X_{\overline{\mathbb{K}}})}$ is a torus, and the type of $\mathcal{X}$ is the identity on $\text{Pic}(X_{\overline{\mathbb{K}}})$. Hence the result follows directly from Proposition 6.1.2.8. $\qquad\square$

Universal torsors have the following universal property; in particular, the existence of universal torsors implies the existence of torsors of arbitrary type.

Proposition 6.1.2.10   *Let X be a variety over $\mathbb{K}$ with finitely generated $\text{Pic}(X_{\overline{\mathbb{K}}})$ and $\Gamma(X_{\overline{\mathbb{K}}}, \mathcal{O}^*) = \overline{\mathbb{K}}^*$ such that a universal torsor $\mathcal{X}$ over X exists. Let H be a quasitorus and $\mathcal{X}' \to X$ an X-torsor under H of any type*

$$\lambda \colon \mathbb{X}(H) \to \text{Pic}(X_{\overline{\mathbb{K}}}).$$

Let $\varphi \colon H_{\text{Pic}(X_{\overline{\mathbb{K}}})} \to H$ be the morphism of quasitori that is dual to λ. Then $\mathcal{X}'$ is the pushforward $\varphi_* \mathcal{X}$ twisted by a $\mathbb{K}$-torsor under H.

Proof   See below [279, Def. 2.3.3]. This follows from Theorem 6.1.2.7 and the fact that the type of the pushforward of a torsor via φ is the composition of the map of characters dual to φ and the type of that torsor, which is the identity here. $\qquad\square$

Over nonclosed fields $\mathbb{K}$, the existence of universal torsors can be characterized as follows, see [100, Sec. 2.2] and [279, Sec. 2.3].

Proposition 6.1.2.11   *Let X be a smooth, geometrically irreducible variety over $\mathbb{K}$, with finitely generated $\text{Pic}(X_{\overline{\mathbb{K}}})$ and $\Gamma(X_{\overline{\mathbb{K}}}, \mathcal{O}^*) = \overline{\mathbb{K}}^*$. Let U be a dense open subset of X such that $\text{Pic}(U_{\overline{\mathbb{K}}}) = 0$. Then the following statements are equivalent.*

(i) *Universal torsors over X exist.*
(ii) *There is a G-equivariant splitting for the natural exact sequence of G-modules*

$$1 \to \overline{\mathbb{K}}^* \to \overline{\mathbb{K}}(X_{\overline{\mathbb{K}}})^* \to \overline{\mathbb{K}}(X_{\overline{\mathbb{K}}})^*/\overline{\mathbb{K}}^* \to 1.$$

(iii) *There is a G-equivariant splitting for the natural exact sequence of G-modules*

$$1 \to \overline{\mathbb{K}}^* \to \Gamma(U_{\overline{\mathbb{K}}}, \mathcal{O}^*) \to \Gamma(U_{\overline{\mathbb{K}}}, \mathcal{O}^*)/\overline{\mathbb{K}}^* \to 1.$$

Remark 6.1.2.12 Let $U \subseteq X$ be as in Proposition 6.1.2.11. Let $\mathrm{Div}_{X_{\overline{\mathbb{K}}} \setminus U_{\overline{\mathbb{K}}}}(X_{\overline{\mathbb{K}}})$ be the group of divisors on $X_{\overline{\mathbb{K}}}$ supported in $X_{\overline{\mathbb{K}}} \setminus U_{\overline{\mathbb{K}}}$. Note that $\mathrm{Pic}(U_{\overline{\mathbb{K}}}) = 0$ is equivalent to the requirement that $\mathrm{Pic}(X_{\overline{\mathbb{K}}})$ is generated by the classes of elements of $\mathrm{Div}_{X_{\overline{\mathbb{K}}} \setminus U_{\overline{\mathbb{K}}}}(X_{\overline{\mathbb{K}}})$. Also note that $\Gamma(U_{\overline{\mathbb{K}}}, \mathcal{O}^*)/\overline{\mathbb{K}}^*$ is free of finite rank because of the exact sequence

$$1 \to \Gamma(U_{\overline{\mathbb{K}}}, \mathcal{O}^*)/\overline{\mathbb{K}}^* \to \mathrm{Div}_{X_{\overline{\mathbb{K}}} \setminus U_{\overline{\mathbb{K}}}}(X_{\overline{\mathbb{K}}}) \to \mathrm{Pic}(X_{\overline{\mathbb{K}}}) \to 0.$$

Torsors under quasitori can be described explicitly locally, see [100, Sec. 2.3] and [279, Sec. 4.3]. For universal torsors, this is as follows.

Theorem 6.1.2.13 *Let X be a smooth, geometrically irreducible variety over $\mathbb{K}$ with $\Gamma(X_{\overline{\mathbb{K}}}, \mathcal{O}^*) = \overline{\mathbb{K}}^*$ and finitely generated $\mathrm{Pic}(X_{\overline{\mathbb{K}}})$, and assume that universal torsors over X exist. Let $j : U \hookrightarrow X$ be an open subset of X with $\mathrm{Pic}(U_{\overline{\mathbb{K}}}) = 0$. Consider the exact sequence of quasitori*

$$1 \to H_{\mathrm{Pic}(X_{\overline{\mathbb{K}}})} \to \mathcal{X}_0 \to R \to 1$$

over $\mathbb{K}$ that is dual to the exact sequence in Remark 6.1.2.12, turning $\mathcal{X}_0$ into an R-torsor under $H_{\mathrm{Pic}(X_{\overline{\mathbb{K}}})}$. Let $\varphi : U \to R$ be induced by the negative of a G-equivariant splitting

$$\widehat{\varphi} : \Gamma(U_{\overline{\mathbb{K}}}, \mathcal{O}^*)/\overline{\mathbb{K}}^* \to \Gamma(U_{\overline{\mathbb{K}}}, \mathcal{O}^*)$$

of the exact sequence in Proposition 6.1.2.11 (iii). Then $\mathcal{X}_0 \times_R U$, constructed via φ, is the restriction to U of a unique universal torsor over X. Conversely, every universal torsor over X has such a description over U.

Example 6.1.2.14 Let $P(z_0) = c(z_0^2 - a)$ be an irreducible polynomial of degree 2 over $\mathbb{K}$, with $c \in \mathbb{K}^*$, where $a \in \mathbb{K}^*$ is not a square. Let $\mathbb{K}$ be a field of characteristic zero, and let $\mathbb{M} \supset \mathbb{K}$ be a field extension of degree 4, with $\mathbb{L} := \mathbb{K}(\sqrt{a}) \subset \mathbb{M}$. Let

$$N_{\mathbb{M}/\mathbb{K}}(z_1, \ldots, z_4) := N_{\mathbb{M}/\mathbb{K}}(z_1 \omega_1 + \cdots + z_4 \omega_4) \in \mathbb{K}[z_1, \ldots, z_4]$$

be a norm form for this field extension, defined via the choice of a basis $\omega_1, \ldots, \omega_4$ of $\mathbb{M}$ over $\mathbb{K}$. Let $X \subset \mathbb{A}_{\mathbb{K}}^5$ be the variety defined by

$$P(z_0) = N_{\mathbb{M}/\mathbb{K}}(z_1, \ldots, z_4).$$

Rational points on X were investigated in [81, 122]. We compute the local description of universal torsors over $U := X \cap \{P(z_0) \neq 0\}$ via Theorem 6.1.2.13. This is a special case of [122, Prop. 2]. We will return to this class of varieties in Examples 6.2.1.3 and 6.2.3.3 below.

First, we check that X is geometrically irreducible and smooth over $\mathbb{K}$, with $\Gamma(X_{\overline{\mathbb{K}}}, \mathcal{O}^*) = \overline{\mathbb{K}}^*$ and $\mathrm{Pic}(X_{\overline{\mathbb{K}}})$ free of rank 3. Over $\overline{\mathbb{K}}$, we have that $X_{\overline{\mathbb{K}}}$ is defined by

$$c(z_0 - \sqrt{a})(z_0 + \sqrt{a}) = u_1 u_2 u_3 u_4,$$

which is clearly irreducible. Because $P(z_0)$ does not have multiple roots, $X_{\overline{\mathbb{K}}}$ is smooth over $\overline{\mathbb{K}}$ and hence X is smooth over $\mathbb{K}$. Because the generic fiber of the projection to the z_0-coordinate $X_{\overline{\mathbb{K}}} \to \mathbb{A}^1_{\overline{\mathbb{K}}}$ is isomorphic to $\mathbb{G}^3_{m, \overline{\mathbb{K}}(z_0)}$, any $f \in \Gamma(X_{\overline{\mathbb{K}}}, \mathcal{O}^*)$ has the form

$$f = g(z_0) u_1^{m_1} \cdots u_4^{m_4}$$

with $g \in \overline{\mathbb{K}}(z_0)$ and $m_1, \ldots, m_4 \in \mathbb{Z}$. Here, $g(z_0)$ cannot have a root or pole outside $\pm\sqrt{a}$, hence $g(z_0) = c'(z_0 - \sqrt{a})^{e^+}(z_0 + \sqrt{a})^{e^-}$ with $e^+, e^- \in \mathbb{Z}$. Hence

$$\mathrm{div}(f) = \sum_{i=1}^{4} (e^+ + m_i) D_i^+ + (e^- + m_i) D_i^-$$

with

$$D_i^+ := \{z_0 = \sqrt{a}, u_i = 0\}, \quad D_i^- := \{z_0 = -\sqrt{a}, u_i = 0\}.$$

Therefore, $f \in \Gamma(X_{\overline{\mathbb{K}}}, \mathcal{O}^*)$ if and only if $e^+ = e^- = -m_1 = \cdots = -m_4$; hence f is a constant in $\overline{\mathbb{K}}^*$ by the description of $X_{\overline{\mathbb{K}}}$ above.

We have $\mathrm{Pic}(U_{\overline{\mathbb{K}}}) = 0$ because $U_{\overline{\mathbb{K}}} \cong (\mathbb{A}^1_{\overline{\mathbb{K}}} \setminus \{\sqrt{a}, -\sqrt{a}\}) \times \mathbb{G}^3_{m, \overline{\mathbb{K}}}$. In the exact sequence in Proposition 6.1.2.11 (iii), the abelian group $\Gamma(U_{\overline{\mathbb{K}}}, \mathcal{O}^*)/\overline{\mathbb{K}}^*$ is free of rank 5, generated by the classes of the functions $z_0 - \sqrt{a}, z_0 + \sqrt{a}, u_1, \ldots, u_4$ with the natural action of $G = \mathrm{Gal}(\overline{\mathbb{K}}/\mathbb{K})$ and the relation

$$[z_0 - \sqrt{a}] + [z_0 + \sqrt{a}] - [u_1] - [u_2] - [u_3] - [u_4] = 0$$

because of the description of $X_{\overline{\mathbb{K}}}$ above. Therefore, the dual torus R in Theorem 6.1.2.13 is isomorphic to the subtorus defined by

$$N_{\mathbb{L}/\mathbb{K}}(\mathbf{z}_1) = N_{\mathbb{M}/\mathbb{K}}(\mathbf{z}_2)$$

in $R_{\mathbb{L}/\mathbb{K}}(\mathbb{G}_{m,\mathbb{L}}) \times R_{\mathbb{M}/\mathbb{K}}(\mathbb{G}_{m,\mathbb{M}})$ with $\mathbb{L}$-coordinate $\mathbf{z}_1$ and $\mathbb{M}$-coordinate $\mathbf{z}_2$.

The sequence in Proposition 6.1.2.11 (iii) is split if and only if we find a G-equivariant splitting

$$\varphi \colon \Gamma(U_{\overline{\mathbb{K}}}, \mathcal{O}^*)/\overline{\mathbb{K}}^* \to \Gamma(U_{\overline{\mathbb{K}}}, \mathcal{O}^*), \quad [z_0 - \sqrt{a}] \mapsto \rho^{-1}(z_0 - \sqrt{a}),$$

$$[u_1] \mapsto \xi^{-1} u_1,$$

with $\rho \in \mathbb{L}^*$ and $\xi \in \mathbb{M}^*$. Because of the relation in $\Gamma(U_{\overline{\mathbb{K}}}, \mathcal{O}^*)/\overline{\mathbb{K}}^*$, such a G-equivariant splitting and hence universal torsors over X exist if and only if there are $\rho \in \mathbb{L}^*$ and $\xi \in \mathbb{M}^*$ satisfying

$$cN_{\mathbb{L}/\mathbb{K}}(\rho) = N_{\mathbb{M}/\mathbb{K}}(\xi).$$

Next, we analyze the other two terms of the exact sequence in Remark 6.1.2.12. The abelian group $\mathrm{Div}_{X_{\overline{\mathbb{K}}} \setminus U_{\overline{\mathbb{K}}}}(X_{\overline{\mathbb{K}}})$ is free of rank 8, with basis $D_1^+, \ldots, D_4^+, D_1^-, \ldots, D_4^-$. The divisor map $\mathrm{div}\colon \Gamma(U_{\overline{\mathbb{K}}}, \mathcal{O}^*)/\overline{\mathbb{K}}^* \to \mathrm{Div}_{X_{\overline{\mathbb{K}}} \setminus U_{\overline{\mathbb{K}}}}(X_{\overline{\mathbb{K}}})$ satisfies

$$[z_0 - \sqrt{a}] \mapsto D_1^+ + \cdots + D_4^+, \quad [z_0 + \sqrt{a}] \mapsto D_1^- + \cdots + D_4^-,$$

$$[u_i] \mapsto D_i^+ + D_i^-.$$

Because of $\mathbb{L} \subset \mathbb{M}$, the action of the Galois group G on our basis of $\mathrm{Div}_{X_{\overline{\mathbb{K}}} \setminus U_{\overline{\mathbb{K}}}}(X_{\overline{\mathbb{K}}})$ has two orbits of order 4, namely (up to reordering of $u_1, \ldots, u_4$) $D_1^+, D_2^+, D_3^-, D_4^-$ and its complement. Therefore, the dual torus $\mathcal{X}_0$ in Theorem 6.1.2.13 is isomorphic to $R_{\mathbb{M}/\mathbb{K}}(\mathbb{G}_{m,\mathbb{M}}) \times R_{\mathbb{M}/\mathbb{K}}(\mathbb{G}_{m,\mathbb{M}})$, with $\mathbb{M}$-coordinates $(\mathbf{x}_1, \mathbf{x}_2)$.

A careful analysis of the map div shows that the induced map on tori can be described as

$$\mathcal{X}_0 \to R, \quad (\mathbf{x}_1, \mathbf{x}_2) \mapsto (N_{\mathbb{M}/\mathbb{L}}(\mathbf{x}_1)\rho(N_{\mathbb{M}/\mathbb{L}}(\mathbf{x}_2)), \mathbf{x}_1 \mathbf{x}_2),$$

where $\sigma \in G$ with $\sigma(\sqrt{a}) = -\sqrt{a}$.

The negative of our splitting φ associated with $(\rho, \xi) \in \mathbb{L}^* \times \mathbb{M}^*$ satisfying $cN_{\mathbb{L}/\mathbb{K}}(\rho) = N_{\mathbb{M}/\mathbb{K}}(\xi)$ induces the map

$$\varphi\colon U \to R, \quad (z_0, z_1, \ldots, z_4) \mapsto (\rho^{-1}(z_0 - \sqrt{a}), \xi^{-1}(z_1\omega_1 + \cdots + z_4\omega_4)).$$

Note that the image is indeed in R because of the equation defining X.

By Theorem 6.1.2.13, $\mathcal{X}_U := \mathcal{X} \times_X U$ is isomorphic to the product $\mathcal{X}_0 \times_R U$ constructed via φ, which is the subvariety of $\mathbb{A}^1_{\mathbb{K}} \times R_{\mathbb{M}/\mathbb{K}}(\mathbb{G}_{m,\mathbb{K}}) \times R_{\mathbb{M}/\mathbb{K}}(\mathbb{G}_{m,\mathbb{K}})$ with coordinates $(z_0, \mathbf{x}_1, \mathbf{x}_2)$ defined by

$$z_0 - \sqrt{a} = \rho \cdot N_{\mathbb{M}/\mathbb{L}}(\mathbf{x}_1) \cdot \sigma(N_{\mathbb{M}/\mathbb{L}}(\mathbf{x}_2)),$$

an equation over $\mathbb{L}$ that can be interpreted as a system of two equations over $\mathbb{K}$. The restriction of $\mathcal{X} \to X$ is

$$\mathcal{X}_U \to U, \quad (z_0, \mathbf{x}_1, \mathbf{x}_2) \mapsto (z_0, \xi \mathbf{x}_1 \mathbf{x}_2).$$

We end this section with two fundamental statements regarding universal torsors and rational points over varieties over an arbitrary field $\mathbb{K}$ of characteristic zero. For further statements specific to number fields, see Section 6.2.3.

Proposition 6.1.2.15 *Let X be a smooth, geometrically irreducible variety over $\mathbb{K}$ with $\Gamma(X_{\overline{\mathbb{K}}}, \mathcal{O}^*) = \overline{\mathbb{K}}^*$ and finitely generated $\mathrm{Pic}(X_{\overline{\mathbb{K}}})$. The existence of universal torsors over X is necessary for the existence of rational points on X.*

Proof See [100, Prop. 2.2.8], [279, Def. 2.3.5]. A $\mathbb{K}$-rational point $p\colon \mathrm{Spec}(\mathbb{K}) \to X$ induces a splitting of the last map in the fundamental exact sequence of Theorem 6.1.2.7, which implies that the type map is surjective. Therefore, the existence of torsors of arbitrary type, and in particular the existence of universal X-torsors, is necessary for the existence of $\mathbb{K}$-rational points on X. $\qquad\square$

In this setting, $e(X) := \partial(\mathrm{id}) \in H^2(\mathbb{K}, H_{\mathrm{Pic}(X_{\overline{\mathbb{K}}})})$ is called the *elementary obstruction* because its vanishing is equivalent to the existence of universal torsors over X and therefore necessary for the existence of rational points on X; see [100, Sec. 2.2] for a systematic discussion. Twisting of torsors leads to:

Proposition 6.1.2.16 *Let H be a quasitorus over $\mathbb{K}$, and X a variety over $\mathbb{K}$. Assume that we have an X-torsor $\pi\colon \mathcal{X} \to X$ under H. For a 1-cocycle σ, let $\pi^\sigma\colon \mathcal{X}^\sigma \to X$ be its twist as in Construction 6.1.2.3. Then we have a disjoint union*

$$X(\mathbb{K}) = \bigcup_{[\sigma] \in H^1(\mathbb{K}, H)} \pi^\sigma(\mathcal{X}^\sigma(\mathbb{K})).$$

The finest partition of $X(\mathbb{K})$ of this kind is obtained using universal torsors.

Proof See [100, (2.7.2), Prop. 2.7.4], [279, Sec. 2.2, p. 22; Sec. 2.3, (2.26)]. Let $x \in X(\mathbb{K})$. Pulling back $\mathcal{X}$ via $x\colon \mathrm{Spec}(\mathbb{K}) \to X$ gives the fiber $\mathcal{X}_x$ of $\pi\colon \mathcal{X} \to X$ over x, which is a $\mathbb{K}$-torsor under H. Let $\tau \in Z^1(\mathbb{K}, H)$ be a representative of the class of $\mathcal{X}_x$ in $H^1(\mathbb{K}, H)$. By Lemma 6.1.1.11, this class is trivial if and only if $\mathcal{X}_x$ has rational points, and in this case $\pi(\mathcal{X}_x) = \{x\}$.

Then $(\mathcal{X}^\sigma)_x$ has class $[\tau] - [\sigma]$. Therefore, $x \in X(\mathbb{K})$ is in the image of $\mathcal{X}^\tau(\mathbb{K})$. By Theorem 6.1.2.7, the disjoint union runs over all isomorphy classes of torsors of the same type. The last statement follows from Proposition 6.1.2.10. $\qquad\square$

6.1.3 Cox rings and characteristic spaces

We extend the constructions of Cox sheaves, Cox rings, and characteristic spaces of Chapter 1 to the case of a not necessarily algebraically closed ground field. Moreover, we compare these concepts to the universal torsors discussed in the preceding section. The setting is the same as in Section 6.1.1. In particular, $\mathbb{K}$ is a field of characteristic zero with algebraic closure $\overline{\mathbb{K}}$ and G is the corresponding Galois group. The constructions and results of this section are part of [121].

A first step is the construction of Cox sheaves and Cox rings. The idea is to perform Construction 1.4.2.1 in an equivariant manner with respect to the action of the Galois group G. For this, the following concept is crucial.

Definition 6.1.3.1 Let X be a variety over $\mathbb{K}$ such that $X_{\overline{\mathbb{K}}}$ is irreducible, normal, has only constant invertible functions and finitely generated divisor class group. Let $K \subseteq \mathrm{WDiv}(X_{\overline{\mathbb{K}}})$ be a G-invariant subgroup mapping onto $\mathrm{Cl}(X_{\overline{\mathbb{K}}})$ and denote by $K^0 \subseteq K$ the subgroup of principal divisors. A *Galois-equivariant identifying character* for K is a G-equivariant homomorphism $\chi \colon K^0 \to \overline{\mathbb{K}}(X_{\overline{\mathbb{K}}})^*$ such that $\mathrm{div}(\chi(E)) = E$ holds for all $E \in K^0$.

Proposition 6.1.3.2 *Let X be a variety over $\mathbb{K}$ such that $X_{\overline{\mathbb{K}}}$ is irreducible, normal, and has only constant invertible functions and finitely generated divisor class group. Then the following statements are equivalent.*

(i) *Every G-invariant subgroup $K \subseteq \mathrm{WDiv}(X_{\overline{\mathbb{K}}})$ mapping onto $\mathrm{Cl}(X_{\overline{\mathbb{K}}})$ admits a Galois-invariant identifying character.*
(ii) *The natural exact sequence $1 \to \overline{\mathbb{K}}^* \to \overline{\mathbb{K}}(X_{\overline{\mathbb{K}}})^* \to \overline{\mathbb{K}}(X_{\overline{\mathbb{K}}})^*/\overline{\mathbb{K}}^* \to 1$ of G-modules admits a G-equivariant splitting.*
(iii) *There exists a G-invariant subgroup $K \subseteq \mathrm{WDiv}(X_{\overline{\mathbb{K}}})$ generated by finitely many prime divisors that maps onto $\mathrm{Cl}(X_{\overline{\mathbb{K}}})$ and admits a Galois-equivariant identifying character.*

Proof Because K^0 is a subgroup of $\mathrm{PDiv}(X_{\overline{\mathbb{K}}})$, which is naturally isomorphic to $\overline{\mathbb{K}}(X_{\overline{\mathbb{K}}})^*/\overline{\mathbb{K}}^*$, a Galois-equivariant identifying character for K is obtained by restricting a G-equivariant splitting of the exact sequence in (ii). Hence (ii) implies (i). Conversely, taking $K := \mathrm{WDiv}(X_{\overline{\mathbb{K}}})$, we have $K^0 = \mathrm{PDiv}(X_{\overline{\mathbb{K}}})$, hence (ii) is a special case of (i).

We show "(iii)$\Rightarrow$(ii)." Removing the singular locus, we may assume that $X_{\overline{\mathbb{K}}}$ is smooth. Let $A_{\overline{\mathbb{K}}} \subseteq X_{\overline{\mathbb{K}}}$ be the union of the supports of all divisors of K. Then $X_{\overline{\mathbb{K}}} \setminus A_{\overline{\mathbb{K}}}$ is defined over $\mathbb{K}$, hence equals $U_{\overline{\mathbb{K}}}$ for an open $U \subseteq X$. Moreover, we have $\mathrm{Cl}(U_{\overline{\mathbb{K}}}) = 0$. Consider the exact sequence

$$1 \to \overline{\mathbb{K}}^* \to \Gamma(U_{\overline{\mathbb{K}}}, \mathcal{O}^*) \to \Gamma(U_{\overline{\mathbb{K}}}, \mathcal{O}^*)/\overline{\mathbb{K}}^* \to 1.$$

By construction, $\Gamma(U_{\overline{\mathbb{K}}}, \mathcal{O}^*)/\overline{\mathbb{K}}^*$ is identified with $K^0 \subseteq K$ and thus the Galois equivariant identifying character $\chi \colon K^0 \to \overline{\mathbb{K}}(X_{\overline{\mathbb{K}}})^*$ gives rise to an equivariant splitting of the above sequence. Proposition 6.1.2.11 then tells us that (ii) holds.

To deduce (iii) from (i), take a finite number of prime divisors whose classes generate $\mathrm{Cl}(X_{\overline{\mathbb{K}}})$. Each of them is defined over a finite extension of $\mathbb{K}$; hence each has a finite orbit under the natural action of G on $\mathrm{WDiv}(X)$. The union of these orbits generates K as in (iii), where the Galois-equivariant identifying character exists by (i). $\qquad\square$

Construction 6.1.3.3 Let X be a variety over $\mathbb{K}$ such that $X_{\overline{\mathbb{K}}}$ is irreducible, normal, and has only constant invertible functions and finitely generated divisor class group. Fix a G-invariant subgroup $K \subseteq \mathrm{WDiv}(X_{\overline{\mathbb{K}}})$ mapping onto $\mathrm{Cl}(X_{\overline{\mathbb{K}}})$ and denote by $K^0 \subseteq K$ the subgroup of principal divisors. Suppose that there is a Galois-equivariant identifying character $\chi \colon K^0 \to \overline{\mathbb{K}}(X_{\overline{\mathbb{K}}})^*$. Denote by $\mathcal{S}_{X_{\overline{\mathbb{K}}}}$ the sheaf of divisorial algebras on $X_{\overline{\mathbb{K}}}$ associated with K and by $\mathcal{I}_{X_{\overline{\mathbb{K}}}}$ the sheaf of ideals of $\mathcal{S}_{X_{\overline{\mathbb{K}}}}$ defined by χ as in 1.4.2.1. Then $\mathcal{S}_{X_{\overline{\mathbb{K}}}}$ is a K-graded G-sheaf as in Section 4.2.3, $\mathcal{I}_{X_{\overline{\mathbb{K}}}}$ is G-invariant and thus $\mathcal{R}_{X_{\overline{\mathbb{K}}}} := \mathcal{S}_{X_{\overline{\mathbb{K}}}}/\mathcal{I}_{X_{\overline{\mathbb{K}}}}$ comes as a $\mathrm{Cl}(X_{\overline{\mathbb{K}}})$-graded G-sheaf; we call it a *Galois-equivariant Cox sheaf* of $X_{\overline{\mathbb{K}}}$. Let $\pi \colon X_{\overline{\mathbb{K}}} \to X$ denote the canonical projection. Then we obtain a *Cox sheaf* and a *Cox ring* of X (associated with χ) by setting

$$\mathcal{R}_X := \pi_*(\mathcal{R}_{X_{\overline{\mathbb{K}}}})^G, \qquad \mathcal{R}_X(X) := \Gamma(X, \mathcal{R}_X) = \mathcal{R}_{X_{\overline{\mathbb{K}}}}(X_{\overline{\mathbb{K}}})^G.$$

Proof Because K is G-invariant, $\mathcal{S}_{X_{\overline{\mathbb{K}}}}$ is a G-sheaf. Because χ is G-equivariant, $\mathcal{I}_{X_{\overline{\mathbb{K}}}}$ is G-invariant. Hence the structure of a K-graded G-sheaf on $\mathcal{S}_{X_{\overline{\mathbb{K}}}}$ induces the same structure on $\mathcal{S}_{X_{\overline{\mathbb{K}}}}/\mathcal{I}_{X_{\overline{\mathbb{K}}}}$. $\qquad\square$

Remark 6.1.3.4 Consider the situation of Construction 6.1.3.3. Note that Cox rings and Cox sheaves of X are generally not unique up to isomorphism, but may depend on the choice of Galois-equivariant identifying character χ.

If the Cox ring $\mathcal{R}_{X_{\overline{\mathbb{K}}}}(X_{\overline{\mathbb{K}}})$ of $X_{\overline{\mathbb{K}}}$ is finitely generated, then also any Cox ring $\mathcal{R}_X(X)$ of X is finitely generated. Similarly, if the Cox sheaf $\mathcal{R}_{X_{\overline{\mathbb{K}}}}$ of $X_{\overline{\mathbb{K}}}$ is locally of finite type, then any Cox sheaf $\mathcal{R}_X$ of X is locally of finite type.

Example 6.1.3.5 Over $\mathbb{K} := \mathbb{R}$, let X be the blow-up of $\mathbb{P}^2_{\mathbb{R}}$ in the points $p_1 := [1, \sqrt{-1}, 0]$ and $p_2 := [1, -\sqrt{-1}, 0]$ that are conjugate under the action of the Galois group $G := \mathrm{Gal}(\mathbb{C}/\mathbb{R})$. Let $p_3 := [0, 0, 1]$. Then $\mathcal{R}_{X_{\mathbb{C}}}(X_{\mathbb{C}}) = \mathbb{C}[T_1, \ldots, T_5]$, where T_i is a section corresponding to the line through p_j, p_k, for any $\{i, j, k\} = \{1, 2, 3\}$, and T_4, T_5 are sections corresponding to the exceptional divisors obtained by blowing up p_1, p_2, such that complex conjugation exchanges T_1 and T_2, and also T_4 and T_5. Then $\mathcal{R}_X(X) = \mathbb{R}[U_1, \ldots, U_5]$ with

$$U_1 := \frac{T_1 + T_2}{2}, \quad U_2 := \frac{T_1 - T_2}{2\sqrt{-1}}, \quad U_3 := T_3, \quad U_4 := \frac{T_4 + T_5}{2}, \quad U_5 := \frac{T_4 - T_5}{2\sqrt{-1}}.$$

Note that $\mathcal{R}_X(X)$ is graded neither by $\mathrm{Cl}(X_{\mathbb{C}})$ nor by $\mathrm{Cl}(X)$, but $\mathcal{R}_X(X) \otimes_{\mathbb{R}} \mathbb{C} \cong \mathcal{R}_{X_{\mathbb{C}}}(X_{\mathbb{C}})$ has a natural $\mathrm{Cl}(X_{\mathbb{C}})$-grading, and that the actions of the Galois group on $\mathcal{R}_X(X) \otimes_{\mathbb{R}} \mathbb{C}$ and $\mathrm{Cl}(X_{\mathbb{C}})$ are compatible.

Example 6.1.3.6 Over $\mathbb{K} := \mathbb{R}$, let $\widetilde{X}$ be the blow-up of the Hirzebruch surface $\mathbb{F}_{2,\mathbb{R}}$ in two points that are conjugate under the action of the Galois group $G := \mathrm{Gal}(\mathbb{C}/\mathbb{R})$, lying on the same fiber on $\mathbb{F}_{2,\mathbb{R}}$, but not lying on the (-2)-curve on $\mathbb{F}_{2,\mathbb{R}}$. Then $\widetilde{X}$ is defined over $\mathbb{R}$. It is a del Pezzo surface of degree 6 with an A_2-singularity. The Cox ring of $\widetilde{X}_{\mathbb{C}}$ is computed in Example 5.4.3.4 as

$$\mathcal{R}_{\widetilde{X}_{\mathbb{C}}}(\widetilde{X}_{\mathbb{C}}) = \mathbb{C}[T_{01}, T_{02}, T_{11}, T_{12}, T_{21}, T_{22}, S]/\langle T_{01}T_{02} + T_{11}T_{12}^2 + T_{21}T_{22}\rangle.$$

The Galois group acts on $\mathcal{R}_{\widetilde{X}_{\mathbb{C}}}(\widetilde{X}_{\mathbb{C}})$ as follows. Complex conjugation exchanges T_{01}, T_{21} and also T_{02}, T_{22}. Then generators of $(\mathcal{R}_{\widetilde{X}_{\mathbb{C}}}(\widetilde{X}_{\mathbb{C}}))^G$ are

$$U_1^+ := \frac{T_{01} + T_{21}}{2}, \ U_1^- := \frac{T_{01} - T_{21}}{2\sqrt{-1}}, \ U_2^+ := \frac{T_{02} + T_{22}}{2}, \ U_2^- := \frac{T_{02} - T_{22}}{2\sqrt{-1}},$$

$$U_{11} := T_{11}, \ U_{12} := T_{12}, \ V := S.$$

This leads to

$$\mathcal{R}_{\widetilde{X}}(\widetilde{X}) = \mathbb{R}[U_1^+, U_1^-, U_2^+, U_2^-, U_{11}, U_{12}, V]/\langle 2U_1^+U_2^+ - 2U_1^-U_2^- + U_{11}U_{12}^2\rangle.$$

As a basic ingredient for the construction of characteristic spaces, we provide a description of quasitorus actions in terms of graded algebras extending Theorem 1.2.2.4 to nonclosed ground fields.

An *affine G-algebra over* $\overline{\mathbb{K}}$ is an affine algebra A over $\overline{\mathbb{K}}$ coming with a continuous action of the Galois group G. By a *graded affine G-algebra over* $\overline{\mathbb{K}}$ we mean an affine G-algebra $A = \bigoplus_M A_w$ over $\overline{\mathbb{K}}$ graded by a G-module M such that $g \cdot A_w = A_{g \cdot w}$ holds for all $g \in G$ and $w \in M$. A *morphism of graded affine G-algebras over* $\overline{\mathbb{K}}$ is a morphism $(\psi, \widetilde{\psi})$ of the underlying graded $\overline{\mathbb{K}}$-algebras such that ψ and $\widetilde{\psi}$ are G-equivariant.

For any M-graded affine G-algebra A over $\overline{\mathbb{K}}$, the action of the quasitorus $H_{\overline{\mathbb{K}}} = \operatorname{Spec} \overline{\mathbb{K}}[M]$ on $X_{\overline{\mathbb{K}}} = \operatorname{Spec} A$ descends to an action of the quasitorus $H = \operatorname{Spec} \overline{\mathbb{K}}[M]^G$ over $\mathbb{K}$ on the affine $\mathbb{K}$-variety $X = \operatorname{Spec} A^G$. Every G-equivariant morphism $(\psi, \widetilde{\psi})$ from an M-graded affine G-algebra A over $\overline{\mathbb{K}}$ to an N-graded affine G-algebra B over $\overline{\mathbb{K}}$ defines a homomorphism $\psi^G \colon A^G \to B^G$ of the $\mathbb{K}$-algebras of invariants and $\widetilde{\psi}$ gives us the $\mathbb{K}$-algebra homomorphism $\overline{\mathbb{K}}[\widetilde{\psi}]^G$. The functorial properties of these assignments are summarized as follows.

Proposition 6.1.3.7 *We have contravariant functors being essentially inverse to each other:*

$$\left\{\begin{array}{l} \textit{graded affine G-algebras} \\ \textit{over } \overline{\mathbb{K}} \end{array}\right\} \longleftrightarrow \left\{\begin{array}{l} \textit{affine } \mathbb{K}\textit{-varieties with} \\ \textit{quasitorus action} \end{array}\right\}$$

$$A \mapsto \operatorname{Spec} A^G,$$

$$(\psi, \widetilde{\psi}) \mapsto (\operatorname{Spec} \psi^G, \operatorname{Spec} \overline{\mathbb{K}}[\widetilde{\psi}]^G),$$

$$\Gamma(X_{\overline{\mathbb{K}}}, \mathcal{O}) \leftmapsto X,$$

$$(\varphi^*, \widetilde{\varphi}^*) \leftmapsto (\varphi, \widetilde{\varphi}).$$

Under these equivalences the graded homomorphisms correspond to the equivariant morphisms.

We are ready for the construction of characteristic spaces. Similarly as before, the idea is to perform Construction 1.6.1.3 equivariantly with respect to the Galois group.

Construction 6.1.3.8 Let X be a variety over $\mathbb{K}$ such that $X_{\overline{\mathbb{K}}}$ is irreducible, normal, and has only constant invertible functions and finitely generated divisor class group. Assume that there is a Galois-equivariant Cox sheaf $\mathcal{R}_{X_{\overline{\mathbb{K}}}}$ on $X_{\overline{\mathbb{K}}}$, and suppose that $\mathcal{R}_{X_{\overline{\mathbb{K}}}}$ is locally of finite type. Then the associated Cox sheaf $\mathcal{R}_X$ over X is locally of finite type. The $\mathbb{K}$-variety $\widehat{X} := \mathrm{Spec}_X \mathcal{R}_X$ comes with an action of the quasitorus $H_X := \mathrm{Spec}\,\overline{\mathbb{K}}[\mathrm{Cl}(X_{\overline{\mathbb{K}}})]^G$ over $\mathbb{K}$ and the canonical $\mathbb{K}$-morphism $p_X \colon \widehat{X} \to X$ is a good quotient for this action. We call H_X the *characteristic quasitorus* of X and $p_X \colon \widehat{X} \to X$ a *characteristic space* over X. We have

$$\mathcal{R}_X = (p_X)_*(\mathcal{O}_{\widehat{X}}), \qquad \mathcal{R}_X(X) = \Gamma(\widehat{X}, \mathcal{O}), \qquad \widehat{X}_{\overline{\mathbb{K}}} = \widehat{X}_{\overline{\mathbb{K}}}.$$

If the Cox ring $\mathcal{R}_X(X)$ is finitely generated, then one has the *total coordinate space* $\overline{X} := \mathrm{Spec}\,\mathcal{R}_X(X)$. It comes with an action of H_X and the canonical map $\widehat{X} \to \overline{X}$ is an open embedding.

Proof Because the Cox sheaf $\mathcal{R}_{X_{\overline{\mathbb{K}}}}$ on $X_{\overline{\mathbb{K}}}$ is constructed via a Galois-equivariant identifying character, the construction of the characteristic space $p_{X_{\overline{\mathbb{K}}}} \colon \widehat{X}_{\overline{\mathbb{K}}} \to X_{\overline{\mathbb{K}}}$ can be performed in a Galois-equivariant manner. Now the statements follow from Proposition 6.1.3.7.   □

Proposition 6.1.3.9 *Let X be a variety over $\mathbb{K}$ such that $X_{\overline{\mathbb{K}}}$ is irreducible, locally factorial, and has only constant invertible functions and finitely generated divisor class group.*

(i) *The characteristic quasitorus H_X of X coincides with the Picard quasitorus of X.*

(ii) *If $p_X \colon \widehat{X} \to X$ is a characteristic space over X, then $p_X \colon \widehat{X} \to X$ is a universal torsor over X.*

(iii) *If $p \colon \mathcal{X} \to X$ is a universal torsor, then $p_*(\mathcal{O}_{\mathcal{X}_{\overline{\mathbb{K}}}})$ is the Cox sheaf of $X_{\overline{\mathbb{K}}}$, $\Gamma(\mathcal{X}_{\overline{\mathbb{K}}}, \mathcal{O}_{\mathcal{X}_{\overline{\mathbb{K}}}})$ is the Cox ring of $X_{\overline{\mathbb{K}}}$, and $p \colon \mathcal{X}_{\overline{\mathbb{K}}} \to X_{\overline{\mathbb{K}}}$ is a characteristic space over $X_{\overline{\mathbb{K}}}$.*

Proof By our assumptions, $\mathrm{Pic}(X_{\overline{\mathbb{K}}})$ can be identified with $\mathrm{Cl}(X_{\overline{\mathbb{K}}})$, and hence (i) holds.

By Corollary 1.6.2.7, $\widehat{X}_{\overline{\mathbb{K}}} \to X_{\overline{\mathbb{K}}}$ is an $X_{\overline{\mathbb{K}}}$-torsor under $(H_{\mathrm{Pic}(X_{\overline{\mathbb{K}}})})_{\overline{\mathbb{K}}}$, and because $\widehat{X}_{\overline{\mathbb{K}}} = \widehat{X}_{\overline{\mathbb{K}}}$, we know that $\widehat{X}$ is an X-torsor under $H_{\mathrm{Pic}(X_{\overline{\mathbb{K}}})}$. That its type as in Definition 6.1.2.5 is the identity map can be checked over $\overline{\mathbb{K}}$, and here we can apply Proposition 1.6.1.7. Therefore, $\widehat{X} \to X$ is a universal X-torsor, and (ii) holds.

Over $\overline{\mathbb{K}}$, characteristic spaces and universal torsors exist and are unique up to isomorphism by Construction 1.6.1.3 resp. Theorem 6.1.2.7. Hence (iii) follows from (ii). □

Remark 6.1.3.10 Let X be a variety over a nonclosed field $\mathbb{K}$ such that $X_{\overline{\mathbb{K}}}$ is irreducible, normal, and has only constant invertible functions and finitely generated divisor class group. One could immitate the construction of Cox rings, Cox sheaves and characteristic spaces with $\mathrm{Cl}(X_{\overline{\mathbb{K}}})$ replaced by the Weil divisor class group $\mathrm{Cl}(X)$ over the ground field $\mathbb{K}$. This would avoid the Galois action, but would generally lead to much smaller rings. For example, for a general smooth cubic surface $X \subset \mathbb{P}^3_{\mathbb{Q}}$ over $\mathbb{Q}$, the class group $\mathrm{Cl}(X)$ is generated by the class of an anticanonical divisor $-K_X$, and hence such a construction would give the anticanonical ring $\bigoplus_{n \geq 0} \Gamma(X, \mathcal{O}(-nK_X))$ instead of a Cox ring, and the affine cone over X instead of a universal torsor over X.

6.2 Existence of rational points

6.2.1 The Hasse principle and weak approximation

Studying rational solutions of Diophantine equations is one of the oldest problems in number theory. In the language of algebraic geometry, it can be rephrased as the question of the existence of rational points on varieties over $\mathbb{Q}$. Finding a general answer seems impossibly hard. However, one can treat some classes of varieties. Two of the most basic questions are whether a variety over $\mathbb{Q}$ satisfies the Hasse principle and weak approximation.

Let $\mathbb{K}$ be a number field, that is, a finite extension of the field $\mathbb{Q}$ of rational numbers. Let $\Omega_{\mathbb{K}}$ be the set of places of $\mathbb{K}$, and, for each $v \in \Omega_{\mathbb{K}}$, let $\mathbb{K}_v$ be the completion of $\mathbb{K}$ with respect to v.

For example, $\Omega_{\mathbb{Q}}$ consists of the archimedean place ∞ and the non-archimedean places p for all prime numbers p. The completions are $\mathbb{Q}_{\infty} := \mathbb{R}$ and the p-adic numbers $\mathbb{Q}_p$ for all prime numbers p.

Let X be a smooth variety over $\mathbb{K}$. A necessary condition for the existence of $\mathbb{K}$-rational points on X clearly is that X is *everywhere locally soluble*, that is, that $X(\mathbb{K}_v)$ is nonempty for every $v \in \Omega_{\mathbb{K}}$. If the converse holds, we say that *the Hasse principle holds on X*.

We say that *weak approximation holds on X* if the diagonal embedding of $X(\mathbb{K})$ is dense in

$$\prod_{v \in \Omega_{\mathbb{K}}} X(\mathbb{K}_v),$$

equipped with the product topology obtained from v-adic metrics $d_v \colon X(\mathbb{K}_v) \times X(\mathbb{K}_v) \to \mathbb{R}$. More concretely, this means that for any system of local points $(x_v) \in \prod_{v \in \Omega_{\mathbb{K}}} X(\mathbb{K}_v)$, any finite set of places $S \subset \Omega_{\mathbb{K}}$ and any $\varepsilon > 0$, there is an $x \in X(\mathbb{K})$ such that for all $v \in S$, we have $d_v(x, x_v) < \varepsilon$.

Example 6.2.1.1   Weak approximation holds for $\mathbb{P}_{\mathbb{K}}^n$ for any $n \geq 1$, essentially by the Chinese remainder theorem.

By the theorem of Hasse–Minkowski, the Hasse principle holds on smooth projective quadrics over arbitrary number fields $\mathbb{K}$. Quadrics $X \subset \mathbb{P}_{\mathbb{K}}^{n-1}$ defined by quadratic forms of rank $n \geq 5$ are locally soluble for all non-archimedean $v \in \Omega_{\mathbb{K}}$, hence in this case, the existence of rational points is equivalent to the requirement that the form is indefinite over the real places of $\mathbb{K}$. See [272] for the case $\mathbb{K} := \mathbb{Q}$. A smooth projective quadric containing a rational point satisfies weak approximation because it is birational to a projective space, as we can see by projecting from the rational point.

By the *Hasse norm principle* [161], the Hasse principle holds for the variety defined in $\mathbb{A}_{\mathbb{K}}^n$ by $N_{\mathbb{L}/\mathbb{K}}(z_1, \dots, z_n) = c$ for $c \in \mathbb{K}^*$, a Galois extension $\mathbb{L} \supseteq \mathbb{K}$ of number fields of degree n with cyclic $\mathrm{Gal}(\mathbb{L}/\mathbb{K})$ and $N_{\mathbb{L}/\mathbb{K}}$ the norm form in n variables obtained via a choice of a basis of $\mathbb{L}$ over $\mathbb{K}$, analogously to the beginning of Example 6.1.2.14. For weak approximation for these varieties, see [302, 6.38].

Whether the Hasse principle holds is invariant under birational maps for smooth, proper varieties over a number field $\mathbb{K}$ because of Nishimura's lemma; see [193, Prop. 6] for a simple proof.

Lemma 6.2.1.2 (Nishimura)   *Let $\mathbb{K}$ be a field. Let X be a proper scheme over $\mathbb{K}$, let X' be a scheme over $\mathbb{K}$ with a smooth rational point, and let $f \colon X' \dashrightarrow X$ be a rational map over $\mathbb{K}$. Then X has a rational point.*

Together with Chow's lemma, Nishimura's lemma reduces the study of the Hasse principle from smooth varieties to the case of smooth, projective varieties. Weak approximation is a birational invariant for smooth varieties over $\mathbb{K}$ by the implicit function theorem.

Example 6.2.1.3   In the setting of Example 6.1.2.14, assume that $\mathbb{K}$ is a number field. We claim that the Hasse principle holds on the varieties $\mathcal{X}_U \subset \mathbb{A}_{\mathbb{K}}^1 \times R_{M/\mathbb{K}}(\mathbb{G}_{m,\mathbb{K}}) \times R_{M/\mathbb{K}}(\mathbb{G}_{m,\mathbb{K}})$ defined by

$$z_0 - \sqrt{a} \, = \, \rho \cdot N_{M/\mathbb{L}}(\mathbf{x}_1) \cdot \sigma(N_{M/\mathbb{L}}(\mathbf{x}_2)).$$

We have $\mathbb{M} = \mathbb{L}(\beta)$ for some $\beta = \sqrt{u + v\sqrt{a}} \in \mathbb{M}$ with $u, v \in \mathbb{K}$, and hence we can write $\mathbf{x}_1 = (y_1 + y_2\sqrt{a}) + (y_3 + y_4\sqrt{a})\beta$. Then we compute that

$$N_{M/\mathbb{L}}(\mathbf{x}_1) = g_0(y_1, \dots, y_4) + g_1(y_1, \dots, y_4)\sqrt{a}$$

with

$$g_0(y_1, \ldots, y_4) = y_1^2 + ay_2^2 - (u(y_3^2 + ay_4^2) + 2avy_3y_4),$$
$$g_1(y_1, \ldots, y_4) = 2y_1y_2 - (2uy_3y_4 + v(y_3^2 + ay_4^2)).$$

Multiplying by $\rho \cdot \sigma(N_{M/L}(\mathbf{x}_2)) \in \mathbb{L}^*$, we see that for any $\mathbf{x}_2 \in \mathbb{M}^*$, the equation defining X has the form

$$z_0 - \sqrt{a} = f_{0,\mathbf{x}_2}(y_1, y_2, y_3, y_4) + f_{1,\mathbf{x}_2}(y_1, y_2, y_3, y_4)\sqrt{a},$$

where $f_{i,\mathbf{x}_2}(y_1, y_2, y_3, y_4) \in \mathbb{K}[y_1, y_2, y_3, y_4]$ are quadratic forms whose coefficients depend on $\mathbf{x}_2$ and that are nontrivial linear combinations of $g_0(y_1, \ldots, y_4)$ and $g_1(y_1, \ldots, y_4)$. Now any such linear combination has rank 4 because for $(\lambda, \mu) \in \mathbb{K}^2 \setminus \{(0, 0)\}$,

$$\lambda g_0(y_1, \ldots, y_4) + \mu g_1(y_1, \ldots, y_4) = q_0(y_1, y_2) + q_1(y_3, y_4)$$

with

$$q_0(y_1, y_2) = \lambda y_1^2 + 2\mu y_1 y_2 + a\lambda y_2^2,$$
$$q_1(y_1, y_2) = -(\lambda u + \mu v)y_3^2 - 2(\lambda av + \mu u)y_3y_4 - a(\lambda u + \mu v)y_4^2,$$

hence their discrimants are

$$\mathrm{disc}(q_0) = \lambda^2 a - \mu^2 \neq 0, \quad \mathrm{disc}(q_1) = -(\lambda^2 a - \mu^2)(v^2 a - u^2) \neq 0,$$

because a is not a square in $\mathbb{K}$.

To prove the Hasse principle on $\mathcal{X}_U$, we apply the fibration method to the projection $p : \mathcal{X}_U \to R_{M/K}(\mathbb{G}_{m,M})$ to the $\mathbf{x}_2$-coordinate. Given local points

$$(z_{0,v}, \mathbf{x}_{1,v}, \mathbf{x}_{2,v}) \in \prod_{v \in \Omega_{\mathbb{K}}} \mathcal{X}_U(\mathbb{K}_v),$$

let $(x_{2,v}) \in \prod_{v \in \Omega_{\mathbb{K}}} R_{M/K}(\mathbb{G}_{m,M})(\mathbb{K}_v)$ be the projection. Because $R_{M/K}(\mathbb{G}_{m,M})$ is rational, it satisfies weak approximation, and hence $(\mathbf{x}_{2,v})$ can be approximated arbitrarily close by a rational point $\mathbf{x}_2 \in R_{M/K}(\mathbb{G}_{m,M})(\mathbb{K})$, in particular at the real places of $\mathbb{K}$. Eliminating the z_0-coordinate, our preceding discussion shows that the fiber over $\mathbf{x}_2$ is defined by

$$-1 = f_{1,\mathbf{x}_2}(y_1, \ldots, y_4),$$

where $f_{1,\mathbf{x}_2}(y_1, \ldots, y_4)$ is a quadric of rank 4. As discussed in Example 6.2.1.1, this implies that the fiber over $\mathbf{x}_2$ has points over every non-archimedean place and every complex place of $\mathbb{K}$. For any real place $v \in \Omega_{\mathbb{K}}$, by assumption we have a real point on the fiber over $\mathbf{x}_{2,v}$; hence $f_{1,\mathbf{x}_{2,v}}(y_1, \ldots, y_4)$ is not negative definite, and because $\mathbf{x}_2$ is very close to $\mathbf{x}_{2,v}$, the same holds for $f_{1,\mathbf{x}_2}(y_1, \ldots, y_4)$, and hence the fiber over $\mathbf{x}_2$ also contains points over any real place of $\mathbb{K}$. By the theorem of Hasse–Minkowski, this implies that this fiber has rational points, and hence $\mathcal{X}_U$ satisfies the Hasse principle.

Similarly, we can show that $\mathcal{X}_U$ satisfies weak approximation, using weak approximation for quadratic forms and the implicit function theorem; see [101, Prop. 3.9] for details.

6.2.2 Brauer–Manin obstructions

While the Hasse principle and weak approximation hold for some classes of varieties, many counterexamples are known, for example, for cubic surfaces [290]. The first systematic attempt to explain these counterexamples was given by Manin, using Grothendieck's Brauer group on varieties; see [214].

The *Brauer group* of a field $\mathbb{K}$ is the Galois cohomology group $\mathrm{Br}(\mathbb{K}) := H^2(\mathbb{K}, \mathbb{G}_{\mathrm{m},\mathbb{K}})$, see [273, Sec. X, Sec. XII], for example. It can be identified with the set of equivalence classes of finite-dimensional central simple algebras over $\mathbb{K}$, where algebras A, A' are equivalent if there are positive integers n, n' such that $A \otimes_{\mathbb{K}} M_{n \times n}(\mathbb{K})$ and $A' \otimes_{\mathbb{K}} M_{n' \times n'}(\mathbb{K})$ are isomorphic $\mathbb{K}$-algebras, where $M_{n \times n}(\mathbb{K})$ is the algebra of $n \times n$-matrices over $\mathbb{K}$.

For example, the Brauer group of $\mathbb{R}$ has two elements, represented by $\mathbb{R}$ itself and *Hamilton's quaternions*, namely the noncommutative $\mathbb{R}$-algebra with basis $1, i, j, k$ satisfying $i^2 = j^2 = k^2 = ijk = -1$.

Finite fields and algebraically closed fields have trivial Brauer groups. For any place v of a number field $\mathbb{K}$, we have the *local invariant map* from class field theory, namely a monomorphism

$$\mathrm{inv}_v \colon \mathrm{Br}(\mathbb{K}_v) \ \to \ \mathbb{Q}/\mathbb{Z}.$$

In fact, inv_v is an isomorphism for non-archimedean v, and $\mathrm{Br}(\mathbb{R}) \cong \mathbb{Z}/2\mathbb{Z}$, while $\mathrm{Br}(\mathbb{C})$ is trivial. For a number field $\mathbb{K}$, a central result of global class field theory is the exact sequence

$$0 \ \to \ \mathrm{Br}(\mathbb{K}) \ \to \ \bigoplus_{v \in \Omega_{\mathbb{K}}} \mathrm{Br}(\mathbb{K}_v) \ \to \ \mathbb{Q}/\mathbb{Z} \ \to \ 0,$$

where the first map is induced by the inclusions $\mathbb{K} \hookrightarrow \mathbb{K}_v$ and the second map is defined as $(A_v) \mapsto \sum_{v \in \Omega_{\mathbb{K}}} \mathrm{inv}_v(A_v)$.

The *(cohomological) Brauer group* of a variety X over a field $\mathbb{K}$ is the étale cohomology group

$$\mathrm{Br}(X) := H^2_{\text{ét}}(X, \mathbb{G}_{\mathrm{m},\mathbb{K}});$$

see [219, Sec. 4], for example. By Grothendieck's purity theorem, we have the following birational invariance of the Brauer group: smooth projective varieties over $\mathbb{K}$ that are birational over $\mathbb{K}$ have isomorphic Brauer groups. Because $\mathrm{Br}(\mathbb{P}^n_{\mathbb{K}}) = \mathrm{Br}(\mathbb{K})$, this implies that the Brauer group is also simply $\mathrm{Br}(\mathbb{K})$ for any smooth projective variety that is rational over $\mathbb{K}$.

Let $\mathbb{K}$ be a number field. Let X be a smooth proper variety over $\mathbb{K}$. For any subset B of $\mathrm{Br}(X)$, we define

$$\left(\prod_{v \in \Omega_{\mathbb{K}}} X(\mathbb{K}_v)\right)^B := \left\{(x_v) \in \prod_{v \in \Omega_{\mathbb{K}}} X(\mathbb{K}_v); \sum_{v \in \Omega_{\mathbb{K}}} \mathrm{inv}_v(A_{x_v}) = 0 \text{ for all } A \in B\right\}.$$

For $A \in \mathrm{Br}(X)$ and $x_v \in X(\mathbb{K}_v)$, we have $A_{x_v} \in \mathrm{Br}(\mathbb{K}_v)$ via the natural map $\mathrm{Br}(X) \to \mathrm{Br}(\mathbb{K}_v)$ induced by x_v. We have

$$X(\mathbb{K}) \subseteq \overline{X(\mathbb{K})} \subseteq \left(\prod_{v \in \Omega_{\mathbb{K}}} X(\mathbb{K}_v)\right)^B \subseteq \prod_{v \in \Omega_{\mathbb{K}}} X(\mathbb{K}_v), \qquad (6.2.1)$$

where $\overline{X(\mathbb{K})}$ is the closure of $X(\mathbb{K})$ in $\prod_{v \in \Omega_{\mathbb{K}}} X(\mathbb{K}_v)$ with the product topology described above.

Definition 6.2.2.1 Let X be a smooth proper variety over a number field $\mathbb{K}$. If $\prod_{v \in \Omega_{\mathbb{K}}} X(\mathbb{K}_v)$ is nonempty but $\left(\prod_{v \in \Omega_{\mathbb{K}}} X(\mathbb{K}_v)\right)^{\mathrm{Br}(X)}$ is empty, the Hasse principle fails for X, and we say that there is a *Brauer–Manin obstruction to the Hasse principle* on X. If nonemptiness of $\left(\prod_{v \in \Omega_{\mathbb{K}}} X(\mathbb{K}_v)\right)^{\mathrm{Br}(X)}$ implies nonemptiness of $X(\mathbb{K})$, we say that the *Brauer–Manin obstruction to the Hasse principle is the only one* on X.

Definition 6.2.2.2 Let X be a smooth proper variety over a number field $\mathbb{K}$. If

$$\left(\prod_{v \in \Omega_{\mathbb{K}}} X(\mathbb{K}_v)\right)^{\mathrm{Br}(X)} \subsetneq \prod_{v \in \Omega_{\mathbb{K}}} X(\mathbb{K}_v),$$

weak approximation fails for X, and we say that there is a *Brauer–Manin obstruction to weak approximation* on X. If $X(\mathbb{K})$ is dense in $\left(\prod_{v \in \Omega_{\mathbb{K}}} X(\mathbb{K}_v)\right)^{\mathrm{Br}(X)}$, we say that the *Brauer–Manin obstruction to weak approximation is the only one* on X.

Let $\mathrm{Br}_0(X)$ be the *constant part* of $\mathrm{Br}(X)$, that is, the image of the natural map $\mathrm{Br}(\mathbb{K}) \to \mathrm{Br}(X)$ induced by the structure map $X \to \mathrm{Spec}(\mathbb{K})$. By the commutative diagram

$$\begin{array}{ccc}
\mathrm{Br}(X) & \longrightarrow & \prod_{v \in \Omega_{\mathbb{K}}} \mathrm{Br}(X_{\mathbb{K}_v}) \\
\uparrow & & \uparrow \quad \searrow^{\sum_{v \in \Omega_{\mathbb{K}}} \mathrm{inv}_v(\cdot)} \\
0 \longrightarrow \mathrm{Br}(\mathbb{K}) \longrightarrow & \bigoplus_{v \in \Omega_{\mathbb{K}}} \mathrm{Br}(\mathbb{K}_v) & \longrightarrow \mathbb{Q}/\mathbb{Z} \longrightarrow 0
\end{array}$$

we have

$$\left(\prod_{v \in \Omega_{\mathbb{K}}} X(\mathbb{K}_v) \right)^{\mathrm{Br}_0(X)} = \prod_{v \in \Omega_{\mathbb{K}}} X(\mathbb{K}_v).$$

Therefore, only the quotient $\mathrm{Br}(X)/\mathrm{Br}_0(X)$ is relevant for Brauer–Manin obstructions.

The *algebraic Brauer group* $\mathrm{Br}_1(X)$ of a variety X over a field $\mathbb{K}$ is the kernel of the natural map $\mathrm{Br}(X) \to \mathrm{Br}(X_{\overline{\mathbb{K}}})$. For example, we have $\mathrm{Br}_1(X) = \mathrm{Br}(X)$ for geometrically rational varieties. For a smooth proper variety X over a number field $\mathbb{K}$, we talk about *algebraic Brauer–Manin obstructions* to the Hasse principle or weak approximation when $\mathrm{Br}(X)$ is replaced by $\mathrm{Br}_1(X)$ in Definitions 6.2.2.1 and 6.2.2.2.

The next result is derived from the Hochschild–Serre spectral sequence; see [279, Cor. 2.3.9].

Proposition 6.2.2.3 *Let X be a variety over a number field $\mathbb{K}$ with $\Gamma(X_{\overline{\mathbb{K}}}, \mathcal{O}^*) = \overline{\mathbb{K}}^*$. We have the exact sequence*

$$0 \to \mathrm{Pic}(X) \to \mathrm{Pic}(X_{\overline{\mathbb{K}}})^G \to \mathrm{Br}(\mathbb{K}) \to \mathrm{Br}_1(X) \to H^1(\mathbb{K}, \mathrm{Pic}(X_{\overline{\mathbb{K}}})) \to 0.$$

This can reduce the computation of $\mathrm{Br}_1(X)/\mathrm{Br}_0(X)$ to the Galois cohomology group $H^1(\mathbb{K}, \mathrm{Pic}(X_{\overline{\mathbb{K}}}))$. It can be described using a universal torsor as follows, see [279, Thm. 4.1.1] for a proof and details on the notation.

Theorem 6.2.2.4 *Let $p \colon X \to \mathrm{Spec}(\mathbb{K})$ be a variety over a field $\mathbb{K}$ of characteristic zero such that $\Gamma(X_{\overline{\mathbb{K}}}, \mathcal{O}^*) = \overline{\mathbb{K}}^*$ and with finitely generated $\mathrm{Pic}(X)$. Assume that a universal X-torsor $\mathcal{X}$ exists. Then any element of $\mathrm{Br}_1(X)/\mathrm{Br}_0(X)$ is represented by the cup product $p^*(\alpha) \cup [\mathcal{X}]$ for some $\alpha \in H^1(\mathbb{K}, \mathrm{Pic}(X_{\overline{\mathbb{K}}}))$.*

6.2.3 Descent and universal torsors

Descent theory allows to study the Hasse principle and weak approximation and their Brauer–Manin obstructions via torsors under quasitori, in particular universal torsors.

Descent theory for torsors under tori over varieties over number fields and its application to rational points were studied by Colliot-Thélène and Sansuc [100] and under quasitori by Skorobogatov [278, 279]. See also [157].

For a quasitorus H over a field $\mathbb{K}$ of characteristic zero, and a variety X over $\mathbb{K}$, we have discussed that the existence of X-torsors $\mathcal{X}$ under H is necessary for the existence of rational points on X, and by Proposition 6.1.2.16 we have the disjoint union

$$X(\mathbb{K}) = \bigcup_{[\sigma] \in H^1(\mathbb{K}, H)} \pi^\sigma(\mathcal{X}^\sigma(\mathbb{K})).$$

If $\mathbb{K}$ is a number field, we have the following finiteness result for this disjoint union; see [99, Prop. 2] and [279, Prop. 5.3.2].

Proposition 6.2.3.1 *Let $\mathbb{K}$ be a number field, H a quasitorus over $\mathbb{K}$, and X a proper variety over $\mathbb{K}$. Assume that we have an X-torsor $\pi : \mathcal{X} \to X$ under H. Then there are only finitely many $[\sigma] \in H^1(\mathbb{K}, H)$ with $\mathcal{X}^\sigma(\mathbb{K}) \neq \emptyset$.*

If H is a quasitorus over a number field $\mathbb{K}$, and X a smooth, proper, geometrically irreducible variety over $\mathbb{K}$, with $\pi : \mathcal{X} \to X$ an X-torsor under H, let

$$\left(\prod_{v \in \Omega_{\mathbb{K}}} X(\mathbb{K}_v) \right)^\pi := \bigcup_{[\sigma] \in H^1(\mathbb{K}, H)} \pi^\sigma \left(\prod_{v \in \Omega_{\mathbb{K}}} \mathcal{X}^\sigma(\mathbb{K}_v) \right).$$

By Proposition 6.1.2.16, this is a subset of $\prod_{v \in \Omega_{\mathbb{K}}} X(\mathbb{K}_v)$ containing $X(\mathbb{K})$; it also contains the closure of $X(\mathbb{K})$. If X is everywhere locally soluble, the emptiness of this set is an obstruction to the Hasse principle, and the inequality

$$\left(\prod_{v \in \Omega_{\mathbb{K}}} X(\mathbb{K}_v) \right)^\pi \subsetneq \prod_{v \in \Omega_{\mathbb{K}}} X(\mathbb{K}_v)$$

is an obstruction to weak approximation. Both are called *descent obstruction* defined by $\pi : \mathcal{X} \to X$, see [279, Sec. 5.3].

For proper X, it turns out that the algebraic Brauer–Manin obstruction is the same as the combination of descent obstructions for all injective types of torsors over X [279, Sec. 6.1], and also the same as the descent obstruction for universal torsors over X:

Theorem 6.2.3.2 *Let X be a smooth, proper, geometrically irreducible variety over a number field $\mathbb{K}$, with finitely generated $\mathrm{Pic}(X_{\overline{\mathbb{K}}})$. If $(\prod_{v \in \Omega_{\mathbb{K}}} X(\mathbb{K}_v))^{\mathrm{Br}_1(X)} \neq \emptyset$, then there is a universal torsor $\pi : \mathcal{X} \to X$, and*

$$\left(\prod_{v \in \Omega_{\mathbb{K}}} X(\mathbb{K}_v) \right)^{\mathrm{Br}_1(X)} = \bigcup_{[\sigma] \in H^1(\mathbb{K}, H_{\mathrm{Pic}(X_{\overline{\mathbb{K}}})})} \pi^\sigma \left(\prod_{v \in \Omega_{\mathbb{K}}} \mathcal{X}^\sigma(\mathbb{K}_v) \right).$$

See [100, Thm. 3.5.1] for the case that $\mathrm{Pic}(X_{\overline{\mathbb{K}}})$ is free, and [278, Thm. 3] for the general case. This result shows, for example, that the existence of universal torsors over X that all satisfy weak approximation implies that the algebraic

Brauer–Manin obstruction to the Hasse principle and weak approximation is the only one on X.

Example 6.2.3.3 Let X be as in Example 6.1.2.14, over a number field $\mathbb{K}$. Then the Brauer–Manin obstruction is the only obstruction to the Hasse principle and weak approximation on any smooth proper model X^c of X. For this, we essentially combine Theorem 6.2.3.2 with Example 6.2.1.3; the fact that we work with universal torsors over X and not over X^c leads to small technical complications; see [122, Prop. 1] for details.

In fact, the Brauer group $\mathrm{Br}(X^c)$ coincides with its constants $\mathrm{Br}_0(X^c)$; see [306, Prop. 1.2(d), Prop. 2.6] for the case of Galois extensions $\mathbb{M}/\mathbb{K}$. Therefore, the Hasse principle and weak approximation hold on X. This leads to another proof of [81, Thm. 1].

On the other hand, our discussion does not apply, for example, to a smooth proper model X^c of $X \subset \mathbb{A}_{\mathbb{Q}}^5$ defined by

$$z_0^2 + 1 = N_{\mathbb{Q}(\sqrt[4]{17})/\mathbb{Q}}(z_1, \ldots, z_4)$$

because $\mathbb{Q}(\sqrt{-1}) \not\subset \mathbb{Q}(\sqrt[4]{17})$. In this case, it turns out that $\mathrm{Br}(X^c)/\mathrm{Br}_0(X^c) \cong \mathbb{Z}/2\mathbb{Z}$ [122, Thm. 4], and there is a Brauer–Manin obstruction to weak approximation on X^c [122, Ex. 1].

6.2.4 Results

In this section, we give an overview of a selection of results regarding the Hasse principle and weak approximation. While we focus on results proved via descent, we also mention several other results and methods. See also [156, 249] for further details and references.

One of the earliest results proved via universal torsors and descent is due to Colliot-Thélène and Sansuc: for smooth proper varieties over a number field $\mathbb{K}$ containing a $\mathbb{K}$-torsor under a torus as a dense open set, the Brauer–Manin obstruction to the Hasse principle and weak approximation is the only one; see [279, Thm. 6.3.1].

The Hasse principle and weak approximation were proved for projective varieties over number fields that are principal homogeneous spaces of semisimple, simply connected affine algebraic groups by Kneser, Harder, and Platonov. This includes the classical result of Hasse and Minkowski for quadrics mentioned in Example 6.2.1.1. For projective varieties that are homogeneous spaces of connected affine algebraic groups with connected stabilizers and of connected and simply connected algebraic groups with finite abelian stabilizers, the Brauer–Manin obstruction to the Hasse principle and weak approximation is the only one. See [57] and the references therein.

By the Hardy–Littlewood circle method, the Hasse principle and weak approximation also hold over number fields $\mathbb{K}$ for nonsingular complete intersections $X \subset \mathbb{P}_{\mathbb{K}}^{n-1}$ of m forms in n variables of degree d with

$$n > m(m+1)(d-1)2^{d-1} + \dim(X^*)$$

[52, 276], where $\dim(X^*) \leq m$ is the dimension of *Birch's singular locus* $X^* \subset X$, which may be nonempty even though X is nonsingular; see [4] for a detailed discussion. For nonsingular cubic (resp. quartic) hypersurfaces $X \subset \mathbb{P}_{\mathbb{Q}}^{n-1}$, $n \geq 9$ (resp. $n \geq 40$) is enough [175, 153].

For Châtelet surfaces, universal torsors were described up to birational equivalence in [99]. The Hasse principle and weak approximation hold for these universal torsors [101, 102] over arbitrary number fields. Hence the Brauer–Manin obstruction to the Hasse principle and weak approximation is the only one on Châtelet surfaces over number fields.

For nonsingular complete intersections $X \subset \mathbb{P}_{\mathbb{K}}^{n-1}$ of two quadratic forms in $n \geq 8$ variables over a number field $\mathbb{K}$, the Hasse principle and weak approximation hold [171]. For $n \geq 9$, this was proved in [101, 102] via the result on Châtelet surfaces above.

The Hasse principle and weak approximation have been widely studied for smooth proper models X^c of varieties $X \subset \mathbb{A}_{\mathbb{K}}^{n+1}$ defined by

$$P(z_0) = N_{\mathbb{L}/\mathbb{K}}(z_1, \ldots, z_n)$$

where $P(z_0) \in \mathbb{K}[z_0]$ is a polynomial and $N_{\mathbb{L}/\mathbb{K}}(z_1, \ldots, z_n)$ is a norm form associated with the extension $\mathbb{L} \supseteq \mathbb{K}$ of number fields of degree n. That the Brauer–Manin obstruction to the Hasse principle and weak approximation on X^c is the only one was proved for several classes of such varieties, in particular:

- Châtelet surfaces, with $[\mathbb{L} : \mathbb{K}] = 2$ and $\deg(P(z_0)) \leq 4$ [101, 102]
- Certain singular cubic hypersurfaces, with $[\mathbb{L} : \mathbb{K}] = 3$ and $\deg(P(z_0)) \leq 3$ [95]
- Arbitrary $\mathbb{L} \supseteq \mathbb{K}$ and $P(z_0)$ split over $\mathbb{K}$ with at most two distinct roots [93, 172, 287]
- $[\mathbb{L} : \mathbb{K}] = 4$ and $\deg(P(z_0)) = 2$ with $P(z_0)$ split over $\mathbb{L}$ and irreducible over $\mathbb{K}$ [81, 122], see Example 6.2.3.3
- Arbitrary $\mathbb{L} \supseteq \mathbb{K} := \mathbb{Q}$ and $\deg(P(z_0)) = 2$ [122]
- Arbitrary $\mathbb{L} \supseteq \mathbb{K} := \mathbb{Q}$ and $P(z_0)$ split over $\mathbb{Q}$ [82, 83, 158]

Most of these results rely on a selection of descent techniques, fibration techniques, classical results on quadratic forms or norm forms, and various analytic techniques, for example, the circle method in [172], sieve methods in [81] or methods of Green and Tao [149] from additive combinatorics in [82, 83, 158].

Under the assumptions of Schinzel's hypothesis (H), which is a far-reaching conjectural generalization of Dirichlet's theorem on primes in arithmetic progressions, or the finiteness of the Tate–Shafarevich group of elliptic curves over number fields, additional results are available, for example, for certain diagonal cubic hypersurfaces in $\mathbb{P}^n_{\mathbb{K}}$ with $n \geq 3$ [91, 291] and certain intersections of two quadrics in $\mathbb{P}^n_{\mathbb{K}}$ with $n \geq 4$ [307].

Beyond descent obstructions for quasitori and algebraic Brauer–Manin obstructions, one may have *transcendental Brauer–Manin obstructions*, coming from elements of $\mathrm{Br}(X)$ not lying in the algebraic Brauer group $\mathrm{Br}_1(X)$, in case of $\mathrm{Br}(X_{\overline{\mathbb{K}}}) \neq 0$; see [154] for a first example.

At least for smooth projective varieties X, one can show that Brauer–Manin obstructions are equivalent to descent obstructions for torsors under the nonabelian groups $\mathrm{PGL}_{n,\mathbb{K}}$ [279, Prop. 5.3.4]; see [155] for more general results in this direction.

The following central conjecture applies in particular to smooth proper geometrically rational varieties; see [91], for example.

Conjecture 6.2.4.1 (Colliot-Thélène) *For any smooth proper rationally connected variety over a number field, the Brauer–Manin obstruction to the Hasse principle and weak approximation is the only one.*

No counterexample to this conjecture is presently known. Skorobogatov [278] gave an example of a (geometrically nonrational) surface X over $\mathbb{Q}$ where the Hasse principle fails even though there is no Brauer–Manin obstruction to the Hasse principle. This can be explained by applying the Brauer–Manin obstruction to finite étale covers of X, which gives the refinement

$$X(\mathbb{K}) \subseteq \overline{X(\mathbb{K})} \subseteq \left(\prod_{v \in \Omega_{\mathbb{K}}} X(\mathbb{K}_v) \right)^{\text{ét,Br}} \subseteq \left(\prod_{v \in \Omega_{\mathbb{K}}} X(\mathbb{K}_v) \right)^{\mathrm{Br}(X)} \subseteq \prod_{v \in \Omega_{\mathbb{K}}} X(\mathbb{K}_v)$$

of (6.2.1), where the *étale Brauer–Manin set* in the middle is empty in Skorobogatov's example and hence is an obstruction to the Hasse principle. Poonen [252] gave an example where

$$\left(\prod_{v \in \Omega_{\mathbb{K}}} X(\mathbb{K}_v) \right)^{\text{ét,Br}} \neq \emptyset$$

while $X(\mathbb{K}) = \emptyset$, so even the étale Brauer–Manin obstruction is insufficient to explain all counterexamples to the Hasse principle. See [94] and the references therein for further examples.

6.3 Distribution of rational points

6.3.1 Heights and Manin's conjecture

For a projective variety X containing infinitely many rational points over a number field $\mathbb{K}$, one may ask for quantitative results describing the density or distribution of rational points. A natural measure of the size of a rational point in projective space is the Weil height. Then we may ask for the asymptotic behavior of the number of rational points of Weil height bounded by a number B on X embedded into projective space, as B tends to infinity. For a large class of varieties, this behavior is precisely predicted by Manin's conjecture. For an introduction to this subject, see also [78, 246, 248].

Definition 6.3.1.1 The *Weil height* of $\mathbf{x} = [x_0, \ldots, x_n] \in \mathbb{P}^n_{\mathbb{K}}(\mathbb{K})$ is defined as

$$H(\mathbf{x}) := \prod_{v \in \Omega_{\mathbb{K}}} \max\{|x_0|_v, \ldots, |x_n|_v\}^{d_v}.$$

Here, for each place $v \in \Omega_{\mathbb{K}}$ extending $w \in \Omega_{\mathbb{Q}}$, the norm $|\cdot|_v : \mathbb{K}_v \to \mathbb{R}$ is normalized such that it extends the usual w-adic absolute value on $\mathbb{Q}$, and $d_v := [\mathbb{K}_v : \mathbb{Q}_w]$.

The Weil height is well-defined because of the product formula $\prod_{v \in \Omega_{\mathbb{K}}} |x|_v^{d_v} = 1$ for any nonzero $x \in \mathbb{K}$. For $\mathbb{K} := \mathbb{Q}$ and coprime integers $x_0, \ldots, x_n$, this is simply

$$H([x_0, \ldots, x_n]) = \max\{|x_0|, \ldots, |x_n|\},$$

where $|\cdot|$ is the usual absolute value on $\mathbb{Z}$.

For a morphism $i : X \to \mathbb{P}^n_{\mathbb{K}}$, the Weil height induces a height H on $X(\mathbb{K})$; we say that H is a height on X with respect to the line bundle $i^*\mathcal{O}_{\mathbb{P}^n_{\mathbb{K}}}(1)$ with its *standard metrization*. More generally, one can define heights with respect to arbitrary *metrized line bundles* on X; see [247, Sec. 2], for example. For a smooth projective variety X over a number field $\mathbb{K}$, an *anticanonical height* H is a height corresponding to a metrization of the anticanonical bundle ω_X^{-1}. If X is a Fano variety with very ample ω_X^{-1}, we can simply take the height induced by the Weil height via an *anticanonical embedding* $i : X \to \mathbb{P}^n_{\mathbb{K}}$ with $\omega_X^{-1} \cong i^*\mathcal{O}_{\mathbb{P}^n_{\mathbb{K}}}(1)$.

Let X be a smooth projective variety over a number field $\mathbb{K}$, with a height function H on its set $X(\mathbb{K})$ of rational points. For any open subset U in X and $B \in \mathbb{R}_{>0}$, we consider the number of rational points of bounded height on U, namely

$$N_{U,H}(B) := \#\{\mathbf{x} \in U(\mathbb{K}); \ H(\mathbf{x}) \leq B\}.$$

The restriction to open subsets is reasonable because sometimes the total number $N_{X,H}(B)$ of rational points of bounded height is dominated by the number of rational points of bounded height on a closed subvariety Z. In this case, $N_{X,H}(B)$ can only capture the geometry and arithmetic of Z, so that we must remove the *accumulating subvariety* Z to get interesting information on X itself.

For Fano varieties with a Zariski-dense set of rational points over a number field, Manin's conjecture makes a precise prediction of the asymptotic behavior of the number of rational points of bounded anticanonical height on sufficiently small dense open subsets; see [32, 138] for its origins. The original heuristic expectation was that for a Fano variety X over a number field $\mathbb{K}$ with a dense set of rational points $X(\mathbb{K})$ and with an anticanonical height function H, there is a constant $c_{X,H} > 0$ such that for all sufficiently small open subsets U of X, we have

$$N_{U,H}(B) \;=\; c_{X,H}\,B(\log B)^{\mathrm{rk}\,\mathrm{Pic}(X)-1}(1 + o(1)), \qquad (6.3.1)$$

as $B \to \infty$; see [138, (0.2)] and [32, Conj. B']. If an asymptotic formula of the shape (6.3.1) holds for such a variety X over a number field $\mathbb{K}$, we say that *Manin's conjecture holds for X*. A conjectural value for the leading constant $c_{X,H}$ for Fano varieties was given by Peyre [243]. For much more general classes of varieties, the Batyrev–Manin conjectures in [32] predict at least upper bounds for the number of rational points of bounded height.

While Manin's conjecture has been proved in many cases, this formulation turns out to be false in general, as the counterexample of [35] shows. On the other hand, we expect that variants of Manin's conjecture hold for some varieties that are in some sense close to being Fano varieties, for example, weak del Pezzo surfaces. Such variants are provided by [32, Conj. C'], [37] and [247, Formule empirique 5.1]; we discuss the latter here, starting with [245, Def. 1.2.1].

Definition 6.3.1.2 An *almost Fano variety* is a smooth, projective, geometrically irreducible variety with $H^1(X, \mathcal{O}_X) = H^2(X, \mathcal{O}_X) = 0$, torsion-free geometric Picard group $\mathrm{Pic}(X_{\overline{\mathbb{K}}})$ and big anticanonical class $-K_X$.

For example, Fano varieties and smooth projective toric varieties are almost Fano varieties [245, Ex. 2.1.3, Ex. 2.1.4]. Not only smooth del Pezzo surfaces, but also weak del Pezzo surfaces are almost Fano varieties; see Remark 6.4.1.11. We must make the notion of accumulating subvarieties precise. The following definition is a variation of [32, Sec. 2.5] and [247, Def. 3.10].

Definition 6.3.1.3 Let H be an anticanonical height function on an almost Fano variety X. A nontrivial irreducible closed subset Z of X is *weakly* *accumulating with respect to* H if for every nonempty open subset W of Z, there is a nonempty open subset U of X with

$$\limsup_{B \to \infty} \frac{N_{W,H}(B)}{N_{U,H}(B)} > 0.$$

If the complement of all subsets that are weakly accumulating with respect to H is an open subset U of X, then $N_{U,H}(B) \sim N_{U',H}(B)$ for all nonempty open $U' \subseteq U$.

One important ingredient for the conjectural description of the leading constant in [37, 243, 247] is the following constant; see [243, Def. 2.4] and [34, Def. 2.4.6].

Definition 6.3.1.4 Let X be an almost Fano variety. Let $\mathrm{Eff}(X)^\vee \subset \mathrm{Pic}(X)^\vee \otimes_\mathbb{Z} \mathbb{R}$ be the dual cone of the effective cone of X. We define

$$\alpha(X) := \mathrm{rk}\,(\mathrm{Pic}(X)) \cdot \mathrm{Vol}\{x \in \mathrm{Eff}(X)^\vee;\ (x, -K_X) \le 1\},$$

where the volume of this polytope is taken with respect to the ordinary Lebesgue measure normalized such that the lattice $\mathrm{Pic}(X)^\vee$ has covolume 1 in $\mathrm{Pic}(X)^\vee \otimes_\mathbb{Z} \mathbb{R}$.

For Fano varieties X with anticanonical height functions H, Peyre [243, Sec. 2.2] defines a *Tamagawa number* $\tau_H(X)$, which is essentially a product of local densities of points at all places $v \in \Omega_\mathbb{K}$ of $\mathbb{K}$, with convergence factors. See [37] for generalizations. For the Tamagawa numbers of almost Fano varieties, see [247, Sec. 4]. Now we are ready to state Peyre's version of Manin's conjecture [247, Formule empirique 5.1].

Conjecture 6.3.1.5 (Manin, Peyre) *Let X be an almost Fano variety over a number field $\mathbb{K}$, with dense set of rational points $X(\mathbb{K})$, finitely generated $\mathrm{Eff}(X_{\overline{\mathbb{K}}})$ and trivial Brauer group $\mathrm{Br}(X_{\overline{\mathbb{K}}})$. Let H be an anticanonical height function, and assume that there is an open subset U of X that is the complement of the weakly accumulating subvarieties on X with respect to H. We say that Manin's conjecture holds for X if there is a positive constant $c_{X,H}$ such that*

$$N_{U,H}(B) = c_{X,H} B (\log B)^{\mathrm{rk}\,\mathrm{Pic}(X)-1}(1 + o(1)).$$

Let $\alpha(X)$ be as in Definition 6.3.1.4, let $\beta(X)$ be the order of the Galois cohomology group $H^1(\mathrm{Gal}(\overline{\mathbb{K}}/\mathbb{K}), \mathrm{Pic}(X_{\overline{\mathbb{K}}}))$, and let $\tau_H(X)$ be the Tamagawa number of X with respect to H. We say that the leading constant is the one predicted by Peyre if furthermore

$$c_{X,H} = \alpha(X)\beta(X)\tau_H(X).$$

We end this section by discussing some progress toward Manin's conjecture; this discussion will be continued in Sections 6.3.2 in the context of universal torsors and in Section 6.4.1 for del Pezzo surfaces.

Let X be a nonsingular complete intersections $X \subset \mathbb{P}_{\mathbb{K}}^{n-1}$ of m forms of degree d with

$$n > m(m + 1)(d - 1)2^{d-1} + \dim(X^*),$$

where $\dim(X^*) \leq m$ is the dimension of Birch's singular locus as in Section 6.2.4. Then Manin's conjecture holds for X. In [209, 243], this is deduced from the circle method results from [52, 276].

Manin's conjecture may be approached via methods of harmonic analysis on adelic points in case of smooth projective varieties over number fields $\mathbb{K}$ with certain structures of algebraic groups, for example:

- Generalized flag varieties G/P, where G is a semisimple connected affine algebraic group over $\mathbb{K}$ and P is a parabolic subgroup defined over $\mathbb{K}$ [138]
- Smooth projective toric varieties over $\mathbb{K}$ [36]
- Smooth projective equivariant compactifications of additive groups $\mathbb{G}_{a,\mathbb{K}}^n$ over $\mathbb{K}$ [88]

Manin's conjecture can be generalized to varieties over function fields of curves over finite fields. Here, analogs of [36, 138] are [59, 61, 201, 250].

6.3.2 Parameterization by universal torsors and Cox rings

Approaching Manin's conjecture via universal torsors was suggested by Salberger and Peyre [244, 263]. For this, one usually needs an explicit description of universal torsors in terms of defining equations. This can be obtained via Cox rings as described in Section 6.1.3; the local description as in Theorem 6.1.2.13 is not enough. The prototype is projective space over $\mathbb{Q}$.

Example 6.3.2.1 Let $X := \mathbb{P}_{\mathbb{Q}}^n$. We have $\omega_{\mathbb{P}_{\mathbb{Q}}^n}^{-1} \cong \mathcal{O}_{\mathbb{P}_{\mathbb{Q}}^n}(n + 1)$, so that an anticanonical embedding is given by the $(n + 1)$-uple embedding of $\mathbb{P}_{\mathbb{Q}}^n$ in $\mathbb{P}_{\mathbb{Q}}^N$ with $N := \binom{2n+1}{n} - 1$. Hence an anticanonical height function is the $(n + 1)$-th power of the Weil height on $\mathbb{P}_{\mathbb{Q}}^n$ from Definition 6.3.1.1, namely

$$H(\mathbf{x}) = (\max\{|x_0|, \ldots, |x_n|\})^{n+1}$$

for $\mathbf{x} = [x_0, \ldots, x_n]$ with coprime integers $x_0, \ldots, x_n$. The counting problem is to estimate

$$N_{\mathbb{P}_{\mathbb{Q}}^n, H}(B) := \#\{\mathbf{x} \in \mathbb{P}_{\mathbb{Q}}^n(\mathbb{Q}); \ H(\mathbf{x}) \leq B\};$$

it turns out that there are no weakly accumulating subsets.

The geometric picture is as follows. Because $\mathbb{P}_{\mathbb{Q}}^n$ is a toric variety described by a fan with $n + 1$ rays, the Cox ring of $\mathbb{P}_{\mathbb{Q}}^n$ is a polynomial ring in $n + 1$ variables. Therefore, the total coordinate space is $\overline{X} = \mathbb{A}_{\mathbb{Q}}^{n+1}$. It contains the characteristic space $\widehat{X} = \mathbb{A}_{\mathbb{Q}}^{n+1} \setminus \{0\}$, with

$$q_X : \widehat{X} \to X, \quad (x_0, \ldots, x_n) \mapsto [x_0, \ldots, x_n].$$

Of course, $\widehat{X} = \mathbb{A}_{\mathbb{Q}}^{n+1} \setminus \{0\}$ is also a universal torsor, see Proposition 6.1.3.9.

Because every $\mathbb{Q}$-rational point in $\mathbb{P}_{\mathbb{Q}}^n$ can be represented uniquely up to sign by coprime integral coordinates and because of the definition of the height function, we have

$$N_{\mathbb{P}_{\mathbb{Q}}^n, H}(B) = \frac{1}{2} \#\{\mathbf{y} \in \mathbb{Z}^{n+1}; \gcd(y_0, \ldots, y_n) = 1, (\max\{|y_0|, \ldots, |y_n|\})^{n+1} \le B\}.$$

This can be interpreted as one half of the number of integral points $\mathbf{y}$ on the integral model $\widehat{\mathfrak{X}} := \mathbb{A}_{\mathbb{Z}}^{n+1} \setminus \{0\}$ of the universal torsor $\widehat{X}$, satisfying $H(q_{\mathfrak{X}}(\mathbf{y})) \le B$, with $q_{\mathfrak{X}} : \widehat{\mathfrak{X}} \to \mathfrak{X} := \mathbb{P}_{\mathbb{Z}}^n$.

Using the simple estimate

$$\#\{y \in \mathbb{Z}; |y| \le C\} = 2\lfloor C \rfloor + 1 = 2C + O(1),$$

this number can be estimated using a Möbius inversion, where the *Möbius function* μ is defined as $\mu(k) = (-1)^n$ if k is the product of n distinct primes, and $\mu(k) = 0$ otherwise:

$$N_{\mathbb{P}_{\mathbb{Q}}^n, H}(B) = \frac{1}{2} \sum_{1 \le k \le B} \mu(k) \#\{\mathbf{y} \in \mathbb{Z}^{n+1} \setminus \{0\}; \max\{|y_0|, \ldots, |y_n|\} \le B^{1/(n+1)}/k\}$$

$$= \frac{(2B^{1/(n+1)})^{n+1}}{2} \sum_{1 \le k \le B} \frac{\mu(k)}{k^{n+1}} + O\left((B^{1/(n+1)})^n \sum_{1 \le k \le B} \frac{|\mu(k)|}{k^n}\right)$$

$$= \frac{2^n}{\zeta(n+1)} B + \begin{cases} O(B^{1-1/(n+1)}), & n \ge 2, \\ O(B^{1/2} \log B), & n = 1. \end{cases}$$

This asymptotic formula agrees with Manin's conjecture because $\mathrm{Pic}(\mathbb{P}_{\mathbb{Q}}^n) \cong \mathbb{Z}$.

To generalize Example 6.3.2.1 to a proof of Manin's conjecture for projective space $\mathbb{P}_{\mathbb{K}}^n$ over other number fields $\mathbb{K}$, major additional difficulties occur. On the one hand, unique factorization may fail in the ring $\mathcal{O}_{\mathbb{K}}$ of integers of $\mathbb{K}$, hence the coordinates $x_0, \ldots, x_n$ of a point $[x_0, \ldots, x_n] \in \mathbb{P}_{\mathbb{K}}^n(\mathbb{K})$ may generate a nonprincipal ideal, and then the point does not have a representative whose coordinates are coprime in the sense that $x_0 \mathcal{O}_{\mathbb{K}} + \cdots + x_n \mathcal{O}_{\mathbb{K}} = \mathcal{O}_{\mathbb{K}}$. On the other hand, $\mathcal{O}_{\mathbb{K}}$ may contain infinitely many units, and hence a rational point on $\mathbb{P}_{\mathbb{K}}^n$ may have infinitely many representatives with coprime coordinates. These difficulties were solved by Schanuel [265], whose results can be interpreted as a proof of Manin's conjecture for projective spaces over arbitrary number fields. By [243, Sec. 6.1], the leading constant in Example 6.3.2.1 and [265] agrees with Peyre's prediction, as in Conjecture 6.3.1.5.

Salberger [263] showed how universal torsors can be used to prove Manin's conjecture for split toric varieties over $\mathbb{Q}$ whose anticanonical sheaf is generated by global sections. See [66] for an improvement of the error term in this result. Subsequently, Manin's conjecture was proved via universal torsor methods for many examples of del Pezzo surfaces over number fields and function fields; see Section 6.4.1 for details. Over function fields, see also [62].

Furthermore, see [67] for Manin's conjecture over $\mathbb{Q}$ for Segre's singular cubic threefold in $\mathbb{P}^5_{\mathbb{Q}}$ defined by

$$x_0^3 + x_1^3 + x_2^3 + x_3^3 + x_4^3 + x_5^3 = x_0 + x_1 + x_2 + x_3 + x_4 + x_5 = 0.$$

The cubic equation

$$x_1 y_2 y_3 + x_2 y_1 y_3 + x_3 y_1 y_2 = 0$$

defines a singular cubic fourfold in $\mathbb{P}^5_{\mathbb{Q}}$; interpreted as bihomogeneous, it also defines a threefold in $\mathbb{P}^2_{\mathbb{Q}} \times_{\mathbb{Q}} \mathbb{P}^2_{\mathbb{Q}}$. See [54, 55] for Manin's conjecture for these varieties.

6.4 Toward Manin's conjecture for del Pezzo surfaces

6.4.1 Classification and results

We state basic background on del Pezzo surfaces over a not necessarily algebraically closed ground field and then briefly survey what is known about Manin's conjecture for these surfaces. The ground field $\mathbb{K}$ is of characteristic zero and $\overline{\mathbb{K}}$ denotes the algebraic closure.

Definition 6.4.1.1 Let $\mathbb{K}$ be a field of characteristic zero.

(i) A *del Pezzo surface over* $\mathbb{K}$ is a geometrically irreducible, normal projective surface over $\mathbb{K}$ with ample anticanonical class.
(ii) A *weak del Pezzo surface over* $\mathbb{K}$ is a geometrically irreducible, smooth projective surface over $\mathbb{K}$ whose anticanonical class is big and nef.

This is a direct generalization of the corresponding notions for the case of an algebraically closed ground field; see Section 5.2. The anticanonical model of a weak del Pezzo surface over $\mathbb{K}$ is defined in the same way as for algebraically closed ground fields and, moreover, part of the characterization given in Theorem 5.2.1.7 extends without changes to the nonclosed case.

Remark 6.4.1.2 The anticanonical model of a weak del Pezzo surface X over $\mathbb{K}$ is a del Pezzo surface Y over $\mathbb{K}$ such that $Y_{\overline{\mathbb{K}}}$ has at most rational double points as singularities. Conversely, for any del Pezzo surface over $\mathbb{K}$ having at most rational double points as singularities, its minimal resolution is a weak del Pezzo surface. Note that each singularity over $\overline{\mathbb{K}}$ may not be defined over $\mathbb{K}$, but that we can blow up singularities that are conjugate under $\mathrm{Gal}(\overline{\mathbb{K}}/\mathbb{K})$ simultaneously to obtain a minimal resolution $\widetilde{X} \to X$ over $\mathbb{K}$ such that its base change $\widetilde{X}_{\overline{\mathbb{K}}} \to X_{\overline{\mathbb{K}}}$ is a minimal resolution over $\overline{\mathbb{K}}$.

Whereas over an algebraically closed ground field all weak del Pezzo surfaces are obtained by blowing up points in almost general position, this holds over a nonclosed field only in the "split case" which is defined as follows.

Definition 6.4.1.3 Let $\mathbb{K}$ be a field of characteristic zero. A weak del Pezzo surface X over $\mathbb{K}$ is *split* if the natural map $\mathrm{Pic}(X) \to \mathrm{Pic}(X_{\overline{\mathbb{K}}})$ is an isomorphism.

Remark 6.4.1.4 Let X be a split weak del Pezzo surface over $\mathbb{K}$. Then X is isomorphic to $\mathbb{P}^1_{\mathbb{K}} \times_{\mathbb{K}} \mathbb{P}^1_{\mathbb{K}}$, the Hirzebruch surface $\mathbb{F}_{2,\mathbb{K}}$ or X is obtained from $\mathbb{P}^2_{\mathbb{K}}$ by a sequence of blow-ups of points in almost general position, each defined over $\mathbb{K}$, that means that we have

$$X = X_r \xrightarrow{\kappa_r} X_{r-1} \to \cdots \to X_1 \xrightarrow{\kappa_1} X_0 = \mathbb{P}^2_{\mathbb{K}}, \qquad (6.4.1)$$

where $r \le 8$ and each $\kappa_i : X_i \to X_{i-1}$ is the blow-up of a rational point of X_{i-1} not lying on a (-2)-curve of X_{i-1}. Conversely, each such blow-up sequence leads to a split weak del Pezzo surface over $\mathbb{K}$.

Remark 6.4.1.5 If X is a weak del Pezzo surface over a field $\mathbb{K}$ of characteristic zero with a nonempty set of rational points $X(\mathbb{K})$, its Picard group $\mathrm{Pic}(X)$ over $\mathbb{K}$ consists of the elements of the geometric Picard group $\mathrm{Pic}(X_{\overline{\mathbb{K}}})$ that are invariant under the natural action of $\mathrm{Gal}(\overline{\mathbb{K}}/\mathbb{K})$; see, for example, [100, proof of Thm. 2.1.2].

Example 6.4.1.6 If X is a *nontrivial Brauer–Severi surface over* $\mathbb{K}$, that is, $X_{\overline{\mathbb{K}}} \cong \mathbb{P}^2_{\overline{\mathbb{K}}}$ and $X(\mathbb{K}) = \emptyset$, then $\mathrm{Gal}(\overline{\mathbb{K}}/\mathbb{K})$ acts trivially on $\mathrm{Pic}(X_{\overline{\mathbb{K}}})$, but $\mathrm{Pic}(X)$ is of index 3 in it, generated by the anticanonical class. In particular, X is a nonsplit smooth del Pezzo surface.

Example 6.4.1.7 Blowing up $\mathbb{P}^2_{\overline{\mathbb{K}}}$ in a set of r points over $\overline{\mathbb{K}}$ in general position on which the Galois group $\mathrm{Gal}(\overline{\mathbb{K}}/\mathbb{K})$ acts nontrivially gives a nonsplit smooth del Pezzo surface that is rational over $\mathbb{K}$. In this case, the action of $\mathrm{Gal}(\overline{\mathbb{K}}/\mathbb{K})$ on $\mathrm{Pic}(X_{\overline{\mathbb{K}}})$ factors through a subgroup of the symmetric group S_r on r elements.

Remark 6.4.1.8 Every weak del Pezzo surface X over $\mathbb{K}$ is geometrically rational, but it is not necessarily rational over $\mathbb{K}$. Furthermore, $\mathrm{Gal}(\overline{\mathbb{K}}/\mathbb{K})$ may act on $\mathrm{Pic}(X_{\overline{\mathbb{K}}})$ through various subgroups of its automorphism group, which is the Weyl group W_{9-d} described in Section 5.2.4 for X of degree $d \le 6$. In fact, for any subgroup H of the Weyl group W_6 associated with the root system E_6, there is a smooth cubic surface over $\mathbb{Q}$ such that $\mathrm{Gal}(\overline{\mathbb{Q}}/\mathbb{Q})$ acts on $\mathrm{Pic}(X_{\overline{\mathbb{K}}})$ via H [135].

As in the case of an algebraically closed ground field, one introduces the degree and the (singularity) type of weak del Pezzo surfaces.

Definition 6.4.1.9 Let X be a weak del Pezzo surface over $\mathbb{K}$ and $X \to Y$ the anticanonical model, that is, X is a minimal resolution of Y.

(i) The *degree* of X and of Y is $\deg(X) := \deg(Y) := (-K_X)^2$.
(ii) The *singularity type* of X and of Y is the configuration of singularities on Y.
(iii) A *line (over $\overline{\mathbb{K}}$)* on X or on Y is a (-1)-curve on $X_{\overline{\mathbb{K}}}$.
(iv) The *type* of X and of Y is the configuration of (-1)- and (-2)-curves on $X_{\overline{\mathbb{K}}}$ together with the degree.

As earlier, we denote the singularity type of X (and Y) in terms of the resolution graphs of the singularities of Y; see Section 5.4.3. So, for example X or Y being of singularity type $2A_1 A_3$ means that Y has precisely three singularities, two with resolution graph A_1 and one with resolution graph A_3.

In degree at most 7, the type is uniquely determined by the degree, the singularity type and the number of lines; in most cases stating the degree and the singularity type is enough. In degree 8, the types are represented by $\mathbb{P}^1_{\mathbb{K}} \times_{\mathbb{K}} \mathbb{P}^1_{\mathbb{K}}$, the Hirzebruch surface $\mathbb{F}_{2,\mathbb{K}}$, and the blow-up $\mathrm{Bl}_1 \mathbb{P}^2_{\mathbb{K}}$ of the projective plane in one rational point. In degree 9, the only type is represented by $\mathbb{P}^2_{\mathbb{K}}$. Note that even over $\overline{\mathbb{K}}$, there may be one, two, or infinitely many isomorphy classes of weak del Pezzo surfaces of the same type.

Remark 6.4.1.10 For a weak del Pezzo surface X over $\mathbb{K}$ of degree $d \geq 3$, the lines in the sense of Definition 6.4.1.9 are precisely the lines over $\overline{\mathbb{K}}$ on an anticanonical model $Y \subset \mathbb{P}^d_{\mathbb{K}}$.

Now we discuss what is known regarding Manin's conjecture for weak del Pezzo surfaces. Conjecture 6.3.1.5 is a version of Manin's conjecture that we may use here because of the following remark.

Remark 6.4.1.11 Let X be a weak del Pezzo surface over a number field $\mathbb{K}$. Then X is an almost Fano variety as in Definition 6.3.1.2, $\mathrm{Br}(X_{\overline{\mathbb{K}}}) = 0$, and the effective cone $\mathrm{Eff}(X_{\overline{\mathbb{K}}})$ is finitely generated. Indeed, because X is a weak del Pezzo surface, $-K_X$ is nef and big, so $H^1(X, \mathcal{O}_X) = 0$ by Kawamata–Viehweg vanishing. By Serre duality, we have $H^2(X, \mathcal{O}_X) \cong H^0(X, \omega_X) = 0$. Because X is geometrically rational and the Brauer group is a birational invariant, we have $\mathrm{Br}(X_{\overline{\mathbb{K}}}) = \mathrm{Br}(\mathbb{P}^2_{\overline{\mathbb{K}}}) = 0$. For the effective cone, see Proposition 5.2.1.10.

Let X be a weak del Pezzo surface over a number field $\mathbb{K}$, corresponding to a smooth or singular del Pezzo surface Y, with $\pi : X \to Y$ contracting precisely the (-2)-curves on X to singularities on Y and being an isomorphism in their complement. Because the singularities on Y are assumed to be rational double points, and π is a minimal resolution, we have $\omega_X^{-1} \cong \pi^* \omega_Y^{-1}$. Hence an anticanonical embedding $Y \subset \mathbb{P}_{\mathbb{K}}^d$ for Y of degree $d \geq 3$ induces an anticanonical height function H on $X(\mathbb{K})$. A line on Y and hence a (-1)-curve on X contains asymptotically cB^2 rational points of height at most B if it is defined over $\mathbb{K}$, and all rational points on a (-2)-curve have the same height. Therefore, we expect that Conjecture 6.3.1.5 holds for each weak del Pezzo surface X with U the complement of its (-1)- and (-2)-curves.

Let us summarize what is known regarding Manin's conjecture for del Pezzo surfaces. By [36, 88], Manin's conjecture is known for toric varieties and equivariant compactifications of additive groups. By [120], this covers all weak del Pezzo surfaces of degree ≥ 6, and some types of degree ≥ 3.

Beyond these results, the following is known. In degree 5, Manin's conjecture holds over $\mathbb{Q}$ for the split smooth del Pezzo surface obtained by blowing up $\mathbb{P}_{\mathbb{K}}^2$ in the rational points

$$[1, 0, 0], \quad [0, 1, 0], \quad [0, 0, 1], \quad [1, 1, 1];$$

the proof uses the description of the universal torsor as an open subset of the affine cone over $G(2, 5)$ defined by five quadratic Plücker equations in $\mathbb{A}_{\mathbb{Q}}^{10}$; see Example 4.1.4.1 and [33, Prop. 4.1]. See [74] for Manin's conjecture for the nonsplit smooth quintic del Pezzo surface over $\mathbb{Q}$ obtained by blowing up

$$[1, 0, 0], \quad [0, 1, 0], \quad [1, 1, i], \quad [1, 1, -i].$$

Furthermore, Manin's conjecture is known over $\mathbb{Q}$ for one singular quintic del Pezzo surface of type A_2, see [112], and one of type A_1, see [38]; see 5.4.4.4 no. 5 and 5.4.4.5 no. 2 or [113, Sec. 3.3] for the Cox rings of these surfaces. All other types in degree 5 are covered by [36, 88].

The following table lists all types of split quartic weak del Pezzo surfaces, together with their singularity types and number of lines in the columns *sg.* and *ln.*; references to the description of their Cox rings or the Cox rings of the corresponding singular del Pezzo surfaces in the column *Cox rings*; minimal numbers of generators and relations in the Cox rings of the weak del Pezzo surfaces in the columns *gn.* and *rl.*; and references to proofs of Manin's conjecture, usually covering precisely one particular split example over $\mathbb{Q}$, in the column *Manin's conjecture*.

sg.	ln.	Cox ring	gn.	rl.	Manin's conjecture
–	16	[281, Cor. 38], Thm. 5.2.3.1	16	20	[70]
A_1	12	[109, 6.4 (i)], [168, Thm. 6.1 (6)]	13	10	
$2A_1$	9	[109, Ex. 6.6], [168, Thm. 6.1 (8)]	11	5	
$2A_1$	8	[109, 6.4 (iii)], 5.4.4.5 (3)	10	2	[71, 73, 75, 124, 208]
A_2	8	[109, 6.4 (iv)], [168, Thm. 6.1 (9)]	10	2	
$3A_1$	6	[113, 3.4], 5.4.4.4 (6)	9	1	[205]
A_1A_2	6	[113, 3.4], 5.4.4.4 (7)	9	1	[205]
A_3	5	[113, 3.4], 5.4.4.4 (8)	9	1	[116]
A_3	4	[109, 6.4 (viii)], [168, Thm. 6.1 (4)]	11	5	[204]
$4A_1$	4	5.4.4.3 (5)	8	–	[36]
$2A_1A_2$	4	5.4.4.3 (7)	8	–	[36]
A_1A_3	3	[113, 3.4], 5.4.4.3 (11)	9	1	[111, 117, 140]
A_4	3	[113, 3.4], 5.4.4.3 (12)	9	1	[80, 117]
D_4	2	[113, 3.4], 5.4.4.3 (13)	9	1	[117, 123]
$2A_1A_3$	2	5.4.4.2 (4)	8	–	[36, 69]
D_5	1	[113, 3.4], 5.4.4.2 (7)	9	1	[68, 88, 117]

Exceptions are [69] for a nonsplit quartic del Pezzo surface of type D_4, [71, 73, 75, 124] for families of nonsplit quartic del Pezzo surfaces of type $2A_1$ that can also be interpreted as families of Châtelet surfaces, [70] for the smooth nonsplit quartic del Pezzo surface $X \subset \mathbb{P}^4_{\mathbb{Q}}$ defined by

$$x_0^2 + x_1^2 + x_2^2 - x_3^2 - 2x_4^2 = x_0x_1 - x_2x_3 = 0,$$

[116, 117] for some quartic del Pezzo surfaces over arbitrary imaginary quadratic fields and [140] for a quartic del Pezzo surface over arbitrary number fields.

For cubic surfaces, the next table lists the classification and the available results. Note that [170] only proves upper and lower bounds of the correct order of magnitude for Cayley's cubic surface of type $4A_1$. For the E_6 cubic surface, see Section 6.5 for a short proof over $\mathbb{Q}$, and [118] for a proof over all imaginary quadratic fields.

Finally, Manin's conjecture is known for a split del Pezzo surface of type E_7 in degree 2 over $\mathbb{Q}$ [25]. See Theorem 5.4.4.2, Theorem 5.4.4.3, [113, Sec. 3.6, 3.7], [168, Thm. 4.1] for Cox rings of several weak and singular del Pezzo surfaces of degrees 1 and 2.

Over function fields, Manin's conjecture was proved for several del Pezzo surfaces [60, 63].

sg.	ln.	Cox ring	gn.	rl.	Manin's conjecture
—	27	[199, Thm. 9.1], Thm. 5.2.3.1	27	81	
A_1	21		22	48	
$2A_1$	16		18	27	
A_2	15		17	21	
$3A_1$	12		15	15	
A_1A_2	11		14	10	
A_3	10		13	6	
$4A_1$	9	[170], [109, Ex. 6.7]	13	9	bounds: [170]
$2A_1A_2$	8		12	5	
A_1A_3	7		11	2	
$2A_2$	7	[109, 6.6], 5.4.4.4 (10)	11	2	
A_4	6		12	5	
D_4	6	[162, (4.2)], [113, 3.5], 5.4.4.4 (9)	10	1	[206]
$2A_1A_3$	5	[113, 3.5], 5.4.4.3 (8)	10	1	
A_12A_2	5	[113, 3.5], 5.4.4.3 (14)	10	1	[203]
A_1A_4	4	[113, 3.5], 5.4.4.3 (15)	10	1	
A_5	3		13	9	
D_5	3	[113, 3.5], 5.4.4.3 (16)	10	1	[79]
$3A_2$	3	5.4.4.2 (5)	9	–	[36], ..., [139]
A_1A_5	2	[113, 3.5], 5.4.4.2 (8)	10	1	[26]
E_6	1	[162, Thm. 3.8], 5.4.4.2 (9)	10	1	[72, 118]

6.4.2 Strategy

Here, we give a detailed explanation of a basic strategy to prove Manin's conjecture for certain weak del Pezzo surfaces. Variations, refinements, and extensions of this strategy were successfully used for example in [72] and in many other cases, each dealing with difficulties not encountered before. In Section 6.5, we apply it to the cubic surface of type E_6.

For simplicity, we assume that X is a split weak del Pezzo surface over $\mathbb{Q}$ of degree $d \geq 3$ with precisely one relation in its Cox ring. Our discussion of the existing results above shows that this is the case that is understood best, but the general strategy is not limited to this case. See [69] for a nonsplit example, [116] for an example over imaginary-quadratic fields, [25] for an example in degree 2, and [204] for an example where the Cox ring has two relations.

Recall Conjecture 6.3.1.5. Because X is split, we have $\beta(X) = 1$, and we expect an asymptotic formula of the shape

$$N_{U,H}(B) = \alpha(X)\omega_\infty \left(\prod_p \omega_p \right) B(\log B)^{9-d}(1 + o(1)),$$

where $\alpha(X)$ is as in Definition 6.3.1.4, ω_∞ is a real density and the product is taken over all primes p, with ω_p a p-adic density with a convergence factor, see [247, Sec. 4].

Our strategy consists of three steps, where we expect the second step to be the major challenge, while the first and third step are expected to be straightforward in any specific case.

Parameterization via Cox rings. Let U be the complement of the (-1)- and (-2)-curves on X. Let $\pi \colon X \to Y \subset \mathbb{P}^d_{\mathbb{K}}$ be an anticanonical map, with corresponding anticanonical height function H on $X(\mathbb{Q})$. We establish a 2^{10-d}-to-1 map between the set $T(B)$ of integral points on an integral model of a characteristic space or universal torsor $\mathcal{X}$ over X of bounded lifted height and the set $\{\mathbf{x} \in U(\mathbb{Q}); H(\mathbf{x}) \le B\}$. Our task is to describe $T(B)$ explicitly as integral points satisfying the *Cox ring relation* for X, *height conditions* and *coprimality conditions*.

Estimating integral points on the universal torsor. We express $\#T(B)$ as a sum over the coordinates on the universal torsor, where one of the coordinates is determined by the relation in the Cox ring. Our task is to estimate this sum by an integral, with an acceptable error term.

Interpretation of the integral. Finally, we must compare this with Manin's predictions. We can rewrite the expected asymptotic formula as an integral that resembles the integral obtained in the previous step. However, the domains of integration differ. Our task is to show that the difference of these integrals is negligible.

6.4.3 Parameterization via Cox rings

Let X be as in Section 6.4.2. Let U be the complement of the (-1)- and (-2)-curves on X. Then X is a blow-up $\kappa \colon X \to \mathbb{P}^2_{\mathbb{Q}}$ in $r := 9 - d$ points over $\mathbb{Q}$ in almost general position; see Remark 6.4.1.4. Because we assume to have precisely one relation in the Cox ring $\mathcal{R}_X(X)$, it is for dimension reasons generated by precisely $r + 4$ sections $\eta_1, \ldots, \eta_{r+4}$ corresponding to divisors $E_1, \ldots, E_{r+4}$, satisfying a unique *Cox ring relation*

$$g(\eta_1, \ldots, \eta_{r+4}) = 0. \tag{6.4.2}$$

Sometimes g is called *torsor equation*. We may assume that all coefficients in g are ± 1 [113, Sec. 2.3]. Furthermore, we assume that the height function H is taken with respect to an anticanonical map $\pi \colon X \to Y \subset \mathbb{P}^d_{\mathbb{Q}}$ satisfying

$$(\pi^*(y_0), \ldots, \pi^*(y_d)) = (\Psi_0(\eta_1, \ldots, \eta_{r+4}), \ldots, \Psi_d(\eta_1, \ldots, \eta_{r+4})),$$

where $\Psi_0, \ldots, \Psi_d$ are monomials in the generators of $\mathcal{R}_X(X)$, with coordinates $y_0, \ldots, y_d$ on $Y \subset \mathbb{P}_{\mathbb{Q}}^d$. We define the *coprimality conditions*

$$\gcd\{\eta_j \, ; \, j \in J\} = 1 \text{ for all minimal } J \subseteq \{1, \ldots, r+4\} \text{ with } \bigcap_{j \in J} E_j = \emptyset$$

(6.4.3)

and the *height condition*

$$\max_{i=0,\ldots,d} \{|\Psi_i(\eta_1, \ldots, \eta_{r+4})|\} \leq B.$$

(6.4.4)

For $i \in \{1, \ldots, r+4\}$, write

$$\mathbb{Z}_{i*} := \begin{cases} \mathbb{Z}_{\neq 0}, & E_i^2 < 0, \\ \mathbb{Z}, & \text{else.} \end{cases}$$

and

$$\mathbb{Z}_*^i := \mathbb{Z}_{1*} \times \cdots \times \mathbb{Z}_{i*}.$$

Claim 6.4.3.1 *Let*

$$T(B) := \{(\eta_1, \ldots, \eta_{r+4}) \in \mathbb{Z}_*^{r+4}; \, (6.4.2), (6.4.3), (6.4.4) \text{ hold}\}.$$

Under the assumptions above, we expect

$$N_{U,H}(B) = \frac{\#T(B)}{2^{r+1}}$$

Remark 6.4.3.2 This should hold because of the following heuristic geometric picture. Note that

$$\overline{\mathfrak{X}} := \mathrm{Spec}(\mathbb{Z}[\eta_1, \ldots, \eta_{r+4}]/\langle g(\eta_1, \ldots, \eta_{r+4})\rangle)$$

is an integral model of the total coordinate space $\overline{X} := \mathrm{Spec}(\mathcal{R}_X(X))$. The coprimality conditions (6.4.3) should explicitly describe integral points on an integral model $\widehat{\mathfrak{X}}$ of the characteristic space $\widehat{X}$ in it. The map

$$\Psi : \widehat{\mathfrak{X}}(\mathbb{Z}) \rightarrow \mathbb{P}_{\mathbb{Z}}^d(\mathbb{Z})$$

$$(\eta_1, \ldots, \eta_{r+4}) \mapsto [\Psi_0(\eta_1, \ldots, \eta_{r+4}), \ldots, \Psi_d(\eta_1, \ldots, \eta_{r+4})]$$

should be the composition of $q_{\mathfrak{X}} : \widehat{\mathfrak{X}} \rightarrow \mathfrak{X}$ and a model of $\pi : X \rightarrow Y$. In fact, we expect that its coordinates $\Psi_i(\eta_1, \ldots, \eta_{r+4})$ are coprime, so that (6.4.4) is the lift of the Weil height on $Y \subset \mathbb{P}_{\mathbb{Q}}^d$. The condition

$$(\eta_1, \ldots, \eta_{r+4}) \in \mathbb{Z}_{1*} \times \cdots \times \mathbb{Z}_{r+4*}$$

ensures that only points not lying above the (-1)- and (-2)-curves on X are taken into account. For each $\mathbf{x} \in U(\mathbb{Q})$, we have a unique rational point on X, giving a unique integral point on $\mathfrak{X}$ because X is projective. Lifting to a universal torsor, this corresponds to an orbit of integral points on $\widehat{\mathfrak{X}}$ under the integral model $\mathbb{G}_{m,\mathbb{Z}}^{r+1}$ of the Picard quasitorus $H_{\text{Pic}(X_{\overline{\mathbb{Q}}})}$, whose rank is rk $\text{Pic}(X) = r + 1$. Therefore, we must divide by $\#\mathbb{G}_{m,\mathbb{Z}}^{r+1}(\mathbb{Z}) = 2^{r+1}$. See [55, Lemma 11] and [140] for examples how this heuristic geometric picture can be turned into a strict proof of Claim 6.4.3.1 in specific cases.

Now we sketch an elementary strategy to prove Claim 6.4.3.1. This is similar to the strategy described over $\mathbb{Q}$ in [123, Sec. 4] and over imaginary quadratic fields in [116, Sec. 4]. The starting point is the birational map $\pi \circ \kappa^{-1} \colon \mathbb{P}_{\mathbb{Q}}^2 \dashrightarrow Y$ with $\kappa = \kappa_1 \circ \cdots \circ \kappa_r$ as in Remark 6.4.1.4. We assume that $E_1, \ldots, E_{r+4}$ are ordered such that E_i is the strict transform of the exceptional divisor of κ_i, for $i = 1, \ldots, r$. Furthermore, we may assume that in $\mathbb{P}_{\mathbb{Q}}^2$, we have

$$\kappa(E_{r+1}) = \{y_0 = 0\}, \qquad \kappa(E_{r+2}) = \{y_1 = 0\},$$

$$\kappa(E_{r+3}) = \{y_2 = 0\}, \qquad \kappa(E_{r+4}) = \{g'(y_0, y_1, y_2) = 0\},$$

for some homogeneous polynomial $g' \in \mathbb{Q}[Y_0, Y_1, Y_2]$ satisfying

$$Y_3 - g'(Y_0, Y_1, Y_2) = g(1, \ldots, 1, Y_0, \ldots, Y_3) \in \mathbb{Q}[Y_0, \ldots, Y_3]; \quad (6.4.5)$$

see [113, Lemma 12]. Using the monomials $\Psi_0, \ldots, \Psi_d$ from above, we define a map

$$\Psi \colon \mathbb{Z}^{r+4} \to \mathbb{Z}^{d+1},$$

$$(\eta_1, \ldots, \eta_{r+4}) \mapsto (\Psi_0(\eta_1, \ldots, \eta_{r+4}), \ldots, \Psi_d(\eta_1, \ldots, \eta_{r+4})).$$

For $i \in \{0, \ldots, r\}$, we consider the coprimality conditions

$$\gcd\{\eta_j; \ j \in J\} = 1 \quad \text{if} \quad \bigcap_{j \in J}(\kappa_{i+1} \circ \cdots \circ \kappa_r)(E_j) = \emptyset \quad (6.4.6)$$

for all $J \subseteq \{1, \ldots, i, r+1, \ldots, r+4\}$; similarly to (6.4.3), these encode the configuration of the strict transforms of the curves E_j on X_i. For $i = 0, \ldots, r$, we claim that Ψ induces a 2^{i+1}-to-1 map from

$$M_i := \{(\eta_1, \ldots, \eta_{r+4}) \in \mathbb{Z}_*^{r+4}; \ \eta_{i+1} = \cdots = \eta_r = 1, (6.4.2), (6.4.6)\} \quad (6.4.7)$$

to $U(\mathbb{Q})$. To obtain Claim 6.4.3.1, it remains to check for $(\eta_1, \ldots, \eta_{r+4}) \in M_r$ using (6.4.3) that $\Psi(\eta_1, \ldots, \eta_{r+4})$ has coprime coordinates so that the height condition $H(\mathbf{x}) \leq B$ on $U(\mathbb{Q})$ lifts to (6.4.4).

Along the following lines, the claim for M_i is easily checked for an explicitly given del Pezzo surface, see Proposition 6.5.3.1, and can be proved in some generality, see [116, Sec. 4]. See [140] for a more abstract approach via Cox rings.

For $i = 0$, this follows essentially from the fact that the complement of the (-1)- and (-2)-curves on X is isomorphic to the complement of the images of the (-1)- and (-2)-curves among $E_{r+1}, \ldots, E_{r+4}$ on $\mathbb{P}^2_{\mathbb{Q}}$ via κ, by parameterizing the rational points on $\mathbb{P}^2_{\mathbb{Q}}$ uniquely up to sign by coprime integers $(\eta_{r+1}, \eta_{r+2}, \eta_{r+3})$, by introducing one additional variable

$$\eta_{r+4} := g'(\eta_{r+1}, \eta_{r+2}, \eta_{r+3})$$

and considering (6.4.5).

To go from $i - 1$ to i, let $(\eta'_1, \ldots, \eta'_{r+4}) \in M_{i-1}$. If $\kappa_i \colon X_i \to X_{i-1}$ is the blow-up of p_i having multiplicity r_j on $(\kappa_i \circ \cdots \circ \kappa_r)(E_j) \subset X_{i-1}$, we define $\eta_i := \pm \gcd\{\eta'_j; r_j > 0\}$ uniquely only up to sign, and

$$\eta_j := \begin{cases} \eta'_j/\eta_i^{r_j}, & j \in \{1, \ldots, i-1, r+1, \ldots, r+4\}, \\ 1, & j \in \{i+1, \ldots, r\}. \end{cases}$$

Then $(\eta_1, \ldots, \eta_{r+4}) \in M_i$. Indeed, (6.4.2) still holds because it is homogeneous with respect to the Pic(X)-grading of the Cox ring and by comparing the degrees of the strict transforms of E_j on X_{i-1} and X_i. Then we must check that the new coprimality conditions (6.4.6) hold. We can check either directly or by considering the Pic(X)-grading that

$$\Psi(\eta_1, \ldots, \eta_{r+4}) = \Psi(\eta'_1, \ldots, \eta'_{r+4}) \in U(\mathbb{Q}).$$

6.4.4 Counting integral points on universal torsors

Once we have established Claim 6.4.3.1, which we expect to be a 2^{r+1}-to-1 map between integral points on a universal torsor and rational points on a del Pezzo surface Y outside its lines, our next step is to estimate $\#T(B)$ by an integral as in Claim 6.4.4.1 below.

We need some more notation. For $i \in \{1, \ldots, r+4\}$, write

$$\mathbb{R}_{i*} := \begin{cases} \mathbb{R}_{\geq 1} \cup \mathbb{R}_{\leq -1}, & E_i^2 < 0, \\ \mathbb{R}, & \text{else} \end{cases}$$

and

$$\mathbb{R}^i_* := \mathbb{R}_{1*} \times \cdots \times \mathbb{R}_{i*}.$$

Let

$$\boldsymbol{\eta} := (\eta_1, \ldots, \eta_{r+3}).$$

By (6.4.5), the Cox ring relation (6.4.2) has the form

$$g(\eta_1, \ldots, \eta_{r+4}) = \eta_{r+4} F(\boldsymbol{\eta}) - G(\boldsymbol{\eta}) = 0 \tag{6.4.8}$$

for some polynomials $F, G \in \mathbb{Z}[\eta_1, \ldots, \eta_{r+3}]$; in fact, F is a monomial in $\eta_1, \ldots, \eta_{r+1}$. We use this to eliminate η_{r+4} from the height condition (6.4.4), giving

$$h(\eta) := \max_{i \in \{0, \ldots, d\}} \{|\Psi_i(\eta, G(\eta)/F(\eta))|\} \leq B.$$

Our task will be to estimate a summation over lattice points in

$$R(B) := \{\eta \in \mathbb{R}_*^{r+3}; \ h(\eta) \leq B\}$$

by an integral.

Claim 6.4.4.1 *Let $T(B)$ be as in Claim 6.4.3.1. For each $i = r + 2, \ldots, 0$, we write $\eta' := (\eta_1, \ldots, \eta_i)$ and $\eta'' := (\eta_{i+1}, \ldots, \eta_{r+3})$, with $\eta = (\eta', \eta'')$, and define*

$$V_i(\eta'; B) := \int_{(\eta', \eta'') \in R(B)} \frac{d\eta''}{F(\eta)}.$$

We expect for $i = r + 2, \ldots, 0$ that

$$\#T(B) = \sum_{\eta' \in \mathbb{Z}_*^i} \vartheta_i(\eta') V_i(\eta'; B) + o(B(\log B)^r)$$

for a suitable explicit arithmetic function $\vartheta_i : \mathbb{Z}_*^i \to [0, 1]$. In particular for $i = 0$, we expect $\vartheta_0 = \prod_p \omega_p \in \mathbb{R}_{>0}$, so

$$\#T(B) = \left(\prod_p \omega_p\right) \int_{\eta \in R(B)} \frac{d\eta}{F(\eta)} + o(B(\log B)^r).$$

Our starting point to prove Claim 6.4.4.1 for $i = r + 2$ is Claim 6.4.3.1. Assume that $G(\eta)$ is linear in η_{r+3} with a coefficient that is coprime to $F(\eta)$; note that $F(\eta)$ is a monomial in $\eta_1, \ldots, \eta_{r+1}$, so that it is independent of η_{r+3} and never vanishes. The basic idea is to regard the torsor equation as a linear congruence, giving

$$\#\{(\eta_{r+3}, \eta_{r+4}) \in \mathbb{Z}_{r+3*} \times \mathbb{Z}; \ \eta_{r+4} F(\eta) - G(\eta) = 0, \ h(\eta) \leq B\}$$

$$= \#\{\eta_{r+3} \in \mathbb{Z}_{r+3*}; \ G(\eta) \equiv 0 \mod F(\eta), \ h(\eta) \leq B\}$$

$$= F(\eta)^{-1} \text{Vol}\{\eta_{r+3} \in \mathbb{R}_{r+3*}; \ h(\eta) \leq B\} + O(1)$$

$$= V_{r+2}(\eta_1, \ldots, \eta_{r+2}; B) + O(1),$$

where the error term $O(1)$ is independent of $\eta_1, \ldots, \eta_{r+2}$ and B because

$$\{\eta_{r+3} \in \mathbb{R}_{r+3*}; \ h(\eta) \leq B\}$$

is a union of finitely many intervals whose number is bounded independently of $\eta_1, \ldots, \eta_{r+2}$ and B. Indeed, $h(\eta) \leq B$ describes a semialgebraic set, see [116, Lemma 3.6] for details.

To apply this to $T(B)$, we must first remove the coprimality conditions on η_{r+3}, η_{r+4} via Möbius inversions; this should give an explicit arithmetic factor $\vartheta_i(\boldsymbol{\eta}')$. If $\mathbb{Z}_{r+4*} = \mathbb{Z}_{\neq 0}$, we get an extra error term in the first step. If η_{r+3} does not appear linearly in $G(\boldsymbol{\eta})$, we introduce an auxiliary variable that causes additional difficulties in the second step; see [111, Prop. 2.4] for a discussion in some generality and Lemma 6.5.4.1 for an example.

The main difficulty in the proof of Claim 6.4.4.1 for $i = r + 2$ is to show that the error term of the first summation over η_{r+3} summed over all $\boldsymbol{\eta}'$ subject to those height conditions that are independent of η_{r+3}, η_{r+4} has order $o(B(\log B)^r)$. This is straightforward in some cases, but especially for varieties of lower degree or with milder singularities, depending on the precise shape of the height conditions, the estimation of the error term may require considerable effort or may even seem out of reach.

Because of the Möbius inversions, the error term is typically a product of $2^{\omega(\eta_j)}$ for suitable η_j appearing in the coprimality conditions; here $\omega(\eta_j)$ is the number of prime divisors of $\eta_j \in \mathbb{Z}_{\neq 0}$. The following result is useful to estimate their sums.

Lemma 6.4.4.2 *For $C \in \mathbb{R}_{\geq 0}$, any $\delta \in \mathbb{R}$, and $t_2 \geq t_1 \in \mathbb{R}_{\geq 1}$, we have*

$$\sum_{t_1 \leq \eta \leq t_2} \frac{(1 + C)^{\omega(\eta)}}{\eta^{\delta}} \ll_{C,\delta} \begin{cases} t_2^{1-\delta}(\log(t_2 + 2))^C, & \delta < 1, \\ \log(t_2 + 2)^{C+1}, & \delta = 1, \\ t_1^{1-\delta}(\log(t_1 + 2))^C \ll_{C,\delta} 1, & \delta > 1. \end{cases}$$

Proof This follows from

$$\sum_{1 \leq \eta \leq t} (1 + C)^{\omega(\eta)} \ll_C t(\log(t + 2))^C$$

and summation by parts. See [111, Ex. 3.3, Lemma 3.4] and [116, Lemma 2.4, Lemma 2.9]. $\square$

To prove Claim 6.4.4.1 recursively for $i = r + 1, \ldots, 0$, we apply partial summation to replace the sum over η_i of the main term of step i by an integral. If ϑ_i is particularly simple, it is enough to apply

$$\sum_{\substack{\eta_i \in \mathbb{Z}_{i*} \\ \gcd(\eta_i, a) = 1}} V_i(\boldsymbol{\eta}', \eta_i; B) = \varphi^*(a) \int_{\eta_i \in \mathbb{R}_{i*}} V_i(\boldsymbol{\eta}', \eta_i; B) \, d\eta_i + O(2^{\omega(a)} \sup_{\eta_i \in \mathbb{R}_{i*}} V_i(\boldsymbol{\eta}', \eta_i; B)),$$

(6.4.9)

using Möbius inversion and partial summation, see [78, Ex. 5.2], and to observe that the integral on the right hand side is just $V_{i-1}(\boldsymbol{\eta}'; B)$ by definition.

Remark 6.4.4.3 Note that in the application of partial summation, it is crucial that there is a constant C such that for any $\boldsymbol{\eta}'$, B, we can split $\mathbb{R}_{i*}$ into at most C intervals on whose interior $V_i(\boldsymbol{\eta}', \eta_i; B)$ is continuously differentiable and monotonic as a function of η_i. In fact, this is proved in [116, Lemma 3.6] using o-minimal structures.

Often, ϑ_i is more complicated, but has the following shape. See [68, Lemma 2] and [111, Sec. 6, Sec. 7] for even more general arithmetic functions.

Definition 6.4.4.4 Let $r \in \mathbb{Z}_{\geq 0}$. For $(\eta_1, \dots, \eta_r) \in \mathbb{Z}^r$ and any prime p, let $I_p(\eta_1, \dots, \eta_r) := \{i; \ p \mid \eta_i\}$. Let $\Theta'_r(C)$ be the set of all functions $\vartheta : \mathbb{Z}^r \to [0, 1]$ such that

$$\vartheta(\eta_1, \dots, \eta_r) = \prod_p \vartheta_p(I_p(\eta_1, \dots, \eta_r))$$

with

$$\vartheta_p(I) \geq \begin{cases} 1 - Cp^{-2}, & \#I = 0, \\ 1 - Cp^{-1}, & \#I = 1. \end{cases}$$

We say that ϑ has *local factors* ϑ_p.

For such arithmetic functions ϑ_i, the estimation (6.4.9) generalizes as follows; the parameter T will be used in the proof of Proposition 6.4.4.7 below.

Proposition 6.4.4.5 *For $i \in \{1, \dots, r+2\}$, assume that there is a $C \in \mathbb{Z}_{\geq 0}$ such that $\vartheta_i \in \Theta'_i(C)$, with local factors $\vartheta_{i,p}$. We define the function $\vartheta_{i-1} \in \Theta'_{i-1}(2C)$ by its local factors*

$$\vartheta_{i-1,p}(I) := \left(1 - \frac{1}{p}\right)\vartheta_{i,p}(I) + \frac{1}{p}\vartheta_{i,p}(I \cup \{i\}).$$

Let $B \in \mathbb{R}_{\geq 2}$. Let V_i be as above, with $V_i(\eta_1, \dots, \eta_i; B) = 0$ for $|\eta_i| > B$. For any $T \in \mathbb{R}_{\geq 0}$, we have

$$\sum_{|\eta_i| \geq T} \vartheta_i(\eta_1, \dots, \eta_i) V_i(\eta_1, \dots, \eta_i; B)$$

$$= \vartheta_{i-1}(\eta_1, \dots, \eta_{i-1}) \int_{|\eta_i| \geq T} V_i(\eta_1, \dots, \eta_i; B)\, d\eta_i$$

$$+ O\left((2C)^{\omega(\eta_1 \cdots \eta_{i-1})} (\log B)^C \sup_{T \leq |\eta_i| \leq B} |V_i(\eta_1, \dots, \eta_i; B)| \right).$$

Proof Let $\vartheta(n) := \vartheta_i(\eta_1, \dots, \eta_{i-1}, n)$. Defining

$$N_p := N_p(\eta_1, \dots, \eta_{i-1}) := \vartheta_p(I_p(\eta_1, \dots, \eta_{i-1}, 1)) = \vartheta_p(I),$$

$$J_p := J_p(\eta_1, \dots, \eta_{i-1}) := \vartheta_p(I_p(\eta_1, \dots, \eta_{i-1}, p)) = \vartheta_p(I \cup \{i\}),$$

with $I := I_p(\eta_1, \dots, \eta_{i-1}, 1)$, we have $\vartheta(n) = \prod_{p \nmid n} N_p \prod_{p \mid n} J_p$.

Because $\vartheta = (\vartheta * \mu) * 1$, we have

$$\sum_{0<n\le t} \vartheta(n) = \sum_{0<d\le t} (\vartheta * \mu)(d) \sum_{0<n\le t/d} 1.$$

Because the inner sum is $t/d + O(1)$, we have

$$\sum_{0<n\le t} \vartheta(n) = t \cdot \sum_{d=1}^{\infty} \frac{(\vartheta * \mu)(d)}{d} + O\left(\sum_{0<d\le t} |(\vartheta * \mu)(d)| + t \cdot \sum_{d>t} \frac{|(\vartheta * \mu)(d)|}{d} \right),$$

under the condition that the infinite sums over d converge, which we will see below.

For squarefree n, we have

$$(\vartheta * \mu)(n) = \sum_{d|n} \mu(d)\vartheta(n/d) = \sum_{d|n} \prod_{\substack{p\nmid n}} N_p \prod_{\substack{p|n \\ p\nmid d}} J_p \prod_{\substack{p|n \\ p|d}} (-N_p)$$

$$= |\mu(n)| \prod_{p\nmid n} N_p \prod_{p|n} (J_p - N_p),$$

and for non-squarefree n, both sides clearly vanish. Therefore,

$$|(\vartheta * \mu)(n)| \le C^{\omega(\eta_1 \cdots \eta_{i-1})} \gcd(\eta_1 \cdots \eta_{i-1}, n)n^{-1}.$$

Hence

$$\sum_{0<n\le t} |(\vartheta * \mu)(n)| \cdot n \ll (2C)^{\omega(\eta_1 \cdots \eta_{i-1})} t(\log t)^{C-1}.$$

This implies that the infinite sums above over d converge absolutely, and that the error term is

$$O((2C)^{\omega(\eta_1 \cdots \eta_{i-1})}(\log(t + 2))^C).$$

For the main term, let $c = \prod_p N_p$, which converges to a positive number because $|N_p - 1| \le Cp^{-2}$ for all $p \nmid \eta_1 \cdots \eta_{i-1}$. The function $B(n) := c^{-1}(\vartheta * \mu)(n)$ is multiplicative, with local factors

$$B(p^{\nu}) = \begin{cases} N_p^{-1}(J_p - N_p), & \nu = 1, \\ 0, & \nu > 1. \end{cases}$$

Absolute convergence implies that we can express the following sum as the Euler product

$$\sum_{n=1}^{\infty} \frac{(\vartheta * \mu)(n)}{n} = \sum_{n=1}^{\infty} cB(n) = c \prod_p \sum_{\nu=0}^{\infty} \frac{B(p^{\nu})}{p^{\nu}} = \prod_p ((1 - p^{-1})N_p + p^{-1}J_p).$$

This proves the claim if $V_i(\eta_1, \ldots, \eta_i; B)$ as a function of η_i is the characteristic function of an interval. The general case follows by partial summation, after splitting $[-B, B]$ into intervals where $V_i(\eta_1, \ldots, \eta_i; B)$ is monotonous and continuously differentiable as a function in η_i; see Remark 6.4.4.3. $\quad\square$

To derive Claim 6.4.4.1 for $i-1$ from Proposition 6.4.4.5, we must sum the error term over $\eta_1, \ldots, \eta_{i-1}$, using Lemma 6.4.4.2. For this, an upper bound on V_i is necessary, which is provided by results such as the following; see [111, Lemma 5.1] for other cases that can be proved similarly.

Lemma 6.4.4.6 *For any $a, b, c \in \mathbb{R}_{\neq 0}$ and $B \in \mathbb{R}_{>0}$, we have*

$$\int_{\left|\frac{a\eta^2 + b\eta'^3}{c}\right| \leq B} \frac{d\eta}{c} \ll \frac{B^{1/2}}{|ac|^{1/2}} \text{ for all } \eta' \in \mathbb{R}, \quad \int_{\left|\frac{a\eta^2 + b\eta'^3}{c}\right| \leq B} \frac{d\eta\, d\eta'}{c} \ll \frac{B^{5/6}}{|a^3 b^2 c|^{1/6}}.$$

Proof We recall the proof from [111, Lemma 5.1]. Without loss of generality, we may assume $a, c > 0$. First, assume that $|\eta'| \leq (2cB/|b|)^{1/3}$. Then $|a\eta^2 + b\eta'^3| \leq cB$ implies $|a\eta^2| \leq 3cB$, so $|\eta| \ll \sqrt{cB/a}$. This gives the first claim for these η', and its integral over η' in this range agrees with the second claim.

Next, suppose $|\eta'| > (2cB/|b|)^{1/3}$; because $b\eta'^3 > 2cB$ is impossible, we have $b\eta'^3 < -2cB$. We have $\sqrt{\frac{-b\eta'^3 - cB}{a}} \leq |\eta| \leq \sqrt{\frac{-b\eta'^3 + cB}{a}}$. This describes two intervals of length $\ll cB/|ab\eta'^3|^{1/2}$. Therefore, the first integral is $\ll B/|ab\eta'^3|^{1/2}$. Under our assumption on η', this gives the first claim, and its integral over η' in this range agrees with the second claim. □

To prove Claim 6.4.4.1 for $i = r + 1$, a successful estimation of the error term of the second summation over η_{r+2} is sometimes straightforward, but tends to be quite difficult for del Pezzo surfaces of lower degrees or with milder singularities, similarly as in the first summation.

To deduce from Claim 6.4.4.1 for $i - 1$ from some $i \in \{1, \ldots, r + 1\}$, estimates such as Lemma 6.4.4.6 can often be applied to show that we are in one of the following situations that can be handled generally:

(i) We have $i = r + 1$, and there exist $k_1, \ldots, k_{r-1} \in \mathbb{R}$, $k_r, k_{r+1} \in \mathbb{R}_{\neq 0}$, $a \in \mathbb{R}_{>0}$ with

$$|V_{r+1}(\eta_1, \ldots, \eta_{r+1}; B)| \ll \frac{B}{|\eta_1 \cdots \eta_{r+1}|} \left(\frac{B}{|\eta_1|^{k_1} \cdots |\eta_{r+1}|^{k_{r+1}}} \right)^{-a}$$

and $V_{r+1}(\eta_1, \ldots, \eta_{r+1}; B) = 0$ unless

$$|\eta_1|^{k_1} \cdots |\eta_{r+1}|^{k_{r+1}} \leq B.$$

(ii) We have $i \leq r$, and

$$|V_i(\eta_1, \ldots, \eta_i; B)| \ll \frac{B(\log B)^{r-i}}{|\eta_1 \cdots \eta_i|}.$$

For $i = r + 1$, we sometimes have alternative estimates that can be handled similarly, see [111, Prop. 3.10].

Proposition 6.4.4.7 *For $i \in \{1, \ldots, r+1\}$, assume $\vartheta_i \in \Theta'_i(C)$ and define $\vartheta_{i-1} \in \Theta'_{i-1}(2C)$ as in Proposition 6.4.4.5. Let V_i satisfy the bounds above, and let*

$$V_{i-1}(\eta_1, \ldots, \eta_{i-1}; B) := \int_{\mathbb{R}} V_i(\eta_1, \ldots, \eta_{i-1}, \eta_i; B) \, d\eta_i.$$

For $B \in \mathbb{R}_{\geq 3}$, we have

$$\sum_{\eta_1, \ldots, \eta_i} \vartheta_i(\eta_1, \ldots, \eta_i) V_i(\eta_1, \ldots, \eta_i; B)$$

$$= \sum_{\eta_1, \ldots, \eta_{i-1}} \vartheta_{i-1}(\eta_1, \ldots, \eta_{i-1}) V_{i-1}(\eta_1, \ldots, \eta_{i-1}; B) + O_{V,C}(B(\log B)^{r-1} \log \log B).$$

Proof We note that

$$\sum_{\eta_1, \ldots, \eta_{i-1}} |V_i(\eta_1, \ldots, \eta_i; B)| \ll \frac{B(\log B)^{r-1}}{|\eta_i|}. \qquad (6.4.10)$$

Indeed, for $i = r + 1$, we apply the first condition of (i) above and Lemma 6.4.4.2 with $\delta := 1 - ak_r$, and the bound on η_r given by the second condition of (i) to the summation over η_r. For any i, we use Lemma 6.4.4.2 with $\delta := 1$ for the summations over the remaining η_j with $j \leq r - 1$.

Let $T := (\log B)^{C + (2C-1)(i-1)+1}$. We proceed in three steps. In the first step, we restrict the summation to $|\eta_i| \geq T$, which is allowed because

$$\sum_{\substack{\eta_1, \ldots, \eta_i \\ |\eta_i| < T}} \vartheta_i(\eta_1, \ldots, \eta_i) V_i(\eta_1, \ldots, \eta_i; B) \ll \sum_{|\eta_i| < T} \frac{B(\log B)^{r-1}}{|\eta_i|}$$

$$\ll_C B(\log B)^{r-1} \log \log B,$$

using $|\vartheta_i(\eta_1, \ldots, \eta_i)| \leq 1$, (6.4.10) and Lemma 6.4.4.2.

In the second step, we apply Proposition 6.4.4.5 to the sum over $|\eta_i| \geq T$. Using

$$V_i(\eta_1, \ldots, \eta_i; B) \ll \frac{B(\log B)^{\max\{r-i, 0\}}}{|\eta_1 \cdots \eta_i|}$$

and Lemma 6.4.4.2, we get the main term

$$\sum_{\eta_1, \ldots, \eta_{i-1}} \vartheta_{i-1}(\eta_1, \ldots, \eta_{i-1}) \int_{|\eta_i| \geq T} V_i(\eta_1, \ldots, \eta_i; B) \, d\eta_i$$

and the error term

$$\ll_{V,C} \sum_{\eta_1, \ldots, \eta_{i-1}} \frac{(2C)^{\omega(\eta_1 \cdots \eta_{i-1})} B(\log B)^{C + \max\{r-i, 0\}}}{|\eta_1 \cdots \eta_{i-1}| T}$$

$$\ll_C B(\log B)^{C + (2C)(i-1) + \max\{r-i, 0\}} T^{-1} \ll B(\log B)^{r-1}.$$

In the third step, we remove the restriction $|\eta_i| \geq T$ on the main term by estimating

$$\sum_{\eta_1,\ldots,\eta_{i-1}} \vartheta_{i-1}(\eta_1,\ldots,\eta_{i-1}) \int_{1\leq|\eta_i|\leq T} V_i(\eta_1,\ldots,\eta_i;B)\,\mathrm{d}\eta_i$$

$$\ll \int_{1\leq|\eta_i|\leq T} \frac{B(\log B)^{r-1}}{|\eta_i|}\,\mathrm{d}\eta_i \ll_C B(\log B)^{r-1}\log\log B,$$

using $|\vartheta_{i-1}(\eta_1,\ldots,\eta_{i-1})| \leq 1$ and (6.4.10).   □

In fact, if we are in the situation to apply Proposition 6.4.4.7 for $i = r+1$, then we can go all the way to $i = 0$ by applying the result repeatedly as in the following Corollary 6.4.4.8. We expect that the arithmetic factor ϑ_0 agrees with $\prod_p \omega_p \in \mathbb{R}_{>0}$, giving Claim 6.4.4.1 in the final case $i = 0$.

Corollary 6.4.4.8 *Let V_{r+1} be as above and $\vartheta_{r+1} \in \Theta'_{r+1}(C)$ with local factors $\vartheta_{r+1,p}$. Then*

$$\sum_{\eta_1,\ldots,\eta_{r+1}} \vartheta_{r+1}(\eta_1,\ldots,\eta_{r+1})V_{r+1}(\eta_1,\ldots,\eta_{r+1};B)$$

$$= \vartheta_0 V_0(B) + O_{V,C}(B(\log B)^{r+1}\log\log B),$$

where

$$\vartheta_0 := \prod_p \sum_{I\subseteq\{1,\ldots,r+1\}} \left(1 - \frac{1}{p}\right)^{r-\#I} \left(\frac{1}{p}\right)^{\#I} \vartheta_{r+1,p}(I).$$

Proof Starting with $\vartheta_{r+1} \in \Theta'_{r+1}(C)$, we define inductively $\vartheta_r,\ldots,\vartheta_0$ as in Proposition 6.4.4.5, with $\vartheta_i \in \Theta'_i(2^{r-i+1}C)$ for $i = r,\ldots,0$, giving ϑ_0 as in our statement. Defining V_{i-1} as before as the integral of V_i over η_i, we have (ii) for V_{i-1} because V_i satisfies the height conditions listed in (i) or (ii) before Proposition 6.4.4.7. An $(r+1)$-fold application of Proposition 6.4.4.7 gives the result.   □

6.4.5 Interpretation of the integral

We redefine our notation

$$\boldsymbol{\eta}' := (\eta_1,\ldots,\eta_r), \quad \boldsymbol{\eta}'' := (\eta_{r+1},\eta_{r+2},\eta_{r+3}),$$

$$\boldsymbol{\eta} := (\boldsymbol{\eta}',\boldsymbol{\eta}'') = (\eta_1,\ldots,\eta_{r+3}).$$

Once Claim 6.4.4.1 is established, it remains to show:

Claim 6.4.5.1 *We expect*

$$\frac{1}{2^{r+1}} \int_{\eta \in R(B)} \frac{d\eta}{F(\eta)} = \alpha(X) \omega_\infty B (\log B)^r + O(B(\log B)^{r-1}).$$

Here, we are guided by the definitions of $\alpha(X)$ as the volume of a polytope in the dual of the effective cone of X, see Definition 6.3.1.4, and of the archimedean density ω_∞, see [243, Sec. 2.2, Sec. 5.4], [247, Sec. 4]. We start with a discussion of $\alpha(X)$ for del Pezzo surfaces. Its value can be determined using the next result, see [119, Thm. 1.2].

Theorem 6.4.5.2 *We have* $\alpha(\mathbb{P}_{\mathbb{K}}^2) = 1/3$, $\alpha(\mathbb{P}_{\mathbb{K}}^1 \times_{\mathbb{K}} \mathbb{P}_{\mathbb{K}}^1) = 1/4$, $\alpha(\mathbb{F}_{2,\mathbb{K}}) = 1/8$, $\alpha(\mathrm{Bl}_1 \mathbb{P}_{\mathbb{K}}^2) = 1/6$. *For any split weak del Pezzo surface X of degree $d \le 7$, we have*

$$\alpha(X) = \sum_{E \in \mathcal{E}} \frac{\alpha(X_E)}{d(9-d)}$$

where $\mathcal{E}$ is the set of (-1)-curves on X, and X_E is the split weak del Pezzo surface of degree $d + 1$ obtained by contracting $E \in \mathcal{E}$.

As every sequence of contractions of (-1)-curves on a split weak del Pezzo surface X of degree $\deg(X) \le 7$ over $\mathbb{K}$ leads eventually to $\mathrm{Bl}_1 \mathbb{P}_{\mathbb{K}}^2$ (the blow-up of $\mathbb{P}_{\mathbb{K}}^2$ in one $\mathbb{K}$-rational point) or $\mathbb{P}_{\mathbb{K}}^1 \times_{\mathbb{K}} \mathbb{P}_{\mathbb{K}}^1$ or the second Hirzebruch surface $\mathbb{F}_{2,\mathbb{K}}$ of degree 8, this allows us to compute $\alpha(X)$ inductively. In the split smooth case, this results in the following table.

$\deg(X)$	7	6	5	4	3	2	1
$\alpha(X)$	$\frac{1}{24}$	$\frac{1}{72}$	$\frac{1}{144}$	$\frac{1}{180}$	$\frac{1}{120}$	$\frac{1}{30}$	1

For weak del Pezzo surfaces that are minimal resolutions of del Pezzo surfaces with ADE-singularities, it is easier to apply the following result, see [119, Thm. 1.3].

Theorem 6.4.5.3 *Let Y be a split singular del Pezzo surface of degree at most 7 with minimal resolution X. Let W be the Weyl group associated with the singularities of Y, that is, the Weyl group of the root system whose simple roots are the (-2)-curves on X. Let X_0 be a split smooth del Pezzo surface of the same degree. Then*

$$\alpha(X) = \frac{\alpha(X_0)}{\#W}.$$

To obtain $\#W$, we must take the product over all singularities of the values that can be found in the following table.

singularity	A_n	D_n	E_6	E_7	E_8
#W	$(n+1)!$	$2^{n-1} \cdot n!$	$2^7 \cdot 3^4 \cdot 5$	$2^{10} \cdot 3^4 \cdot 5 \cdot 7$	$2^{14} \cdot 3^5 \cdot 5^2 \cdot 7$

For example, for the split cubic surface X of type E_6 that we will treat in Section 6.5, we have $\alpha(X) = \frac{1}{120 \cdot 2^7 \cdot 3^4 \cdot 5} = \frac{1}{6220800}$.

For nonsplit smooth del Pezzo surfaces X over $\mathbb{K}$, the value of $\alpha(X)$ depends on the action of $G := \mathrm{Gal}(\overline{\mathbb{K}}/\mathbb{K})$ on the set of (-1)-curves on $X_{\overline{\mathbb{K}}}$, via a subgroup of a Weyl group. See [115] for details. For nonsplit singular del Pezzo surfaces, we have [119, Thm. 7.4], which is similar to Theorem 6.4.5.3, but the action of G on the (-2)-curves of $X_{\overline{\mathbb{K}}}$ must be taken into account, leading additionally to root systems B_n, C_n, F_4, G_2; see [119, Sec. 6, Sec. 7] for details.

To prove Claim 6.4.5.1, it is usually unnecessary to know the value of $\alpha(X)$. Instead, an explicit description of the polytope appearing in Definition 6.3.1.4 is more useful.

Suppose that the (-1)- and (-2)-curves on X are $E_1, \ldots, E_{r+1+s}$, for some $s \geq 0$. By our assumption in Section 6.4.3, $E_1, \ldots, E_{r+1}$ are a basis of $\mathrm{Pic}(X)$. Expressing $-K_X$ and $E_{r+2}, \ldots, E_{r+1+s}$ in terms of this basis, we have

$$[-K_X] = \sum_{j=1}^{r+1} c_j [E_j]$$

and, for $i = 1, \ldots, s$,

$$[E_{r+1+i}] = \sum_{j=1}^{r+1} b_{i,j} [E_j]$$

for some $b_{i,j}, c_j \in \mathbb{Z}$. Assume that $c_{r+1} > 0$.

Lemma 6.4.5.4 *Define, for $j = 1, \ldots, r$ and $i = 1, \ldots, s$,*

$$a_{0,j} := c_j, \qquad\qquad a_{i,j} := b_{i,r+1} c_j - b_{i,j} c_{r+1},$$
$$A_0 := 1, \qquad\qquad A_i := b_{i,r+1}.$$

Then

$$\alpha(X)(\log B)^r = \frac{1}{2^r c_{r+1}} \int_{R_1'(B)} \frac{1}{|\eta_1 \cdots \eta_r|} \, d\boldsymbol{\eta}'$$

with a domain of integration

$$R_1'(B) := \left\{ \boldsymbol{\eta}' \in \mathbb{R}^r; \ |\eta_1|, \ldots, |\eta_r| \geq 1, \ \prod_{j=1}^{r} |\eta_j|^{a_{i,j}} \leq B^{A_i} \ (i \in \{0, \ldots, s\}) \right\}.$$

Proof By Definition 6.3.1.4, using the basis $E_1, \ldots, E_{r+1}$ of $\mathrm{Pic}(X)$, we have

$$\alpha(X) = (r+1) \cdot \mathrm{Vol}\left\{ (t'_1, \ldots, t'_{r+1}) \in \mathbb{R}^{r+1}_{\geq 0};\ \sum_{j=1}^{r+1} b_{i,j} t'_j \geq 0\ (i = 1, \ldots, s),\ \sum_{j=1}^{r+1} c_j t'_j \leq 1 \right\}.$$

We make a linear change of variables $(t_1, \ldots, t_r, t_{r+1}) = (t'_1, \ldots, t'_r, c_1 t'_1 + \cdots + c_{r+1} t'_{r+1})$, with Jacobian c_{r+1}. This transforms the polytope in the previous formula into a pyramid whose base is $R_0 \times \{1\}$ in the hyperplane $\{t_{r+1} = 1\}$ in $\mathbb{R}^{r+1}$, and whose apex is the origin, where

$$R_0 := \left\{ (t_1, \ldots, t_r) \in \mathbb{R}^r_{\geq 0};\ \sum_{j=1}^{r} a_{i,j} t_j \leq A_i\ (i \in \{0, \ldots, s\}) \right\}.$$

This pyramid has volume $(r+1)^{-1} \mathrm{Vol}\, R_0$ because its height is 1 and its dimension is $r+1$. Writing $\mathrm{Vol}\, R_0$ as an integral, we get

$$\alpha(X) = c_{r+1}^{-1} \int_{(t_1, \ldots, t_r) \in R_0} dt_r \cdots dt_1,$$

where the factor c_{r+1}^{-1} appears because of our change of coordinates. Now the result is derived using the change of coordinates $\eta_i = B^{t_i}$ for $i \in \{1, \ldots, r\}$ and the observation that allowing arbitrary signs for $\eta_1, \ldots, \eta_r$ leads to a factor 2^{-r}. $\qquad\square$

Let

$$R'_2(\boldsymbol{\eta}'; B) := \{\boldsymbol{\eta}'' \in \mathbb{R}^3;\ h(\boldsymbol{\eta}', \boldsymbol{\eta}'') \leq B\}.$$

We expect that

$$\int_{\boldsymbol{\eta}'' \in R'_2(\boldsymbol{\eta}';B)} \frac{d\boldsymbol{\eta}''}{F(\boldsymbol{\eta}', \boldsymbol{\eta}'')} = \frac{2\omega_\infty B}{c_{r+1} |\eta_1 \cdots \eta_r|} \tag{6.4.11}$$

can be proved using the substitution

$$x_j = B^{-1} \Psi_j \left(\boldsymbol{\eta}, \frac{G(\boldsymbol{\eta})}{F(\boldsymbol{\eta})} \right),$$

with F, G from (6.4.8), for $j \in \{0, \ldots, d\}$, if we express ω_∞ as an integral over $|x_0|, \ldots, |x_d| \leq 1$, see [243, Sec. 5.4].

Let

$$R'(B) := \{\boldsymbol{\eta} = (\boldsymbol{\eta}', \boldsymbol{\eta}'') \in \mathbb{R}^{r+3};\ \boldsymbol{\eta}' \in R'_1(B), \boldsymbol{\eta}'' \in R'_2(\boldsymbol{\eta}'; B)\}$$

$$= \left\{ \boldsymbol{\eta} \in \mathbb{R}^{r+3};\ \begin{array}{l} |\eta_1|, \ldots, |\eta_r| \geq 1,\ h(\boldsymbol{\eta}) \leq B, \\[2mm] \displaystyle\prod_{j=1}^{r} |\eta_j|^{a_{i,j}} \leq B^{A_i}\ (i \in \{0, \ldots, s\}) \end{array} \right\}.$$

If (6.4.11) holds, with Lemma 6.4.5.4, we have

$$\frac{1}{2^{r+1}} \int_{R'(B)} \frac{d\eta}{F(\eta)} = \frac{\omega_\infty B}{2^r c_{r+1}} \int_{R_1'(B)} \frac{1}{|\eta_1 \cdots \eta_r|} \, d\eta' = \alpha(X) \omega_\infty B (\log B)^r.$$

To establish Claim 6.4.5.1, it remains to show that replacing the domain of integration $R(B)$ from Claim 6.4.4.1 by $R'(B)$ results in a negligible error, namely

$$\int_{R(B)} \frac{d\eta}{F(\eta)} = \int_{R'(B)} \frac{d\eta}{F(\eta)} + O(B(\log B)^{r-1}). \qquad (6.4.12)$$

It may be necessary to split the proof of this into several steps whose order is crucial. We expect the following steps:

(i) Restrict $\eta_1, \ldots, \eta_r$ to $\prod_{j=1}^r |\eta_j|^{a_{i,j}} \le B^{A_i}$ for $i = 0, \ldots, s$. Some of these conditions may be trivially true because of $h(\eta) \le B$, while we have to do some work for others.

(ii) Allow $|\eta_i| < 1$ for $i \ge r+1$ with $E_i^2 < 0$. We expect that we must use the conditions $\prod_{j=1}^r |\eta_j|^{a_{i,j}} \le B^{A_i}$ that we introduced in the previous step.

Once we have proved (6.4.11) and (6.4.12), Claim 6.4.5.1 follows. Together with Claim 6.4.3.1 and Claim 6.4.4.1, this implies Manin's conjecture.

6.5 Manin's conjecture for a singular cubic surface

6.5.1 Statement of the result

In this section, we apply the strategy and auxiliary results from Section 6.4 to prove Manin's conjecture as stated in Conjecture 6.3.1.5 for the weak del Pezzo surface X of degree 3 over $\mathbb{Q}$ that is the minimal resulstion of the cubic surface $Y \subset \mathbb{P}_{\mathbb{Q}}^3$ of singularity type E_6 given as hypersurface in the projective space by

$$y_0 y_2^2 + y_0^2 y_3 + y_1^3 = 0.$$

The surface $Y_{\overline{\mathbb{Q}}}$ comes with a $\overline{\mathbb{Q}}^*$-action and is discussed in the series of examples starting with 3.4.3.3, see in particular 3.4.4.2 and, finally, 5.4.3.5, where we show that Y is of singularity type E_6 and compute the Cox ring of its minimal resolution. Here is the main result of this section.

Theorem 6.5.1.1 *Let $Y \subset \mathbb{P}_{\mathbb{Q}}^3$ be the singular cubic surface defined above and let $U \subset Y$ be the complement of the line $\{y_0 = y_1 = 0\}$ on Y. Then*

$$N_{U,H}(B) = \frac{1}{6220800} \prod_p \left(1 - \frac{1}{p}\right)^7 \left(1 + \frac{7}{p} + \frac{1}{p^2}\right) \omega_\infty B (\log B)^6 (1 + o(1))$$

holds with

$$\omega_\infty = \frac{3}{2} \int_{|y_0^3|, |y_0^2 y_1|, |y_0^2 y_2|, |y_0 y_2^2 + y_1^3| \leq 1} dy_0 \, dy_1 \, dy_2.$$

See [72] for a version of this result with a much stronger estimation of the error term. Our treatment of the first two summations is similar to the first proof of Manin's conjecture for this cubic surface in [114].

6.5.2 Geometry and Cox ring

To prepare for the parameterization of rational points on Y, we discuss its geometry and its Cox ring. Recall from [84] that up to isomorphy Y is uniquely determined as a cubic surface of singularity type E_6 in the projective space. The singular point of Y is $[0, 0, 0, 1]$ and $\{y_0 = y_1 = 0\}$ is the only line on Y. The minimal resolution $X \to Y$ is obtained as a blow-up of $\mathbb{P}^2_{\mathbb{Q}}$ in six points in almost general position, that is, as a series

$$X = X_6 \xrightarrow{\kappa_6} X_5 \to \dots \to X_1 \xrightarrow{\kappa_1} X_0 = \mathbb{P}^2_{\mathbb{K}},$$

where each $\kappa_i : X_i \to X_{i-1}$ describes the blow up of a rational point. We make this more concrete. Let x_0, x_1, x_2 be the coordinates on $X_0 = \mathbb{P}^2_{\mathbb{Q}}$ and consider the curves

$$E_7^{(0)} := \{x_0 = 0\}, \ E_8^{(0)} := \{x_1 = 0\},$$

$$E_9^{(0)} := \{x_2 = 0\}, \ E_{10}^{(0)} := \{-x_0 x_2^2 - x_1^3 = 0\}.$$

For any curve $E_j^{(i-1)}$ on X_{i-1}, denote by $E_j^{(i)}$ the strict transform of $E_j^{(i-1)}$ under $\kappa_i : X_i \to X_{i-1}$, where $i = 1, \dots, 6$. Then

- κ_1 is the blow-up of $E_7^{(0)} \cap E_8^{(0)} \cap E_{10}^{(0)}$, giving $E_1^{(1)}$.
- κ_2 is the blow-up of $E_1^{(1)} \cap E_7^{(1)} \cap E_{10}^{(1)}$, giving $E_2^{(2)}$.
- κ_3 is the blow-up of $E_2^{(2)} \cap E_7^{(2)} \cap E_{10}^{(2)}$, giving $E_3^{(3)}$.
- κ_4 is the blow-up of $E_3^{(3)} \cap E_{10}^{(3)}$, giving $E_4^{(4)}$.
- κ_5 is the blow-up of $E_4^{(4)} \cap E_{10}^{(4)}$, giving $E_5^{(5)}$.
- κ_6 is the blow-up of $E_5^{(5)} \cap E_{10}^{(5)}$, giving $E_6^{(6)}$.

For $j = 1, \dots, 10$, let $E_j := E_j^{(6)}$ be the strict transforms of the exceptional divisors $E_1^{(1)}, \dots, E_6^{(6)}$ of $\kappa_1, \dots, \kappa_6$ and the four curves $E_7^{(0)}, \dots, E_{10}^{(0)}$ in $\mathbb{P}^2_{\mathbb{Q}}$. Then $E_1, \dots, E_5, E_7$ are (-2)-curves coming from the singularity on Y, and E_6 is a (-1)-curve, the strict transform of the line on Y. Their configuration is encoded in the following *extended Dynkin diagram*.

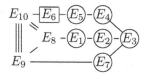

Write $\kappa\colon X \to \mathbb{P}^2_{\mathbb{Q}}$ for the composition of all the κ_i and $\pi\colon X \to Y$ for the minimal resolution. Then, on its area of definition, the rational map $\pi \circ \kappa^{-1}\colon \mathbb{P}^2_{\mathbb{Q}} \dashrightarrow Y$ coincides with

$$\psi\colon \mathbb{P}^2_{\mathbb{Q}} \dashrightarrow Y, \qquad [y_0, y_1, y_2] \mapsto [y_0^3, y_0^2 y_1, y_0^2 y_2, -y_0 y_2^2 - y_1^3].$$

As mentioned above, we determined the Cox ring of $X_{\overline{\mathbb{Q}}}$ in 5.4.3.5; see also 4.4.1.8 for another approach to the Cox ring of X. The explicit construction of X by a sequence of blow-ups of $\mathbb{P}^2_{\mathbb{Q}}$ allows moreover a Cox ring computation via toric ambient modifications in the sense of Section 4.1.3. Consider the embedding

$$\mathbb{P}^2_{\mathbb{K}} \to \mathbb{P}_{1,1,1,3}, \qquad [x_0, x_1, x_2] \mapsto [x_0, x_1, x_2, -x_0 x_2^2 - x_1^3].$$

Then the divisors $E_j^{(0)}$ are intersections of $\mathbb{P}^2_{\mathbb{K}}$ with the toric divisors of $\mathbb{P}_{1,1,1,3}$. The toric modification $Z \to \mathbb{P}_{1,1,1,3}$ defined by the blow-up sequence specified before satisfies the assumptions of Theorem 4.1.3.5. Thus, the Cox ring of $X_{\overline{\mathbb{Q}}}$ is obtained by pulling back the defining equation of $\mathbb{P}^2_{\mathbb{K}}$ in $\mathbb{P}_{1,1,1,3}$:

$$\mathcal{R}(X_{\overline{\mathbb{Q}}}) = \overline{\mathbb{Q}}[\eta_1, \ldots, \eta_{10}]/\langle \eta_1^2 \eta_2 \eta_8^3 + \eta_7 \eta_9^2 + \eta_4 \eta_5^2 \eta_6^3 \eta_{10}\rangle,$$

where we denote by $\eta_1, \ldots, \eta_{10}$ the generators of the Cox ring corresponding to the curves $E_1, \ldots, E_{10}$ on $X_{\overline{\mathbb{Q}}}$. The total coordinate space $\overline{X}_{\overline{\mathbb{Q}}} \subseteq \mathbb{A}^{10}$ is the zero set of

$$g(\eta_1, \ldots, \eta_{10}) := \eta_1^2 \eta_2 \eta_8^3 + \eta_7 \eta_9^2 + \eta_4 \eta_5^2 \eta_6^3 \eta_{10}.$$

This Cox ring was first computed in [162]. Because X is split and $E_1, \ldots, E_{10}$ are invariant under the Galois group $G := \mathrm{Gal}(\overline{\mathbb{Q}}/\mathbb{Q})$, Construction 6.1.3.3 shows that a Cox ring over $\mathbb{Q}$ is given by

$$\mathcal{R}(X) = \mathbb{Q}[\eta_1, \ldots, \eta_{10}]/\langle \eta_1^2 \eta_2 \eta_8^3 + \eta_7 \eta_9^2 + \eta_4 \eta_5^2 \eta_6^3 \eta_{10}\rangle.$$

6.5.3 Parameterization via Cox rings

Our first step in the proof of Manin's conjecture is the following parameterization of rational points outside the line on Y by integral points on its characteristic space, or universal torsor. The following result confirms Claim 6.4.3.1 in our case.

Proposition 6.5.3.1 *Let $T(B)$ be the set of all $(\eta_1, \ldots, \eta_{10}) \in \mathbb{Z}_{\neq 0}^7 \times \mathbb{Z}^3$ satisfying the Cox ring relation $g(\eta_1, \ldots, \eta_{10}) = 0$, the height condition $H(\Psi(\eta_1, \ldots, \eta_{10})) \leq B$ and the coprimality conditions*

$$\gcd(\eta_j, \eta_{j'}) = 1 \text{ if } E_j \cdot E_{j'} = 0 \tag{6.5.1}$$

for all $j \neq j' \in \{1, \ldots, 10\}$, encoded in the extended Dynkin diagram. Then the map

$$\Psi \colon \mathbb{Z}^{10} \to \mathbb{Z}^4$$

$$(\eta_1, \ldots, \eta_{10}) \mapsto (\eta_1^2 \eta_2^4 \eta_3^6 \eta_4^5 \eta_5^4 \eta_6^3 \eta_7^3, \, \eta_1^2 \eta_2^3 \eta_3^4 \eta_4^3 \eta_5^2 \eta_6 \eta_7^2 \eta_8, \, \eta_1 \eta_2^2 \eta_3^3 \eta_4^2 \eta_5 \eta_7^2 \eta_9, \, \eta_{10})$$

induces a 2^7-to-1 map

$$T(B) \to \{\mathbf{x} \in U(\mathbb{Q}); \ H(\mathbf{x}) \leq B\}.$$

Proof For $i = 0, \ldots, 6$, we claim that Ψ induces a 2^{i+1}-to-1 map from

$$M_i := \left\{ \begin{array}{l} (\eta_1, \ldots, \eta_{10}) \in \mathbb{Z}_{\neq 0}^7 \times \mathbb{Z}^3; \ \eta_{i+1} = \cdots = \eta_6 = 1, \ g(\eta_1, \ldots, \eta_{10}) = 0, \\ \gcd\{\eta_j; \ j \in J\} = 1 \text{ for all } J \subseteq \{1, \ldots, i, 7, \ldots, 10\} \text{ with } \bigcap_{j \in J} E_j^{(i)} = \varnothing \end{array} \right\}$$

as in (6.4.7) to $U(\mathbb{Q})$.

For $i = 0$, we note that ψ induces a 2-to-1 map from

$$\{(y_0, y_1, y_2) \in \mathbb{Z}^3; \ y_0 \neq 0, \ \gcd(y_0, y_1, y_2) = 0\} \to U(\mathbb{Q}),$$

using $U = \{x_0 \neq 0\} \cap Y$. Introducing new variables

$$(\eta_1, \ldots, \eta_{10}) := (1, 1, 1, 1, 1, 1, y_0, y_1, y_2, -y_0 y_2^2 - y_1^3)$$

and observing that it makes no difference whether we require $\gcd(\eta_7, \eta_9, \eta_{10}) = 1$ in addition to $\gcd(\eta_7, \eta_8, \eta_9) = 1$, we get the 2-to-1 map from $M_0 \to U(\mathbb{Q})$.

For $i = 1$, we replace η_1 by $\pm \gcd(\eta_7, \eta_8, \eta_{10})$ and $(\eta_7, \eta_8, \eta_{10})$ by $(\eta_7/\eta_1, \eta_8/\eta_1, \eta_{10}/\eta_1)$ and keep all other η_j unchanged. Note the similarity to our description of $\kappa_1 \colon X_1 \to X_0$. We check directly that the new $(\eta_1, \ldots, \eta_{10})$ still satisfy $g(\eta_1, \ldots, \eta_{10}) = 0$, and that Ψ is changed by a factor η_1 in each component, giving the same map to $U \subset \mathbb{P}_{\mathbb{Q}}^3$. The new coprimality conditions are $\gcd(\eta_1, \eta_9) = 1$, which holds because $\gcd(\eta_7, \ldots, \eta_{10}) = 1$ in the previous step, and $\gcd(\eta_7, \eta_8) = 1$, because otherwise $g(\eta_1, \ldots, \eta_{10}) = 0$ would contradict $\gcd(\eta_7, \eta_8, \eta_{10}) = 1$, which holds by construction. For a converse, given $(\eta_1, 1, \ldots, 1, \eta_7, \ldots, \eta_{10}) \in M_1$, we have $(1, \ldots, 1, \eta_1 \eta_7, \eta_1 \eta_8, \eta_9, \eta_1 \eta_{10}) \in M_0$, which is clearly inverse to the above.

For $i = 2$, in view of $\kappa_2 \colon X_2 \to X_1$, given

$$(\eta_1, 1, 1, 1, 1, 1, \eta_7, \eta_8, \eta_9, \eta_{10}) \in M_1,$$

we define $\eta_2 := \pm \gcd(\eta_1, \eta_7, \eta_{10})$ and obtain

$$(\eta_1/\eta_2, \eta_2, 1, 1, 1, 1, \eta_7/\eta_2, \eta_8, \eta_9, \eta_{10}/\eta_2) \in M_2.$$

Again we check that $g(\eta_1, \ldots, \eta_{10}) = 0$ and Ψ are essentially unchanged and that the result satisfies the new coprimality conditions encoded by the curves $E_j^{(2)}$ on X_2.

For $i = 3, \ldots, 6$, we continue analogously. In the final step, we note that it is enough to consider coprimality conditions just for all minimal J with $\bigcap_{j \in J} E_j = \emptyset$, and in view of the extended Dynkin diagram and the fact that $E_8 \cap E_9 \cap E_{10} \neq \emptyset$, these are precisely the sets $J = \{j, j'\}$ with $j \neq j'$ and $E_j \cdot E_{j'} = 0$. $\qquad \square$

6.5.4 Counting integral points on universal torsors

The second step in our proof of Manin's conjecture for the E_6 cubic surface is to estimate the number of integral points in our parameterization via Cox rings from the previous step, using analytical methods.

For $(\eta_1, \ldots, \eta_{10})$ satisfying the coprimality conditions (6.5.1), we observe that the coordinates of $\Psi(\eta_1, \ldots, \eta_{10})$ are coprime. Eliminating η_{10} using the Cox ring relation $g(\eta_1, \ldots, \eta_{10}) = 0$, this implies that $H(\Psi(\eta_1, \ldots, \eta_{10})) \leq B$ if and only if

$$h(\eta_1, \ldots, \eta_9) := \max \left\{ \begin{array}{l} |\eta_1^2 \eta_2^4 \eta_3^6 \eta_4^5 \eta_5^4 \eta_6^3 \eta_7^3|, \; |\eta_1^2 \eta_2^3 \eta_3^4 \eta_4^3 \eta_5^2 \eta_6 \eta_7^2 \eta_8|, \\[2mm] |\eta_1 \eta_2^2 \eta_3^3 \eta_4^2 \eta_5 \eta_7^2 \eta_9|, \; \left| \dfrac{\eta_1^2 \eta_2 \eta_8^3 + \eta_7 \eta_9^2}{\eta_4 \eta_5^2 \eta_6^3} \right| \end{array} \right\} \leq B.$$

For $q \in \mathbb{Z}_{\neq 0}$, we define $\varphi^*(q) := \frac{\varphi(q)}{q} = \prod_{p \mid q} \left(1 - \frac{1}{p}\right)$, where φ is the Euler function. The following result of the first summation over η_9, with dependent coordinate η_{10}, is Claim 6.4.4.1 for $i = 8$.

Lemma 6.5.4.1 *Write $\eta' := (\eta_1, \ldots, \eta_8)$. Let*

$$V_8(\eta'; B) := \int_{h(\eta', \eta_9) \leq B} \frac{d\eta_9}{|\eta_4 \eta_5^2 \eta_6^3|}$$

and

$$\vartheta_8(\eta') := \sum_{\substack{k \mid \eta_3 \eta_4 \eta_5 \\ \gcd(k, \eta_2 \eta_7) = 1}} \frac{\mu(k) \varphi^*(\eta_1 \eta_2 \eta_3)}{k \varphi^*(\gcd(\eta_3, k\eta_4))} \sum_{\substack{1 \leq \rho \leq k\eta_4 \eta_5^2 \eta_6^3 \\ \gcd(\rho, k\eta_4 \eta_5 \eta_6) = 1 \\ \eta_2 \eta_8 + \eta_7 \rho^2 \equiv 0 \mod k\eta_4 \eta_5^2 \eta_6^3}} 1$$

if η' satisfies the coprimality conditions (6.5.1) for all $j \neq j' \in \{1, \ldots, 8\}$, and $\vartheta_8(\eta') := 0$ otherwise. Then

$$N_{U,H}(B) = 2^{-7} \sum_{\eta' \in \mathbb{Z}_{\neq 0}^7 \times \mathbb{Z}} \vartheta_8(\eta') V_8(\eta'; B) + O(B(\log B)^2).$$

Proof We must apply the first step of the strategy from Section 6.4.3. This is done in some generality in [111, Prop. 2.4], which gives our result directly, but nevertheless, we carry out the details here. For $B \geq 2$ and $\eta' \in \mathbb{Z}_{\neq 0}^7 \times \mathbb{Z}$ satisfying the coprimality conditions among $\eta_1, \ldots, \eta_8$ from (6.5.1), we must estimate the number N_1 of all $(\eta_9, \eta_{10}) \in \mathbb{Z}^2$ satisfying the Cox ring relation $g(\eta_1, \ldots, \eta_{10}) = 0$, the height condition $h(\eta_1, \ldots, \eta_9) \leq B$ and

$$\gcd(\eta_9, \eta_1 \cdots \eta_6) = \gcd(\eta_{10}, \eta_1 \cdots \eta_5 \eta_7) = 1.$$

In view of $g(\eta_1, \ldots, \eta_{10}) = 0$ and the other coprimality conditions, these may be simplified to $\gcd(\eta_9, \eta_1\eta_2\eta_3) = \gcd(\eta_{10}, \eta_3\eta_4\eta_5) = 1$. By Möbius inversion, we have

$$N_1 = \sum_{k | \eta_3\eta_4\eta_5} \mu(k) \# \left\{ (\eta_9, \eta_{10}') \in \mathbb{Z}^2; \begin{array}{l} \eta_1^2\eta_2\eta_8^3 + \eta_7\eta_9^2 + \eta_4\eta_5^2\eta_6^3 k\eta_{10}' = 0, \\ \gcd(\eta_9, \eta_1\eta_2\eta_3) = 1, \ h(\eta) \leq B \end{array} \right\}$$

$$= \sum_{\substack{k | \eta_3\eta_4\eta_5 \\ \gcd(k, \eta_1\eta_2\eta_7\eta_8)=1}} \mu(k) \# \left\{ \eta_9 \in \mathbb{Z}; \begin{array}{l} \eta_1^2\eta_2\eta_8^3 + \eta_7\eta_9^2 \equiv 0 \mod k\eta_4\eta_5^2\eta_6^3, \\ \gcd(\eta_9, \eta_1\eta_2\eta_3) = 1, \ h(\eta) \leq B \end{array} \right\}$$

because the Cox ring relation $g(\eta_1, \ldots, \eta_9, k\eta_{10}') = 0$ determines η_{10}' uniquely if the congruence holds. Note that we could restrict the summation over k by the coprimality condition because for all other k, there is no η_9 satisfying the Cox ring relation, resp. the congruence, because of the coprimality conditions on $\eta_1, \ldots, \eta_9$.

Because η_9 does not appear linearly in the Cox ring relation $g(\eta_1, \ldots, \eta_{10})$, we must do the following intermediate step. Using $\gcd(k\eta_4\eta_5^2\eta_6^3, \eta_1\eta_2\eta_7\eta_8) = 1$, we see that, for each η_9 satisfying the congruence above, there is a unique $\rho \in \{1, \ldots, k\eta_4\eta_5^2\eta_6^3\}$ such that

$$\gcd(\rho, k\eta_4\eta_5\eta_6) = 1, \quad \eta_2\eta_8 + \eta_7\rho^2 \equiv 0 \mod k\eta_4\eta_5^2\eta_6^3,$$

$$\eta_9 \equiv \rho\eta_1\eta_8 \mod k\eta_4\eta_5^2\eta_6^3.$$

Using this and removing the coprimality condition on η_9 by Möbius inversion, we have

$$N_1 = \sum_{\substack{k | \eta_3\eta_4\eta_5 \\ \gcd(k, \eta_1\eta_2\eta_7\eta_8)=1}} \mu(k) \sum_{\substack{1 \leq \rho \leq k\eta_4\eta_5^2\eta_6^3 \\ \gcd(\rho, k\eta_4\eta_5\eta_6)=1 \\ \eta_2\eta_8 + \eta_7\rho^2 \equiv 0 \mod k\eta_4\eta_5^2\eta_6^3}} \sum_{\substack{k' | \eta_1\eta_2\eta_3 \\ \gcd(k', k\eta_4\eta_5\eta_6)=1}} \mu(k') N_1(k, k', \rho)$$

with

$$N_1(k, k', \rho) = \#\left\{\eta'_9 \in \mathbb{Z};\ k'\eta'_9 \equiv \rho\eta_1\eta_8 \mod k\eta_4\eta_5^2\eta_6^3,\ h(\eta', k'\eta'_9) \leq B\right\}.$$

Here, we could restrict the summation over k' by a coprimality condition because of the congruence and the condition $\gcd(\rho\eta_1\eta_8, k\eta_4\eta_5\eta_6) = 1$.

Because the height condition $h(\eta', k'\eta'_9) \leq B$ is equivalent to the requirement that η'_9 lies in intervals whose number is bounded uniformly with respect to η' and B, as discussed in Remark 6.4.4.3, we have

$$N_1(k, k', \rho) = \int_{h(\eta', k'\eta'_9) \leq B} \frac{d\eta'_9}{k\eta_4\eta_5^2\eta_6^3} + O(1).$$

The change of variables $\eta_9 := k'\eta'_9$ transforms the main term into $\vartheta_8(\eta')V_8(\eta'; B)$.

We sum the error term of $N_1(k, k', \rho)$ over k, k', ρ and the remaining variables η'. Both $N_1(k, k', \rho)$ and the integral and hence the error term vanish unless $|\eta_1^2\eta_2^3\eta_3^4\eta_4^3\eta_5^2\eta_6\eta_7^2\eta_8| \leq B$ and $|\eta_1|, \ldots, |\eta_7| \leq B$. Therefore, we can restrict the sum of $O(1)$ over k, k', ρ and $\eta_1, \ldots, \eta_8 \in \mathbb{Z}_{\neq 0}^7 \times \mathbb{Z}$ to these ranges and get a total error term

$$\ll \sum_{\eta_1, \ldots, \eta_8} 2^{\omega(\eta_1\eta_2\eta_3)+\omega(\eta_3\eta_4\eta_5)+\omega(\eta_3\eta_4\eta_5\eta_6)}$$

$$\ll \sum_{\eta_1, \ldots, \eta_7} \frac{2^{\omega(\eta_1\eta_2\eta_3)+\omega(\eta_3\eta_4\eta_5)+\omega(\eta_3\eta_4\eta_5\eta_6)} B}{|\eta_1^2\eta_2^3\eta_3^4\eta_4^3\eta_5^2\eta_6\eta_7^2|} \ll B(\log B)^2. \qquad \square$$

To establish Claim 6.4.4.1 for $i = 7$, we perform the second summation over η_8, replacing the sum in the main term of Lemma 6.5.4.1 by an integral. For the estimation of the error term of the second summation in Lemma 6.5.4.3, we will use the following auxiliary result.

Lemma 6.5.4.2 *Let $a, q \in \mathbb{Z}$ with $q > 0$ and $\gcd(a, q) = 1$. Let $t_1 \leq t_2 \in \mathbb{R}$. Then*

$$\sum_{\substack{1 \leq \rho \leq q \\ \gcd(\rho, q)=1}} \#\{\eta \in \mathbb{Z}; t_1 < \eta \leq t_2, \eta \equiv a\rho^2 \mod q\} = \varphi^*(q)(t_2 - t_1) + O_\varepsilon(q^{1/2+\varepsilon}).$$

Proof For the proof via exponential sums, see [68, Lemma 3, Lemma 5]. $\square$

Note that Lemma 6.5.4.2 is trivially true if we replace the error term $O_\varepsilon(q^{1/2+\varepsilon})$ by $O(q)$, but that this trivial bound on the error term is not sufficient for the proof of Lemma 6.5.4.3.

Lemma 6.5.4.3 *Write $\eta'' = (\eta_1, \ldots, \eta_7)$ and let*

$$V_7(\eta''; B) := \int_{\mathbb{R}} V_8(\eta'', \eta_8; B) \, d\eta_8$$

and $\vartheta_7 \in \Theta_7'(3)$ with local factors

$$\vartheta_{7,p}(I) := \begin{cases} 1, & I = \emptyset, \\ 1 - \frac{1}{p}, & I = \{1\}, \{6\}, \{7\}, \\ \left(1 - \frac{1}{p}\right)^2, & I = \{2\}, \{4\}, \{5\}, \{1,2\}, \{2,3\}, \\ & \{3,4\}, \{4,5\}, \{5,6\}, \{3,7\}, \\ \left(1 - \frac{1}{p}\right)\left(1 - \frac{2}{p}\right), & I = \{3\}, \\ 0, & \text{otherwise.} \end{cases}$$

Then

$$N_{U,H}(B) = 2^{-7} \sum_{\eta'' \in \mathbb{Z}_{\neq 0}^7} \vartheta_7(\eta'') V_7(\eta''; B) + O(B(\log B)^2).$$

Proof Starting from Lemma 6.5.4.1, we change the order of summation over ρ, η_8 and remove the coprimality condition $\gcd(\eta_8, \eta_2 \cdots \eta_7) = 1$ by Möbius inversion, writing $\eta_8 = k_8 \eta_8'$. Here we can restrict the sum over k_8 to $\gcd(k_8, k\eta_4 \eta_5^2 \eta_6^3) = 1$ because otherwise no η_8' satisfies the congruence $\eta_2 k_8 \eta_8' + \eta_7 \rho^2 \equiv 0 \mod k\eta_4 \eta_5^2 \eta_6^3$. Hence

$$N_{U,H}(B) = 2^{-7} \sum_{\eta'' \in \mathbb{Z}_{\neq 0}^7} \sum_{\substack{k | \eta_3 \eta_4 \eta_5 \\ \gcd(k, \eta_2 \eta_7) = 1}} \frac{\mu(k)\varphi^*(\eta_1 \eta_2 \eta_3)}{k\varphi^*(\gcd(\eta_3, k\eta_4))} \sum_{\substack{k_8 | \eta_2 \cdots \eta_7 \\ \gcd(k_8, k\eta_4 \eta_5^2 \eta_6^3) = 1}} A$$

with

$$A := \sum_{\substack{1 \leq \rho \leq k\eta_4 \eta_5^2 \eta_6^3 \\ \gcd(\rho, k\eta_4 \eta_5^2 \eta_6^3) = 1}} \sum_{\substack{\eta_8' \in \mathbb{Z} \\ \eta_2 k_8 \eta_8' + \eta_7 \rho^2 \equiv 0 \mod k\eta_4 \eta_5^2 \eta_6^3}} V_8(\eta'', k_8 \eta_8'; B).$$

Combining Lemma 6.5.4.2 with partial summation, we have

$$A = \frac{\varphi^*(k\eta_4 \eta_5^2 \eta_6^3)}{k_8} V_7(\eta''; B) + O_\varepsilon \left((k\eta_4 \eta_5^2 \eta_6^3)^{1/2+\varepsilon} \sup_{\eta_8} V_8(\eta'', \eta_8; B)\right).$$

For the main term, we observe that

$$\vartheta_7(\eta'') = \sum_{\substack{k | \eta_3 \eta_4 \eta_5 \\ \gcd(k, \eta_2 \eta_7) = 1}} \frac{\mu(k)\varphi^*(\eta_1 \eta_2 \eta_3)\varphi^*(k\eta_4 \eta_5^2 \eta_6^3)\varphi^*(\eta_2 \cdots \eta_7)}{k\varphi^*(\gcd(\eta_3, k\eta_4))\varphi^*(\gcd(\eta_2 \cdots \eta_7, k\eta_4 \eta_5^2 \eta_6^3))}.$$

We estimate the error term of A summed over k, k_8 and η'' as

$$\ll \sum_{\eta''} \frac{2^{\omega(\eta_3 \eta_4 \eta_5) + \omega(\eta_2 \cdots \eta_7)} |\eta_4 \eta_5^2 \eta_6^3|^{1/2+\varepsilon} B^{1/2}}{|\eta_4 \eta_5^2 \eta_6^3 \eta_7|^{1/2}}$$

$$\ll \sum_{\eta_2, \dots, \eta_7} \frac{2^{\omega(\eta_3 \eta_4 \eta_5) + \omega(\eta_2 \cdots \eta_6)} B}{|\eta_2|^2 |\eta_3|^3 |\eta_4|^{5/2-\varepsilon} |\eta_5|^{2-2\varepsilon} |\eta_6|^{3/2-3\varepsilon} |\eta_7|^2} \ll B,$$

using Lemma 6.4.4.6 to estimate sup $V_8(\eta'', \eta_8; B)$, the first height condition $|\eta_1^2 \eta_2^4 \eta_3^6 \eta_4^5 \eta_5^4 \eta_6^3 \eta_7^3| \le B$ for the sum over η_1, and Lemma 6.4.4.2. $\square$

Finally, we prove Claim 6.4.4.1 for $i = 0$.

Lemma 6.5.4.4 *Let*

$$V_0(B) := \int_{|\eta_1|,\dots,|\eta_7| \ge 1} V_7(\eta_1, \dots, \eta_7; B) \, d\eta_7 \dots d\eta_1$$

and

$$\vartheta_0 := \prod_p \left(1 - \frac{1}{p}\right)^7 \left(1 + \frac{7}{p} + \frac{1}{p^2}\right).$$

Then

$$N_{U,H}(B) = 2^{-7} \vartheta_0 V_0(B) + O(B(\log B)^5 \log \log B).$$

Proof Applying Lemma 6.4.4.6 with the last part of our height condition $h(\eta_1, \dots, \eta_9) \le B$ gives

$$V_2(\eta_1, \dots, \eta_7; B) \ll \frac{B}{|\eta_1 \cdots \eta_7|} \left(\frac{B}{|\eta_1^2 \eta_2^4 \eta_3^6 \eta_4^5 \eta_5^4 \eta_6^3 \eta_7^3|}\right)^{-1/6}.$$

By the first part of the height condition, we have $V_2(\eta_1, \dots, \eta_7; B) = 0$ unless $|\eta_1^2 \eta_2^4 \eta_3^6 \eta_4^5 \eta_5^4 \eta_6^3 \eta_7^3| \le B$ and $1 \le |\eta_i| \le B$ for $i = 1, \dots, 7$. Therefore, we may apply Corollary 6.4.4.8. $\square$

6.5.5 Interpretation of the integral

To complete the proof of Manin's conjecture for our E_6 cubic surface, we must show that the integral $V_0(B)$ in the outcome of our counting of points on universal torsors as in Lemma 6.5.4.4 grows asymptotically as a constant multiple of $B(\log B)^6$. To this end, we prove Claim 6.4.5.1 for our cubic surface.

Lemma 6.5.5.1 *We have*

$$2^{-7} V_0(B) = 6220800^{-1} \omega_\infty B(\log B)^6 + O(B(\log B)^5).$$

Proof The substitutions $x_0 = y_0^3$, $x_1 = y_0^2 y_1$, $x_2 = y_0^2 y_1$ and

$$x_0 = \eta_1^2 \eta_2^4 \eta_3^6 \eta_4^5 \eta_5^4 \eta_6^3 \eta_7^3 / B, \quad x_1 = \eta_1^2 \eta_2^3 \eta_3^4 \eta_4^3 \eta_5^2 \eta_6 \eta_7^2 \eta_8 / B, \quad x_2 = \eta_1 \eta_2^2 \eta_3^3 \eta_4^2 \eta_5 \eta_7^2 \eta_9 / B$$

show that

$$\omega_\infty = \frac{1}{2} \int_{|x_0|,|x_1|,|x_2|,|(x_0 x_2^2 + x_1^3)/x_0^2| \le 1} \frac{dx_0\, dx_1\, dx_2}{|x_0^2|} = \frac{3|\eta_1 \ldots \eta_6|}{2B} \int \frac{d\eta_7\, d\eta_8\, d\eta_9}{|\eta_4 \eta_5^2 \eta_6^3|},$$

confirming (6.4.11), where the latter integral is over the domain $h(\eta_1, \ldots, \eta_9) \le B$.

Now we could apply Lemma 6.4.5.4 and Theorem 6.4.5.3, but here it is equally straightforward to compute directly, using the substitutions $\eta_i = B^{t_i}$ for $i = 1, \ldots, 6$, that

$$\frac{(\log B)^6}{6220800} = \frac{(\log B)^6}{3} \int_{\substack{2t_1 + 4t_2 + 6t_3 + 5t_4 + 4t_5 + 3t_6 \le 1 \\ t_1, \ldots, t_6 \ge 0}} dt_1 \ldots dt_6$$

$$= \frac{1}{3 \cdot 2^6} \int_{\substack{|\eta_1^2 \eta_2^4 \eta_3^6 \eta_4^5 \eta_5^4 \eta_6^3| \le B \\ |\eta_1|, \ldots, |\eta_6| \ge 1}} \frac{d\eta_1 \cdots d\eta_6}{|\eta_1 \cdots \eta_6|},$$

because the integral over $t_1, \ldots, t_6$ is the easily computable volume of a scaled six-dimensional simplex.

Combining these shows that $6220800^{-1} \omega_\infty B (\log B)^6 = V_0'(B)$ where $V_0'(B)$ is defined as $V_0(B)$ with $|\eta_7| \ge 1$ replaced by $|\eta_1^2 \eta_2^4 \eta_3^6 \eta_4^5 \eta_5^4 \eta_6^3| \le B$. Because $h(\eta_1, \ldots, \eta_9) \le B$ and $|\eta_7| \ge 1$ imply that $|\eta_1^2 \eta_2^4 \eta_3^6 \eta_4^5 \eta_5^4 \eta_6^3| \le B$, we can estimate $|V_3(B) - V_3'(B)|$ as

$$\ll \int_{\substack{|\eta_1|, \ldots, |\eta_6| \ge 1 \\ |\eta_1^2 \eta_2^4 \eta_3^6 \eta_4^5 \eta_5^4 \eta_6^3| \le B \\ 0 \le |\eta_7| \le 1}} \frac{B}{|\eta_1 \cdots \eta_7|} \left(\frac{B}{|\eta_1^2 \eta_2^4 \eta_3^6 \eta_4^5 \eta_5^4 \eta_6^3 \eta_7^3|} \right)^{-1/6} \ll B(\log B)^5,$$

confirming (6.4.12). □

In view of Lemma 6.5.4.4, this completes the proof of Manin's conjecture for the E_6 cubic surface as in Theorem 6.5.1.1, with a total error term of order $B(\log B)^5 \log \log B$.

Exercises to Chapter 6

Exercise 6.1 (Rational points)   Let $\mathbb{K}$ be a field of characteristic zero, with algebraic closure $\overline{\mathbb{K}}$. Given polynomials $f_1, \ldots, f_r \in \mathbb{K}[T_1, \ldots, T_n]$, consider the $\mathbb{K}$-algebra

$$A := \mathbb{K}[T_1, \ldots, T_n]/\langle f_1, \ldots, f_r \rangle$$

and the associated $\mathbb{K}$-variety $X := \operatorname{Spec} A$. Show that there are canonical bijections

$$X(\mathbb{K}) \cong \{x \in X;\ x \text{ closed in } X,\ \mathbb{K}(x) = \mathbb{K}\} \cong V(\overline{\mathbb{K}}^n; f_1, \ldots, f_r) \cap \mathbb{K}^n.$$

Exercise 6.2 (Rational points of algebraic groups) Let H be an algebraic group over a field $\mathbb{K}$ of characteristic zero. Show that the structure of H as an algebraic group induces the structure of an abstract group on the set $H(\mathbb{K})$ of rational points on H.

Exercise 6.3 (Split tori) Let H be a torus over a field $\mathbb{K}$ of characteristic zero. Show that there is a finite Galois extension $\mathbb{L} \supseteq \mathbb{K}$ such that $H_{\mathbb{L}} \cong \mathbb{G}_{m,\mathbb{L}}^n$ holds.

Exercise 6.4 (Norm 1 tori) Let $\mathbb{L} \supset \mathbb{K}$ be a finite extension of fields of characteristic zero, with $n := [\mathbb{L} : \mathbb{K}]$. Let $N_{\mathbb{L}/\mathbb{K}}$ be a norm form associated with this extension, and let $H \subset \mathbb{A}_{\mathbb{K}}^n$ be defined by $N_{\mathbb{L}/\mathbb{K}}(z_1, \ldots, z_n) = 1$. Show that H is a torus over $\mathbb{K}$, compute the Galois module $\mathbb{X}(H)$, and compare to the Weil restriction $R_{\mathbb{L}/\mathbb{K}}(\mathbb{G}_{m,\mathbb{L}})$.

Exercise 6.5 (Local description of universal torsors) Let $\mathbb{K}$ be a number field, and let $\mathbb{L}$ and $\mathbb{M}$ be finite extensions of $\mathbb{K}$, with $l := [\mathbb{L} : \mathbb{K}]$ and $m := [\mathbb{M} : \mathbb{K}]$. Let $X \subset \mathbb{A}_{\mathbb{K}}^{l+m}$ be defined by

$$N_{\mathbb{L}/\mathbb{K}}(z_1, \ldots, z_l) = N_{\mathbb{M}/\mathbb{K}}(z_{l+1}, \ldots, z_{l+m}).$$

Determine local equations for universal torsors over X, and deduce that the Brauer–Manin obstruction to the Hasse principle and weak approximation is the only one on smooth proper models of X.

Exercise 6.6 (Cox rings over nonclosed fields) In the situation of Constructions 6.1.3.3 and 6.1.3.8 over a field $\mathbb{K}$ of characteristic zero that is not algebraically closed, show that the Cox sheaves, Cox rings, and characteristic spaces are generally not unique up to isomorphism, but may depend on the choice of a Galois-equivariant identifying character χ.

Exercise 6.7 (Weak approximation on $\mathbb{P}_{\mathbb{Q}}^n$) Use the Chinese remainder theorem to show that $\mathbb{P}_{\mathbb{Q}}^n$ satisfies weak approximation.

Exercise 6.8 (Cox rings and universal torsors) Let X be a geometrically integral and normal variety over a field $\mathbb{K}$ of characteristic zero, and $p : \mathcal{X} \to X$ a universal torsor. Show that $p_*(\mathcal{O}_{\mathcal{X}})$ is a Cox sheaf on X, that $\Gamma(\mathcal{X}, \mathcal{O}_{\mathcal{X}})$ is a Cox ring of X, and that $p : \mathcal{X} \to X$ is a characteristic space over X.

Exercise 6.9 (Manin's conjecture for $\mathbb{P}_{\mathbb{K}}^n$) Prove Manin's conjecture for $\mathbb{P}_{\mathbb{K}}^n$, where $\mathbb{K}$ is the field $\mathbb{Q}(\sqrt{-1})$ of Gaussian rational numbers or any other imaginary quadratic number field.

Exercise 6.10* (Cox rings of nonsplit del Pezzo surfaces) Construct examples of Cox rings of nonsplit weak del Pezzo surfaces over a field $\mathbb{K}$ of characteristic zero that is not algebraically closed.

Exercise 6.11* (Picard quotient presentation over nonclosed fields) Let X be geometrically irreducible variety over a field $\mathbb{K}$ of characteristic zero with finitely generated $\mathrm{Pic}(X_{\overline{\mathbb{K}}})$. Generalize Exercise 4.8 analogously to Constructions 6.1.3.3 and 6.1.3.8. Under which conditions do we obtain a Picard quotient presentation $\widetilde{X} \to X$ over $\mathbb{K}$ that is a universal torsor over X?

Exercise 6.12* (Manin's conjecture over $\mathbb{Q}$ for a family of intrinsic quadrics) For $n \geq 3$, let X_n be the smooth projective variety constructed in [62, Sec. 4] over $\mathbb{Q}$, with Cox ring

$$\mathcal{R}_X(X) \cong \mathbb{Q}[S_0, S_1, \ldots, S_n, T_1, \ldots, T_n]/(S_1 T_1 + \cdots + S_n T_n).$$

Because X_n is an equivariant compactification of $\mathbb{G}_{a,\mathbb{Q}}^{n-1}$, Manin's conjecture for X_n holds by techniques from harmonic analysis [88]. Give a new proof of Manin's conjecture for X_n over $\mathbb{Q}$ using the universal torsor method.

Exercise 6.13* Let $\mathbb{K}$ be a number field. Prove Manin's conjecture for some nonsplit del Pezzo surfaces over $\mathbb{K}$, using a parameterization of rational points via Cox rings constructed in Exercise 6.10.

Exercise 6.14* (Manin's conjecture for $\mathbb{K}^*$-surfaces) Let $\mathbb{K}$ be a number field. Prove Manin's conjecture for del Pezzo $\mathbb{K}^*$-surfaces with at most rational double points as singularities using the $\mathbb{K}^*$-action. Consider also wider classes of singular del Pezzo $\mathbb{K}^*$-surfaces and higher dimensional almost Fano varieties with a torus action of complexity 1.

Bibliography

[1] A. A'Campo-Neuen. Note on a counterexample to Hilbert's fourteenth problem given by P. Roberts. *Indag. Math. (N.S.)*, 5(3):253–257, 1994.

[2] A. A'Campo-Neuen and J. Hausen. Examples and counterexamples for existence of categorical quotients. *J. Algebra*, 231(1):67–85, 2000.

[3] A. A'Campo-Neuen and J. Hausen. Toric prevarieties and subtorus actions. *Geom. Dedicata*, 87(1–3):35–64, 2001.

[4] A. G. Aleksandrov and B. Z. Moroz. Complete intersections in relation to a paper of B. J. Birch. *Bull. London Math. Soc.*, 34(2):149–154, 2002.

[5] V. Alexeev and V. V. Nikulin. Del Pezzo and K3 surfaces. MSJ Memoirs, Vol. 15. Mathematical Society of Japan, Tokyo, 2006.

[6] D. F. Anderson. Graded Krull domains. *Comm. Algebra*, 7(1):79–106, 1979.

[7] E. Arbarello, M. Cornalba, P. A. Griffiths, and J. Harris. *Geometry of algebraic curves. Vol. I*, Grundlehren der Mathematischen Wissenschaften, Vol. 267. Springer-Verlag, New York, 1985.

[8] M. Artebani, A. Garbagnati, and A. Laface. Cox rings of extremal rational elliptic surfaces. arXiv:1302.4361, 2013.

[9] M. Artebani, J. Hausen, and A. Laface. On Cox rings of K3 surfaces. *Compos. Math.*, 146(4):964–998, 2010.

[10] M. Artebani and A. Laface. Cox rings of surfaces and the anticanonical Iitaka dimension. *Adv. Math.*, 226(6):5252–5267, 2011.

[11] M. Artebani and A. Laface. Hypersurfaces in Mori dream spaces. *J. Algebra*, 371:26–37, 2012.

[12] M. Artin. Some numerical criteria for contractability of curves on algebraic surfaces. *Amer. J. Math.*, 84:485–496, 1962.

[13] I. V. Arzhantsev. On the factoriality of Cox rings. *Mat. Zametki*, 85(5):643–651, 2009.

[14] I. V. Arzhantsev. Projective embeddings with a small boundary for homogeneous spaces. *Izv. Ross. Akad. Nauk Ser. Mat.*, 73(3):5–22, 2009.

[15] I. V. Arzhantsev, D. Celik, and J. Hausen. Factorial algebraic group actions and categorical quotients. *J. Algebra*, 387:87–98, 2013.

[16] I. V. Arzhantsev and S. A. Gaïfullin. Cox rings, semigroups, and automorphisms of affine varieties. *Mat. Sb.*, 201(1):3–24, 2010.

[17] I. V. Arzhantsev and S. A. Gaïfullin. Homogeneous toric varieties. *J. Lie Theory*, 20(2):283–293, 2010.

[18] I. V. Arzhantsev and J. Hausen. On embeddings of homogeneous spaces with small boundary. *J. Algebra*, 304(2):950–988, 2006.

[19] I. V. Arzhantsev and J. Hausen. On the multiplication map of a multigraded algebra. *Math. Res. Lett.*, 14(1):129–136, 2007.

[20] I. V. Arzhantsev and J. Hausen. Geometric invariant theory via Cox rings. *J. Pure Appl. Algebra*, 213(1):154–172, 2009.

[21] I. V. Arzhantsev, J. Hausen, E. Herppich, and A. Liendo. The automorphism group of a variety with torus action of complexity one. *Mosc. Math. J.*, 14(3):429–471, 2014.

[22] M. F. Atiyah and I. G. Macdonald. *Introduction to commutative algebra*. Addison-Wesley, Reading, MA-London-Don Mills, Ont., 1969.

[23] M. Audin. *The topology of torus actions on symplectic manifolds*. Progress in Mathematics, Vol. 93. Birkhäuser Verlag, Basel, 1991.

[24] R. Avdeev. An epimorphic subgroup arising from Roberts' counterexample. *Indag. Math. (N.S.)*, 23(1–2):10–18, 2012.

[25] S. Baier and T. D. Browning. Inhomogeneous cubic congruences and rational points on del Pezzo surfaces. *J. reine angew. Math.*, 680:69–151, 2013.

[26] S. Baier and U. Derenthal. Quadratic congruences on average and rational points on cubic surfaces. arXiv:1205.0373, 2012.

[27] H. Bäker. Good quotients of Mori dream spaces. *Proc. Amer. Math. Soc.*, 139(9):3135–3139, 2011.

[28] H. Bäker, J. Hausen, and S. Keicher. On Chow quotients of torus actions. arXiv:1203.3759, 2012.

[29] W. P. Barth, K. Hulek, C. A. M. Peters, and A. Van de Ven. *Compact complex surfaces*, Ergebnisse der Mathematik und ihrer Grenzgebiete (3), Vol. 4, 2nd ed. Springer-Verlag, Berlin, 2004.

[30] V. V. Batyrev. Quantum cohomology rings of toric manifolds. *Astérisque*, (218):9–34, 1993. Journées de Géométrie Algébrique d'Orsay (Orsay, 1992).

[31] V. V. Batyrev and F. Haddad. On the geometry of SL(2)-equivariant flips. *Mosc. Math. J.*, 8(4):621–646, 846, 2008.

[32] V. V. Batyrev and Yu. I. Manin. Sur le nombre des points rationnels de hauteur borné des variétés algébriques. *Math. Ann.*, 286(1–3):27–43, 1990.

[33] V. V. Batyrev and O. N. Popov. The Cox ring of a del Pezzo surface. In *Arithmetic of higher-dimensional algebraic varieties* (Palo Alto, CA, 2002). Progress in Mathematics, Vol. 226, pp. 85–103. Birkhäuser Boston, Boston, 2004.

[34] V. V. Batyrev and Yu. Tschinkel. Rational points of bounded height on compactifications of anisotropic tori. *Int. Math. Res. Not. IMRN*, (12):591–635, 1995.

[35] V. V. Batyrev and Yu. Tschinkel. Rational points on some Fano cubic bundles. *C. R. Acad. Sci. Paris Sér. I Math.*, 323(1):41–46, 1996.

[36] V. V. Batyrev and Yu. Tschinkel. Manin's conjecture for toric varieties. *J. Algebraic Geom.*, 7(1):15–53, 1998.

[37] V. V. Batyrev and Yu. Tschinkel. Tamagawa numbers of polarized algebraic varieties. *Astérisque*, (251):299–340, 1998. Nombre et répartition de points de hauteur bornée (Paris, 1996).

[38] S. Bauer. Die Manin-Vermutung für eine del-Pezzo-Fläche, Master's thesis, Universität München, 2013.

[39] A. Beauville. *Complex algebraic surfaces*. London Mathematical Society Student Texts, Vol. 34, 2nd ed. Cambridge University Press, Cambridge, 1996.

[40] B. Bechtold. Factorially graded rings and Cox rings. *J. Algebra*, 369:351–359, 2012.

[41] O. Benoist. Quasi-projectivity of Normal Varieties. *Int. Math. Res. Not. IMRN*, (17):3878–3885, 2013.

[42] F. Berchtold and J. Hausen. Homogeneous coordinates for algebraic varieties. *J. Algebra*, 266(2):636–670, 2003.

[43] F. Berchtold and J. Hausen. Bunches of cones in the divisor class group—a new combinatorial language for toric varieties. *Int. Math. Res. Not. IMRN*, (6):261–302, 2004.

[44] F. Berchtold and J. Hausen. Cox rings and combinatorics. *Trans. Amer. Math. Soc.*, 359(3):1205–1252 (electronic), 2007.

[45] A. Białynicki-Birula. Finiteness of the number of maximal open subsets with good quotients. *Transform. Groups*, 3(4):301–319, 1998.

[46] A. Białynicki-Birula, J. B. Carrell, and W. M. McGovern. *Algebraic quotients. Torus actions and cohomology. The adjoint representation and the adjoint action.* Invariant Theory and Algebraic Transformation Groups, II. Encyclopaedia of Mathematical Sciences, Vol. 131. Springer-Verlag, Berlin, 2002.

[47] A. Białynicki-Birula and J. Święcicka. Three theorems on existence of good quotients. *Math. Ann.*, 307(1):143–149, 1997.

[48] A. Białynicki-Birula and J. Święcicka. A recipe for finding open subsets of vector spaces with a good quotient. *Colloq. Math.*, 77(1):97–114, 1998.

[49] F. Bien and A. Borel. Sous-groupes épimorphiques de groupes linéaires algébriques. I. *C. R. Acad. Sci. Paris Sér. I Math.*, 315(6):649–653, 1992.

[50] F. Bien and A. Borel. Sous-groupes épimorphiques de groupes linéaires algébriques. II. *C. R. Acad. Sci. Paris Sér. I Math.*, 315(13):1341–1346, 1992.

[51] F. Bien, A. Borel, and J. Kollár. Rationally connected homogeneous spaces. *Invent. Math.*, 124(1–3):103–127, 1996.

[52] B. J. Birch. Forms in many variables. *Proc. Roy. Soc. Ser. A*, 265:245–263, 1961/1962.

[53] C. Birkar, P. Cascini, C. D. Hacon, and J. McKernan. Existence of minimal models for varieties of log general type. *J. Amer. Math. Soc.*, 23(2):405–468, 2010.

[54] V. Blomer and J. Brüdern. The density of rational points on a certain threefold. In *Contributions in analytic and algebraic number theory*. Springer Proceedings in Mathematics, Vol. 9, pp. 1–15. Springer-Verlag, New York, 2012.

[55] V. Blomer, J. Brüdern, and P. Salberger. On a senary cubic form. *Proc. Lond. Math. Soc. (3)*, 108(4):911–964, 2014.

[56] A. Borel. *Linear algebraic groups*. Graduate Texts in Mathematics, Vol. 126, 2nd ed. Springer-Verlag, New York, 1991.

[57] M. Borovoi. The Brauer-Manin obstructions for homogeneous spaces with connected or abelian stabilizer. *J. reine angew. Math.*, 473:181–194, 1996.

[58] N. Bourbaki. *Commutative algebra*. Chapters 1–7. Elements of Mathematics (Berlin). Springer-Verlag, Berlin, 1998.

[59] D. Bourqui. Fonction zêta des hauteurs des variétés toriques déployées dans le cas fonctionnel. *J. reine angew. Math.*, 562:171–199, 2003.

[60] D. Bourqui. Comptage de courbes sur le plan projectif éclaté en trois points alignés. *Ann. Inst. Fourier (Grenoble)*, 59(5):1847–1895, 2009.

[61] D. Bourqui. Fonction zêta des hauteurs des variétés toriques non déployées. *Mem. Amer. Math. Soc.*, 211(994):viii+151, 2011.

[62] D. Bourqui. La conjecture de Manin géométrique pour une famille de quadriques intrinsèques. *Manuscripta Math.*, 135(1–2):1–41, 2011.

[63] D. Bourqui. Exemples de comptage de courbes sur les surfaces. *Math. Ann.*, 357(4):1291–1327, 2013.

[64] J.-F. Boutot. Singularités rationnelles et quotients par les groupes réductifs. *Invent. Math.*, 88(1):65–68, 1987.

[65] H. Brenner and S. Schröer. Ample families, multihomogeneous spectra, and algebraization of formal schemes. *Pacific J. Math.*, 208(2):209–230, 2003.

[66] R. de la Bretèche. Compter des points d'une variété torique. *J. Number Theory*, 87(2):315–331, 2001.

[67] R. de la Bretèche. Répartition des points rationnels sur la cubique de Segre. *Proc. Lond. Math. Soc. (3)*, 95(1):69–155, 2007.

[68] R. de la Bretèche and T. D. Browning. On Manin's conjecture for singular del Pezzo surfaces of degree 4. I. *Michigan Math. J.*, 55(1):51–80, 2007.

[69] R. de la Bretèche and T. D. Browning. On Manin's conjecture for singular del Pezzo surfaces of degree four. II. *Math. Proc. Cambridge Philos. Soc.*, 143(3):579–605, 2007.

[70] R. de la Bretèche and T. D. Browning. Manin's conjecture for quartic del Pezzo surfaces with a conic fibration. *Duke Math. J.*, 160(1):1–69, 2011.

[71] R. de la Bretèche and T. D. Browning. Binary forms as sums of two squares and Châtelet surfaces. *Israel J. Math.*, 191(2):973–1012, 2012.

[72] R. de la Bretèche, T. D. Browning, and U. Derenthal. On Manin's conjecture for a certain singular cubic surface. *Ann. Sci. École Norm. Sup. (4)*, 40(1):1–50, 2007.

[73] R. de la Bretèche, T. D. Browning, and E. Peyre. On Manin's conjecture for a family of Châtelet surfaces. *Ann. of Math. (2)*, 175(1):297–343, 2012.

[74] R. de la Bretèche and É. Fouvry. L'éclaté du plan projectif en quatre points dont deux conjugués. *J. reine angew. Math.*, 576:63–122, 2004.

[75] R. de la Bretèche and G. Tenenbaum. Sur la conjecture de Manin pour certaines surfaces de Châtelet. *J. Inst. Math. Jussieu*, 12(4):759–819, 2013.

[76] M. Brion. The total coordinate ring of a wonderful variety. *J. Algebra*, 313(1):61–99, 2007.

[77] M. Brion and S. Kumar. *Frobenius splitting methods in geometry and representation theory*. Progress in Mathematics, Vol. 231. Birkhäuser Boston, Boston, 2005.

[78] T. D. Browning. *Quantitative arithmetic of projective varieties*. Progress in Mathematics, Vol. 277. Birkhäuser Verlag, Basel, 2009.

[79] T. D. Browning and U. Derenthal. Manin's conjecture for a cubic surface with D_5 singularity. *Int. Math. Res. Not. IMRN*, (14):2620–2647, 2009.

[80] T. D. Browning and U. Derenthal. Manin's conjecture for a quartic del Pezzo surface with A_4 singularity. *Ann. Inst. Fourier (Grenoble)*, 59(3):1231–1265, 2009.

[81] T. D. Browning and D. R. Heath-Brown. Quadratic polynomials represented by norm forms. *Geom. Funct. Anal.*, 22(5):1124–1190, 2012.

[82] T. D. Browning and L. Matthiesen. Norm forms for arbitrary number fields as products of linear polynomials. arXiv:1307.7641, 2013.

[83] T. D. Browning, L. Matthiesen, and A. N. Skorobogatov. Rational points on pencils of conics and quadrics with many degenerate fibres. *Ann. of Math.* 180(1):381–402, 2014.

[84] J. W. Bruce and C. T. C. Wall. On the classification of cubic surfaces. *J. London Math. Soc. (2)*, 19(2):245–256, 1979.

[85] W. Bruns and J. Gubeladze. Polytopal linear groups. *J. Algebra*, 218(2):715–737, 1999.

[86] A.-M. Castravet and J. Tevelev. Hilbert's 14th problem and Cox rings. *Compos. Math.*, 142(6):1479–1498, 2006.

[87] D. Celik. A categorical quotient in the category of dense constructible subsets. *Colloq. Math.*, 116(2):147–151, 2009.

[88] A. Chambert-Loir and Yu. Tschinkel. On the distribution of points of bounded height on equivariant compactifications of vector groups. *Invent. Math.*, 148(2):421–452, 2002.

[89] R. Chirivì, P. Littelmann, and A. Maffei. Equations defining symmetric varieties and affine Grassmannians. *Int. Math. Res. Not. IMRN*, (2):291–347, 2009.

[90] R. Chirivì and A. Maffei. The ring of sections of a complete symmetric variety. *J. Algebra*, 261(2):310–326, 2003.

[91] J.-L. Colliot-Thélène. Points rationnels sur les fibrations. In *Higher dimensional varieties and rational points* (Budapest, 2001), Bolyai Society Mathematical Studies, Vol. 12, pp. 171–221. Springer-Verlag, Berlin, 2003.

[92] J.-L. Colliot-Thélène. Lectures on linear algebraic groups, Morning Side Centre, Beijing, http://www.math.u-psud.fr/~colliot,2007.

[93] J.-L. Colliot-Thélène, D. Harari, and A. N. Skorobogatov. Valeurs d'un polynôme à une variable représentées par une norme. In *Number theory and algebraic geometry*. London Mathematical Society Lecture Note Series, Vol. 303, pp. 69–89. Cambridge University Press, Cambridge, 2003.

[94] J.-L. Colliot-Théléne, A. Pál, and A. N. Skorobogatov. Pathologies of the Brauer-Manin obstruction. arXiv:1310.5055, 2013.

[95] J.-L. Colliot-Thélène and P. Salberger. Arithmetic on some singular cubic hypersurfaces. *Proc. London Math. Soc. (3)*, 58(3):519–549, 1989.

[96] J.-L. Colliot-Thélène and J.-J. Sansuc. Torseurs sous des groupes de type multiplicatif; applications à l'étude des points rationnels de certaines variétés algébriques. *C. R. Acad. Sci. Paris Sér. A-B*, 282(18):Aii, A1113–A1116, 1976.

[97] J.-L. Colliot-Thélène and J.-J. Sansuc. La descente sur une variété rationnelle définie sur un corps de nombres. *C. R. Acad. Sci. Paris Sér. A-B*, 284(19):A1215–A1218, 1977.

[98] J.-L. Colliot-Thélène and J.-J. Sansuc. Variétés de première descente attachées aux variétés rationnelles. *C. R. Acad. Sci. Paris Sér. A-B*, 284(16):A967–A970, 1977.

[99] J.-L. Colliot-Thélène and J.-J. Sansuc. La descente sur les variétés rationnelles. In *Journées de géometrie algébrique d'Angers, Juillet 1979/Algebraic geometry, Angers, 1979*, pp. 223–237. Sijthoff & Noordhoff, Alphen aan den Rijn, 1980.

[100] J.-L. Colliot-Thélène and J.-J. Sansuc. La descente sur les variétés rationnelles. II. *Duke Math. J.*, 54(2):375–492, 1987.

[101] J.-L. Colliot-Thélène, J.-J. Sansuc, and P. Swinnerton-Dyer. Intersections of two quadrics and Châtelet surfaces. I. *J. reine angew. Math.*, 373:37–107, 1987.

[102] J.-L. Colliot-Thélène, J.-J. Sansuc, and P. Swinnerton-Dyer. Intersections of two quadrics and Châtelet surfaces. II. *J. reine angew. Math.*, 374:72–168, 1987.

[103] F. R. Cossec and I. V. Dolgachev. *Enriques surfaces. I.* Progress in Mathematics, Vol. 76. Birkhäuser Boston, Boston, 1989.

[104] D. A. Cox. The homogeneous coordinate ring of a toric variety. *J. Algebraic Geom.*, 4(1):17–50, 1995.

[105] D. A. Cox, J. B. Little, and H. K. Schenck. *Toric varieties*. Graduate Studies in Mathematics, Vol. 124. American Mathematical Society, Providence, RI, 2011.

[106] C. De Concini and C. Procesi. Complete symmetric varieties. In *Invariant theory* (Montecatini, 1982). Lecture Notes in Mathematics, Vol. 996, pp. 1–44. Springer-Verlag, Berlin, 1983.

[107] J. A. De Loera, J. Rambau, and F. Santos. *Triangulations.* Algorithms and Computation in Mathematics, Vol. 25. Springer-Verlag, Berlin, 2010. Structures for algorithms and applications.

[108] M. Demazure. Sous-groupes algébriques de rang maximum du groupe de Cremona. *Ann. Sci. École Norm. Sup. (4)*, 3:507–588, 1970.

[109] U. Derenthal. *Geometry of universal torsors*. PhD thesis, Universität Göttingen, 2006.

[110] U. Derenthal. Universal torsors of del Pezzo surfaces and homogeneous spaces. *Adv. Math.*, 213(2):849–864, 2007.

[111] U. Derenthal. Counting integral points on universal torsors. *Int. Math. Res. Not. IMRN*, (14):2648–2699, 2009.

[112] U. Derenthal. Manin's conjecture for a quintic del Pezzo surface with A_2 singularity. arXiv:0710.1583, 2007.

[113] U. Derenthal. Singular Del Pezzo surfaces whose universal torsors are hypersurfaces. *Proc. Lond. Math. Soc. (3)*, 108(3):638–681, 2014.

[114] U. Derenthal. Manin's conjecture for a certain singular cubic surface. arXiv:math.NT/0504016, 2005.

[115] U. Derenthal, A.-S. Elsenhans, and J. Jahnel. On the factor alpha in Peyre's constant. *Math. Comp.*, 83(286):965–977, 2014.

[116] U. Derenthal and C. Frei. Counting imaginary quadratic points via universal torsors. *Compositio Math.*, in press, arXiv:1302.6151, 2013.

[117] U. Derenthal and C. Frei. Counting imaginary quadratic points via universal torsors, II. *Math. Proc. Cambridge Philos. Soc.*, 156(3):383–407, 2014.

[118] U. Derenthal and C. Frei. On Manin's conjecture for a certain singular cubic surface over imaginary quadratic fields. *Int. Math. Res. Not. IMRN*, in press, arXiv:1311.2809, 2013.

[119] U. Derenthal, M. Joyce, and Z. Teitler. The nef cone volume of generalized del Pezzo surfaces. *Algebra Number Theory*, 2(2):157–182, 2008.

[120] U. Derenthal and D. Loughran. Singular del Pezzo surfaces that are equivariant compactifications. *Zap. Nauchn. Sem. S.-Peterburg. Otdel. Mat. Inst. Steklov. (POMI)*, 377(Issledovaniya po Teorii Chisel. 10):26–43, 241, 2010.

[121] U. Derenthal and M. Pieropan. Cox rings over nonclosed fields, preprint, 2014.

[122] U. Derenthal, A. Smeets, and D. Wei. Universal torsors and values of quadratic polynomials represented by norms. *Math. Ann.*, in press, arXiv:1202.3567, 2012.

[123] U. Derenthal and Yu. Tschinkel. Universal torsors over del Pezzo surfaces and rational points. In *Equidistribution in number theory, an introduction.* NATO Science Series II: Mathematics, Physics and Chemistry, Vol. 237, pp. 169–196. Springer-Verlag, Dordrecht, 2007.

[124] K. Destagnol. La conjecture de Manin sur les surfaces de Châtelet, Master's thesis, Université Paris 7 Diderot, 2013.

[125] I. V. Dolgachev. Newton polyhedra and factorial rings. *J. Pure Appl. Algebra*, 18(3):253–258, 1980.

[126] I. V. Dolgachev. On automorphisms of Enriques surfaces. *Invent. Math.*, 76(1):163–177, 1984.

[127] I. V. Dolgachev. *Classical algebraic geometry*. Cambridge University Press, Cambridge, 2012. A modern view.

[128] I. V. Dolgachev and Yi Hu. Variation of geometric invariant theory quotients. *Inst. Hautes Études Sci. Publ. Math.*, (87):5–56, 1998.

[129] I. V. Dolgachev and De-Qi Zhang. Coble rational surfaces. *Amer. J. Math.*, 123(1):79–114, 2001.

[130] F. Donzelli. Algebraic density property of Danilov-Gizatullin surfaces. *Math. Z.*, 272(3–4):1187–1194, 2012.

[131] J. J. Duistermaat. *Discrete integrable systems*. Springer Monographs in Mathematics. Springer-Verlag, New York, 2010. QRT maps and elliptic surfaces.

[132] D. Eisenbud. *Commutative algebra*. Graduate Texts in Mathematics, Vol. 150. Springer-Verlag, New York, 1995.

[133] D. Eisenbud and S. Popescu. The projective geometry of the Gale transform. *J. Algebra*, 230(1):127–173, 2000.

[134] E. J. Elizondo, K. Kurano, and K. Watanabe. The total coordinate ring of a normal projective variety. *J. Algebra*, 276(2):625–637, 2004.

[135] A.-S. Elsenhans and J. Jahnel. Moduli spaces and the inverse Galois problem for cubic surfaces. *Trans. Amer. Math. Soc.*, in press, arXiv:1209.5591, 2012.

[136] G. Ewald. Polygons with hidden vertices. *Beiträge Algebra Geom.*, 42(2):439–442, 2001.

[137] H. Flenner, S. Kaliman, and M. Zaidenberg. On the Danilov-Gizatullin isomorphism theorem. *Enseign. Math. (2)*, 55(3-4):275–283, 2009.

[138] J. Franke, Yu. I. Manin, and Yu. Tschinkel. Rational points of bounded height on Fano varieties. *Invent. Math.*, 95(2):421–435, 1989.

[139] C. Frei. Counting rational points over number fields on a singular cubic surface. *Algebra Number Theory*, 7(6):1451–1479, 2013.

[140] C. Frei and M. Pieropan. O-minimality on twisted universal torsors and Manin's conjecture over number fields. arXiv:1312.6603, 2013.

[141] W. Fulton. *Introduction to toric varieties*. Annals of Mathematics Studies, Vol. 131. Princeton University Press, Princeton, NJ, 1993.

[142] W. Fulton. *Intersection theory*. Ergebnisse der Mathematik und ihrer Grenzgebiete (3), Vol. 2, 2nd ed. Springer-Verlag, Berlin, 1998.

[143] W. Fulton and J. Harris. *Representation theory*. Graduate Texts in Mathematics, Vol. 129. Springer-Verlag, New York, 1991.

[144] G. Gagliardi. The Cox ring of a spherical embedding. *J. Algebra*, 397:548–569, 2014.

[145] S. A. Gaĭfullin. Affine toric SL(2)-embeddings. *Mat. Sb.*, 199(3):3–24, 2008.

[146] C. Galindo and F. Monserrat. The total coordinate ring of a smooth projective surface. *J. Algebra*, 284(1):91–101, 2005.

[147] M. H. Gizatullin and V. I. Danilov. Automorphisms of affine surfaces. II. *Izv. Akad. Nauk SSSR Ser. Mat.*, 41(1):54–103, 231, 1977.

[148] Y. Gongyo, S. Okawa, A. Sannai, and S. Takagi. Characterization of varieties of Fano type via singularities of Cox rings. *J. Algebraic Geom.*, in press, arXiv:1201.1133, 2012.

[149] B. Green and T. Tao. Linear equations in primes. *Ann. Math. (2)*, 171(3):1753–1850, 2010.

[150] F. D. Grosshans. *Algebraic homogeneous spaces and invariant theory*. Lecture Notes in Mathematics, Vol. 1673. Springer-Verlag, Berlin, 1997.

[151] A. Grothendieck. Éléments de géométrie algébrique. II. Étude globale élémentaire de quelques classes de morphismes. *Inst. Hautes Études Sci. Publ. Math.*, (8):222, 1961.

[152] N. Guay. Embeddings of symmetric varieties. *Transform. Groups*, 6(4):333–352, 2001.

[153] M. Hanselmann. *Rational points on quartic hypersurfaces*. PhD thesis, Ludwig-Maximilians-Universität München, 2012.

[154] D. Harari. Obstructions de Manin transcendantes. In *Number theory* (Paris, 1993–1994), London Mathematical Society Lecture Note Series, Vol. 235, pp. 75–87. Cambridge University Press, Cambridge, 1996.

[155] D. Harari. Groupes algébriques et points rationnels. *Math. Ann.*, 322(4):811–826, 2002.

[156] D. Harari. Weak approximation on algebraic varieties. In *Arithmetic of higher-dimensional algebraic varieties* (Palo Alto, CA, 2002), Progress in Mathematics, Vol. 226, pp. 43–60. Birkhäuser Boston, Boston, 2004.

[157] D. Harari and A. N. Skorobogatov. Descent theory for open varieties. In *Torsors, étale homotopy and applications to rational points*. London Mathematical Society Lecture Note Series, Vol. 405, pp. 250–279. Cambridge University Press, Cambridge, 2013.

[158] Y. Harpaz, A. N. Skorobogatov, and O. Wittenberg. The Hardy–Littlewood conjecture and rational points. *Compositio Math.*, in press, arXiv:1304.3333, 2013.

[159] R. Hartshorne. *Ample subvarieties of algebraic varieties*. Notes written in collaboration with C. Musili. Lecture Notes in Mathematics, Vol. 156. Springer-Verlag, Berlin, 1970.

[160] R. Hartshorne. *Algebraic geometry*. Graduate Texts in Mathematics, Vol. 52. Springer-Verlag, New York, 1977.

[161] H. Hasse. Die Normenresttheorie relativ-Abelscher Zahlkörper als Klassenkörpertheorie im Kleinen. *J. reine angew. Math.*, 162:145–154, 1930.

[162] B. Hassett and Yu. Tschinkel. Universal torsors and Cox rings. In *Arithmetic of higher-dimensional algebraic varieties* (Palo Alto, CA, 2002). Progress in Mathematics, Vol. 226, pp. 149–173. Birkhäuser Boston, Boston, 2004.

[163] J. Hausen. Equivariant embeddings into smooth toric varieties. *Canad. J. Math.*, 54(3):554–570, 2002.

[164] J. Hausen. Geometric invariant theory based on Weil divisors. *Compos. Math.*, 140(6):1518–1536, 2004.

[165] J. Hausen. Cox rings and combinatorics. II. *Mosc. Math. J.*, 8(4):711–757, 847, 2008.

[166] J. Hausen and E. Herppich. Factorially graded rings of complexity one. In *Torsors, étale homotopy and applications to rational points*. London Mathematical Society Lecture Note Series, Vol. 405, pp. 414–428. Cambridge University Press, Cambridge, 2013.

[167] J. Hausen, E. Herppich, and H. Süß. Multigraded factorial rings and Fano varieties with torus action. *Doc. Math.*, 16:71–109, 2011.

[168] J. Hausen, S. Keicher, and A. Laface. Computing Cox rings. arXiv:1305.4343, 2013.

[169] J. Hausen and H. Süß. The Cox ring of an algebraic variety with torus action. *Adv. Math.*, 225(2):977–1012, 2010.

[170] D. R. Heath-Brown. The density of rational points on Cayley's cubic surface. In *Proceedings of the Session in Analytic Number Theory and Diophantine Equations*. Bonner Mathematische Schriften, Vol. 360, pp. 33, Bonn, 2003. Universität Bonn.

[171] D. R. Heath-Brown. Zeros of pairs of quadratic forms. arXiv:1304.3894, 2013.

[172] D. R. Heath-Brown and A. Skorobogatov. Rational solutions of certain equations involving norms. *Acta Math.*, 189(2):161–177, 2002.

[173] H. Hironaka. Triangulations of algebraic sets. In *Algebraic geometry* (Humboldt State Univ., Arcata, Calif., 1974). Proc. Sympos. Pure Math., Vol. 29, pp. 165–185. Amer. Math. Soc., Providence, RI, 1975.

[174] M. Hochster and J. L. Roberts. Rings of invariants of reductive groups acting on regular rings are Cohen-Macaulay. *Adv. Math.*, 13:115–175, 1974.

[175] C. Hooley. On nonary cubic forms. *J. reine angew. Math.*, 386:32–98, 1988.

[176] Y. Hu and S. Keel. Mori dream spaces and GIT. *Michigan Math. J.*, 48:331–348, 2000.

[177] E. Huggenberger. *Fano varieties with a torus action of complexity one*. PhD thesis, Universität Tübingen, 2013.

[178] J. E. Humphreys. *Linear algebraic groups*. Graduate Texts in Mathematics, Vol. 21. Springer-Verlag, New York, 1975.

[179] D. Hwang and J. Park. Redundant blow-ups and Cox rings of rational surfaces. arXiv:1303.2274, 2013.

[180] M.-N. Ishida. Graded factorial rings of dimension 3 of a restricted type. *J. Math. Kyoto Univ.*, 17(3):441–456, 1977.

[181] S.-Y. Jow. A Lefschetz hyperplane theorem for Mori dream spaces. *Math. Z.*, 268(1-2):197–209, 2011.

[182] T. Kajiwara. The functor of a toric variety with enough invariant effective Cartier divisors. *Tohoku Math. J. (2)*, 50(1):139–157, 1998.

[183] Y. Kawamata and S. Okawa. Mori dream spaces of Calabi-Yau type and the log canonicity of the Cox rings. *J. reine angew. Math.*, in press, arXiv:1202.2696, 2012.

[184] A. D. King. Moduli of representations of finite-dimensional algebras. *Quart. J. Math. Oxford Ser. (2)*, 45(180):515–530, 1994.

[185] S. L. Kleiman. Toward a numerical theory of ampleness. *Ann. of Math. (2)*, 84:293–344, 1966.

[186] P. Kleinschmidt. A classification of toric varieties with few generators. *Aequationes Math.*, 35(2-3):254–266, 1988.

[187] F. Knop. The Luna-Vust theory of spherical embeddings. In *Proceedings of the Hyderabad Conference on Algebraic Groups* (Hyderabad, 1989), pp. 225–249, Madras, 1991. Manoj Prakashan.

[188] F. Knop. Über Hilberts vierzehntes Problem für Varietäten mit Kompliziertheit eins. *Math. Z.*, 213(1):33–36, 1993.

[189] F. Knop, H. Kraft, D. Luna, and T. Vust. Local properties of algebraic group actions. In *Algebraische Transformationsgruppen und Invariantentheorie*. DMV Seminar, Vol. 13, pp. 63–75. Birkhäuser, Basel, 1989.

[190] F. Knop, H. Kraft, and T. Vust. The Picard group of a G-variety. In *Algebraische Transformationsgruppen und Invariantentheorie*. DMV Seminar, Vol. 13, pp. 77–87. Birkhäuser, Basel, 1989.

[191] M. Koitabashi. Automorphism groups of generic rational surfaces. *J. Algebra*, 116(1):130–142, 1988.

[192] J. Kollár and S. Mori. *Birational geometry of algebraic varieties*. Cambridge Tracts in Mathematics, Vol. 134. Cambridge University Press, Cambridge, 1998. With the collaboration of C. H. Clemens and A. Corti.

[193] J. Kollár and E. Szabó. Fixed points of group actions and rational maps. arXiv:math/9905053, 1999.

[194] S. Kondō. Enriques surfaces with finite automorphism groups. *Jpn. J. Math.*, 12(2):191–282, 1986.

[195] S. Kondō. Algebraic $K3$ surfaces with finite automorphism groups. *Nagoya Math. J.*, 116:1–15, 1989.

[196] M. Koras and P. Russell. Linearization problems. In *Algebraic group actions and quotients*, pp. 91–107. Hindawi, Cairo, 2004.

[197] H. Kraft. *Geometrische Methoden in der Invariantentheorie*. Aspects of Mathematics, D1. Friedr. Vieweg & Sohn, Braunschweig, 1984.

[198] A. Laface and D. Testa. Nef and semiample divisors on rational surfaces. In *Torsors, étale homotopy and applications to rational points*. London Mathematical Society Lecture Note Series, Vol. 405, pp. 429–446. Cambridge University Press, Cambridge, 2013.

[199] A. Laface and M. Velasco. Picard-graded Betti numbers and the defining ideals of Cox rings. *J. Algebra*, 322(2):353–372, 2009.

[200] A. Laface and M. Velasco. A survey on Cox rings. *Geom. Dedicata*, 139:269–287, 2009.

[201] K. F. Lai and K. M. Yeung. Rational points in flag varieties over function fields. *J. Number Theory*, 95(2):142–149, 2002.

[202] R. Lazarsfeld. *Positivity in algebraic geometry. I*. Ergebnisse der Mathematik und ihrer Grenzgebiete (3), Vol. 48. Springer-Verlag, Berlin, 2004.

[203] P. Le Boudec. Manin's conjecture for a cubic surface with $2A_2 + A_1$ singularity type. *Math. Proc. Cambridge Philos. Soc.*, 153(3):419–455, 2012.

[204] P. Le Boudec. Manin's conjecture for a quartic del Pezzo surface with A_3 singularity and four lines. *Monatsh. Math.*, 167(3-4):481–502, 2012.

[205] P. Le Boudec. Manin's conjecture for two quartic del Pezzo surfaces with $3A_1$ and $A_1 + A_2$ singularity types. *Acta Arith.*, 151(2):109–163, 2012.

[206] P. Le Boudec. Affine congruences and rational points on a certain cubic surface. *Algebra Number Theory*, in press, arXiv:1207.2685, 2012.

[207] J. Lipman. Rational singularities, with applications to algebraic surfaces and unique factorization. *Inst. Hautes Études Sci. Publ. Math.*, (36):195–279, 1969.

[208] D. Loughran. Manin's conjecture for a singular quartic del Pezzo surface. *J. Lond. Math. Soc. (2)*, 86(2):558–584, 2012.

[209] D. Loughran. Rational points of bounded height and the Weil restriction. *Israel J. Math.*, in press, arXiv:1210.1792, 2012.

[210] D. Luna. Slices étales. In *Sur les groupes algébriques*, pages 81–105. Bull. Soc. Math. France, Paris, Mémoire 33. Soc. Math. France, Paris, 1973.

[211] D. Luna. Toute variété magnifique est sphérique. *Transform. Groups*, 1(3):249–258, 1996.

[212] D. Luna and T. Vust. Plongements d'espaces homogènes. *Comment. Math. Helv.*, 58(2):186–245, 1983.

[213] D. Maclagan and B. Sturmfels. *Introduction to tropical geometry.* In preparation.

[214] Yu. I. Manin. *Cubic forms.* North-Holland Mathematical Library, Vol. 4, 2nd ed. North-Holland Publishing Co., Amsterdam, 1986.

[215] K. Matsuki. *Introduction to the Mori program.* Universitext. Springer-Verlag, New York, 2002.

[216] H. Matsumura. *Commutative ring theory.* Cambridge Studies in Advanced Mathematics, Vol. 8, 2nd ed. Cambridge University Press, Cambridge, 1989.

[217] J. McKernan. Mori dream spaces. *Jpn. J. Math.*, 5(1):127–151, 2010.

[218] E. Miller and B. Sturmfels. *Combinatorial commutative algebra.* Graduate Texts in Mathematics, Vol. 227. Springer-Verlag, New York, 2005.

[219] J. S. Milne. *Étale cohomology.* Princeton Mathematical Series, Vol. 33. Princeton University Press, Princeton, NJ, 1980.

[220] R. Miranda and U. Persson. On extremal rational elliptic surfaces. *Math. Z.*, 193(4):537–558, 1986.

[221] S. Mori. Graded factorial domains. *Jpn. J. Math.*, 3(2):223–238, 1977.

[222] D. R. Morrison. On $K3$ surfaces with large Picard number. *Invent. Math.*, 75(1):105–121, 1984.

[223] S. Mukai. Geometric realization of T-shaped root systems and counterexamples to Hilbert's fourteenth problem. In *Algebraic transformation groups and algebraic varieties.* Encyclopaedia Mathematical Sciences, Vol. 132, pp. 123–129. Springer-Verlag, Berlin, 2004.

[224] D. Mumford. *The red book of varieties and schemes.* Lecture Notes in Mathematics, Vol. 1358, expanded edition. Springer-Verlag, Berlin, 1999.

[225] D. Mumford, J. Fogarty, and F. Kirwan. *Geometric invariant theory.* Ergebnisse der Mathematik und ihrer Grenzgebiete (2), Vol. 34, 3rd ed. Springer-Verlag, Berlin, 1994.

[226] I. M. Musson. Differential operators on toric varieties. *J. Pure Appl. Algebra*, 95(3):303–315, 1994.

[227] M. Mustaţă. Vanishing theorems on toric varieties. *Tohoku Math. J. (2)*, 54(3):451–470, 2002.

[228] M. Nagata. On the 14-th problem of Hilbert. *Amer. J. Math.*, 81:766–772, 1959.

[229] N. Nakayama. Classification of log del Pezzo surfaces of index two. *J. Math. Sci. Univ. Tokyo*, 14(3):293–498, 2007.

[230] V. V. Nikulin. Quotient-groups of groups of automorphisms of hyperbolic forms of subgroups generated by 2-reflections. *Dokl. Akad. Nauk SSSR*, 248(6):1307–1309, 1979.

[231] V. V. Nikulin. $K3$ surfaces with a finite group of automorphisms and a Picard group of rank three. *Trudy Mat. Inst. Steklov.*, 165:119–142, 1984. Algebraic geometry and its applications.

[232] V. V. Nikulin. A remark on algebraic surfaces with polyhedral Mori cone. *Nagoya Math. J.*, 157:73–92, 2000.

[233] B. Nill. Complete toric varieties with reductive automorphism group. *Math. Z.*, 252(4):767–786, 2006.

[234] T. Oda. *Convex bodies and algebraic geometry.* Ergebnisse der Mathematik und ihrer Grenzgebiete (3), Vol. 15. Springer-Verlag, Berlin, 1988.

[235] T. Oda and H. S. Park. Linear Gale transforms and Gel'fand-Kapranov-Zelevinskij decompositions. *Tohoku Math. J. (2)*, 43(3):375–399, 1991.

[236] A. L. Onishchik and È. B. Vinberg. *Lie groups and algebraic groups*. Springer Series in Soviet Mathematics. Springer-Verlag, Berlin, 1990.

[237] P. Orlik and P. Wagreich. Isolated singularities of algebraic surfaces with C* action. *Ann. of Math. (2)*, 93:205–228, 1971.

[238] P. Orlik and P. Wagreich. Algebraic surfaces with k^*-action. *Acta Math.*, 138(1–2):43–81, 1977.

[239] J. C. Ottem. On the Cox ring of $\mathbf{P}^2$ blown up in points on a line. *Math. Scand.*, 109(1):22–30, 2011.

[240] J. C. Ottem. Cox rings of K3 surfaces with Picard number 2. *J. Pure Appl. Algebra*, 217(4):709–715, 2013.

[241] T. E. Panov. Toric Kempf–Ness sets. *Proc. Steklov Inst. Math.*, 263:159–172, 2008.

[242] H. S. Park. The Chow rings and GKZ-decompositions for $\mathbf{Q}$-factorial toric varieties. *Tohoku Math. J. (2)*, 45(1):109–145, 1993.

[243] E. Peyre. Hauteurs et mesures de Tamagawa sur les variétés de Fano. *Duke Math. J.*, 79(1):101–218, 1995.

[244] E. Peyre. Terme principal de la fonction zêta des hauteurs et torseurs universels. *Astérisque*, (251):259–298, 1998. Nombre et répartition de points de hauteur bornée (Paris, 1996).

[245] E. Peyre. Torseurs universels et méthode du cercle. In *Rational points on algebraic varieties*. Progress in Mathematics, Vol. 199, pp. 221–274. Birkhäuser, Basel, 2001.

[246] E. Peyre. Points de hauteur bornée et géométrie des variétés (d'après Y. Manin et al.). *Astérisque*, (282):Exp. No. 891, ix, 323–344, 2002. Séminaire Bourbaki, Vol. 2000/2001.

[247] E. Peyre. Points de hauteur bornée, topologie adélique et mesures de Tamagawa. *J. Théor. Nombres Bordeaux*, 15(1):319–349, 2003. Les XXIIèmes Journées Arithmetiques (Lille, 2001).

[248] E. Peyre. Counting points on varieties using universal torsors. In *Arithmetic of higher-dimensional algebraic varieties* (Palo Alto, CA, 2002). Progress in Mathematics, Vol. 226, pp. 61–81. Birkhäuser Boston, Boston, 2004.

[249] E. Peyre. Obstructions au principe de Hasse et à l'approximation faible. *Astérisque*, (299):Exp. No. 931, viii, 165–193, 2005. Séminaire Bourbaki. Vol. 2003/2004.

[250] E. Peyre. Points de hauteur bornée sur les variétés de drapeaux en caractéristique finie. *Acta Arith.*, 152(2):185–216, 2012.

[251] I. I. Pjateckiĭ-Šapiro and I. R. Šafarevič. Torelli's theorem for algebraic surfaces of type K3. *Izv. Akad. Nauk SSSR Ser. Mat.*, 35:530–572, 1971.

[252] B. Poonen. Insufficiency of the Brauer-Manin obstruction applied to étale covers. *Ann. of Math. (2)*, 171(3):2157–2169, 2010.

[253] B. Poonen. Rational points on varieties, http://www-math.mit.edu/~poonen/papers/Qpoints.pdf, 2008.

[254] O. N. Popov. The Cox ring of a Del Pezzo surface has rational singularities. arXiv:math.AG/0402154, 2004.

[255] V. L. Popov. Quasihomogeneous affine algebraic varieties of the group SL(2). *Izv. Akad. Nauk SSSR Ser. Mat*, 37:792–832, 1973.

[256] G. V. Ravindra and V. Srinivas. The Grothendieck-Lefschetz theorem for normal projective varieties. *J. Algebraic Geom.*, 15(3):563–590, 2006.

[257] G. V. Ravindra and V. Srinivas. The Noether-Lefschetz theorem for the divisor class group. *J. Algebra*, 322(9):3373–3391, 2009.

[258] L. Renner. The cone of semi-simple monoids with the same factorial hull. arXiv:math.AG/0603222, 2006.

[259] A. Rittatore. Algebraic monoids and group embeddings. *Transform. Groups*, 3(4):375–396, 1998.

[260] A. Rittatore. Very flat reductive monoids. *Publ. Mat. Urug.*, 9:93–121 (2002), 2001.

[261] P. Roberts. An infinitely generated symbolic blow-up in a power series ring and a new counterexample to Hilbert's fourteenth problem. *J. Algebra*, 132(2):461–473, 1990.

[262] B. Saint-Donat. Projective models of $K - 3$ surfaces. *Amer. J. Math.*, 96:602–639, 1974.

[263] P. Salberger. Tamagawa measures on universal torsors and points of bounded height on Fano varieties. *Astérisque*, (251):91–258, 1998. Nombre et répartition de points de hauteur bornée (Paris, 1996).

[264] P. Samuel. *Lectures on unique factorization domains*. Notes by M. Pavman Murthy. Tata Institute of Fundamental Research Lectures on Mathematics, No. 30. Tata Institute of Fundamental Research, Bombay, 1964.

[265] S. Schanuel. Heights in number fields. *Bull. Soc. Math. France*, 107(4):433–449, 1979.

[266] G. Scheja and U. Storch. Zur Konstruktion faktorieller graduierter Integritätsbereiche. *Arch. Math. (Basel)*, 42(1):45–52, 1984.

[267] V. V. Serganova. Torsors and representation theory of reductive groups. In *Torsors, étale homotopy and applications to rational points*. London Mathematical Society Lecture Note Series, Vol. 405, pp. 75–119. Cambridge University Press, Cambridge, 2013.

[268] V. V. Serganova and A. N. Skorobogatov. Del Pezzo surfaces and representation theory. *Algebra Number Theory*, 1(4):393–419, 2007.

[269] V. V. Serganova and A. N. Skorobogatov. On the equations for universal torsors over del Pezzo surfaces. *J. Inst. Math. Jussieu*, 9(1):203–223, 2010.

[270] V. V. Serganova and A. N. Skorobogatov. Adjoint representation of E_8 and del Pezzo surfaces of degree 1. *Ann. Inst. Fourier (Grenoble)*, 61(6):2337–2360 (2012), 2011.

[271] J.-P. Serre. Géométrie algébrique et géométrie analytique. *Ann. Inst. Fourier, Grenoble*, 6:1–42, 1955–1956.

[272] J.-P. Serre. *A course in arithmetic*. Springer-Verlag, New York, 1973.

[273] J.-P. Serre. *Local fields*. Graduate Texts in Mathematics, Vol. 67. Springer-Verlag, New York, 1979.

[274] J.-P. Serre. *Galois cohomology*. Springer Monographs in Mathematics. Springer-Verlag, Berlin, English edition, 2002.

[275] V. V. Shokurov. A nonvanishing theorem. *Izv. Akad. Nauk SSSR Ser. Mat.*, 49(3):635–651, 1985.

[276] C. M. Skinner. Forms over number fields and weak approximation. *Compositio Math.*, 106(1):11–29, 1997.

[277] A. N. Skorobogatov. On a theorem of Enriques-Swinnerton-Dyer. *Ann. Fac. Sci. Toulouse Math. (6)*, 2(3):429–440, 1993.

[278] A. N. Skorobogatov. Beyond the Manin obstruction. *Invent. Math.*, 135(2):399–424, 1999.

[279] A. N. Skorobogatov. *Torsors and rational points*. Cambridge Tracts in Mathematics, Vol. 144. Cambridge University Press, Cambridge, 2001.

[280] T. A. Springer. *Linear algebraic groups*, 2nd ed. Modern Birkhäuser Classics. Birkhäuser Boston, Boston, 2009.

[281] M. Stillman, D. Testa, and M. Velasco. Gröbner bases, monomial group actions, and the Cox rings of del Pezzo surfaces. *J. Algebra*, 316(2):777–801, 2007.

[282] B. Sturmfels. *Gröbner bases and convex polytopes*. University Lecture Series, Vol. 8. American Mathematical Society, Providence, RI, 1996.

[283] B. Sturmfels and M. Velasco. Blow-ups of $\mathbb{P}^{n-3}$ at n points and spinor varieties. *J. Commut. Algebra*, 2(2):223–244, 2010.

[284] B. Sturmfels and Z. Xu. Sagbi bases of Cox-Nagata rings. *J. Eur. Math. Soc. (JEMS)*, 12(2):429–459, 2010.

[285] H. Sumihiro. Equivariant completion. *J. Math. Kyoto Univ.*, 14:1–28, 1974.

[286] H. Süß. Canonical divisors on T-varieties. arXiv:0811.0626, 2008.

[287] M. Swarbrick Jones. A Note On a Theorem of Heath-Brown and Skorobogatov. *Q. J. Math.* 64(4):1239–1251, 2013.

[288] J. Święcicka. Quotients of toric varieties by actions of subtori. *Colloq. Math.*, 82(1):105–116, 1999.

[289] J. Święcicka. A combinatorial construction of sets with good quotients by an action of a reductive group. *Colloq. Math.*, 87(1):85–102, 2001.

[290] P. Swinnerton-Dyer. Two special cubic surfaces. *Mathematika*, 9:54–56, 1962.

[291] P. Swinnerton-Dyer. The solubility of diagonal cubic surfaces. *Ann. Sci. École Norm. Sup. (4)*, 34(6):891–912, 2001.

[292] D. Testa, A. Várilly-Alvarado, and M. Velasco. Cox rings of degree one del Pezzo surfaces. *Algebra Number Theory*, 3(7):729–761, 2009.

[293] D. Testa, A. Várilly-Alvarado, and M. Velasco. Big rational surfaces. *Math. Ann.*, 351(1):95–107, 2011.

[294] J. Tevelev. Compactifications of subvarieties of tori. *Amer. J. Math.*, 129(4):1087–1104, 2007.

[295] M. Thaddeus. Geometric invariant theory and flips. *J. Amer. Math. Soc.*, 9(3):691–723, 1996.

[296] D. A. Timashev. *Homogeneous spaces and equivariant embeddings*. Invariant Theory and Algebraic Transformation Groups, 8. Encyclopaedia of Mathematical Sciences, Vol. 138. Springer-Verlag, Heidelberg, 2011.

[297] B. Totaro. The cone conjecture for Calabi-Yau pairs in dimension 2. *Duke Math. J.*, 154(2):241–263, 2010.

[298] È. B. Vinberg. Complexity of actions of reductive groups. *Funktsional. Anal. i Prilozhen.*, 20(1):1–13, 96, 1986.

[299] È. B. Vinberg. On reductive algebraic semigroups. In *Lie groups and Lie algebras: E. B. Dynkin's Seminar*. American Mathematical Society Translations Series 2, Vol. 169, pp. 145–182. American Mathematical Society, Providence, RI, 1995.

[300] È. B. Vinberg. Classification of 2-reflective hyperbolic lattices of rank 4. *Tr. Mosk. Mat. Obs.*, 68:44–76, 2007.

[301] È. B. Vinberg and V. L. Popov. Invariant theory. In *Algebraic Geometry, IV*. Encyclopedia of Mathematical Sciences, Vol. 55, pp. 123–278. Springer-Verlag, Berlin, 1994.

[302] V. E. Voskresenskiĭ. *Algebraicheskie tory*. Izdat. "Nauka," Moscow, 1977.

[303] V. E. Voskresenskiĭ. *Algebraic groups and their birational invariants*. Translations of Mathematical Monographs, Vol. 179. American Mathematical Society, Providence, RI, 1998.

[304] W. C. Waterhouse. *Introduction to affine group schemes*. Graduate Texts in Mathematics, Vol. 66. Springer-Verlag, New York, 1979.

[305] R. Wazir. Arithmetic on elliptic threefolds. *Compos. Math.*, 140(3):567–580, 2004.

[306] D. Wei. On the equation $N_{K/k}(\Xi) = P(t)$. *Proc. London Math. Soc. (3)*, in press, arXiv:1202.4115, 2012.

[307] O. Wittenberg. *Intersections de deux quadriques et pinceaux de courbes de genre 1/Intersections of two quadrics and pencils of curves of genus 1*. Lecture Notes in Mathematics, Vol. 1901. Springer-Verlag, Berlin, 2007.

[308] J. Włodarczyk. Embeddings in toric varieties and prevarieties. *J. Algebraic Geom.*, 2(4):705–726, 1993.

[309] J. Włodarczyk. Maximal quasiprojective subsets and the Kleiman-Chevalley quasiprojectivity criterion. *J. Math. Sci. Univ. Tokyo*, 6(1):41–47, 1999.

[310] D.-Q. Zhang. Quotients of $K3$ surfaces modulo involutions. *Jpn. J. Math.*, 24(2):335–366, 1998.

[311] V. S. Zhgun. On embeddings of universal torsors over del Pezzo surfaces into cones over flag varieties. *Izv. Ross. Akad. Nauk Ser. Mat.*, 74(5):3–44, 2010.

Index